Student Solutions Manual

Technical Calculus with Analytic Geometry

FIFTH EDITION

Peter Kuhfittig
Milwaukee School of Engineering

Prepared by

Christopher Schroeder
Morehead State University

Australia • Brazil • Japan • Korea • Mexico • Singapore • Spain • United Kingdom • United States

ISBN-13: 978-1-285-05257-1
ISBN-10: 1-285-05257-9

Brooks/Cole
20 Channel Center Street
Boston, MA 02210
USA

Cengage Learning is a leading provider of customized learning solutions with office locations around the globe, including Singapore, the United Kingdom, Australia, Mexico, Brazil, and Japan. Locate your local office at: **www.cengage.com/global**

Cengage Learning products are represented in Canada by Nelson Education, Ltd.

To learn more about Brooks/Cole, visit **www.cengage.com/brookscole**

Purchase any of our products at your local college store or at our preferred online store **www.cengagebrain.com**

Printed in the United States of America
1 2 3 4 5 6 7 17 16 15 14 13

Contents

Chapter 1

Introduction to Analytic Geometry

1.1 The Cartesian Coordinate System

1. Let $(x_2, y_2) = (2, 4)$ and $(x_1, y_1) = (5, 2)$. From the distance formula

$$d = \sqrt{(x_2 - x_1)^2 + (y_2 - y_1)^2}$$

we get

$$d = \sqrt{(2-5)^2 + (4-2)^2} = \sqrt{(-3)^2 + 2^2} = \sqrt{9+4} = \sqrt{13}.$$

3. Let $(x_2, y_2) = (-3, -6)$ and $(x_1, y_1) = (5, -2)$. Then

$$d = \sqrt{(-3-5)^2 + [-6-(-2)]^2} = \sqrt{(-8)^2 + (-4)^2} = \sqrt{64+16} = \sqrt{16 \cdot 5} = 4\sqrt{5}.$$

5. Let $(x_2, y_2) = (\sqrt{3}, 4)$ and $(x_1, y_1) = (0, 2)$. Then

$$d = \sqrt{(\sqrt{3} - 0)^2 + (4-2)^2} = \sqrt{3+4} = \sqrt{7}.$$

7. $d = \sqrt{[1-(-1)]^2 + (-\sqrt{2} - 0)^2} = \sqrt{4+2} = \sqrt{6}$

9. $d = \sqrt{[-9-(-11)]^2 + (-1-1)^2} = \sqrt{2^2 + (-2)^2} = \sqrt{8} = \sqrt{2 \cdot 4} = 2\sqrt{2}$

11b. x/y is negative whenever x and y have opposite signs: quadrants II and IV.

13a. Any point on the y-axis has coordinates of the form $(0, y)$.

15. Let $A = (-2, -5), B = (-4, 1)$ and $C = (5, 4)$; then $AB = \sqrt{[-2-(-4)]^2 + (-5-1)^2} = \sqrt{40}, AC = \sqrt{(-2-5)^2 + (-5-4)^2} = 130$, and $BC = \sqrt{(-4-5)^2 + (1-4)^2} = \sqrt{90}$. Since $(AB)^2 + (BC)^2 = (AC)^2$, the triangle must be a right triangle.

17. The points $(12, 0), (-4, 8)$ and $(-1, -13)$ are all $5\sqrt{5}$ units from $(1, -2)$.

19. Distance from $(-1, -1)$ to $(2, 8)$:

$$\sqrt{(-1-2)^2 + (-1-8)^2} = \sqrt{9+81} = \sqrt{90} = \sqrt{9 \cdot 10} = 3\sqrt{10}$$

Distance from $(2, 8)$ to $(5, 17)$:

$$\sqrt{(5-2)^2 + (17-8)^2} = \sqrt{90} = 3\sqrt{10}$$

Distance from $(-1,-1)$ to $(5,17)$:

$$\sqrt{6^2+18^2}=\sqrt{360}=6\sqrt{10}$$

Total distance $6\sqrt{10}=3\sqrt{10}+3\sqrt{10}$, the sum of the other two distances.

21. Distance from (x,y) to y-axis: x units
Distance from (x,y) to $(2,0)$: $\sqrt{(x-2)^2+(y-0)^2}=\sqrt{(x-2)^2+y^2}$

By assumption,

$$\begin{aligned}\sqrt{(x-2)^2+y^2} &= x \\ (x-2)^2+y^2 &= x^2 \qquad \text{squaring both sides} \\ x^2-4x+4+y^2 &= x^2 \\ y^2-4x+4 &= 0\end{aligned}$$

23. Let $(x_1,y_1)=(-2,6)$ and $(x_2,y_2)=(2,-4)$. Then from the midpoint formula

$$\left(\frac{x_1+x_2}{2},\frac{y_1+y_2}{2}\right)$$

we get

$$\left(\frac{-2+2}{2},\frac{6+(-4)}{2}\right)=(0,1).$$

25. Let $(x_1,y_1)=(5,0)$ and $(x_2,y_2)=(9,4)$. Then from the midpoint formula

$$\left(\frac{x_1+x_2}{2},\frac{y_1+y_2}{2}\right)$$

we get

$$\left(\frac{5+9}{2},\frac{0+4}{2}\right)=(7,2).$$

27. The center is the midpoint: $\left(\frac{-2+6}{2},\frac{-1+11}{2}\right)=(2,5)$.

1.2 The Slope

1. Let $(x_2,y_2)=(1,7)$ and $(x_1,y_1)=(2,6)$. Then, by formula (1.4),

$$m=\frac{y_2-y_1}{x_2-x_1}$$

we get

$$m=\frac{7-6}{1-2}=\frac{1}{-1}=-1.$$

3. Let $(x_1,y_1)=(0,2)$ and $(x_2,y_2)=(-4,-4)$. Then

$$m=\frac{-4-2}{-4-0}=\frac{-6}{-4}=\frac{3}{2}.$$

5. Let $(x_2, y_2) = (7, 8)$ and $(x_1, y_1) = (-3, -4)$. Then

$$m = \frac{8-(-4)}{7-(-3)} = \frac{8+4}{7+3} = \frac{12}{10} = \frac{6}{5}.$$

7. $m = \dfrac{43-(-1)}{-1-1} = \dfrac{44}{-2} = -22$

9. $m = \dfrac{-5-4}{3-3} = \dfrac{-9}{0}$ (undefined)

11. $m = \dfrac{-3-(-3)}{9-5} = \dfrac{0}{4} = 0$

13. $m = \dfrac{3-2}{12-(-2)} = \dfrac{1}{14}$

15. See answer section of book.

17. Slope of given line is $\dfrac{1-(-5)}{-7-6} = \dfrac{6}{-13} = -\dfrac{6}{13}$. Slope of perpendicular is given by the negative reciprocal and is therefore $-\dfrac{1}{(-6/13)} = \dfrac{13}{6}$.

19. Slope of line through $(-4, 6)$ and $(6, 10)$: $\dfrac{6-10}{-4-6} = \dfrac{-4}{-10} = \dfrac{2}{5}$
Slope of line through $(6, 10)$ and $(10, 0)$: $\dfrac{10-0}{6-10} = \dfrac{10}{-4} = -\dfrac{5}{2}$
Since the slopes are negative reciprocals, the lines are perpendicular.

21. Slope of line through $(0, -3)$ and $(-2, 3)$: $\dfrac{-3-3}{0-(-2)} = \dfrac{-6}{2} = -3$
Slope of line through $(7, 6)$ and $(9, 0)$: $\dfrac{6-0}{7-9} = \dfrac{6}{-2} = -3$
Slope of line through $(-2, 3)$ and $(7, 6)$: $\dfrac{3-6}{-2-7} = \dfrac{-3}{-9} = \dfrac{1}{3}$
Slope of line through $(0, -3)$ and $(9, 0)$: $\dfrac{-3-0}{0-9} = \dfrac{1}{3}$
Since -3 and $\frac{1}{3}$ are negative reciprocals, adjacent sides are perpendicular and opposite sides are parallel.

23. Midpoint: $\left(\dfrac{-3+9}{2}, \dfrac{-2+0}{2}\right) = (3, -1)$
Slope of line through $(5, 6)$ and $(3, -1)$: $\dfrac{6-(-1)}{5-3} = \dfrac{7}{2}$

25. $\tan\theta = \dfrac{\text{rise}}{\text{run}} = \dfrac{10.0\,\text{ft}}{160\,\text{ft}} = 0.0625$
$\theta = 3.6^\circ$

27. Slope of line through $(-1,-1)$ and $(3,-5)$: $\dfrac{-1-(-5)}{-1-3} = \dfrac{4}{-4} = -1$

Slope of line through $(x, 2)$ and $(4,-6)$: $\dfrac{2+6}{x-4} = \dfrac{8}{x-4}$

Since the two slopes must be equal, we have:

$$\begin{aligned} \frac{8}{x-4} &= -1 \\ 8 &= -x+4 \qquad \text{multiplying both sides by } x-4 \\ x &= -4 \end{aligned}$$

1.3 The Straight Line

1. Since $(x_1, y_1) = (-7, 2)$ and $m = 1/2$, we get

$$\begin{aligned} y-2 &= \tfrac{1}{2}(x+7) \qquad y-y_1 = m(x-x_1) \\ 2y-4 &= x+7 \qquad \text{clearing fractions} \\ 0 &= x-2y+7+4 \\ x-2y+11 &= 0 \end{aligned}$$

3.

$$\begin{aligned} y-y_1 &= m(x-x_1) \\ y+4 &= 3(x-3) \qquad (x_1,y_1)=(3,-4);\ m=3 \\ y+4 &= 3x-9 \\ 3x-y-13 &= 0 \end{aligned}$$

5.

$$\begin{aligned} y-y_1 &= m(x-x_1) \\ y-0 &= -\tfrac{1}{3}(x-0) \qquad (x_1,y_1)=(0,0);\ m=-1/3 \\ 3y &= -x \\ x+3y &= 0 \end{aligned}$$

7. The line $y = 1 = 0x + 1$ has slope 0.

$$\begin{aligned} y-y_1 &= m(x-x_1) \\ y-0 &= 0(x+4) \qquad (x_1,y_1)=(-4,0);\ m=0 \\ y &= 0 \qquad x\text{-axis} \end{aligned}$$

9. First determine the slope using $m = \dfrac{y_2-y_1}{x_2-x_1}$ to get $m = \dfrac{4-(-6)}{-3-3} = \dfrac{10}{-6} = -\dfrac{5}{3}$.

Then let $(x_1, y_1) = (-3, 4)$ to get

$$\begin{aligned} y-4 &= -\tfrac{5}{3}(x+3) \qquad y-y_1 = m(x-x_1) \\ 3y-12 &= -5x-15 \qquad \text{multiplying by 3} \\ 5x+3y+3 &= 0 \end{aligned}$$

11. $m = \dfrac{-4-0}{9-5} = -1$

$$\begin{aligned} y-0 &= -1(x-5) \qquad \text{choosing } (x_1,y_1)=(5,0) \\ x+y-5 &= 0 \end{aligned}$$

13.

$$\begin{aligned} 6x+2y &= 5 \\ 2y &= -6x+5 \\ y &= -3x+\tfrac{5}{2} \qquad y=mx+b \end{aligned}$$

$m = -3$, y-intercept $= \frac{5}{2}$; see graph in answer section of book.

15. Since $2x = 3y$, $y = \frac{2}{3}x$. From the form $y = mx + b$, $m = \frac{2}{3}$ and $b = 0$. The line passes through the origin and has slope $\frac{2}{3}$. See graph in answer section of book.

17. $$\begin{aligned} 2y - 7 &= 0 \\ y &= 0x + \tfrac{7}{2} \qquad y = mx + b \end{aligned}$$
$m = 0$, y-intercept $= \frac{7}{2}$; see graph in answer section of book.

19. $$\begin{aligned} 2x - 3y &= 1 \\ -3y &= -2x + 1 \\ y &= \tfrac{2}{3}x - \tfrac{1}{3} \end{aligned} \qquad \begin{aligned} 4x - 6y + 3 &= 0 \\ -6y &= -4x - 3 \\ y &= \tfrac{4}{6}x + \tfrac{3}{6} \\ y &= \tfrac{2}{3}x + \tfrac{1}{2} \end{aligned}$$
From the form $y = mx + b$, $m = \frac{2}{3}$ in both cases, so that the lines are parallel.

21. $$\begin{aligned} 3x - 4y &= 1 \\ -4y &= -3x + 1 \\ y &= \tfrac{3}{4}x - \tfrac{1}{4} \end{aligned} \qquad \begin{aligned} 3y - 4x &= 3 \\ 3y &= 4x + 3 \\ y &= \tfrac{4}{3}x + 1 \end{aligned}$$
The lines are neither parallel nor perpendicular.

23. $$\begin{aligned} x + 3y &= 5 \\ 3y &= -x + 5 \\ y &= -\tfrac{1}{3}x + \tfrac{5}{3} \end{aligned} \qquad \begin{aligned} y - 3x - 2 &= 0 \\ y &= 3x + 2 \end{aligned}$$
The slopes are $-\frac{1}{3}$ and 3, respectively. Since the slopes are negative reciprocals, the lines are perpendicular.

25. $$\begin{aligned} 3x - 5y &= 6 \\ -5y &= -3x + 6 \\ y &= \tfrac{3}{5}x - \tfrac{6}{5} \end{aligned} \qquad \begin{aligned} 9x - 15y &= 4 \\ -15y &= -9x + 4 \\ y &= \tfrac{-9}{-15}x + \tfrac{4}{-15} \\ y &= \tfrac{3}{5}x - \tfrac{4}{15} \end{aligned}$$
From the form $y = mx + b$, the slope m is $\frac{3}{5}$ in both cases; so the lines are parallel.

27. $$\begin{aligned} 2y - 3x &= 6 \\ 2y &= 3x + 6 \\ y &= \tfrac{3}{2}x + 3 \end{aligned} \qquad \begin{aligned} 6y + 4x &= 5 \\ 6y &= -4x + 5 \\ y &= \tfrac{-4}{6}x + \tfrac{5}{6} \\ y &= -\tfrac{2}{3}x - \tfrac{5}{6} \end{aligned}$$
The respective slopes are $\frac{3}{2}$ and $-\frac{2}{3}$. Since the slopes are negative reciprocals, the lines are perpendicular.

29. $$\begin{aligned} 3y - 2x - 12 &= 0 \\ 3y &= 2x + 12 \\ y &= \tfrac{2}{3}x + 4 \end{aligned} \qquad \begin{aligned} 2x + 3y - 4 &= 0 \\ 3y &= -2x + 4 \\ y &= -\tfrac{2}{3}x + \tfrac{4}{3} \end{aligned}$$
The lines are neither parallel nor pependicular.

31.
$$\begin{aligned}
3x+4y &= 5 && \text{(given line)}\\
y &= -\tfrac{3}{4}x+\tfrac{5}{4} && \text{slope} = -\tfrac{3}{4}\\
y-y_1 &= m(x-x_1) && \text{point-slope form}\\
y-1 &= -\tfrac{3}{4}(x+2) && \text{point: } (-2,1)\\
4y-4 &= -3x-6\\
3x+4y+2 &= 0
\end{aligned}$$

33. To find the coordinates of the point of intersection, solve the equations simultaneously:

$$\begin{aligned}
2x - 4y &= 1\\
3x + 4y &= 4\\
\hline
5x &= 5 && \text{adding}\\
x &= 1
\end{aligned}$$

From the second equation, $3(1)+4y=4$, and $y=\frac{1}{4}$. So the point of intersection is $(1,\frac{1}{4})$.
From the equation $5x+7y+3=0$, we get

$$\begin{aligned}
7y &= -5x-3\\
y &= -\tfrac{5}{7}x-\tfrac{3}{7} && \text{slope} = -5/7
\end{aligned}$$

Thus $(x_1,y_1)=(1,\frac{1}{4})$ and $m=-\frac{5}{7}$. The desired line is $y-\frac{1}{4}=-\frac{5}{7}(x-1)$. To clear fractions, we multiply both sides by 28:

$$\begin{aligned}
28y-7 &= -20(x-1)\\
28y-7 &= -20x+20\\
20x+28y-27 &= 0
\end{aligned}$$

35. See graph in answer section of book.

37. From $F=kx$, we get $3=k\cdot\frac{1}{2}$. Thus $k=6$ and $F=6x$.

39.
$$\begin{aligned}
F &= mC+b\\
212 &= m(100)+b && F=212,\ C=100\\
32 &= m(0)+b && F=32,\ C=0\\
\hline
b &= 32 && \text{second equation}\\
212 &= m(100)+32 && \text{substituting into first equation}\\
\hline
m &= \frac{180}{100}=\frac{9}{5}
\end{aligned}$$

Solution: $F=\frac{9}{5}C+32$

41.
$$\begin{aligned}
R &= aT+b\\
51 &= a\cdot 100+b && R=51,\ T=100\\
54 &= a\cdot 400+b && R=54,\ T=400\\
\hline
-3 &= -300a && \text{subtracting}\\
a &= \frac{-3}{-300}=0.01
\end{aligned}$$

From the first equation, $51=a\cdot 100+b$, we get

$$\begin{aligned}
51 &= (0.01)(100)+b && (a=0.01)\\
b &= 50
\end{aligned}$$

So the formula $R=aT+b$ becomes $R=0.01T+50$.

1.4 Curve Sketching

3. Intercepts. If $x = 0$, then $y = -9$. If $y = 0$, then

$$\begin{aligned} 0 &= x^2 - 9 \\ x^2 &= 9 \qquad \text{solving for } x \\ x &= \pm 3 \qquad x = 3 \text{ and } x = -3. \end{aligned}$$

Symmetry. If x is replaced by $-x$, we get $y = (-x)^2 - 9$, which reduces to the given equation $y = x^2 - 9$. The graph is therefore symmetric with respect to the y-axis. There is no other type of symmetry.

Asymptotes. Since the equation is not in the form of a quotient with a variable in the denominator, there are no asymptotes.

Extent. y is defined for all x.

Graph.

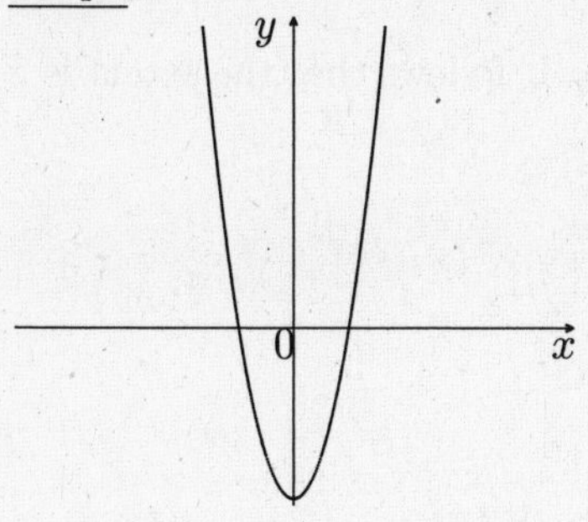

5. Intercepts. If $x = 0$, then $y = 1$. If $y = 0$, then

$$\begin{aligned} 0 &= 1 - x^2 \\ x^2 &= 1 \qquad \text{solving for } x \\ x &= \pm 1 \qquad x = 1 \text{ and } x = -1. \end{aligned}$$

Symmetry. If x is replaced by $-x$, we get $y = 1 - (-x)^2$, which reduces to the given equation $y = 1 - x^2$. The graph is therefore symmetric with respect to the y-axis. There is no other type of symmetry.

Asymptotes. Since the equation is not in the form of a quotient with a variable in the denominator, there are no asymptotes.

Extent. y is defined for all x.

Graph.

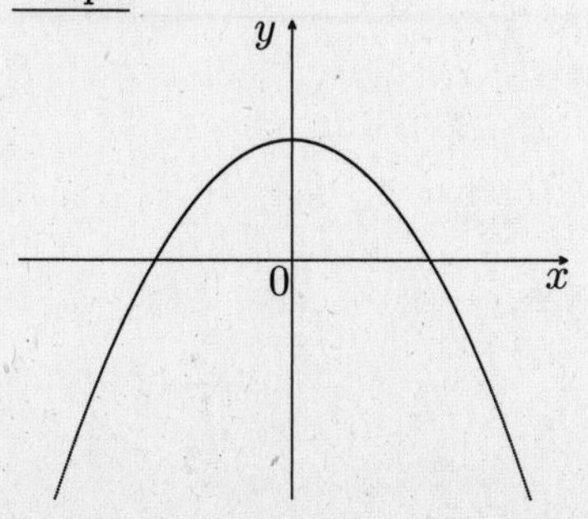

7. Intercepts. If $x = 0$, then $y = 0$, and if $y = 0$, then $x = 0$. So the only intercept is the origin.
   Symmetry. If we replace x by $-x$, we get $y^2 = -x$, which does not reduce to the given equation. So there is no symmetry with respect to the y-axis.
   If y is replaced by $-y$, we get $(-y)^2 = x$, which reduces to $y^2 = x$, the given equation. It follows that the graph is symmetric with respect to the x-axis.
   To check for symmetry with respect to the origin, we replace x by $-x$ and y by $-y$: $(-y)^2 = -x$. The resulting equation, $y^2 = -x$, does not reduce to the given equation. So there is no symmetry with respect to the origin.
   Asymptotes. Since the equation is not in the form of a fraction with a variable in the denominator, there are no asymptotes.
   Extent. Solving the equation for y in terms of x, we get

$$y = \pm\sqrt{x}.$$

   Note that to avoid imaginary values, x cannot be negative. It follows that the extent is $x \geq 0$.
   Graph.

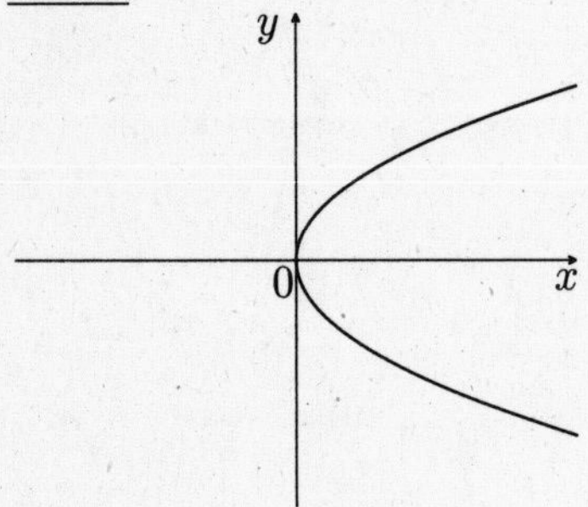

9. Intercepts. If $x = 0$, then $y = \pm 1$. If $y = 0$, then $x = -1$.
   Symmetry. If we replace x by $-x$ we get $y^2 = -x + 1$, which does not reduce to the given equation. So there is no symmetry with respect to the y-axis.
   If y is replaced by $-y$, we get $(-y)^2 = x + 1$, which reduces to $y^2 = x + 1$, the given equation. It follows that the graph is symmetric with respect to the x-axis.
   The graph is not symmetric with respect to the origin.
   Asymptotes. Since the equation is not in the form of a fraction with a variable in the denominator, there are no asymptotes.
   Extent. Solving the equation for y, we get $y = \pm\sqrt{x+1}$. To avoid imaginary values, we must have $x + 1 \geq 0$ or $x \geq -1$. Therefore the extent is $x \geq -1$.
   Graph.

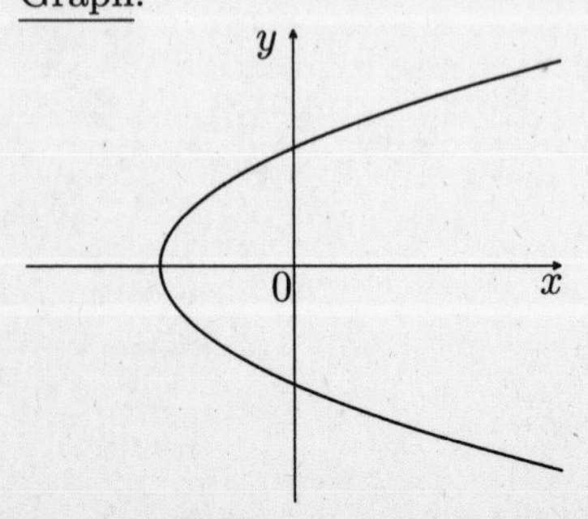

11. Intercepts. If $x = 0$, then $y = (0-3)(0+5) = -15$. If $y = 0$, then

$$0 = (x-3)(x+5)$$
$$x - 3 = 0 \qquad x + 5 = 0$$
$$x = 3 \qquad x = -5.$$

Symmetry. If x is replaced by $-x$, we get $y = (-x-3)(-x+5)$, which does not reduce to the given equation. So there is no symmetry with respect to the y-axis. Similarly, there is no other type of symmetry.

Asymptotes. Since the equation is not in the form of a quotient with a variable in the denominator, there are no asymptotes.

Extent. y is defined for all x.

Graph.

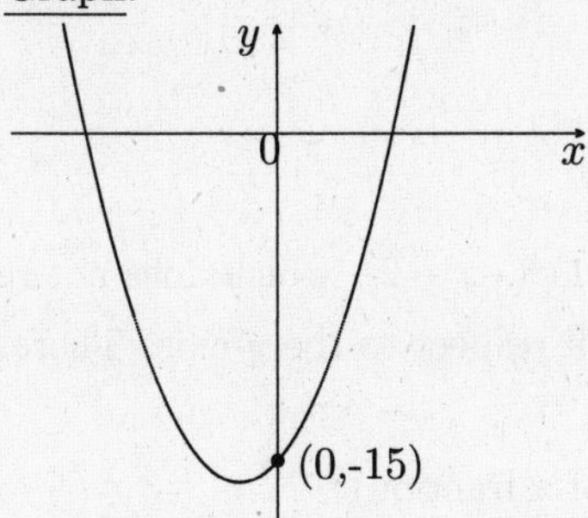

13. Intercepts. If $x = 0$, then $y = 0$. If $y = 0$, then

$$0 = x(x+3)(x-2)$$
$$x = 0, -3, 2.$$

Symmetry. If x is replaced by $-x$, we get $y = -x(-x+3)(-x-2)$, which does not reduce to the given equation. So the graph is not symmetric with respect to the y-axis. There is no other type of symmetry.

Asymptotes. Since the equation is not in the form of a quotient with a variable in the denominator, there are no asymptotes.

Extent. y is defined for all x.

Graph.

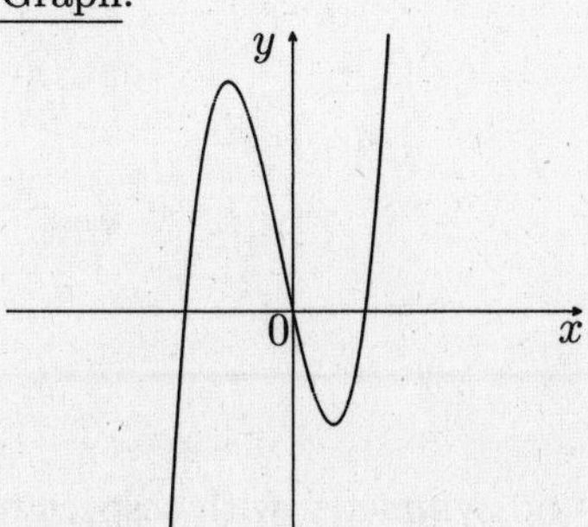

15. Intercepts. If $x = 0$, $y = 0$; if $y = 0$, then

$$x(x-1)(x-2)^2 = 0$$
$$x = 0, 1, 2.$$

Symmetry. If x is replaced by $-x$, we get $y = -x(-x-1)(-x-2)^2$, which does not reduce to the given equation. So there is no symmetry with respect to the y-axis. Similarly, there is no other type of symmetry.

Asymptotes. None (the equation does not have the form of a fraction).

Extent. y is defined for all x.

Graph.

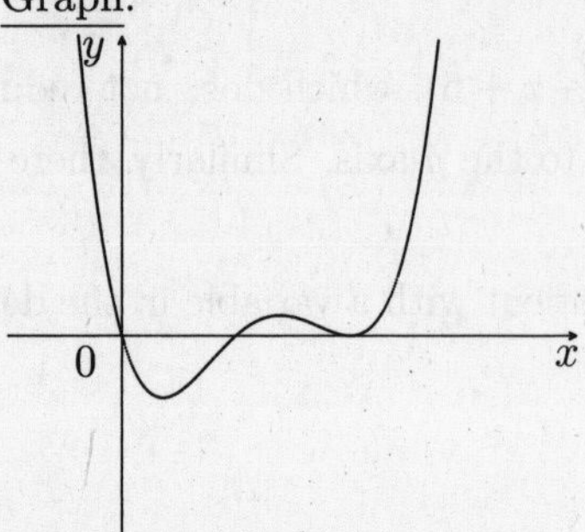

17. Intercepts. If $x = 0$, then $y = 0$. If $y = 0$, then

$$\begin{aligned} 0 &= x(x-1)^2(x-2) \\ x &= 0, 1, 2. \end{aligned}$$

Symmetry. If x is replaced with $-x$, we get $y = -x(-x-1)^2(-x-2)$, which does not reduce to the given equation. Therefore there is no symmetry with respect to the y-axis. There is no other type of symmetry.

Asymptotes. None (the equation does not have the form of a fraction).

Extent. y is defined for all x.

Graph.

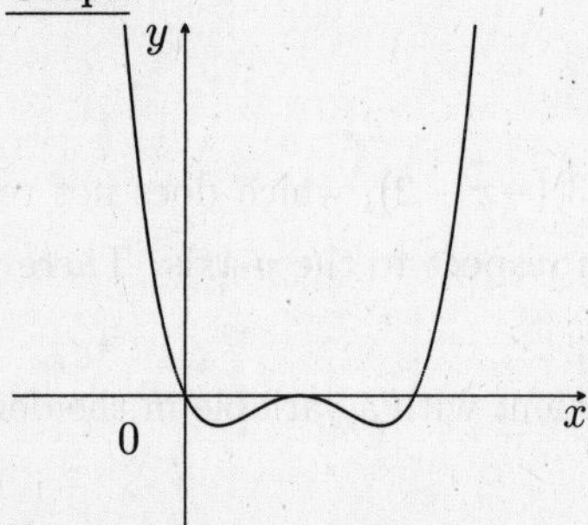

19. Intercepts. If $x = 0$, $y = 1$; if $y = 0$, we have

$$0 = \frac{2}{x+2}.$$

This equation has no solution.

Symmetry. Replacing x by $-x$, we get

$$y = \frac{2}{-x+2}$$

which does not reduce to the given equation. So there is no symmetry with respect to the y-axis. Similarly, there is no other type of symmetry.

Asymptotes. Setting the denominator equal to 0, we get

$$x + 2 = 0 \text{ or } x = -2.$$

It follows that $x = -2$ is a vertical asymptote. Also, as x gets large, y approaches 0. So the x-axis is a horizontal asymptote.

Extent. To avoid division by 0, x cannot be equal to -2. So the extent is all x except $x = -2$.

Graph.

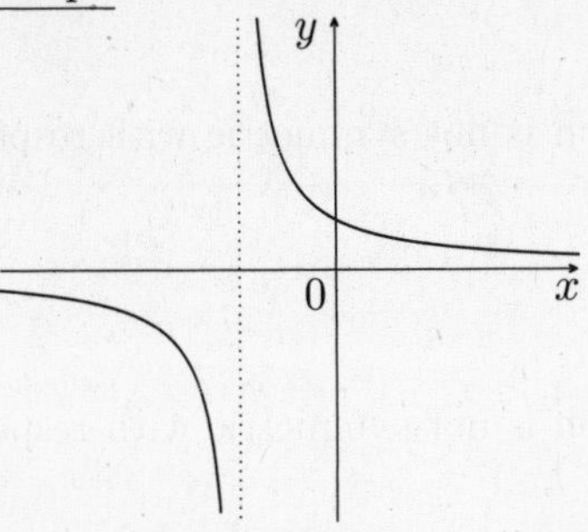

21. Intercepts. If $x = 0$, then $y = 2$. If $y = 0$, then

$$\frac{2}{(x-1)^2} = 0.$$

This equation has no solution.

Symmetry. Replacing x by $-x$, we get

$$y = \frac{2}{(-x-1)^2}$$

which does not reduce to the given equation. There are no other types of symmetry.

Asymptotes. Setting the denominator equal to 0 gives

$$(x-1)^2 = 0 \text{ or } x = 1.$$

It follows that $x = 1$ is a vertical asymptote. Also, as x gets large, y approaches 0. So the x-axis is a horizontal asymptote.

Extent. To avoid division by 0, x cannot be equal to 1. So the extent is the set of all x except $x = 1$.

Graph.

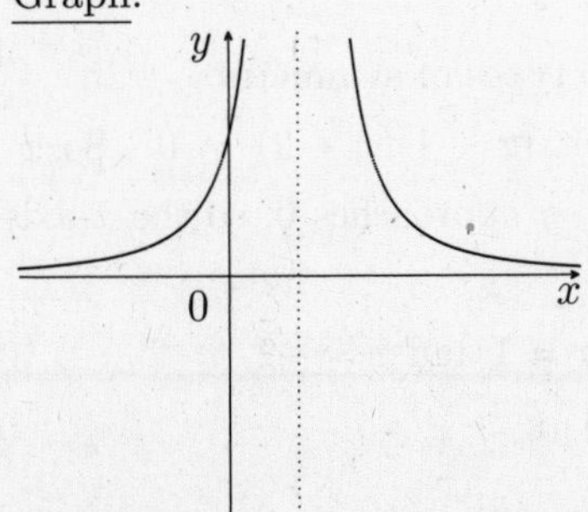

23. Intercepts. If $x = 0$, then $y = 0$. If $y = 0$, then

$$0 = \frac{x^2}{x-1}.$$

The only solution is $x = 0$.

Symmetry. Replacing x by $-x$ yields

$$y = \frac{(-x)^2}{-x-1} = \frac{x^2}{-x-1}$$

which is not the same as the given equation. So the graph is not symmetric with respect to the y-axis. Replacing y by $-y$, we have

$$-y = \frac{x^2}{x-1}$$

which does not reduce to the given equation. So the graph is not symmetric with respect to the x-axis.

Similarly, there is no symmetry with respect to the origin.

Asymptotes. Setting the denominator equal to 0, we get $x - 1 = 0$, or $x = 1$. So $x = 1$ is a vertical asymptote. There are no horizontal asymptotes.

(Observation: for very large x the 1 in the denominator becomes insignificant. So the graph gets ever closer to $y = \frac{x^2}{x} = x$; the line $y = x$ is a slant asymptote.)

Extent. To avoid division by 0, x cannot be equal to 1. So the extent is all x except $x = 1$.

Graph.

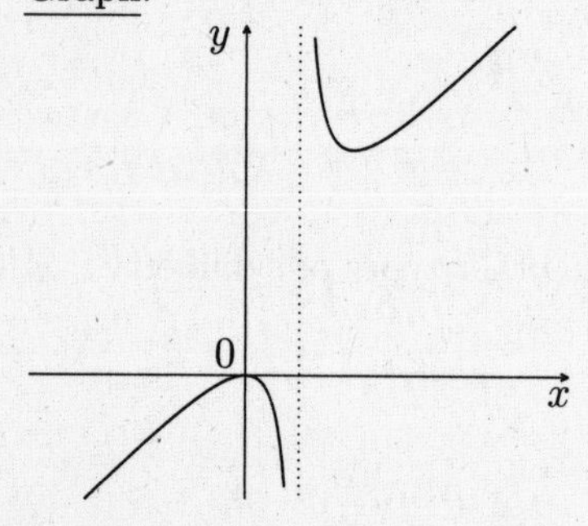

25. Intercepts. If $x = 0$, then $y = -1/2$. If $y = 0$, then

$$0 = \frac{x+1}{(x-1)(x+2)}.$$

The only solution is $x = -1$.

Symmetry. Replacing x by $-x$ yields

$$y = \frac{-x+1}{(-x-1)(-x+2)}$$

which is not the same as the given equation. There are no types of symmetry.

Asymptotes. Setting the denominator equal to 0, we get $(x-1)(x+2) = 0$. So $x = 1$ and $x = -2$ are the vertical asymptotes. As x gets large, y approaches 0, so the x-axis is a horizontal asymptote.

Extent. To avoid division by 0, the extent is all x except $x = 1$ and $x = -2$.

Graph.

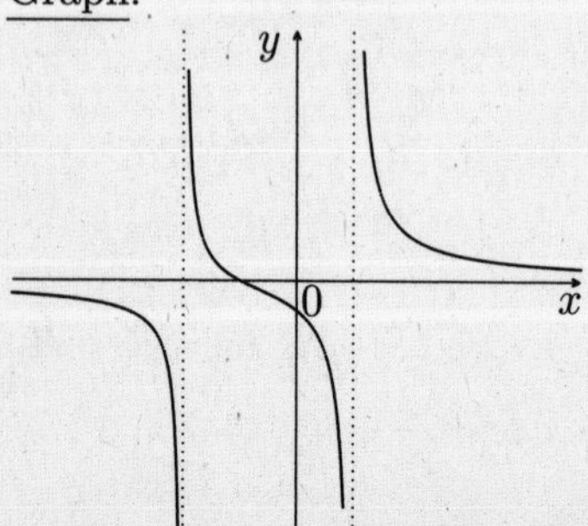

27. Intercepts. If $x = 0$, then $y = 4$. If $y = 0$, then

$$\begin{aligned} \frac{x^2-4}{x^2-1} &= 0 \\ x^2 - 4 &= 0 && \text{multiplying by } x^2 - 1 \\ x &= \pm 2. && \text{solution} \end{aligned}$$

Symmetry. Replacing x by $-x$ reduces to the given equation. So there is symmetry with respect to the y-axis. There is no other type of symmetry.
Asymptotes. Vertical: setting the denominator equal to 0, we have

$$x^2 - 1 = 0 \text{ or } x = \pm 1.$$

Horizontal: dividing numerator and denominator by x^2, the equation becomes

$$y = \frac{1 - \frac{4}{x^2}}{1 - \frac{1}{x^2}}.$$

As x gets large, y approaches 1. So $y = 1$ is a horizontal asymptote.
Extent. All x except $x = \pm 1$ (to avoid division by 0).
Graph.

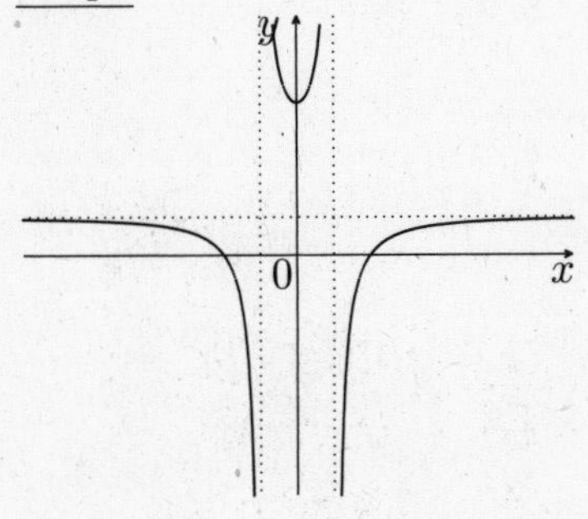

29. Intercepts. If $x = 0$, then $y^2 = \frac{-4}{-1} = 4$, or $y = \pm 2$. If $y = 0$, then

$$0 = \frac{x^2-4}{x^2-1}$$

which is possible only if $x^2 - 4 = 0$, or $x = \pm 2$.
Symmetry. The even powers on x and y tell us that if x is replaced by $-x$ and y is replaced by $-y$, the resulting equation will reduce to the given equation. The graph is therefore symmetric with respect to both axes and the origin.
Asymptotes. Vertical: setting the denominator equal to 0, we get

$$x^2 - 1 = 0 \text{ or } x = \pm 1.$$

Horizontal: dividing numerator and denominator by x^2, we get

$$y^2 = \frac{1 - \frac{4}{x^2}}{1 - \frac{1}{x^2}}.$$

The right side approaches 1 as x gets large. Thus y^2 approaches 1, so that $y = \pm 1$ are the horizontal asymptotes.

Extent. From

$$y = \pm\sqrt{\frac{x^2-4}{x^2-1}}$$

we conclude that

$$\frac{x^2-4}{x^2-1} = \frac{(x-2)(x+2)}{(x-1)(x+1)} \geq 0.$$

Since the signs change only at $x = 2, -2, 1,$ and -1, we need to use arbitrary "test values" between these points. The results are summarized in the following chart.

	test values	$x-2$	$x-1$	$x+1$	$x+2$	$\frac{(x-2)(x+2)}{(x-1)(x+1)}$
$x > 2$	3	+	+	+	+	+
$1 < x < 2$	3/2	−	+	+	+	−
$-1 < x < 1$	0	−	−	+	+	+
$-2 < x < -1$	−3/2	−	−	−	+	−
$x < -2$	−3	−	−	−	−	+

Note that the fraction is positive only when $x > 2$, $-1 < x < 1$ and $x < -2$. Since $y = 0$ when $x = \pm 2$, the extent is $x \geq 2$, $-1 < x < 1$, $x \leq -2$.

Graph.

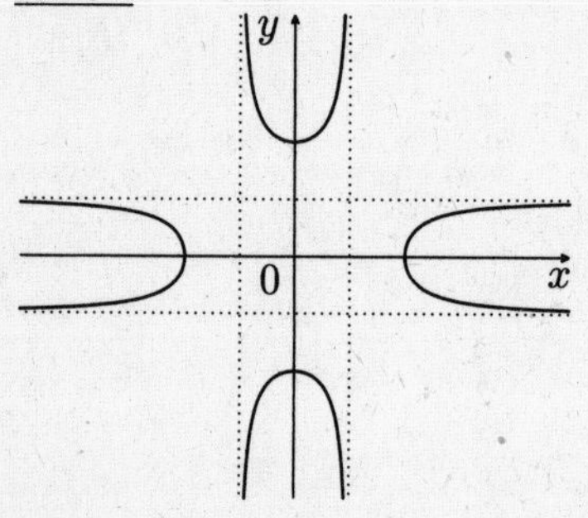

31. Intercepts. If $x = 0$, $y^2 = (-3)(5) = -15$, or $y = \pm\sqrt{15}\,\mathrm{j}$, which is a pure imaginary number. If $y = 0$,

$$\begin{aligned}(x-3)(x+5) &= 0\\ x &= 3, -5.\end{aligned}$$

Symmetry. Replacing y by $-y$, we get $(-y)^2 = (x-3)(x+5)$, which reduces to the given equation. Hence the graph is symmetric with respect to the x-axis.

Asymptotes. None (no fractions).

Extent. From $y = \pm\sqrt{(x-3)(x+5)}$, we conclude that $(x-3)(x+5) \geq 0$. If $x \geq 3$, $(x-3)(x+5) \geq 0$. If $x \leq -5$, $(x-3)(x+5) \geq 0$, since both factors are negative (or zero). If $-5 < x < 3$, $(x-3)(x+5) < 0$. [For example, if $x = 0$, we get $(-3)(5) = -15$.] These observations are summarized in the following chart.

	test values	$x-3$	$x+5$	$(x-3)(x+5)$
$x > 3$	4	+	+	+
$-5 < x < 3$	0	−	+	−
$x < -5$	−6	−	−	+

Extent: $x \leq -5$, $x \geq 3$

Graph.

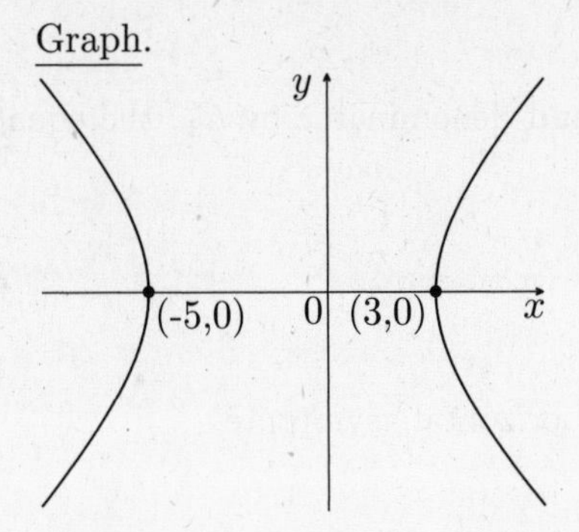

33. Intercepts. If $x = 0,\ y = 0$; if $y = 0,\ x = 0$.
Symmetry. Replacing y by $-y$ leaves the equation unchanged. So there is symmetry with respect to the x-axis. There is no other type of symmetry.
Asymptotes. Vertical: setting the denominator equal to 0, we get

$$(x-3)(x-2) = 0 \text{ or } x = 3, 2.$$

Horizontal: as x gets large, y approaches 0 (x-axis).
Extent. From

$$y = \pm\sqrt{\frac{x}{(x-3)(x-2)}}$$

we conclude that

$$\frac{x}{(x-3)(x-3)} \geq 0.$$

Since signs change only at $x = 0, 2$ and 3, we need to use "test values"between these points. The results are summarized in the following chart.

	test values	x	$x-2$	$x-3$	$\frac{x}{(x-3)(x-2)}$
$x < 0$	-1	$-$	$-$	$-$	$-$
$0 < x < 2$	1	$+$	$-$	$-$	$+$
$2 < x < 3$	5/2	$+$	$+$	$-$	$-$
$x > 3$	4	$+$	$+$	$+$	$+$

So the inequality is satisfied for $0 < x < 2$ and $x > 3$. In addition, $y = 0$ when $x = 0$. So the extent is $0 \leq x < 2$ and $x > 3$.
Graph.

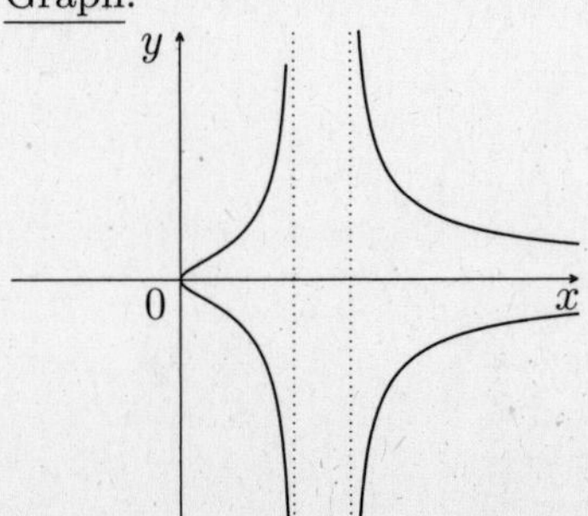

35. $C = \dfrac{10^{-2}C_1}{C_1 + 10^{-2}}, \; C_1 \geq 0$

The only intercept is the origin. Dividing numerator and denominator by C_1, the equation becomes

$$C = \frac{10^{-2}}{1 + 10^{-2}/C_1}.$$

As C_1 gets large, C approaches 10^{-2}; so $C = 10^{-2}$ is a horizontal asymptote.

See graph in answer section of book.

37. Intercepts. If $t = 0$, $S = 0$; if $S = 0$, we get

$$\begin{aligned} 0 &= 60t - 5t^2 \\ 0 &= 5t(12 - t) \end{aligned}$$

or $t = 0, 12$.

Symmetry. None.

Asymptotes. None.

Extent. $t \geq 0$ by assumption.

Graph. See graph in answer section of book.

39. Extent $L \geq 0$.

See graph in answer section of book.

1.5 Discussion of Curves with Graphing Utilities

Graphs are from the answer section of the book.

1. If $y = 0$, then

$$x^2(x-1)(x-2) = 0.$$

Setting each factor equal to 0, we get

$$x = 0, 1, 2.$$

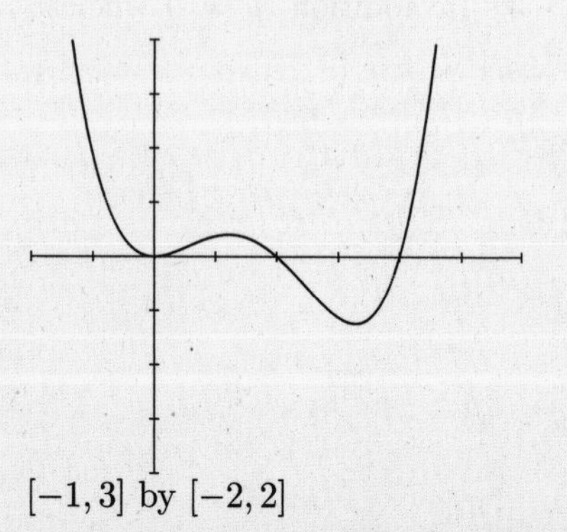

$[-1, 3]$ by $[-2, 2]$

5. $$\begin{aligned} x^4 - 2x^3 &= 0 \\ x^3(x-2) &= 0 \\ x &= 0, 2 \end{aligned}$$

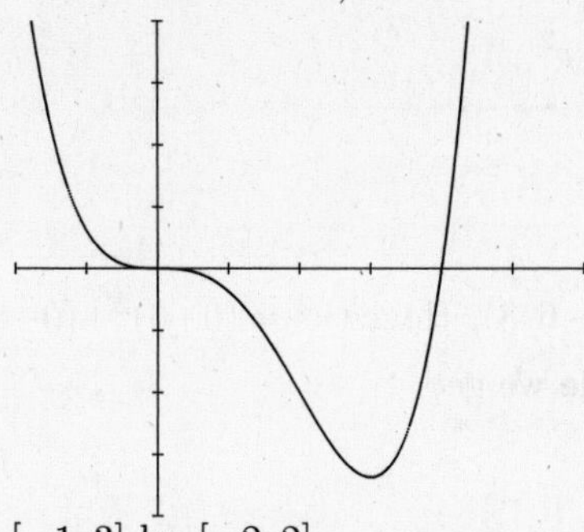

$[-1, 3]$ by $[-2, 2]$

9. Domain: $x \geq 0$ (to avoid imaginary values).
Vertical asymptotes: None. (The denominator is always positive, that is, $1 + \sqrt{x} \neq 0$.)

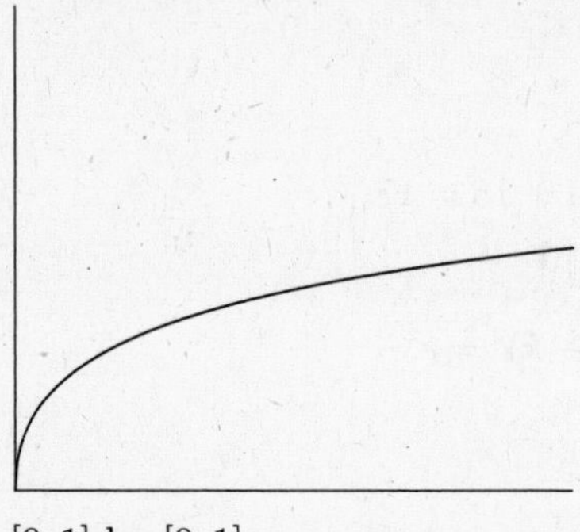

$[0, 1]$ by $[0, 1]$

13. To find the vertical asymptotes, we set the denominator equal to 0:
$$\begin{aligned} 2x^2 - 3 &= 0 \\ 2x^2 &= 3 \\ x^2 &= \frac{3}{2} \cdot \frac{2}{2} = \frac{6}{4} \\ x &= \pm\frac{\sqrt{6}}{2}. \end{aligned}$$
Domain: y is defined for all x except $x = \pm\frac{\sqrt{6}}{2}$.

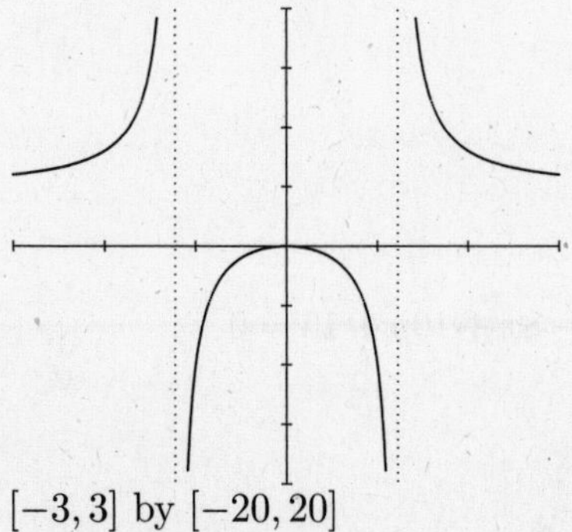

$[-3, 3]$ by $[-20, 20]$

17. See graph in answer section of book.

21. See graph in answer section of book.

1.7 The Circle

1. Since $(h, k) = (0, 0)$ and $r = 4$, we get from the form

$$(x-h)^2 + (y-k)^2 = r^2$$

the equation

$$x^2 + y^2 = 16.$$

3. The radius of the circle is the distance from the origin to $(-6, 8)$. Hence $r^2 = (0+6)^2 + (0-8)^2 = 100$. From the standard form of the equation of the circle we get

$$\begin{aligned} (x-0)^2 + (y-0)^2 &= 100 \qquad \text{center: } (0,0) \\ x^2 + y^2 &= 100. \end{aligned}$$

5.
$$\begin{aligned} (x-h)^2 + (y-k)^2 &= r^2 \\ (x+2)^2 + (y-5)^2 &= 1^2 \\ x^2 + y^2 + 4x - 10y + 28 &= 0 \end{aligned}$$

7. The radius is the distance from $(-1, -4)$ to the origin:

$$r^2 = (-1-0)^2 + (-4-0)^2 = 1 + 16 = 17.$$

Hence,

$$\begin{aligned} (x+1)^2 + (y+4)^2 &= 17 \qquad (x-h)^2 + (y-k)^2 = r^2 \\ x^2 + 2x + 1 + y^2 + 8y + 16 &= 17 \\ x^2 + y^2 + 2x + 8y &= 0. \end{aligned}$$

9. Diameter: distance from $(-2, -6)$ to $(1, 5)$. Hence

$$r = \frac{1}{2}\sqrt{(-2-1)^2 + (-6-5)^2} = \frac{1}{2}\sqrt{9+121} = \frac{1}{2}\sqrt{130}$$

and thus,

$$r^2 = \frac{1}{4}(130) = \frac{65}{2}.$$

Center: midpoint of the line segment, whose coordinates are

$$\left(\frac{-2+1}{2}, \frac{-6+5}{2}\right) = \left(-\frac{1}{2}, -\frac{1}{2}\right).$$

Thus

$$\begin{aligned} (x+\tfrac{1}{2})^2 + (y+\tfrac{1}{2})^2 &= \tfrac{65}{2} \\ x^2 + x + \tfrac{1}{4} + y^2 + y + \tfrac{1}{4} &= \tfrac{65}{2} \\ x^2 + y^2 + x + y - 32 &= 0. \end{aligned}$$

11. Since $r = 5$, we get

$$(x-4)^2 + (y+5)^2 = 25 \text{ or } x^2 + y^2 - 8x + 10y + 16 = 0$$

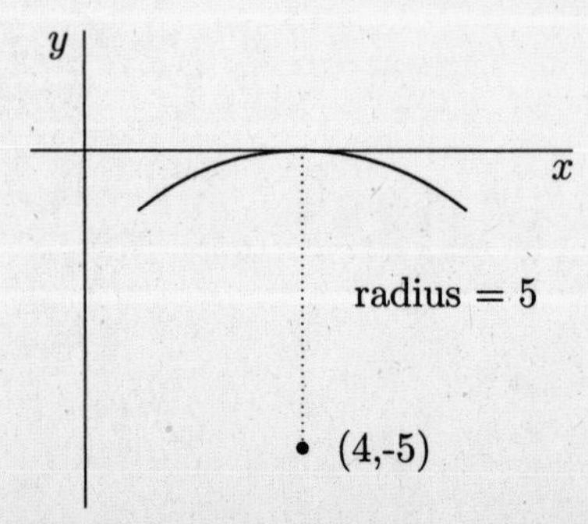

13. From the diagram, $r^2 = 1^2 + 1^2 = 2$; so $x^2 + y^2 = 2$.

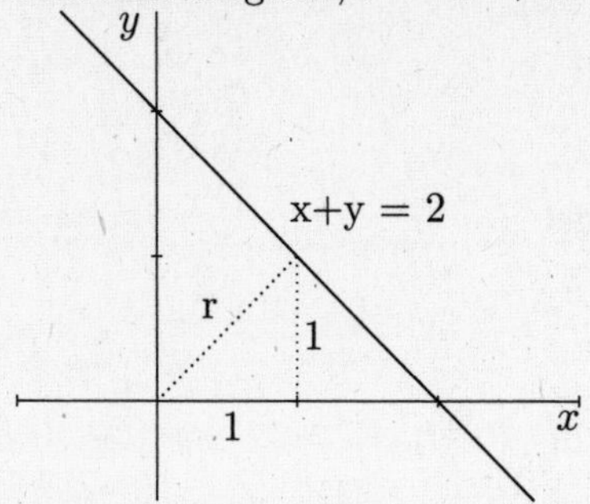

15.
$$\begin{aligned} x^2 + y^2 - 2x - 2y - 2 &= 0 \\ x^2 - 2x + y^2 - 2y &= 2 \end{aligned}$$

We now add to each side the square of one-half the coefficient of x:

$$\begin{aligned} \left[\tfrac{1}{2}(-2)\right]^2 &= 1 \\ x^2 - 2x + \underline{1} + y^2 - 2y &= 2 + \underline{1}. \end{aligned}$$

Similarly, we add 1 (the square of one-half the coefficient of y):

$$\begin{aligned} (x^2 - 2x + 1) + (y^2 - 2y + \underline{1}) &= 2 + 1 + \underline{1} \\ (x-1)^2 + (y-1)^2 &= 4. \end{aligned}$$

Center: $(h, k) = (1, 1)$; radius: $\sqrt{4} = 2$.

17.
$$\begin{aligned} x^2 + y^2 + 4x - 8y + 4 &= 0 \\ x^2 + 4x \quad + y^2 - 8y &= -4 \end{aligned}$$

Since

$$\left(\frac{1}{2} \cdot 4\right)^2 = 4 \text{ and } \left[\frac{1}{2}(-8)\right]^2 = 16$$

we get

$$\begin{aligned} (x^2 + 4x + \underline{4}) + (y^2 - 8y + \underline{16}) &= -4 + \underline{4} + \underline{16} \\ (x+2)^2 + (y-4)^2 &= 16. \end{aligned}$$

The equation can be written

$$[x - (-2)]^2 + (y - 4)^2 = 4^2.$$

It follows that

$$(h, k) = (-2, 4) \text{ and } r = 4.$$

19.
$$\begin{aligned} x^2 + y^2 - 4x + y + \tfrac{9}{4} &= 0 \\ x^2 - 4x + y^2 + y &= -\tfrac{9}{4} \end{aligned}$$

We add to each side the square of one-half the coefficient of x: $\left[\frac{1}{2}(-4)\right]^2 = 4$. This gives

$$x^2 - 4x + \underline{4} + y^2 + y = -\frac{9}{4} + \underline{4}.$$

Similarly, we add the square of one-half the coefficient of y: $\left[\frac{1}{2} \cdot 1\right]^2 = \frac{1}{4}$. This gives

$$\begin{aligned} x^2 - 4x + 4 + y^2 + y + \underline{\tfrac{1}{4}} &= -\tfrac{9}{4} + 4 + \underline{\tfrac{1}{4}} \\ (x-2)^2 + (y + \tfrac{1}{2})^2 &= -\tfrac{8}{4} + 4 \\ (x-2)^2 + (y + \tfrac{1}{2})^2 &= 2. \end{aligned}$$

Center: $(2, -\frac{1}{2})$; radius: $\sqrt{2}$.

21.
$$\begin{aligned} 4x^2+4y^2-8x-12y+9 &= 0 \\ x^2+y^2-2x-3y+\tfrac{9}{4} &= 0 \qquad \text{dividing by 4} \\ x^2-2x \;+y^2-3y &= -\tfrac{9}{4} \end{aligned}$$

Add to each side: $\left[\frac{1}{2}(-2)\right]^2 = 1$ and $\left[\frac{1}{2}(-3)\right]^2 = \frac{9}{4}$ to get

$$\begin{aligned} (x^2-2x+1)+(y^2-3y+\tfrac{9}{4}) &= -\tfrac{9}{4}+1+\tfrac{9}{4} \\ (x-1)^2+(y-\tfrac{3}{2})^2 &= 1. \end{aligned}$$

Center: $(1, \frac{3}{2})$; radius: 1.

23.
$$\begin{aligned} x^2+y^2+4x-2y-4 &= 0 \\ x^2+4x \;+y^2-2y &= 4 \end{aligned}$$

Note that

$$\left(\frac{1}{2}\cdot 4\right)^2 = 4 \text{ and } \left[\frac{1}{2}(-2)\right]^2 = 1.$$

Adding 4 and 1, respectively, we get

$$\begin{aligned} (x^2+4x+4)+(y^2-2y+1) &= 4+4+1 \\ (x+2)^2+(y-1)^2 &= 9. \end{aligned}$$

The equation can be written

$$[x-(-2)]^2+(y-1)^2 = 3^2.$$

So the center is $(-2, 1)$ and the radius is 3.

25.
$$\begin{aligned} x^2+y^2-x-2y+\tfrac{1}{4} &= 0 \\ x^2-x \;+\; y^2-2y &= -\tfrac{1}{4} \end{aligned}$$

Add to each side: $\left[\frac{1}{2}(-1)\right]^2 = \frac{1}{4}$ and $\left[\frac{1}{2}(-2)\right]^2 = 1$ to get

$$\begin{aligned} (x^2-x+\tfrac{1}{4})+(y^2-2y+1) &= -\tfrac{1}{4}+\tfrac{1}{4}+1 \\ (x-\tfrac{1}{2})^2+(y-1)^2 &= 1. \end{aligned}$$

Center: $(\frac{1}{2}, 1)$; radius: 1.

27.
$$\begin{aligned} x^2+y^2-4x+y+\tfrac{9}{4} &= 0 \\ x^2-4x \;+y^2+y &= -\tfrac{9}{4} \end{aligned}$$

Note that

$$\left[\frac{1}{2}(-4)\right]^2 = 4 \text{ and } \left(\frac{1}{2}\cdot 1\right)^2 = \frac{1}{4}.$$

Adding 4 and $\frac{1}{4}$, respectively, we get

$$\begin{aligned} (x^2-4x+4)+(y^2+y+\tfrac{1}{4}) &= -\tfrac{9}{4}+4+\tfrac{1}{4} \\ (x-2)^2+(y+\tfrac{1}{2})^2 &= 2. \end{aligned}$$

The equation can be written

$$(x-2)^2+\left[y-\left(-\frac{1}{2}\right)\right]^2 = (\sqrt{2})^2.$$

Center: $(2, -\frac{1}{2})$; radius: $\sqrt{2}$.

29.
$$\begin{aligned} 4x^2+4y^2+12x+16y+5 &= 0 \\ x^2+y^2+3x+4y+\tfrac{5}{4} &= 0 \\ x^2+3x \quad +y^2+4y &= -\tfrac{5}{4} \end{aligned}$$

Add to each side: $\left[\frac{1}{2}\cdot 3\right]^2 = \frac{9}{4}$ and $\left(\frac{1}{2}\cdot 4\right)^2 = 4$ to get

$$\begin{aligned} x^2+3x+\tfrac{9}{4}+y^2+4y+4 &= -\tfrac{5}{4}+\tfrac{9}{4}+4 \\ (x+\tfrac{3}{2})^2+(y+2)^2 &= 5. \end{aligned}$$

Center: $(-\frac{3}{2},-2)$; radius: $\sqrt{5}$.

31.
$$\begin{aligned} 4x^2+4y^2-20x-4y+26 &= 0 \\ x^2+y^2-5x-y+\tfrac{26}{4} &= 0 \qquad \text{dividing by 4} \\ x^2-5x \quad +y^2-y &= -\tfrac{26}{4} \\ x^2-5x+\tfrac{25}{4}+y^2-y+\tfrac{1}{4} &= -\tfrac{26}{4}+\tfrac{25}{4}+\tfrac{1}{4} \\ (x-\tfrac{5}{2})^2+(y-\tfrac{1}{2})^2 &= 0 \end{aligned}$$

Locus is the single point $(\frac{5}{2},\frac{1}{2})$.

33.
$$\begin{aligned} x^2+y^2-6x+8y+25 &= 0 \\ x^2-6x \quad +y^2+8y &= -25 \\ (x^2-6x+9)+(y^2+8y+16) &= -25+9+16 \\ (x-3)^2+(y+4)^2 &= 0 \end{aligned}$$

Locus is the single point $(3,-4)$.

35.
$$\begin{aligned} x^2+y^2-6x-8y+30 &= 0 \\ x^2-6x \quad +y^2-8y &= -30 \\ x^2-6x+9+y^2-8y+16 &= -30+9+16 \\ (x-3)^2+(y-4)^2 &= -5 \qquad \text{(imaginary circle)} \end{aligned}$$

37.
$$\begin{aligned} x^2+y^2-x+4y+\tfrac{17}{4} &= 0 \\ x^2-x \quad +y^2+4y &= -\tfrac{17}{4} \end{aligned}$$

We add to each side $\left[\frac{1}{2}(-1)\right]^2$ and $\left[\frac{1}{2}(4)\right]^2$:

$$\begin{aligned} x^2-x+\tfrac{1}{4}+y^2+4y+4 &= -\tfrac{17}{4}+\tfrac{1}{4}+4 \\ \left(x-\tfrac{1}{2}\right)^2+(y+2)^2 &= 0 \qquad \text{(point circle)} \end{aligned}$$

39. $x^2+y^2=(2.00)^2=4.00$; $x^2+y^2=(3.40)^2=11.6$

41. The radius is $22,300+4000=$26,300 mi.

43. $(h,k)=(0,0)$ and $r=\frac{3}{2}$ ft: $x^2+y^2=\frac{9}{4}$ and $y=\sqrt{\frac{9}{4}-x^2}$

1.8 The Parabola

1. Since the focus is on the x-axis, the form is $y^2=4px$. Since the focus is at $(3,0)$, $p=3$ (positive). Thus $y^2=4(3)x$, or $y^2=12x$.

3. Since the focus is on the y-axis, the form is $x^2=4py$. Since the focus is at $(0,-5)$, $p=-5$ (negative). Thus $x^2=4(-5)y$, or $x^2=-20y$.

5. Since the focus is on the x-axis, the form is $y^2 = 4px$. The focus is on the left side of the origin, at $(-4, 0)$. So $p = -4$ (negative). It follows that $y^2 = 4(-4)x$, or $y^2 = -16x$.

7. Since the directrix is $x = -1$, the focus is at $(1, 0)$. So the form is $y^2 = 4px$ with $p = 1$, and the equation is $y^2 = 4x$.

9. Since the directrix is $x = 2$, the focus is at $(-2, 0)$. So the form is $y^2 = 4px$ with $p = -2$. Thus $y^2 = -8x$.

11. Form: $y^2 = 4px$. Substituting the coordinates of the point $(-2, -4)$, we get

$$(-4)^2 = 4p(-2) \text{ and } 4p = -8.$$

Thus $y^2 = -8x$.

13. The form is either

$$y^2 = 4px \text{ or } x^2 = 4py.$$

Substituting the coordinates of the point $(1, 1)$, we get

$$1^2 = 4p \cdot 1 \text{ or } 1^2 = 4p \cdot 1.$$

In either case, $p = \frac{1}{4}$. So the equations are $y^2 = x$ and $x^2 = y$.

15. The form is either $y^2 = 4px$ or $x^2 = 4py$. Substituting the coordinates of the point $(-2, 4)$, we get

$$4^2 = 4p(-2) \quad \text{or} \quad (-2)^2 = 4p(4)$$

The respective values of p are -2 and $\frac{1}{4}$; so the equations are $y^2 = -8x$ and $x^2 = y$.

17. From $x^2 = 12y$, we have $x^2 = 4(3y)$. Thus $p = 3$ and the focus is at $(0, 3)$.

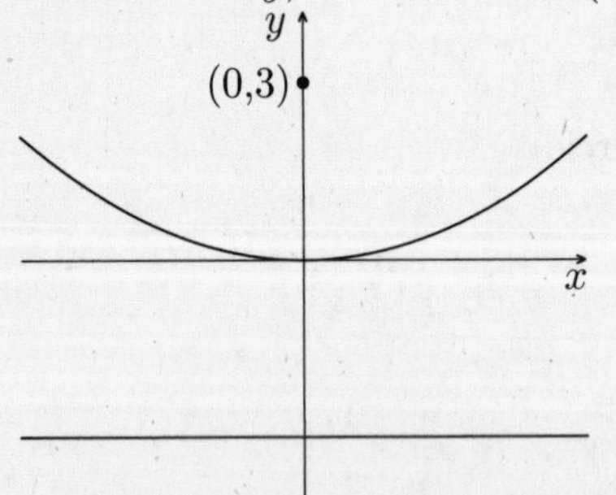

19. From $x^2 = -8y$, we have $x^2 = 4(-2)y$. So $p = -2$ and the focus is at $(0, -2)$.

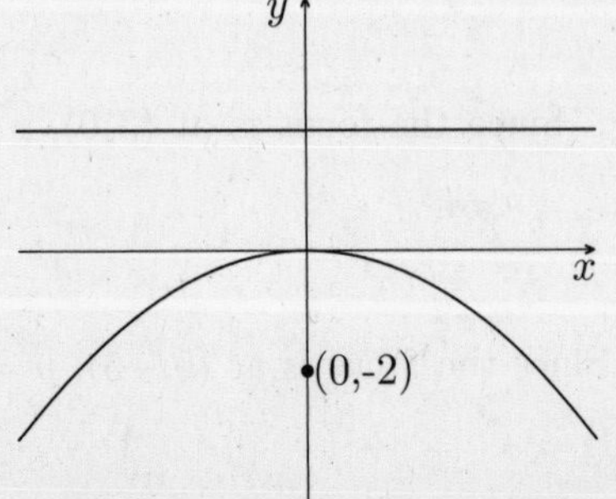

21. $y^2 = 24x = 4(6)x$; $p = 6$ and the focus is at $(6, 0)$.

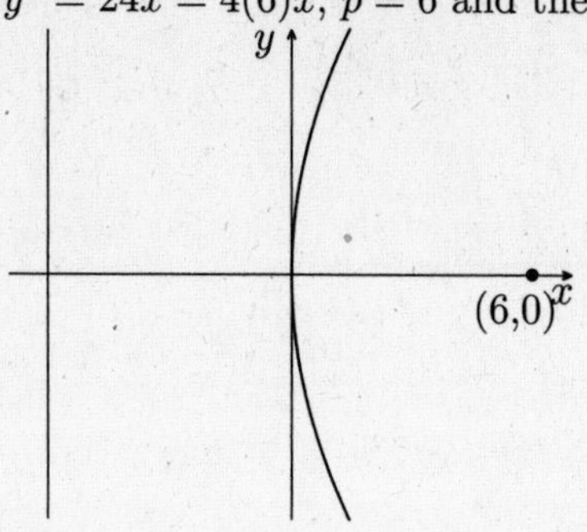

23. From $y^2 = -4x$, $y^2 = 4(-1)x$. So $p = -1$ and the focus is at $(-1, 0)$.

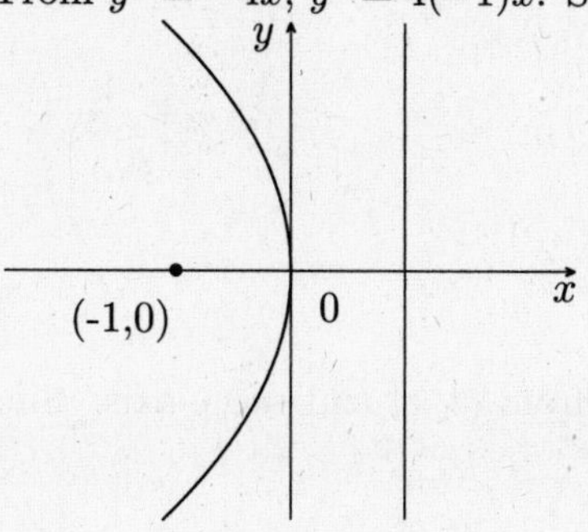

25. $x^2 = 4y = 4(1)y$; $p = 1$ and the focus is at $(0, 1)$.

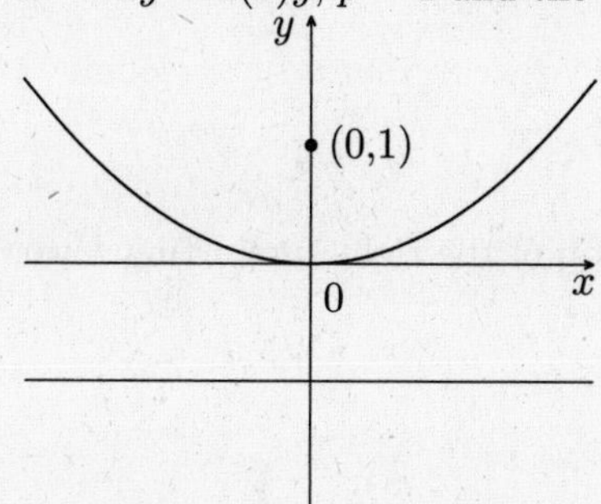

27. From $y^2 = 9x$, $y^2 = 4(\frac{9}{4})x$ (inserting 4). So $p = \frac{9}{4}$ and the focus is at $\left(\frac{9}{4}, 0\right)$.

29. $y^2 = -x = 4(-\frac{1}{4})x$; $p = -\frac{1}{4}$ and the focus is at $\left(-\frac{1}{4}, 0\right)$.

31.
$$\begin{aligned} 3y^2 + 2x &= 0 \\ y^2 &= -\tfrac{2}{3}x \\ y^2 &= 4(-\tfrac{2}{3} \cdot \tfrac{1}{4})x \\ y^2 &= 4(-\tfrac{1}{6})x \end{aligned}$$

So the focus is at $\left(-\frac{1}{6}, 0\right)$.

33. $x^2 = 4(3)y$; $p = 3$. Focus: $(0, 3)$; directrix: $y = -3$.

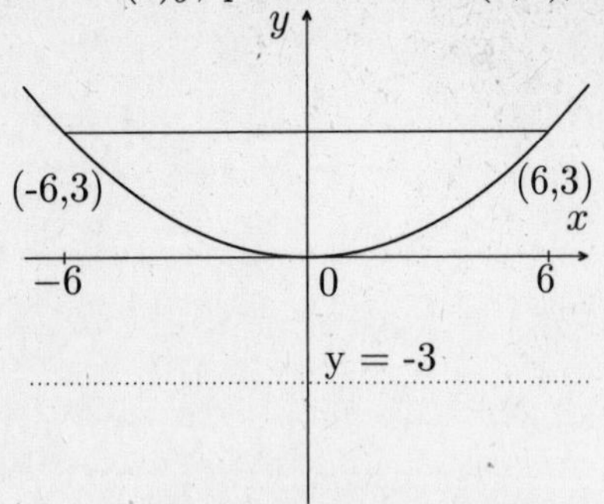

Observe that the points $(6, 3)$ and $(-6, 3)$ lie on the curve because the distance to the focus must be equal to the distance to the directrix.

Circle:
$$\begin{aligned}(x-h)^2 + (y-k)^2 &= r^2\\ (x-0)^2 + (y-3)^2 &= 6^2\\ x^2 + y^2 - 6y + 9 &= 36\\ x^2 + y^2 - 6y - 27 &= 0.\end{aligned}$$

35. We need to find the locus of points (x, y) equidistant from $(4, 1)$ and the y-axis. Since the distance from (x, y) to the y-axis is x units, we get

$$\begin{aligned}\sqrt{(x-4)^2 + (y-1)^2} &= x\\ (x-4)^2 + (y-1)^2 &= x^2\\ x^2 - 8x + 16 + y^2 - 2y + 1 &= x^2\\ y^2 - 2y - 8x + 17 &= 0.\end{aligned}$$

37. If the origin is the lowest point on the cable, then the top of the right supporting tower is at $(100, 70)$.

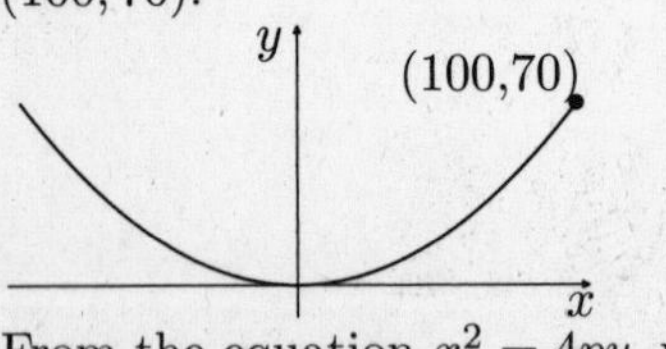

From the equation $x^2 = 4py$, we get

$$\begin{aligned}(100)^2 &= 4p(70)\\ 4p &= \frac{10,000}{70} = \frac{1000}{7}.\end{aligned}$$

The equation is therefore

$$x^2 = \frac{1000}{7}y.$$

To find the length of the cable 30 m from the center, we let $x = 30$:

$$30^2 = \frac{1000}{7}y \text{ and } y = \frac{6300}{1000} = 6.3.$$

So the length of the cable is $20 + 6.3 = 26.3$ m.

39.

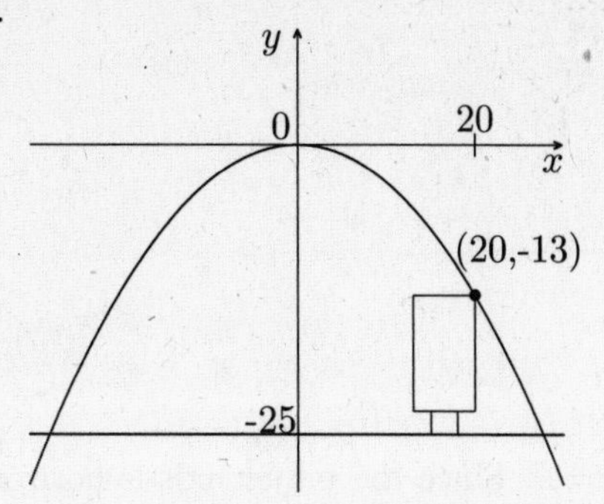

The required minimum clearance of 12 ft yields the point $(20, -13)$ in the figure.

$$x^2 = 4py$$

$$20^2 = 4p(-13) \quad \text{or} \quad 4p = \frac{20^2}{-13}$$

Equation: $x^2 = -\frac{20^2}{13}y$.

When $y = -25$,

$$x^2 = -\frac{20^2}{13}(-25)$$

$$x = \sqrt{\frac{20^2 \cdot 25}{13}} = \frac{20 \cdot 5}{\sqrt{13}} = \frac{100}{\sqrt{13}}$$

$$2x = \frac{200}{\sqrt{13}} \approx 55.5\,\text{ft}$$

41. We place the vertex of the parabola at the origin, so that one point on the parabola is $(3, -3)$ (from the given dimensions). Substituting in the equation $x^2 = 4py$, we get

$$3^2 = 4p(-3)$$

$$4p = -3.$$

The equation is therefore seen to be $x^2 = -3y$.

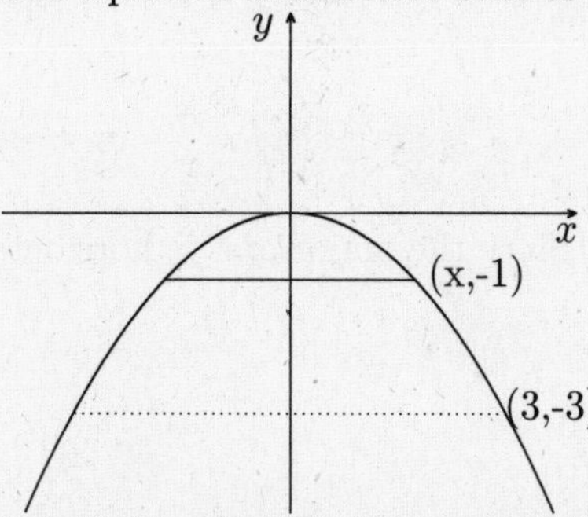

The right end of the beam 2 m above the base is at $(x, -1)$. To find x, let $y = -1$:

$$x^2 = -3(-1) = 3$$

$$x = \pm\sqrt{3}.$$

Hence the length of the beam is $2|x| = 2\sqrt{3}\,\text{m}$.

43. $x^2 = 4py$. From Figure 1.55, we see that the point $(4, 1)$ lies on the curve: $4^2 = 4p(1)$. So $p = 4\,\text{ft}$.

1.9 The Ellipse

1. The equation is

$$\frac{x^2}{25} + \frac{y^2}{16} = 1.$$

So by (1.16), $a^2 = 25$ and $b^2 = 16$; thus $a = 5$ and $b = 4$. Since the major axis is horizontal, the vertices are at $(\pm 5, 0)$. From $b^2 = a^2 - c^2$,

$$\begin{aligned} 16 &= 25 - c^2 \\ c^2 &= 9 \\ c &= \pm 3. \end{aligned}$$

The foci are therefore at $(\pm 3, 0)$, on the major axis. Finally, the length of the semi-minor axis is equal to $b = 4$.

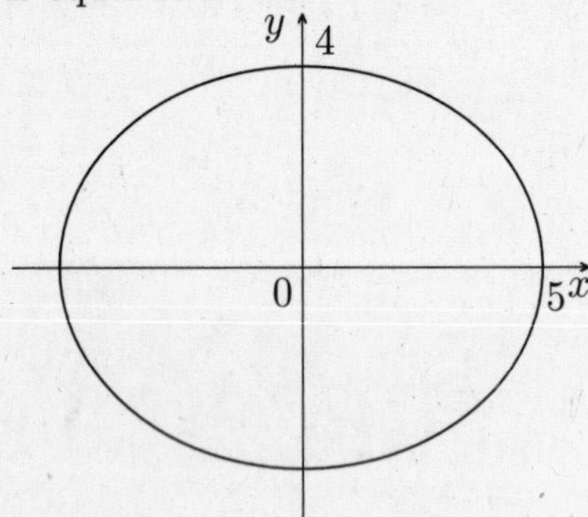

3. The equation is

$$\frac{x^2}{9} + \frac{y^2}{4} = 1.$$

So by (1.16), $a^2 = 9$ and $b^2 = 4$; thus $a = 3$ and $b = 2$. Since the major axis is horizontal, the vertices are at $(\pm 3, 0)$. From $b^2 = a^2 - c^2$,

$$\begin{aligned} 4 &= 9 - c^2 \\ c^2 &= 5 \\ c &= \pm\sqrt{5}. \end{aligned}$$

The foci are therefore at $\left(\pm\sqrt{5}, 0\right)$, on the major axis. Finally, the length of the semi-minor axis is equal to $b = 2$.

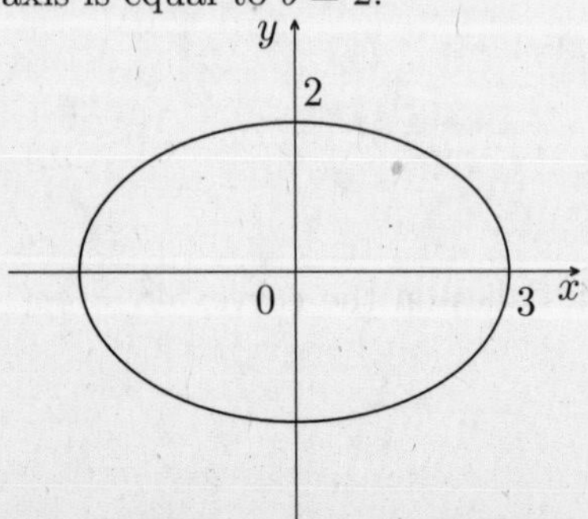

5. The equation is

$$\frac{x^2}{16} + y^2 = 1.$$

By (1.16), $a = 4$ and $b = 1$. From $b^2 = a^2 - c^2$,

$$\begin{aligned} 1 &= 16 - c^2 \\ c^2 &= 15 \\ c &= \pm\sqrt{15}. \end{aligned}$$

Since the major axis is horizontal, the vertices and foci lie on the x-axis. The vertices are therefore at $(\pm 4, 0)$ and the foci are at $(\pm\sqrt{15}, 0)$. The length of the semi-minor axis is $b = 1$.

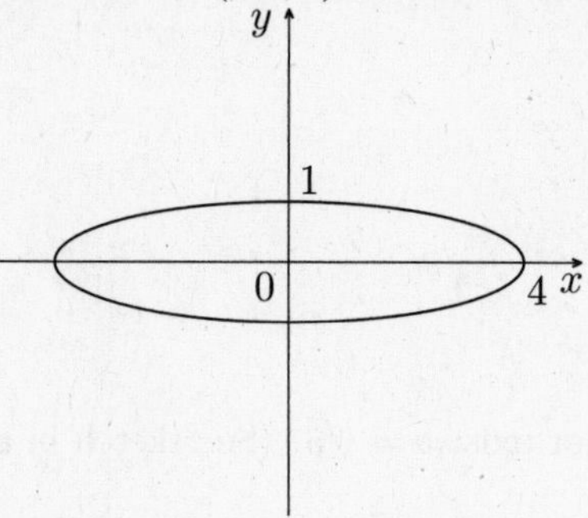

7. $$\begin{aligned} 16x^2 + 9y^2 &= 144 \\ \frac{x^2}{9} + \frac{y^2}{16} &= 1 \end{aligned}$$

So by (1.17), $a^2 = 16$ and $b^2 = 9$; so $a = 4$ and $b = 3$. Since the major axis is vertical, the vertices are at $(0, \pm 4)$. From $b^2 = a^2 - c^2$

$$\begin{aligned} 9 &= 16 - c^2 \\ c &= \pm\sqrt{7}. \end{aligned}$$

The foci are therefore at $(0, \pm\sqrt{7})$. Semi-minor axis: $b = 3$.

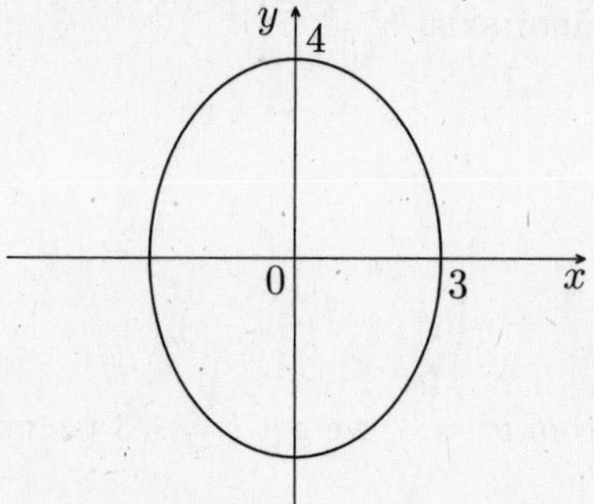

9. $$\begin{aligned} 5x^2 + 2y^2 &= 20 \\ \frac{5x^2}{20} + \frac{2y^2}{20} &= 1 \\ \frac{x^2}{4} + \frac{y^2}{10} &= 1 \end{aligned}$$

By (1.17), $a = \sqrt{10}$ and $b = 2$. Since the major axis is vertical, the vertices are at $(0, \pm\sqrt{10})$. From $b^2 = a^2 - c^2$

$$\begin{aligned} 4 &= 10 - c^2 \\ c &= \pm\sqrt{6}. \end{aligned}$$

The foci, also on the major axis, are therefore at $(0, \pm\sqrt{6})$, while the length of the semi-minor axis is $b = 2$. (See sketch in answer section of book.)

11. $$\begin{aligned} 5x^2 + y^2 &= 5 \\ \frac{x^2}{1} + \frac{y^2}{5} &= 1 \qquad \text{major axis vertical} \end{aligned}$$

Vertices: $(0, \pm\sqrt{5})$, foci: $(0, \pm 2)$. Length of semi-minor axis: $b = 1$. (See sketch in answer section of book.)

13. $$\begin{aligned} x^2 + 2y^2 &= 6 \\ \frac{x^2}{6} + \frac{2y^2}{6} &= 1 \\ \frac{x^2}{6} + \frac{y^2}{3} &= 1 \qquad \text{major axis horizontal} \end{aligned}$$

Thus $a = \sqrt{6}$ and $b = \sqrt{3}$. From $b^2 = a^2 - c^2$,

$$\begin{aligned} 3 &= 6 - c^2 \\ c &= \pm\sqrt{3}. \end{aligned}$$

Vertices: $(\pm\sqrt{6}, 0)$; foci: $(\pm\sqrt{3}, 0)$. Length of semi-minor axis: $b = \sqrt{3}$. (See sketch in answer section of book.)

15. $$\begin{aligned} 15x^2 + 7y^2 &= 105 \\ \frac{x^2}{7} + \frac{y^2}{15} &= 1 \qquad \text{major axis vertical} \end{aligned}$$

Thus $a = \sqrt{15}$ and $b = \sqrt{7}$. From $b^2 = a^2 - c^2$

$$\begin{aligned} 7 &= 15 - c^2 \\ c &= \pm\sqrt{8} = \pm 2\sqrt{2}. \end{aligned}$$

Vertices: $(0, \pm\sqrt{15})$, foci: $(0, \pm 2\sqrt{2})$. Length of semi-minor axis: $b = \sqrt{7}$.

17. $$\begin{aligned} 3x^2 + 4y^2 &= 12 \\ \frac{x^2}{4} + \frac{y^2}{3} &= 1 \qquad \text{dividing by 12} \end{aligned}$$

Since $a^2 = 4, a = \pm 2$, so the vertices are at $(\pm 2, 0)$. From $b^2 = 3$, we get $b = \sqrt{3}$ (semiminor axis). $b^2 = a^2 - c^2$ yields $c = \pm 1$.

23. Since the foci $(\pm 2, 0)$ lie along the major axis, the major axis is horizontal. So the form of the equation is

$$\frac{x^2}{a^2} + \frac{y^2}{b^2} = 1.$$

Since the vertices are at $(\pm 3, 0)$, $a = 3$. From $b^2 = a^2 - c^2$ (with $c = 2$), we get $b^2 = 9 - 4 = 5$. So the equation is

$$\frac{x^2}{9} + \frac{y^2}{5} = 1.$$

25. Since the foci $(0, \pm 2)$ lie on the major axis, the major axis is vertical. So by (1.17) the form of the equation is

$$\frac{x^2}{b^2} + \frac{y^2}{a^2} = 1.$$

Since the length of the major axis is 8, $a = 4$. From $b^2 = a^2 - c^2$ with $c = 2$, $b^2 = 16 - 4 = 12$. Hence

$$\frac{x^2}{12} + \frac{y^2}{16} = 1 \quad \text{or} \quad 4x^2 + 3y^2 = 48.$$

27. Since the foci are at $(0, \pm 3)$, $c = 3$, and the major axis is vertical. By (1.17)

$$\frac{x^2}{b^2} + \frac{y^2}{a^2} = 1.$$

Since the length of the minor axis is 6, $b = 3$. From $b^2 = a^2 - c^2$, $9 = a^2 - 9$ and $a^2 = 18$.

$$\text{Equation: } \frac{x^2}{9} + \frac{y^2}{18} = 1 \quad \text{or} \quad 2x^2 + y^2 = 18.$$

29. Since the vertices and foci are on the y-axis, the form of the equation is, by (1.17),

$$\frac{x^2}{b^2} + \frac{y^2}{a^2} = 1.$$

From $b^2 = a^2 - c^2$ with $a = 8$ and $c = 5$, $b^2 = 64 - 25 = 39$. Hence

$$\frac{x^2}{39} + \frac{y^2}{64} = 1.$$

31. Form: $\dfrac{x^2}{a^2} + \dfrac{y^2}{b^2} = 1$; $c = 2\sqrt{3}$, $b = 2$. From $b^2 = a^2 - c^2$, $4 = a^2 - (2\sqrt{3})^2$ and $a^2 = 16$.

$$\text{Equation: } \frac{x^2}{16} + \frac{y^2}{4} = 1 \quad \text{or} \quad x^2 + 4y^2 = 16.$$

33. From the form

$$\frac{x^2}{a^2} + \frac{y^2}{b^2} = 1$$

we get (since $b = 2$)

$$\frac{x^2}{a^2} + \frac{y^2}{4} = 1$$

Substituting the coordinates of the point $(3, 1)$ yields

$$\frac{9}{a^2} + \frac{1}{4} = 1 \quad \text{and} \quad \frac{9}{a^2} = \frac{3}{4}.$$

So

$$\frac{a^2}{9} = \frac{4}{3} \quad \text{and} \quad a^2 = 12.$$

Equation:

$$\frac{x^2}{12} + \frac{y^2}{4} = 1 \quad \text{or} \quad x^2 + 3y^2 = 12$$

35. From the original derivation of the ellipse, $2a = 16$ and $a = 8$. Since the foci are at $(\pm 6, 0)$, $c = 6$. Thus $b^2 = a^2 - c^2 = 64 - 36 = 28$.
By (1.16) the equation is

$$\frac{x^2}{64} + \frac{y^2}{28} = 1.$$

37. $9x^2 + 5y^2 = 45$ or $\dfrac{x^2}{5} + \dfrac{y^2}{9} = 1$; $a = 3$; $b = \sqrt{5}$. From $b^2 = a^2 - c^2$, $5 = 9 - c^2$ and $c = 2$. Thus

$$e = \frac{c}{a} = \frac{2}{3}.$$

39. We want the center of the ellipse to be at the origin with the center of the earth at one of the foci. Study the following diagram:

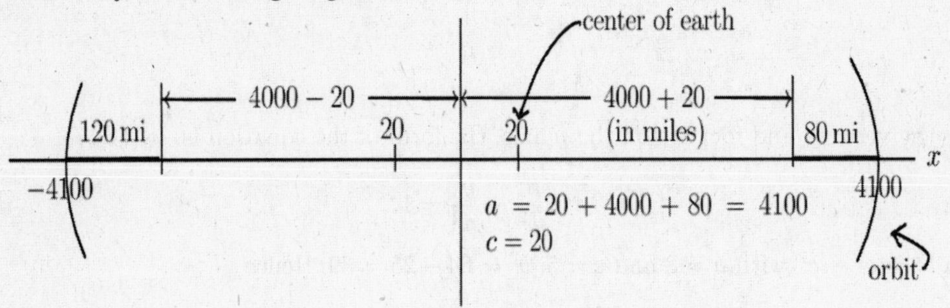

$$b^2 = 4100^2 - 20^2 = 16,809,600.$$

41. Let $A =$ the maximum distance and $P =$ the minimum distance as shown.

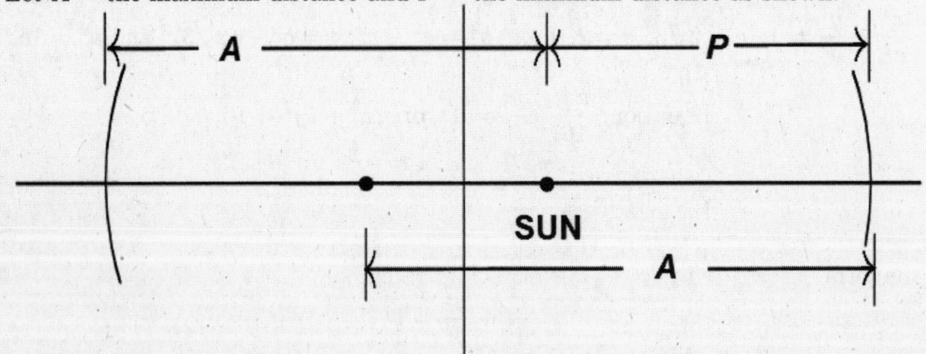

A is also the distance from the left focus to the right vertex. So $A - P$ is the distance between the foci. Therefore $\frac{1}{2}(A - P)$ is the distance from the center to the sun (the focus), or $c = \frac{1}{2}(A - P)$.
Now $a = c + P = \frac{1}{2}(A - P) + P = \frac{1}{2}(A + P)$. So

$$e = \frac{c}{a} = \frac{\frac{1}{2}(A - P)}{\frac{1}{2}(A + P)} = \frac{A - P}{A + P}.$$

In our problem

$$e = \frac{3.285 \times 10^9 - 5.48 \times 10^7}{3.285 \times 10^9 + 5.48 \times 10^7} = 0.967.$$

43. Since $a = 2$ and $b = \frac{3}{2}$, we get

$$\frac{x^2}{4} + \frac{y^2}{\frac{9}{4}} = 1 \quad \text{or} \quad \frac{x^2}{4} + \frac{4y^2}{9} = 1,$$

and $9x^2 + 16y^2 = 36$.

45. Placing the center at the origin, the vertices are at $(\pm 6, 0)$. The road extends from $(-4, 0)$ to $(4, 0)$. Since the clearance is 4 m, the point $(4, 4)$ lies on the ellipse, as shown.

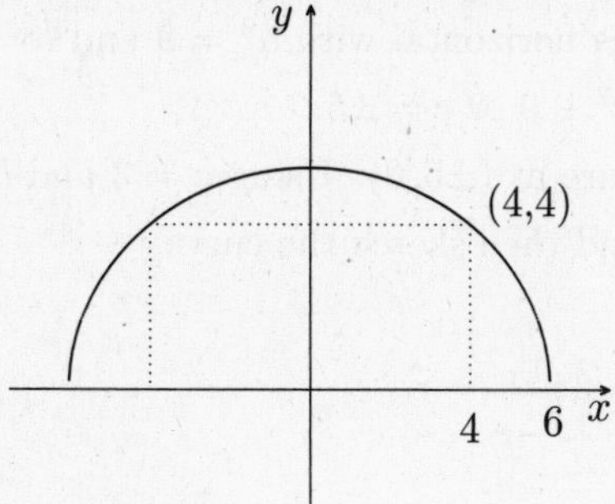

By (1.16),

$$\begin{aligned} \frac{x^2}{a^2} + \frac{y^2}{b^2} &= 1 \\ \frac{x^2}{36} + \frac{y^2}{b^2} &= 1. \end{aligned}$$

To find b, we substitute the coordinates of $(4, 4)$ in the equation:

$$\begin{aligned} \frac{16}{36} + \frac{16}{b^2} &= 1 \\ \frac{16}{b^2} &= \frac{36}{36} - \frac{16}{36} = \frac{20}{36} = \frac{5}{9} \\ \frac{b^2}{16} &= \frac{9}{5} \\ b^2 &= \frac{(9)(16)}{5} \\ b &= \frac{(3)(4)}{\sqrt{5}} = \frac{12}{\sqrt{5}} = \frac{12\sqrt{5}}{5}. \end{aligned}$$

So the height of the arch is $\dfrac{12\sqrt{5}}{5} = 5.4\,\text{m}$ to two significant digits.

1.10 The Hyperbola

1. Comparing the given equation,

$$\frac{x^2}{16} - \frac{y^2}{9} = 1$$

to form (1.22), we see that the transverse axis is horizontal, with $a^2 = 16$ and $b^2 = 9$. So $a = 4$ and $b = 3$. From $b^2 = c^2 - a^2$, we get

$$\begin{aligned} 9 &= c^2 - 16 \\ c &= \pm 5. \end{aligned}$$

So the vertices are at $(\pm 4, 0)$ and the foci are at $(\pm 5, 0)$. Using $a = 4$ and $b = 3$, we draw the auxiliary rectangle and sketch the curve:

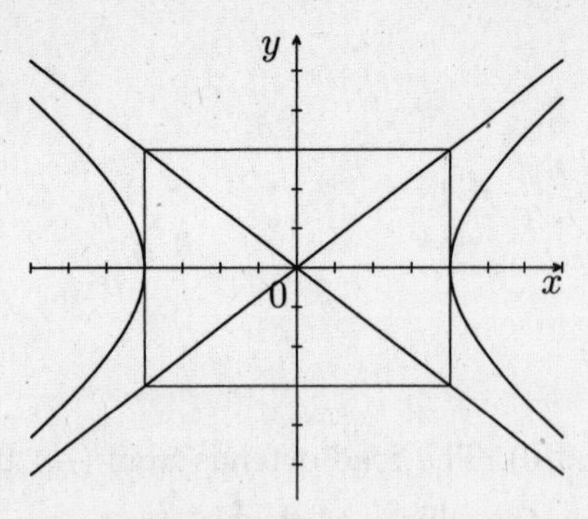

3. $\frac{x^2}{9} - \frac{y^2}{16} = 1$; by Equation (1.22), the transverse axis is horizontal with $a^2 = 9$ and $b^2 = 16$. So $a = 3$ and $b = 4$. From $b^2 = c^2 - a^2$, we have $16 = c^2 - 9$ or $c = \pm 5$.
It follows that the vertices are at $(\pm 3, 0)$ and the foci are at $(\pm 5, 0)$. Using $a = 3$ and $b = 4$, we draw the auxiliary rectangle and the asymptotes, and then sketch the curve.

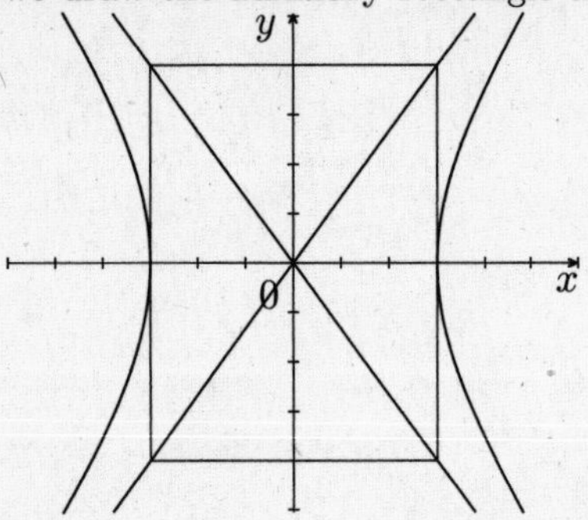

5. By (1.23), $a = 2$ and $b = 2$, transverse axis vertical along the y-axis. From $b^2 = c^2 - a^2$, $4 = c^2 - 4$ and $c = \pm\sqrt{8} = \pm 2\sqrt{2}$. So the vertices are at $(0, \pm 2)$ and the foci are at $(0, \pm 2\sqrt{2})$. Using $a = 2$ and $b = 2$, we draw the auxiliary rectangle and sketch the curve:

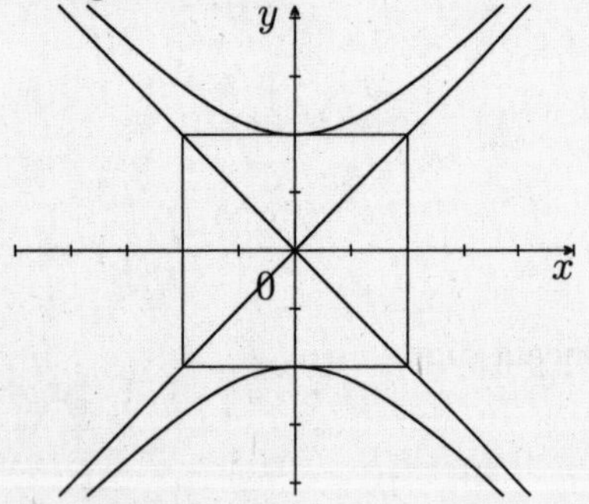

7. $x^2 - \frac{y^2}{5} = 1$ transverse axis horizontal
$a^2 = 1$ and $b^2 = 5$; so $a = 1$ and $b = \sqrt{5}$. From $b^2 = c^2 - a^2$, $5 = c^2 - 1$ and $c = \pm\sqrt{6}$. Vertices: $(\pm 1, 0)$; foci: $(\pm\sqrt{6}, 0)$. Using $a = 1$ and $b = \sqrt{5}$, we draw the auxiliary rectangle and sketch the curve.

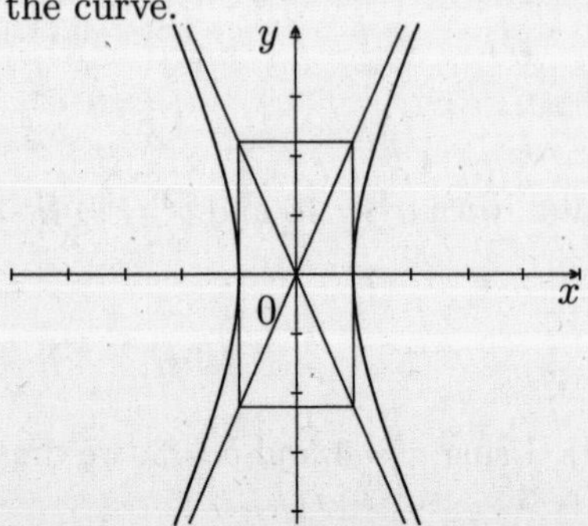

9.
$$\begin{aligned} 2y^2 - 3x^2 &= 24 \\ \frac{2y^2}{24} - \frac{3x^2}{24} &= 1 \\ \frac{y^2}{12} - \frac{x^2}{8} &= 1 \end{aligned}$$

By (1.23), $a = \sqrt{12} = 2\sqrt{3}$ and $b = \sqrt{8} = 2\sqrt{2}$. From $b^2 = c^2 - a^2$, $8 = c^2 - 12$, so that $c = \pm\sqrt{20} = \pm 2\sqrt{5}$. Since the transverse axis lies along the y-axis, the vertices are at $(0, \pm 2\sqrt{3})$ and the foci at $(0, \pm 2\sqrt{5})$. Using $a = 2\sqrt{3}$ and $b = 2\sqrt{2}$, we draw the auxiliary rectangle and sketch the curve:

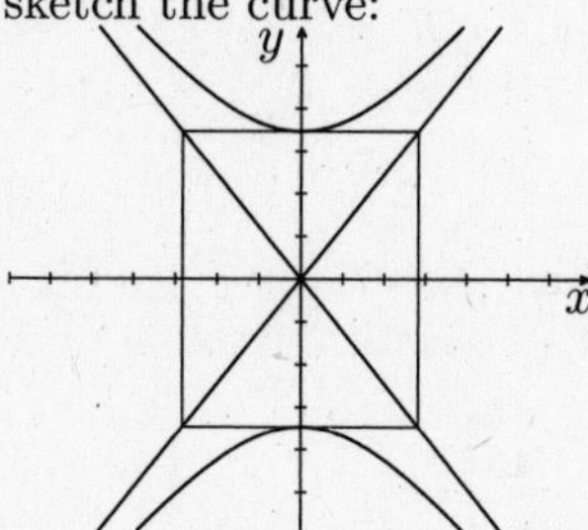

11. $3y^2 - 2x^2 = 6$ or $\dfrac{y^2}{2} - \dfrac{x^2}{3} = 1$

By (1.23) the transverse axis is vertical with $a = \sqrt{2}$ and $b = \sqrt{3}$. From $b^2 = c^2 - a^2$, $3 = c^2 - 2$ or $c = \pm\sqrt{5}$.

So the vertices are at $(0, \pm\sqrt{2})$ and the foci at $(0, \pm\sqrt{5})$. Using $a = \sqrt{2}$ and $b = \sqrt{3}$, we draw the auxiliary rectangle and sketch the curve.

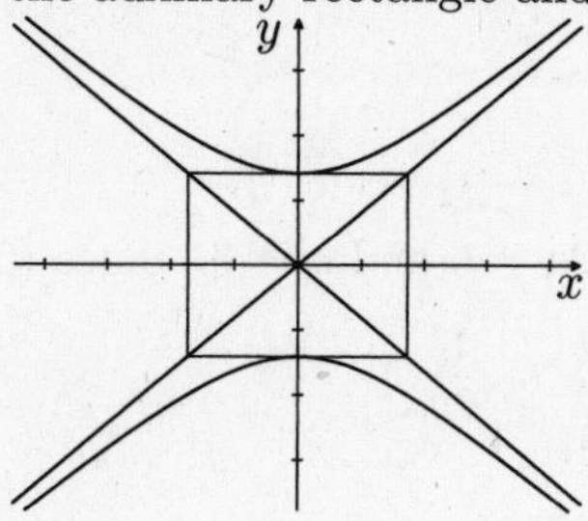

13. Since the foci (and hence the vertices) lie on the x-axis, the transverse axis is horizontal. By (1.22),

$$\frac{x^2}{a^2} - \frac{y^2}{b^2} = 1.$$

Since the length of the transverse axis is 4, $a = 2$, and since the length of the conjugate axis is 2, $b = 1$. It follows that

$$\frac{x^2}{4} - \frac{y^2}{1} = 1 \quad \text{and} \quad x^2 - 4y^2 = 4.$$

15. Since the foci (and hence the vertices) lie on the y-axis, the form is

$$\frac{y^2}{a^2} - \frac{x^2}{b^2} = 1.$$

Since the length of the transverse axis is 8, $a = 4$, while $c = 6$. From $b^2 = c^2 - a^2$, we get $b^2 = 36 - 16 = 20$. So the equation is

$$\frac{y^2}{16} - \frac{x^2}{20} = 1$$

17. Since the vertices lie on the x-axis, the form is

$$\frac{x^2}{a^2} - \frac{y^2}{b^2} = 1.$$

The conjugate axis of length 8 implies that $b = 4$ and the position of the vertices imply that $a = 4$.

$$\text{Equation: } \frac{x^2}{16} - \frac{y^2}{16} = 1 \text{ or } x^2 - y^2 = 16$$

19. Since the vertices are on the y-axis, the form is

$$\frac{y^2}{a^2} - \frac{x^2}{b^2} = 1.$$

Since $a = 5$ and $c = 7$, we get $b^2 = c^2 - a^2 = 49 - 25 = 24$. Thus

$$\frac{y^2}{25} - \frac{x^2}{24} = 1.$$

21. Since the foci are on the x-axis, the form is

$$\frac{x^2}{a^2} - \frac{y^2}{b^2} = 1.$$

Since the length of the conjugate axis is 10, $b = 5$, while $c = 6$. From $b^2 = c^2 - a^2$, we get $a^2 = c^2 - b^2 = 36 - 25 = 11$. So the equation is

$$\frac{x^2}{11} - \frac{y^2}{25} = 1.$$

23. By the original derivation of the equation of the hyperbola, $2a = 6$ and $a = 3$. Since $(0, \pm 5)$ are the foci, $c = 5$. Thus $b^2 = 25 - 9 = 16$. By (1.23)

$$\begin{aligned} \frac{y^2}{a^2} - \frac{x^2}{b^2} &= 1 \\ \frac{y^2}{9} - \frac{x^2}{16} &= 1 \quad \text{or} \quad 16y^2 - 9x^2 = 144. \end{aligned}$$

25. By (1.23), the equation has the form $\dfrac{y^2}{a^2} - \dfrac{x^2}{b^2} = 1$. Since $a = 12$, we have

$$\frac{y^2}{144} - \frac{x^2}{b^2} = 1.$$

To find b, we substitute the coordinates of $(-1, 13)$ in the last equation:

$$\begin{aligned} \frac{169}{144} - \frac{1}{b^2} &= 1 \\ -\frac{1}{b^2} &= \frac{144}{144} - \frac{169}{144} = -\frac{25}{144}. \end{aligned}$$

Thus $b^2 = \dfrac{144}{25}$. The equation is

$$\frac{y^2}{144} - \frac{x^2}{144/25} = 1 \quad \text{or} \quad \frac{y^2}{144} - \frac{25x^2}{144} = 1.$$

27.
$$\begin{aligned} pV &= k \\ (12)(3.0) &= k \qquad V = 3.0\,\text{m}^3,\ p = 12\,\text{Pa} \end{aligned}$$

So $pV = 36$. (See graph in answer section of book.)

1.11 Translation of Axes; Standard Equations of the Conics

1. Circle, center at $(1,2)$, $r = \sqrt{3}$.

3. $$\begin{aligned}(y+3)^2 &= 8(x-2)\\ (y+3)^2 &= 4(2)(x-2) \qquad p = +2\end{aligned}$$
Vertex at $(2,-3)$, focus at $(2+2,-3) = (4,-3)$.

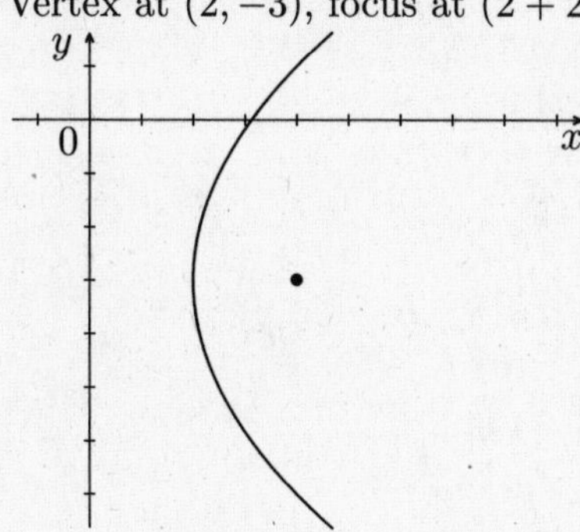

5. $$\begin{aligned}2x^2 - 3y^2 + 8x - 12y + 14 &= 0\\ 2x^2 + 8x - 3y^2 - 12y + 14 &= 0\\ 2(x^2 + 4x \quad) - 3(y^2 + 4y \quad) &= -14 \qquad \text{factoring 2 and } -3\end{aligned}$$
Note that the square of one-half the coefficient of x and y is $\left(\frac{1}{2}\cdot 4\right)^2 = 4$.
Inserting these values inside the parentheses and balancing the equation, we get
$$\begin{aligned}2(x^2 + 4x + \underline{4}) - 3(y^2 + 4y + \underline{4}) &= -14 + 2\cdot\underline{4} - 3\cdot\underline{4}\\ 2(x+2)^2 - 3(y+2)^2 &= -18\\ \frac{3(y+2)^2}{18} - \frac{2(x+2)^2}{18} &= 1\\ \frac{(y+2)^2}{6} - \frac{(x+2)^2}{9} &= 1.\end{aligned}$$
The equation represents a hyperbola with transverse axis vertical. Center: $(-2,-2)$, $a = \sqrt{6}$, $b = 3$.

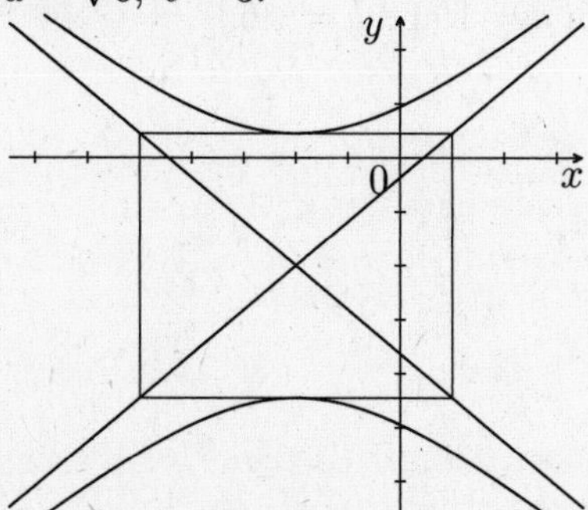

7. $$\begin{aligned}16x^2 + 4y^2 + 64x - 12y + 57 &= 0\\ 16x^2 + 64x + 4y^2 - 12y + 57 &= 0\\ 16(x^2 + 4x \quad) + 4(y^2 - 3y \quad) &= -57 \qquad \text{factoring 16 and 4}\end{aligned}$$
Note that
$$\left(\frac{1}{2}\cdot 4\right)^2 = 4 \quad\text{and}\quad \left[\frac{1}{2}(-3)\right]^2 = \frac{9}{4}.$$
Inserting these values inside the parentheses and balancing the equation, we get
$$\begin{aligned}16(x^2 + 4x + 4) + 4(y^2 - 3y + \tfrac{9}{4}) &= -57 + 16\cdot 4 + 4(\tfrac{9}{4})\\ 16(x+2)^2 + 4(y - \tfrac{3}{2})^2 &= 16\\ \frac{(x+2)^2}{1} + \frac{(y-3/2)^2}{4} &= 1.\end{aligned}$$

The equation represents an ellipse with major axis vertical. Center: $(-2, \frac{3}{2})$, $a = 2$, $b = 1$.

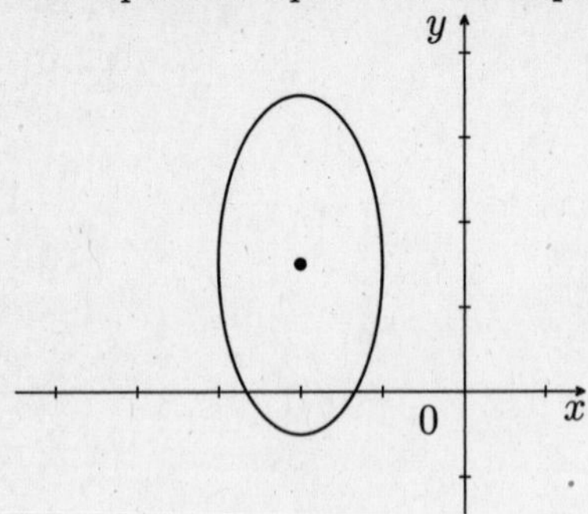

9.
$$\begin{aligned} x^2 + y^2 + 2x - 2y + 2 &= 0 \\ x^2 + 2x \quad + y^2 - 2y &= -2 \\ (x^2 + 2x + 1) + (y^2 - 2y + 1) &= -2 + 1 + 1 \\ (x+1)^2 + (y-1)^2 &= 0 \end{aligned}$$

Point: $(-1, 1)$.

11.
$$\begin{aligned} 2x^2 - 12y^2 + 60y - 63 &= 0 \\ 2x^2 - 12(y^2 - 5y \quad) &= 63 \end{aligned}$$

The square of one-half the coefficient of y is $\left[\frac{1}{2}(-5)\right]^2 = \frac{25}{4}$. Inserting this number <u>inside</u> the parentheses, we get

$$\begin{aligned} 2x^2 - 12(y^2 - 5y + \tfrac{25}{4}) &= 63 - 12(\tfrac{25}{4}) = -12 \\ x^2 - 6(y^2 - 5y + \tfrac{25}{4}) &= -6 \\ x^2 - 6(y - \tfrac{5}{2})^2 &= -6 \\ (y - \tfrac{5}{2})^2 - \frac{x^2}{6} &= 1. \end{aligned}$$

Hyperbola, center at $(0, \frac{5}{2})$, transverse axis vertical with $a = 1$ and $b = \sqrt{6}$.

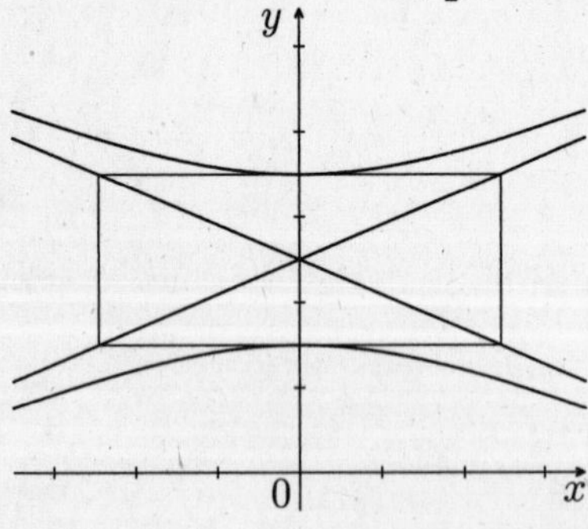

13.
$$\begin{aligned} 64x^2 + 64y^2 - 16x - 96y - 27 &= 0 \\ 64x^2 - 16x + 64y^2 - 96y - 27 &= 0 \\ 64(x^2 - \tfrac{x}{4} \quad) + 64(y^2 - \tfrac{3y}{2} \quad) &= 27 \\ 64(x^2 - \tfrac{x}{4} + \tfrac{1}{64}) + 64(y^2 - \tfrac{3y}{2} + \tfrac{9}{16}) &= 27 + 1 + 36 \\ 64(x - \tfrac{1}{8})^2 + 64(y - \tfrac{3}{4})^2 &= 64 \\ (x - \tfrac{1}{8})^2 + (y - \tfrac{3}{4})^2 &= 1. \end{aligned}$$

Circle of radius 1 centered at $\left(\frac{1}{8}, \frac{3}{4}\right)$.

15.
$$\begin{aligned} 3x^2 + y^2 - 18x + 2y + 29 &= 0 \\ 3x^2 - 18x + y^2 + 2y &= -29 \\ 3(x^2 - 6x \quad) + (y^2 + 2y \quad) &= -29 \end{aligned}$$

Observe that $\left[\frac{1}{2}(-6)\right]^2 = 9$ and $\left(\frac{1}{2} \cdot 2\right)^2 = 1$. Adding these values inside the parentheses and balancing the equation, we get

$$\begin{aligned} 3(x^2 - 6x + 9) + (y^2 + 2y + 1) &= -29 + 3 \cdot 9 + 1 \\ 3(x-3)^2 + (y+1)^2 &= -1 \end{aligned}$$

which is an imaginary locus.

17.
$$\begin{aligned} x^2 + 2x - 12y + 25 &= 0 \\ x^2 + 2x &= 12y - 25 \end{aligned}$$

We add to each side of the equation the square of one-half the coefficient of x, $\left[\frac{1}{2} \cdot 2\right]^2 = 1$:

$$\begin{aligned} x^2 + 2x + \underline{1} &= 12y - 25 + \underline{1} \\ (x+1)^2 &= 12y - 24 \\ (x+1)^2 &= 12(y-2) \\ (x+1)^2 &= 4 \cdot 3(y-2). \qquad p = +3 \end{aligned}$$

Vertex at $(-1, 2)$, focus at $(-1, 5)$.

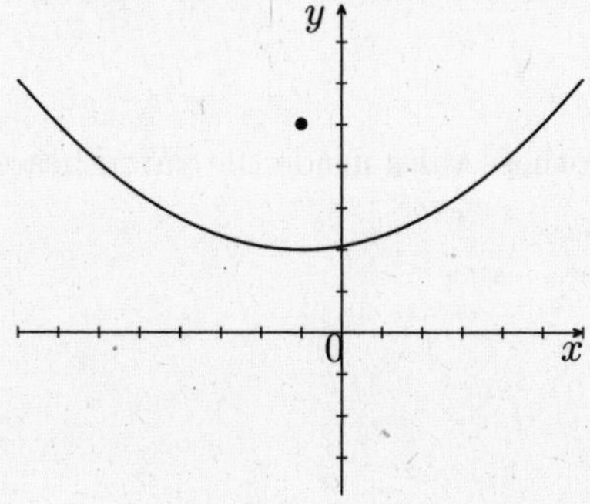

19.
$$\begin{aligned} x^2 + 2y^2 + 6x - 4y + 9 &= 0 \\ x^2 + 6x + 2y^2 - 4y &= -9 \\ (x^2 + 6x \quad) + 2(y^2 - 2y \quad) &= -9 \end{aligned}$$

Adding $\left[\frac{1}{2}(6)\right]^2 = 9$ and $\left[\frac{1}{2}(-2)\right]^2 = 1$ inside the parentheses and balancing the equation, we have

$$\begin{aligned} (x^2 + 6x + 9) + 2(y^2 - 2y + 1) &= -9 + 9 + 2 \\ (x+3)^2 + 2(y-1)^2 &= 2 \\ \frac{(x+3)^2}{2} + (y-1)^2 &= 1. \end{aligned}$$

Ellipse, center at $(-3, 1)$ with $a = \sqrt{2}$ and $b = 1$.

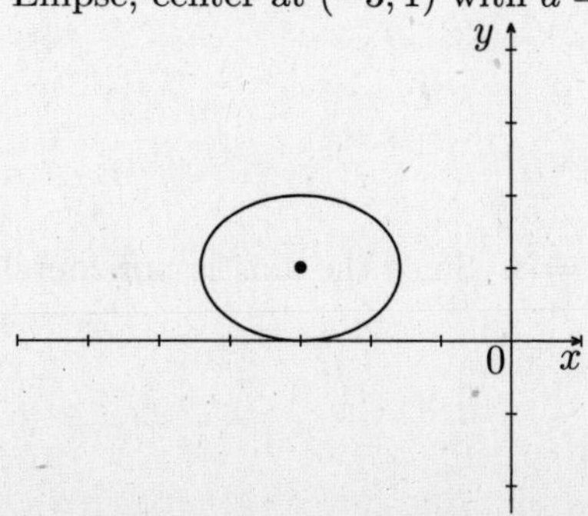

21.
$$\begin{aligned} x^2+4x+4y+16 &= 0 \\ x^2+4x &= -4y-16 \end{aligned}$$

We add to each side $\left[\frac{1}{2}\cdot 4\right]^2=4$:

$$\begin{aligned} x^2+4x+\underline{4} &= -4y-16+\underline{4} \\ (x+2)^2 &= -4y-12 \\ (x+2)^2 &= -4(y+3) \\ (x+2)^2 &= 4(-1)(y+3). \qquad p=-1 \end{aligned}$$

Vertex at $(-2,-3)$, focus at $(-2,-4)$.

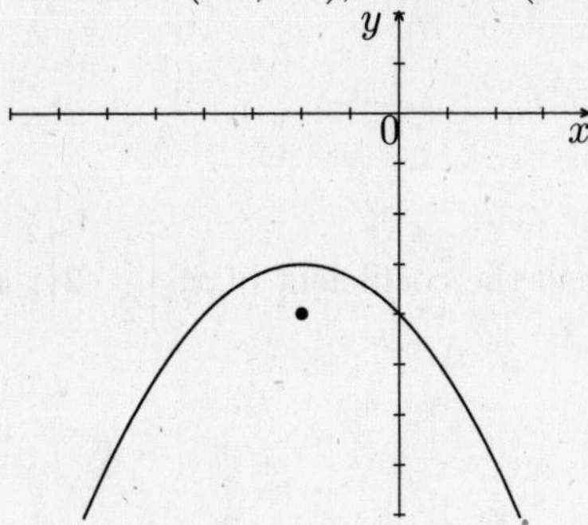

23.
$$\begin{aligned} x^2+2y^2-4x+12y+14 &= 0 \\ x^2-4x \quad +2y^2+12y &= -14 \\ (x^2-4x \quad)+2(y^2+6y \quad) &= -14 \qquad \text{factoring} \end{aligned}$$

Observe that $\left[\frac{1}{2}(-4)\right]^2=4$ and $\left[\frac{1}{2}(6)\right]^2=9$. Inserting these vales inside the parentheses and balancing the equation, we get

$$\begin{aligned} (x^2-4x+\underline{4})+2(y^2+6y+\underline{9}) &= -14+\underline{4}+2\cdot\underline{9} \\ (x-2)^2+2(y+3)^2 &= 8 \\ \frac{(x-2)^2}{8}+\frac{(y+3)^2}{4} &= 1. \end{aligned}$$

Ellipse, center at $(2,-3)$.

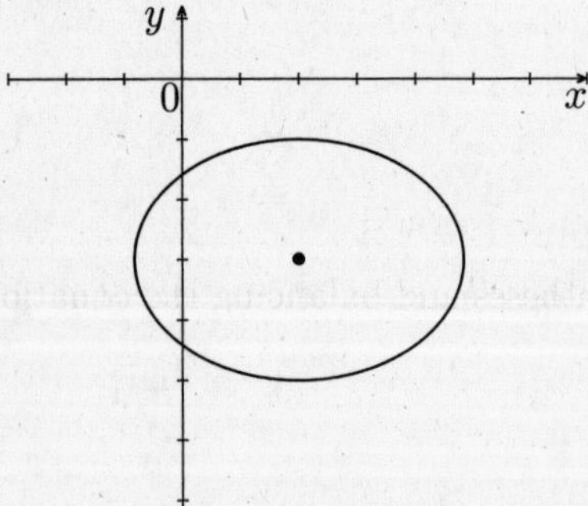

25.

vertex focus

Distance from vertex to focus: $3-(-1)=4$. Thus $p=4$. Since the axis is horizontal, the form of the equation is $(y-k)^2=4p(x-h)^2$. Thus

$$\begin{aligned} (y-2)^2 &= 4\cdot 4(x+1) \qquad (h,k)=(-1,2),\ p=4 \\ (y-2)^2 &= 16(x+1). \end{aligned}$$

27.

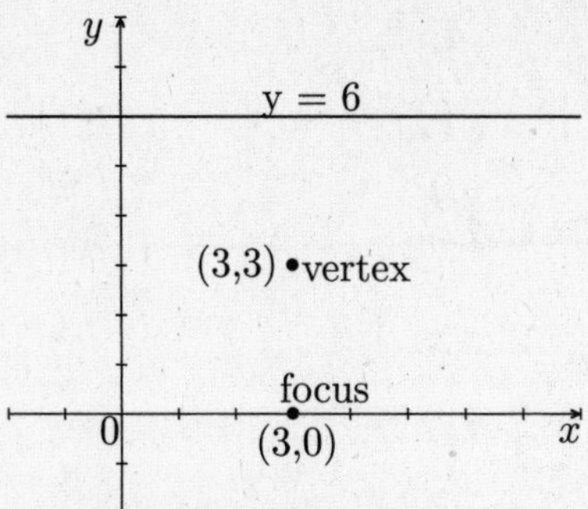

The distance from the vertex to the focus is 3, with the focus below the vertex; so $p = -3$.

Since the axis is vertical, the form of the equation is

$$\begin{aligned}(x-h)^2 &= 4p(y-k)\\(x-3)^2 &= 4(-3)(y-3)\\(x-3)^2 &= -12(y-3)\end{aligned}$$

29.

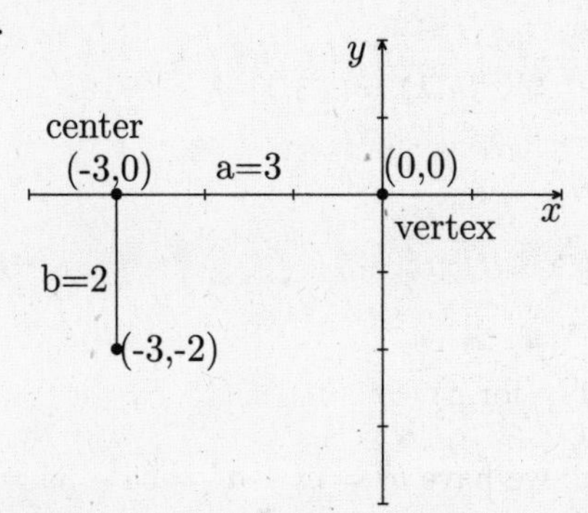

$$\frac{(x-h)^2}{a^2} + \frac{(y-k)^2}{b^2} = 1 \qquad \text{major axis horizontal}$$

Since the center is at $(-3, 0)$, we get for the equation

$$\frac{(x+3)^2}{9} + \frac{y^2}{4} = 1. \qquad a = 3,\ b = 2$$

31.

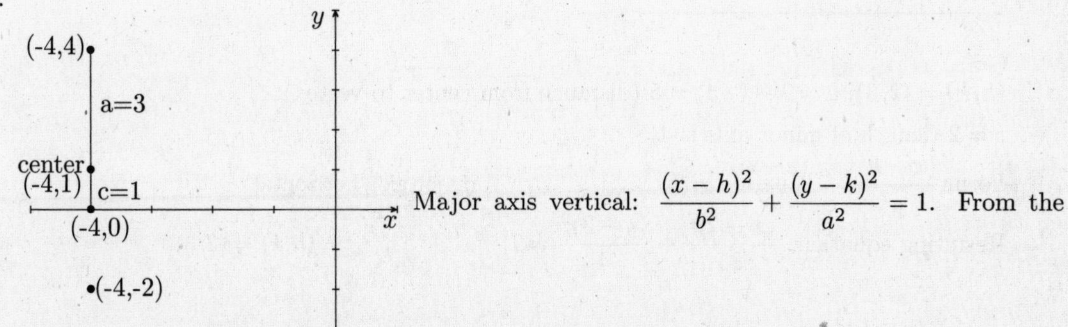

Major axis vertical: $\dfrac{(x-h)^2}{b^2} + \dfrac{(y-k)^2}{a^2} = 1$. From the diagram: $a = 3$ and $c = 1$; so $b^2 = a^2 - c^2 = 9 - 1 = 8$; center: $(-4, 1)$.

$$\text{Equation: } \frac{(x+4)^2}{8} + \frac{(y-1)^2}{9} = 1$$

33.

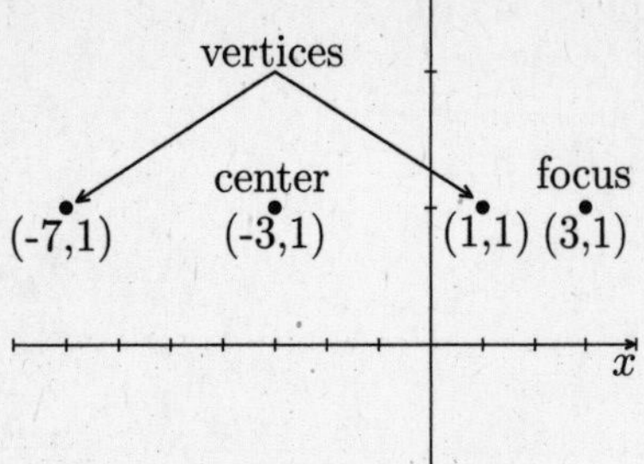

Distance between vertices is 8, so that $a = 4$. Center: $(-3, 1)$ (point midway between vertices). Distance from center to one focus is 6, so that $c = 6$. The transverse axis is horizontal, resulting in the form

$$\frac{(x-h)^2}{a^2} - \frac{(y-k)^2}{b^2} = 1.$$

Since $b^2 = c^2 - a^2 = 36 - 16 = 20$, we get

$$\frac{(x+3)^2}{16} - \frac{(y-1)^2}{20} = 1. \qquad (h,k) = (-3,1)$$

35. Transverse axis vertical:

$$\frac{(y-k)^2}{a^2} - \frac{(x-h)^2}{b^2} = 1 \quad \text{(form)}$$

Distance from center to focus: 8, so $c = 8$. Since $a = 6$, we have $b^2 = c^2 - a^2 = 64 - 36 = 28$.

$$\text{Equation: } \frac{(y+4)^2}{36} - \frac{(x-3)^2}{28} = 1$$

37.

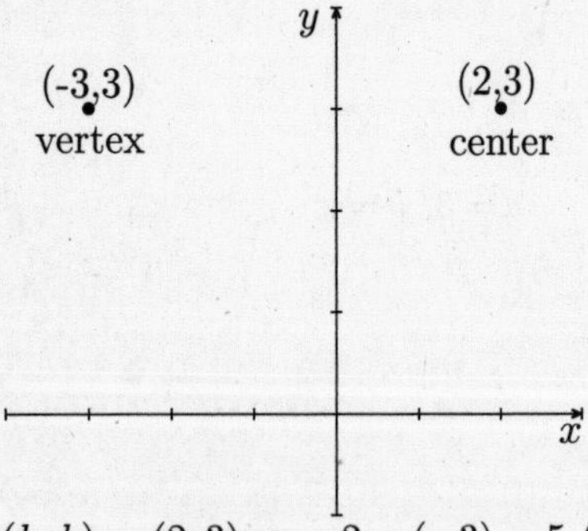

$(h,k) = (2,3)$; $a = 2 - (-3) = 5$ (distance from center to vertex);
$b = 2$ (length of minor axis is 4).

Form: $\dfrac{(x-h)^2}{a^2} + \dfrac{(y-k)^2}{b^2} = 1.$ major axis horizontal

Resulting equation: $\dfrac{(x-2)^2}{25} + \dfrac{(y-3)^2}{4} = 1.$ $(h,k) = (2,3)$

39. Form: $\dfrac{(x-h)^2}{a^2} - \dfrac{(y-k)^2}{b^2} = 1.$
Distance from center to vertex is 2, so that $a = 2$. From $x - 2y = 1$, $y = \frac{1}{2}x - \frac{1}{2}$. So the slope m of one of the asymptotes is $\frac{1}{2}$. But $m = \frac{b}{a}$. Thus $\frac{1}{2} = \frac{b}{a} = \frac{b}{2}$ or $b = 1$. The equation is

$$\frac{(x-1)^2}{4} - \frac{y^2}{1} = 1. \qquad (h,k) = (1,0)$$

41.

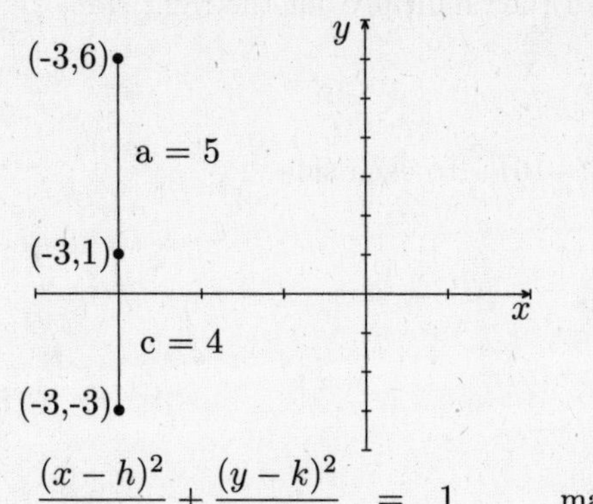

$$b^2 = a^2 - c^2 = 25 - 16 = 9$$

$$\frac{(x-h)^2}{b^2} + \frac{(y-k)^2}{a^2} = 1 \qquad \text{major axis vertical}$$

$$\frac{(x+3)^2}{9} + \frac{(y-1)^2}{25} = 1$$

43. Distance from vertex to focus: $4 - 1 = 3$. Since the focus is to the left of the vertex, $p = -3$.

Form:

$$(y-k)^2 = 4p(x-h) \qquad \text{axis horizontal}$$

$$(y-k)^2 = -12(x-h). \qquad p = -3$$

Equation: $(y+2)^2 = -12(x-4)$. $\qquad (h,k) = (4,-2)$

45.

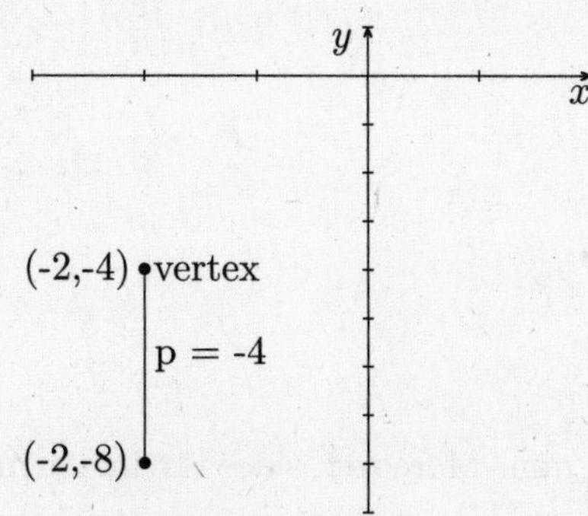

Since the vertex is midway between the focus and directrix, its coordinates are $(-2,-4)$. Since $p = -4$, we have

$$(x+2)^2 = 4(-4)(y+4)$$

$$(x+2)^2 = -16(y+4).$$

47.

y
(-1,3) • vertex
(-1,1) • center
x
(-1,2) • focus

$(h,k) = (-1,1)$; $a = 3 - 1 = 2$ (distance from center to vertex);

$c = 1 - (-2) = 3$ (distance from center to focus); $b^2 = c^2 - a^2 = 9 - 4 = 5$.

Form: $\dfrac{(y-k)^2}{a^2} - \dfrac{(x-h)^2}{b^2} = 1.$ $\qquad$ transverse axis vertical

Equation: $\dfrac{(y-1)^2}{4} - \dfrac{(x+1)^2}{5} = 1.$ $\qquad (h,k) = (-1,1)$

49. Multiply both sides of the given equation by $\frac{81}{5}$ and then multiply out the right side:

$$\begin{aligned} \tfrac{81}{5}y &= -x^2 + 16x + 17 \\ x^2 - 16x &= -\tfrac{81}{5}y + 17 \\ x^2 - 16x + 64 &= -\tfrac{81}{5}y + 17 + 64 \qquad \text{add } \left[\tfrac{1}{2}(-16)\right]^2 \text{ to each side} \\ (x-8)^2 &= -\tfrac{81}{5}(y-5) \end{aligned}$$

vertex: $(8, 5)$; maximum height $= 5$ units

Chapter 1 Review

1. Slope of line segment joining $(3, 10)$ and $(7, 4)$: $-\dfrac{3}{2}$.

   Slope of line segment joining $(4, 2)$ and $(7, 4)$: $\dfrac{2}{3}$.

   Since the slopes are negative reciprocals, the line segments are perpendicular.

3. $C = \frac{5}{9}(F - 32)$

   By assumption $C = F$,

$$\begin{aligned} F &= \tfrac{5}{9}(F-32) \\ F &= \tfrac{5}{9}F - \tfrac{160}{9} \\ F - \tfrac{5}{9}F &= -\tfrac{160}{9} \\ \tfrac{4}{9}F &= -\tfrac{160}{9} \\ F &= -40^\circ. \end{aligned}$$

5. Slope of line segment joining $(-1, 5)$ and $(3, 9)$: 1.

   Slope of line segment joining $(3, 1)$ and $(7, 5)$: 1.

   Slope of line segment joining $(-1, 5)$ and $(3, 1)$: -1.

   Slope of line segment joining $(3, 9)$ and $(7, 5)$: -1.

   Since opposite sides are parallel, the figure is a parallelogram. Moreover, since the line segment joining $(-1, 5)$ and $(3, 1)$ is perpendicular to the line segment joining $(-1, 5)$ and $(3, 9)$, the figure must be a rectangle. Finally:

   Length of line segment joining $(-1, 5)$ and $(3, 1) = 4\sqrt{2}$.

   Length of line segment joining $(-1, 5)$ and $(3, 9) = 4\sqrt{2}$.

   Thus the figure is a square.

7.
$$\begin{aligned} 3x + y &= 3 \\ y &= -3x + 3 \qquad y = mx + b \end{aligned}$$

   Since $m = -3$, we get

$$\begin{aligned} y - 5 &= -3(x+1) \qquad y - y_1 = m(x - x_1) \\ 3x + y - 2 &= 0. \end{aligned}$$

9. $r^2 = (1-0)^2 + (-2-0)^2 = 1 + 4 = 5$. We now get

   $(x-1)^2 + (y+2)^2 = 5$ or $x^2 + y^2 - 2x + 4y = 0$.

11.
$$\begin{aligned} x^2 + y^2 + 2x + 2y &= 0 \\ x^2 + 2x \quad + y^2 + 2y &= 0 \\ (x^2 + 2x + 1) + (y^2 + 2y + 1) &= 1 + 1 \\ (x+1)^2 + (y+1)^2 &= 2 \end{aligned}$$

   Center: $(-1, -1)$; $r = \sqrt{2}$.

13. Ellipse, major axis vertical, $a = 4$, $b = 3$. From $b^2 = a^2 - c^2$,

$$\begin{aligned} 9 &= 16 - c^2 \\ c &= \pm\sqrt{7}. \end{aligned}$$

Vertices: $(0, \pm 4)$; foci: $(0, \pm\sqrt{7})$. (See sketch in answer section of book.)

15. $\dfrac{y^2}{4} - \dfrac{x^2}{7} = 1$

Hyperbola, transverse axis vertical with $a = 2$ and $b = \sqrt{7}$. From $b^2 = c^2 - a^2$, $7 = c^2 - 4$ and $c = \pm\sqrt{11}$. So the vertices are at $(0, \pm 2)$ and the foci are at $(0, \pm\sqrt{11})$. Using $a = 2$ and $b = \sqrt{7}$, we draw the auxiliary rectangle and sketch the curve.

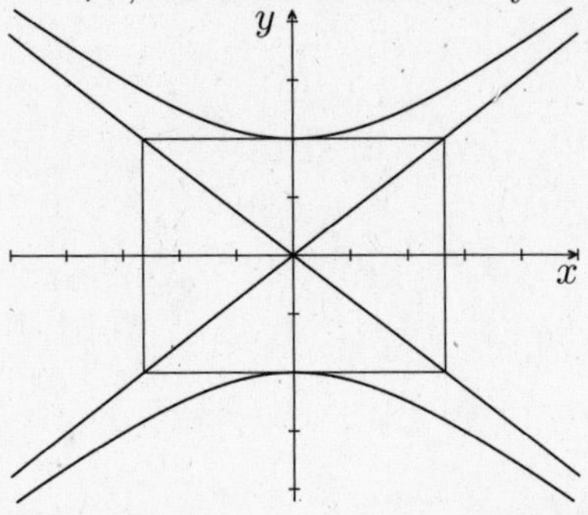

17. Parabola, axis horizontal. From $y^2 = -3x$, we have

$$y^2 = 4\left(-\frac{3}{4}\right)x. \qquad \text{inserting 4}$$

Thus, $p = -\dfrac{3}{4}$, placing the focus at $\left(-\dfrac{3}{4}, 0\right)$. (See sketch in answer section of book.)

19.
$$\begin{aligned} y^2 + 6y + 4x + 1 &= 0 \\ y^2 + 6y &= -4x - 1 \end{aligned}$$

Adding $\left[\frac{1}{2}(6)\right]^2 = 9$ to each side,

$$\begin{aligned} y^2 + 6y + 9 &= -4x - 1 + 9 \\ (y+3)^2 &= -4x + 8 \\ (y+3)^2 &= 4(-1)(x-2). \qquad p = -1 \end{aligned}$$

Parabola, vertex at $(2, -3)$, focus at $(2 - 1, -3) = (1, -3)$.

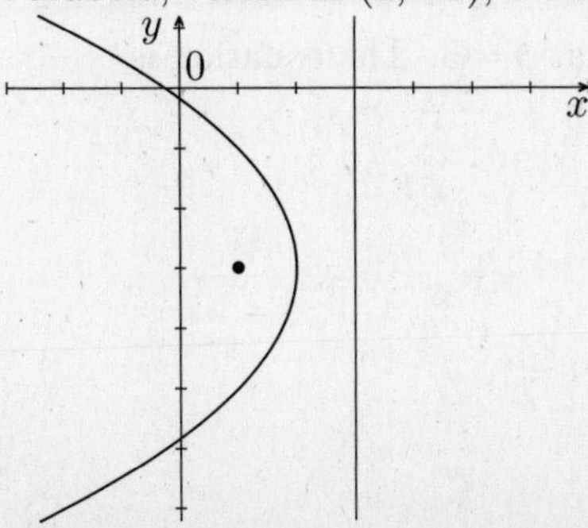

21.
$$\begin{aligned} 16x^2 - 64x + 9y^2 + 18y &= 71 \\ 16(x^2 - 4x \qquad) + 9(y^2 + 2y \qquad) &= 71 \qquad \text{factoring 16 and 9} \end{aligned}$$

Note that $\left[\frac{1}{2}(-4)\right]^2 = 4$ and $\left[\frac{1}{2}(2)\right]^2 = 1$. Inserting these values inside the parentheses and balancing the equation, we get

$$\begin{aligned} 16(x^2 - 4x + \underline{4}) + 9(y^2 + 2y + \underline{1}) &= 71 + 16 \cdot \underline{4} + 9 \cdot \underline{1} \\ 16(x-2)^2 + 9(y+1)^2 &= 144 \\ \frac{(x-2)^2}{9} + \frac{(y+1)^2}{16} &= 1. \qquad \text{dividing by 144} \end{aligned}$$

Ellipse, center at $(2, -1)$, major axis vertical, $a = 4$, $b = 3$.

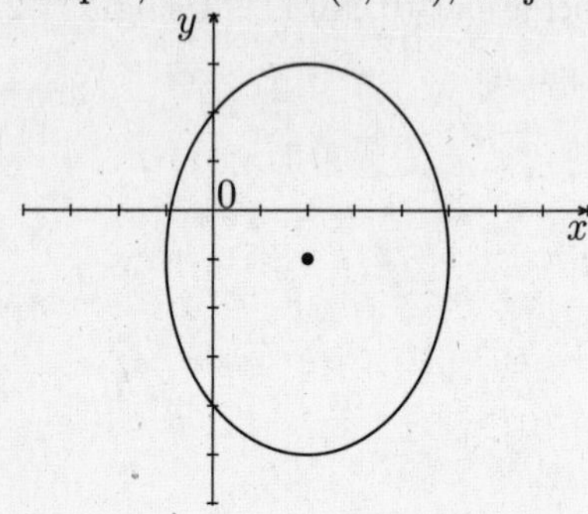

23.
$$\begin{aligned} x^2 - y^2 - 4x + 8y - 21 &= 0 \\ x^2 - 4x - y^2 + 8y &= 21 \\ (x^2 - 4x \qquad) - (y^2 - 8y \qquad) &= 21 \end{aligned}$$

Adding $\left[\frac{1}{2}(-4)\right]^2 = 4$ and $\left[\frac{1}{2}(-8)\right]^2 = 16$ inside the parentheses and balancing the equation we get

$$\begin{aligned} (x^2 - 4x + 4) - (y^2 - 8y + 16) &= 21 + 4 - 1(16) \\ (x-2)^2 - (y-4)^2 &= 9 \\ \frac{(x-2)^2}{9} - \frac{(y-4)^2}{9} &= 1. \end{aligned}$$

Hyperbola, center at $(2, 4)$.

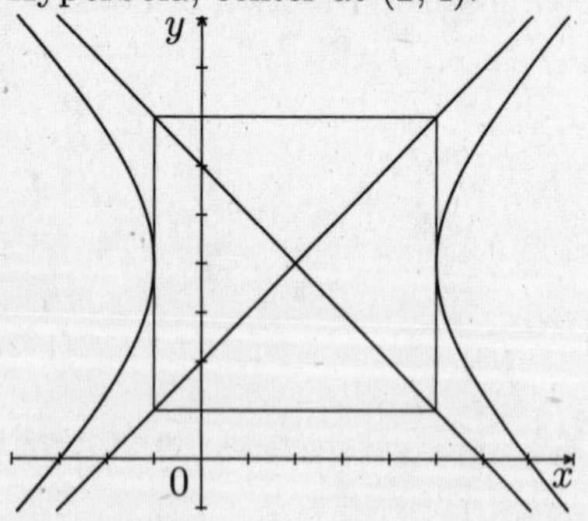

25. Form:

$$(x-h)^2 = 4p(y-k). \qquad \text{axis vertical}$$

Distance from vertex $(1, 3)$ to directrix $y = 0$ is 3, so that $p = 3$. The equation is

$$(x-1)^2 = 4(3)(y-3) \qquad (h,k) = (1,3),\ p = 3$$

or

$$(x-1)^2 = 12(y-3).$$

27. Form: $\frac{x^2}{b^2} + \frac{y^2}{a^2} = 1$; $a = 4$; $c = 3$; $b^2 = a^2 - c^2 = 16 - 9 = 7$.

Equation: $\frac{x^2}{7} + \frac{y^2}{16} = 1$.

29.

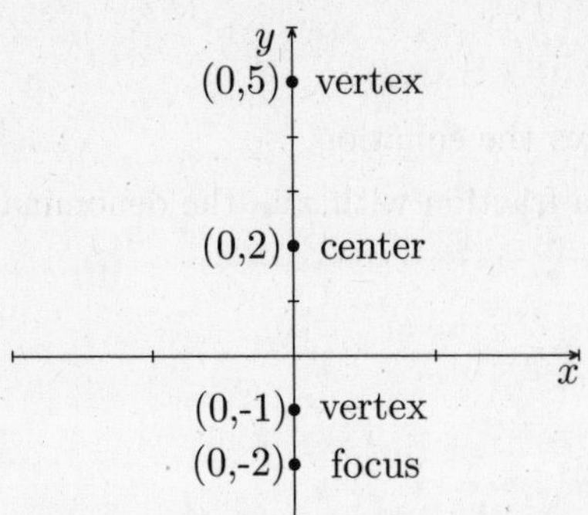

Center: $(0, 2)$ (midway between vertices). Distance from center to vertex is 3, so that $a = 3$. Distance from center to focus is 4, so that $c = 4$. Thus $b^2 = c^2 - a^2 = 16 - 9 = 7$. Since the transverse axis is vertical, the form is

$$\frac{(y-k)^2}{a^2} - \frac{(x-h)^2}{b^2} = 1,$$

and the equation is

$$\frac{(y-2)^2}{9} - \frac{x^2}{7} = 1. \qquad (h, k) = (0, 2)$$

31.

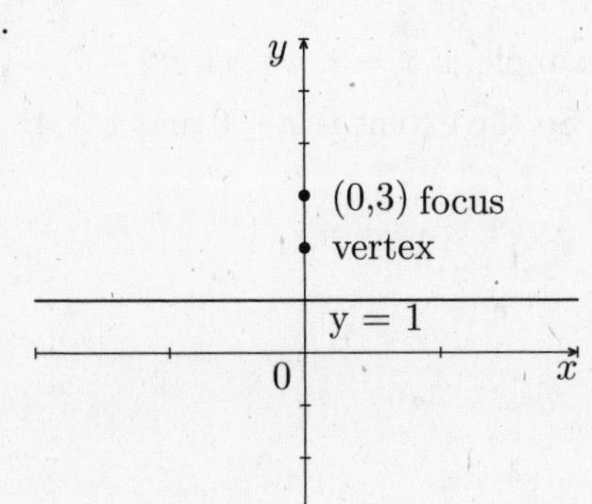

Since the vertex is midway between the focus and directrix, its coordinates are $(0, 2)$. It follows that $p = +1$; so the equation is

$$\begin{aligned} (x-0)^2 &= 4(1)(y-2) \\ x^2 &= 4(y-2). \end{aligned}$$

33.

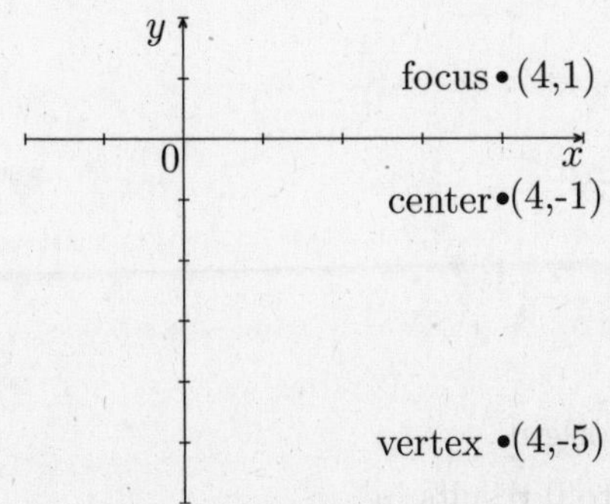

$c = 2$ (distance from center to focus); $a = 4$ (distance from center to vertex);

$b^2 = a^2 - c^2 = 16 - 4 = 12$. Form: $\dfrac{(x-h)^2}{b^2} + \dfrac{(y-k)^2}{a^2} = 1$. major axis vertical

Equation: $\dfrac{(x-4)^2}{12} + \dfrac{(y+1)^2}{16} = 1$. $(h, k) = (4, -1)$

35. $y = (x+1)^3$

Intercepts. If $x = 0$, then $y = 1$. If $y = 0$, then $x = -1$.

Symmetry. None: replacing x by $-x$ or y by $-y$ changes the equation.

Asymptotes. None: the equation is not in the form of a fraction with x in the denominator.

Extent. All x and all y.

Graph.

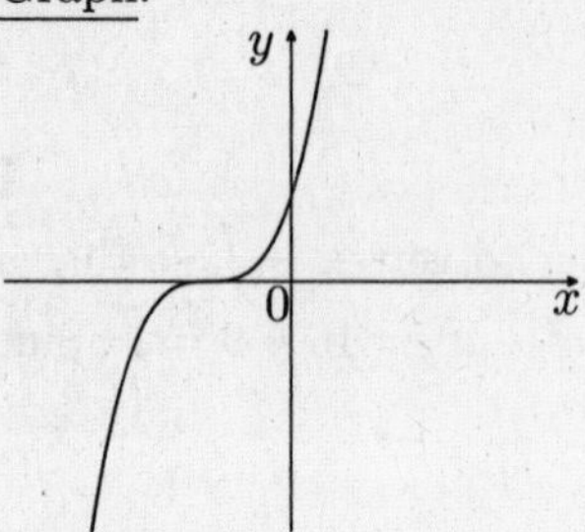

37. Intercepts. If $x = 0$, then $y = 0$; if $y = 0$, then $x(x-4) = 0$ and thus $x = 0, 4$.

Symmetry. Replacing y by $-y$, we get $(-y)^2 = x(x-4)$, which reduces to the given equation. So the curve is symmetric with respect to the x-axis.

Asymptotes. None (equation is not in the form of a fraction).

Extent. Solving for y we have

$$y = \pm\sqrt{x(x-4)}.$$

If $x > 4$, $x(x-4) > 0$. If $0 < x < 4$, $x(x-4) < 0$ [for example, if $x = 2$, we get $2(2-4) = -4$]. If $x < 0$, $x(x-4) > 0$, since both factors are negative. So the extent is $x \leq 0$ and $x \geq 4$.

Graph.

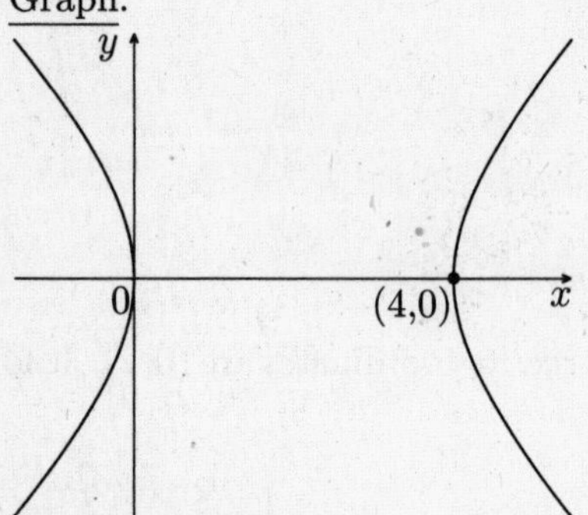

39. $y = \dfrac{x}{x^2-4}$

Intercept. $(0,0)$

Symmetry. Replacing x by $-x$ and y by $-y$ we get

$$-y = \frac{-x}{(-x)^2 - 4}$$

which reduces to

$$y = \frac{x}{x^2-4}.$$

The graph is therefore symmetric with respect to the origin.

Asymptotes. Vertical: setting the denominator equal to 0 results in

$$x^2 - 4 = 0 \quad \text{and} \quad x = \pm 2.$$

If x gets large, y approaches 0, so that the x-axis is a horizontal asymptote.

Extent. All x except $x = 2$ and $x = -2$.

Graph.

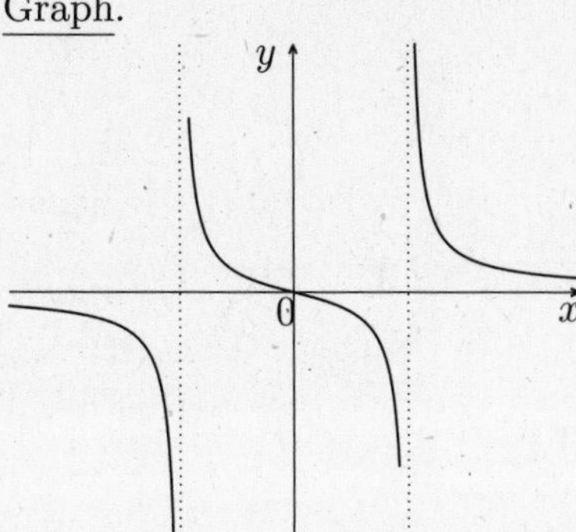

41. Placing the vertex at the origin, one point on the parabola is $(0.90, 0.60)$, as shown in the figure. The form is $y^2 = 4px$. To find p, we substitute the coordinates of the point in the equation:

$$\begin{aligned} (0.60)^2 &= 4p(0.90) \\ p &= \frac{(0.60)^2}{(4)(0.90)} = 0.10. \end{aligned}$$

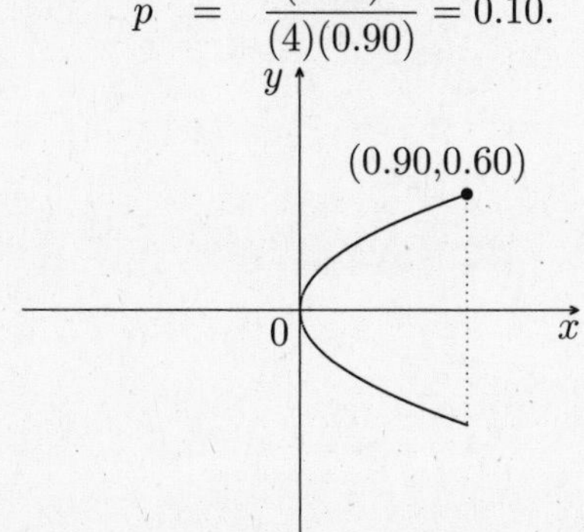

To be at the focus, the light must be placed 0.10 feet from the vertex.

43.

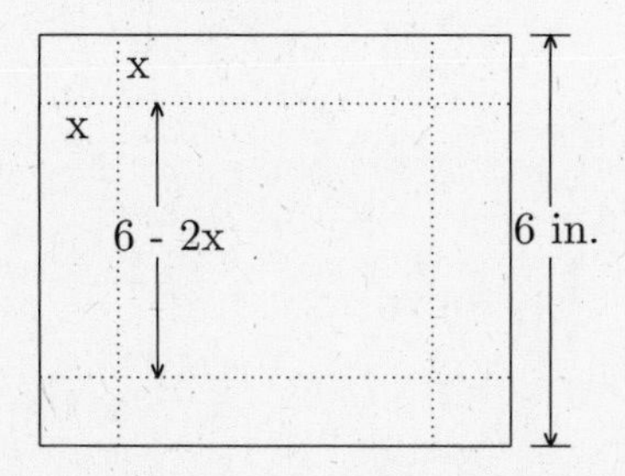

Volume=length×width×height.

$$\begin{aligned} V &= (6-2x)(6-2x)x \\ &= x(6-2x)^2 \end{aligned}$$

See graph in answer section of book.

45. See Exercise 41, Section 1.9:

$$e = \frac{A-P}{A+P};$$

$P = 4000 + 119 = 4119\,\text{mi}$; $A = 4000 + 122,000 = 126,000\,\text{mi}$;

$$e = \frac{126,000 - 4119}{126,000 + 4119} = 0.94.$$

Chapter 2

Introduction to Calculus: The Derivative

2.1 Functions and Intervals

13. The two half-circles have a total length of $2\pi r$. So $L = 200 + 2\pi r$.

15. From the given equation $y = x + 2$, we see that y is defined for all x. From $x = y - 2$ it follows that x is defined for all y.

17. To avoid imaginary values, we must have $x - 4 \geq 0$ or $x \geq 4$. We can also say that the domain is the interval $[4, \infty)$. When $x = 4$, then $y = 0$. By definition of principal square root, y cannot be negative. So the range is $y \geq 0$ or $[0, \infty)$.

19. For the y-values to be real, the radicand must be positive or zero:

$$\begin{aligned} 4 - x^2 &\geq 0 & \\ 4 &\geq x^2 & \text{adding } x^2 \text{ to both sides} \\ x^2 &\leq 4 & \text{same inequality} \\ -2 \leq x &\leq 2 & \text{domain} \end{aligned}$$

If $x = 0$, $y = \sqrt{4} = 2$; $y = 2$ is the largest possible value of y (if $x \neq 0$, then $y < 2$). If $x = \pm 2$, then $y = 0$. So the range is $0 \leq y \leq 2$.

21. $y = x\sqrt{x - 3}$. To avoid imaginary values, we require that $x - 3 \geq 0$ or $x \geq 3$.

23. For $\sqrt{x - 2}$ to be real, we must have $x \geq 2$. So the domain is $x \geq 2$. For the x-values under discussion, $y \geq 0$, which is the range.

25. $y = \dfrac{1}{x - 1}$. Domain: all x except $x = 1$ (to avoid division by 0).
Range: all y except $y = 0$ because the x-axis ($y = 0$) is a horizontal asymptote. This can also be seen by solving the equation for x in terms of y:

$$x = \frac{y + 1}{y}$$

y cannot be equal to zero.

27. For the y-values to be real, the radicand must be positive or zero:

$$\begin{aligned} 1-x^2 &\geq 0 & \\ 1 &\geq x^2 & \text{adding } x^2 \text{ to both sides} \\ x^2 &\leq 1 & \\ -1 \leq x &\leq 1. & \text{domain} \end{aligned}$$

If $x = 0$, then $y = 1 + \sqrt{1} = 2$; $y = 2$ is the largest possible value. (Whenever $x \neq 0$, $y < 2$.) If $x = \pm 1$, then $y = 1$. So the range is the closed interval $[1, 2]$.

29. $f(2) = 2^3 = 8$; $f(-2) = (-2)^3 = -8$

31.
$$\begin{aligned} f(x) &= 2x & \\ f(0) &= 2 \cdot 0 = 0 & x = 0 \\ f(6) &= 2 \cdot 6 = 12 & x = 6 \end{aligned}$$

33. $h(x) = x^2 + 2x$. Leaving a blank space for x, we get $h(\ ) = (\ )^2 + 2(\ )$.
Now fill in the blanks:

$$\begin{aligned} h(1) &= 1^2 + 2(1) = 3 \\ h(3) &= 3^2 + 2(3) = 15 \end{aligned}$$

35. $\phi(x) = \dfrac{1}{x}$; $\phi(3) = \dfrac{1}{3}$ and $\phi(a) = \dfrac{1}{a}$

37. $G(z) = \sqrt{z^2 - 1}$. Leaving a blank space for z, we get $G(\ ) = \sqrt{(\ )^2 - 1}$.
Now fill in the blanks:

$$\begin{aligned} G(a^2) &= \sqrt{(a^2)^2 - 1} = \sqrt{a^4 - 1} \\ G(x-1) &= \sqrt{(x-1)^2 - 1} = \sqrt{x^2 - 2x} \end{aligned}$$

39. $f(x) = 1 - x^2$; $f(\ ) = 1 - (\ )^2$. Filling in the blanks:

$$f(x + \Delta x) = 1 - (x + \Delta x)^2 = 1 - x^2 - 2x\Delta x - (\Delta x)^2$$
$$f(x - \Delta x) = 1 - (x - \Delta x)^2 = 1 - x^2 + 2x\Delta x - (\Delta x)^2$$

41. $f(x) = x^2$; $g(x) = x + 1$; $f(\ ) = (\ )^2$; $g(\ ) = (\ ) + 1$.
Filling in the blanks:

$$\begin{aligned} (a)\quad f(g(x)) &= (x+1)^2 & g(x) = x + 1 \\ &= x^2 + 2x + 1 & \\ (b)\quad g(f(x)) &= (x^2) + 1 & f(x) = x^2 \\ &= x^2 + 1 & \\ (c)\quad f(f(x)) &= (x^2)^2 & f(x) = x^2 \\ &= x^4 & \end{aligned}$$

43. $f(x) = x^3$ and $g(x) = \sqrt[3]{x}$
$f(g(x)) = \left(\sqrt[3]{x}\right)^3 = x$; $g(f(x)) = \sqrt[3]{x^3} = x$

45. See graph in answer section of book.

2.2 Limits

1. Set your calculator in the radian mode and evaluate $\dfrac{\tan x}{x}$ for values near 0:

x:	0.1	0.05	0.01	0.001
$\frac{\tan x}{x}$	1.003	1.0008	1.00003	1.0000003

The values approach 1.

5. Set your calculator in the radian mode and evaluate $\sec x - \tan x$ for values near $\frac{\pi}{2} = 1.570796$.

x :	1.5 ($< \frac{\pi}{2}$)	1.6 ($> \frac{\pi}{2}$)	1.57 ($< \frac{\pi}{2}$)	1.571 ($> \frac{\pi}{2}$)	1.5705 ($< \frac{\pi}{2}$)
$\sec x - \tan x$:	0.04	-0.01	4.0×10^{-4}	-1.0×10^{-4}	1.5×10^{-4}

Based on these calculations, the limit appears to be 0.

7. Since

$$\frac{x^2 + 4x}{x} = \frac{x(x+4)}{x} = x + 4,$$

the two functions are

$$y = \frac{x^2 + 4x}{x} \quad \text{and} \quad y = x + 4.$$

As $x \to 0$, $y \to 4$.

9. Since

$$\frac{x^2 - 3x - 4}{x - 4} = \frac{(x-4)(x+1)}{x-4} = x + 1,$$

the two functions are

$$y = \frac{x^2 - 3x - 4}{x - 4} \quad \text{and} \quad y = x + 1.$$

As $x \to 4$, $y \to 5$.

11. The function $y = 4x$ is defined for all x. As x approaches 0, y approaches 0.

13. The function $y = x^3 + 2x - 4$ is defined for all x. As x approaches 1, y approaches -1.

15. Since the function $f(x) = \dfrac{x^2 - 4}{x + 2}$ is defined at $x = 2$, we obtain the limit by inspection: as $x \to 2$

$$\frac{x^2 - 4}{x + 2} \to 0$$

17. Since the function $f(x) = \dfrac{x^2 - 4}{x - 2}$ is undefined at $x = 2$, we evaluate the limit as follows:

$$\lim_{x \to 2} \frac{x^2 - 4}{x - 2} = \lim_{x \to 2} \frac{(x-2)(x+2)}{x - 2} = \lim_{x \to 2} (x + 2) = 4$$

(Since x approaches 2 but is never equal to 2, division by 0 has been avoided.)

19. The function is undefined at $x = 0$:

$$\lim_{x \to 0} \frac{x(x-2)}{2x} = \lim_{x \to 0} \frac{(x-2)}{2} = -1$$

21. The function is undefined at $x = 2$:

$$\lim_{x \to 2} \frac{x^2 - 4x + 4}{x - 2} = \lim_{x \to 2} \frac{(x-2)^2}{x - 2} = \lim_{x \to 2} (x - 2) = 0$$

23. The function is undefined at $x = 2$:

$$\lim_{x \to 2} \frac{x^2 + x - 6}{x - 2} = \lim_{x \to 2} \frac{(x-2)(x+3)}{x - 2} = \lim_{x \to 2} (x + 3) = 5$$

25. The function is defined at $x = 4$. So we obtain the limit by inspection: as $x \to 4$

$$\frac{x^2 + x - 8}{x + 4} \to \frac{4^2 + 4 - 8}{4 + 4} = \frac{12}{8} = \frac{3}{2}$$

27. $\lim_{x\to 1} \dfrac{x^3 - 2x^2 + x}{x - 1} = \lim_{x\to 1} \dfrac{x(x^2 - 2x + 1)}{x} = \lim_{x\to 1}(x^2 - 2x + 1) = 0$

29. $\lim_{r\to 3} \dfrac{r^2 - 9}{3 - r} = \lim_{r\to 3} \dfrac{(r-3)(r+3)}{(-1)(r-3)} = \lim_{r\to 3} \dfrac{r+3}{-1} = -6$

31. $\lim_{x\to 5} \dfrac{25 - x^2}{5 - x} = \lim_{x\to 5} \dfrac{(5-x)(5+x)}{5 - x} = \lim_{x\to 5}(5 + x) = 10$

33. Since the function $f(x) = \dfrac{4x^2 - 2}{x}$ is defined at $x = 1$, we obtain the limit by inspection: as $x \to 1$

$$\frac{4x^2 - 2}{1} \to 2$$

35. Dividing numerator and denominator by x^3, the highest power of x, we obtain

$$\lim_{x\to\infty} \frac{4 - 2/x^2 + 1/x^3}{5 + 3/x - 1/x^2} = \frac{4 - 0 + 0}{5 + 0 - 0} = \frac{4}{5}$$

37. Dividing numerator and denominator by x^2, the highest power of x, we get

$$\lim_{x\to\infty} \frac{3 + 2/x}{4 - 3/x^2} = \frac{3 + 0}{4 - 0} = \frac{3}{4}$$

39. Dividing numerator and denominator by y^2, the highest power of y, we obtain

$$\lim_{y\to\infty} \frac{1 - 4y}{y^2 + 1} = \lim_{y\to\infty} \frac{1/y^2 - 4/y}{1 + 1/y^2} = \frac{0}{1} = 0$$

(The conclusion also follows from the fact that the denominator is of higher degree than the numerator.)

41. Dividing numerator and denominator by t^2, we get

$$\lim_{t\to\infty} \frac{\frac{1}{t} - 3}{\frac{7}{t^2} - \frac{2}{t} - 9} = \frac{-3}{-9} = \frac{1}{3}$$

43. Rationalize the numerator:

$$\lim_{x\to\infty} \frac{\sqrt{x^2+4} - x}{1} \frac{\sqrt{x^2+4} + x}{\sqrt{x^2+4} + x} = \lim_{x\to\infty} \frac{\left(\sqrt{x^2+4}\right)^2 - x^2}{\sqrt{x^2+4} + x}$$

$$= \lim_{x\to\infty} \frac{x^2 + 4 - x^2}{\sqrt{x^2+4} + x} = \lim_{x\to\infty} \frac{4}{\sqrt{x^2+4} + x} = 0$$

45. Rationalize the numerator:

$$\lim_{x\to\infty} \left(x - \sqrt{x^2 - 9}\right) = \lim_{x\to\infty} \frac{x - \sqrt{x^2-9}}{1} \frac{x + \sqrt{x^2-9}}{x + \sqrt{x^2+9}} = \lim_{x\to\infty} \frac{x^2 - \left(\sqrt{x^2-9}\right)^2}{x + \sqrt{x^2+9}}$$

$$= \lim_{x\to\infty} \frac{x^2 - (x^2 - 9)}{x + \sqrt{x^2 - 9}} = \lim_{x\to\infty} \frac{9}{x + \sqrt{x^2 - 9}} = 0$$

47. $\lim_{x\to 3^+} \sqrt{x - 3} = 0$

49. (a) As $x \to 0$ from the right (through positive values), $1/x$ gets large beyond all bounds. So

$$\lim_{x\to 0+} 3^{1/x} = \infty. \quad \text{(no limit)}$$

(b) As $x \to 0$ from the left (through negative values), $1/x$ becomes large and negative. As a result

$$\lim_{x \to 0-} 3^{1/x} = 0.$$

51. By the way the function is defined, its value at $x = 2$ is -1, that is, $f(2) = -1$. But this value does not match the right-hand limit: $\lim_{x \to 2+} f(x) = 1$.
So the condition $\lim_{x \to a} f(x) = f(a)$ has been violated.

53. $\lim_{x \to -1} \frac{x^2 - 1}{x + 1} = \lim_{x \to -1} \frac{(x-1)(x+1)}{x+1} = -2$
But $f(x)$ is undefined at $x = -1$, that is, $f(-1)$ does not exist, even though the limit exists. This violates the condition

$$\lim_{x \to -1} f(x) = f(-1).$$

2.4 The Derivative by the Four-Step Process

1. To find $f(x + \Delta x)$, write $f(\ \ ) = 2(\ \ ) + 1$ and fill in the blanks.
   Step 1. $y + \Delta y = f(x + \Delta x) = 2(x + \Delta x) + 1$
   Step 2. $\Delta y = 2(x + \Delta x) + 1 - (2x + 1) = 2x + 2\Delta x + 1 - 2x - 1 = 2\Delta x$
   Step 3. $\frac{\Delta y}{\Delta x} = \frac{2\Delta x}{\Delta x} = 2$
   Step 4. $f'(x) = \lim_{\Delta x \to 0} \frac{\Delta y}{\Delta x} = \lim_{\Delta x \to 0} 2 = 2$

3. To find $f(x + \Delta x)$, write $f(\ \ ) = 2 - 3(\ \ )$ and fill in the blanks.
   Step 1. $y + \Delta y = f(x + \Delta x) = 2 - 3(x + \Delta x)$
   Step 2. $\Delta y = 2 - 3(x + \Delta x) - (2 - 3x) = 2 - 3x - 3\Delta x - 2 + 3x = -3\Delta x$
   Step 3. $\frac{\Delta y}{\Delta x} = \frac{-3\Delta x}{\Delta x} = -3$
   Step 4. $f'(x) = \lim_{\Delta x \to 0} \frac{\Delta y}{\Delta x} = \lim_{\Delta x \to 0} (-3) = -3$

5. To find $f(x + \Delta x)$, write $f(\ \ ) = (\ \ )^2 + 1$ and fill in the blanks.
   Step 1. $y + \Delta y = f(x + \Delta x) = (x + \Delta x)^2 + 1$
   Step 2. $\Delta y = (x + \Delta x)^2 + 1 - (x^2 + 1) = x^2 + 2x\Delta x + (\Delta x)^2 + 1 - x^2 - 1 = 2x\Delta x + (\Delta x)^2$
   Step 3. $\frac{\Delta y}{\Delta x} = \frac{2x\Delta x + (\Delta x)^2}{\Delta x} = \frac{\Delta x(2x + \Delta x)}{\Delta x} = 2x + \Delta x$
   Step 4. $f'(x) = \lim_{\Delta x \to 0} \frac{\Delta y}{\Delta x} = \lim_{\Delta x \to 0} (2x + \Delta x) = 2x$

7. To find $f(x + \Delta x)$, write $f(\ \ ) = (\ \ )^2 - 2(\ \ )$ and fill in the blanks.
   Step 1. $y + \Delta y = f(x + \Delta x) = (x + \Delta x)^2 - 2(x + \Delta x)$
   Step 2.
   $$\begin{aligned} \Delta y &= (x + \Delta x)^2 - 2(x + \Delta x) - (x^2 - 2x) \\ &= x^2 + 2x\Delta x + (\Delta x)^2 - 2x - 2\Delta x - x^2 + 2x \\ &= 2x\Delta x + (\Delta x)^2 - 2\Delta x \end{aligned}$$
   Step 3. $\frac{\Delta y}{\Delta x} = \frac{2x\Delta x + (\Delta x)^2 - 2\Delta x}{\Delta x} = \frac{\Delta x(2x + \Delta x - 2)}{\Delta x} = 2x + \Delta x - 2$
   Step 4. $f'(x) = \lim_{\Delta x \to 0} \frac{\Delta y}{\Delta x} = \lim_{\Delta x \to 0} (2x + \Delta x - 2) = 2x - 2$

9. $f(\ ) = (\ )^3 - 2(\ )$

$$\begin{aligned}
\underline{\text{Step 1.}}\quad y+\Delta y &= f(x+\Delta x) = (x+\Delta x)^3 - 2(x+\Delta x) \\
\underline{\text{Step 2.}}\quad \Delta y &= (x+\Delta x)^3 - 2(x+\Delta x) - (x^3 - 2x) \\
&= x^3 + 3x^2\Delta x + 3x(\Delta x)^2 + (\Delta x)^3 - 2x - 2\Delta x - x^3 + 2x \\
&= 3x^2\Delta x + 3x(\Delta x)^2 + (\Delta x)^3 - 2\Delta x \\
\underline{\text{Step 3.}}\quad \frac{\Delta y}{\Delta x} &= \frac{\Delta x[3x^2 + 3x\Delta x + (\Delta x)^2 - 2]}{\Delta x} \\
&= 3x^2 + 3x\Delta x + (\Delta x)^2 - 2 \\
\underline{\text{Step 4.}}\quad f'(x) &= \lim_{\Delta x\to 0}[3x^2 + 3x\Delta x + (\Delta x)^2 - 2] \\
&= 3x^2 - 2
\end{aligned}$$

11. $\underline{\text{Step 1.}}\ y+\Delta y = \dfrac{1}{x+\Delta x}$

$$\begin{aligned}
\underline{\text{Step 2.}}\quad \Delta y &= \frac{1}{x+\Delta x} - \frac{1}{x} \\
&= \frac{1}{x+\Delta x}\cdot\frac{x}{x} - \frac{1}{x}\cdot\frac{x+\Delta x}{x+\Delta x} \\
&= \frac{x - x - \Delta x}{x(x+\Delta x)} = \frac{-\Delta x}{x(x+\Delta x)}
\end{aligned}$$

$\underline{\text{Step 3.}}\ \dfrac{\Delta y}{\Delta x} = -\dfrac{\Delta x}{x(x+\Delta x)}\cdot\dfrac{1}{\Delta x} = -\dfrac{1}{x(x+\Delta x)}$

$\underline{\text{Step 4.}}\ f'(x) = \lim\limits_{\Delta x\to 0}\dfrac{\Delta y}{\Delta x} = \lim\limits_{\Delta x\to 0}\left[-\dfrac{1}{x(x+\Delta x)}\right] = -\dfrac{1}{x^2}$

13. $\underline{\text{Step 1.}}\ y+\Delta y = \dfrac{1}{x+\Delta x+1}$

$$\begin{aligned}
\underline{\text{Step 2.}}\quad \Delta y &= \frac{1}{x+\Delta x+1} - \frac{1}{x+1} \\
&= \frac{1}{x+\Delta x+1}\cdot\frac{x+1}{x+1} - \frac{1}{x+1}\cdot\frac{x+\Delta x+1}{x+\Delta x+1} \\
&= \frac{(x+1)-(x+\Delta x+1)}{(x+\Delta x+1)(x+1)} = \frac{-\Delta x}{(x+\Delta x+1)(x+1)}
\end{aligned}$$

$\underline{\text{Step 3.}}\ \dfrac{\Delta y}{\Delta x} = \dfrac{-\Delta x}{(x+\Delta x+1)(x+1)}\cdot\dfrac{1}{\Delta x} = \dfrac{-1}{(x+\Delta x+1)(x+1)}$

$\underline{\text{Step 4.}}\ f'(x) = \lim\limits_{\Delta x\to 0}\dfrac{\Delta y}{\Delta x} = \lim\limits_{\Delta x\to 0}\dfrac{-1}{(x+\Delta x+1)(x+1)} = -\dfrac{1}{(x+1)^2}$

15. $\underline{\text{Step 1.}}\ y+\Delta y = \dfrac{1}{1-(x+\Delta x)^2}$

$\underline{\text{Step 2.}}\ \Delta y = \dfrac{1}{1-(x+\Delta x)^2} - \dfrac{1}{1-x^2} = \dfrac{(1-x^2)-[1-(x+\Delta x)^2]}{[1-(x+\Delta x)^2](1-x^2)} = \dfrac{2x\Delta x + (\Delta x)^2}{[1-(x+\Delta x)^2](1-x^2)}$

$\underline{\text{Step 3.}}\ \dfrac{\Delta y}{\Delta x} = \dfrac{\Delta x(2x+\Delta x)}{[1-(x+\Delta x)^2](1-x^2)}\cdot\dfrac{1}{\Delta x} = \dfrac{2x+\Delta x}{[1-(x+\Delta x)^2](1-x^2)}$

$\underline{\text{Step 4.}}\ f'(x) = \lim\limits_{\Delta x\to 0}\dfrac{\Delta y}{\Delta x} = \lim\limits_{\Delta x\to 0}\dfrac{2x+\Delta x}{[1-(x+\Delta x)^2](1-x^2)} = \dfrac{2x}{(1-x^2)^2}$

17. Step 1. $y + \Delta y = f(x + \Delta x) = \sqrt{x + \Delta x}$

Step 2. $\Delta y = \sqrt{x + \Delta x} - \sqrt{x}$

As in Example 4, we rationalize the numerator by multiplying both numerator and denominator by the quantity $\sqrt{x + \Delta x} - \sqrt{x}$:

$$\begin{aligned} \Delta y &= \frac{\sqrt{x + \Delta x} - \sqrt{x}}{1} \cdot \frac{\sqrt{x + \Delta x} + \sqrt{x}}{\sqrt{x + \Delta x} + \sqrt{x}} \\ &= \frac{(\sqrt{x + \Delta x})^2 - (\sqrt{x})^2}{\sqrt{x + \Delta x} + \sqrt{x}} = \frac{x + \Delta x - x}{\sqrt{x + \Delta x} + \sqrt{x}} \\ &= \frac{\Delta x}{\sqrt{x + \Delta x} + \sqrt{x}} \end{aligned}$$

Step 3. $\dfrac{\Delta y}{\Delta x} = \dfrac{\Delta x}{\sqrt{x + \Delta x} + \sqrt{x}} \cdot \dfrac{1}{\Delta x} = \dfrac{1}{\sqrt{x + \Delta x} + \sqrt{x}}$

Step 4. $f'(x) = \lim\limits_{\Delta x \to 0} \dfrac{1}{\sqrt{x + \Delta x} + \sqrt{x}} = \dfrac{1}{\sqrt{x} + \sqrt{x}} = \dfrac{1}{2\sqrt{x}}$

19. Step 1. $y + \Delta y = \sqrt{1 - (x + \Delta x)}$

Step 2. $\Delta y = \sqrt{1 - (x + \Delta x)} - \sqrt{1 - x}$

Step 3.

$$\begin{aligned} \frac{\Delta y}{\Delta x} &= \frac{\sqrt{1 - (x + \Delta x)} - \sqrt{1 - x}}{\Delta x} \cdot \frac{\sqrt{1 - (x + \Delta x)} + \sqrt{1 - x}}{\sqrt{1 - (x + \Delta x)} + \sqrt{1 - x}} \\ &= \frac{[1 - (x + \Delta x)] - (1 - x)}{\Delta x[\sqrt{1 - (x + \Delta x)} + \sqrt{1 - x}]} = \frac{-1}{\sqrt{1 - (x + \Delta x)} + \sqrt{1 - x}} \end{aligned}$$

Step 4. $f'(x) = \lim\limits_{\Delta x \to 0} \dfrac{\Delta y}{\Delta x} = \lim\limits_{\Delta x \to 0} \dfrac{-1}{\sqrt{1 - (x + \Delta x)} + \sqrt{1 - x}} = -\dfrac{1}{2\sqrt{1 - x}}$

21. Similar to Exercise 5: $f'(x) = 2x$. Thus $f'(-1) = 2(-1) = -2$; $f'(2) = 4$; $f'(3) = 6$.

23. Step 1. $y + \Delta y = \dfrac{1}{\sqrt{x + \Delta x}}$

Step 2. $\Delta y = \dfrac{1}{\sqrt{x + \Delta x}} - \dfrac{1}{\sqrt{x}}$

Step 3.

$$\begin{aligned} \frac{\Delta y}{\Delta x} &= \frac{1}{\Delta x} \cdot \frac{\sqrt{x} - \sqrt{x + \Delta x}}{\sqrt{x + \Delta x}\sqrt{x}} \cdot \frac{\sqrt{x} + \sqrt{x + \Delta x}}{\sqrt{x} + \sqrt{x + \Delta x}} \\ &= \frac{1}{\Delta x} \cdot \frac{x - (x + \Delta x)}{\sqrt{x + \Delta x}\sqrt{x}(\sqrt{x} + \sqrt{x + \Delta x})} \\ &= \frac{-1}{\sqrt{x + \Delta x}\sqrt{x}(\sqrt{x} + \sqrt{x + \Delta x})} \end{aligned}$$

Step 4. $f'(x) = \lim\limits_{\Delta x \to 0} \dfrac{-1}{\sqrt{x + \Delta x}\sqrt{x}(\sqrt{x} + \sqrt{x + \Delta x}} = \dfrac{-1}{\sqrt{x}\sqrt{x}(\sqrt{x} + \sqrt{x})} = -\dfrac{1}{2x\sqrt{x}}$

$$f'(4) = -\frac{1}{2 \cdot 4\sqrt{4}} = -\frac{1}{16};\ f'(9) = -\frac{1}{2 \cdot 9\sqrt{9}} = -\frac{1}{54}$$

25. First we need to find the slope of the tangent line at the given point. From $f(\ ) = (\ )^2 - 1$:

$$\begin{aligned}
\underline{\text{Step 1.}}\quad y+\Delta y &= f(x+\Delta x) = (x+\Delta x)^2 - 1\\
\underline{\text{Step 2.}}\quad \Delta y &= (x+\Delta x)^2 - 1 - (x^2-1)\\
&= x^2 + 2x\Delta x + (\Delta x)^2 - 1 - x^2 + 1\\
&= 2x\Delta x + (\Delta x)^2\\
&= \Delta x(2x+\Delta x)\\
\underline{\text{Step 3.}}\quad \frac{\Delta y}{\Delta x} &= \frac{\Delta x(2x+\Delta x)}{\Delta x} = 2x + \Delta x\\
\underline{\text{Step 4.}}\quad f'(x) &= \lim_{\Delta x\to 0}(2x+\Delta x) = 2x
\end{aligned}$$

So $f'(x) = 2x$ and $f'(2) = 4$. Using the point-slope form $y - y_1 = m(x - x_1)$, we have

$$\begin{aligned}
y - 3 &= 4(x-2) \qquad (x_1, y_1) = (2,3);\ m = 4\\
y &= 4x - 5.
\end{aligned}$$

$$\begin{aligned}
27.\quad f'(x) &= \lim_{\Delta x\to 0}\frac{\frac{2}{x+\Delta x} - \frac{2}{x}}{\Delta x}\\
&= \lim_{\Delta x\to 0}\frac{1}{\Delta x}\cdot\frac{2x - 2x - 2\Delta x}{(x+\Delta x)x}\\
&= \lim_{\Delta x\to 0}\frac{-2}{(x+\Delta x)x} = \frac{-2}{x^2}
\end{aligned}$$

Thus $f'(x) = \dfrac{-2}{x^2}$ and $f'(2) = -\dfrac{2}{2^2} = -\dfrac{1}{2}$. Using the point-slope form $y - y_1 = m(x - x_1)$, we have

$$\begin{aligned}
y - 1 &= -\tfrac{1}{2}(x-2) \qquad (x_1, y_1) = (2,1)\\
y &= \tfrac{1}{2}(4-x).
\end{aligned}$$

2.5 Derivatives of Polynomials

$$\begin{aligned}
9.\quad y &= 5x^3 - 7x^2 + 2\\
y' &= 5(3x^2) - 7(2x) + 0\\
&= 15x^2 - 14x
\end{aligned}$$

$$\begin{aligned}
11.\quad y &= 7x^3 - x^2 - x + 2\\
y' &= 7(3x^2) - 2x - 1 + 0 = 21x^2 - 2x - 1
\end{aligned}$$

$$\begin{aligned}
13.\quad y &= \tfrac{1}{4}x^4 + \tfrac{1}{2}x^2 + x\\
y' &= \tfrac{1}{4}(4x^3) + \tfrac{1}{2}(2x) + 1\\
&= x^3 + x + 1
\end{aligned}$$

15. $y = \frac{x^2}{2} - \frac{x^3}{3} + 4^2 = \frac{1}{2}x^2 - \frac{1}{3}x^3 + 16$

$y' = \frac{1}{2}(2x) - \frac{1}{3}(3x^2) + 0 = x - x^2$

$$\begin{aligned}
17.\quad y &= 7x^9 - 5x^7 - 3x^2 - \sqrt{2}\\
y' &= 7(9x^8) - 5(7x^6) - 3(2x) + 0 \qquad \text{note that } \sqrt{2} \text{ is a constant}\\
&= 63x^8 - 35x^6 - 6x
\end{aligned}$$

19.
$$\begin{aligned} y &= \sqrt{2}t^3 - \sqrt{5}t + \sqrt{2} + \pi^2 \\ y' &= \sqrt{2}\,(3t^2) - \sqrt{5}(1) + 0 \\ &= 3\sqrt{2}t^2 - \sqrt{5} \end{aligned}$$

21.
$$\begin{aligned} y &= \tfrac{1}{6}R^6 + \tfrac{1}{5}R^4 - \sqrt{3} \\ \frac{dy}{dR} &= \tfrac{1}{6}(6R^5) + \tfrac{1}{5}(4R^3) - 0 \\ &= R^5 + \tfrac{4}{5}R^3 \end{aligned}$$

23. $f(x) = 4 - x^2$; $f'(x) = -2x$; $f'(2) = -4$. Slope of tangent line: -4; slope of normal line: $\frac{1}{4}$; point: $(2, 0)$.

$$\begin{aligned} y - y_1 &= m(x - x_1) \\ y - 0 &= -4(x - 2) \qquad \text{tangent line} \\ y - 0 &= \tfrac{1}{4}(x - 2) \qquad \text{normal line} \end{aligned}$$

(See graphs in answer section in the back of the book.)

25. $f(x) = \frac{1}{3}x^3 + x$; $f'(x) = x^2 + 1$; $f'(1) = 1^2 + 1 = 2$. Slope of tangent line: 2; slope of normal line: $-\frac{1}{2}$; point: $(1, \frac{4}{3})$.

$$\begin{aligned} y - y_1 &= m(x - x_1) \\ y - \tfrac{4}{3} &= 2(x - 1) \qquad \text{tangent line} \\ y - \tfrac{4}{3} &= -\tfrac{1}{2}(x - 1) \qquad \text{normal line} \end{aligned}$$

(Simplified forms and graphs are in the answer section in the back of the book.)

27. $y = cu$ where u is a function of x.

Step 1. $y + \Delta y = c(u + \Delta u)$

Step 2. $\Delta y = c(u + \Delta u) - cu = c\Delta u$

Step 3. $\dfrac{\Delta y}{\Delta x} = c\dfrac{\Delta u}{\Delta x}$

Step 4. $\dfrac{dy}{dx} = \lim\limits_{\Delta x \to 0} \dfrac{\Delta y}{\Delta x} = \lim\limits_{\Delta x \to 0} c\dfrac{\Delta u}{\Delta x} = c\dfrac{du}{dx}$

2.6 Instantaneous Rates of Change

1.
$$\begin{aligned} s &= 2t^2 \\ v &= \frac{ds}{dt} = 2(2t) = 4t \\ a &= \frac{dv}{dt} = 4 \end{aligned}$$

$v = 0$ when $4t = 0$ or $t = 0$.

3.
$$\begin{aligned} s &= t^2 - 2t + 1 \\ v &= \frac{ds}{dt} = 2t - 2 \\ a &= \frac{dv}{dt} = 2 \end{aligned}$$

$v = 0$ when $2t - 2 = 0$ or $t = 1$.

5. $s = 2t - t^2$

a. $v = \dfrac{ds}{dt} = 2 - 2t$

b. $v = 0$ when $2 - 2t = 0$, or $t = 1$. When $t = 1$, $s = 2(1) - 1^2 = 1$; so the particle is momentarily at rest at $s = 1$.

c. When $t < 1$, then $v > 0$; so the particle is moving to the right before coming to rest.

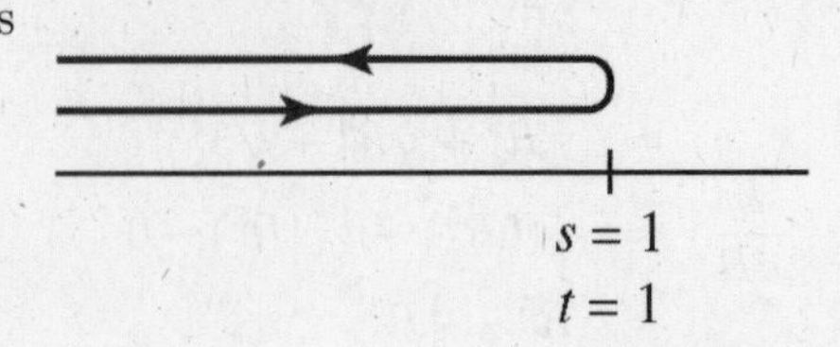

7. $s = t^3 - 3t^2$

a. $v = \dfrac{ds}{dt} = 3t^2 - 6t$

b. $v = 0$ when
$$\begin{aligned} 3t^2 - 6t &= 0 \\ 3t(t-2) &= 0 \\ t &= 0, 2 \end{aligned}$$

When $t = 0$, $s = 0$ and when $t = 2$, $s = -4$; so the particle is at rest at $s = 0$ and $s = -4$.

c. From $v = 3t(t-2)$, we deduce that $v > 0$ when $t > 2$ or when $t < 0$ (particle is moving to the right). The particle moves to the left in the time interval $0 < t < 2$.

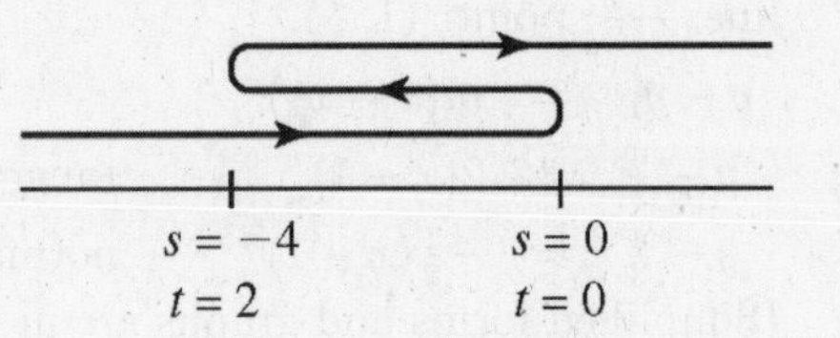

9. $s = 2 - 2t^3 + 9t^2 - 12t$

a. $v = \dfrac{ds}{dt} = -6t^2 + 18t - 12$

b. $v = 0$ when
$$\begin{aligned} -6t^2 + 18t - 12 &= 0 \\ -6\left(t^2 - 3t + 2\right) &= 0 \\ -6(t-1)(t-2) &= 0 \\ t &= 1, 2 \end{aligned}$$

When $t = 1$, $s = -3$ and when $t = 2$, $s = -2$.

c. From $v = -6(t-1)(t-2)$, it follows that $v < 0$ whenever $t > 2$ or $t < 1$ (particle moving to the left). The particle moves to the right in the time interval $1 < t < 2$.

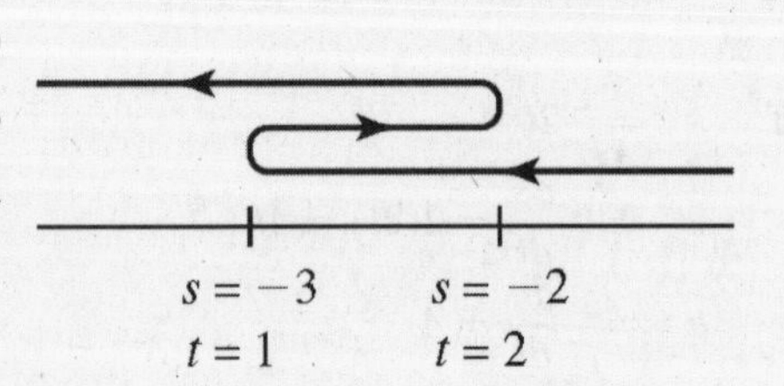

11.
$$\begin{aligned} s &= 50t^2 \\ v &= \frac{ds}{dt} = 50(2t) = 100t\Big|_{t=10} = 1000\,\text{m/s} \end{aligned}$$

13.
$$\begin{aligned} y &= 50t - 0.83t^2 \\ v &= \frac{dy}{dt} = 50 - 0.83(2t)\Big|_{t=10\,\text{s}} = 33.4\,\text{m/s} \end{aligned}$$

15. If x is the length of the side, then $A = x^2$;

$$\frac{dA}{dx} = 2x\Big|_{x=2} = 4\,\text{cm}^2 \text{ per cm.}$$

17. The instantaneous rate of change of R with respect to T is $\frac{dR}{dT}$:

$$\begin{aligned} R &= 20.0 + 0.520T + 0.00973T^2 \\ \frac{dR}{dT} &= 0 + 0.520(1) + 0.00973(2T) \\ &= 0.520 + (0.00973)(2)T\Big|_{T=125} \\ &= 0.520 + (0.00973)(2)(125) \\ &= 2.95\,\frac{\Omega}{^\circ\text{C}}. \end{aligned}$$

19. The instantaneous rate of change is the derivative of E with respect to T:

$$\frac{dE}{dT} = 0.0042\left(4T^3\right)\Big|_{T=8.0} = 8.6\,\text{J}/^\circ\text{C}$$

21.
$$\begin{aligned} \theta &= 3t^2 \\ \omega &= \frac{d\theta}{dt} = 6t\Big|_{t=1} = 6\,\text{rad/s} \\ \alpha &= \frac{d\omega}{dt} = 6\,\text{rad/s}^2 \end{aligned}$$

23. $V = L\frac{di}{dt} = 5.00 \times 10^{-3}\frac{d}{dt}(3.00t) = 0.0150\,V$

25. $P = \frac{dW}{dt} = \frac{d}{dt}\left(10t^3 + 3t^4\right) = 30t^2 + 12t^3\Big|_{t=2} = 216\,\text{J/s} = 216\,\text{W (watts)}$

27. $V = 6.4P^{-1/2}$; $\frac{dV}{dP} = 6.4\left(-\frac{1}{2}\right)P^{-3/2}$

$$\beta = -\frac{1}{V}\frac{dV}{dP} = -\frac{1}{6.4P^{-1/2}}(6.4)\left(-\frac{1}{2}\right)P^{-3/2} = \frac{1}{2P}$$

29. $C(x) = 500 + 2.5x + 0.001x^2, N = 500$

The average cost is $\frac{C(500)}{500}$ and the marginal cost is $C'(500)$:

$$\frac{C(500)}{500} = \frac{500 + 2.5(500) + 0.001(500)^2}{500} = \$4.00$$

$C'(x) = 2.5 + 0.001(2x)$; when $x = 500, C' = \$3.50$.

31. $C(x) = 2000 + 2x + 0.01x^2 + 0.0005x^3, N = 80$

Average cost: $\frac{C(80)}{80} = \frac{2000 + 2(80) + 0.01(80)^2 + 0.0005(80)^3}{80} = \$31.00$

Marginal cost: $C'(x) = 2 + 0.01(2x) + 0.0005(3x^2)\Big|_{x=80} = \$13.20$

2.7 Differentiation Formulas

Group A

1. $y = 4x^4 - 4x^2 + 8$; $y' = 4(4x^3) - 4(2x) + 0 = 16x^3 - 8x$

3. $y = x^{-1}$; $y' = -1x^{-2} = -\dfrac{1}{x^2}$

5. $y = x^5 - 3x^{-3} + 2x^{-2}$. By (2.4),
   $y' = 5x^4 - 3(-3)x^{-4} + 2(-2)x^{-3} = 5x^4 + 9x^{-4} - 4x^{-3}$.

7. $y = x^{5/2} + x^{-1/2}$; $y' = \dfrac{5}{2}x^{3/2} - \dfrac{1}{2}x^{-3/2} = \dfrac{5}{2}x^{3/2} - \dfrac{1}{2x^{3/2}}$

9. $y = \sqrt{2}x - \sqrt{3}x^2 + \pi^2$; $y' = \sqrt{2} - 2\sqrt{3}x$
   (Recall that the derivative of a constant is 0.)

11. $y = (2x^2 - 3)^4$. By the generalized power rule,
$$\begin{aligned} y' &= 4(2x^2-3)^3 \frac{d}{dx}(2x^2-3) \\ &= 4(2x^2-3)^3(4x) = 16x(2x^2-3)^3. \end{aligned}$$

13. $y = \frac{1}{4}\left(4 - 3x^2\right)^4$. By the generalized power rule
$$\begin{aligned} y' &= \tfrac{1}{4} \cdot 4(4-3x^2)^3(-6x) \\ &= -6x(4-3x^2)^3. \end{aligned}$$

15. $y = 2(x^6 + 1)^{10}$. By the generalized power rule,
$$y' = 2 \cdot 10\left(x^6+1\right)^9\left(6x^5\right) = 120x^5\left(x^6+1\right)^9$$

17. $y = \dfrac{2}{\sqrt{x^3 - 3x}}$. We avoid the quotient rule by writing the function as follows:
   $y = 2(x^3 - 3x)^{-1/2}$. By the generalized power rule,
$$\begin{aligned} y' &= 2\left(-\frac{1}{2}\right)(x^3-3x)^{-3/2}\frac{d}{dx}(x^3-3x) \\ &= -(x^3-3x)^{-3/2}(3x^2-3) \\ &= -\frac{3x^2-3}{(x^3-3x)^{3/2}} = -\frac{3(x^2-1)}{(x^3-3x)^{3/2}}. \end{aligned}$$

19. $v = \sqrt[3]{t^3 - 3} = (t^3 - 3)^{1/3}$. By the power rule,
$$\begin{aligned} \frac{dv}{dt} &= \frac{1}{3}(t^3-3)^{-2/3}\frac{d}{dt}(t^3-3) = \frac{1}{3}(t^3-3)^{-2/3}(3t^2) \\ &= \frac{t^2}{(t^3-3)^{2/3}}. \end{aligned}$$

21. $y = \dfrac{4}{\sqrt{2x^2 - 1}}$. To avoid having to use the quotient rule, we rewrite the function as follows:
   $y = 4\left(2x^2 - 1\right)^{-1/2}$; by the generalized power rule
$$y' = 4\left(-\frac{1}{2}\right)\left(2x^2-1\right)^{-3/2}(4x) = -\frac{8x}{\left(2x^2-1\right)^{3/2}}.$$

23. $y = 4\left(x - x^2\right)^{-1/4}$. By the generalized power rule,

$$\begin{aligned} y' &= 4\left(-\frac{1}{4}\right)\left(x - x^2\right)^{-5/4}\frac{d}{dx}\left(x - x^2\right) \\ &= -\left(x - x^2\right)^{-5/4}(1 - 2x) \\ &= -\frac{1 - 2x}{\left(x - x^2\right)^{5/4}} = \frac{2x - 1}{\left(x - x^2\right)^{5/4}}. \end{aligned}$$

25. $y = x^3(x+1)^2$. By the product rule:

$$\begin{aligned} y' &= x^3\frac{d}{dx}(x+1)^2 + (x+1)^2\frac{d}{dx}(x^3) \\ &= x^3 \cdot 2(x+1) + (x+1)^2(3x^2) \end{aligned}$$

Note that the terms on the right have common factor $x^2(x+1)$. So

$$\begin{aligned} y' &= x^2(x+1)[x \cdot 2 + (x+1)(3)] \\ &= x^2(x+1)(5x+3). \end{aligned}$$

27. $y = 2x^4(x+2)^2$. By the product rule,

$$\begin{aligned} y' &= 2x^4\frac{d}{dx}(x+2)^2 + (x+2)^2\frac{d}{dx}(2x^4) \\ &= 2x^4 \cdot 2(x+2) + (x+2)^2(8x^3) \\ &= 4x^3(x+2)[x + 2(x+2)] && \text{common factor: } 4x^3(x+2) \\ &= 4x^3(x+2)(3x+4). \end{aligned}$$

29. $y = x^2(x^2-5)^2$

By the product rule,

$$y' = x^2\frac{d}{dx}(x^2-5)^2 + (x^2-5)^2\frac{d}{dx}(x^2).$$

Now by the generalized power rule,

$$y' = x^2 \cdot 2(x^2-5)(2x) + (x^2-5)^2(2x).$$

Factoring $2x(x^2-5)$, we get

$$\begin{aligned} y' &= 2x(x^2-5)[x^2 \cdot 2 + (x^2-5)] \\ &= 2x(x^2-5)(3x^2-5). \end{aligned}$$

31. $y = 4x(x+5)^4$. By the product rule,

$$\begin{aligned} y' &= 4x\frac{d}{dx}(x+5)^4 + (x+5)^4\frac{d}{dx}(4x) \\ &= 4x \cdot 4(x+5)^3(1) + (x+5)^4(4) \\ &= 4(x+5)^3[4x + (x+5)] && \text{common factor: } 4(x+5)^3 \\ &= 4(x+5)^3(5x+5) \\ &= 20(x+5)^3(x+1). && \text{common factor: } 5 \end{aligned}$$

33. $Q = 2r^3(r-2)^3$; by the product rule

$$\frac{dQ}{dr} = 2r^3\frac{d}{dr}(r-2)^3 + (r-2)^3\frac{d}{dr}\left(2r^3\right) = 2r^3 \cdot 3(r-2)^2(1) + (r-2)^3\left(6r^2\right).$$

Factoring $6r^2(r-2)^2$, we get

$$\begin{aligned} \frac{dQ}{dr} &= 6r^2(r-2)^2[r + (r-2)] = 6r^2(r-2)^2(2r-2) \\ &= 6r^2(r-2)^2 \cdot 2(r-1) = 12r^2(r-2)^2(r-1). \end{aligned}$$

35. $y = \dfrac{x}{x-1}$. By the quotient rule,

$$\begin{aligned} y' &= \frac{(x-1)\frac{d}{dx}(x) - x\frac{d}{dx}(x-1)}{(x-1)^2} = \frac{x-1-x}{(x-1)^2} \\ &= \frac{-1}{(x-1)^2}. \end{aligned}$$

37. $P = \dfrac{t-2}{t^2+4}$. By the quotient rule:

$$\begin{aligned} \frac{dP}{dt} &= \frac{(t^2+4)\frac{d}{dt}(t-2) - (t-2)\frac{d}{dt}(t^2+4)}{(t^2+4)^2} \\ &= \frac{(t^2+4)(1) - (t-2)(2t)}{(t^2+4)^2} \\ &= \frac{t^2+4-2t^2+4t}{(t^2+4)^2} = \frac{-t^2+4t+4}{(t^2+4)^2}. \end{aligned}$$

39. $R = \dfrac{s^2-3}{s-2}$. By the quotient rule,

$$\frac{dR}{ds} = \frac{(s-2)(2s)-(s^2-3)(1)}{(s-2)^2} = \frac{s^2-4s+3}{(s-2)^2}.$$

41. $y = \dfrac{x^3+2x}{x^2-8}$; by the quotient rule,

$$\begin{aligned} y' &= \frac{(x^2-8)\frac{d}{dx}(x^3+2x) - (x^3+2x)\frac{d}{dx}(x^2-8)}{(x^2-8)^2} \\ &= \frac{(x^2-8)(3x^2+2) - (x^3+2x)(2x)}{(x^2-8)^2}. \end{aligned}$$

After collecting terms,

$$y' = \frac{x^4-26x^2-16}{(x^2-8)^2}.$$

43. $y = x^2\sqrt{x+1} = x^2(x+1)^{1/2}$. By the product rule,

$$\begin{aligned} y' &= x^2\frac{d}{dx}(x+1)^{1/2} + (x+1)^{1/2}\frac{d}{dx}(x^2) \\ &= x^2 \cdot \tfrac{1}{2}(x+1)^{-1/2} + (x+1)^{1/2} \cdot 2x \\ &= \frac{x^2}{2(x+1)^{1/2}} + 2x(x+1)^{1/2}. \end{aligned}$$

The common denominator is $2(x+1)^{1/2}$:

$$\begin{aligned} y' &= \frac{x^2}{2(x+1)^{1/2}} + \frac{2x(x+1)^{1/2}}{1} \cdot \frac{2(x+1)^{1/2}}{2(x+1)^{1/2}} \\ &= \frac{x^2+4x(x+1)}{2(x+1)^{1/2}} \qquad a^{1/2} \cdot a^{1/2} = a \\ &= \frac{5x^2+4x}{2\sqrt{x+1}}. \end{aligned}$$

45. $y = 4x\sqrt{x^2+2} = 4x\left(x^2+2\right)^{1/2}$; by the product rule,

$$\begin{aligned} y' &= 4x\frac{d}{dx}\left(x^2+2\right)^{1/2} + \left(x^2+2\right)^{1/2}\frac{d}{dx}(4x) \\ &= 4x \cdot \frac{1}{2}\left(x^2+2\right)^{-1/2}(2x) + \left(x^2+2\right)^{1/2}(4) \\ &= \frac{4x^2}{\left(x^2+2\right)^{1/2}} + 4\left(x^2+2\right)^{1/2} \end{aligned}$$

Since the common denominator is $\left(x^2+2\right)^{1/2}$, we need to write the second fraction as

$$\frac{4\left(x^2+2\right)^{1/2}}{1}\cdot\frac{\left(x^2+2\right)^{1/2}}{\left(x^2+2\right)^{1/2}}$$

The result is

$$\begin{aligned} y' &= \frac{4x^2}{\left(x^2+2\right)^{1/2}}+\frac{4\left(x^2+2\right)^{1/2}}{1}\cdot\frac{\left(x^2+2\right)^{1/2}}{\left(x^2+2\right)^{1/2}} \\ &= \frac{4x^2+4\left(x^2+2\right)^{1/2}\left(x^2+2\right)^{1/2}}{\left(x^2+2\right)^{1/2}} \\ &= \frac{4x^2+4\left(x^2+2\right)}{\left(x^2+2\right)^{1/2}} && A^{1/2}\cdot A^{1/2}=A \\ &= \frac{8x^2+8}{\left(x^2+2\right)^{1/2}} \\ &= \frac{8\left(x^2+1\right)}{\sqrt{x^2+2}} \end{aligned}$$

47. $y=x\sqrt{x^2-1}=x\left(x^2-1\right)^{1/2}$. By the product rule,

$$\begin{aligned} y' &= x\frac{d}{dx}\left(x^2-1\right)^{1/2}+\left(x^2-1\right)^{1/2}\frac{d}{dx}(x) \\ &= x\left(\frac{1}{2}\right)\left(x^2-1\right)^{-1/2}(2x)+\left(x^2+1\right)^{1/2} && \text{generalized power rule} \\ &= \frac{x^2}{\sqrt{x^2-1}}+\sqrt{x^2-1} \end{aligned}$$

since the common denominator is $\sqrt{x^2-1}$, we need to write the second term as

$$\frac{\sqrt{x^2-1}}{1}\cdot\frac{\sqrt{x^2-1}}{\sqrt{x^2-1}}.$$

So

$$\begin{aligned} y' &= \frac{x^2}{\sqrt{x^2-1}}+\frac{\sqrt{x^2-1}}{1}\cdot\frac{\sqrt{x^2-1}}{\sqrt{x^2-1}} \\ &= \frac{x^2+\left(x^2-1\right)}{\sqrt{x^2-1}}=\frac{2x^2-1}{\sqrt{x^2-1}}. && \sqrt{A}\sqrt{A}=A \end{aligned}$$

49. $y=2x(1-x)^{1/2}$. By the product rule,

$$\begin{aligned} y' &= 2x\cdot\tfrac{1}{2}(1-x)^{-1/2}(-1)+2(1-x)^{1/2} \\ &= \frac{-x}{(1-x)^{1/2}}+\frac{2(1-x)^{1/2}}{1}\cdot\frac{(1-x)^{1/2}}{(1-x)^{1/2}} \\ &= \frac{-x+2(1-x)}{(1-x)^{1/2}}=\frac{2-3x}{\sqrt{1-x}}. \end{aligned}$$

51. $T=\theta^3(\theta+7)^{1/2}$. By the product rule:

$$\begin{aligned} \frac{dT}{d\theta} &= \theta^3\frac{d}{d\theta}(\theta+7)^{1/2}+(\theta+7)^{1/2}\frac{d}{d\theta}(\theta^3) \\ &= \theta^3\cdot\tfrac{1}{2}(\theta+7)^{-1/2}+(\theta+7)^{1/2}(3\theta^2) \\ &= \frac{\theta^3}{2(\theta+7)^{1/2}}+3\theta^2(\theta+7)^{1/2}. \end{aligned}$$

Since the common denominator is $2(\theta+7)^{1/2}$, we need to write the second fraction as

$$\frac{3\theta^2(\theta+7)^{1/2}}{1}\cdot\frac{2(\theta+7)^{1/2}}{2(\theta+7)^{1/2}}.$$

So

$$\begin{aligned}\frac{dT}{d\theta} &= \frac{\theta^3}{2(\theta+7)^{1/2}} + \frac{3\theta^2(\theta+7)^{1/2}}{1}\cdot\frac{2(\theta+7)^{1/2}}{2(\theta+7)^{1/2}}\\ &= \frac{\theta^3 + 3\theta^2(\theta+7)^{1/2}\cdot 2(\theta+7)^{1/2}}{2(\theta+7)^{1/2}}\\ &= \frac{\theta^3+6\theta^2(\theta+7)}{2(\theta+7)^{1/2}}\\ &= \frac{\theta^3+6\theta^3+42\theta^2}{2(\theta+7)^{1/2}} = \frac{7\theta^3+42\theta^2}{2(\theta+7)^{1/2}}\\ &= \frac{7\theta^2(\theta+6)}{2\sqrt{\theta+7}}.\end{aligned}$$

Group B

1. $y = \dfrac{\sqrt{x}}{x-4} = \dfrac{x^{1/2}}{x-4}$. By the quotient rule,

$$\begin{aligned} y' &= \frac{(x-4)\frac{d}{dx}(x^{1/2}) - x^{1/2}\frac{d}{dx}(x-4)}{(x-4)^2}\\ &= \frac{(x-4)\cdot\frac{1}{2}x^{-1/2} - x^{1/2}}{(x-4)^2} = \frac{\frac{x-4}{2x^{1/2}} - x^{1/2}}{(x-4)^2}\\ &= \frac{\frac{x-4}{2x^{1/2}} - \frac{x^{1/2}}{1}\cdot\frac{2x^{1/2}}{2x^{1/2}}}{(x-4)^2} \qquad \text{common denominator: } 2x^{1/2}\\ &= \frac{\frac{x-4-2x}{2x^{1/2}}}{(x-4)^2} = \frac{-x-4}{2x^{1/2}}\cdot\frac{1}{(x-4)^2}\\ &= -\frac{x+4}{2(x-4)^2\sqrt{x}}\end{aligned}$$

3. $y = \dfrac{x^2}{\sqrt{x+1}} = \dfrac{x^2}{(x+1)^{1/2}}$. By the quotient rule:

$$\begin{aligned} y' &= \frac{(x+1)^{1/2}\frac{d}{dx}(x^2) - x^2\frac{d}{dx}(x+1)^{1/2}}{[(x+1)^{1/2}]^2}\\ &= \frac{(x+1)^{1/2}(2x) - x^2(\frac{1}{2})(x+1)^{-1/2}}{x+1} = \frac{2x\sqrt{x+1} - \frac{x^2}{2\sqrt{x+1}}}{x+1}.\end{aligned}$$

To combine the expressions on top, note that the common denominator is $2\sqrt{x+1}$:

$$\begin{aligned} &= \frac{\frac{2x\sqrt{x+1}}{1}\cdot\frac{2\sqrt{x+1}}{2\sqrt{x+1}} - \frac{x^2}{2\sqrt{x+1}}}{x+1} = \frac{\frac{4x(x+1)}{2\sqrt{x+1}} - \frac{x^2}{2\sqrt{x+1}}}{x+1}\\ &= \frac{\frac{4x^2+4x-x^2}{2\sqrt{x+1}}}{x+1} = \frac{3x^2+4x}{2\sqrt{x+1}}\cdot\frac{1}{x+1} = \frac{3x^2+4x}{2(x+1)\sqrt{x+1}}.\end{aligned}$$

The expression

$$\frac{2x\sqrt{x+1} - \frac{x^2}{2\sqrt{x+1}}}{x+1}$$

can also be simplified by clearing fractions: multiply numerator and denominator by $2\sqrt{x+1}$ to reduce the complex fraction to an ordinary fraction. Thus

$$\begin{aligned} y' &= \frac{2x\sqrt{x+1} - \frac{x^2}{2\sqrt{x+1}}}{x+1}\cdot\frac{2\sqrt{x+1}}{2\sqrt{x+1}}\\ &= \frac{2x\sqrt{x+1}\cdot 2\sqrt{x+1} - x^2}{(x+1)\cdot 2\sqrt{x+1}}\\ &= \frac{4x(x+1)-x^2}{2(x+1)\sqrt{x+1}} = \frac{3x^2+4x}{2(x+1)\sqrt{x+1}}.\end{aligned}$$

5. $y = \dfrac{(x^2-1)^{1/2}}{x^2}$. By the quotient rule,

$$\begin{aligned} y' &= \frac{x^2\frac{d}{dx}(x^2-1)^{1/2} - (x^2-1)^{1/2}\frac{d}{dx}(x^2)}{x^4} \\ &= \frac{x^2 \cdot \frac{1}{2}(x^2-1)^{-1/2}(2x) - (x^2-1)^{1/2}(2x)}{x^4} \\ &= \frac{\frac{x^3}{(x^2-1)^{1/2}} - 2x(x^2-1)^{1/2}}{x^4}. \end{aligned}$$

To clear fractions, we multiply numerator and denominator by $(x^2-1)^{1/2}$:

$$\begin{aligned} y' &= \frac{\frac{x^3}{(x^2-1)^{1/2}} - 2x(x^2-1)^{1/2}}{x^4} \cdot \frac{(x^2-1)^{1/2}}{(x^2-1)^{1/2}} \\ &= \frac{\frac{x^3}{(x^2-1)^{1/2}}(x^2-1)^{1/2} - 2x(x^2-1)^{1/2}(x^2-1)^{1/2}}{x^4(x^2-1)^{1/2}} \\ &= \frac{x^3 - 2x(x^2-1)}{x^4(x^2-1)^{1/2}} = \frac{2x - x^3}{x^4\sqrt{x^2-1}} \\ &= \frac{2-x^2}{x^3\sqrt{x^2-1}}. \end{aligned}$$

7. $y = \dfrac{x^2\sqrt{x}}{x^2+3} = \dfrac{x^2 x^{1/2}}{x^2+3} = \dfrac{x^{5/2}}{x^2+3}$. By the quotient rule:

$$\begin{aligned} y' &= \frac{(x^2+3)(\frac{5}{2})x^{3/2} - x^{5/2}(2x)}{(x^2+3)^2} \\ &= \frac{x^{3/2}[(\frac{5}{2})(x^2+3) - x(2x)]}{(x^2+3)^2}. \qquad \text{common factor } x^{3/2} \end{aligned}$$

Now we multiply numerator and denominator by 2 (to clear fractions):

$$\begin{aligned} y' &= \frac{x^{3/2}[\frac{5}{2}(x^2+3) - x(2x)]}{(x^2+3)^2} \cdot \frac{2}{2} \\ &= \frac{x^{3/2}[5(x^2+3) - 4x^2]}{2(x^2+3)^2} \\ &= \frac{x\sqrt{x}(5x^2+15-4x^2)}{2(x^2+3)^2} \qquad x^{3/2} = xx^{1/2} = x\sqrt{x} \\ &= \frac{x\sqrt{x}(x^2+15)}{2(x^2+3)^2}. \end{aligned}$$

9. $R = (t-1)\sqrt{t-2} = (t-1)(t-2)^{1/2}$. By the product rule,

$$\begin{aligned} \frac{dR}{dt} &= (t-1) \cdot \frac{1}{2}(t-2)^{-1/2} + (t-2)^{1/2} \cdot 1 \\ &= \frac{t-1}{2(t-2)^{1/2}} + \frac{(t-2)^{1/2}}{1} \cdot \frac{2(t-2)^{1/2}}{2(t-2)^{1/2}} \\ &= \frac{t-1+2(t-2)}{2(t-2)^{1/2}} = \frac{3t-5}{2\sqrt{t-2}}. \end{aligned}$$

11. $y = \dfrac{x\sqrt{x-1}}{2x+3} = \dfrac{x(x-1)^{1/2}}{2x+3}$. By the quotient rule:

$$\begin{aligned}
y' &= \frac{(2x+3)\frac{d}{dx}[x(x-1)^{1/2}] - x(x-1)^{1/2}\frac{d}{dx}(2x+3)}{(2x+3)^2} \\
&= \frac{(2x+3)[x(\frac{1}{2})(x-1)^{-1/2} + (x-1)^{1/2}] - x(x-1)^{1/2}(2)}{(2x+3)^2} \\
&= \frac{(2x+3)[\frac{x}{2\sqrt{x-1}} + \sqrt{x-1}] - 2x\sqrt{x-1}}{(2x+3)^2} \cdot \frac{2\sqrt{x-1}}{2\sqrt{x-1}} \\
&= \frac{(2x+3)[\frac{x}{2\sqrt{x-1}} + \sqrt{x-1}]\cdot 2\sqrt{x-1} - 2x\sqrt{x-1}\cdot 2\sqrt{x-1}}{(2x+3)^2\cdot 2\sqrt{x-1}} \\
&= \frac{(2x+3)[x+2(x-1)] - 4x(\sqrt{x-1})^2}{2(2x+3)^2\sqrt{x-1}} \\
&= \frac{(2x+3)(3x-2) - 4x(x-1)}{2(2x+3)^2\sqrt{x-1}} \\
&= \frac{6x^2+5x-6-4x^2+4x}{2(2x+3)^2\sqrt{x-1}} = \frac{2x^2+9x-6}{2(2x+3)^2\sqrt{x-1}}.
\end{aligned}$$

13. $y = \dfrac{x^2(x-5)^{1/2}}{(x+3)^{1/2}}$. By the quotient and product rules,

$$\begin{aligned}
y' &= \frac{(x+3)^{1/2}\frac{d}{dx}[x^2(x-5)^{1/2}] - x^2(x-5)^{1/2}\frac{d}{dx}(x+3)^{1/2}}{[(x+3)^{1/2}]^2} \\
&= \frac{(x+3)^{1/2}[x^2\cdot\frac{1}{2}(x-5)^{-1/2} + 2x(x-5)^{1/2}] - x^2(x-5)^{1/2}\cdot\frac{1}{2}(x+3)^{-1/2}}{x+3}.
\end{aligned}$$

To start clearing fractions, multiply numerator and denominator by $2(x-5)^{1/2}$:

$$y' = \frac{(x+3)^{1/2}[x^2+4x(x-5)] - x^2(x-5)(x+3)^{-1/2}}{2(x-5)^{1/2}(x+3)}.$$

Now multiply numerator and denominator by $(x+3)^{1/2}$:

$$y' = \frac{(x+3)(5x^2-20x) - x^2(x-5)}{2(x-5)^{1/2}(x+3)^{3/2}}.$$

After collecting terms, we get

$$y' = \frac{4x^3-60x}{2(x+3)^{3/2}(x-5)^{1/2}} = \frac{2x^3-30x}{(x+3)^{3/2}(x-5)^{1/2}}.$$

15. By the product rule,

$$\begin{aligned}
y' &= (x^3-1)(2x+3) + (3x^2)(x^2+3x+2)\Big|_{x=1} \\
&= 0 + (3)(1+3+2) = 18.
\end{aligned}$$

17. $Z = \sqrt{16+X^2}$; $\dfrac{dZ}{dX} = \dfrac{1}{2}(16+X^2)^{-1/2}(2X) = \dfrac{X}{\sqrt{16+X^2}}$.

19.
$$\begin{aligned}
i &= \frac{dq}{dt} = \frac{d}{dt}\frac{t}{t^2+4} = \frac{(t^2+4)(1) - t(2t)}{(t^2+4)^2} \\
&= \frac{t^2+4-2t^2}{(t^2+4)^2} = \frac{4-t^2}{(t^2+4)^2} = 0
\end{aligned}$$

Hence

$$\begin{aligned}
4-t^2 &= 0 \\
t &= \pm 2.
\end{aligned}$$

Taking only the positive root, $t = 2\,\text{s}$.

2.8 Implicit Differentiation

1. $$\begin{aligned} 2x+3y &= 3 \\ 2+3\frac{dy}{dx} &= 0 \qquad \text{derivative of } y \text{ is } \frac{dy}{dx} \\ \frac{dy}{dx} &= -\frac{2}{3} \end{aligned}$$

3. $$\begin{aligned} x^2-y^2 &= 2 \\ 2x-2y\frac{dy}{dx} &= 0 \\ -2y\frac{dy}{dx} &= -2x \\ \frac{dy}{dx} &= \frac{-2x}{-2y}=\frac{x}{y} \end{aligned}$$

5. $$\begin{aligned} 2x^2-3y^2 &= 1 \\ 4x-6y\frac{dy}{dx} &= 0 \qquad \text{formula (2.11)} \\ -6y\frac{dy}{dx} &= -4x \\ \frac{dy}{dx} &= \frac{-4x}{-6y}=\frac{2x}{3y} \end{aligned}$$

7. $$\begin{aligned} x-5x^2-6y^3 &= 0 \\ 1-10x-18y^2\frac{dy}{dx} &= 0 \\ -18y^2\frac{dy}{dx} &= -1+10x \\ \frac{dy}{dx} &= \frac{1-10x}{18y^2} \end{aligned}$$

9. $$\begin{aligned} \frac{1}{a^2}x^2+\frac{1}{b^2}y^2 &= 1 \\ \frac{1}{a^2}(2x)+\frac{1}{b^2}(2y)\frac{dy}{dx} &= 0 \\ \frac{2y}{b^2}\frac{dy}{dx} &= -\frac{2x}{a^2} \\ \frac{dy}{dx} &= -\frac{2x}{a^2}\frac{b^2}{2y} \\ &= -\frac{b^2x}{a^2y} \end{aligned}$$

11. $$\begin{aligned} 2xy &= 6 \\ 2y+2x\frac{dy}{dx} &= 0 \qquad \frac{d}{dx}(y)=\frac{dy}{dx};\ \text{product rule} \\ y+x\frac{dy}{dx} &= 0 \\ \frac{dy}{dx} &= -\frac{y}{x} \end{aligned}$$

13.

$$\begin{aligned} x^2 y &= 3 \\ x^2\frac{dy}{dx} + y\cdot 2x &= 0 \qquad \frac{d}{dx}(y) = \frac{dy}{dx} \\ x^2\frac{dy}{dx} &= -2xy \\ \frac{dy}{dx} &= -\frac{2xy}{x^2} = -\frac{2y}{x} \end{aligned}$$

15.

$$\begin{aligned} x^2 + xy^2 + 2 &= 0 \\ 2x + x\cdot 2y\frac{dy}{dx} + 1\cdot y^2 + 0 &= 0 \\ 2xy\frac{dy}{dx} &= -\left(2x + y^2\right) \\ \frac{dy}{dx} &= -\frac{2x+y^2}{2xy} \end{aligned}$$

17.

$$\begin{aligned} xy - 2y^3 &= 4 \\ x\frac{dy}{dx} + 1\cdot y - 6y^2\frac{dy}{dx} &= 0 \\ \left(x - 6y^2\right)\frac{dy}{dx} &= -y \\ \frac{dy}{dx} &= \frac{-y}{x-6y^2} = \frac{y}{6y^2 - x} \end{aligned}$$

19.

$$\begin{aligned} x^2 + x^2y^2 + x &= 0 \\ 2x + x^2\frac{d}{dx}y^2 + y^2\frac{d}{dx}x^2 + 1 &= 0 \qquad \text{product rule} \\ 2x + x^2\cdot 2y\frac{dy}{dx} + y^2\cdot 2x + 1 &= 0 \\ 2x^2y\frac{dy}{dx} &= -(2x + 2xy^2 + 1) \\ \frac{dy}{dx} &= -\frac{2x + 2xy^2 + 1}{2x^2y} \end{aligned}$$

21.

$$\begin{aligned} x^3 - 4x^2y^2 + y^2 &= 1 \\ 3x^2 - 4x^2\frac{d}{dx}(y^2) + y^2\frac{d}{dx}(-4x^2) + 2y\frac{dy}{dx} &= 0 \qquad \text{product rule} \\ 3x^2 - 4x^2(2y)\frac{dy}{dx} + y^2(-8x) + 2y\frac{dy}{dx} &= 0 \\ -8x^2y\frac{dy}{dx} + 2y\frac{dy}{dx} &= -3x^2 + 8xy^2 \\ (-8x^2y + 2y)\frac{dy}{dx} &= 8xy^2 - 3x^2 \\ \frac{dy}{dx} &= \frac{8xy^2 - 3x^2}{2y - 8x^2y} \end{aligned}$$

23.

$$\begin{aligned}
5x^2y^3 - y^4 &= 2x^3 \\
5\left(x^2 \cdot 3y^2\frac{dy}{dx} + y^3 \cdot 2x\right) - 4y^3\frac{dy}{dx} &= 6x^2 \qquad \text{product rule} \\
15x^2y^2\frac{dy}{dx} - 4y^3\frac{dy}{dx} &= 6x^2 - 10xy^3 \\
(15x^2y^2 - 4y^3)\frac{dy}{dx} &= 6x^2 - 10xy^3 \\
\frac{dy}{dx} &= \frac{6x^2 - 10xy^3}{15x^2y^2 - 4y^3}
\end{aligned}$$

25.

$$\begin{aligned}
x^4y^4 - 3y^2 + 5x &= 6 \\
x^4\frac{d}{dx}(y^4) + y^4\frac{d}{dx}(x^4) - 6y\frac{dy}{dx} + 5 &= 0 \\
x^4(4y^3)\frac{dy}{dx} + y^4(4x^3) - 6y\frac{dy}{dx} + 5 &= 0 \\
4x^4y^3\frac{dy}{dx} - 6y\frac{dy}{dx} &= -4x^3y^4 - 5 \\
(4x^4y^3 - 6y)\frac{dy}{dx} &= -(4x^3y^4 + 5) \\
\frac{dy}{dx} &= -\frac{4x^3y^4 + 5}{4x^4y^3 - 6y}
\end{aligned}$$

27.

$$\begin{aligned}
x^2 + 4y^2 &= 5 \\
2x + 8y\frac{dy}{dx} &= 0 \\
\frac{dy}{dx} &= -\frac{x}{4y}\bigg|_{(1,-1)} = \frac{1}{4}
\end{aligned}$$

Slope of tangent line: $\frac{1}{4}$; slope of normal line: -4; point: $(1, -1)$.

$$\begin{aligned}
y - y_1 &= m(x - x_1) \\
y + 1 &= \tfrac{1}{4}(x - 1) \\
y &= \tfrac{1}{4}(x - 5) \qquad \text{tangent line} \\
y + 1 &= -4(x - 1) \\
y &= -4x + 3 \qquad \text{normal line}
\end{aligned}$$

29.

$$\begin{aligned}
y^2 &= -4x \\
2y\frac{dy}{dx} &= -4 \\
\frac{dy}{dx} &= -\frac{2}{y}\bigg|_{(-1,-2)} = -\frac{2}{-2} = 1
\end{aligned}$$

Slope of tangent line: 1; slope of normal line: -1; point: $(-1, -2)$.

$$\begin{aligned}
y - y_1 &= m(x - x_1) \\
y + 2 &= 1(x + 1) \\
y &= x - 1 \qquad \text{tangent line} \\
y + 2 &= -1(x + 1) \\
y &= -x - 3 \qquad \text{normal line}
\end{aligned}$$

31.
$$\begin{aligned}
x^2 - 2y^2 &= 2 \\
2x - 4y\frac{dy}{dx} &= 0 \\
\frac{dy}{dx} &= \left.\frac{x}{2y}\right|_{(-2,1)} = -1
\end{aligned}$$

Slope of tangent line: -1; slope of normal line: 1; point: $(-2, 1)$.

$$\begin{aligned}
y - y_1 &= m(x - x_1) \\
y - 1 &= -1(x + 2) \\
y &= -x - 1 && \text{tangent line} \\
y - 1 &= 1(x + 2) \\
y &= x + 3 && \text{normal line}
\end{aligned}$$

33.
$$\begin{aligned}
2x^2 + y^2 &= 17 \\
4x + 2y\frac{dy}{dx} &= 0 \\
\frac{dy}{dx} &= -\left.\frac{2x}{y}\right|_{(-2,-3)} = -\frac{2(-2)}{-3} = -\frac{4}{3}
\end{aligned}$$

Slope of tangent line: $-\frac{4}{3}$; slope of normal line: $\frac{3}{4}$; point: $(-2, -3)$.

$$\begin{aligned}
y - y_1 &= m(x - x_1) \\
y + 3 &= -\tfrac{4}{3}(x + 2) && \text{tangent line} \\
y + 3 &= \tfrac{3}{4}(x + 2) && \text{normal line}
\end{aligned}$$

(Simplified forms and graphs are in the answer section in the back of the book.)

2.9 Higher Derivatives

1.
$$\begin{aligned}
y &= 5x^4 + 5x^3 - 3x + 1 \\
y' &= 20x^3 + 15x^2 - 3 \\
y'' &= 60x^2 + 30x
\end{aligned}$$

3.
$$\begin{aligned}
f(x) &= \sqrt{2x - 3} &&= (2x - 3)^{1/2} \\
f'(x) &= \frac{1}{2}(2x - 3)^{-1/2}(2) &&= (2x - 3)^{-1/2} \\
f''(x) &= -\frac{1}{2}(2x - 3)^{-3/2}(2) &&= -\frac{1}{(2x - 3)^{3/2}}
\end{aligned}$$

5.
$$\begin{aligned}
y &= x^6 - 2x^5 - x^4 \\
\frac{dy}{dx} &= 6x^5 - 10x^4 - 4x^3 \\
\frac{d^2y}{dx^2} &= 30x^4 - 40x^3 - 12x^2 \\
\frac{d^3y}{dx^3} &= 120x^3 - 120x^2 - 24x
\end{aligned}$$

7.
$$\begin{aligned} f(x) &= \frac{x}{x+1} \\ f'(x) &= \frac{(x+1)\cdot 1 - x\cdot 1}{(x+1)^2} \qquad \text{quotient rule} \\ &= \frac{1}{(x+1)^2} = (x+1)^{-2} \\ f''(x) &= -2(x+1)^{-3} \\ f'''(x) &= 6(x+1)^{-4} \\ f^{(4)}(x) &= -24(x+1)^{-5} = -\frac{24}{(x+1)^5} \end{aligned}$$

9.
$$\begin{aligned} y &= \frac{1}{\sqrt{x}} = x^{-1/2} \\ y' &= -\frac{1}{2}x^{-3/2} \\ y'' &= \left(-\frac{1}{2}\right)\left(-\frac{3}{2}\right)y^{-5/2} \\ y''' &= \left(-\frac{1}{2}\right)\left(-\frac{3}{2}\right)\left(-\frac{5}{2}\right)y^{-7/2} \\ y^{(4)} &= \left(-\frac{1}{2}\right)\left(-\frac{3}{2}\right)\left(-\frac{5}{2}\right)\left(-\frac{7}{2}\right)y^{-9/2} = \frac{105}{16y^{9/2}} \end{aligned}$$

11.
$$\begin{aligned} y &= \frac{3+2x}{3-2x} \\ \frac{dy}{dx} &= \frac{(3-2x)(2)-(3+2x)(-2)}{(3-2x)^2} = \frac{6-4x+6+4x}{(3-2x)^2} = \frac{12}{(3-2x)^2} = 12(3-2x)^{-2} \\ \frac{d^2y}{dx^2} &= 12(-2)(3-2x)^{-3}(-2) = \frac{48}{(3-2x)^3} \end{aligned}$$

Chapter 2 Review

1.
$$\begin{aligned} f(x) &= x^2-1 \\ f(0) &= 0^2-1=-1 \qquad x=0 \\ f(1) &= 1^2-1=0 \qquad x=1 \\ f(\sqrt{2}) &= (\sqrt{2})^2-1=1 \qquad x=\sqrt{2} \end{aligned}$$

3. $f(0)=0$ and $f(\frac{1}{2})=0$ since $0 \le x < 1$
$f(\frac{5}{2})=2;$ since $x > 2$
$f(x)$ is not defined for $x=1$.

5. (a) To avoid imaginary values, we must have $x \ge 1$. The range is $y \ge 0$, since y cannot be negative.

(b) The cube root of any real number is a real number. So y is defined for all x.

7. $\lim_{x\to 4}\frac{16-x^2}{4-x} = \lim_{x\to 4}\frac{(4-x)(4+x)}{4-x} = \lim_{x\to 4}(4+x) = 8$

9.
$$\begin{aligned} \lim_{x\to 0}\frac{x^3-x^2+3x}{x} &= \lim_{x\to 0}\frac{x(x^2-x+3)}{x} \qquad \text{factoring } x \\ &= \lim_{x\to 0}(x^2-x+3)=3 \end{aligned}$$

11. (a) $f(x)=1-x^2$ is defined at $x=2$. So we obtain the limit by inspection:
$\lim_{x\to 2}(1-x^2)=1-2^2=-3$.

(b) $\lim\limits_{x\to 1} \dfrac{x^2-5x+4}{x-1} = \lim\limits_{x\to 1} \dfrac{(x-1)(x-4)}{x-1} = \lim\limits_{x\to 1}(x-4) = -3$

13. Since the function is not defined at $x = 1$, we find the limit by rationalizing the numerator:

$$\begin{aligned} \lim_{x\to 1} \frac{\sqrt{x}-1}{x-1} &= \lim_{x\to 1} \frac{\sqrt{x}-1}{x-1}\cdot\frac{\sqrt{x}+1}{\sqrt{x}+1} = \lim_{x\to 1}\frac{x-1}{(x-1)(\sqrt{x}+1)} \\ &= \lim_{x\to 1}\frac{1}{\sqrt{x}+1} = \frac{1}{2}. \end{aligned}$$

15. $\lim\limits_{x\to\infty} \dfrac{2-3x-2x^2}{1+10x-x^2} = \lim\limits_{x\to\infty} \dfrac{2/x^2-3/x-2}{1/x^2+10/x-1} = \dfrac{-2}{-1} = 2$

17. By inspection, $\sqrt{x-4} \to 0$ as $x \to 4$ from the right. (The restriction 4+ is necessary to avoid imaginary values.)

19. $f(x) = x - 3x^2$; $f(\ \ ) = (\ \ ) - 3(\ \ )^2$

Step 1. $y + \Delta y = (x+\Delta x) - 3(x+\Delta x)^2$

Step 2.
$$\begin{aligned} \Delta y &= (x+\Delta x) - 3(x+\Delta x)^2 - (x - 3x^2) \\ &= x + \Delta x - 3[x^2 + 2x\Delta x + (\Delta x)^2] - x + 3x^2 \\ &= \Delta x - 6x\Delta x - 3(\Delta x)^2 \end{aligned}$$

Step 3. $\dfrac{\Delta y}{\Delta x} = \dfrac{\Delta x - 6x\Delta x - 3(\Delta x)^2}{\Delta x} = \dfrac{\Delta x(1-6x-3\Delta x)}{\Delta x} = 1 - 6x - 3\Delta x$

Step 4. $f'(x) = \lim\limits_{\Delta x\to 0} \dfrac{\Delta y}{\Delta x} = \lim\limits_{\Delta x\to 0}(1-6x-3\Delta x) = 1 - 6x$

21. (a) $f(\ \ ) = \dfrac{1}{4-(\ \ )}$

Step 1. $y + \Delta y = f(x+\Delta x) = \dfrac{1}{4-(x+\Delta x)}$

Step 2. $\Delta y = \dfrac{1}{4-(x+\Delta x)} - \dfrac{1}{4-x}$

Note that the common denominator is $[4-(x+\Delta x)](4-x)$. We now get

$$\begin{aligned} \Delta y &= \frac{4-x}{[4-(x+\Delta x)](4-x)} - \frac{4-(x+\Delta x)}{[4-(x+\Delta x)](4-x)} \\ &= \frac{4-x-4+(x+\Delta x)}{[4-(x+\Delta x)](4-x)} = \frac{\Delta x}{[4-(x+\Delta x)](4-x)}. \end{aligned}$$

Step 3.
$$\begin{aligned} \frac{\Delta y}{\Delta x} &= \frac{\Delta x}{[4-(x+\Delta x)](4-x)}\cdot\frac{1}{\Delta x} \\ &= \frac{1}{[4-(x+\Delta x)](4-x)} \end{aligned}$$

Step 4.
$$\begin{aligned} f'(x) &= \lim_{\Delta x\to 0}\frac{1}{[4-(x+\Delta x)](4-x)} \\ &= \frac{1}{(4-x)^2} \end{aligned}$$

(b) $f(\ \ ) = \sqrt{\ \ }$

Step 1. $y + \Delta y = f(x+\Delta x) = \sqrt{x+\Delta x}$

Step 2. $\Delta y = \sqrt{x+\Delta x} - \sqrt{x}$

Rationalizing the numerator,

$$\begin{aligned} \Delta y &= \frac{\sqrt{x+\Delta x}-\sqrt{x}}{1}\cdot\frac{\sqrt{x+\Delta x}+\sqrt{x}}{\sqrt{x+\Delta x}+\sqrt{x}} \\ &= \frac{(\sqrt{x+\Delta x})^2-(\sqrt{x})^2}{\sqrt{x+\Delta x}+\sqrt{x}} = \frac{x+\Delta x-x}{\sqrt{x+\Delta x}+\sqrt{x}} = \frac{\Delta x}{\sqrt{x+\Delta x}+\sqrt{x}}. \end{aligned}$$

Step 3. $\dfrac{\Delta y}{\Delta x} = \dfrac{\Delta x}{\sqrt{x+\Delta x}+\sqrt{x}} \cdot \dfrac{1}{\Delta x} = \dfrac{1}{\sqrt{x+\Delta x}+\sqrt{x}}$

Step 4. $f'(x) = \lim\limits_{\Delta x \to 0} \dfrac{1}{\sqrt{x+\Delta x}+\sqrt{x}} = \dfrac{1}{\sqrt{x}+\sqrt{x}} = \dfrac{1}{2\sqrt{x}}$

23. $y = (x^3-2)^4$. By the generalized power rule,
$y' = 4(x^3-2)^3 \dfrac{d}{dx}(x^3-2) = 4(x^3-2)^3(3x^2) = 12x^2(x^3-2)^3$.

25. $y = \dfrac{x-4}{x+1}$. By the quotient rule, $y' = \dfrac{(x+1)-(x-4)}{(x+1)^2} = \dfrac{5}{(x+1)^2}$.

27. $y = \dfrac{x^2}{(4-x^2)^{1/2}}$. By the quotient rule,

$$\begin{aligned} y' &= \frac{(4-x^2)^{1/2}(2x) - x^2 \cdot \frac{1}{2}(4-x^2)^{-1/2}(-2x)}{[(4-x^2)^{1/2}]^2} \\ &= \frac{2x(4-x^2)^{1/2} + x^3(4-x^2)^{-1/2}}{4-x^2}. \end{aligned}$$

Now multiply numerator and denominator by $(4-x^2)^{1/2}$:

$$y' = \frac{2x(4-x^2)+x^3}{(4-x^2)^{3/2}} = \frac{8x-x^3}{(4-x^2)^{3/2}}.$$

29. $y = x\sqrt{4-x^2} = x(4-x^2)^{1/2}$. By the product rule,

$$\begin{aligned} y' &= x\frac{d}{dx}(4-x^2)^{1/2} + (4-x^2)^{1/2}\frac{d}{dx}(x) \\ &= x \cdot \tfrac{1}{2}(4-x^2)^{-1/2}(-2x) + (4-x^2)^{1/2} \\ &= \frac{-x^2}{(4-x^2)^{1/2}} + (4-x^2)^{1/2} \\ &= \frac{-x^2}{(4-x^2)^{1/2}} + \frac{(4-x^2)^{1/2}}{1} \cdot \frac{(4-x^2)^{1/2}}{(4-x^2)^{1/2}} \\ &= \frac{-x^2+(4-x^2)}{(4-x^2)^{1/2}} \\ &= \frac{4-2x^2}{\sqrt{4-x^2}}. \end{aligned}$$

31.

$$\begin{aligned} x^2y + xy^2 + y^3 &= 1 \\ x^2\frac{d}{dx}(y) + y\frac{d}{dx}(x^2) + x\frac{d}{dx}(y^2) + y^2\frac{d}{dx}(x) + \frac{d}{dx}(y^3) &= 0 \\ x^2\frac{dy}{dx} + 2xy + 2xy\frac{dy}{dx} + y^2 + 3y^2\frac{dy}{dx} &= 0 \\ x^2\frac{dy}{dx} + 2xy\frac{dy}{dx} + 3y^2\frac{dy}{dx} &= -(2xy+y^2) \\ (x^2+2xy+3y^2)\frac{dy}{dx} &= -(2xy+y^2) \\ \frac{dy}{dx} &= -\frac{2xy+y^2}{x^2+2xy+3y^2} \end{aligned}$$

33.

$$\begin{aligned} y^3 - xy &= 4 \\ 3y^2y' - xy' - y &= 0 \\ y' &= \frac{y}{3y^2-x} \end{aligned}$$

35.
$$\begin{aligned}
8x^2+4xy+5y^2+28x-2y+20 &= 0\\
16x+(4x\frac{dy}{dx}+4y)+10y\frac{dy}{dx}+28-2\frac{dy}{dx} &= 0\\
(4x+10y-2)\frac{dy}{dx} &= -16x-4y-28\\
\frac{dy}{dx} &= \left.\frac{-16x-4y-28}{4x+10y-2}\right|_{(-1,\,6/5)}\\
&= \frac{-16(-1)-4(6/5)-28}{4(-1)+10(6/5)-2}\cdot\frac{5}{5}\\
&= \frac{80-24-140}{-20+60-10}=\frac{-84}{30}=-\frac{14}{5}
\end{aligned}$$

37. (a)
$$\begin{aligned}
f(x) &= \frac{x}{\sqrt{x-1}}=\frac{x}{(x-1)^{1/2}}\\
f'(x) &= \frac{(x-1)^{1/2}-x(\frac{1}{2})(x-1)^{-1/2}}{[(x-1)^{1/2}]^2}\\
&= \frac{\sqrt{x-1}-\frac{x}{2\sqrt{x-1}}}{x-1}\cdot\frac{2\sqrt{x-1}}{2\sqrt{x-1}}\\
&= \frac{2(x-1)-x}{2(x-1)\sqrt{x-1}}=\frac{x-2}{2(x-1)\sqrt{x-1}}\\
f'(2) &= 0
\end{aligned}$$

Thus $f'(x)=0$ when $x=2$.

(b)
$$\begin{aligned}
f(x) &= 2x^3-6x^2+4\\
f'(x) &= 6x^2-12x\\
f''(x) &= 12x-12\\
f''(1) &= 0
\end{aligned}$$

Thus $f''(x)=0$ when $x=1$.

39. $\dfrac{dR}{dr}=k(-2r^{-3})=-\dfrac{2k}{r^3}$

41. $V=IR=(4.12+0.020t)(0.010t^2)$. By the product rule, $\dfrac{dV}{dt}=(4.12+0.020t)(0.020t)+(0.020)(0.010t^2)$. Letting $t=2.5\,\text{s}$, we get

$$\frac{dV}{dt}=0.21\,\frac{\text{V}}{\text{s}}.$$

43.
$$\begin{aligned}
C'(y) &= \left.\frac{1}{2}-\frac{1}{250}y\right|_{y=50}\\
&= \$0.30 \text{ per widget} = 30 \text{ cents per widget.}
\end{aligned}$$

45. $F=\dfrac{1}{r^2}=r^{-2}$. Thus

$$\frac{dF}{dr}=-2r^{-3}=\left.-\frac{2}{r^3}\right|_{r=5.02}=-\frac{2}{(5.02)^3}=-0.0158\,\text{dyne/cm}.$$

(The negative sign indicates that F decreases as r increases.)

Chapter 3

Applications of the Derivative

3.1 The First-Derivative Test

1. $y = -x^2 - 2x$; $y' = -2x - 2 = 0$; $x = -1$, the critical value.
   To the left of $x = -1$, y' is positive (for example, the test value $x = -2$ yields $y' = 2$). So the function is increasing. To the right of $x = -1$, y' is negative (for example, the test value $x = 0$ yields $y' = -2$). So the function is decreasing.

3. $y = \frac{1}{4}x^3 - 3x + 2$;
   $$\begin{aligned} y' = \tfrac{3}{4}x^2 - 3 &= 0 \\ \tfrac{3}{4}x^2 &= 3 \\ x^2 &= 4 \\ x &= \pm 2, \text{ the critical values.} \end{aligned}$$
   To see where y' is negative and positive, we substitute certain test values.

	test values	$y' = \frac{3}{4}x^2 - 3$	
$x < -2$	-3	$+$	increasing
$-2 < x < 2$	0	$-$	decreasing
$x > 2$	3	$+$	increasing

We conclude that the function is increasing on $(-\infty, -2]$, decreasing on $[-2, 2]$, and increasing on $[2, \infty)$.

5. $y = x^2 - 2x + 1$
   $$\begin{aligned} y' &= 2x - 2 \\ 2x - 2 &= 0 \\ x &= 1 \quad \text{critical value} \end{aligned}$$
   Substituting $x = 1$ in $y = x^2 - 2x + 1$, we get $y = 0$. So $(1, 0)$ is the critical point.
   If $x < 1$, $y' < 0$; if $x > 1$, $y' > 0$. Since the function is decreasing to the left of $x = 1$ and increasing to the right, $(1, 0)$ is a minimum point.

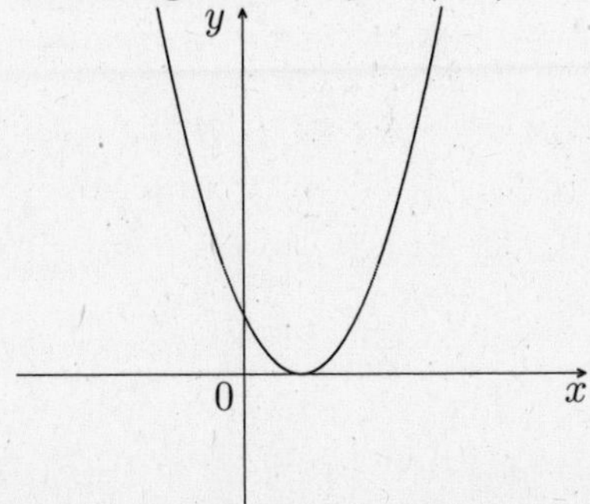

7. $y = 8 - 2x - x^2$;

$$\begin{aligned} y' = -2 - 2x &= 0 \\ x &= -1 \text{ (critical value).} \end{aligned}$$

If $x = -1$, $y = 8 - 2(-1) - (-1)^2 = 9$; so $(-1, 9)$ is the critical point. To the left of $x = -1$, $y' > 0$. (For example, for the test value $x = -2$, $y' = 2$.) So the function is increasing. Similarly, to the right of $x = -1$, the function is decreasing. It follows that the critical point $(-1, 9)$ is a maximum.

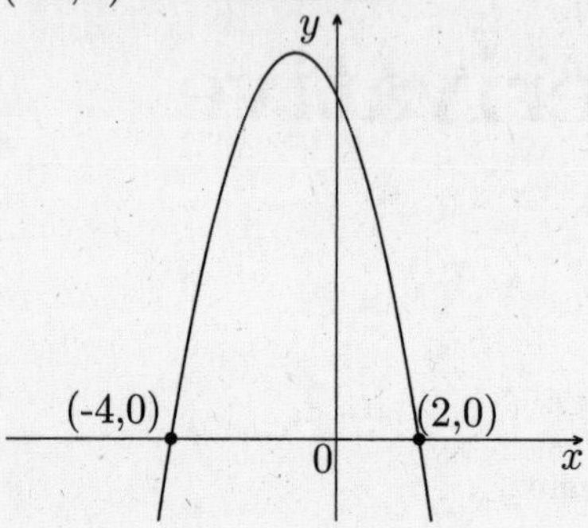

9. $y = 2x^3 + 3x^2 - 12x + 6$; $y' = 6x^2 + 6x - 12$

$$\begin{aligned} 6x^2 + 6x - 12 &= 0 \\ 6(x^2 + x - 2) &= 0 \\ 6(x+2)(x-1) &= 0 \qquad x = -2, 1 \end{aligned}$$

If $x = -2$, $y = 26$; if $x = 1$, $y = -1$. So $(-2, 26)$ and $(1, -1)$ are the critical points. To see where y' is positive or negative, we substitute certain convenient values between the critical values.

	test values	$x+2$	$x-1$	$y' = 6(x+2)(x-1)$
$x < -2$	-3	$-$	$-$	$+$
$-2 < x < 1$	0	$+$	$-$	$-$
$x > 1$	2	$+$	$+$	$+$

Summary:

If $x < -2$, $y' > 0$, so that $f(x)$ is increasing.

If $-2 < x < 1$, $y' < 0$ so that $f(x)$ is decreasing.

If $x > 1$, $y' > 0$, so that $f(x)$ is increasing.

We conclude that $(-2, 26)$ is a maximum point and $(1, -1)$ is a minimum point.

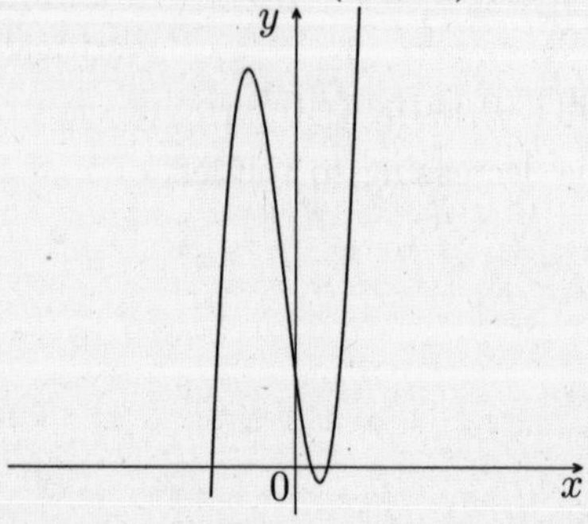

11. $y = -x^3 + 6x^2 - 9x - 5$; $y' = -3x^2 + 12x - 9$

$$\begin{aligned} -3x^2 + 12x - 9 &= 0 \\ -3(x^2 - 4x + 3) &= 0 \\ -3(x-1)(x-3) &= 0 \\ x &= 1, 3 \text{ (critical values)} \end{aligned}$$

If $x = 1$, $y = -9$ and if $x = 3$, $y = -5$. So $(1, -9)$ and $(3, -5)$ are the critical points.

Now we use test values to determine the sign of y'.

	test values	$x - 1$	$x - 3$	$y' = -3(x-1)(x-3)$
$x < 1$	0	$-$	$-$	$-$
$1 < x < 3$	2	$+$	$-$	$+$
$x > 3$	4	$+$	$+$	$-$

Summary:

If $x < 1$, $y' < 0$, so that $f(x)$ is decreasing.

If $1 < x < 3$, $y' > 0$, so that $f(x)$ is increasing.

If $x > 3$, $y' < 0$, so that $f(x)$ is decreasing.

We conclude that $(1, -9)$ is a minimum and $(3, -5)$ is a maximum point.

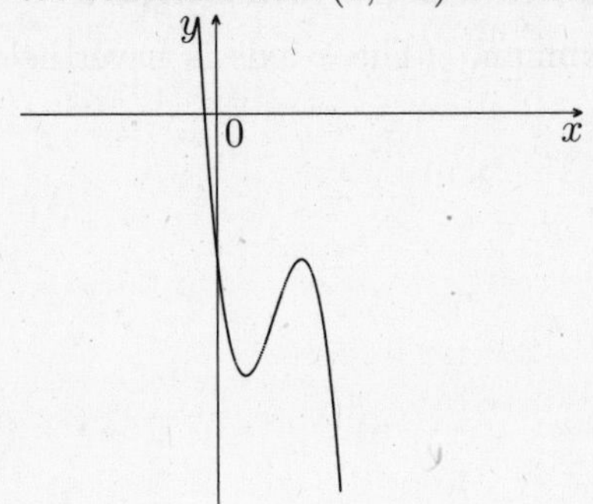

13. $y = x^4 - 2x^2 - 2$; $y' = 4x^3 - 4x$

$$\begin{aligned} 4x^3 - 4x &= 0 \\ 4x(x^2 - 1) &= 0 \\ 4x(x+1)(x-1) &= 0 \\ x &= -1,\ 0,\ 1 \qquad \text{(critical values).} \end{aligned}$$

To see where $y' < 0$ and where $y' > 0$, we substitute certain test values:

	test values	$4x$	$x + 1$	$x - 1$	$y' = 4x(x+1)(x-1)$
$x < -1$	-2	$-$	$-$	$-$	$-$
$-1 < x < 0$	$-1/2$	$-$	$+$	$-$	$+$
$0 < x < 1$	$1/2$	$+$	$+$	$-$	$-$
$x > 1$	2	$+$	$+$	$+$	$+$

Summary:

If $x < -1$, $y' < 0$: $f(x)$ is decreasing

If $-1 < x < 0$, $y' > 0$: $f(x)$ is increasing

If $0 < x < 1$, $y' < 0$: $f(x)$ is decreasing

If $x > 1$, $y' > 0$: $f(x)$ is increasing

It follows that $(\pm 1, -3)$ are minimum points and $(0, -2)$ is a maximum point.

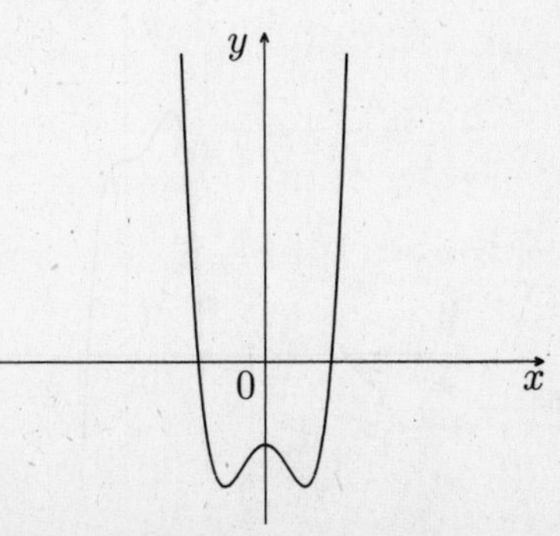

15. $y = x^4 + \frac{4}{3}x^3$; $y' = 4x^3 + 4x^2$

$$\begin{aligned} 4x^3 + 4x^2 &= 0 \\ 4x^2(x+1) &= 0 \\ x &= 0, -1 \text{ (critical values).} \end{aligned}$$

Critical points: $(0,0)$ and $(-1, -\frac{1}{3})$.

	test values	$4x^2$	$x+1$	$y' = 4x^2(x+1)$
$x < -1$	-2	+	−	−
$-1 < x < 0$	$-1/2$	+	+	+
$x > 0$	1	+	+	+

To the left of $x = -1$, $f(x)$ is decreasing. To the right of $x = -1$, $f(x)$ is increasing. So $(-1, -\frac{1}{3})$ is a minimum point. The function continues to increase to the right of $x = 0$. Consequently, $(0,0)$ is neither a minimum nor a maximum. (The x-axis is nevertheless a horizontal tangent.)

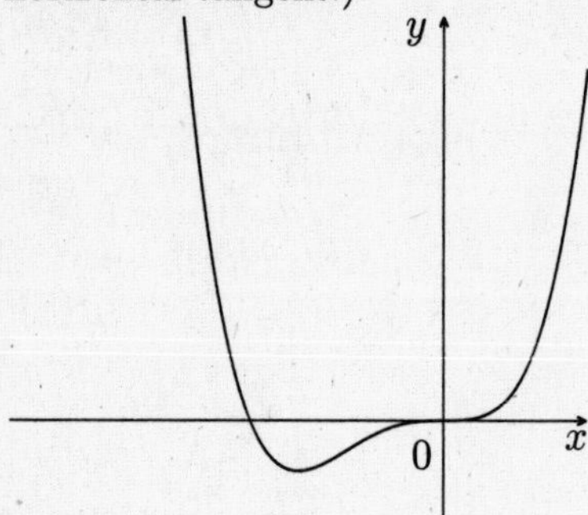

17. $y = 4 - 4x^3 - 3x^4$; $y' = -12x^2 - 12x^3$

$$\begin{aligned} -12x^2 - 12x^3 &= 0 \\ -12x^2(1+x) &= 0 \\ x &= -1,\ 0 \end{aligned}$$

Critical points: $(-1, 5)$ and $(0, 4)$.

	test values	$-12x^2$	$1+x$	$y' = -12x^2(1+x)$
$x < -1$	-2	−	−	+
$-1 < x < 0$	$-1/2$	−	+	−
$x > 0$	1	−	+	−

Observe that to the left of $x = -1$, the function is increasing, and to the right of $x = -1$, the function is decreasing. It follows that $(-1, 5)$ is a maximum.

Since the function is decreasing to the right of $x = -1$, it is decreasing to the left and right of $x = 0$. So the point $(0, 4)$ is neither a minimum nor a maximum.

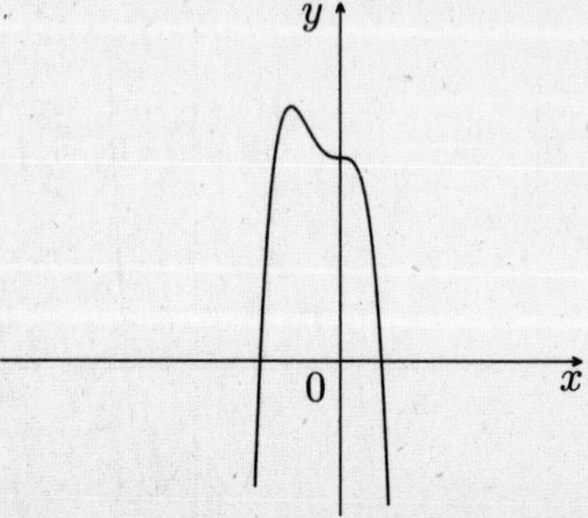

19. $y = 2\sqrt[3]{x} = 2x^{1/3}$; $y' = 2\left(\frac{1}{3}\right)x^{-2/3} = \dfrac{2}{3x^{2/3}} \neq 0$.
y' cannot be 0, but y' is undefined (∞) at $x = 0$, indicating a vertical tangent at $(0,0)$. So $(0,0)$ is a critical point. Now observe that

$$y' = \frac{2}{3(x^{1/3})^2} > 0 \text{ for all } x \neq 0,$$

so that f is an increasing function. We conclude that $(0,0)$ is neither a minimum nor a maximum point.

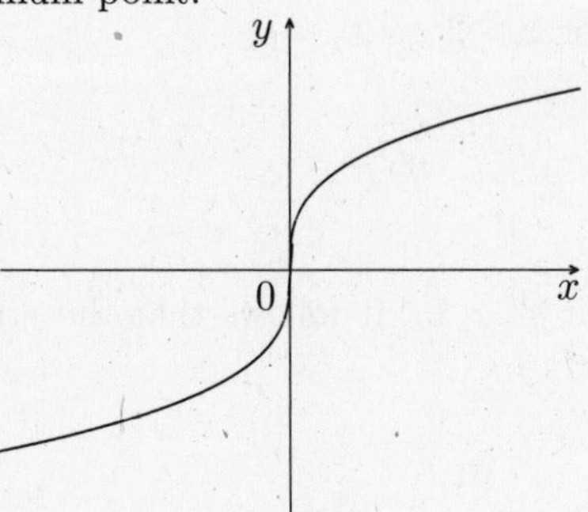

21. $y = 4\sqrt{x} - 2x = 4x^{1/2} - 2x$; $\qquad y' = 2x^{-1/2} - 2 = 0$
$x = 1$

Domain: $[0,\infty)$. To the right if $(0,0)$, $y' > 0$, so that y is increasing. This makes $(0,0)$ an endpoint and therefore a minimum. At $x = 1$, y' changes from positive to negative. So $(1,2)$ is a maximum point.

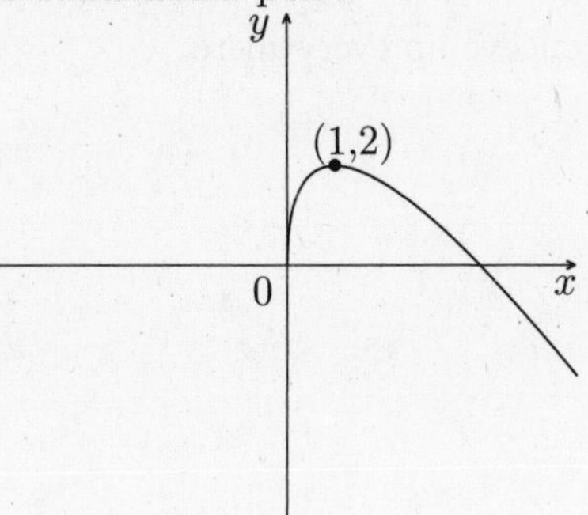

23. $s = 160t - 16t^2$; $\qquad \dfrac{ds}{dt} = 160 - 32t = 0$

$$t = \frac{160}{32} = 5\,\text{s}$$

When $t = 5$, $s = 400$ ft. At $t = 5$, $\dfrac{ds}{dt}$ changes from positive to negative, making $s = 400$ ft a maximum value.

25. Using the quotient rule, we find that $\dfrac{dP}{dR} = -\dfrac{16(R-2)}{(R+2)^3}$. $R = 2\,\Omega$ is the critical value. Since P is increasing to the left of $R = 2$ and decreasing to the right, $R = 2\,\Omega$ does correspond to a maximum.

3.2 The Second-Derivative Test

1. $y = 2x - x^2$; $y' = 2 - 2x$; $y'' = -2$. Since y'' is strictly negative, the graph is everywhere concave down.

3. $y = x^3 - 12x + 6$; $y' = 3x^2 - 12$; $y'' = 6x$. If $x < 0$, then $y'' < 0$ and if $x > 0$, then $y'' > 0$. We conclude that the graph is concave down on $(-\infty, 0]$ and concave up on $[0,\infty)$.

5. $$\begin{aligned} y &= x^4 - 4x^3 + 8x - 5; \\ y' &= 4x^3 - 12x^2 + 8; \\ y'' &= 12x^2 - 24x = 12x(x-2) = 0. \end{aligned}$$
So $y'' = 0$ when $x = 0$ and $x = 2$. These are the only values for which y'' is zero; y'' is therefore different from zero everywhere else, and we may use arbitrary test values:
$x = -1 \;:\; y'' = 12(-1)(-3) > 0$ concave up for $x < 0$
$x = 1 \;:\; y'' = 12(1)(-1) < 0$ concave down on $[0, 2]$
$x = 3 \;:\; y'' = 12(3)(1) > 0$ concave up for $x > 2$

7. $$\begin{aligned} y &= x^3 - 9x^2 + 27x - 27; \\ y' &= 3x^2 - 18x + 27; \\ y'' &= 6x - 18 = 0 \text{ when } x = 3. \end{aligned}$$
Observe that if $x < 3$, then $y'' < 0$ and if $x > 3$, then $y'' > 0$. It follows that the graph is concave down on $(-\infty, 3]$ and concave up on $[3, \infty)$.

9. $f(x) = 2x^2 - 4x$; $f'(x) = 4x - 4$; $f''(x) = 4$.
Step 1. Critical points: $$\begin{aligned} f'(x) = 4x - 4 &= 0 \\ x &= 1. \end{aligned}$$
Substituting $x = 1$ in the given equation $y = 2x^2 - 4x$ yields $y = -2$; thus $(1, -2)$ is the critical point.
Step 2. Test of critical point: Since $f''(x) = 4$ for all x, $f''(1) = 4 > 0$. Thus $(1, -2)$ is a minimum.
Step 3. Concavity: Since $f''(x) = 4 > 0$, the graph is concave up everywhere.

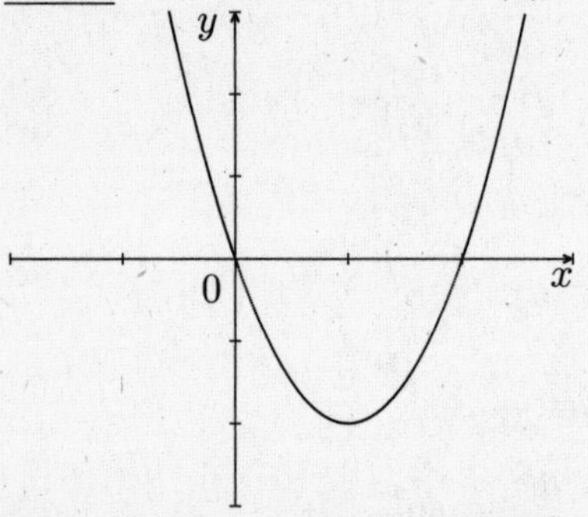

11. $f(x) = 6x - 6x^2$; $f'(x) = 6 - 12x$; $f''(x) = -12$.
Step 1. Critical points: $$\begin{aligned} f'(x) = 6 - 12x &= 0 \\ x &= \frac{1}{2}. \end{aligned}$$
Substituting $x = 1/2$ in the given equation yields $y = 3/2$. So $(\frac{1}{2}, \frac{3}{2})$ is the critical point.
Step 2. Test of critical point: Since $f''(x) = -12$ for all x, $f''(\frac{1}{2}) = -12 < 0$. So $(\frac{1}{2}, \frac{3}{2})$ is a maximum point.
Step 3. Concavity: Since $f''(x) < 0$ for all x, the graph is concave down everywhere.

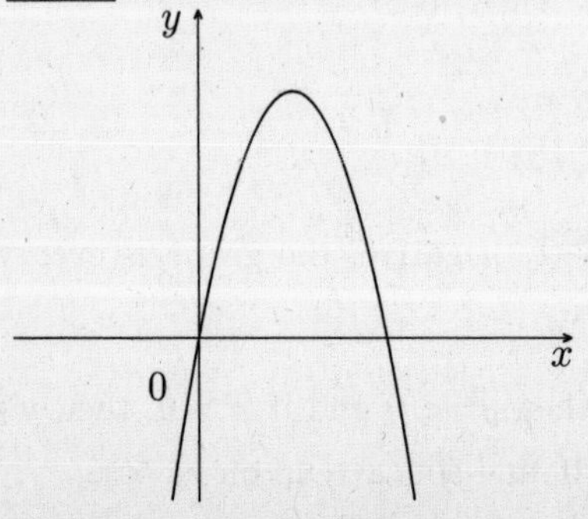

13. $f(x) = -4 - 3x - \frac{1}{2}x^2$; $f'(x) = -3 - x$; $f''(x) = -1$.

Step 1. Critical points: $f'(x) = -3 - x = 0$

$$x = -3.$$

Substituting $x = -3$ in $y = -4 - 3x - \frac{1}{2}x^2$, we get $y = \frac{1}{2}$. So $(-3, \frac{1}{2})$ is a critical point.

Step 2. Test of critical point: Since $f''(x) = -1$ for all x, $f''(-3) = -1 < 0$. So $(-3, \frac{1}{2})$ is a maximum.

Step 3. Concavity: Since $f''(x) = -1$, the graph is concave down everywhere.

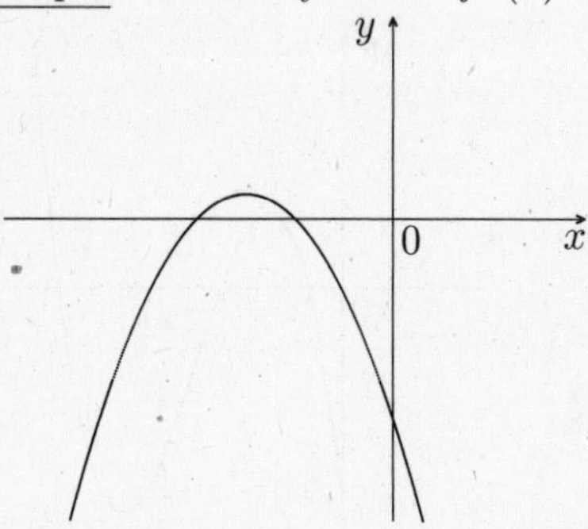

15. $f(x) = 2x^3 - 6x + 1$; $f'(x) = 6x^2 - 6$; $f''(x) = 12x$.

Step 1. Critical points: $f'(x) = 6x^2 - 6 = 0$

$$x = \pm 1.$$

Substituting these values in $y = 2x^3 - 6x + 1$, we get $(1, -3)$ and $(-1, 5)$ for the critical points.

Step 2. Test of critical points: $f''(-1) = -12 < 0$; $(-1, 5)$ is a maximum.

$f''(1) = 12 > 0$; $(1, -3)$ is a minimum.

Step 3. Concavity: We need to determine where $y'' < 0$ and where $y'' > 0$. To this end, we first determine where $y'' = 0$: $f''(x) = 12x = 0$ when $x = 0$; the point is $(0, 1)$.

	test values	$y'' = 12x$
$x < 0$	-1	$-$
$x > 0$	1	$+$

So $f(x)$ is concave down to the left of $(0, 1)$ and concave up to the right of $(0, 1)$.

Step 4. Inflection point: Since the concavity changes, $(0, 1)$ is an inflection point.

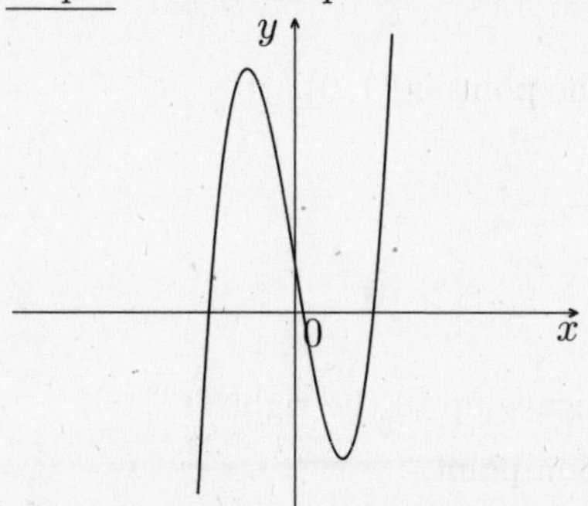

17: $f(x) = x^3 - 6x^2 + 9x - 3$; $f'(x) = 3x^2 - 12x + 9$; $f''(x) = 6x - 12$.

Step 1. Critical points:

$$\begin{aligned} f'(x) &= 3x^2 - 12x + 9 = 0 \\ 3(x^2 - 4x + 3) &= 0 \\ 3(x - 3)(x - 1) &= 0 \\ x &= 1, 3. \end{aligned}$$

Substituting these values in $y = x^3 - 6x^2 + 9x - 3$, we see that $(1, 1)$ and $(3, -3)$ are the critical points.

Step 2. Test of critical points:
$f''(1) = 6 - 12 = -6 < 0$; thus $(1, 1)$ is a maximum.
$f''(3) = 18 - 12 = 6 > 0$; thus $(3, -3)$ is a minimum.
Step 3. Concavity: We need to determine where $y'' < 0$ and where $y'' > 0$. To this end, we first determine where $y'' = 0$: $f''(x) = 6x - 12 = 0$ when $x = 2$. If $x = 2$, then $y = -1$; the point is $(2, -1)$.

	test values	$y'' = 6x - 12$
$x < 2$	1	$-$
$x > 2$	3	$+$

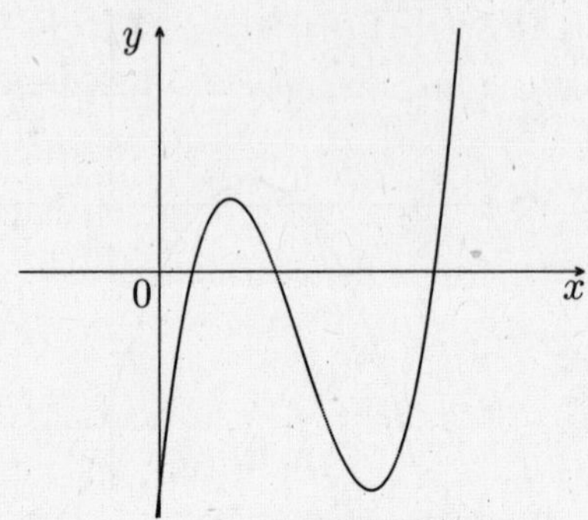

So $f(x)$ is concave down to the left of $(2, -1)$ and $f(x)$ is concave up to the right of $(2, -1)$.
Step 4. Inflection point: Since the concavity changes, the point $(2, -1)$ is an inflection point.

19. $f(x) = x^3 - 3x^2 - 9x + 11$; $f'(x) = 3x^2 - 6x - 9$; $f''(x) = 6x - 6$.
Step 1. Critical points:

$$\begin{aligned} f'(x) = 3x^2 - 6x - 9 &= 0 \\ 3(x^2 - 2x - 3) &= 0 \\ 3(x+1)(x-3) &= 0 \\ x &= -1, 3. \end{aligned}$$

Substituting these values in $y = x^3 - 3x^2 - 9x + 11$, we get the following critical points: $(-1, 16)$ and $(3, -16)$.
Step 2. Test of critical points:
$f''(-1) = 6(-1) - 6 = -12 < 0$; $(-1, 16)$ is a maximum.
$f''(3) = 6(3) - 6 = 12 > 0$; $(3, -16)$ is a minimum.
Step 3. Concavity: $f''(x) = 6x - 6 = 0$ when $x = 1$. The point is $(1, 0)$.

	test values	$f''(x) = 6x - 6$	
$x < 1$	0	$-$	concave down
$x > 1$	2	$+$	concave up

The graph is concave down to the left of $x = 1$ and concave up to the right of $x = 1$.
Step 4. Since the concavity changes, $(1, 0)$ is an inflection point.

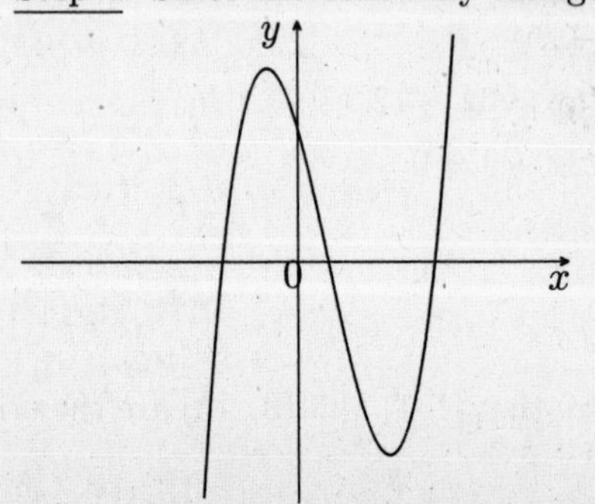

21. $f(x) = 2 + 3x - x^3$; $f'(x) = 3 - 3x^2$; $f''(x) = -6x$.

Step 1. Critical points:
$$\begin{aligned} f'(x) &= 3 - 3x^2 = 0 \\ 3(1 - x^2) &= 0 \\ x &= \pm 1. \end{aligned}$$

Substituting these values in $y = 2 + 3x - x^3$, we get the following critical points: $(-1, 0)$ and $(1, 4)$.

Step 2. Test of critical points:

$f''(-1) = -6(-1) = 6 > 0$; $(-1, 0)$ is a minimum.

$f''(1) = -6(1) = -6 < 0$; $(1, 4)$ is maximum.

Step 3. Concavity: $f''(x) = -6x$ is positive for x less than 0 (concave up) and negative for x greater than 0 (concave down). When $x = 0$, $y = 2$.

Step 4. Since the concavity changes at $x = 0$, the point $(0, 2)$ is an inflection point.

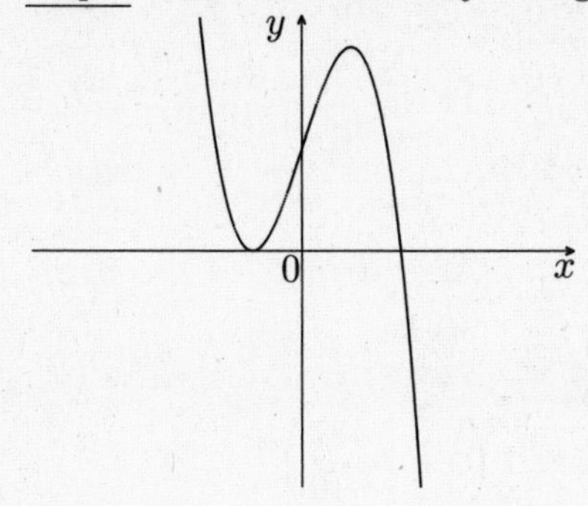

23. $f(x) = x^4 - \frac{4}{3}x^3 + 2$; $f'(x) = 4x^3 - 4x^2$; $f''(x) = 12x^2 - 8x$.

Step 1. Critical points:
$$\begin{aligned} f'(x) = 4x^3 - 4x^2 &= 0 \\ 4x^2(x - 1) &= 0 \\ x &= 0, 1. \end{aligned}$$

Substituting these values in $y = x^4 - \frac{4}{3}x^3 + 2$, we get $(0, 2)$ and $\left(1, \frac{5}{3}\right)$.

Step 2. Test of critical points:

$f''(1) > 0$; $\left(1, \frac{5}{3}\right)$ is a minimum.

$f''(0) = 0$; the test fails.

Using the first-derivative test, $f'\left(-\frac{1}{2}\right) < 0$ and $f'\left(\frac{1}{2}\right) < 0$. So $(0, 2)$ is neither a minimum nor a maximum.

Step 3. Concavity: We need to determine where $y'' < 0$ and where $y'' > 0$. To this end we first determine where $f''(x) = 0$:
$$\begin{aligned} f''(x) = 12x^2 - 8x &= 0 \\ 4x(3x - 2) &= 0 \\ x &= 0, \frac{2}{3}. \end{aligned}$$

Substituting in $y = x^4 - \frac{4}{3}x^3 + 2$, we find that the points are $(0, 2)$ and $\left(\frac{2}{3}, \frac{146}{81}\right) \approx \left(\frac{2}{3}, 1.8\right)$.

	test values	$4x$	$3x - 2$	$y'' = 4x(3x - 2)$
$x < 0$	-1	$-$	$-$	$+$
$0 < x < 2/3$	$1/2$	$+$	$-$	$-$
$x > 2/3$	1	$+$	$+$	$+$

Summary:
$f(x)$ is concave up for $x < 0$.
$f(x)$ is concave down for $0 < x < 2/3$.
$f(x)$ is concave up for $x > 2/3$.
Step 4. Inflection points: Since the concavity changes, $(0, 2)$ and $\left(\frac{2}{3}, 1.8\right)$ are inflection points.

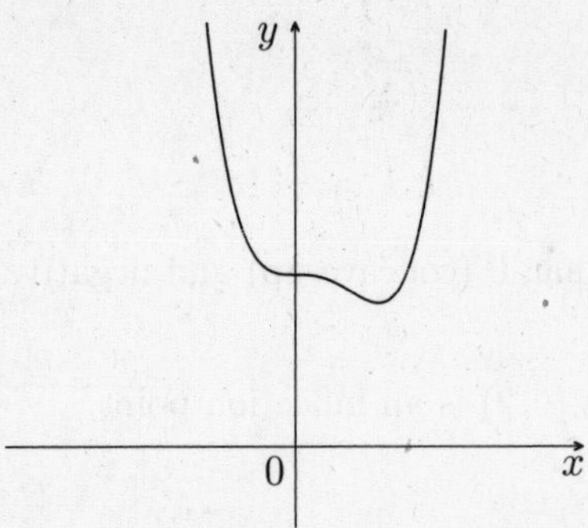

25. $f(x) = x^4 + 2x^3$; $f'(x) = 4x^3 + 6x^2$; $f''(x) = 12x^2 + 12x$.
Step 1. Critical points:
$$\begin{aligned} f'(x) = 4x^3 + 6x^2 &= 0 \\ 2x^2(2x+3) &= 0 \\ x &= 0, -\frac{3}{2}. \end{aligned}$$
Substituting these values in $y = x^4 + 2x^3$, we get $(0, 0)$ and $\left(-\frac{3}{2}, -\frac{27}{16}\right)$.
Step 2. Test of critical points:
$f''(-\frac{3}{2}) > 0$; $\left(-\frac{3}{2}, -\frac{27}{16}\right)$ is a minimum.
$f''(0) = 0$; the test fails.
Using the first derivative test, $f'\left(-\frac{1}{2}\right) > 0$ and $f'\left(\frac{1}{2}\right) > 0$. So $(0, 0)$ is neither a minimum nor a maximum.
Step 3. Concavity: We need to determine where $y'' < 0$ and where $y'' > 0$. To this end we find those values of x for which $f''(x) = 0$:
$$\begin{aligned} f''(x) = 12x^2 + 12x &= 0 \\ 12x(x+1) &= 0 \\ x &= 0, -1. \end{aligned}$$
Substituting in $y = x^4 + 2x^3$, we find that the points are $(0, 0)$ and $(-1, -1)$.

	test values	$12x$	$x+1$	$y'' = 12x(x+1)$
$x < -1$	-2	$-$	$-$	$+$
$-1 < x < 0$	$-1/2$	$-$	$+$	$-$
$x > 0$	$1/2$	$+$	$+$	$+$

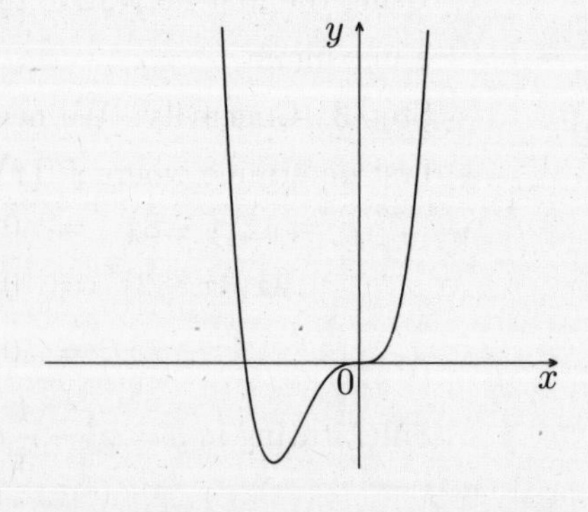

Summary:
$f(x)$ is concave up for $x < -1$.
$f(x)$ is concave down for $-1 < x < 0$.
$f(x)$ is concave up for $x > 0$.
Step 4. Inflection points: Since the concavity changes at each of the points, $(0, 0)$ and $(-1, -1)$ are inflection points.

27. $f(x) = (x-1)^4$; $f'(x) = 4(x-1)^3$; $f''(x) = 12(x-1)^2$.

Step 1. Critical points: $f'(x) = 4(x-1)^3 = 0$

$$x = 1.$$

Substituting in $y = (x-1)^4$, we get $(1, 0)$ for the critical point.

Step 2. Test of critical point:

$f''(1) = 0$; the test fails.

Returning to the first-derivative test, observe that $y' < 0$ for $x < 1$ and $y' > 0$ for $x > 1$. The point $(1, 0)$ is a minimum.

Step 3. Concavity: $f''(x) = 12(x-1)^2 > 0$ for all $x \neq 1$.

The graph is concave up everywhere.

Step 4. There are no inflection points.

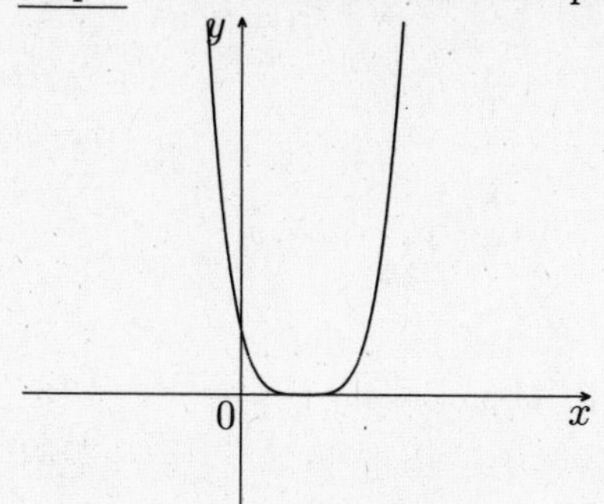

29. $f(x) = \dfrac{1}{(x-2)^3} = (x-2)^{-3}$; $f'(x) = -3(x-2)^{-4}$; $f''(x) = 12(x-2)^{-5}$

Vertical asymptote: $x = 2$; horizontal asymptote: x-axis; no horizontal tangents: $y' = \dfrac{-3}{(x-2)^4} \neq 0$

(Observe that the slope is always negative.)

Concavity: $f''(x) = \dfrac{12}{(x-2)^5} \neq 0$

However, $y'' < 0$ whenever $x < 2$ and $y'' > 0$ whenever $x > 2$. So the graph is concave down to the left of the asymptote and concave up to the right. (There are no inflection points.)

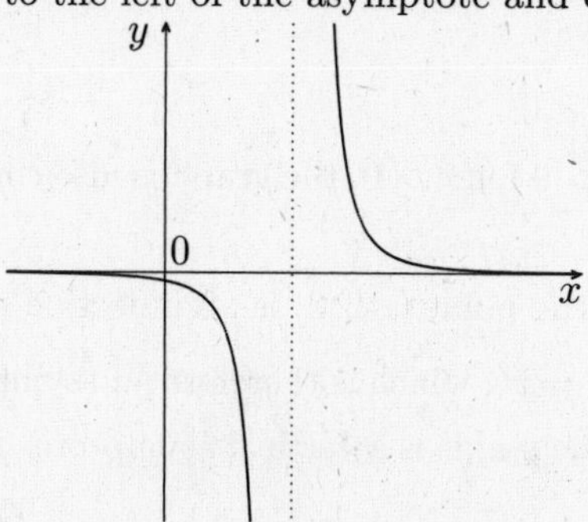

31. $f(x) = \dfrac{x-1}{x+1}$. Vertical asymptote: $x = -1$. Horizontal asymptote: $y = 1$ since

$$\lim_{x\to\infty} \frac{x-1}{x+1} = \lim_{x\to\infty} \frac{1-1/x}{1+1/x} = 1.$$

$$f'(x) = \frac{2}{(x+1)^2};\ f''(x) = -\frac{4}{(x+1)^3}$$

Step 1. Critical points: Since $f'(x) = \dfrac{2}{(x+1)^2} > 0$, there are no critical points. In fact, the graph is strictly increasing.

Step 2. Does not apply.

Step 3. Concavity: $f''(x) = -\dfrac{4}{(x+1)^3}$
$f''(x) > 0$ for $x < -1$ (concave up)
$f''(x) < 0$ for $x > -1$ (concave down)
Step 4. Inflection points: none. (Observe that the concavity changes at the vertical asymptote.)

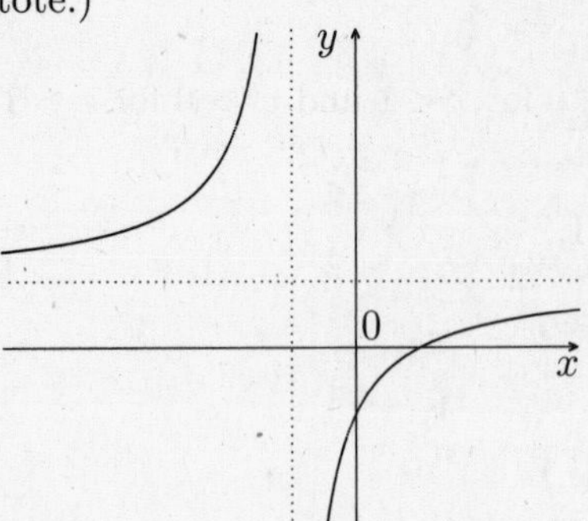

33. $f(x) = x^2 + \dfrac{8}{x}$; $f'(x) = 2x - \dfrac{8}{x^2}$; $f''(x) = 2 + \dfrac{16}{x^3}$.
Step 1. Critical points:

$$f'(x) = 2x - \frac{8}{x^2} = \frac{2x^3 - 8}{x^2} = 0, \text{ or } x = \sqrt[3]{4}.$$

Step 2. Test of critical points: $f''(\sqrt[3]{4}) > 0$ (minimum).
Step 3. Concavity: We need to determine where $y'' < 0$ and $y'' > 0$. To this end we first determine those values of x for which $f''(x) = 0$:

$$\begin{aligned} f''(x) = 2 + \frac{16}{x^3} = \frac{2x^3 + 16}{x^3} &= 0 \\ 2x^3 + 16 &= 0 \\ x^3 &= -8 \\ x &= -2. \end{aligned}$$

From $y = x^2 + \dfrac{8}{x}$, the point is $(-2, 0)$.
If $x = -3$, $y'' > 0$; so $f(x)$ is concave up for $x < -2$.
If $x = -1$, $y'' < 0$; so $f(x)$ is concave down for $-2 < x < 0$.
We see that the concavity changes at $(-2, 0)$. Since $y'' > 0$ for $x > 0$, the graph is also concave up for $x > 0$.
Step 4. Inflection points: Since the concavity changes, the point $(-2, 0)$ is an inflection point.
Step 5. Other: If x gets large, $y = x^2 + \dfrac{8}{x}$ approaches $y = x^2$, which is therefore an asymptotic curve. Since the denominator of $f(x)$ is 0 when $x = 0$, the y-axis is a vertical asymptote. (Note that the concavity changes at the asymptote.)

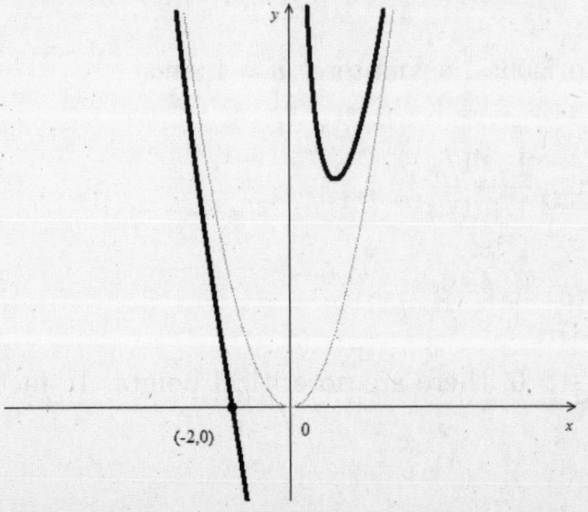

35. $f(x) = \dfrac{x+1}{x-2}$. Vertical asymptote: $x = 2$. Horizontal asymptote: $y = 1$ since

$$\lim_{x\to\infty} \frac{x+1}{x-2} = \lim_{x\to\infty} \frac{1+1/x}{1-2/x} = 1.$$

$f'(x) = \dfrac{-3}{(x-2)^2}$; $f''(x) = \dfrac{6}{(x-2)^3}$.

Step 1. Critical points: Since $f'(x) = \dfrac{-3}{(x-2)^2} < 0$ for all $x \neq 2$, there are no critical points. Moreover, $f(x)$ is always decreasing.

Step 2. (Does not apply.)

Step 3. Concavity: $f''(x) = \dfrac{6}{(x-2)^3}$.

$f''(x) > 0$ for $x > 2$ (concave up).

$f''(x) < 0$ for $x < 2$ (concave down).

Step 4. Inflection points: None. (Note that the concavity changes at $x = 2$, the vertical asymptote.)

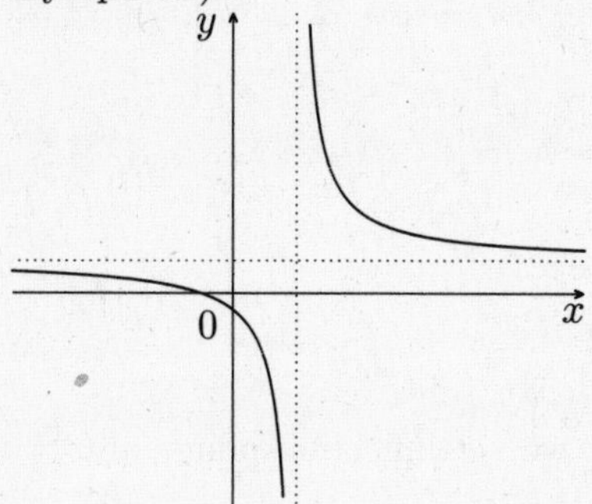

37. $y = \dfrac{2x}{(x+1)^2}$. There is a vertical asymptote at $x = -1$.

$f'(x) = -\dfrac{2(x-1)}{(x+1)^3}$; $f''(x) = \dfrac{4(x-2)}{(x+1)^4}$.

Step 1. Critical points: $f'(x) = 0$ when $x = 1$. The critical point is $\left(1, \dfrac{1}{2}\right)$.

Step 2. Test of critical point: $f''(1) < 0$; $\left(1, \dfrac{1}{2}\right)$ is a maximum.

Step 3. Concavity: $y'' = 0$ when $x = 2$. Now observe that for $x < 2$, $y'' < 0$. So the graph is concave down to the left of the asymptote and also concave down on $(-1, 2]$. For $x > 2$, $y'' > 0$. The graph is concave up on $[2, \infty)$.

Step 4. Because the concavity changes, $\left(2, \dfrac{4}{9}\right)$ is an inflection point.

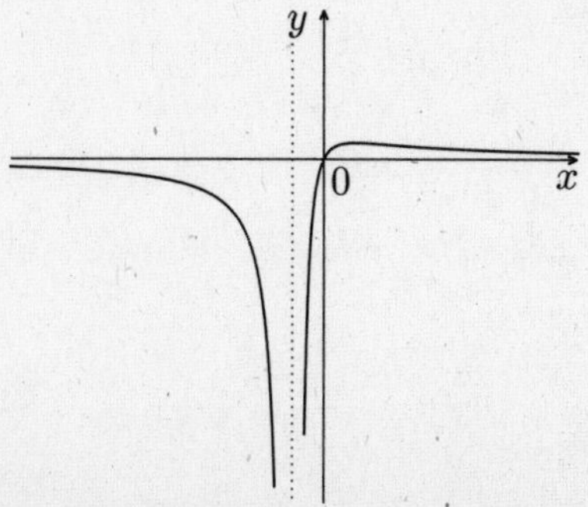

39. $f(x) = \dfrac{6x}{x^2+3}$; $f'(x) = \dfrac{6(3-x^2)}{(x^2+3)^2}$; $f''(x) = \dfrac{12x(x^2-9)}{(x^2+3)^3}$.

Step 1. Critical points: $6(3-x^2) = 0$ when $x = -\sqrt{3}, \sqrt{3}$.

Step 2. Test of critical points:

$f''(-\sqrt{3}) > 0$ (minimum).

$f''(\sqrt{3}) < 0$ (maximum).

Step 3. Concavity: We need to determine where $y'' < 0$ and where $y'' > 0$. To this end we first determine where $y'' = 0$:

$$\begin{aligned} 12x(x^2-9) &= 0 \\ x &= 0, \pm 3. \end{aligned}$$

	test values	$12x$	x^2-9	$y'' = \dfrac{12x(x^2-9)}{(x^2+3)^3}$
$x < -3$	-4	$-$	$+$	$-$
$-3 < x < 0$	-1	$-$	$-$	$+$
$0 < x < 3$	1	$+$	$-$	$-$
$x > 3$	4	$+$	$+$	$+$

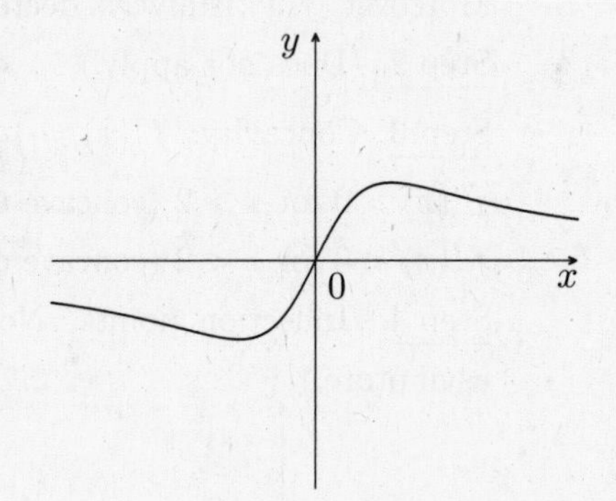

Summary:

$f(x)$ is concave down for $x < -3$.

$f(x)$ is concave up for $-3 < x < 0$.

$f(x)$ is concave down for $0 < x < 3$.

$f(x)$ is concave up for $x > 3$.

Step 4. Inflection points: Since the concavity changes, we get inflection points at $x = 0$ and $x = \pm 3$.

3.3 Exploring with Graphing Utilities

1. $y = x^5 + x^3 + 1$

$$\begin{aligned} y' = 5x^4 + 3x^2 &= 0 \\ x^2(5x^2+3) &= 0 \\ x = 0,\ 5x^2+3 &> 0 \end{aligned}$$

Critical point: $(0, 1)$.

$$\begin{aligned} y'' = 20x^3 + 6x &= 0 \\ x(20x^2+6) &= 0 \\ x = 0,\ 20x^2+6 &> 0 \end{aligned}$$

Inflection point: $(0, 1)$.

$[-2, 2]$ by $[-1, 3]$

3. $y = 4x^2 - \frac{4}{5}x^5 + 2$

$$\begin{aligned} y' = 8x - 4x^4 &= 0 \\ 4x(2 - x^3) &= 0 \\ x &= 0, \sqrt[3]{2} \end{aligned}$$

Critical points: $(0, 2)$ and $(\sqrt[3]{2}, 5.8)$.

$$\begin{aligned} y'' = 8 - 16x^3 &= 0 \\ 1 - 2x^3 &= 0 \\ x &= \frac{1}{\sqrt[3]{2}} \end{aligned}$$

Inflection point: $\left(\frac{1}{\sqrt[3]{2}}, 4.27\right)$.

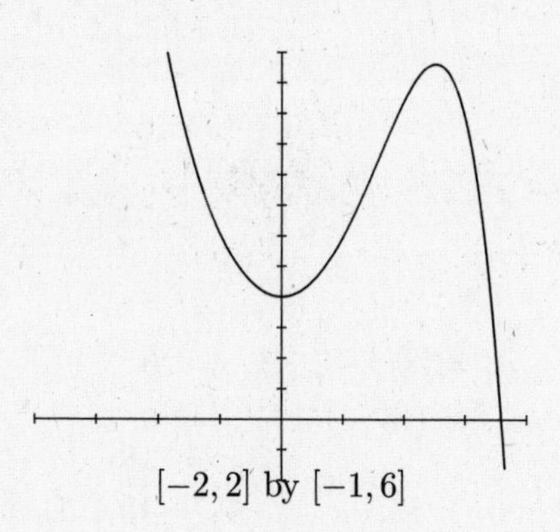

$[-2, 2]$ by $[-1, 6]$

5. $y = 3.0x^2 + 1.2x^5 - 1.0$

$$\begin{aligned} y' = 6x + 1.2(5x^4) &= 0 \\ 6x + 6x^4 &= 0 \\ 6x(1 + x^3) &= 0 \\ x &= 0,\ -1 \end{aligned}$$

$$\begin{aligned} y'' = 6 + 24x^3 &= 0 \\ 6(1 + 4x^3) &= 0 \\ 4x^3 &= -1 \\ x^3 &= -\frac{1}{4} \\ x &= -\frac{1}{\sqrt[3]{4}} \approx -0.63 \end{aligned}$$

If $x = 0$, $y = -1.0$.

If $x = -1$, $y = 3.0(-1)^2 + 1.2(-1)^5 - 1.0 = 0.80$.

If $x = -\frac{1}{\sqrt[3]{4}}$, $y = 3.0\left(-\frac{1}{\sqrt[3]{4}}\right)^2 + 1.2\left(-\frac{1}{\sqrt[3]{4}}\right)^5 - 1.0 = 0.071$.

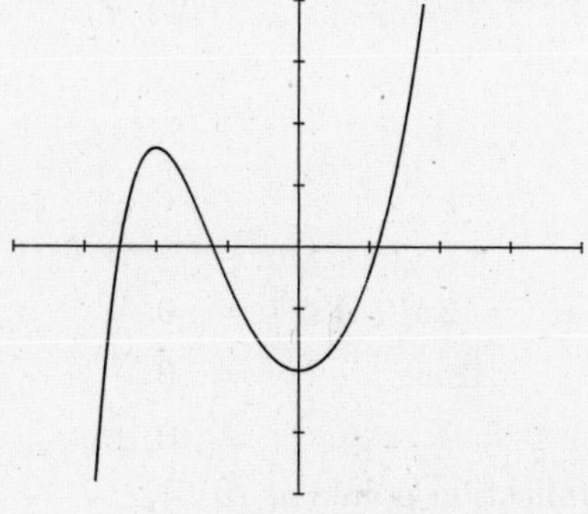

$[-2, 2]$ by $[-2, 2]$

7. $y = 3.7x^6 + 2.4x^4 + 1.5$

$$\begin{aligned} y' = 22.2x^5 + 9.6x^3 &= 0 \\ x^3(22.2x^2 + 9.6) &= 0 \end{aligned}$$

$x = 0$ is the only real solution. By the graph, $(0, 1.5)$ is a minimum.

$$\begin{aligned} y'' = 111x^4 + 28.8x^2 &= 0 \\ x^2(111x^2 + 28.8) &= 0 \end{aligned}$$

$x = 0$ is the only real solution. Even though $y'' = 0$ at $x = 0$, $(0, 1.5)$ is not an inflection point.

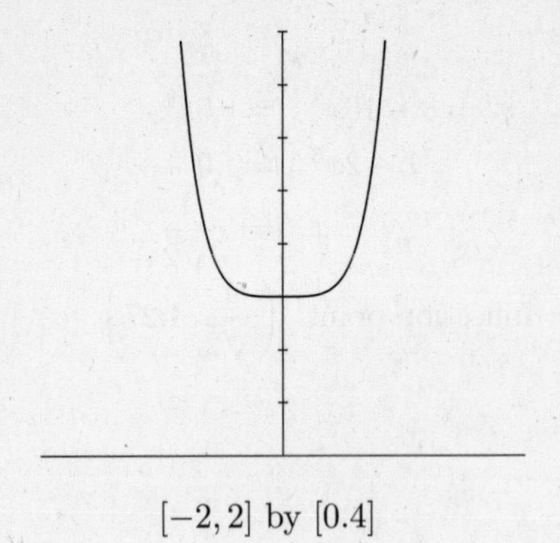

[−2, 2] by [0.4]

9. $y = 1.5x^4 - 0.50x^6 + 0.20$

$$\begin{aligned} y' = 1.5(4x^3) - 0.50(6x^5) &= 0 \\ 6x^3 - 3x^5 &= 0 \\ 3x^3(2 - x^2) &= 0 \\ x &= 0,\ \pm\sqrt{2} \end{aligned}$$

$$\begin{aligned} y'' = 18x^2 - 15x^4 &= 0 \\ x^2(18 - 15x^2) &= 0 \\ x = 0,\ x^2 &= \frac{18}{15} \\ x &= \pm\sqrt{\frac{18}{15}} \approx \pm 1.1 \end{aligned}$$

If $x = 0$, $y = 0.20$.
If $x = \pm\sqrt{2}$, $y = 1.5(\pm\sqrt{2})^4 - 0.50(\pm\sqrt{2})^6 + 0.20 = 2.2$.
If $x = \pm\sqrt{\frac{18}{15}}$, $y = 1.5\left(\pm\sqrt{\frac{18}{15}}\right)^4 - 0.50\left(\pm\sqrt{\frac{18}{15}}\right)^6 + 0.20 = 1.5$.

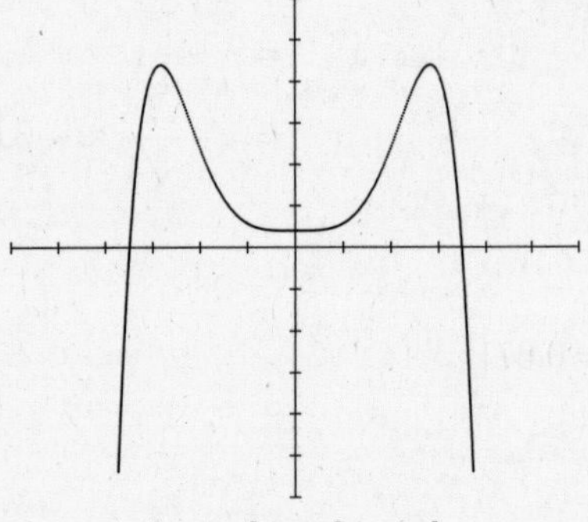

[−3, 3] by [−3, 3]

11. $y = \frac{3}{5}x^5 - 2x^3$

$$\begin{aligned} y' = 3x^4 - 6x^2 &= 0 \\ 3x^2(x^2 - 2) &= 0 \\ x &= 0, \pm\sqrt{2} \end{aligned}$$

(Critical points.)

$$\begin{aligned} y'' = 12x^3 - 12x &= 0 \\ 12x(x^2 - 1) &= 0 \\ x &= 0, \pm 1 \end{aligned}$$

Inflection points at $(0, 0)$, $(-1, 1.4)$, $(1, -1.4)$.

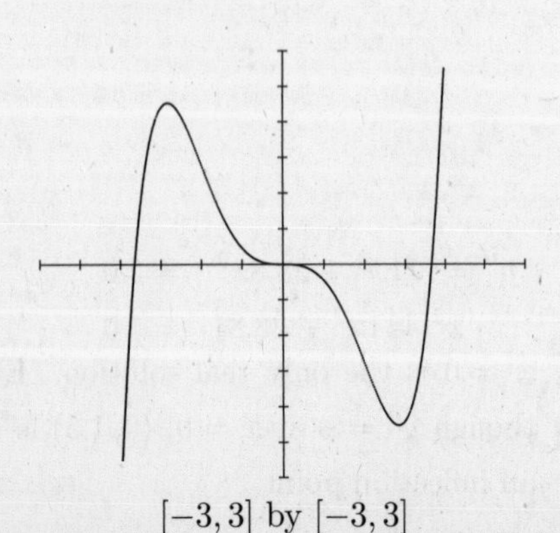

[−3, 3] by [−3, 3]

13. $y = \frac{1}{7}x^7 - \frac{1}{5}x^5$

$$\begin{aligned} y' = x^6 - x^4 &= 0 \\ x^4(x^2-1) &= 0 \\ x &= 0, \pm 1 \end{aligned}$$

$$\begin{aligned} y'' = 6x^5 - 4x^3 &= 0 \\ x^3(6x^2-4) &= 0 \\ x = 0,\ x^2 &= \frac{2}{3} \\ x &= \pm\frac{\sqrt{6}}{3} \end{aligned}$$

If $x = 0$, $y = 0$.
If $x = 1$, $y = \frac{1}{7} - \frac{1}{5} = -\frac{2}{35}$.
If $x = -1$, $y = -\frac{1}{7} + \frac{1}{5} = \frac{2}{35}$.
If $x = \frac{\sqrt{6}}{3}$, $y = \frac{1}{7}\left(\frac{\sqrt{6}}{3}\right)^7 - \frac{1}{5}\left(\frac{\sqrt{6}}{3}\right)^5 \approx -0.038$.

$[-2, 2]$ by $[-0.1, 0.1]$

15. $y = 1.2x + \frac{4.0}{\sqrt{x}}$

Vertical asymptote: y-axis.

$$\begin{aligned} y' = 1.2 - \frac{2}{x^{3/2}} &= 0 \\ 1.2x^{3/2} - 2 &= 0 \\ x^{3/2} &= \frac{2}{1.2} \\ x &= 1.41 \end{aligned}$$

Thus $(1.41, 5.06)$ is a minimum. (See graph in answer section of book.)

17. By the quotient rule,

$$\begin{aligned} \frac{dy}{dx} &= \frac{(x^2-1)\cdot 1 - x(2x)}{(x^2-1)^2} = \frac{x^2-1-2x^2}{(x^2-1)^2} \\ &= \frac{-x^2-1}{(x^2-1)^2} = -\frac{x^2+1}{(x^2-1)^2} \neq 0. \end{aligned}$$

In fact, $y' < 0$ for all x, so that the slope of the tangent line is negative everywhere. (See graph in answer section of book.)

19. $y = \frac{x^3}{x^2-3}$. By the quotient rule,

$$\begin{aligned} y' = \frac{x^2(x^2-9)}{(x^2-3)^2} &= 0 \\ x &= 0, \pm 3 \text{ (critical points)}. \end{aligned}$$

Vertical asymptotes: $x = \pm\sqrt{3}$. (See graph in answer section of book.)

21. $$\begin{aligned} y &= 0.50x^3 + 2.34x^{-1} \\ y' &= 0.50(3x^2) - (2.34)x^{-2} \\ &= 1.50x^2 - \frac{2.34}{x^2} = 0 \end{aligned}$$
Multiplying both sides of the equation by x^2, we get
$$\begin{aligned} 1.50x^4 - 2.34 &= 0 \\ x &= \pm\sqrt[4]{\frac{2.34}{1.50}} \approx \pm 1.12 \end{aligned}$$
If $x = 1.12$, $y = 0.50(1.12)^3 + \dfrac{2.34}{1.12} = 2.79$.
If $x = -1.12$, $y = -2.79$.
(See graph in answer section of book.)

23. $y = 2x - \dfrac{1}{x}$
Vertical asymptote: y-axis.
$y' = 2 + \dfrac{1}{x^2} = 0$ has no real solutions. (See graph in answer section of book.)

3.4 Applications of Minima and Maxima

1. Find the critical value:
$$\begin{aligned} \frac{dP}{di} &= 0 + 12.8 - 6.40i = 0 \\ i &= \frac{12.8}{6.40} = 2.0\,\text{A}. \end{aligned}$$
Since $\dfrac{d^2P}{di^2} = -6.40 < 0$, $i = 2.0$ leads to a maximum. Thus
$P = 4.50 + 12.8(2.0) - 3.20(2.0)^2 = 17.3\,\text{W}$.

3. $q(t) = \dfrac{t}{t^2+1}$; by the quotient rule, $q'(t) = \dfrac{1-t^2}{(t^2+1)^2} = 0$, whence $t = \pm 1$.
Consider the critical value $t = 1$: for t slightly less than 1, $q' > 0$; for t slightly more than 1, $q' < 0$. So t does correspond to a maximum. When $t = 1\,\text{s}$,
$$q(1) = \frac{1}{1^2+1} = \frac{1}{2}\,\text{C}.$$

5. $d(x) = 1.5 \times 10^{-6}x^2(40-x)^2$; by the product rule,
$$\begin{aligned} d'(x) &= 1.5 \times 10^{-6}[x^2 \cdot 2(40-x)(-1) + (40-x)^2 \cdot 2x] \\ &= 1.5 \times 10^{-6}(2x)(40-x)[-x + (40-x)] \\ &= 1.5 \times 10^{-6}(2x)(40-x)(40-2x) = 0 \end{aligned}$$
when $x = 0$, 40, and 20. At $x = 0$ and $x = 40$, $d(x) = 0$ (no deflection). At $x = 20$, $d(x)$ is positive. The value $x = 20\,\text{in.}$ leads to a maximum by the first-derivative test: $d'(19) > 0$ and $d'(21) < 0$.

7. Let $x =$ the first number; then $60 - x =$ the second number. For the product P we have
$P(x) = x(60-x) = 60x - x^2$.
$$\begin{aligned} \frac{dP}{dx} = 60 - 2x &= 0 \qquad \frac{d^2P}{dx^2} = -2 \\ x &= 30 \qquad \text{maximum} \\ 60 - x &= 30 \end{aligned}$$

9. If x is the length, then $4 - x$ is the width.

$$\begin{aligned} \text{Area} = A &= x(4-x) \\ A &= 4x - x^2 \\ \frac{dA}{dx} &= 4 - 2x = 0 \text{ and } x = 2\,\text{cm} \\ \frac{d^2A}{dx^2} &= -2 < 0 \quad \text{(maximum)} \end{aligned}$$

Answer: square with side 2 cm.

11. From the given figure, $A = xy$. To eliminate one variable, observe that

$$2x + 3y = 600, \text{ so that } y = \frac{1}{3}(600 - 2x)$$

and

$$A = x \cdot \frac{1}{3}(600 - 2x) = \frac{1}{3}\left(600x - 2x^2\right)$$

$$\frac{dA}{dx} = \frac{1}{3}(600 - 4x) = 0;\ x = 150\,\text{m and } y = \frac{1}{3}(600 - 300) = 100\,\text{m}.$$

13.

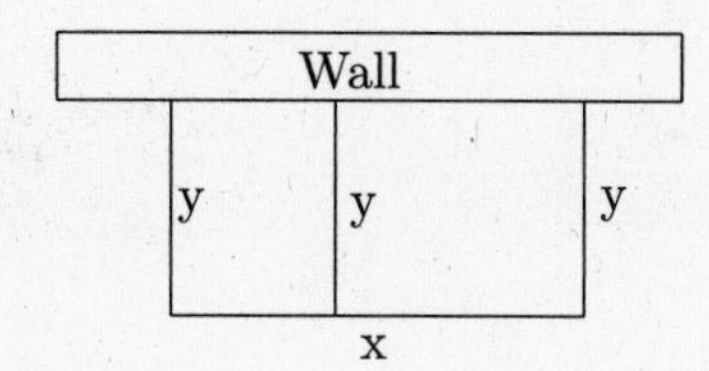

Let L be the total length; then $L = x + 3y$. To eliminate x, observe that $1200 = xy$. So

$$x = \frac{1200}{y} \text{ and } L = \frac{1200}{y} + 3y = 1200y^{-1} + 3y$$

$$\begin{aligned} \frac{dL}{dy} = -1(1200)y^{-2} + 3 &= 0 \\ -400y^{-2} &= -1 \\ y^2 &= 400 \\ y &= 20\,\text{ft} \qquad x = \frac{1200}{20} = 60\,\text{ft} \end{aligned}$$

15.

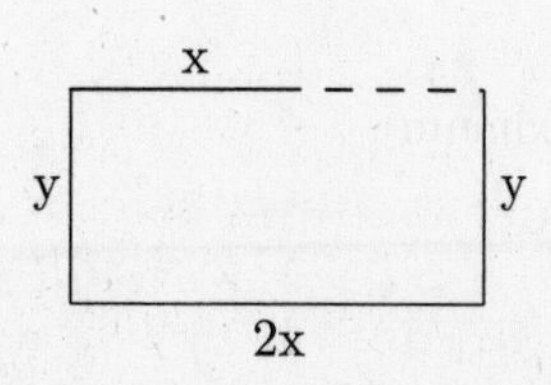

Cost $= C = 3x + 2y$; area $= A = 2xy$ and $y = \dfrac{A}{2x}$. Now eliminate y from the cost:

$$\begin{aligned} C &= 3x + 2\left(\frac{A}{2x}\right) \\ C &= 3x + Ax^{-1} \end{aligned}$$

$$\begin{aligned} \frac{dC}{dx} = 3 - Ax^{-2} &= 0 \\ 3x^2 &= A \\ x &= \sqrt{\frac{A}{3}} \end{aligned}$$

$$y = \frac{A}{2x} = \frac{A}{2\sqrt{\frac{A}{3}}} = \frac{A}{2} \cdot \frac{\sqrt{3}}{\sqrt{A}} = \frac{\sqrt{A}\sqrt{3}}{2} = \frac{1}{2}\sqrt{3A}$$

Length $= 2x = \dfrac{2\sqrt{A}}{\sqrt{3}} = \dfrac{2}{3}\sqrt{3A}$; Dimensions: $\dfrac{2}{3}\sqrt{3A} \times \dfrac{1}{2}\sqrt{3A}$

17. Let x = width and y = depth. Then S, the strength of the beam, is $S = kxy^2$, k a constant. This is the quantity to be maximized. To eliminate y, note that from the Pythagorean theorem,

$$\begin{aligned} x^2 + y^2 &= d^2 \\ y^2 &= 9 - x^2. \qquad \text{since } d = 3\,\text{ft} \end{aligned}$$

So

$$\begin{aligned} S &= kx(9 - x^2) \\ &= k(9x - x^3) \\ S' &= k(9 - 3x^2) = 0 \\ 9 - 3x^2 &= 0 \\ x^2 &= 3, \\ y^2 &= 9 - 3 = 6. \end{aligned}$$

So $x = \sqrt{3}$ ft and $y = \sqrt{6}$ ft.

19. From Figure 3.41,

$$\begin{aligned} V(x) &= (12 - 2x)(12 - 2x)x \qquad \text{length} \times \text{width} \times \text{height} \\ V(x) &= x(12 - 2x)^2 \\ V'(x) &= x \cdot 2(12 - 2x)(-2) + (12 - 2x)^2 \cdot 1 \\ &= (12 - 2x)[-4x + (12 - 2x)] \\ &= (12 - 2x)(12 - 6x) = 0 \\ x &= 6, 2. \end{aligned}$$

The only meaningful solution is $x = 2$ cm.

$$V''(x) = -96 + 24x\Big|_{x=2} < 0 \text{ (maximum).}$$

21.

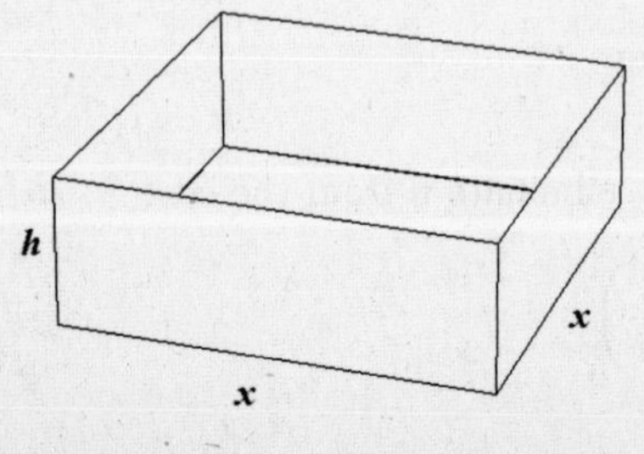

Surface area: $S = 2x^2 + 4xh$. To eliminate h:

$$\begin{aligned}
32 &= x^2h \text{ and } h = \frac{32}{x^2} \\
S &= 2x^2 + 4x \cdot \frac{32}{x^2} = 2x^2 + 128x^{-1} \\
\frac{dS}{dx} &= 4x - 128x^{-2} = 0 \\
x - 32x^{-2} &= 0 \\
x^3 - 32 &= 0 \\
x &= \sqrt[3]{32} = \sqrt[3]{8 \cdot 4} = 2\sqrt[3]{4} \\
h &= \frac{32}{x^2} = \frac{32}{\left[32^{1/3}\right]^2} = \frac{32}{32^{2/3}} = 32^{1/3}
\end{aligned}$$

So $h = x$ and we get a cube with side $2\sqrt[3]{4}$ cm.

23.

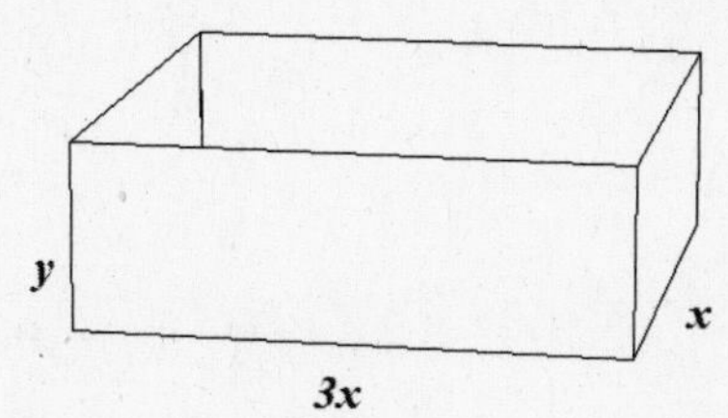

Quantity to be minimized:

$$\begin{aligned}
S &= \text{area of base} + \text{area of sides} \\
&= (3x)(x) + 2(3xy + xy) \\
&= 3x^2 + 8xy.
\end{aligned}$$

To eliminate y:

$$486 = 3x \cdot x \cdot y \quad \text{or} \quad y = \frac{162}{x^2}.$$

It follows that $S = 3x^2 + 8x \cdot \dfrac{162}{x^2} = 3x^2 + 1296x^{-1}$.

$$\begin{aligned}
S' = 6x - 1296x^{-2} &= 0 \\
x - 216x^{-2} &= 0 \\
x^3 &= 216 \\
x &= 6 \text{ in.} \\
3x &= 18 \text{ in.} \\
y &= \frac{162}{6^2} = 4.5 \text{ in.}
\end{aligned}$$

25. From Figure 3.43, the quantity to be maximized is the volume

$$V = x^2y.$$

The combined length and perimeter of a cross-section is $4x + y = 108$, which can be used to eliminate one of the variables: $y = 108 - 4x$ and

$$\begin{aligned}
V &= x^2(108 - 4x) \\
&= 108x^2 - 4x^3. \\
\frac{dV}{dx} &= 216x - 12x^2 \\
&= 12x(18 - x) = 0 \\
x &= 0,\ 18, \\
y &= 108 - 4(18) = 36.
\end{aligned}$$

27.

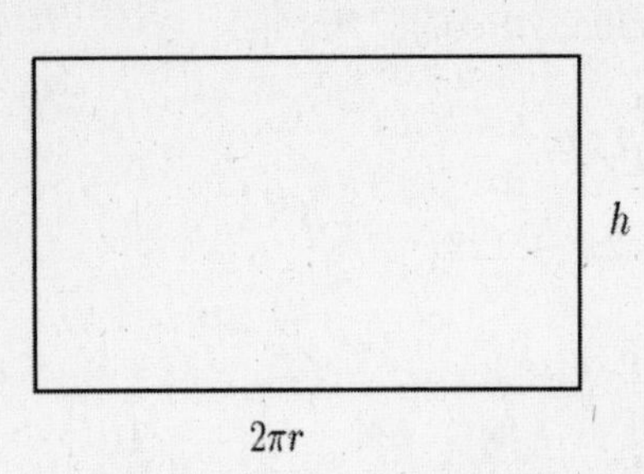

Area of the side of a cylinder $= 2\pi \cdot r \cdot h$; total area of top and bottom $= 2\pi \cdot r^2$

$$S = 2\pi \cdot r^2 + 2\pi \cdot rh$$

To eliminate h, we use $V = \pi \cdot r^2 h$ or $h = \dfrac{V}{\pi \cdot r^2}$. Substituting,

$$\begin{aligned}
S &= 2\pi \cdot r^2 + 2\pi \cdot r \cdot \frac{V}{\pi \cdot r^2} = 2\pi \cdot r^2 + 2Vr^{-1} \\
\frac{dS}{dr} &= 4\pi \cdot r - 2Vr^{-2} = 0 \\
2\pi \cdot r - \frac{V}{r^2} &= 0 \\
2\pi \cdot r^3 = V \quad &\text{and} \quad r^3 = \frac{V}{2\pi}
\end{aligned}$$

29. Let (x, y) be a point on the curve. Then

$$d^2 = (x-1)^2 + (y-2)^2.$$

Since $y = \frac{1}{4}x^2$,

$$\begin{aligned}
d^2 &= (x-1)^2 + (\tfrac{1}{4}x^2 - 2)^2 \\
&= x^2 - 2x + 1 + \tfrac{1}{16}x^4 - x^2 + 4 \\
&= \tfrac{1}{16}x^4 - 2x + 5.
\end{aligned}$$

Critical value:

$$\begin{aligned}
\tfrac{1}{16}(4x^3) - 2 &= 0 \\
4x^3 &= 32 \\
x^3 &= 8 \text{ and } x = 2, \\
y &= \tfrac{1}{4}(2)^2 = 1.
\end{aligned}$$

31. Let x be the piece to be bent into the shape of a circle, leaving $50 - x$ for the square. Circumference of a circle: $C = 2\pi \cdot r$. So $2\pi \cdot r = x$ and $r = \dfrac{x}{2\pi}$.

Area of a circle: $\pi \cdot r^2 = \pi \cdot \dfrac{x^2}{4\pi^2} = \dfrac{1}{4\pi}x^2$. Length of a side of the square: $\dfrac{1}{4}(50 - x)$. Area of the square: $\dfrac{1}{16}(50 - x)^2$. Let $A =$ the total area; then

$$\begin{aligned}
A &= \frac{1}{4\pi}x^2 + \frac{1}{16}(50 - x)^2 \\
\frac{dA}{dx} &= \frac{1}{2\pi}x + \frac{1}{8}(50 - x)(-1) = 0 \\
8x &= 2\pi(50 - x) \\
4x &= 50\pi - \pi \cdot x \\
x &= \frac{50\pi}{4 + \pi}
\end{aligned}$$

So the wire is cut in such a way that $\dfrac{50\pi}{4 + \pi}$ cm is used for the circle.

33. Quantity to be minimized: $A = (x+2)(y+4)$. To eliminate x or y:

$$xy = 72 \quad \text{or} \quad y = \frac{72}{x}.$$

So $A = (x+2)\left(\frac{72}{x}+4\right) = 4x + 144x^{-1} + 80$.

$$\begin{aligned} A' = 4 - 144x^{-2} &= 0 \\ x^2 - 36 &= 0 \\ x &= 6\,\text{in. and } y = 12\,\text{in.} \end{aligned}$$

So the width of the poster is $6 + 2 = 8$ in. and the height is $12 + 4 = 16$ in.

35.

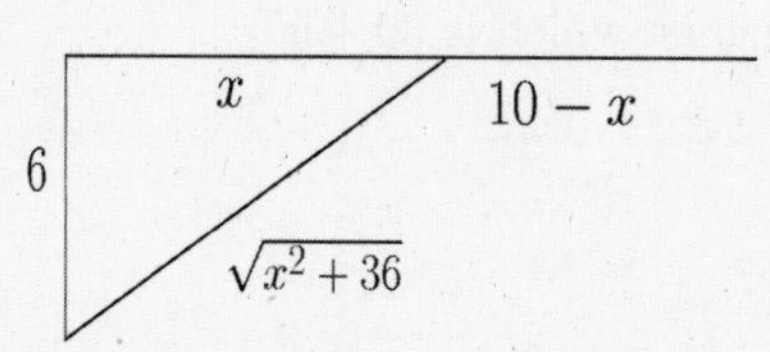

Since distance = rate × time, time = $\frac{\text{distance}}{\text{rate}}$. From the figure, total time T is

$$T = \frac{\sqrt{x^2+36}}{4} + \frac{10-x}{5}.$$

$$\begin{aligned} \frac{dT}{dx} = \frac{1}{4}\cdot\frac{1}{2}(x^2+36)^{-1/2}(2x) - \frac{1}{5} &= 0 \\ \frac{1}{4}\cdot\frac{x}{\sqrt{x^2+36}} &= \frac{1}{5} \\ \frac{x}{\sqrt{x^2+36}} &= \frac{4}{5} \\ \frac{x^2}{x^2+36} &= \frac{16}{25} \\ 25x^2 &= 16x^2 + 16\cdot 36 \\ 9x^2 &= 16\cdot 36 \\ 3x &= 4\cdot 6 \\ x &= 8\,\text{km} \end{aligned}$$

37. Let h and r be the height and radius of the cylinder and h_1 and r_1 the height and radius of the cone, so that h_1 and r_1 are fixed quantities.

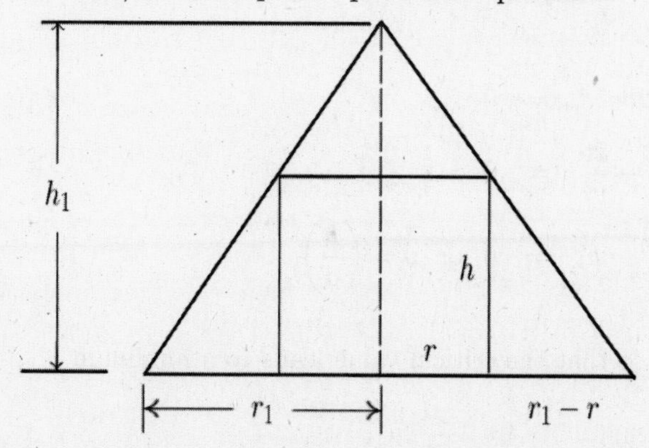

Quantity to be maximized: $V = \pi r^2 h$. To eliminate h, note that by similar triangles we get

$$\frac{h}{h_1} = \frac{r_1 - r}{r_1} \quad \text{or} \quad h = \frac{h_1(r_1 - r)}{r_1}. \qquad (1)$$

Substituting in V:

$$\begin{aligned} V &= \pi r^2 \frac{h_1(r_1 - r)}{r_1} \\ &= \frac{\pi h_1}{r_1}(r_1 r^2 - r^3). \qquad h_1 \text{ and } r_1 \text{ constants} \\ \frac{dV}{dr} &= \frac{\pi h_1}{r_1}(2r_1 r - 3r^2) = 0 \\ r &= \frac{2}{3} r_1 \end{aligned}$$

By (1), $h = \frac{h_1}{r_1}\left(r_1 - \frac{2}{3}r_1\right) = \frac{1}{3}h_1$.

So the height of the cylinder is one-third the height of the cone.

39. Let $x =$ the number of passengers above 30. Then $400 - 10x =$ price per ticket.

Intake: $I =$ price per ticket times the number of passengers, which is $30 + x$.

$$\begin{aligned} I &= (400 - 10x)(30 + x) \\ &= -10x^2 + 100x + 12{,}000 \\ I' &= -20x + 100 = 0 \text{ or } x = 5 \end{aligned}$$

So $30 + x = 35$ passengers.

41. Let $M(x) =$ the marginal cost, that is,

$$\begin{aligned} M(x) = C'(x) &= 3(2.0 \times 10^{-6})x^2 - 0.0030x + 2.5 \\ &= 6.0 \times 10^{-6}x^2 - 0.0030x + 2.5. \end{aligned}$$

Then

$$\begin{aligned} M'(x) &= 2(6.0 \times 10^{-6})x - 0.0030 = 0 \\ x &= \frac{0.0030}{2(6.0 \times 10^{-6})} = 250 \text{ units.} \end{aligned}$$

Since $M''(x) = 2(6.0 \times 10^{-6}) > 0$, $x = 250$ corresponds to the minimum.

43. $c(x) = \frac{C(x)}{x}$; critical value:

$$\begin{aligned} c'(x) = \frac{xC'(x) - C(x) \cdot 1}{x^2} &= 0 \\ xC'(x) - C(x) &= 0 \\ C'(x) &= \frac{C(x)}{x} \\ \text{or } C'(x) &= c(x) \end{aligned}$$

45. $D = av^2 + \frac{b}{v^2} = av^2 + bv^{-2}$

$$\begin{aligned} \frac{dD}{dv} = 2av - 2bv^{-3} &= 0 \\ 2av - \frac{2b}{v^3} &= 0 \\ 2(av^4 - b) &= 0 \text{ or } v = \left(\frac{b}{a}\right)^{1/4} \end{aligned}$$

$\frac{d^2D}{dv^2} = 2a + \frac{6b}{v^4} > 0$ for all nonzero v, so that the critical value leads to a minimum.

3.5 Related Rates

1. $\frac{dy}{dt} = 2x\frac{dx}{dt} = 2(2)(1) = 4$

3. $x^2 + y^2 = 25$

$$\begin{aligned} 2x\frac{dx}{dt} + 2y\frac{dy}{dt} &= 0 \\ x\frac{dx}{dt} + y\frac{dy}{dt} &= 0 \\ x\frac{dx}{dt} + y(3) &= 0 \qquad \frac{dy}{dt} = 3 \\ \frac{dx}{dt} &= -\frac{3y}{x} \end{aligned}$$

If $y = 3$, then $x = \sqrt{25 - 3^2} = 4$. Substituting $y = 3$ and $x = 4$, we get

$$\frac{dx}{dt} = -\frac{3(3)}{4} = -\frac{9}{4}.$$

5. $I = \frac{E}{R} = \frac{100}{R} = 100R^{-1}$. Given $\frac{dR}{dt} = 2$, find $\frac{dI}{dt}$ when $R = 10$.

$$\frac{dI}{dt} = 100(-1)R^{-2}\frac{dR}{dt} = -\frac{100}{R^2}\frac{dR}{dt} = -\frac{(100)(2)}{R^2}\Big|_{R=10} = -2\,\text{A/s}$$

7. Given: $\frac{di}{dt} = 0.20\,\text{A/s}$, find: $\frac{dP}{dt}$ when $i = 2.0\,\text{A}$. $P = 100i^2$ (since $R = 100\,\Omega$).
$\frac{dP}{dt} = 100\left(2i\frac{di}{dt}\right) = 200i(0.20) = 40i$. At the instant when $i = 2.0$, we have

$$\frac{dP}{di} = 40i\Big|_{i=2.0} = 80\,\text{W/s}.$$

9. $A = \pi r^2$. Given: $\frac{dr}{dt} = 0.25\,\text{mm/min}$. Find $\frac{dA}{dt}$ when $r = 10\,\text{mm}$.

$$\frac{dA}{dt} = 2\pi r\frac{dr}{dt} = 2\pi r(0.25)\Big|_{r=10} = 15.7\,\text{mm}^2/\text{min}$$

11. $A = \pi r^2$. Given: $\frac{dr}{dt} = 1.9\,\text{ft/s}$; find $\frac{dA}{dt}$ at the end of 10 s, when $r = 1.9\,\text{ft/s} \times 10\,\text{s} = 19\,\text{ft}$.

$$\frac{dA}{dt} = 2\pi r\frac{dr}{dt} = 2\pi r(1.9)\Big|_{r=19} = 227\,\text{ft}^2/\text{s}$$

13 . $V = \frac{4}{3}\pi r^3$. Given: $\frac{dr}{dt} = 1.0\,\text{in./s}$; find $\frac{dV}{dt}$ at the end of 5.0 s, when $r = 1.0\,\text{in./s} \times 5.0\,\text{s} = 5\,\text{in.}$

$$\frac{dV}{dt} = 4\pi r^2\frac{dr}{dt} = 4\pi r^2(1.0)\Big|_{r=5} = 314\,\text{in.}^3/s$$

15. In the diagram we label all quantities that change and all quantities that remain fixed.

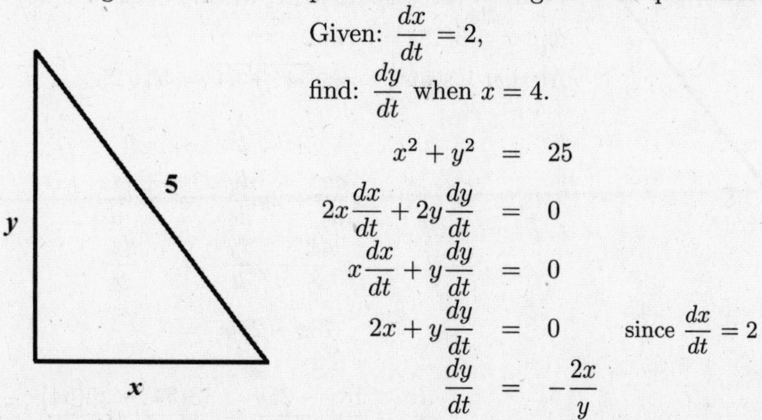

Given: $\frac{dx}{dt} = 2$,

find: $\frac{dy}{dt}$ when $x = 4$.

$$\begin{aligned} x^2 + y^2 &= 25 \\ 2x\frac{dx}{dt} + 2y\frac{dy}{dt} &= 0 \\ x\frac{dx}{dt} + y\frac{dy}{dt} &= 0 \\ 2x + y\frac{dy}{dt} &= 0 \qquad \text{since } \frac{dx}{dt} = 2 \\ \frac{dy}{dt} &= -\frac{2x}{y} \end{aligned}$$

At the instant when $x = 4$, we have $y = 3$. Thus $\frac{dy}{dt} = -\frac{(2)(4)}{3} = -\frac{8}{3}\,\text{m/min}$.

17. Given: $\frac{dx}{dt} = 11.8\,\text{ft/s}$, find: $\frac{dz}{dt}$ when $x = 151$ ft.

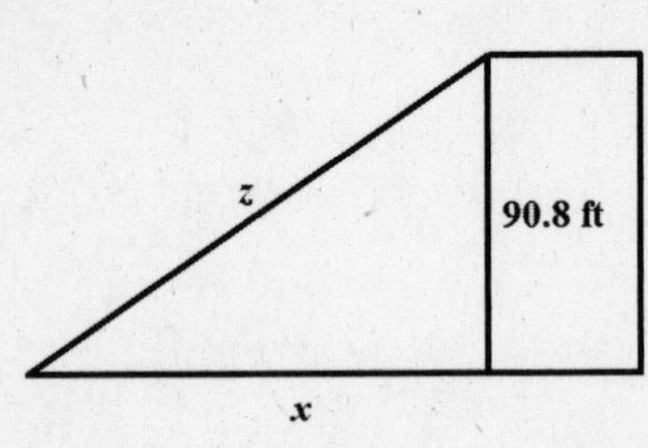

$$\begin{aligned} z^2 &= x^2 + (90.8)^2 \\ 2z\frac{dz}{dt} &= 2x\frac{dx}{dt} + 0 \\ z\frac{dz}{dt} &= x\frac{dx}{dt} = x(11.8) \\ \frac{dz}{dt} &= \frac{11.8x}{z}. \end{aligned}$$

At the instant when $x = 151$, $z = \sqrt{151^2 + (90.8)^2} = 176.2$ ft. Substituting these values,

$$\frac{dz}{dt} = \frac{11.8(151)}{176.2} = 10.1\,\text{ft/s}.$$

19. In the diagram, we label all quantities that change and all quantities that remain fixed.

Given: $\frac{dx}{dt} = 350$,

find: $\frac{ds}{dt}$ when $x = 3000$.

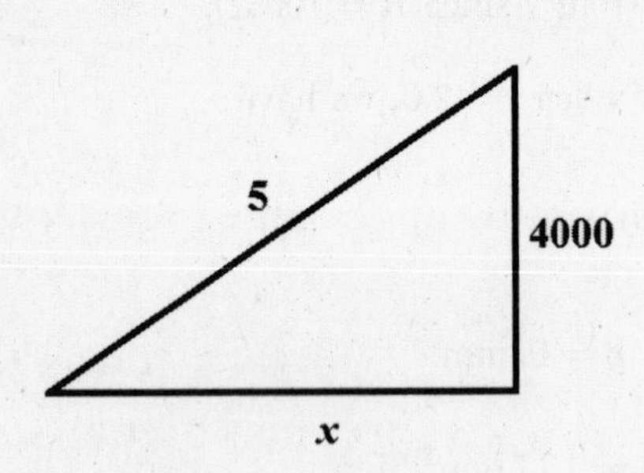

$$\begin{aligned} x^2 + 4000^2 &= s^2 \\ 2x\frac{dx}{dt} &= 2s\frac{ds}{dt} \\ x\frac{dx}{dt} &= s\frac{ds}{dt} \\ 350x &= s\frac{ds}{dt} \\ \frac{ds}{dt} &= \frac{350x}{s} \end{aligned}$$

At the instant when $x = 3000$, we have $s = 5000$. Thus

$$\frac{ds}{dt} = \frac{(350)(3000)}{5000} = \frac{(350)(3)}{5} = (70)(3) = 210\,\text{km/h}.$$

21. Given: $\frac{dy}{dt} = -25\,\text{km/h}$; $\frac{dx}{dt} = 18\,\text{km/h}$.

Find $\frac{dz}{dt}$ three hours later, when $y = 129 - 3(25) = 54$ km and $x = 3(18) = 54$ km.

At that instant, $z = \sqrt{54^2 + 54^2} = 54\sqrt{2}$.

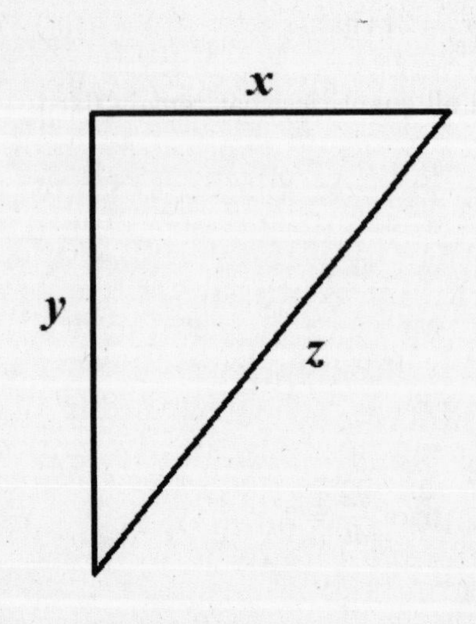

$$\begin{aligned} x^2 + y^2 &= z^2 \\ 2x\frac{dx}{dt} + 2y\frac{dy}{dt} &= 2z\frac{dz}{dt} \\ x\frac{dx}{dt} + y\frac{dy}{dt} &= z\frac{dz}{dt} \\ 18x - 25y &= z\frac{dz}{dt} \end{aligned}$$

$$\frac{dz}{dt} = \frac{18x - 25y}{z} = \frac{18(54) - 25(54)}{54\sqrt{2}}$$

$$= -\frac{7}{\sqrt{2}} = -\frac{7\sqrt{2}}{2} \approx -4.9\,\text{km/h}$$

23.

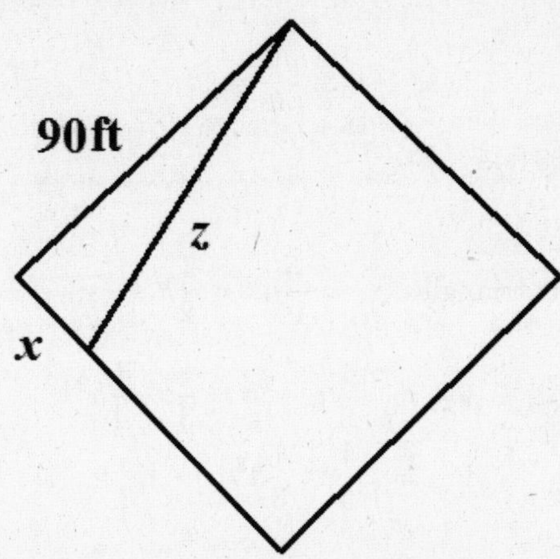

Given: $\dfrac{dx}{dt} = 24\,\text{ft/s}$, find: $\dfrac{dz}{dt}$ when $x = 90 - 25 = 65\,\text{ft}$.

$$\begin{aligned} z^2 &= x^2 + 90^2 \\ 2z\frac{dz}{dt} &= 2x\frac{dx}{dt} + 0 \\ z\frac{dz}{dt} &= x\frac{dx}{dt} = x(24) \\ \frac{dz}{dt} &= \frac{24x}{z}. \end{aligned}$$

When $x = 65$, then $z = \sqrt{65^2 + 90^2} = 111.0\,\text{ft}$. So $\dfrac{dz}{dt} = \dfrac{24(65)}{111.0} = 14\,\text{ft/s}$

25. Given $\dfrac{dx}{dt} = -2$, find $\dfrac{dy}{dt}$ when $x = 16$ and $y = 4$. From $y^2 = x$, we have

$$\begin{aligned} 2y\frac{dy}{dt} = \frac{dx}{dt} &= -2 \\ \frac{dy}{dt} &= -\frac{1}{y}\bigg|_{y=4} = -\frac{1}{4}\,\frac{\text{unit}}{\text{min}}. \end{aligned}$$

27.

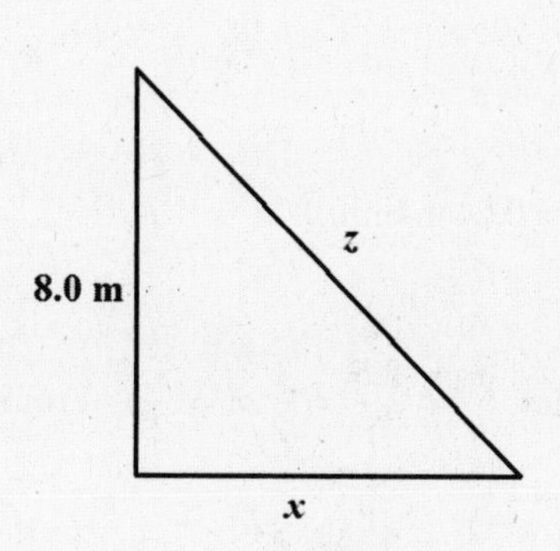

Given: $\dfrac{dz}{dt} = -2.0\,\text{m/min}$ (decreasing quantity), find: $\dfrac{dx}{dt}$ when $x = 12.0\,\text{m}$.

$$\begin{aligned} z^2 &= x^2 + 8.0^2 \\ 2z\frac{dz}{dt} &= 2x\frac{dx}{dt} + 0 \\ z\frac{dz}{dt} &= x\frac{dx}{dt} \\ z(-2.0) &= x\frac{dx}{dt} \\ \frac{dx}{dt} &= -\frac{2.0z}{x} \end{aligned}$$

At the instant when $x = 12.0$, $z = \sqrt{(12.0)^2 + (8.0)^2} = \sqrt{208} = \sqrt{16 \cdot 13} = 4\sqrt{13}$.
$\dfrac{dx}{dt} = -\dfrac{2.0(4\sqrt{13})}{12.0} = -\dfrac{2\sqrt{13}}{3} \approx -2.4\,\text{m/min}$. (decreasing)

29.

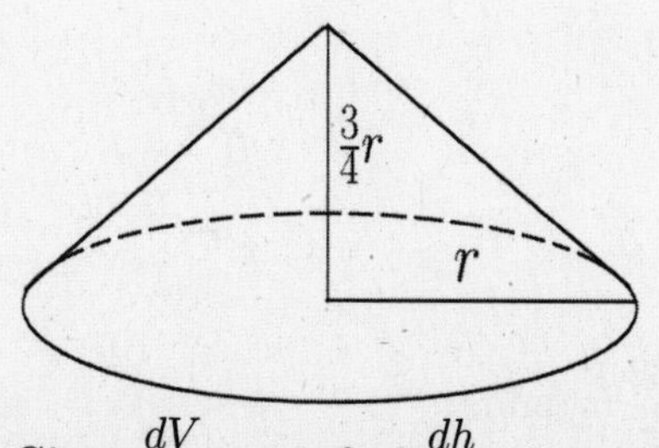

Since $h = \dfrac{3}{4}r$, $r = \dfrac{4}{3}h$.

$$\begin{aligned} V &= \frac{1}{3}\pi r^2 h = \frac{1}{3}\pi\left(\frac{4}{3}h\right)^2 h \\ V &= \frac{16\pi}{27}h^3 \end{aligned}$$

Given $\dfrac{dV}{dt} = 12$, find $\dfrac{dh}{dt}$ when $h = 6.0$.

$$\begin{aligned}\frac{dV}{dt} &= \frac{16\pi}{27}(3h^2)\frac{dh}{dt}\\ 12 &= \frac{16\pi}{9}h^2\frac{dh}{dt} \qquad \text{since } \frac{dV}{dt}=12\\ \frac{dh}{dt} &= \frac{108}{16\pi h^2}\bigg|_{h=6.0} = \frac{108}{16\pi(36)} = \frac{3}{16\pi} = 0.060\,\text{ft/s}.\end{aligned}$$

31.

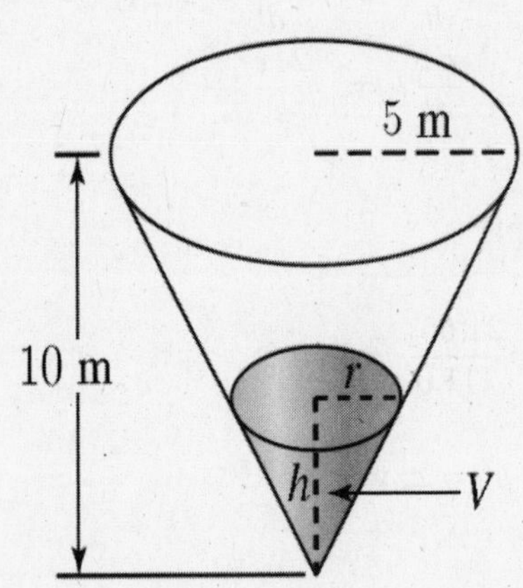

By similar triangles, $\frac{r}{h}=\frac{5}{10}, r=\frac{1}{2}h.$

$$\begin{aligned}V &= \frac{1}{3}\pi\cdot r^2h = \frac{1}{3}\pi\cdot\frac{1}{4}h^2h\\ V &= \frac{1}{4}\pi\cdot\frac{1}{3}h^3\end{aligned}$$

Given: $\frac{dV}{dt}=3.0\,\text{m}^3/\text{min}$. Find: $\frac{dh}{dt}$ when $h=4.0\,\text{m}$.

$$\begin{aligned}\frac{dV}{dt} &= \frac{1}{4}\pi h^2\frac{dh}{dt}\\ 3.0 &= \frac{1}{4}\pi h^2\frac{dh}{dt}\\ \frac{dh}{dt} &= \frac{12}{\pi h^2}\bigg|_{h=4.0} = \frac{12}{16\pi} = \frac{3}{4\pi} \approx 0.24\,\text{m/min}.\end{aligned}$$

33.

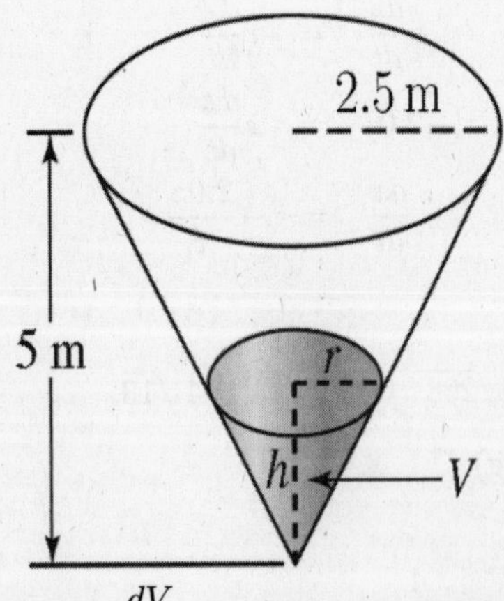

By similar triangles, $\frac{r}{h}=\frac{2.5}{5}=\frac{1}{2}, r=\frac{1}{2}h.$ From Exercise 31, $V=\frac{1}{4}\pi\cdot\frac{1}{3}h^3$.

Given: $\frac{dV}{dt}=2.0\,\text{m}^3/\text{min}$. Find: $\frac{dh}{dt}$ when $h=\frac{1}{2}\,\text{m}$.

$$\begin{aligned}\frac{dV}{dt} &= \frac{1}{4}\pi\cdot h^2\frac{dh}{dt}\\ 2.0 &= \frac{1}{4}\pi\cdot h^2\frac{dh}{dt}\\ \frac{dh}{dt} &= \frac{8.0}{\pi\cdot h^2}\bigg|_{h=\frac{1}{2}} = \frac{32}{\pi}\,\text{m/min}.\end{aligned}$$

35.

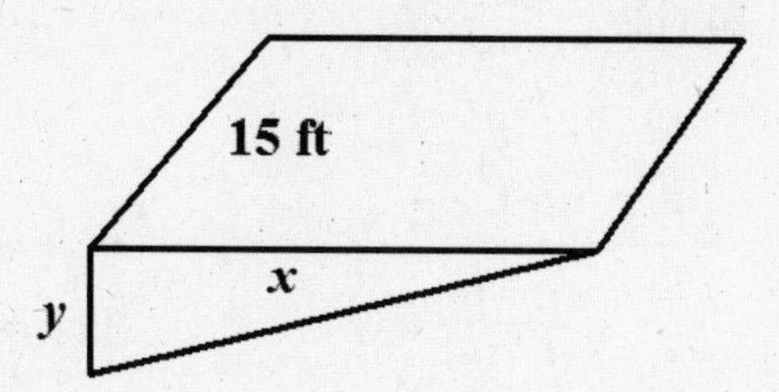

Referring to Figure 3.63 and the diagram, the volume is equal to the area of the triangle times 15 ft, the width of the pool. Using similar triangles,

$$\frac{x}{y} = \frac{50}{12} \quad \text{or} \quad x = \frac{25}{6}y.$$

$$\begin{aligned} V &= \frac{1}{2}y\left(\frac{25}{6}y\right)\cdot 15 = \frac{125}{4}y^2 \\ \frac{dV}{dt} &= \frac{125}{2}y\frac{dy}{dt} \\ 120 &= \frac{125}{2}y\frac{dy}{dt} \qquad\qquad \frac{dV}{dt} = 120\,\text{ft}^3/\text{min} \\ \frac{dy}{dt} &= \left.\frac{240}{125y}\right|_{y=8} = \frac{240}{(125)(8)} = \frac{6}{25}\,\text{ft/min}. \end{aligned}$$

3.6 Differentials

1. Since $\frac{dy}{dx} = 3x^2 - 1$, $dy = (3x^2 - 1)dx$.

3. By the quotient rule, $\frac{dy}{dx} = -\frac{1}{(x-1)^2}$; hence $dy = -\frac{dx}{(x-1)^2}$.

5. If $x = 2$, $y = 2^2 - 2 = 2$. If $x = 2.1$, $y = (2.1)^2 - 2.1 = 2.31$. Hence $\Delta y = 2.31 - 2 = 0.31$. From $y = x^2 - x$, we get the differential $dy = (2x - 1)dx$. So if $x = 2$ and $dx = 0.1$, $dy = (4 - 1)(0.1) = 0.3$.

7. $A = x^2$, where x is the length of a side. Given: $x = 6.00\,\text{cm}$, $dx = 0.02\,\text{cm}$.
$\Delta A \approx dA = 2x dx = 2(6.00)(0.02) = 0.24\,\text{cm}^2$.
Percentage error: $\frac{dA}{A} \times 100 = \frac{0.24}{(6.00)^2} \times 100 = 0.67\%$.

9. $V = s^3$, where s is the length of a side. $s = 5.00\,\text{in.}$ and $ds = \pm 0.01\,\text{in.}$
$\Delta V \approx dV = 3s^2 ds = 3(5.00)^2(\pm 0.01) = \pm 0.75\,\text{in.}^3$
Percentage error: $\frac{dV}{V} \times 100 = \frac{0.75}{(5.00)^3} \times 100 = 0.6\%$.

11. $V = \frac{4}{3}\pi r^3$ and $S = 4\pi r^2$. Given: $r = 10.00\,\text{cm}$, $dr = \pm 1\,\text{mm} = \pm 0.1\,\text{cm}$.
$\Delta V \approx dV = 4\pi r^2 dr = 4\pi(10.00)^2(\pm 0.1) = \pm 125.66\,\text{cm}^3$.
Percentage error: $\frac{dV}{V} \times 100 = \frac{125.66}{(4/3)\pi(10.00)^3} \times 100 = 3\%$.
$\Delta S \approx dS = 8\pi r dr = 8\pi(10.00)(\pm 0.1) = \pm 25.13\,\text{cm}^2$.
Percentage error: $\frac{dS}{S} \times 100 = \frac{25.13}{4\pi(10.00)^2} \times 100 = 2\%$.

13. $T = 2\pi\sqrt{\frac{L}{10}} = \frac{2\pi}{\sqrt{10}}L^{1/2}$. $L = 2.0\,\text{m}$ and $dL = \pm 0.1\,\text{m}$.

$$\begin{aligned}\Delta T \approx dT &= \frac{2\pi}{\sqrt{10}} \cdot \frac{1}{2}L^{-1/2}dL \\ &= \frac{\pi}{\sqrt{10}}\frac{1}{\sqrt{L}}dL \\ &= \frac{\pi}{\sqrt{10}}\frac{1}{\sqrt{2.0}}(\pm 0.1) = \pm 0.07\,\text{s}.\end{aligned}$$

Percentage error: $\frac{dT}{T} \times 100 = \frac{0.07}{2\pi\sqrt{\frac{2.0}{10}}} \times 100 = 2.5\%$.

15. $P = 10.0i^2$. Given: $i = 2.1\,\text{A}$, $di = (2.2 - 2.1) = 0.1\,\text{A}$.
$\Delta P \approx dP = 20.0i\,di = 20.0(2.1)(0.1) = 4.2\,\text{W}$.

17. $R = 60.0 + 0.020T^2, T = 50.0°\text{F}, dT = \pm 2.0°\text{F}$

$$dR = [0 + 0.020(2T)]dR = 0.020(2)(50.0)(\pm 2.0) = \pm 4.0\,\Omega$$

$$\frac{dR}{R} \times 100 = \frac{4.0}{60.0 + 0.020(50.0)^2} \times 100 = 3.6\%$$

19. $A = \pi r^2$; $\Delta A \approx dA = 2\pi r dr$. The geometric interpretation of this formula can be seen from the following diagram:

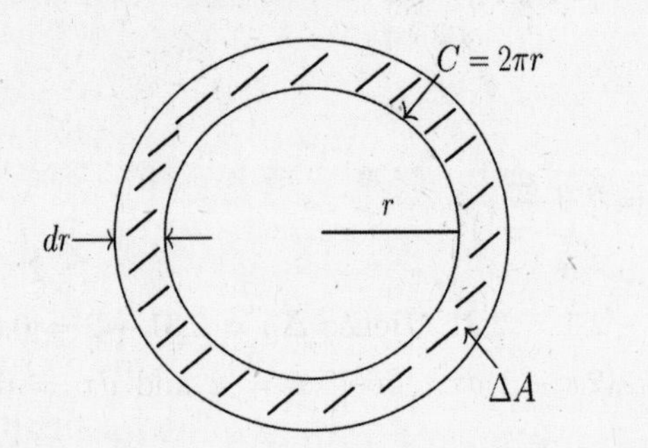

dr | dA | 2πr

Area $= 2\pi r\, dr = dA \approx \Delta A$

21. $V = \frac{4}{3}\pi r^3$; $dV = 4\pi r^2\, dr = 4\pi r^3 \frac{dr}{r} = 4\pi r^3(\pm 0.02)$ since $\frac{dr}{r} = \pm 0.02$.

$$\frac{dV}{V} \times 100 = \frac{4\pi r^3(\pm 0.02)}{\frac{4}{3}\pi r^3} \times 100 = \pm 0.06 = \pm 6\%$$

Chapter 3 Review

1. $y = (x-2)^{1/2}$, $y' = \frac{1}{2}(x-2)^{-1/2} = \frac{1}{2\sqrt{x-2}}\Big|_{x=3} = \frac{1}{2}$

Tangent line:
$$\begin{aligned} y - 1 &= \tfrac{1}{2}(x-3) \qquad m = \tfrac{1}{2} \\ 2y - 2 &= x - 3 \\ x - 2y - 1 &= 0 \end{aligned}$$

Normal line:
$$\begin{aligned} y - 1 &= -2(x-3) \qquad m = -\frac{1}{1/2} = -2 \\ y - 1 &= -2x + 6 \\ 2x + y - 7 &= 0 \end{aligned}$$

3. $f(x) = x^2 - 4x + 3$; $f'(x) = 2x - 4$; $f''(x) = 2$.

Step 1. Critical points:
$$f'(x) = 2x - 4 = 0$$
$$x = 2.$$
Substituting in $y = x^2 - 4x + 3$, we get $(2, -1)$ for the critical point.

Step 2. Test of critical point: $f''(2) = 2 > 0$. The critical point is a minimum.

Step 3. Concavity: $y'' = 2 > 0$; the graph is concave up everywhere.

Step 4. There are no inflection points.

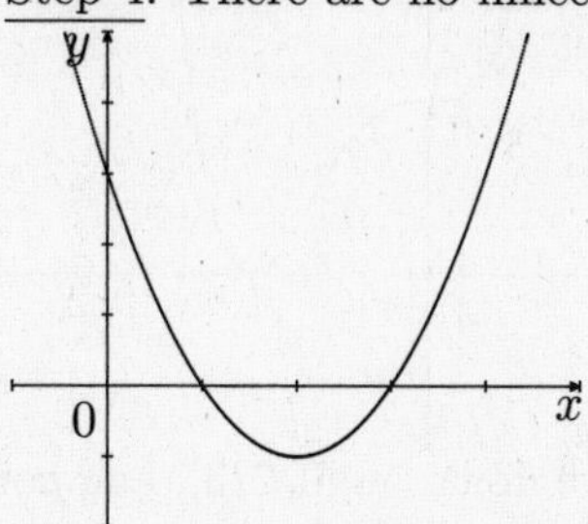

5. $f(x) = -x^3 + 12x + 2$; $f'(x) = -3x^2 + 12$; $f''(x) = -6x$.

Step 1. Critical points:
$$f'(x) = -3x^2 + 12 = 0$$
$$x = \pm 2.$$
The points are $(2, 18)$ and $(-2, -14)$.

Step 2. Test of critical points:

$f''(2) = -12 < 0$ (maximum).

$f''(-2) = 12 > 0$ (minimum).

Step 3. Concavity: $f''(x) = -6x = 0$ when $x = 0$.

If $x < 0$, $y'' > 0$ (concave up).

If $x > 0$, $y'' < 0$ (concave down).

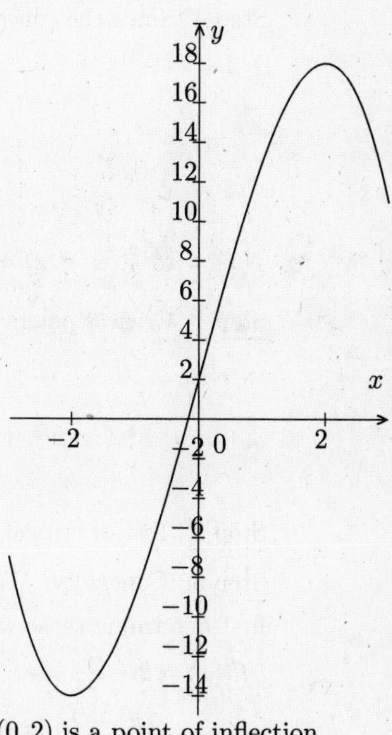

Step 4. Inflection points: Since the concavity changes, the point $(0, 2)$ is a point of inflection.

7. $f(x) = 3x^4 - 4x^3 + 1$; $f'(x) = 12x^3 - 12x^2$; $f''(x) = 36x^2 - 24x$.

Step 1. Critical points:
$$f'(x) = 12x^3 - 12x^2 = 0$$
$$12x^2(x - 1) = 0$$
$$x = 0, 1$$
Substituting in the given function, we find that $(0, 1)$ and $(1, 0)$ are the critical points.

Step 2. Test of critical points:

$f''(1) = 36 - 24 > 0$; $(1, 0)$ is a minimum.

$f''(0) = 0$; the test fails.

By the first-derivative test, using $x = -1/2$ and $x = 1/2$ for the test values, we see that $f(x)$ is a decreasing function in the neighborhood of $x = 0$. So $(0, 1)$ is neither a minimum nor a maximum.

Step 3. Concavity:
$$\begin{aligned} f''(x) = 36x^2 - 24x &= 0 \\ 12x(3x-2) &= 0 \\ x &= 0, \frac{2}{3}. \end{aligned}$$

	test values	$12x$	$3x-2$	$y'' = 12x(3x-2)$
$x < 0$	-1	$-$	$-$	$+$
$0 < x < 2/3$	$1/2$	$+$	$-$	$-$
$x > 2/3$	1	$+$	$+$	$+$

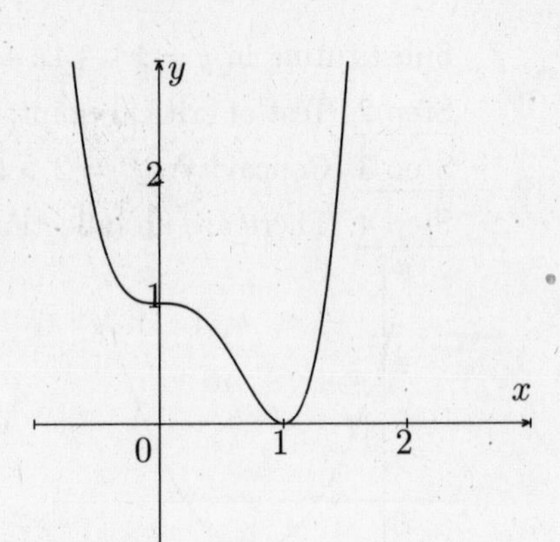

We conclude that the graph is concave up on $(-\infty, 0]$, concave down on $[0, 2/3]$, and concave up on $[2/3, \infty)$.

Step 4. Since the concavity changes, the points $(0, 1)$ and $\left(\frac{2}{3}, \frac{11}{27}\right)$ are inflection points.

9. $f(x) = x^2 - \frac{1}{x}$; $f'(x) = 2x + \frac{1}{x^2}$; $f''(x) = 2 - \frac{2}{x^3}$.

Step 1. Critical points:
$$\begin{aligned} f'(x) &= 2x + \frac{1}{x^2} = 0 \\ 2x^3 + 1 &= 0 \\ 2x^3 &= -1 \\ x &= -\frac{1}{\sqrt[3]{2}} \end{aligned}$$

Step 2. Test of critical points: $f''\left(-\frac{1}{\sqrt[3]{2}}\right) > 0$ (minimum).

Step 3. Concavity: We need to determine where $y'' < 0$ and where $y'' > 0$. To this end we first determine those values of x for which $f''(x) = 0$:
$$\begin{aligned} f''(x) = 2 - \frac{2}{x^3} &= 0 \\ 2x^3 - 2 &= 0 \\ x &= 1. \end{aligned}$$
(The concavity may also change at the vertical asymptote $x = 0$.)

	test values	$y'' = 2 - \frac{2}{x^3}$
$x > 1$	2	$+$
$0 < x < 1$	$1/2$	$-$
$x < 0$	-1	$+$

Summary:
If $x > 1$, $f(x)$ is concave up.
If $0 < x < 1$, $f(x)$ is concave down.
If $x < 0$, $f(x)$ is concave up.

Step 4. Inflection points: Because of the change in concavity, $(1, 0)$ is a point of inflection. (Note that the concavity also changes at $x = 0$, the vertical asymptote.)

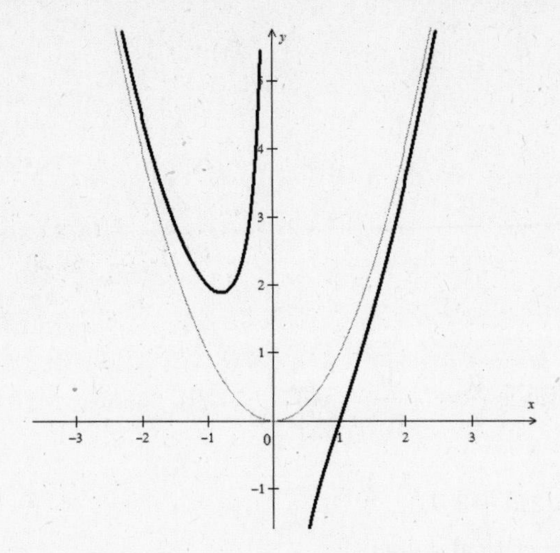

11. $v = kr^2(a-r) = k(ar^2 - r^3)$.

$$\begin{aligned} \frac{dv}{dr} = k(2ar - 3r^2) &= 0 \\ kr(2a-3r) &= 0 \\ r &= \frac{2a}{3}. \qquad \text{positive root} \end{aligned}$$

$$\frac{d^2v}{dr^2} = k(2a - 6r)\Big|_{r=2a/3} < 0.$$

So the critical value $r = 2a/3$ leads to a maximum.

13.

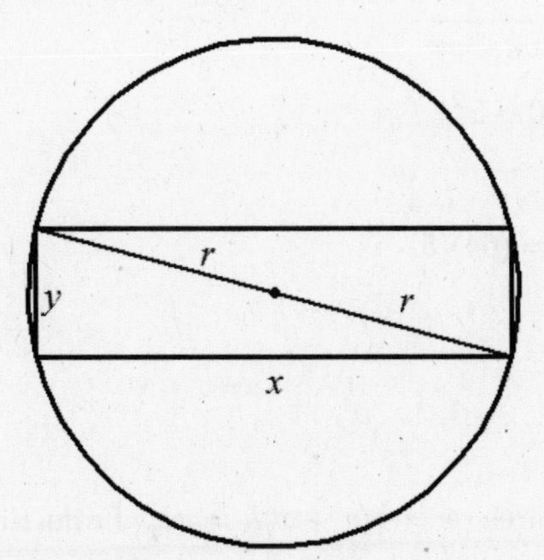

Denote the radius of the log by r, the width of the beam by x, and the depth by y.

Quantity to be maximized: $S = kxy^3$. To eliminate x, note that

$$\begin{aligned} x^2 + y^2 &= 4r^2 \\ x &= \sqrt{4r^2 - y^2}. \end{aligned}$$

Substituting in the expression for S, we get

$$\begin{aligned} S &= ky^3(4r^2 - y^2)^{1/2} \\ S' &= ky^3 \cdot \tfrac{1}{2}(4r^2 - y^2)^{-1/2}(-2y) + k(3y^2)(4r^2 - y^2)^{1/2} \\ &= k\left[\frac{-y^4}{\sqrt{4r^2 - y^2}} + 3y^2\sqrt{4r^2 - y^2}\right] = 0. \end{aligned}$$

Multiplying by $\frac{1}{k}\sqrt{4r^2-y^2}$, we get

$$\begin{aligned}
-y^4+3y^2(4r^2-y^2) &= 0\\
-y^4+12r^2y^2-3y^4 &= 0\\
12r^2y^2-4y^4 &= 0\\
4y^2(3r^2-y^2) &= 0\\
y^2 &= 3r^2\\
y &= \sqrt{3}r.
\end{aligned}$$

Hence $x=\sqrt{4r^2-3r^2}=r$. So the desired ratio is $\frac{y}{x}=\frac{\sqrt{3}r}{r}=\sqrt{3}$.

15. Marginal cost: $C'(x)=0.00025\left(3x^2\right)-0.015(2x)+3.5$

To obtain the minimum for $C'(x)$, we find the critical value:

$$\frac{d}{dx}C'(x)=0.00025(6x)-0.015(2)=0 \text{ and } x=20 \text{ units}$$

17.

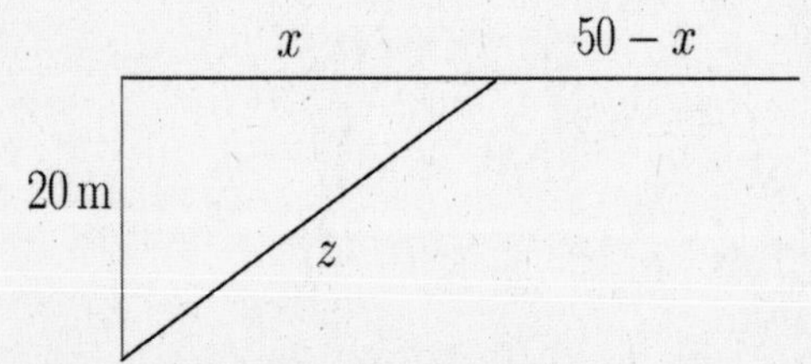

$$z=\sqrt{400+x^2}$$

Quantity to be minimized:

$$\begin{aligned}
C &= 45\sqrt{400+x^2}+30(50-x)\\
C' &= 45\cdot\tfrac{1}{2}(400+x^2)^{-1/2}(2x)-30=0
\end{aligned}$$

$$\begin{aligned}
\frac{45x}{\sqrt{400+x^2}} &= 30\\
\frac{3x}{\sqrt{400+x^2}} &= 2\\
3x &= 2\sqrt{400+x^2}\\
9x^2 &= 4(400+x^2)\\
5x^2 &= 1600\\
x^2 &= 320=(64)(5)\\
x &= 8\sqrt{5}
\end{aligned}$$

So the distance on land is $(50-8\sqrt{5})$ m.

19. The quantity to be maximized is the cross-sectional area $A=\frac{1}{2}(x+x)h=xh$. From Figure 3.65, $x=\sqrt{8^2-h^2}$, and $A=h(64-h^2)^{1/2}$.

$$\begin{aligned}
\frac{dA}{dh} &= h\cdot\tfrac{1}{2}(64-h^2)^{-1/2}(-2h)+(64-h^2)^{1/2}\\
&= \frac{-h^2}{\sqrt{64-h^2}}+\sqrt{64-h^2}=0.
\end{aligned}$$

Multiplying by $\sqrt{64-h^2}$, we get

$$\begin{aligned}
-h^2+64-h^2 &= 0\\
h^2 &= 32\\
h &= \sqrt{32}=\sqrt{16\cdot 2}=4\sqrt{2}\text{ in.}
\end{aligned}$$

21. $R = 25 + 0.020T^2$

Given: $\frac{dT}{dt} = 1.6\,\frac{°\text{C}}{\text{min}}$, find: $\frac{dR}{dt}$ when $T = 50.0\,°\text{C}$.

$$\begin{aligned}\frac{dR}{dt} &= 0 + 0.020(2T)\frac{dT}{dt} \\ &= 0.020(2T)(1.6) \\ &= 0.020(2T)(1.6)\Big|_{T=50.0} = 3.2\,\frac{\Omega}{\text{min}}\end{aligned}$$

23. Given: $\frac{dE}{dt} = 3.00\,\text{V/s}$, find: $\frac{dI}{dt}$. $E = \frac{0.1138I}{(0.060)^2}$.

$$\begin{aligned}\frac{dE}{dt} &= \frac{0.1138}{(0.060)^2}\frac{dI}{dt} \\ 3.00 &= \frac{0.1138}{(0.060)^2}\frac{dI}{dt} \\ \frac{dI}{dt} &= \frac{3.00(0.060)^2}{0.1138} = 0.095\,\text{A/s}.\end{aligned}$$

25.

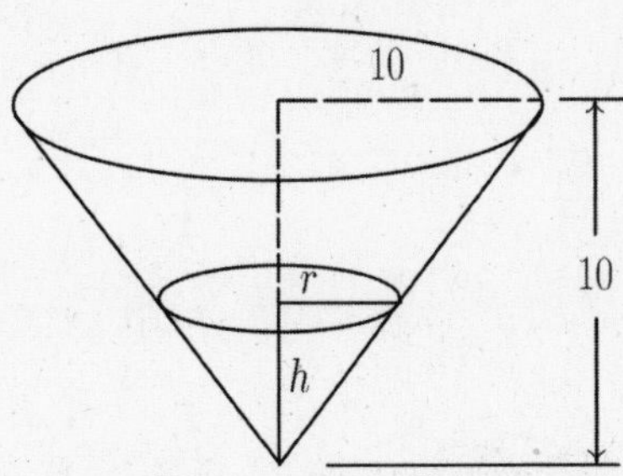

$\frac{r}{h} = \frac{10}{10} = 1$, or $r = h$

$V = \frac{1}{3}\pi r^2 h = \frac{1}{3}\pi h^3$

Given $\frac{dV}{dt} = 9$, find $\frac{dh}{dt}$ when $h = 3$.

$$\begin{aligned}\frac{dV}{dt} &= \pi h^2\frac{dh}{dt} \\ 9 &= \pi h^2\frac{dh}{dt} \\ \frac{dh}{dt} &= \frac{9}{\pi h^2}\Big|_{h=3} = \frac{1}{\pi}\,\frac{\text{m}}{\text{min}}\end{aligned}$$

27. First we need to determine the constant k:

$k = PV^{1.4} = 300(40)^{1.4} = 52481.38$.

$P = 52481.38V^{-1.4}$.

Given: $\frac{dV}{dt} = -80\,\text{in.}^3/\text{s}$, find: $\frac{dP}{dt}$ when $V = 40\,\text{in.}^3$

$$\begin{aligned}\frac{dP}{dt} &= 52481.38(-1.4)V^{-2.4}\frac{dV}{dt} \\ &= 52481.38(-1.4)V^{-2.4}(-80).\end{aligned}$$

When $V = 40$, $\frac{dP}{dt} = 840$ (lb/in.^{2}) per second.

29. $V = s^3$, $s = 10.00\,\text{cm}$, $ds = \pm 0.02\,\text{cm}$.

$\Delta V \approx dV = 3s^2 ds = 3(10.00)^2(\pm 0.02) = \pm 6\,\text{cm}^3$.

Percentage error: $\frac{dV}{V} \times 100 = \frac{6}{(10.00)^3} \times 100 = 0.6\%$.

31. $A = \pi r^2$.

$\Delta A \approx dA = 2\pi r\,dr = 2\pi(5.00)(\pm 0.02) = \pm 0.63\,\text{in.}^2$

Percentage error: $\frac{dA}{A} \times 100 = \frac{0.63}{\pi(5.00)^2} \times 100 = 0.8\%$.

Chapter 4

The Integral

4.1 Antiderivatives

1. By (4.3), $F(x) = 3x + C$.

3. $f(x) = 1 - 3x^2$. By (4.3) and (4.2), $F(x) = x - 3\dfrac{x^3}{3} + C = x - x^3 + C$.

5. $f(x) = 2x^3 - 3x^2 + x$. By (4.2), $F(x) = 2\dfrac{x^4}{4} - 3\dfrac{x^3}{3} + \dfrac{x^2}{2} + C = \dfrac{1}{2}x^4 - x^3 + \dfrac{1}{2}x^2 + C$.

7. Since 3 is a constant, $F(x) = \dfrac{x^4}{4} - 3\dfrac{x^3}{3} + 3x + C = \dfrac{1}{4}x^4 - x^3 + 3x + C$.

9. $f(x) = x^5 - 6x^4 + 2x^3$
$$F(x) = \frac{x^6}{6} - 6\frac{x^5}{5} + 2\frac{x^4}{4} + C = \frac{1}{6}x^6 - \frac{6}{5}x^5 + \frac{1}{2}x^4 + C$$

11. $f(x) = 3x^{-4} - 2x^{-3} + x^{-2} + 4$
$$F(x) = 3\frac{x^{-3}}{-3} - 2\frac{x^{-2}}{-2} + \frac{x^{-1}}{-1} + 4x + C = -x^{-3} + x^{-2} - x^{-1} + 4x + C$$

13. $f(x) = \dfrac{1}{x^2} - 2 = x^{-2} - 2$. By (4.1) and (4.3), $F(x) = \dfrac{x^{-1}}{-1} - 2x + C = -\dfrac{1}{x} - 2x + C$.

15. $f(x) = \dfrac{2}{3x^2} + \dfrac{5}{4x^3} + \sqrt{x} = \dfrac{2}{3}x^{-2} + \dfrac{5}{4}x^{-3} + x^{1/2}$.
$$F(x) = \frac{2}{3}\frac{x^{-1}}{-1} + \frac{5}{4}\frac{x^{-2}}{-2} + \frac{x^{3/2}}{3/2} + C = -\frac{2}{3x} - \frac{5}{8x^2} + \frac{2}{3}x^{3/2} + C.$$

4.2 The Area Problem

1. Subdivide the interval $[0,1]$ into n equal parts, each of length $1/n$. Thus $\Delta x_i = 1/n$. Choosing the right endpoint of each subinterval for x_i, we get

$$x_1 = 1\cdot\frac{1}{n},\ x_2 = 2\cdot\frac{1}{n},\ x_3 = 3\cdot\frac{1}{n},\ldots,\ x_i = i\cdot\frac{1}{n},\ldots,\ x_n = n\cdot\frac{1}{n} = 1.$$

Since $f(x) = x$, we get, for the corresponding altitudes,

$$f(x_1) = 1\cdot\frac{1}{n},\ f(x_2) = 2\cdot\frac{1}{n},\ldots,\ f(x_i) = i\cdot\frac{1}{n},\ldots,\ f(x_n) = n\cdot\frac{1}{n}.$$

Since $\Delta x_i = 1/n$, the sum of the areas is given by

$$\sum_{i=1}^{n} f(x_i)\Delta x_i = \sum_{i=1}^{n} i\cdot\frac{1}{n}\cdot\frac{1}{n} = \frac{1}{n^2}\sum_{i=1}^{n} i = \frac{1}{n^2}\cdot\frac{n(n+1)}{2}$$

by formula A. So the exact area is

$$\int_0^1 x\,dx = \lim_{n\to\infty}\frac{n(n+1)}{2n^2} = \lim_{n\to\infty}\frac{n^2+n}{2n^2} = \lim_{n\to\infty}\frac{1+1/n}{2} = \frac{1}{2}.$$

3. Subdivide the interval $[0,1]$ into n equal parts, each of length $1/n$. Thus $\Delta x_i = 1/n$. Choosing the right endpoint of each subinterval for x_i, we get

$$x_1 = 1\cdot\frac{1}{n},\ x_2 = 2\cdot\frac{1}{n},\ x_3 = 3\cdot\frac{1}{n},\ldots,\ x_i = i\cdot\frac{1}{n},\ldots,\ x_n = n\cdot\frac{1}{n} = 1.$$

Since $f(x) = x^2$, we get, for the corresponding altitudes,

$$f(x_1) = \left(1\cdot\frac{1}{n}\right)^2,\ f(x_2) = \left(2\cdot\frac{1}{n}\right)^2,\ldots,\ f(x_i) = \left(i\cdot\frac{1}{n}\right)^2,\ldots,\ f(x_n) = \left(n\cdot\frac{1}{n}\right)^2 = 1.$$

Since $\Delta x_i = 1/n$, the sum of the areas is given by

$$\sum_{i=1}^{n} f(x_i)\Delta x_i = \sum_{i=1}^{n}\left(i\cdot\frac{1}{n}\right)^2\frac{1}{n} = \frac{1}{n^3}\sum_{i=1}^{n} i^2 = \frac{1}{n^3}\frac{n(n+1)(2n+1)}{6}$$

by Formula B. So the exact area is

$$\int_0^1 x^2\,dx = \lim_{n\to\infty}\frac{n(n+1)(2n+1)}{6n^3} = \lim_{n\to\infty}\frac{2n^3+3n^2+n}{6n^3} = \lim_{n\to\infty}\frac{2+3/n+1/n^2}{6} = \frac{2}{6} = \frac{1}{3}.$$

5. Subdividing the interval $[0,2]$ into n equal parts, we get $\Delta x_i = 2/n$. Choosing the right endpoint in each subinterval for x_i again, we have

$$x_1 = 1\cdot\frac{2}{n},\ x_2 = 2\cdot\frac{2}{n},\ldots,\ x_i = i\cdot\frac{2}{n},\ldots,\ x_n = n\cdot\frac{2}{n} = 2.$$

For the altitudes, we get from $f(x) = 3x^2$

$$f(x_1) = 3(1)^2\frac{4}{n^2},\ f(x_2) = 3(2)^2\frac{4}{n^2},\ldots,\ f(x_i) = 3(i)^2\frac{4}{n^2},\ldots,\ f(x_n) = 3(n)^2\frac{4}{n^2}.$$

Since $\Delta x_i = 2/n$, the sum of the areas is

$$\sum_{i=1}^{n} f(x_i)\Delta x_i = \sum_{i=1}^{n} 3i^2\frac{4}{n^2}\cdot\frac{2}{n} = \frac{24}{n^3}\sum_{i=1}^{n} i^2 = \frac{24}{n^3}\cdot\frac{n(n+1)(2n+1)}{6} = \frac{4n(n+1)(2n+1)}{n^3}$$

by Formula B. Hence

$$\int_0^2 3x^2\,dx = \lim_{n\to\infty}\frac{4n(n+1)(2n+1)}{n^3} = \lim_{n\to\infty}\frac{8n^3+12n^2+4n}{n^3} = \lim_{n\to\infty}\frac{8+12/n+4/n^2}{1} = 8.$$

4.3 The Fundamental Theorem of Calculus

1. Partial solution:

(5) $\int_0^2 3x^2\,dx = 3\frac{x^3}{3}\Big|_0^2 = x^3\Big|_0^2 = 2^3 - 0^3 = 8$

(6) $\int_0^2 (1+2x)\,dx = x + 2\frac{x^2}{2}\Big|_0^2 = x + x^2\Big|_0^2 = (2+2^2) - 0 = 6$

3. $\int_0^2 \frac{1}{2}x\,dx = \frac{1}{2}\frac{x^2}{2}\Big|_0^2 = \frac{2^2}{4} - \frac{0^2}{4} = 1$

5. $\int_0^4 2x^2\,dx = 2\frac{x^3}{3}\Big|_0^4 = \frac{2}{3}(4^3) - \frac{2}{3}(0^3) = \frac{128}{3}$

7. $\int_0^1 (x^3+1)\,dx = \frac{1}{4}x^4 + x\Big|_0^1 = \left(\frac{1}{4}+1\right) - (0) = \frac{5}{4}$

9. For the function $y = 4 - x^2$, the x-intercepts are $x = \pm 2$.

$$\begin{aligned} A &= \int_{-2}^2 (4-x^2)\,dx = 4x - \frac{x^3}{3}\Big|_{-2}^2 = \left(8 - \frac{8}{3}\right) - \left(-8 + \frac{8}{3}\right) \\ &= 2\left(8 - \frac{8}{3}\right) = 2\left(\frac{24}{3} - \frac{8}{3}\right) = 2\cdot\frac{16}{3} = \frac{32}{3} \end{aligned}$$

11. $\int_1^3 \frac{1}{x^3}\,dx = \int_1^3 x^{-3}\,dx = \frac{x^{-2}}{-2}\Big|_1^3 = -\frac{1}{2x^2}\Big|_1^3 = -\frac{1}{2\cdot 3^2} + \frac{1}{2\cdot 1^2} = -\frac{1}{18} + \frac{1}{2} = \frac{8}{18} = \frac{4}{9}$

13. The x-intercepts are 0 and 2:

$$\int_0^2 (4x - 2x^2)\,dx = 2x^2 - \frac{2}{3}x^3\Big|_0^2 = 8 - \frac{16}{3} = \frac{8}{3}$$

4.5 Basic Integration Formulas

1. $\int \sqrt{x}\,dx = \int x^{1/2}\,dx = \frac{x^{3/2}}{3/2} + C = \frac{2}{3}x^{3/2} + C$

3. $\int \left(\frac{1}{x^3} - \frac{3}{x^2}\right)\,dx = \int (x^{-3} - 3x^{-2})\,dx = \frac{x^{-2}}{-2} - 3\frac{x^{-1}}{-1} + C = -\frac{1}{2x^2} + \frac{3}{x} + C$

5. $\int (2\sqrt{x} - 3x^2 + 1)\,dx = 2\frac{x^{3/2}}{3/2} - 3\frac{x^3}{3} + x + C = \frac{4}{3}x^{3/2} - x^3 + x + C$

7. $\int (x^2-4)^3(2x)\,dx$. Let $u = x^2 - 4$; then $du = 2x\,dx$.

$\int u^3\,du = \frac{1}{4}u^4 + C = \frac{1}{4}(x^2-4)^4 + C$

9. $\int (2x^2-3)^3(4x)\,dx$. Let $u = 2x^2 - 3$; then $du = 4x\,dx$.

$\int u^3\,du = \frac{1}{4}u^4 + C = \frac{1}{4}(2x^2-3)^4 + C$

11. $\int (2-x^2)^4 x\,dx$. Let $u = 2-x^2$; then $du = -2x\,dx$.

$$\begin{aligned}\int (2-x^2)^4 x\,dx &= -\frac{1}{2}\int (2-x^2)^4(-2x)\,dx = -\frac{1}{2}\int u^4\,du \\ &= -\frac{1}{2}\frac{u^5}{5} + C = -\frac{1}{10}(2-x^2)^5 + C\end{aligned}$$

13. $\int \frac{x\,dx}{(x^2-1)^2} = \int (x^2-1)^{-2}x\,dx$. Let $u = x^2-1$; then $du = 2x\,dx$.

$$\frac{1}{2}\int (x^2-1)^{-2}(2x)\,dx = \frac{1}{2}\int u^{-2}\,du = \frac{1}{2}\frac{u^{-1}}{-1} + C = -\frac{1}{2(x^2-1)} + C$$

15. $\int \frac{dt}{\sqrt{1-t}} = \int (1-t)^{-1/2}\,dt$. Let $u = 1-t$; then $du = -dt$.

$$-\int (1-t)^{-1/2}(-dt) = -\int u^{-1/2}\,du = -\frac{u^{1/2}}{1/2} + C = -2\sqrt{1-t} + C$$

17. $\int \frac{x\,dx}{\sqrt{1-x^2}} = \int (1-x^2)^{-1/2}x\,dx$. Let $u = 1-x^2$; then $du = -2x\,dx$.

$$\begin{aligned}\int (1-x^2)^{-1/2}x\,dx &= -\frac{1}{2}\int (1-x^2)^{-1/2}(-2x\,dx) = -\frac{1}{2}\int u^{-1/2}\,du \\ &= -\frac{1}{2}\frac{u^{1/2}}{1/2} + C = -u^{1/2} + C = -\sqrt{1-x^2} + C\end{aligned}$$

19. $\int (x^2+1)^2\,dx$. If we try letting $u = x^2+1$, then $du = 2x\,dx$ and no substitution can be made. As noted in Example 4, the integral is not of the proper form. Consequently, we must multiply out the binomial and integrate term by term using (4.8):

$$\int (x^2+1)^2\,dx = \int (x^4+2x^2+1)\,dx = \frac{1}{5}x^5 + \frac{2}{3}x^3 + x + C$$

21. $\int (1-x^2)^2 x\,dx$. Let $u = 1-x^2$; then $du = -2x\,dx$.

$$\int (1-x^2)^2 x\,dx = -\frac{1}{2}\int (1-x^2)^2(-2x)\,dx = -\frac{1}{2}\int u^2\,du = -\frac{1}{2}\frac{u^3}{3} + C = -\frac{1}{6}(1-x^2)^3 + C$$

23. As in Exercise 19, we must multiply out the integrand:

$$\int (1+\sqrt{x})^2\,dx = \int (1+2x^{1/2}+x)\,dx = x + 2\frac{x^{3/2}}{3/2} + \frac{x^2}{2} + C = x + \frac{4}{3}x^{3/2} + \frac{1}{2}x^2 + C$$

25. $\int (x^3+1)^2(3x)\,dx$. If we let $u = x^3+1$, then $du = 3x^2\,dx$, which does not match $3x\,dx$. So we need to multiply out the integrand and integrate term by term.

$$\begin{aligned}\int (x^6+2x^3+1)(3x)\,dx &= 3\int (x^7+2x^4+x)\,dx \\ &= 3\left(\frac{x^8}{8} + 2\frac{x^5}{5} + \frac{x^2}{2}\right) + C \\ &= \frac{3}{8}x^8 + \frac{6}{5}x^5 + \frac{3}{2}x^2 + C\end{aligned}$$

27. $\int (x^3+1)^3(5x^2)\,dx$. Let $u = x^3+1$; then $du = 3x^2\,dx$.

$$\begin{aligned}\int (x^3+1)^3(5x^2)\,dx &= 5\int (x^3+1)^3x^2\,dx = \frac{5}{3}\int (x^3+1)(3x^2)\,dx \\ &= \frac{5}{3}\int u^3\,du = \frac{5}{3}\frac{u^4}{4} + C = \frac{5}{12}(x^3+1)^4 + C\end{aligned}$$

29. $\int (4V^3-1)^2(12V)\,dV$. If we let $u = 4V^3-1$, then $du = 12V^2\,dV$, which does not match $12V\,dV$. So we need to multiply out the binomial and integrate term by term.

$$\begin{aligned}\int (16V^6-8V^3+1)(12V)\,dV &= 12\int (16V^7-8V^4+V)\,dV\\ &= 12\left(16\frac{V^8}{8}-8\frac{V^5}{5}+\frac{V^2}{2}\right)+C\\ &= 24V^8-\frac{96}{5}V^5+6V^2+C\end{aligned}$$

31. $\int (1+x^3)^4(3x^2)\,dx$. Let $u = 1+x^3$; then $du = 3x^2\,dx$.
$\int u^4\,du = \frac{1}{5}u^5+C = \frac{1}{5}(1+x^3)^5+C$

33. $\int (1+x^3)^4(x^2)\,dx$. Let $u = 1+x^3$; then $du = 3x^2\,dx$.
$\frac{1}{3}\int (1+x^3)^4(3x^2)\,dx = \frac{1}{3}\int u^4\,du = \frac{1}{3}\frac{u^5}{5}+C = \frac{1}{15}(1+x^3)^5+C$

35. $\int (x^4+2)^2(4x^3)\,dx = \frac{1}{3}(x^4+2)^3+C$ $\qquad u = x^4+2$, $du = 4x^3\,dx$

37. As in Exercise 25, if $u = x^3+1$, then $du = 3x^2\,dx$, which cannot be made to match $x\,dx$ in the integral. Instead, we multiply out the integrand and integrate term by term:

$$\int (x^3+1)^2x\,dx = \int (x^6+2x^3+1)x\,dx = \int (x^7+2x^4+x)\,dx = \frac{1}{8}x^8+\frac{2}{5}x^5+\frac{1}{2}x^2+C$$

39. $\int (3-t^4)^2t^3\,dt$. Let $u = 3-t^4$; then $du = -4t^3\,dt$.
$-\frac{1}{4}\int (3-t^4)^2(-4t^3)\,dt = -\frac{1}{4}\int u^2\,du = -\frac{1}{4}\frac{u^3}{3}+C = -\frac{1}{12}(3-t^4)^3+C$

41. $\int_0^1 (1-x)\,dx = x-\frac{1}{2}x^2\Big|_0^1 = \left[1-\frac{1}{2}(1)^2\right]-\left[0-\frac{1}{2}(0)^2\right] = \frac{1}{2}$

43. $\int_1^8 \sqrt[3]{x}\,dx = \int_1^8 x^{1/3}\,dx = \frac{x^{4/3}}{4/3}\Big|_1^8 = \frac{3}{4}x^{4/3}\Big|_1^8 = \frac{3}{4}(8^{4/3}-1^{4/3}) = \frac{3}{4}(16-1) = \frac{45}{4}$

45. $\int_0^1 \sqrt{1-x}\,dx$; $\qquad u = 1-x$, $du = -dx$
Lower limit: if $x = 0$, then $u = 1-x = 1-0 = 1$
Upper limit: if $x = 1$, then $u = 1-x = 1-1 = 0$

$$\int_0^1 (1-x)^{1/2}\,dx = -\int_0^1 (1-x)^{1/2}(-dx) = -\int_1^0 u^{1/2}\,du = -\frac{2}{3}u^{3/2}\Big|_1^0 = 0-\left(-\frac{2}{3}\right) = \frac{2}{3}$$

Alternatively, we can find the indefinite integral first and substitute $x = 0$ and $x = 1$:

$$\begin{aligned}\int_0^1 (1-x)^{1/2}\,dx &= -\int_0^1 (1-x)^{1/2}(-dx) \qquad u = 1-x;\ du = -dx\\ &= -\frac{2}{3}(1-x)^{3/2}\Big|_0^1 = 0-\left(-\frac{2}{3}\cdot 1\right) = \frac{2}{3}\end{aligned}$$

47. $$\begin{aligned}\int_0^1 (x^2-1)^2\,dx &= \int_0^1 (x^4-2x^2+1)\,dx = \frac{1}{5}x^5 - \frac{2}{3}x^3 + x\Big|_0^1 \\ &= \left(\frac{1}{5}-\frac{2}{3}+1\right) - 0 = \frac{3}{15} - \frac{10}{15} + \frac{15}{15} = \frac{8}{15}\end{aligned}$$

49. $\displaystyle\int_2^7 \frac{dt}{\sqrt{t+2}} = \int_2^7 (t+2)^{-1/2}\,dt; \qquad u = t+2,\ du = dt$

Lower limit: if $t=2$, then $u = t+2 = 2+2 = 4$

Upper limit: if $t=7$, then $u = t+2 = 7+2 = 9$

$$\int_2^7 (t+2)^{-1/2}\,dt = \int_4^9 u^{-1/2}\,du = 2u^{1/2}\Big|_4^9 = 2\sqrt{u}\Big|_4^9 = 2\sqrt{9} - 2\sqrt{4} = 6 - 4 = 2$$

Alternatively, we can find the indefinite integral first and substitute $t=2$ and $t=7$:

$$\begin{aligned}\int_2^7 (t+2)^{-1/2}\,dt &= 2(t+2)^{1/2}\Big|_2^7 \qquad u = t+2,\ du = dt \\ &= 2(9)^{1/2} - 2(4)^{1/2} = 2\end{aligned}$$

51. $\displaystyle\int_{-4}^0 \sqrt{1-2x}\,dx = \int_{-4}^0 (1-2x)^{1/2}\,dx$. Let $u = 1-2x$; then $du = -2\,dx$.

Lower limit: if $x = -4$, then $u = 9$

Upper limit: if $x = 0$, then $u = 1$

$$-\frac{1}{2}\int_{-4}^0 (1-2x)^{1/2}(-2)\,dx = -\frac{1}{2}\int_9^1 u^{1/2}\,du = -\frac{1}{2}\cdot\frac{2}{3}u^{3/2}\Big|_9^1 = -\frac{1}{3}(1-9^{3/2}) = -\frac{1}{3}(1-27) = \frac{26}{3}$$

53. $\displaystyle\int_4^9 \frac{1+\sqrt{x}}{\sqrt{x}}\,dx = \int_4^9 \frac{1+x^{1/2}}{x^{1/2}}\,dx = \int_4^9 \left(\frac{1}{x^{1/2}} + \frac{x^{1/2}}{x^{1/2}}\right)dx = \int_4^9 (x^{-1/2}+1)\,dx$

$\displaystyle = 2x^{1/2} + x\Big|_4^9 = [2(9)^{1/2}+9] - [2(4)^{1/2}+4] = (6+9) - (4+4) = 15 - 8 = 7$

55. $\displaystyle\int_1^2 \theta\sqrt{4-\theta^2}\,d\theta = \int_1^2 (4-\theta^2)^{1/2}\theta\,d\theta$. Let $u = 4-\theta^2$; then $du = -2\theta\,d\theta$.

Lower limit: if $\theta = 1$, then $u = 3$

Upper limit: if $\theta = 2$, then $u = 0$

$$-\frac{1}{2}\int_1^2 (4-\theta^2)^{1/2}(-2\theta)\,d\theta = -\frac{1}{2}\int_3^0 u^{1/2}\,du = -\frac{1}{2}\cdot\frac{2}{3}u^{3/2}\Big|_3^0 = -\frac{1}{3}(0-3^{3/2}) = 3^{1/2} = \sqrt{3}$$

4.6 Area Between Curves

1.

$$\int_0^1 y\,dx = \int_0^1 2x\,dx = x^2\Big|_0^1 = 1$$

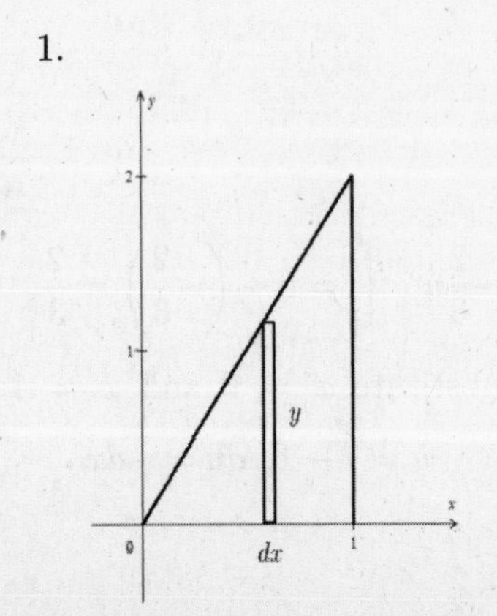

3.

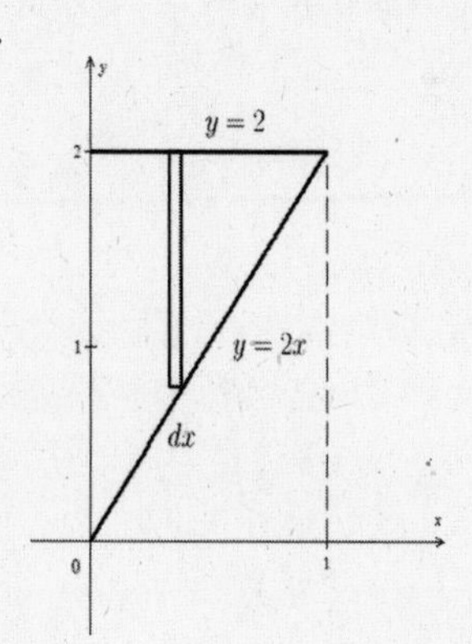

$$\int_0^1 (2 - 2x)\,dx = 2x - x^2\Big|_0^1 = 2 - 1 = 1$$

5.

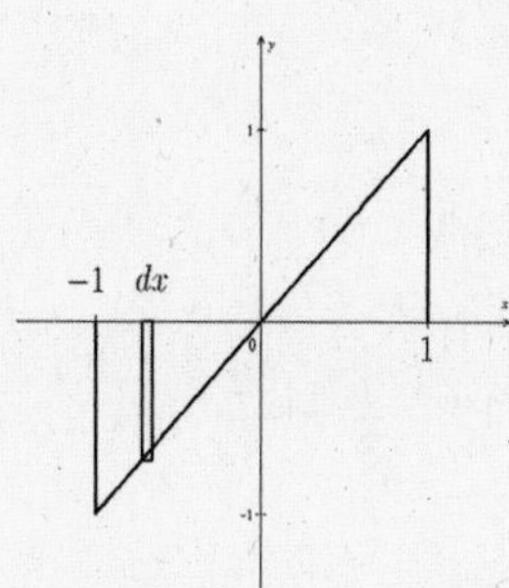

For the region on the left,

$$\int_{-1}^0 (0 - x)\,dx = -\frac{x^2}{2}\Big|_{-1}^0$$

$$= 0 - \left(-\frac{1}{2}\right) = \frac{1}{2}.$$

For the region on the right: $\displaystyle\int_0^1 x\,dx = \frac{x^2}{2}\Big|_0^1 = \frac{1}{2}$ for a total of $\dfrac{1}{2} + \dfrac{1}{2} = 1$.

7.

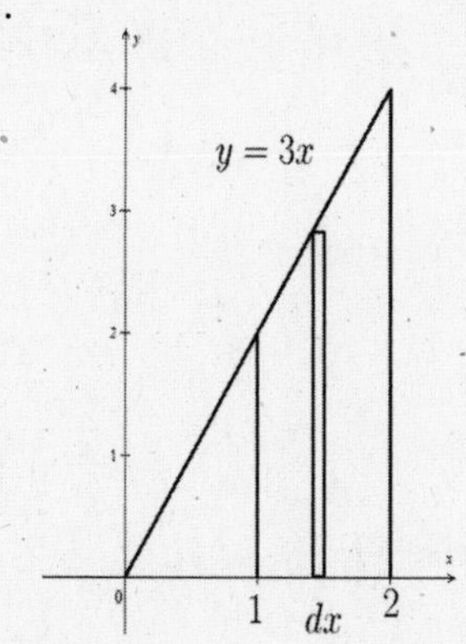

$$\int_1^2 3x\,dx = \frac{3}{2}x^2\Big|_1^2 = \frac{3}{2}(4 - 1) = \frac{9}{2}$$

9.

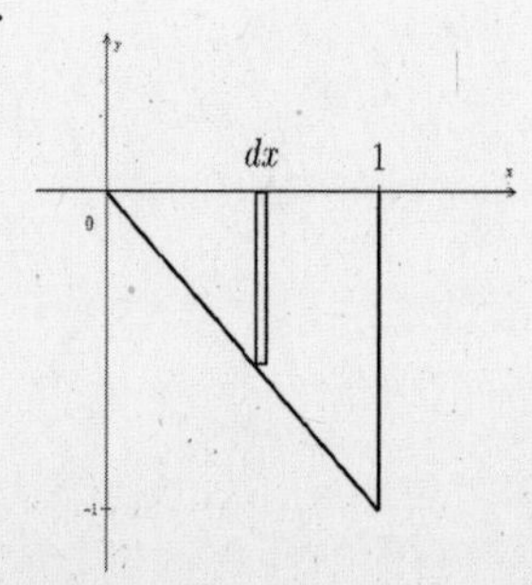

$$\int_0^1 [0 - (-x)\,dx] = \int_0^1 x\,dx = \frac{1}{2}$$

11.

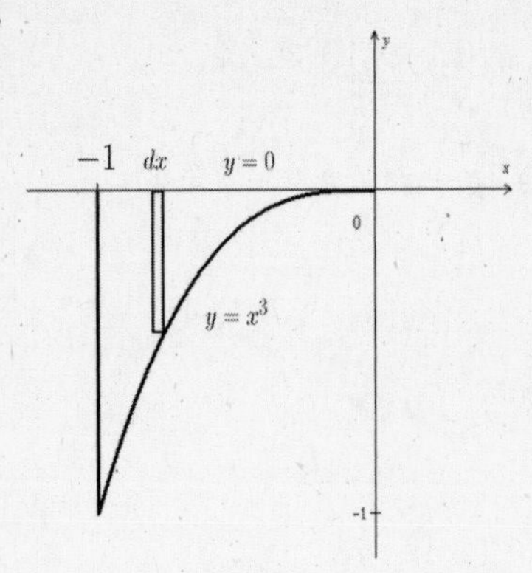

$$\int_{-1}^{0} (0 - x^3)\,dx = -\frac{x^4}{4}\Big|_{-1}^{0} = 0 + \frac{1}{4} = \frac{1}{4}$$

13.

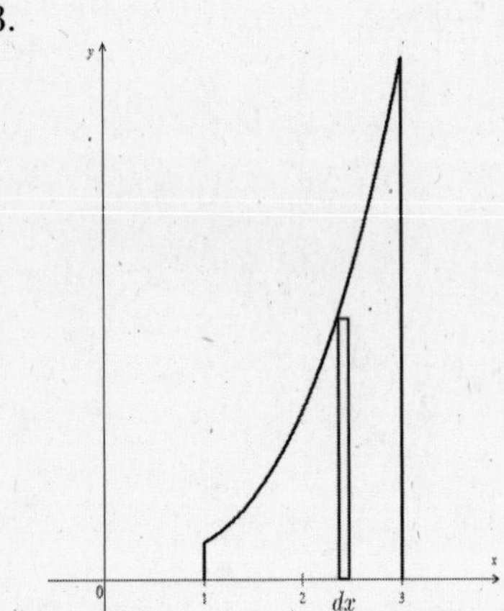

$$\int_{1}^{3} (x^2 + 1)\,dx = \frac{1}{3}x^3 + x\Big|_{1}^{3}$$
$$= (9 + 3) - \left(\frac{1}{3} + 1\right) = 12 - \frac{4}{3} = \frac{32}{3}$$

15.

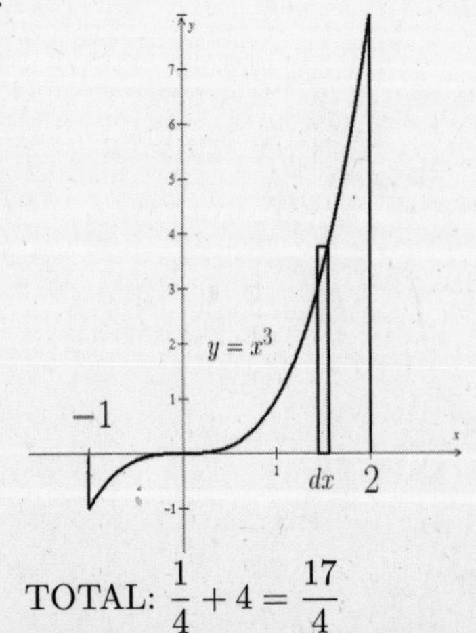

Left region: $\frac{1}{4}$ by Exercise 11

Right region: $\int_{0}^{2} x^3\,dx = \frac{1}{4}x^4\Big|_{0}^{2} = 4$

TOTAL: $\frac{1}{4} + 4 = \frac{17}{4}$

17.

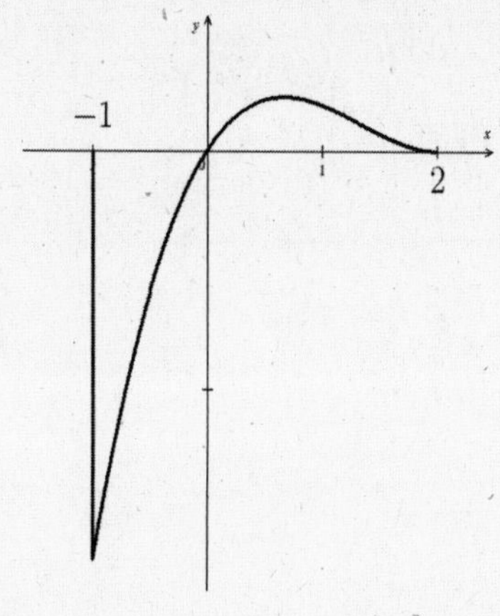

Left region: $\int_{-1}^{0}[0-x(x-2)^2]\,dx$

$=\int_{-1}^{0}(-x^3+4x^2-4x)\,dx$

$=-\frac{1}{4}x^4+\frac{4}{3}x^3-2x^2\Big|_{-1}^{0}$

$=0-\left(-\frac{1}{4}-\frac{4}{3}-2\right)=\frac{43}{12}$

Right region: $\int_{0}^{2}x(x-2)^2\,dx$

$=\int_{0}^{2}(x^3-4x^2+4x)\,dx=\frac{1}{4}x^4-\frac{4}{3}x^3+2x^2\Big|_{0}^{2}$

$=4-\frac{32}{3}+8=\frac{36}{3}-\frac{32}{3}=\frac{4}{3}$

TOTAL: $\frac{43}{12}+\frac{4}{3}=\frac{43}{12}+\frac{16}{12}=\frac{59}{12}$

19.

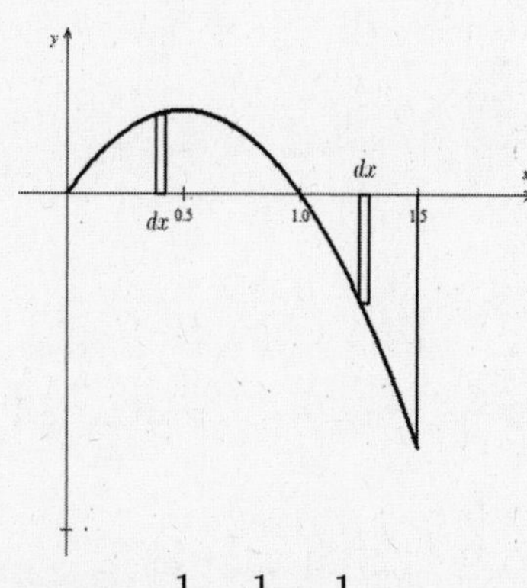

Left region: $\int_{0}^{1}\left(x-x^2\right)\,dx$

$=\frac{1}{2}x^2-\frac{1}{3}x^3\Big|_{0}^{1}=\frac{1}{2}-\frac{1}{3}=\frac{1}{6}$

Right region: $\int_{1}^{3/2}\left[0-\left(x-x^2\right)\right]\,dx$

$=-\frac{1}{2}x^2+\frac{1}{3}x^3\Big|_{1}^{3/2}$

$=-\frac{1}{2}\left(\frac{3}{2}\right)^2+\frac{1}{3}\left(\frac{3}{2}\right)^3-\left(-\frac{1}{2}+\frac{1}{3}\right)=\frac{1}{6}$

TOTAL: $\frac{1}{6}+\frac{1}{6}=\frac{1}{3}$

21.

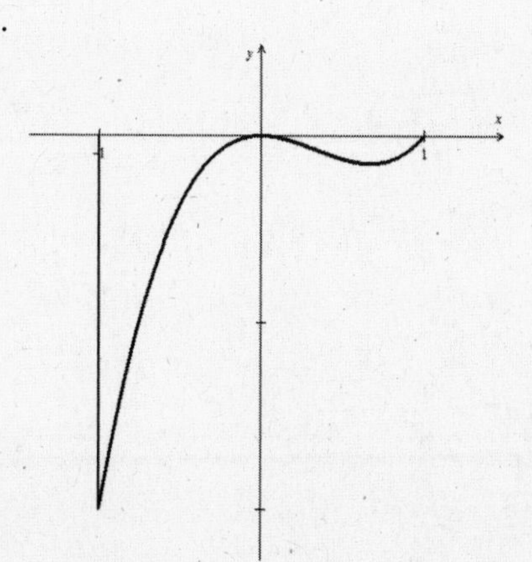

The region lies below the x-axis:

$$\int_{-1}^{1}\left[0-x^2(x-1)\right]\,dx=\int_{-1}^{1}\left(-x^3+x^2\right)\,dx$$

$$=-\frac{1}{4}x^4+\frac{1}{3}x^3\Big|_{-1}^{1}=\left(-\frac{1}{4}+\frac{1}{3}\right)-\left(-\frac{1}{4}-\frac{1}{3}\right)=\frac{2}{3}.$$

23. Here the typical element is drawn horizontally.

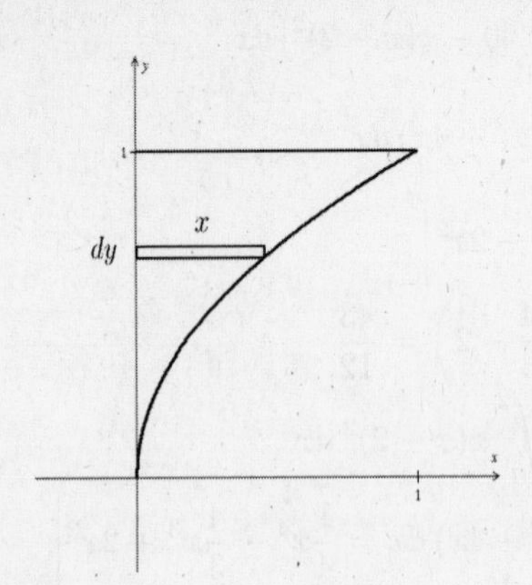

$$\int_0^1 x\,dy = \int_0^1 y^2\,dy = \frac{1}{3}y^3\Big|_0^1 = \frac{1}{3}$$

25.

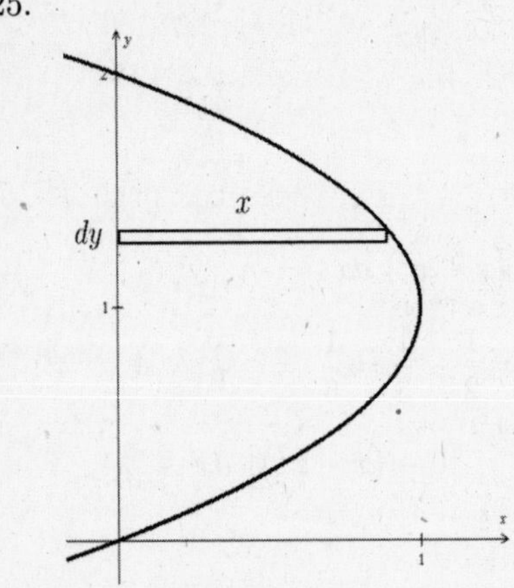

$$\int_0^2 (2y - y^2)\,dy = y^2 - \frac{1}{3}y^3\Big|_0^2$$
$$= 4 - \frac{8}{3} - 0 = \frac{4}{3}$$

27.

$$\int_0^1 (\sqrt{x} - x)\,dx = \frac{2}{3}x^{3/2} - \frac{1}{2}x^2\Big|_0^1 = \frac{1}{6}$$

29.

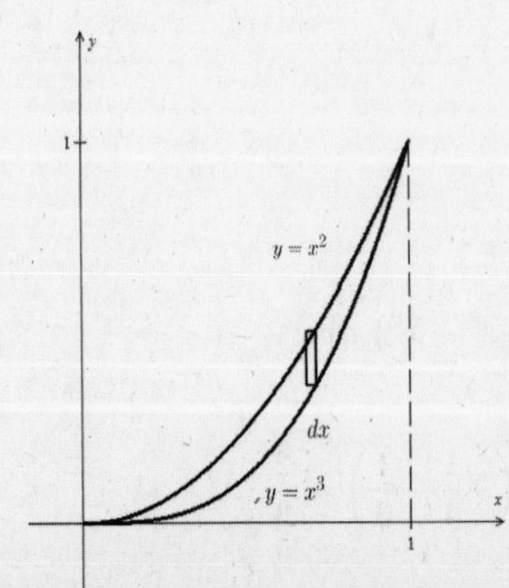

$$\int_0^1 (x^2 - x^3)\,dx = \frac{x^3}{3} - \frac{x^4}{4}\Big|_0^1$$
$$= \frac{1}{3} - \frac{1}{4} = \frac{1}{12}$$

31. To see where the curves intersect, we need to solve the equations simultaneously:

$$\begin{aligned} y &= 2x \\ y &= 8 - x^2 \\ \hline 0 &= 2x - 8 + x^2 \qquad \text{(subtracting)} \\ (x+4)(x-2) &= 0 \\ x &= -4, 2 \end{aligned}$$

Substituting in $y = 2x$, we obtain the points of intersection: $(-4, -8)$ and $(2, 4)$. The length of the typical element is $8 - x^2 - 2x$

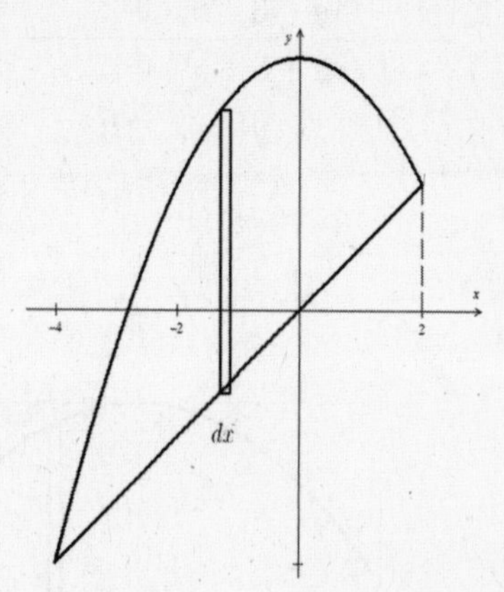

Thus,

$$\begin{aligned} \int_{-4}^{2} \left(8 - x^2 - 2x\right) dx &= 8x - \frac{1}{3}x^3 - x^2 \Big|_{-4}^{2} \\ &= \left(16 - \frac{8}{3} - 4\right) - \left(-32 + \frac{64}{3} - 16\right) \\ &= 60 - \frac{72}{3} = 36 \end{aligned}$$

33. To see where the curves intersect, we need to solve the equations simultaneously:

$$\begin{aligned} y^2 &= x - 1 \\ y &= x - 3 \\ \hline y^2 - y &= 2 \qquad \text{(subtracting)} \\ y^2 - y - 2 &= 0 \\ (y-2)(y+1) &= 0 \\ y &= -1, 2 \end{aligned}$$

The points are $(2, -1)$ and $(5, 2)$.

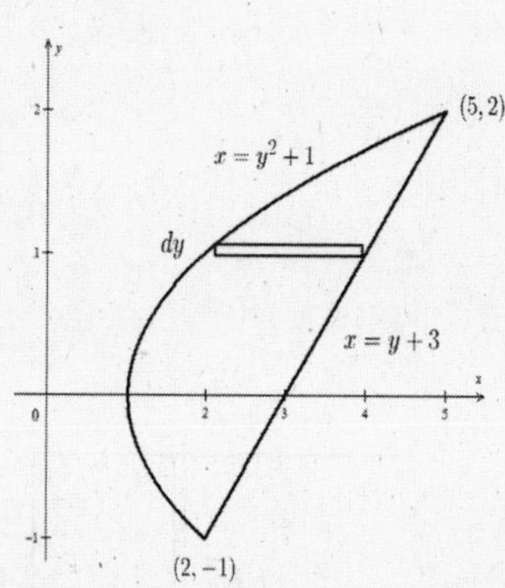

Note that in the resulting region the typical element should be drawn sideways. Solving the given equations for x in terms of y, we get

$$x = y^2 + 1 \text{ and } x = y + 3,$$

respectively. The length of the typical element is now seen to be: $(y+3) - (y^2+1)$. Thus

$$\begin{aligned} \int_{-1}^{2} [(y+3) - (y^2+1)]\, dy &= \int_{-1}^{2} (y + 3 - y^2 - 1)\, dy \\ &= \int_{-1}^{2} (-y^2 + y + 2)\, dy \\ &= -\frac{1}{3}y^3 + \frac{1}{2}y^2 + 2y \Big|_{-1}^{2} \\ &= \left(-\frac{8}{3} + 2 + 4\right) - \left(\frac{1}{3} + \frac{1}{2} - 2\right) \\ &= -\frac{8}{3} + 2 + 4 - \frac{1}{3} - \frac{1}{2} + 2 \\ &= -\frac{8}{3} - \frac{1}{3} - \frac{1}{2} + 8 = \frac{9}{2}. \end{aligned}$$

35.
$$
\begin{aligned}
x^2 + 4y &= 0 \\
x^2 - 4y - 8 &= 0 \\
\hline
8y + 8 &= 0 \\
y &= -1; \\
x^2 &= 4,\ x = \pm 2
\end{aligned}
\qquad
\begin{aligned}
y &= -\tfrac{1}{4}x^2 \\
y &= \tfrac{1}{4}(x^2 - 8)
\end{aligned}
$$

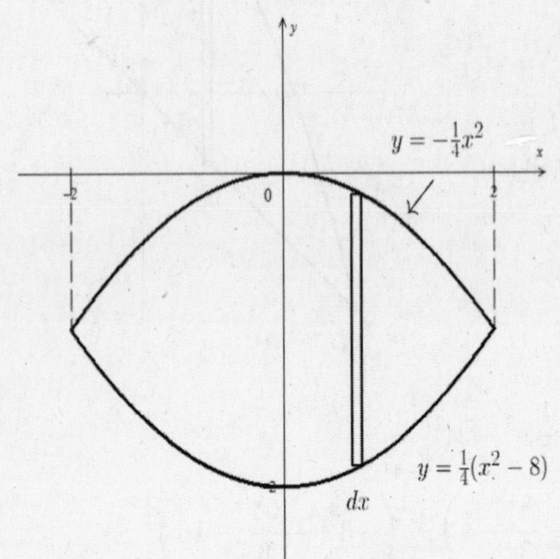

$$
\begin{aligned}
\int_{-2}^{2} \left[-\frac{1}{4}x^2 - \frac{1}{4}(x^2 - 8) \right] dx &= \frac{1}{4}\int_{-2}^{2} (-x^2 - x^2 + 8)\, dx \\
&= \frac{1}{4}\int_{-2}^{2} (8 - 2x^2)\, dx = \frac{1}{4}\left(8x - \frac{2}{3}x^3\right)\Big|_{-2}^{2} \\
&= \frac{1}{4}\left(16 - \frac{16}{3}\right) - \frac{1}{4}\left(-16 + \frac{16}{3}\right) = \frac{1}{2}\left(16 - \frac{16}{3}\right) \\
&= \frac{1}{2} \cdot \frac{48 - 16}{3} = \frac{16}{3}
\end{aligned}
$$

37.

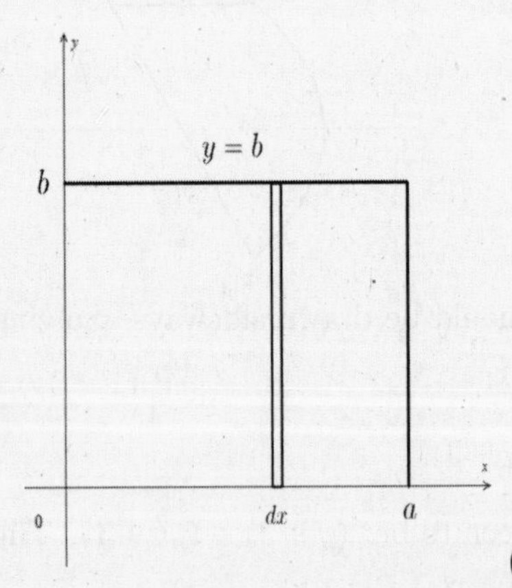

(b) Since the function is $y = b$, $A = \int_0^a b\, dx = bx\Big|_0^a = ab$.

39. We can see from the graph in the Answer Section that the limits of integration are -2 and 4. Alternatively, we can solve the equations simultaneously:

$$
\begin{aligned}
y &= \tfrac{1}{2}x^2 \\
y &= x + 4 \\
\hline
0 &= \tfrac{1}{2}x^2 - x - 4 \qquad \text{(subtraction)} \\
x^2 - 2x - 8 &= 0 \\
(x + 2)(x - 4) &= 0 \\
x &= -2, 4.
\end{aligned}
$$

Using the integration capability of a graphing utility, we get

$$\int_{-2}^{4} \left(x + 4 - \frac{1}{2}x^2\right) dx = 18.$$

41. We can see from the graph in the Answer Section that the limits of integration are -1 and 2. Alternatively, solving the equations simultaneously,

$$\begin{aligned} y &= 2 - x^2 \\ y &= -x \\ \hline -x &= 2 - x^2 \qquad \text{(substitution)} \\ x^2 - x - 2 &= 0 \\ (x+1)(x-2) &= 0 \\ x &= -1, 2. \end{aligned}$$

Using the integration capability of a graphing utility, we get $A = \displaystyle\int_{-1}^{2} (2 - x^2 + x)\, dx = \frac{9}{2}$.

43. We can see from the graph in the Answer Section that the limits of integration are -2 and 2. As a check:
If $x = -2$, then $y = 0$ for both equations; if $x = 2$, then $y = 8$ for both equations.

$$\int_{-2}^{2} [(2x + 4) - (x^2 + 2x)]\, dx = \frac{32}{3}$$

45. From the graph in the Answer Section, the limits of integration are 0 and 4. Alternatively,

$$\begin{aligned} y &= x^2 - 4x + 2 \\ y &= 2 + 4x - x^2 \\ \hline x^2 - 4x + 2 &= 2 + 4x - x^2 \qquad \text{(substitution)} \\ 2x^2 - 8x &= 0 \\ 2x(x - 4) &= 0 \\ x &= 0, 4. \end{aligned}$$

Using the integration capability of a graphing utility, we get

$$A = \int_0^4 [(2 + 4x - x^2) - (x^2 - 4x + 2)]\, dx = \int_0^4 (8x - 2x^2)\, dx = \frac{64}{3}.$$

47. According to the graphs in the Answer Section, the graphs intersect at $(\pm 1, 1)$ and $(\pm 2, 4)$. Using the integration capability of a graphing utility, we can find the sum of the areas of the regions on the right and double the result:

$$2\left(\int_0^1 [(x^4 - 4x^2 + 4) - x^2]\, dx + \int_1^2 [x^2 - (x^4 - 4x^2 + 4)]\, dx\right) = 2\left(\frac{38}{15} + \frac{22}{15}\right) = 2 \cdot \frac{60}{15} = 8.$$

4.7 Improper Integrals

1.
$$\begin{aligned} \int_1^{\infty} \frac{3}{x^4}\, dx &= \lim_{b\to\infty} \int_1^b 3x^{-4}\, dx = \lim_{b\to\infty} \left(-x^{-3}\right)\Big|_1^b \\ &= \lim_{b\to\infty} \left(-\frac{1}{b^3} + 1\right) = 0 + 1 = 1 \end{aligned}$$

3.
$$\begin{aligned}
\int_{-\infty}^{0} \frac{3}{(x-1)^4}\,dx &= \lim_{b\to-\infty} \int_{b}^{0} 3(x-1)^{-4}\,dx \qquad u = x-1,\ du = dx \\
&= \lim_{b\to-\infty} \left[-(x-1)^{-3}\right]\Big|_b^0 \\
&= \lim_{b\to-\infty} \left(-\frac{1}{(x-1)^3}\right)\Big|_b^0 \\
&= \lim_{b\to-\infty} \left(1+\frac{1}{(b-1)^3}\right) = 1+0 = 1
\end{aligned}$$

5.
$$\begin{aligned}
\int_{2}^{\infty} \frac{x}{(x^2+5)^2}\,dx &= \lim_{b\to\infty} \int_{2}^{b} (x^2+5)^{-2}x\,dx \\
&= \lim_{b\to\infty} \frac{1}{2}\int_{2}^{b} (x^2+4)^{-2}(2x\,dx) \qquad u = x^2+5,\ du = 2x\,dx \\
&= \lim_{b\to\infty} \frac{1}{2}\frac{(x^2+5)^{-1}}{-1}\Big|_2^b \\
&= \lim_{b\to\infty} \left(-\frac{1}{2(x^2+5)}\right)\Big|_2^b \\
&= \lim_{b\to\infty} \left[-\frac{1}{2(b^2+5)}+\frac{1}{18}\right] = \frac{1}{18}
\end{aligned}$$

7.
$$\begin{aligned}
\int_{-\infty}^{0} \frac{dt}{(2t-5)^3} &= \lim_{b\to-\infty} \int_{b}^{0} (2t-5)^{-3}\,dt \\
&= \lim_{b\to-\infty} \frac{1}{2}\int_{b}^{0} (2t-5)^{-3}(2\,dt) \qquad u = 2t-5,\ du = 2\,dt \\
&= \lim_{b\to-\infty} \frac{1}{2}\frac{(2t-5)^{-2}}{-2}\Big|_b^0 \\
&= \lim_{b\to-\infty} -\frac{1}{4(2t-5)^2}\Big|_b^0 \\
&= \lim_{b\to-\infty} \left(-\frac{1}{100}+\frac{1}{4(2b-5)^2}\right) = -\frac{1}{100}
\end{aligned}$$

9.

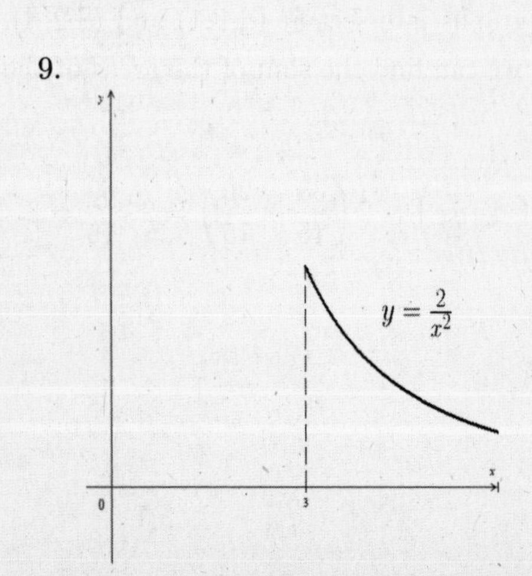

$$\begin{aligned}
\int_{3}^{\infty} \frac{2}{x^2}\,dx &= \lim_{b\to\infty} \int_{3}^{b} 2x^{-2}\,dx \\
&= \lim_{b\to\infty} 2\frac{x^{-1}}{-1}\Big|_3^b = \lim_{b\to\infty}\left(-\frac{2}{x}\right)\Big|_3^b \\
&= \lim_{b\to\infty} \left(-\frac{2}{b}+\frac{2}{3}\right) = \frac{2}{3}
\end{aligned}$$

11.

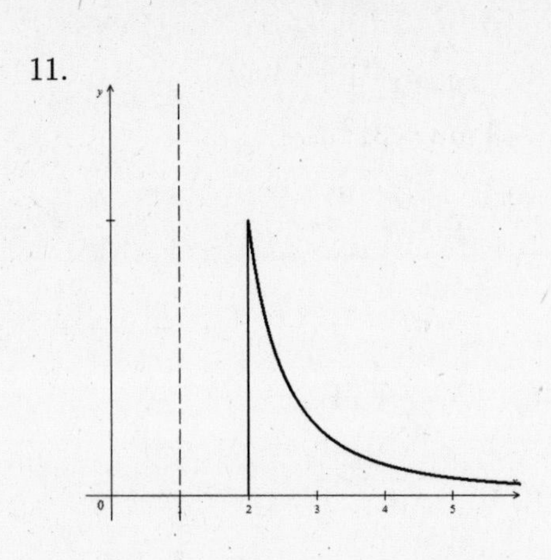

$$\begin{aligned}\int_2^\infty \frac{1}{(x-1)^2}\,dx &= \lim_{b\to\infty}\int_2^b (x-1)^{-2}\,dx \\ &= \lim_{b\to\infty} \left.\frac{-1}{x-1}\right|_2^b \\ &= \lim_{b\to\infty}\left(\frac{-1}{b-1}+1\right) \\ &= 0+1=1\end{aligned}$$

$u = x-1$
$du = dx$

13.

$y = \frac{2}{x^3}$

$$\begin{aligned}\int_1^\infty \frac{2}{x^3}\,dx &= \lim_{b\to\infty}\int_1^b 2x^{-3}\,dx \\ &= \lim_{b\to\infty}\left.\left(2\frac{x^{-2}}{-2}\right)\right|_1^b \\ &= \lim_{b\to\infty}\left.\left(-\frac{1}{x^2}\right)\right|_1^b \\ &= \lim_{b\to\infty}\left(-\frac{1}{b^2}+1\right)=1\end{aligned}$$

15.

$$\begin{aligned}\int_0^\infty \frac{1}{(x+1)^{3/2}}\,dx &= \lim_{b\to\infty}\int_0^b (x+1)^{-3/2}\,dx \\ &= \lim_{b\to\infty}\left.\frac{(x+1)^{-1/2}}{-1/2}\right|_0^b \\ &= \lim_{b\to\infty}\left.\left(\frac{-2}{(x+1)^{1/2}}\right)\right|_0^b \\ &= \lim_{b\to\infty}\left(\frac{-2}{(b+1)^{1/2}}+\frac{2}{1}\right) \\ &= 0+2=2\end{aligned}$$

$u = x+1$
$du = dx$

17.

$$\begin{aligned}A &= \int_{-\infty}^0 \left[0-\frac{1}{(2x-3)^3}\right]dx \\ &= \lim_{b\to-\infty}\int_b^0 [-(2x-3)^{-3}]\,dx \\ &= \lim_{b\to-\infty}\frac{1}{2}\int_b^0 [-(2x-3)^{-3}]2\,dx \\ &= \lim_{b\to-\infty}\frac{1}{2}\left.\left(-\frac{(2x-3)^{-2}}{-2}\right)\right|_b^0 \\ &= \lim_{b\to-\infty}\left.\frac{1}{4}\frac{1}{(2x-3)^2}\right|_b^0 \\ &= \lim_{b\to-\infty}\left(\frac{1}{4}\cdot\frac{1}{9}-\frac{1}{4}\cdot\frac{1}{(2b-3)^2}\right)=\frac{1}{36}\end{aligned}$$

$u = 2x-3$
$du = 2\,dx$

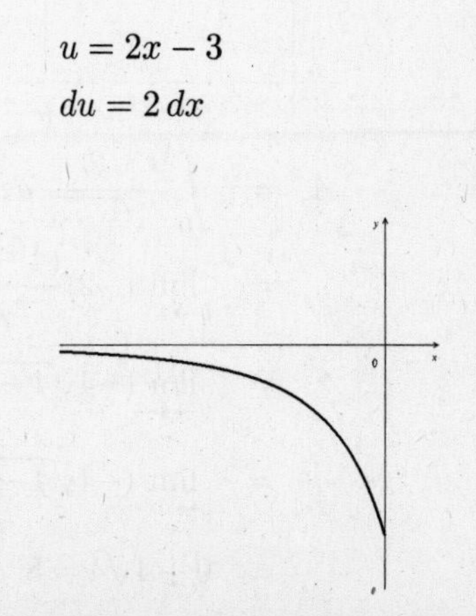

19.
$$\begin{aligned}
\int_{-\infty}^{-2} \frac{3x^2\,dx}{(x^3+1)^2} &= \lim_{b\to-\infty} \int_b^{-2} (x^3+1)^{-2}(3x^2)\,dx \\
&= \lim_{b\to-\infty} \left.\frac{-1}{x^3+1}\right|_b^{-2} \\
&= \lim_{b\to-\infty} \left(\frac{-1}{-8+1} + \frac{1}{b^3+1}\right) \\
&= \frac{1}{7} + 0 = \frac{1}{7}
\end{aligned}$$

$u = x^3 + 1$
$du = 3x^2\,dx$

21.

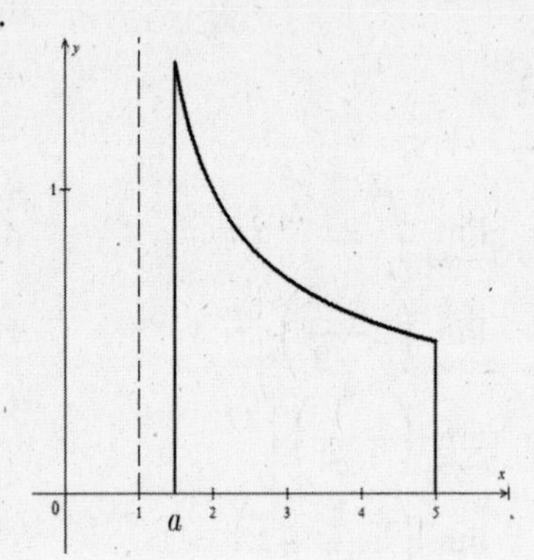

$$\begin{aligned}
\int_1^5 \frac{dx}{\sqrt{x-1}} &= \lim_{a\to1^+} \int_a^5 (x-1)^{-1/2}\,dx \\
&= \lim_{a\to1^+} \left. 2(x-1)^{1/2}\right|_a^5 \\
&= \lim_{a\to1^+} (4 - 2\sqrt{a-1}) = 4
\end{aligned}$$

$u = x - 1;\ du = dx$

23.

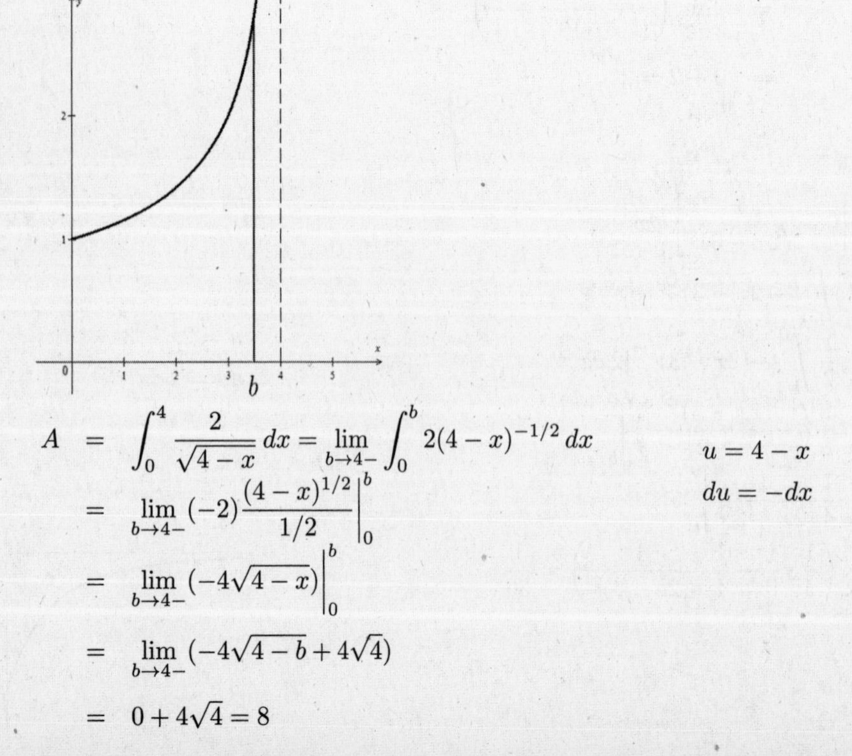

$$\begin{aligned}
A &= \int_0^4 \frac{2}{\sqrt{4-x}}\,dx = \lim_{b\to4-} \int_0^b 2(4-x)^{-1/2}\,dx \\
&= \lim_{b\to4-} (-2)\left.\frac{(4-x)^{1/2}}{1/2}\right|_0^b \\
&= \lim_{b\to4-} \left.(-4\sqrt{4-x})\right|_0^b \\
&= \lim_{b\to4-} (-4\sqrt{4-b} + 4\sqrt{4}) \\
&= 0 + 4\sqrt{4} = 8
\end{aligned}$$

$u = 4 - x$
$du = -dx$

25.

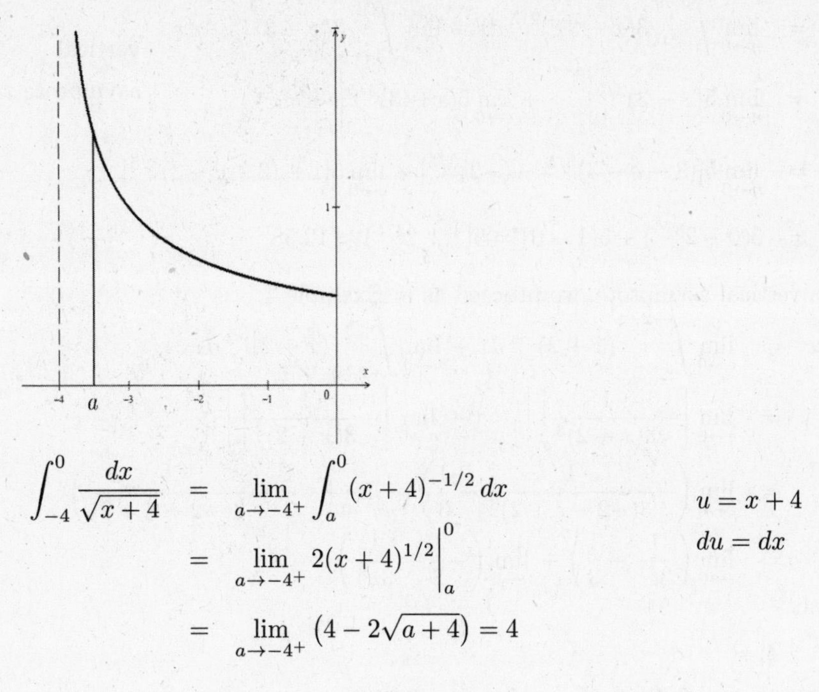

$$\begin{aligned}\int_{-4}^{0}\frac{dx}{\sqrt{x+4}} &= \lim_{a\to-4^+}\int_a^0 (x+4)^{-1/2}\,dx \qquad u=x+4 \\ &= \lim_{a\to-4^+} 2(x+4)^{1/2}\Big|_a^0 \qquad du=dx \\ &= \lim_{a\to-4^+}\left(4-2\sqrt{a+4}\right)=4\end{aligned}$$

27. $\displaystyle\int_1^\infty \frac{1}{\sqrt{x}}\,dx = \lim_{b\to\infty}\int_1^b x^{-1/2}\,dx = \lim_{b\to\infty} 2x^{1/2}\Big|_1^b = \lim_{b\to\infty}(2\sqrt{b}-2)=\infty$

29. Since the y-axis is a vertical asymptote, we need to split the integral as follows:

$$\begin{aligned}\int_{-1}^{2}\frac{dx}{\sqrt[3]{x}} &= \lim_{\eta\to0-}\int_{-1}^{\eta} x^{-1/3}\,dx + \lim_{\epsilon\to0+}\int_{\epsilon}^{2} x^{-1/3}\,dx \\ &= \lim_{\eta\to0-}\frac{3}{2}x^{2/3}\Big|_{-1}^{\eta} + \lim_{\epsilon\to0+}\frac{3}{2}x^{2/3}\Big|_{\epsilon}^{2} \\ &= \lim_{\eta\to0-}\frac{3}{2}[\eta^{2/3}-(-1)^{2/3}] + \lim_{\epsilon\to0+}\frac{3}{2}(2^{2/3}-\epsilon^{2/3}) \\ &= \frac{3}{2}(0-1)+\frac{3}{2}(2^{2/3}-0) \\ &= \frac{3}{2}(2^{2/3}-1)\approx 0.881\end{aligned}$$

31. Since $x=3$ is a vertical asymptote, we proceed as in Example 4:

$$\begin{aligned}\int_0^4 \frac{x}{(9-x^2)^2}\,dx &= \int_0^3 \frac{x}{(9-x^2)^2}\,dx + \int_3^4 \frac{x}{(9-x^2)^2}\,dx \\ &= \lim_{\epsilon\to0}\int_0^{3-\epsilon}(9-x^2)^{-2}x\,dx + \lim_{\eta\to0}\int_{3+\eta}^{4}(9-x^2)^{-2}x\,dx.\end{aligned}$$

Let $u=9-x^2$; then $du=-2x\,dx$:

$$\begin{aligned}&= \lim_{\epsilon\to0}\left(-\frac{1}{2}\frac{(9-x^2)^{-1}}{-1}\right)\Big|_0^{3-\epsilon} + \lim_{\eta\to0}\left(-\frac{1}{2}\frac{(9-x^2)^{-1}}{-1}\right)\Big|_{3+\eta}^{4} \\ &= \lim_{\epsilon\to0}\frac{1}{2(9-x^2)}\Big|_0^{3-\epsilon} + \lim_{\eta\to0}\frac{1}{2(9-x^2)}\Big|_{3+\eta}^{4} \\ &= \lim_{\epsilon\to0}\left\{\frac{1}{2[9-(3-\epsilon)^2]}-\frac{1}{18}\right\} + \lim_{\eta\to0}\left\{\frac{1}{2(-7)}-\frac{1}{2[9-(3+\eta)^2]}\right\}.\end{aligned}$$

Neither limit exists.

33.
$$\begin{aligned}
\int_1^4 \frac{3\,dx}{(x-3)^{2/5}} &= \lim_{\eta\to 0}\int_1^{3-\eta} 3(x-3)^{-2/5}\,dx + \lim_{\epsilon\to 0}\int_{3+\epsilon}^{4} 3(x-3)^{-2/5}\,dx \\
&= \lim_{\eta\to 0} 5(x-3)^{3/5}\Big|_1^{3-\eta} + \lim_{\epsilon\to 0} 5(x-3)^{3/5}\Big|_{3+\epsilon}^{4} \\
&= \lim_{\eta\to 0} 5[(3-\eta-3)^{3/5} - (-2)^{3/5}] + \lim_{\epsilon\to 0} 5[1-(3+\epsilon-3)^{3/5}] \\
&= 5(0+2^{3/5}) + 5(1-0) = 5(1+2^{3/5}) \approx 12.58
\end{aligned}$$

vertical asymptote $x = 3$

35. Since $x = -2$ is a vertical asymptote, we proceed as in Example 4:
$$\begin{aligned}
\int_{-3}^{-1} \frac{1}{(x+2)^4}\,dx &= \lim_{\epsilon\to 0}\int_{-3}^{-2-\epsilon} (x+2)^{-4}\,dx + \lim_{\eta\to 0}\int_{-2+\eta}^{-1} (x+2)^{-4}\,dx \\
&= \lim_{\epsilon\to 0}\left[-\frac{1}{3(x+2)^3}\right]\Bigg|_{-3}^{-2-\epsilon} + \lim_{\eta\to 0}\left[-\frac{1}{3(x+2)^3}\right]\Bigg|_{-2+\eta}^{-1} \\
&= \lim_{\epsilon\to 0}\left(-\frac{1}{3(-2-\epsilon+2)^3} - \frac{1}{3}\right) + \lim_{\eta\to 0}\left(-\frac{1}{3} + \frac{1}{3(-2+\eta+2)}\right) \\
&= \lim_{\epsilon\to 0}\left(\frac{1}{3\epsilon^3} - \frac{1}{3}\right) + \lim_{\eta\to 0}\left(-\frac{1}{3} + \frac{1}{3\eta^3}\right).
\end{aligned}$$
Neither limit exists.

4.8 The Constant of Integration

1. From $\frac{dy}{dx} = 3x$, we get $y = \frac{3}{2}x^2 + C$. Substituting $(0,1)$ in the equation, we get $1 = 0 + C$ or $C = 1$. The resulting function is
$$y = \frac{3}{2}x^2 + 1.$$

3. From $\frac{dy}{dx} = 6x^2 + 1$, we get $y = 2x^3 + x + C$. Substituting $(-1,1)$ in the equation, we get $1 = 2(-1)^3 + (-1) + C$ or $C = 4$. The resulting function is
$$y = 2x^3 + x + 4.$$

5. From $\frac{dy}{dx} = 3x^2 + 2$, we get $y = x^3 + 2x + C$. Substituting $(1,0)$ in the equation, we get $0 = 3 + C$, or $C = -3$. It follows that
$$y = x^3 + 2x - 3.$$

7.

+

$g = -10\,\mathrm{m/s}^2$

$s = 0$ $\quad v_0 = +15\,\mathrm{m/s}$

Taking the upward direction as positive, we get $g = -10\,\mathrm{m/s}^2$ and $v_0 = +15\,\mathrm{m/s}$, since the stone is moving in the upward direction initially. Integrating, we have $v = -10t + C$. If $t = 0$, $v = +15$, so that $15 = 0 + C$.

Thus
$$v = -10t + 15.$$

To see how long the stone takes to reach its highest point, we set v equal to 0:
$$0 = -10t + 15 \quad\text{and}\quad t = 1.5\,\mathrm{s}.$$

So the entire trip takes $2(1.5) = 3\,\mathrm{s}$.

9.

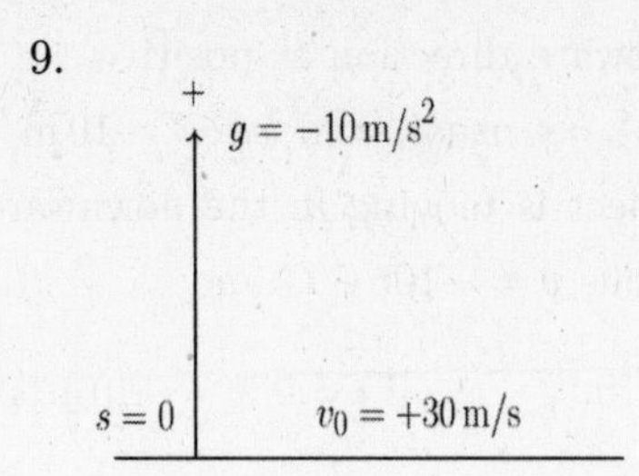

Taking the upward direction as positive (as was done in the examples), we get $g = -10\,\text{m/s}^2$ and $v_0 = +30\,\text{m/s}$, since the ball is moving in the positive direction initially. Integrating, we have

$$v = -10t + C.$$

If $t = 0$, $v = +30$, so that $30 = 0 + C$.

Thus

$$v = -10t + 30.$$

To find s, we integrate v:

$$s = -5t^2 + 30t + k.$$

If the origin is on the ground, as in the figure, $s = 0$ when $t = 0$. Thus $0 = 0 + 0 + k$ and

$$s = -5t^2 + 30t.$$

To find how high the ball rises, we first determine how long it takes to reach the highest point by setting v to 0:

$$v = -10t + 30 = 0, \text{ whence } t = 3.$$

Thus

$$s = -5t^2 + 30t\Big|_{t=3} = -45 + 90 = 45.$$

In the coordinate system chosen, 45 corresponds to 45 m above the ground.

11.

+ $g = -10\,\text{m/s}^2$

$s = 125\,\text{m}$

$v_0 = 0$

$s = 0$

Taking the upward direction to be positive, $g = -10\,\text{m/s}^2$. Since the initial velocity is 0,

$$v = -10t. \qquad v = 0 \text{ when } t = 0$$

Integrating v, we have

$$s = -5t^2 + k.$$

Since the origin ($s = 0$) is on the ground, $s = 125\,\text{m}$ initially (when $t = 0$). So

$$125 = 0 + k$$

and

$$s = -5t^2 + 125. \qquad k = 125$$

To see how long the object takes to reach the ground, we let $s = 0$:

$$\begin{aligned} 0 &= -5t^2 + 125 \\ t^2 &= 25 \text{ and } t = 5\,\text{s}. \end{aligned}$$

13.

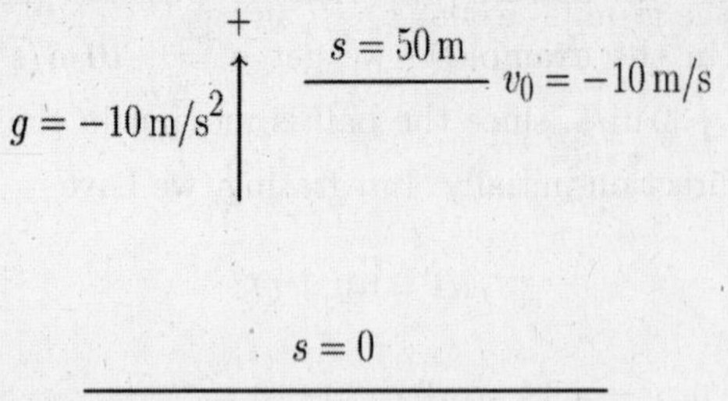

$+$ $s = 50\,\text{m}$ $v_0 = -10\,\text{m/s}$

$g = -10\,\text{m/s}^2$

$s = 0$

Taking the upward direction as positive, $g = -10\,\text{m/s}^2$, as usual, and $v_0 = -10\,\text{m/s}$ (since the object is moving in the downward direction). Thus $v = -10t + C$ and

$$v = -10t - 10. \qquad \text{if } t = 0,\ v = -10\,\text{m/s}$$

Hence

$$s = -5t^2 - 10t + k. \qquad \text{integrating } v$$

Now observe that when $t = 0$, $s = 50\,\text{m}$. So

$$50 = 0 + 0 + k$$

and

$$s = -5t^2 - 10t + 50.$$

To see how long it takes the object to reach the ground, we let $s = 0$ and solve for t:

$$\begin{aligned} 0 &= -5t^2 - 10t + 50 \\ t^2 + 2t - 10 &= 0 \qquad \text{dividing by } -5 \\ t &= \frac{-2 \pm \sqrt{4+40}}{2} = \frac{-2 \pm 2\sqrt{11}}{2} = -1 \pm \sqrt{11}. \end{aligned}$$

Taking the positive root, $t = -1 + \sqrt{11} \approx 2.3\,\text{s}$. Finally, when $t = 2.3$,

$$v = -10(2.3) - 10 = -33.$$

We conclude that the velocity is $33\,\text{m/s}$ in the downward direction.

15.

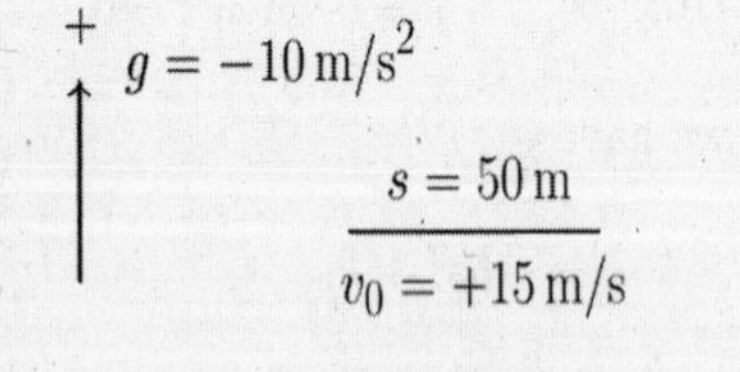

$+$ $g = -10\,\text{m/s}^2$

$s = 50\,\text{m}$ $v_0 = +15\,\text{m/s}$

$s = 0$

Taking the upward direction to be positive, $g = -10\,\text{m/s}^2$ and $v_0 = +15\,\text{m/s}$ (upward direction).

$$\begin{aligned} v &= -10t + C \\ +15 &= 0 + C \qquad \text{when } t = 0,\ v = +15\,\text{m/s} \\ v &= -10t + 15. \end{aligned}$$

So

$$s = -5t^2 + 15t + k.$$

When $t = 0$, $s = 50\,\text{m}$:

$$50 = 0 + k$$

and

$$s = -5t^2 + 15t + 50.$$

To see how long it takes for the rock to reach the ground, we let $s = 0$ and solve for t:

$$\begin{aligned} 0 &= -5t^2 + 15t + 50 \\ t^2 - 3t - 10 &= 0 \\ (t-5)(t+2) &= 0 \\ t &= 5\,\text{s}. \qquad \text{(positive root)} \end{aligned}$$

Finally, when $t = 5$, $v = -10(5) + 15 = -35\,\text{m/s}$. We conclude that the velocity is $35\,\text{m/s}$ in the downward direction.

17. Taking the upward direction as positive, $g = -10\,\text{m/s}^2$ and $v_0 = +10\,\text{m/s}$ (since the object is moving in the upward direction). Thus $v = -10t + C$, and

$$v = -10t + 10. \qquad \text{if } t = 0,\ v = +10\,\text{m/s}$$

So

$$s = -5t^2 + 10t + k \qquad \text{integrating } v$$

and

$$s = -5t^2 + 10t + 40. \qquad \text{if } t = 0,\ s = 40\,\text{m}$$

To see how long the object travels before striking the ground, we let $s = 0$:

$$\begin{aligned} 0 &= -5t^2 + 10t + 40 \\ t^2 - 2t - 8 &= 0 \qquad \text{dividing by -5} \\ (t-4)(t+2) &= 0 \\ t &= +4. \qquad \text{positive root} \end{aligned}$$

From $v = -10t + 10$, we get $v = -30\,\text{m/s}$, that is, the object is moving in the downward direction.

19. The simplest way to solve this problem is to take the bottom of the pit to be the origin ($s = 0$). However, to be consistent with our general method, let $s = 0$ at ground level, so that $s = -20\,\text{m}$ initially, with $v_0 = +25\,\text{m/s}$.

$$\begin{aligned} v &= -10t + 25 \\ s &= -5t^2 + 25t + k \\ -20 &= 0 + k \end{aligned}$$

and

$$s = -5t^2 + 25t - 20.$$

To find how long it takes for the stone to reach the top of the pit (ground level), we let $s = 0$ and solve for t:

$$\begin{aligned} 0 &= -5t^2 + 25t - 20 \\ t^2 - 5t + 4 &= 0 \\ (t-1)(t-4) &= 0 \\ t &= 1, 4. \end{aligned}$$

Answer: $t = 1\,\mathrm{s}$ to reach the top. (The stone actually overshoots and reaches the top of the pit again after $4\,\mathrm{s}$ on the way down.)

21. If we let the direction of motion be the positive direction, then $a = -7$ and $v_0 = 28$. For convenience, we let $s = 0$ be the point at which the car starts to decelerate. We now get

$$\begin{aligned} v &= -7t + 28, \\ s &= -\frac{7}{2}t^2 + 28t + 0. \qquad k = 0 \text{ (since } s = 0 \text{ when } t = 0) \end{aligned}$$

If we let $v = 0$, we find that the car stops in four seconds. From

$$s = -\frac{7}{2}t^2 + 28t\Big|_{t=4} = 56$$

we see that the car stops in $56\,\mathrm{m}$.

23. The easiest way to solve this problem is to assume that the keys are dropped from rest at $s = 15\,\mathrm{m}$ and determine the speed when it hits the ground. Alternatively, let v_0 be the unknown initial velocity. Then

$$\begin{aligned} v &= -10t + v_0 \\ s &= -5t^2 + v_0 t + k \end{aligned}$$

and

$$(1)\quad s = -5t^2 + v_0 t. \qquad s = 0 \text{ when } t = 0$$

By assumption, the object reaches the dorm window when $v = 0$ and $s = 15$. If $v = 0$,

$$0 = -10t + v_0 \quad \text{and} \quad t = \frac{v_0}{10}.$$

Substituting in Equation (1),

$$\begin{aligned} 15 &= -5\left(\frac{v_0}{10}\right)^2 + v_0\left(\frac{v_0}{10}\right) \\ 15 &= \left(-\frac{5}{10^2} + \frac{1}{10}\right)v_0^2 \\ v_0^2 &= 3 \cdot 10^2 \\ v_0 &= 10\sqrt{3}\,\mathrm{m/s}. \end{aligned}$$

25.
$$\begin{aligned} a &= \frac{t}{(t^2+1)^2} \\ v &= \int \frac{t}{(t^2+1)^2}\,dt = \int (t^2+1)^{-2} t\,dt \qquad u = t^2+1;\ du = 2t\,dt \\ &= \frac{1}{2}\int (t^2+1)^{-2}(2t\,dt) \\ &= \frac{1}{2}\frac{(t^2+1)^{-1}}{-1} + C = -\frac{1}{2(t^2+1)} + C \end{aligned}$$

Since $v = 10$ when $t = 0$, we get from

$$\begin{aligned} v &= -\frac{1}{2(t^2+1)} + C, \\ 10 &= -\frac{1}{2(0+1)} + C \quad \text{or } C = \frac{21}{2}. \end{aligned}$$

So

$$\begin{aligned} v &= -\frac{1}{2(t^2+1)} + \frac{21}{2} \\ &= \frac{1}{2}\left(21 - \frac{1}{t^2+1}\right). \end{aligned}$$

27. Since $i = \frac{dq}{dt}$, $q = \int i\,dt = \int \left(3t^2+2t+1\right)\,dt = t^3+t^2+t+C$. To compute C, observe that $q = 4.5\,\mathrm{C}$ when $t = 0$ (initial charge). So $4.5 = 0 + C$ and $C = 4.5\,\mathrm{C}$.

(a) $q(t) = t^3 + t^2 + t + 4.5$

(b) $q(4.00) = 88.5\,\mathrm{C}$

29. Given: $i = 0.010t + 0.020$ and $q = 0.040\,\mathrm{C}$ when $t = 0$.
$q = \int i\,dt = \int (0.010t + 0.020)\,dt = 0.010\left(\frac{t^2}{2}\right) + 0.020t + C$
Substituting $t = 0$ and $q = 0.040$, we get

$$0.040 = 0 + C \quad \text{or} \quad C = 0.040,$$

so that

$$q(t) = 0.010\left(\frac{t^2}{2}\right) + 0.020t + 0.040 \quad \text{and} \quad q(3.0) = 0.145\,\mathrm{C}.$$

31. From $v = L\frac{di}{dt}$, we have

$$i = \frac{1}{L}\int v\,dt = \frac{1}{7.5}\int \sqrt{t+9.0}\,dt = \frac{1}{7.5}\left(\frac{2}{3}\right)(t+9.0)^{3/2} + C.$$

We are given that $i = 1.5\,\mathrm{A}$ when $t = 0$:

$$1.5 = \frac{1}{7.5}\left(\frac{2}{3}\right)(9.0)^{3/2} + C,$$

which yields $C = -0.9$. We now get

$$i = \frac{1}{7.5}\left(\frac{2}{3}\right)(t+9.0)^{3/2} - 0.9\Big|_{t=4.0} = 3.3\,\mathrm{A}.$$

33. $v = \frac{q}{C}$ and $q = \int i\,dt$; so

$$q = \int 0.030t^{1/3}\,dt = 0.030\left(\frac{3}{4}\right)t^{4/3} + k \quad \text{and} \quad v = \frac{0.030\left(\frac{3}{4}\right)t^{4/3} + k}{C}.$$

Given: when $t=0$, $v=150$ and when $t=1, v=300$. Substituting,

$$150=\frac{0+k}{C} \quad\text{and}\quad 300=\frac{0.030\left(\frac{3}{4}\right)+k}{C}.$$

From the first equation, $k=150C$; the second equation now yields

$$300=\frac{0.030\left(\frac{3}{4}\right)+150C}{C}.$$

Solving, $C=1.5\times10^{-4}\,\text{F}$.

35. Integrate the given expression:
$$\begin{aligned} R &= 0.00060\frac{T^3}{3}+0.0080\frac{T^2}{2}+0.14T+C \\ R &= 0.00020T^3+0.0040T^2+0.14T+C. \end{aligned}$$
Let $R=50$ and $T=0$:
$$50=0+C \quad\text{or}\quad C=50.$$
$$R=0.00020T^3+0.0040T^2+0.14T+50\Big|_{T=20}=56\,\Omega$$

37.
$$\begin{aligned} P &= 3(t+1)^{1/2} \\ W &= \int P\,dt=\int 3(t+1)^{1/2}\,dt=2(t+1)^{3/2}+C \\ 5 &= 2+C \text{ and } C=3 \qquad\qquad W=5\text{ when }t=0 \\ W &= 2(t+1)^{3/2}+3. \end{aligned}$$
When $t=8\,\text{s}$, $W=2(8+1)^{3/2}+3=57\,\text{J}$.

4.9 Numerical Integration

1. If $n=6$, we have $h=\dfrac{b-a}{6}=\dfrac{1}{3}$. Thus $x_0=1,\ x_1=\dfrac{4}{3},\ x_2=\dfrac{5}{3},\ x_3=\dfrac{6}{3},\ x_4=\dfrac{7}{3}$, $x_5=\dfrac{8}{3},\ x_6=\dfrac{9}{3}=3$.
By the trapezoidal rule:
$$\int_1^3 x^2\,dx\approx\frac{1}{3}\left[\frac{1}{2}(1)^2+\left(\frac{4}{3}\right)^2+\left(\frac{5}{3}\right)^2+\left(\frac{6}{3}\right)^2+\left(\frac{7}{3}\right)^2+\left(\frac{8}{3}\right)^2+\frac{1}{2}\left(\frac{9}{3}\right)^2\right]=8.704.$$
By Simpson's rule:
$$\int_1^3 x^2\,dx\approx\frac{1/3}{3}\left[1^2+4\left(\frac{4}{3}\right)^2+2\left(\frac{5}{3}\right)^2+4\left(\frac{6}{3}\right)^2+2\left(\frac{7}{3}\right)^2+4\left(\frac{8}{3}\right)^2+\left(\frac{9}{3}\right)^2\right]=8.667.$$
By direct integration:
$$\int_1^3 x^2\,dx=\frac{1}{3}x^3\Big|_1^3=\frac{26}{3}.$$

3. Since $n=4$, we have $h=\dfrac{b-a}{4}=\dfrac{1}{4}$. Thus $x_0=0,\ x_1=\dfrac{1}{4},\ x_2=\dfrac{1}{2},\ x_3=\dfrac{3}{4},\ x_4=1$.
By the trapezoidal rule:
$$\int_0^1\frac{1}{1+x^2}\,dx\approx\frac{1}{4}\left[\frac{1}{2}\frac{1}{1+0^2}+\frac{1}{1+(1/4)^2}+\frac{1}{1+(1/2)^2}+\frac{1}{1+(3/4)^2}+\frac{1}{2}\frac{1}{1+1^2}\right]=0.783.$$
By Simpson's rule:
$$\int_0^1\frac{1}{1+x^2}\,dx\approx\frac{1/4}{3}\left[\frac{1}{1+0^2}+4\frac{1}{1+(1/4)^2}+2\frac{1}{1+(1/2)^2}+4\frac{1}{1+(3/4)^2}+\frac{1}{1+1^2}\right]=0.785.$$

5. If $n = 4$, we have $h = \dfrac{b-a}{4} = \dfrac{1}{2}$. Thus $x_0 = 0,\ x_1 = \dfrac{1}{2},\ x_2 = 1,\ x_3 = \dfrac{3}{2},\ x_4 = 2$.
By the trapezoidal rule:

$$\int_0^2 \sqrt{1+x}\,dx \approx \frac{1}{2}\left[\frac{1}{2}\sqrt{1} + \sqrt{\frac{3}{2}} + \sqrt{2} + \sqrt{\frac{5}{2}} + \frac{1}{2}\sqrt{3}\right] = 2.793.$$

By Simpson's rule:

$$\begin{aligned}\int_0^2 \sqrt{1+x}\,dx &\approx \frac{1/2}{3}\left[\sqrt{1} + 4\sqrt{\frac{3}{2}} + 2\sqrt{2} + 4\sqrt{\frac{5}{2}} + \sqrt{3}\right] \\ &= \frac{1}{6}\left[1 + 4\sqrt{1.5} + 2\sqrt{2} + 4\sqrt{2.5} + \sqrt{3}\right] = 2.797\end{aligned}$$

7. Since $n = 5$, $h = \dfrac{3-0}{5} = \dfrac{3}{5}$.
Thus $x_0 = 0,\ x_1 = \dfrac{3}{5},\ x_2 = \dfrac{6}{5},\ x_3 = \dfrac{9}{5},\ x_4 = \dfrac{12}{5},\ x_5 = \dfrac{15}{5} = 3$.

$$\int_0^3 \sqrt{x}\,dx \approx \frac{3}{5}\left(0 + \sqrt{\frac{3}{5}} + \sqrt{\frac{6}{5}} + \sqrt{\frac{9}{5}} + \sqrt{\frac{12}{5}} + \frac{1}{2}\sqrt{3}\right) = 3.376.$$

9. $\displaystyle\int_{-1}^{2} \frac{dx}{x^3+2}$ $\qquad h = \dfrac{2-(-1)}{12} = 0.25$
$x_0 = -1,\ x_1 = -1 + 0.25 = -0.75,\ x_2 = -0.75 + 0.25 = -0.5,\ x_3 = -0.5 + 0.25 = -0.25$, etc.
The function values are listed next:

$$\begin{aligned}
f(x_0) &= 1/(-1+2) = 1 \\
f(x_1) &= 1/[(-0.75)^3 + 2] = 0.63366 \\
f(x_2) &= 1/[(-0.5)^3 + 2] = 0.53333 \\
f(x_3) &= 1/[(-0.25)^3 + 2] = 0.50394 \\
f(x_4) &= 1/(0+2) = 0.5 \\
f(x_5) &= 1/(0.25^3 + 2) = 0.49612 \\
f(x_6) &= 1/(0.5^3 + 2) = 0.47059 \\
f(x_7) &= 1/(0.75^3 + 2) = 0.41290 \\
f(x_8) &= 1/(1+2) = 0.33333 \\
f(x_9) &= 1/(1.25^3 + 2) = 0.25296 \\
f(x_{10}) &= 1/(1.5^3 + 2) = 0.18605 \\
f(x_{11}) &= 1/(1.75^3 + 2) = 0.13588 \\
f(x_{12}) &= 1/(2^3 + 2) = 0.1
\end{aligned}$$

By the trapezoidal rule:

$$\int_{-1}^{2} \frac{dx}{x^3+2} \approx 0.25\left[\frac{1}{2}(1) + 0.63366 + 0.53333 + \ \dots\ + \frac{1}{2}(0.1)\right] = 1.252.$$

11. Since $n = 8$, $h = \dfrac{6-0}{8} = \dfrac{3}{4}$. Thus $x_0 = 0,\ x_1 = \dfrac{3}{4},\ x_2 = \dfrac{6}{4} = \dfrac{3}{2},\ x_3 = \dfrac{9}{4},\ x_4 = \dfrac{12}{4} = 3,$
$x_5 = \dfrac{15}{4},\ x_6 = \dfrac{18}{4} = \dfrac{9}{2},\ x_7 = \dfrac{21}{4},\ x_8 = \dfrac{24}{4} = 6.$
The function values are listed next:

$$f(x) = \frac{\sqrt{x}+3}{x+7}$$

$$\begin{aligned}
f(x_0) &= \frac{3}{7} = 0.4286 \\
f(x_1) &= \frac{\sqrt{3/4}+3}{3/4+7} = 0.4988 \\
f(x_2) &= \frac{\sqrt{3/2}+3}{3/2+7} = 0.4970 \\
f(x_3) &= \frac{\sqrt{9/4}+3}{9/4+7} = 0.4865 \\
f(x_4) &= \frac{\sqrt{3}+3}{3+7} = 0.4732 \\
f(x_5) &= \frac{\sqrt{15/4}+3}{15/4+7} = 0.4592 \\
f(x_6) &= \frac{\sqrt{9/2}+3}{9/2+7} = 0.4453 \\
f(x_7) &= \frac{\sqrt{21/4}+3}{21/4+7} = 0.4319 \\
f(x_8) &= \frac{\sqrt{6}+3}{6+7} = 0.4192.
\end{aligned}$$

$$\int_0^6 \frac{\sqrt{x}+3}{x+7}\,dx \approx \frac{3}{4}\left[\frac{1}{2}(0.4286) + 0.4988 + 0.4970 + \cdots + \frac{1}{2}(0.4192)\right] = 2.787.$$

13. Since $n = 6$, we get $h = \frac{b-a}{6} = \frac{1}{2}$. Thus $x_0 = 1,\ x_1 = \frac{3}{2},\ x_2 = 2,\ x_3 = \frac{5}{2}$, $x_4 = 3,\ x_5 = \frac{7}{2},\ x_6 = 4$.

By Simpson's rule:

$$\begin{aligned}
\int_1^4 \sqrt{1+x^2}\,dx &\approx \frac{1/2}{3}\Big[\sqrt{1+1^2} + 4\sqrt{1+(3/2)^2} + 2\sqrt{1+2^2} + 4\sqrt{1+(5/2)^2} \\
&\quad +2\sqrt{1+3^2} + 4\sqrt{1+(7/2)^2} + \sqrt{1+4^2}\Big] = 8.146.
\end{aligned}$$

15. $h = \frac{2-0}{6} = \frac{1}{3};\ x_0 = 0,\ x_1 = \frac{1}{3},\ x_2 = \frac{2}{3},\ x_3 = 1,\ x_4 = \frac{4}{3},\ x_5 = \frac{5}{3},\ x_6 = 2.$

$$\begin{aligned}
\int_0^2 \sqrt{1+x^4}\,dx &\approx \frac{1/3}{3}\left[\sqrt{1} + 4\sqrt{1+\left(\frac{1}{3}\right)^4} + 2\sqrt{1+\left(\frac{2}{3}\right)^4}\right. \\
&\quad + \left. 4\sqrt{1+1^4} + 2\sqrt{1+\left(\frac{4}{3}\right)^4} + 4\sqrt{1+\left(\frac{5}{3}\right)^4} + \sqrt{1+2^4}\right] \\
&= 3.6535.
\end{aligned}$$

17. From the table, $h = 1$. Using the given y-values, we get, by Simpson's rule:

$$\frac{1}{3}[1.3 + 4(1.9) + 2(3.2) + 4(3.8) + 2(4.7) + 4(6.8) + 2(10.2) + 4(15.6) + 20.3] = 56.7.$$

19. Since $h = 10$, we get by the trapezoidal rule the approximate area

$$10\left[\frac{1}{2}(8.0) + 12.5 + 20.3 + 27.9 + 26.1 + 19.7 + 15.4 + \frac{1}{2}(6.6)\right] = 1292\,\text{ft}^2.$$

21. From the table, $h = 0.2$. By the trapezoidal rule, the approximate value of the integral is

$$0.2\left(\frac{1}{2}\cdot 0.10 + 0.40 + 0.46 + 0.57 + 0.64 + 0.54 + 0.43 + \frac{1}{2}\cdot 0.31\right) = 0.65\,\text{C}.$$

Chapter 4 Review

1. Subdivide the interval $[0, 3]$ into n equal parts, each of length $3/n$. Thus $\Delta x_i = 3/n$. Choosing the right endpoint of each subinterval for x, we get

$$x_1 = 1 \cdot \frac{3}{n},\ x_2 = 2 \cdot \frac{3}{n},\ \ldots,\ x_i = i \cdot \frac{3}{n},\ \ldots,\ x_n = n \cdot \frac{3}{n} = 3.$$

For the altitudes, we get from $f(x) = 3x^2$

$$f(x_1) = 3(1)^2\frac{9}{n^2},\ f(x_2) = 3(2)^2\frac{9}{n^2},\ \ldots,\ f(x_i) = 3(i)^2\frac{9}{n^2},\ \ldots,\ f(x_n) = 3(n)^2\frac{9}{n^2}.$$

Since $\Delta x_i = 3/n$, the sum of the areas is

$$\sum_{i=1}^{n} f(x_i)\Delta x_i = \sum_{i=1}^{n} 3i^2 \frac{9}{n^2}\frac{3}{n} = \frac{81}{n^3}\frac{n(n+1)(2n+1)}{6} = \frac{27n(n+1)(2n+1)}{2n^3},$$

by Formula B. Hence

$$\begin{aligned}\int_0^3 3x^2\,dx &= \lim_{n\to\infty}\frac{27n(n+1)(2n+1)}{2n^3} = \lim_{n\to\infty}\frac{54n^3+81n^2+27n}{2n^3}\\ &= \lim_{n\to\infty}\frac{54+81/n+27/n^2}{2} = 27.\end{aligned}$$

3. $$\begin{aligned}\int (3\sqrt{x} - x^{-4} + 1)\,dx &= \int (3x^{1/2} - x^{-4} + 1)\,dx\\ &= 3\frac{x^{3/2}}{3/2} - \frac{x^{-3}}{-3} + x + C\\ &= 2x^{3/2} + \frac{1}{3}x^{-3} + x + C\end{aligned}$$

Since $x^{3/2} = x^{1+1/2} = x^1x^{1/2} = x\sqrt{x}$, the answer can also be written

$$2x\sqrt{x} + \frac{1}{3}x^{-3} + x + C.$$

5. $\int (1-x^2)^5 x\,dx$. Let $u = 1 - x^2$; then $du = -2x\,dx$.

$$\begin{aligned}\int (1-x^2)^5 x\,dx &= -\frac{1}{2}\int (1-x^2)^5(-2x)\,dx = -\frac{1}{2}\int u^5\,du\\ &= -\frac{1}{2}\frac{u^6}{6} + C = -\frac{1}{12}(1-x^2)^6 + C.\end{aligned}$$

7. $\int \frac{3x\,dx}{\sqrt{x^2-2}} = 3\int (x^2-2)^{-1/2}x\,dx$. Let $u = x^2 - 2$; then $du = 2x\,dx$. Thus

$$\begin{aligned}3\int (x^2-2)^{-1/2}x\,dx &= \frac{3}{2}\int (x^2-2)^{-1/2}(2x)\,dx = \frac{3}{2}\int u^{-1/2}\,du\\ &= \frac{3}{2}\frac{u^{1/2}}{1/2} + C = 3\sqrt{x^2-2} + C.\end{aligned}$$

9. $\int (x^3+1)^2(3x)\,dx$. If $u = x^3 + 1$, then $du = 3x^2\,dx$, which is different from $3x\,dx$. Since the substitution cannot be made, we multiply out the integrand and integrate term by term:

$$\int (x^6 + 2x^3 + 1)(3x)\,dx = \int (3x^7 + 6x^4 + 3x)\,dx = \frac{3}{8}x^8 + \frac{6}{5}x^5 + \frac{3}{2}x^2 + C.$$

11. $$\begin{aligned}\int (\sqrt{x}-1)^2\,dx &= \int (x - 2x^{1/2} + 1)\,dx = \frac{1}{2}x^2 - 2\frac{x^{3/2}}{3/2} + x + C\\ &= \frac{1}{2}x^2 - \frac{4}{3}x^{1+1/2} + x + C = \frac{1}{2}x^2 - \frac{4}{3}x\sqrt{x} + x + C.\end{aligned}$$

13. $\int (x-2)\sqrt{x^2-4x}\,dx = \int (x^2-4x)^{1/2}(x-2)\,dx.$

Let $u = x^2 - 4x$; then $du = (2x-4)\,dx = 2(x-2)\,dx$.

$$\begin{aligned}\int (x^2-4x)^{1/2}(x-2)\,dx &= \frac{1}{2}\int (x^2-4x)^{1/2}2(x-2)\,dx \\ &= \frac{1}{2}\int u^{1/2}\,du = \frac{1}{2}\frac{u^{3/2}}{3/2} + C \\ &= \frac{1}{3}(x^2-4x)^{3/2} + C.\end{aligned}$$

15.

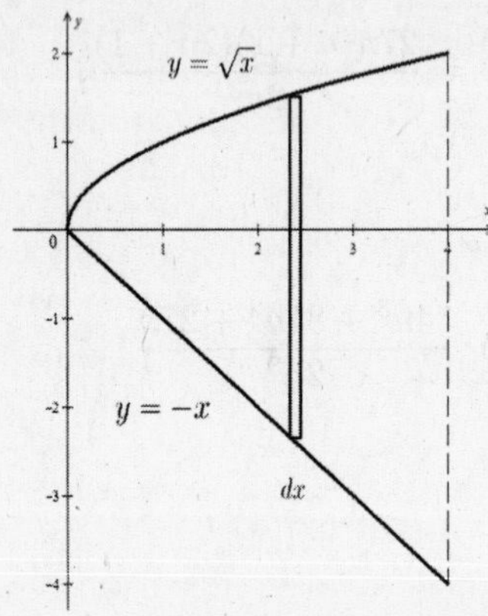

$$\begin{aligned}A &= \int_0^4 [\sqrt{x} - (-x)]\,dx = \int_0^4 (x^{1/2}+x)\,dx = \frac{2}{3}x^{3/2} + \frac{1}{2}x^2\Big|_0^4 \\ &= \frac{2}{3}(4)^{3/2} + \frac{1}{2}(4)^2 - 0 = \frac{16}{3} + 8 = \frac{16}{3} + \frac{24}{3} = \frac{40}{3}.\end{aligned}$$

17.

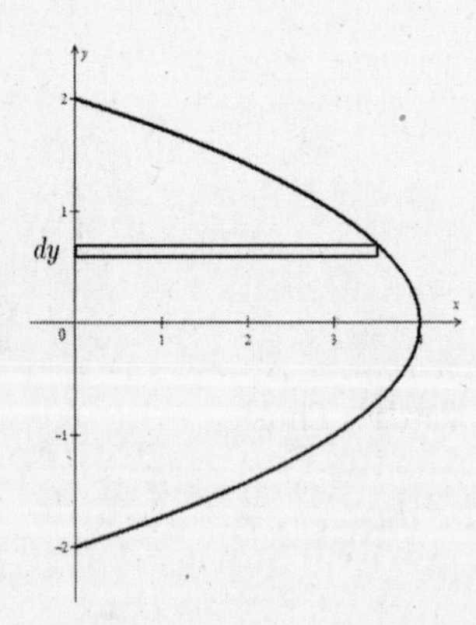

$$\begin{aligned}\int_{-2}^{2} x\,dy &= \int_{-2}^{2} (4-y^2)\,dy \\ &= 4y - \frac{y^3}{3}\Big|_{-2}^{2} \\ &= \left(8 - \frac{8}{3}\right) - \left(-8 + \frac{8}{3}\right) \\ &= 16 - \frac{16}{3} = \frac{32}{3}.\end{aligned}$$

19.
$$\begin{aligned}x &= y^2 - 4y + 4 \\ x &= 4 - y \\ \hline 0 &= y^2 - 3y \qquad \text{(subtracting)} \\ y(y-3) &= 0 \\ y &= 0, 3\end{aligned}$$

The points of intersection are $(4,0)$ and $(1,3)$.

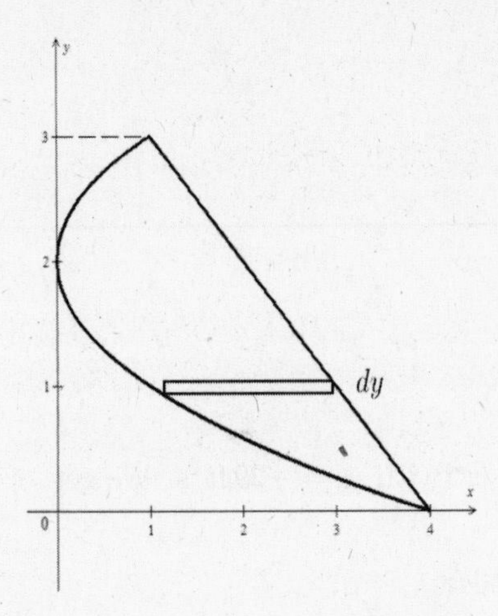

The typical element is drawn horizontally.

$$\begin{aligned} A &= \int_0^3 [(4-y)-(y^2-4y+4)]\,dy = \int_0^3 (-y^2+3y)\,dy = -\frac{1}{3}y^3+\frac{3}{2}y^2\Big|_0^3 \\ &= -\frac{1}{3}\cdot 27+\frac{3}{2}\cdot 9 = -9+\frac{27}{2} = \frac{-18}{2}+\frac{27}{2} = \frac{9}{2}. \end{aligned}$$

21. $$\begin{aligned} \int_{-\infty}^0 \frac{2x}{(x^2+4)^2}\,dx &= \lim_{b\to-\infty}\int_b^0 (x^2+4)^{-2}(2x)\,dx \\ &= \lim_{b\to-\infty}\frac{(x^2+4)^{-1}}{-1}\Big|_b^0 \qquad u = x^2+4,\quad du = 2x\,dx \\ &= \lim_{b\to-\infty}\left(-\frac{1}{x^2+4}\right)\Big|_b^0 \\ &= \lim_{b\to-\infty}\left(-\frac{1}{4}+\frac{1}{b^2+4}\right) = -\frac{1}{4}. \end{aligned}$$

23. $\int_{-1}^1 \frac{3}{4x^2}\,dx$. Since the y-axis is a vertical asymptote, we need to split the integral as follows:

$$\int_{-1}^0 \frac{3}{4x^2}\,dx + \int_0^1 \frac{3}{4x^2}\,dx.$$

Consider the second integral:

$$\int_0^1 \frac{3}{4}x^{-2}\,dx = \lim_{\epsilon\to0+}\int_\epsilon^1 \frac{3}{4}x^{-2}\,dx = \lim_{\epsilon\to0+}\left(-\frac{3}{4x}\right)\Big|_\epsilon^1 = \lim_{\epsilon\to0+}\left(-\frac{3}{4}+\frac{3}{4\epsilon}\right).$$

This limit does not exist $(+\infty)$.

25. $$\begin{aligned} \int_2^\infty \frac{1}{x^3}\,dx &= \lim_{b\to\infty}\int_2^b x^{-3}\,dx = \lim_{b\to\infty}\frac{x^{-2}}{-2}\Big|_2^b = \lim_{b\to\infty}\left(-\frac{1}{2x^2}\right)\Big|_2^b \\ &= \lim_{b\to\infty}\left(-\frac{1}{2b^2}+\frac{1}{8}\right) = \frac{1}{8}. \end{aligned}$$

27. $\int_0^1 \frac{1}{\sqrt{x}}\,dx = \lim_{\epsilon\to0+}\int_\epsilon^1 x^{-1/2}\,dx = \lim_{\epsilon\to0+} 2x^{1/2}\Big|_\epsilon^1 = \lim_{\epsilon\to0+}(2-2\sqrt{\epsilon}) = 2.$

29. $q = \int i\,dt = \int 3.08t^{1/2}\,dt = 3.08\left(\frac{2}{3}\right)t^{3/2}+C$. If $t=0$, $q=0$, so that $C=0$. Thus $q = 3.08\left(\frac{2}{3}\right)t^{3/2}\big|_{t=1.75} = 4.75\,\mathrm{C}$.

31.

+

$\uparrow$ $g = -32\,\text{ft/s}^2$

$s = 121\,\text{ft}$

$v_0 = +29\,\text{ft/s}$

$s = 0$

Taking the upward direction to be positive, $g = -32\,\text{ft/s}^2$ with $v_0 = +29\,\text{ft/s}$. We get

$$\begin{aligned} v &= -32t + C \\ +29 &= 0 + C && \text{when } t = 0,\ v = 29\,\text{ft/s} \\ v &= -32t + 29 \\ s &= -16t^2 + 29t + k \\ 121 &= 0 + k && \text{when } t = 0,\ s = 121\,\text{ft} \\ s &= -16t^2 + 29t + 121. \end{aligned}$$

To see how long the stone takes to reach the ground, we let $s = 0$ and solve for t:

$$\begin{aligned} 0 &= -16t^2 + 29t + 121 \\ t &= \frac{-29 \pm \sqrt{29^2 + 4(16)(121)}}{2(-16)} = 3.8\,\text{s}. \end{aligned}$$

33. If the direction of motion is the positive direction, then $a = -20\,\text{ft/s}^2$. Let $s = 0$ be the point at which the brakes are first applied. Then

$$\begin{aligned} v &= -20t + v_0, && \text{unknown initial velocity} \\ s &= -10t^2 + v_0 t + k, && \text{integrating } v \\ s &= -10t^2 + v_0 t. && \text{since } k = 0 \end{aligned}$$

At the instant when the minivan comes to a stop, we have $v = 0$ and $s = 180\,\text{ft}$. From the first equation, $0 = -20t + v_0$ and $t = \dfrac{v_0}{20}$.

Substituting in the third equation,

$$\begin{aligned} 180 &= -10\left(\frac{v_0}{20}\right)^2 + v_0\frac{v_0}{20} \\ 180 &= -10\frac{v_0^2}{20^2} + \frac{v_0^2}{20}\cdot\frac{20}{20} \\ 180 &= \frac{-10v_0^2 + 20v_0^2}{20^2} = \frac{10v_0^2}{20^2}. \\ v_0^2 &= \frac{180(20)^2}{10} \\ v_0 &\approx 85\,\text{ft/s}. \end{aligned}$$

35. $h = \dfrac{2-(-1)}{6} = \dfrac{1}{2}$; $x_0 = -1$, $x_1 = -\dfrac{1}{2}$, $x_2 = 0$, $x_3 = \dfrac{1}{2}$, $x_4 = 1$, $x_5 = \dfrac{3}{2}$, $x_6 = 2$.

$$\begin{aligned} \int_{-1}^{2} \frac{1}{x+3}\,dx &= \frac{1/2}{3}\left[\frac{1}{-1+3} + 4\frac{1}{-1/2+3} + 2\frac{1}{0+3} + 4\frac{1}{1/2+3} + 2\frac{1}{1+3} + 4\frac{1}{3/2+3} + \frac{1}{2+3}\right] \\ &= 0.916. \end{aligned}$$

Chapter 5

Applications of the Integral

5.1 Means and Root Mean Squares

1. $f_{\text{av}} = \dfrac{1}{16-1}\displaystyle\int_1^{16} x^{1/2}\,dx = \frac{1}{15}\cdot\frac{2}{3}x^{3/2}\Big|_1^{16} = \frac{2}{45}(16^{3/2}-1) = \frac{2}{45}(64-1) = \frac{2(63)}{45} = \frac{14}{5}$

3. $f_{\text{av}} = \dfrac{1}{2-(-2)}\displaystyle\int_{-2}^{2} x\sqrt{x^2+1}\,dx$. Let $u = x^2+1$, then $du = 2x\,dx$.

$$f_{\text{av}} = \frac{1}{4}\int_{-2}^{2}(x^2+1)^{1/2}\,dx = \frac{1}{4}\cdot\frac{1}{2}\int_{-2}^{2}(x^2+1)^{1/2}(2x)\,dx = \frac{1}{8}\frac{(x^2+1)^{3/2}}{3/2}\Big|_{-2}^{2} = 0$$

5. $f_{\text{av}} = \dfrac{1}{2-0}\displaystyle\int_0^2 (2x-x^2)\,dx = \frac{1}{2}\left(x^2-\frac{1}{3}x^3\right)\Big|_0^2 = \frac{1}{2}\left(4-\frac{8}{3}\right) = \frac{2}{3}$

7. $$\begin{aligned} f_{\text{av}} &= \frac{1}{9-4}\int_4^9 (\sqrt{x}+2x)\;dx = \frac{1}{5}\left(\frac{2}{3}x^{3/2}+x^2\right)\Big|_4^9 \\ &= \frac{1}{5}\left[\frac{2}{3}(9)^{3/2}+9^2\right]-\frac{1}{5}\left[\frac{2}{3}(4)^{3/2}+4^2\right] \\ &= \frac{1}{5}\left(18+81-\frac{16}{3}-16\right) = \frac{233}{15} \end{aligned}$$

9. $f_{\text{rms}}^2 = \dfrac{1}{2-1}\displaystyle\int_1^2\left(\frac{1}{x}\right)^2 dx = \int_1^2 x^{-2}\,dx = \frac{x^{-1}}{-1}\Big|_1^2 = -\frac{1}{x}\Big|_1^2 = -\frac{1}{2}+1 = \frac{1}{2}$

Thus $f_{\text{rms}} = \sqrt{\dfrac{1}{2}} = \dfrac{\sqrt{2}}{2}$.

11. $$\begin{aligned} f_{\text{rms}}^2 &= \frac{1}{1-0}\int_0^1[\sqrt{x}(x^2+1)]^2\,dx = \int_0^1(x^5+2x^3+x)\,dx = \frac{1}{6}x^6+\frac{1}{2}x^4+\frac{1}{2}x^2\Big|_0^1 \\ &= \frac{1}{6}+\frac{1}{2}+\frac{1}{2} = \frac{7}{6} \end{aligned}$$

$f_{\text{rms}} = \sqrt{\dfrac{7}{6}} = \dfrac{\sqrt{42}}{6}$

13. Taking the upward direction as positive, we have $g = -10\,\text{m/s}^2$. Also, when $t=0$, we have $v=0$ and $s=180$. Thus

$$\begin{aligned} v &= -10t && \text{since } v_0 = 0 \\ s &= -5t^2+k \\ s &= -5t^2+180 && \text{since } s = 180 \text{ when } t = 0 \end{aligned}$$

Now let $s = 0$ and solve for t: $0 = -5t^2 + 180$ or $t = 6\,\text{s}$.
So it takes the object 6 s to reach the ground. We now get

$$v_{\text{av}} = \frac{1}{6-0}\int_0^6 (-10t)\,dt = -\frac{1}{6}(5t^2)\Big|_0^6 = -30\,\text{m/s},$$

so the average velocity is 30 m/s in the downward direction. (This agrees with the usual notion of average speed: 180 m in 6 s results in an average speed of 180 m/6 s = 30 m/s.)

15. Mean current $= \dfrac{1}{4.0-0.0}\displaystyle\int_{0.0}^{4.0} (1.0t + 1.0\sqrt{t})\,dt = 3.3\,\text{A}$

17.
$$\begin{aligned} i_{\text{rms}}^2 &= \frac{1}{3-0}\int_0^3 (1-t^2)^2\,dt \qquad \text{(leaving out final zeros)} \\ &= \frac{1}{3}\int_0^3 (1-2t^2+t^4)\,dt = \frac{1}{3}\left(t - \frac{2}{3}t^3 + \frac{1}{5}t^5\right)\Big|_0^3 \\ &= \frac{1}{3}\left(3 - 18 + \frac{243}{5}\right) = \frac{1}{3}\,\frac{-75+243}{5} = \frac{168}{15} \end{aligned}$$

Since $R = 5$, we get $P = i_{\text{rms}}^2 R = \dfrac{168}{15}(5) = 56\,\text{W}$ to two significant digits.

5.2 Volumes of Revolution: Disk and Washer Methods

1.

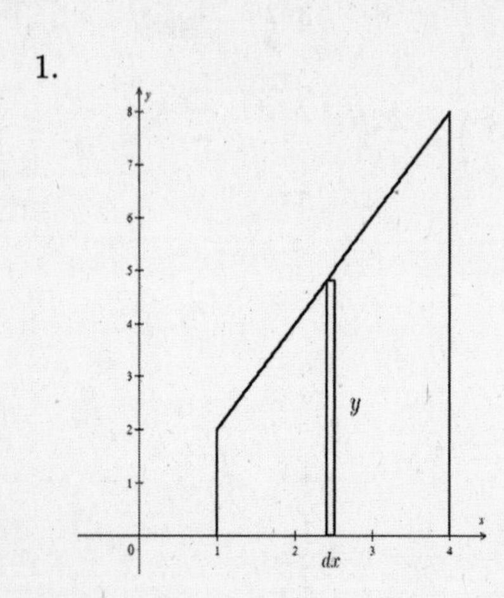

Volume of disk: $\pi(\text{radius})^2 \cdot \text{thickness} = \pi y^2\,dx$

$$\begin{aligned} V &= \int_1^4 \pi(2x)^2\,dx = \pi\int_1^4 4x^2\,dx \\ &= \pi\left(\frac{4}{3}\right)x^3\Big|_1^4 = \frac{4\pi}{3}(64-1) \\ &= \frac{252\pi}{3} = 84\pi \end{aligned}$$

3. $V = \displaystyle\int_1^3 \pi(\sqrt{x^2+1})^2\,dx = \pi\int_1^3 (x^2+1)\,dx = \frac{32\pi}{3}$

5.

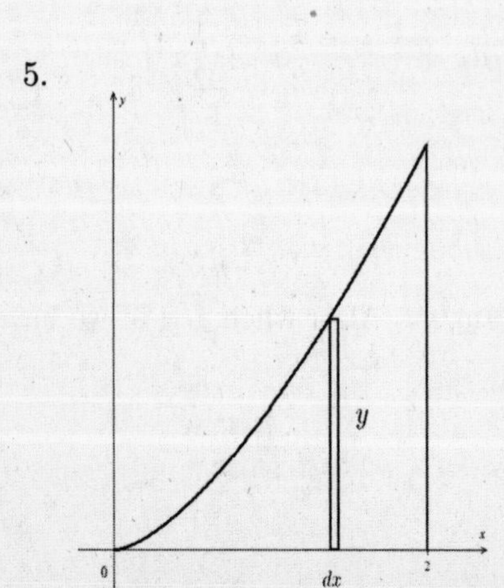

$$\begin{aligned} V &= \pi\int_0^2 (x^{3/2})^2\,dx \\ &= \pi\int_0^2 x^3\,dx \\ &= \pi\left(\frac{1}{4}x^4\right)\Big|_0^2 = 4\pi \end{aligned}$$

7.

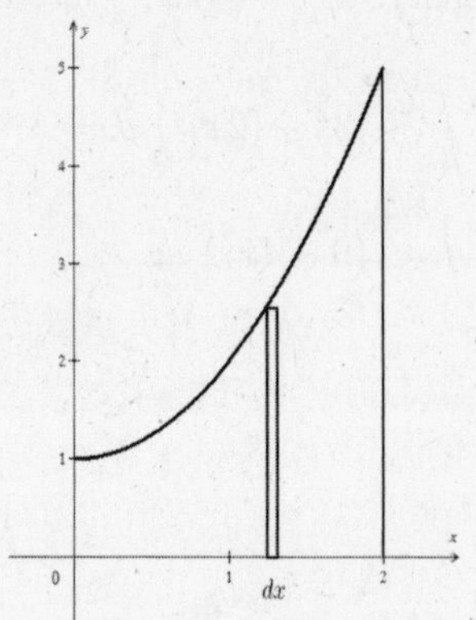

Volume of disk: $\pi(\text{radius})^2 \cdot (\text{thickness}) = \pi y^2\, dx$

Integrating (summing) from $x = 0$ to $x = 2$, we get

$$\begin{aligned} V &= \int_0^2 \pi(x^2+1)^2\, dx = \pi \int_0^2 (x^4 + 2x^2 + 1)\, dx \\ &= \pi\left(\frac{1}{5}x^5 + \frac{2}{3}x^3 + x\right)\Bigg|_0^2 = \pi\left(\frac{32}{5} + \frac{16}{3} + 2\right) = \frac{206\pi}{15}. \end{aligned}$$

9.

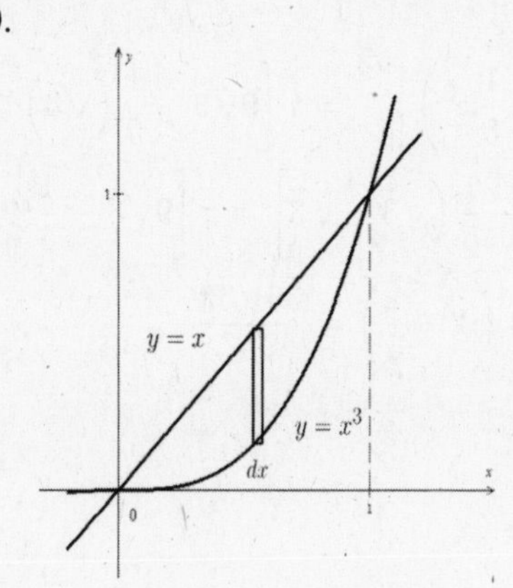

Volume of typical washer: $\pi(y_2^2 - y_1^2)\, dx$

$$\begin{aligned} V &= \int_0^1 \pi[x^2 - (x^3)^2]\, dx \\ &= \pi\left(\frac{1}{3}x^3 - \frac{1}{7}x^7\right)\Bigg|_0^1 \\ &= \pi\left(\frac{1}{3} - \frac{1}{7}\right) = \frac{4\pi}{21} \end{aligned}$$

11.

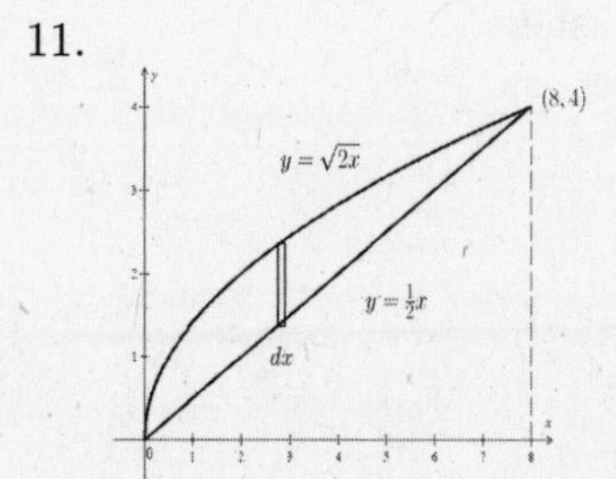

Volume of typical washer: $\pi\left[\left(\sqrt{2x}\right)^2 - \left(\frac{1}{2}x\right)^2\right] dx$

$$\begin{aligned} V &= \pi\int_0^8 \left(2x - \frac{1}{4}x^2\right) dx = \pi\left(x^2 - \frac{x^3}{12}\right)\Bigg|_0^8 \\ &= \pi\left(8^2 - \frac{8^3}{12}\right) = \pi(8^2)\left(1 - \frac{8}{12}\right) = \frac{64\pi}{3} \end{aligned}$$

13.

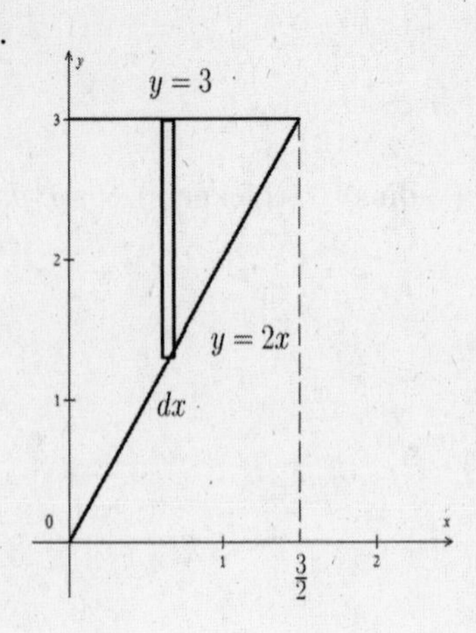

Volume of typical washer: $\pi\left[3^2-(2x)^2\right]\,dx$

$$\begin{aligned} V &= \pi\int_0^{3/2}\left[3^2-(2x)^2\right]\,dx \\ &= \pi\int_0^{3/2}\left(9-4x^2\right)\,dx \\ &= 9\pi \end{aligned}$$

15. If $y=x^2$, then $x=\sqrt{y}$.

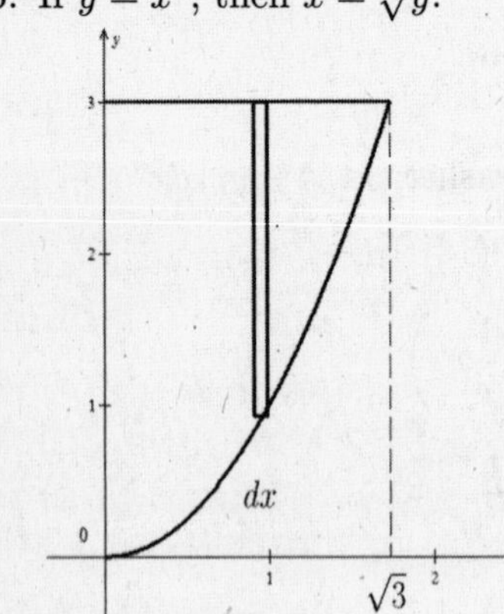

Volume of typical washer: $\pi\left[3^2-\left(x^2\right)^2\right]\,dx$

$$\begin{aligned} V &= \pi\int_0^{\sqrt{3}}\left[3^2-\left(x^2\right)^2\right]\,dx = \pi\int_0^{\sqrt{3}}\left(9-x^4\right)\,dx \\ &= \pi\left(9x-\frac{1}{5}x^5\right)\Big|_0^{\sqrt{3}} = \pi\left[9\sqrt{3}-\frac{1}{5}\left(\sqrt{3}\right)^5\right] \\ &= \pi\left[9\sqrt{3}-\frac{1}{5}\left(\sqrt{3}\right)^4\sqrt{3}\right] = \pi\left[9\sqrt{3}-\frac{1}{5}(9)\sqrt{3}\right] \\ &= \pi\left(9\sqrt{3}\right)\left(1-\frac{1}{5}\right) = \frac{36\sqrt{3}\pi}{5} \end{aligned}$$

17.

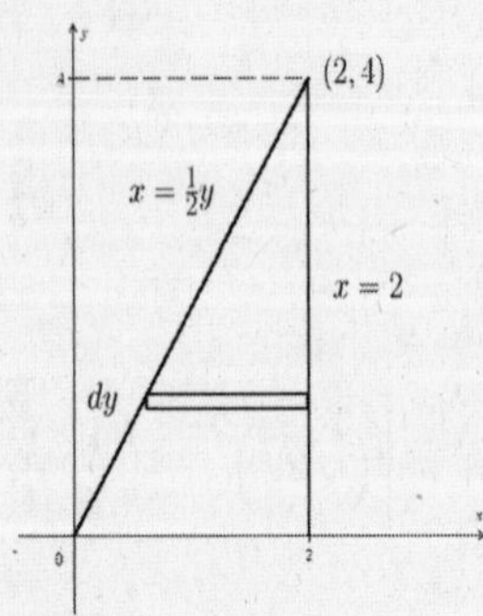

Note that the typical element has to be drawn horizontally to generate washers.

Volume of typical washer: $\pi(x_2^2-x_1^2)\,dy=\pi\left[2^2-\left(\frac{1}{2}y\right)^2\right]$ formula(5.4)

Integrating (summing) from $y=0$ to $y=4$, we get

$$\begin{aligned} V &= \int_0^4\pi\left[2^2-\left(\frac{1}{2}y\right)^2\right]\,dy = \pi\int_0^4\left(4-\frac{1}{4}y^2\right)\,dy = \pi\left(4y-\frac{1}{12}y^3\right)\Big|_0^4 \\ &= \pi\left(4^2-\frac{1}{12}4^3\right) = \pi\left(4^2-\frac{1}{3}\cdot 4^2\right) = 4^2\pi\left(1-\frac{1}{3}\right) = 16\pi\cdot\frac{2}{3} = \frac{32\pi}{3}. \end{aligned}$$

19. Since $y = \frac{1}{2}x$, we have $x = 2y$.

Volume of typical washer:
$\pi\left(x_2^2 - x_1^2\right) dy = \pi[4^2 - (2y)^2]\, dy$

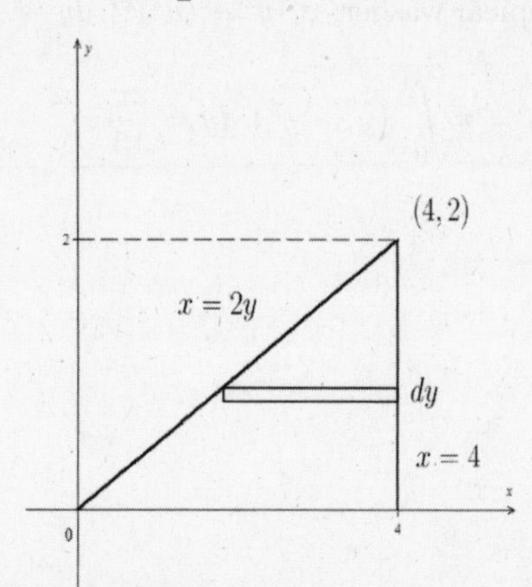

Summing from $y = 0$ to $y = 2$, we get

$$\begin{aligned} V &= \int_0^2 \pi[4^2 - (2y)^2]\, dy = \pi \int_0^2 (16 - 4y^2)\, dy = \pi\left(16y - \frac{4}{3}y^3\right)\Big|_0^2 \\ &= \pi\left(16 \cdot 2 - \frac{4}{3} \cdot 2^3\right) = \pi\left(32 - \frac{1}{3} \cdot 32\right) = 32\pi\left(1 - \frac{1}{3}\right) = 32\pi\left(\frac{2}{3}\right) = \frac{64\pi}{3}. \end{aligned}$$

21.
$$\begin{aligned} y &= x \\ y &= 2 - x \\ \hline 0 &= 2x - 2 \\ x &= 1, y = 1 \end{aligned}$$

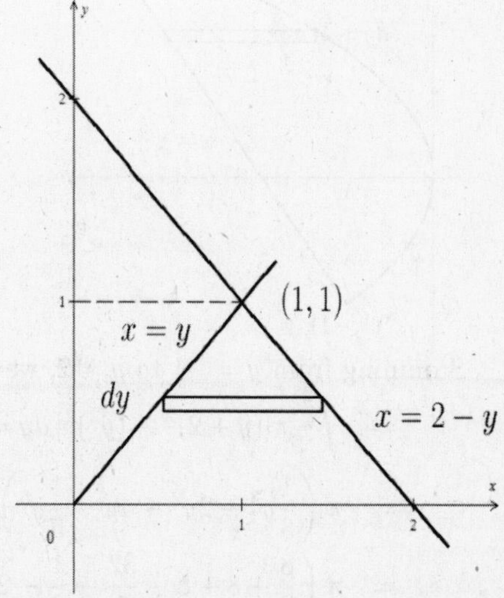

The lines intersect at $(1, 1)$. The typical element has to be drawn horizontally to generate washers. Volume of typical washer: $\pi(x_2^2 - x_1^2)\, dy$

$V = \int_0^1 \pi[(2 - y)^2 - y^2]\, dy = \pi \int_0^1 (4 - 4y)\, dy = 2\pi$

23. Solving each equation for x, we get $x = y$ and $x = y^2$.

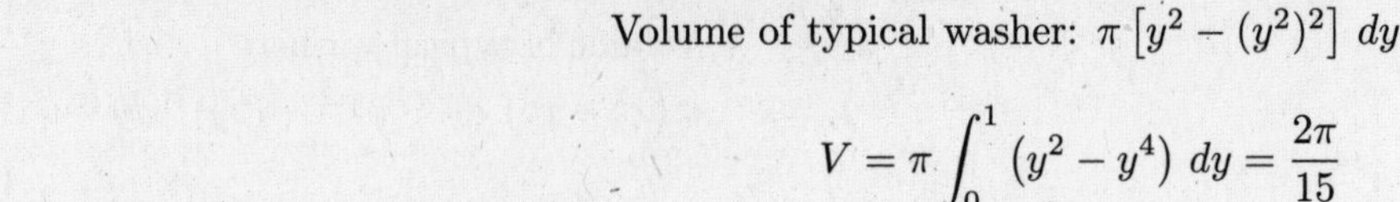

Volume of typical washer: $\pi\left[y^2 - (y^2)^2\right]\,dy$

$$V = \pi \int_0^1 \left(y^2 - y^4\right)\,dy = \frac{2\pi}{15}$$

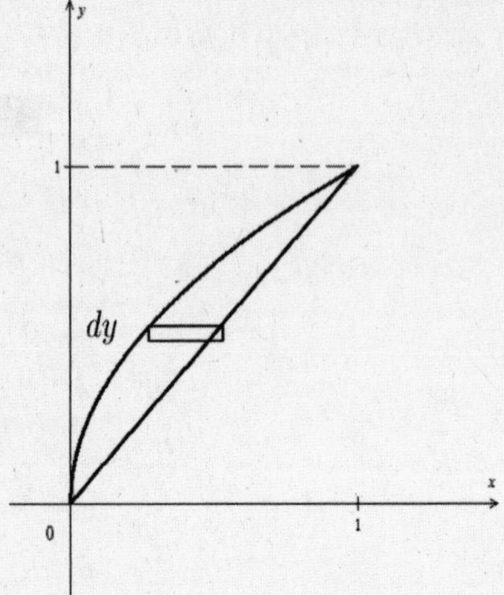

25. To find the points of intersection, we need to solve the equations simultaneously:

$$\begin{aligned} x &= y^2 \\ x &= y + 2 \\ \hline 0 &= y^2 - y - 2 \qquad \text{(subtracting)} \\ (y+1)(y-2) &= 0 \\ y &= -1, 2 \end{aligned}$$

The points of intersection are $(1, -1)$ and $(4, 2)$.

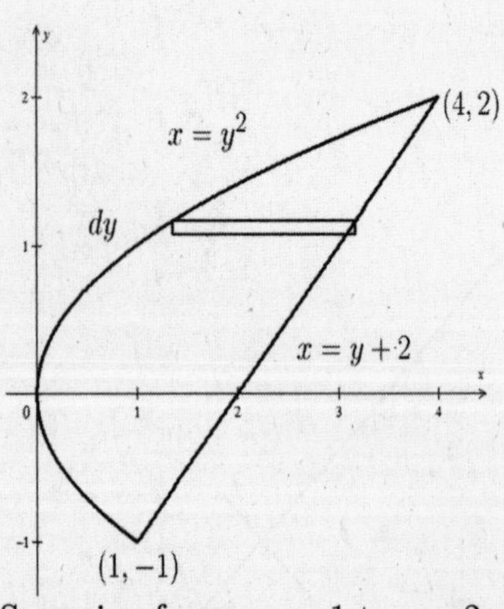

Note that the typical element has to be drawn horizontally. Now recall that the volume of a typical washer is $\pi(x_2^2 - x_1^2)\,dy = \pi[(y+2)^2 - (y^2)^2]\,dy$.

Summing from $y = -1$ to $y = 2$, we get

$$\begin{aligned} V &= \int_{-1}^{2} \pi[(y+2)^2 - (y^2)^2]\,dy = \pi \int_{-1}^{2} (y^2 + 4y + 4 - y^4)\,dy \\ &= \pi\left(\frac{1}{3}y^3 + 2y^2 + 4y - \frac{1}{5}y^5\right)\Big|_{-1}^{2} = \pi\left[\left(\frac{8}{3} + 8 + 8 - \frac{32}{5}\right) - \left(-\frac{1}{3} + 2 - 4 + \frac{1}{5}\right)\right] \\ &= \pi\left(\frac{8}{3} + 8 + 8 - \frac{32}{5} + \frac{1}{3} - 2 + 4 - \frac{1}{5}\right) = \pi\left(\frac{9}{3} - \frac{33}{5} + 18\right) \\ &= \pi\left(21 - \frac{33}{5}\right) = \pi\frac{105 - 33}{5} = \frac{72\pi}{5}. \end{aligned}$$

27.

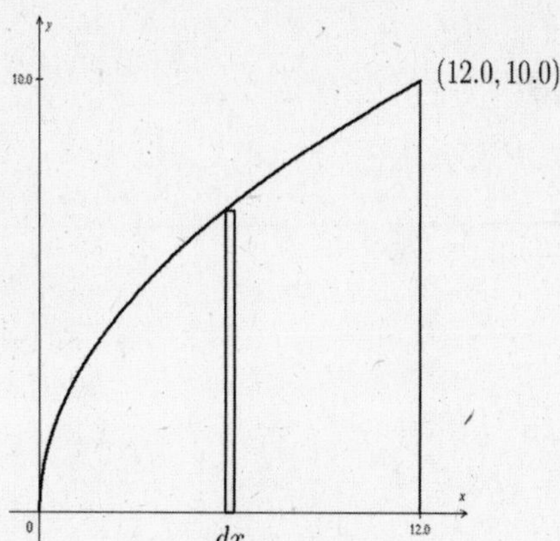

Assume that the equation of the parabola has the form $y^2 = 4px$ (axis horizontal). From the given information, the point (12.0, 10.0) lies on the curve:

$$10.0^2 = 4p(12.0) \quad \text{or} \quad 4p = \frac{10.0^2}{12.0} = \frac{25}{3}$$

Equation: $y^2 = \frac{25}{3}x$. Volume of disk: $\pi(\text{radius})^2 \cdot (\text{thickness}) = \pi y^2\,dx = \pi\left(\frac{25}{3}x\right)\,dx$.

$$V = \int_0^{12.0} \pi\left(\frac{25}{3}x\right)\,dx = \frac{25\pi}{3}\frac{x^2}{2}\bigg|_0^{12.0} = 600\pi \approx 1880\,\text{cm}^3$$

29. By turning the tank on the side, we can find the volume of the water by rotating the region in the figure about the x-axis. (Since the radius is 12 ft, note that the equation of the circle is $x^2 + y^2 = 12^2$.)

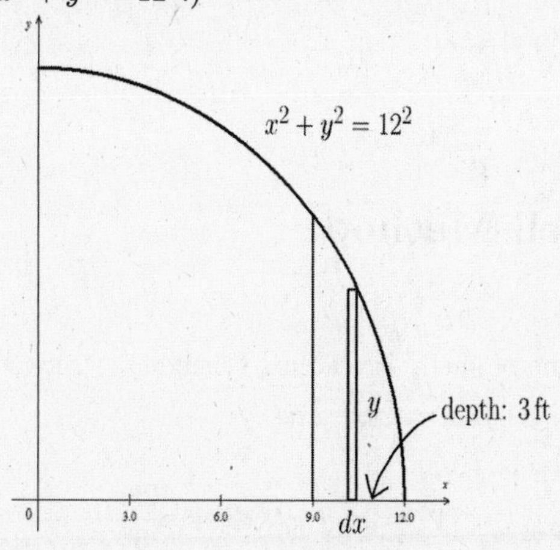

From $x^2 + y^2 = 144$, we get $y^2 = 144 - x^2$. Volume of typical disk: $\pi y^2\,dx = \pi(144 - x^2)\,dx$.

Integrating from $x = 9$ to $x = 12$,

$$\begin{aligned} V &= \pi\int_9^{12}(144 - x^2)\,dx = \pi\left(144x - \frac{1}{3}x^3\right)\bigg|_9^{12} \\ &= \pi\left[\left(144\cdot 12 - \frac{1}{3}(12)^3\right) - \left(144\cdot 9 - \frac{1}{3}\cdot 9^3\right)\right] \\ &= \pi[(1728 - 576) - (1296 - 243)] = 99\pi \approx 310\,\text{ft}^3. \end{aligned}$$

31.

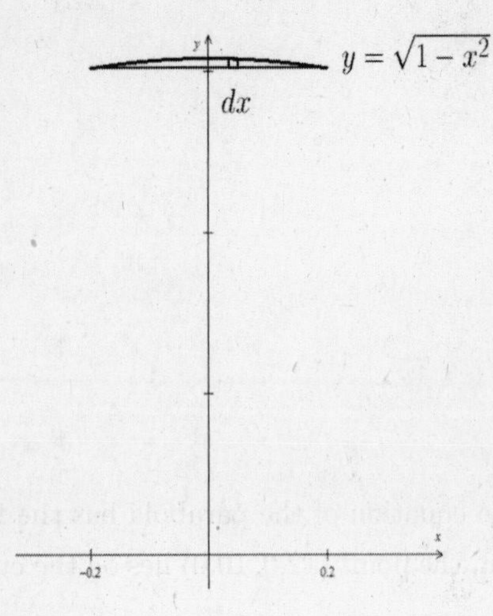

From $x^2 + y^2 = 1, y = \sqrt{1-x^2}$; if $x = 0.2$, then $y = \sqrt{1-(0.2)^2} = 0.979796$.

Volume of washer: $\pi\left[\left(\sqrt{1-x^2}\right)^2 - (0.979796)^2\right]\,dx$

Integrating from -0.2 to 0.2, we get $0.03351\,\text{cm}^3 \approx (0.32\,\text{cm})^3$, equivalent to a cube with side $0.32\,\text{cm}$.

33. $y = \dfrac{1}{x^{3/4}} = x^{-3/4}$; Volume of typical disk: $\pi y^2\,dx = \pi(x^{-3/4})^2\,dx = \pi x^{-3/2}\,dx$

$$\begin{aligned} V &= \pi\int_4^\infty x^{-3/2}\,dx = \lim_{b\to\infty} \pi\int_4^b x^{-3/2}\,dx = \lim_{b\to\infty} \pi(-2)x^{-1/2}\Big|_4^b \\ &= \lim_{b\to\infty} \frac{-2\pi}{\sqrt{x}}\Big|_4^b = \lim_{b\to\infty}\left[\frac{-2\pi}{\sqrt{b}} + \frac{2\pi}{\sqrt{4}}\right] = \frac{2\pi}{2} = \pi \\ A &= \int_4^\infty x^{-3/4}\,dx = \lim_{b\to\infty}\int_4^b x^{-3/4}\,dx = \lim_{b\to\infty} 4x^{1/4}\Big|_4^b \\ &= \lim_{b\to\infty}(4b^{1/4} - 4(4)^{1/4}) = \infty \text{ (does not exist)} \end{aligned}$$

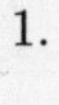

5.3 Volumes of Revolution: Shell Method

1.

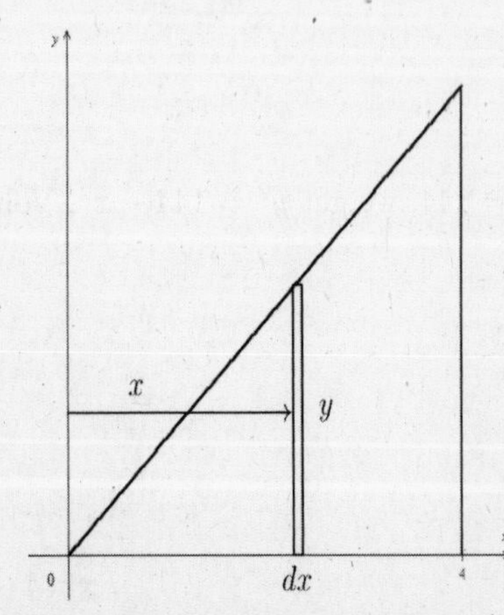

Volume of shell: $2\pi(\text{radius})\cdot(\text{height})\cdot(\text{thickness})$
$= 2\pi\cdot x\cdot y\,dx = 2\pi x\cdot x\,dx$

$$V = 2\pi\int_0^4 x^2\,dx = \frac{128\pi}{3}$$

3.

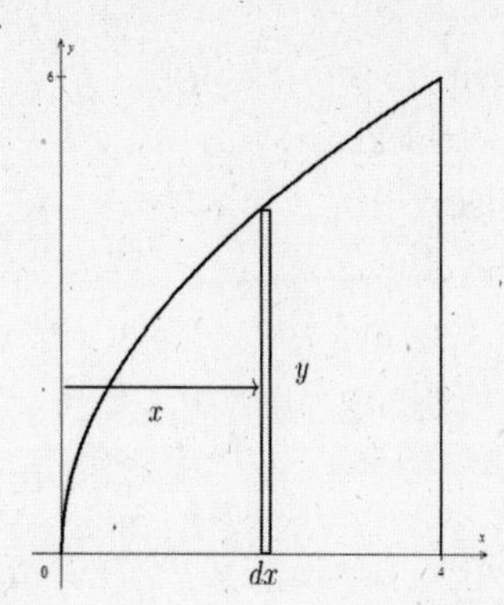

Volume of shell: $2\pi(\text{radius}) \cdot (\text{height}) \cdot (\text{thickness})$
$= 2\pi x \cdot 3\sqrt{x}\, dx$

$$\begin{aligned} V &= 2\pi \int_0^4 x \cdot 3\sqrt{x}\, dx = 6\pi \int_0^4 x^{3/2}\, dx \\ &= 6\pi \left(\frac{2}{5}\right) x^{5/2}\Big|_0^4 = 6\pi \left(\frac{2}{5}\right)(32) = \frac{384\pi}{5} \end{aligned}$$

5.

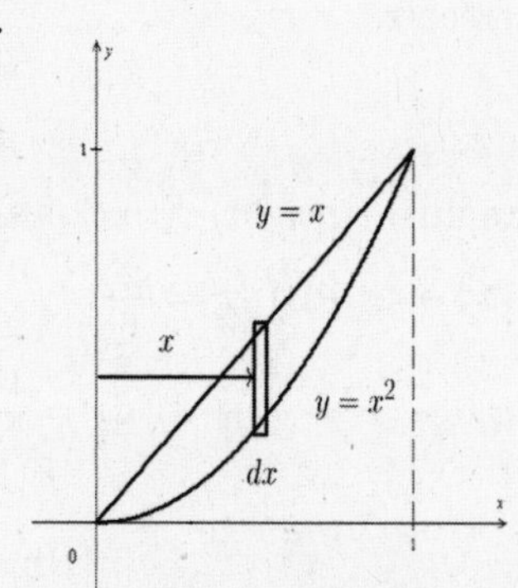

Volume of shell: $2\pi(\text{radius}) \cdot (\text{height}) \cdot (\text{thickness})$
$= 2\pi x(x - x^2)\, dx$

Integrating (summing) from $x = 0$ to $x = 1$, we get

$$V = 2\pi \int_0^1 x(x - x^2)\, dx = 2\pi \left(\frac{1}{3}x^3 - \frac{1}{4}x^4\right)\Big|_0^1 = 2\pi\left(\frac{1}{3} - \frac{1}{4}\right) = 2\pi\left(\frac{1}{12}\right) = \frac{\pi}{6}.$$

7.

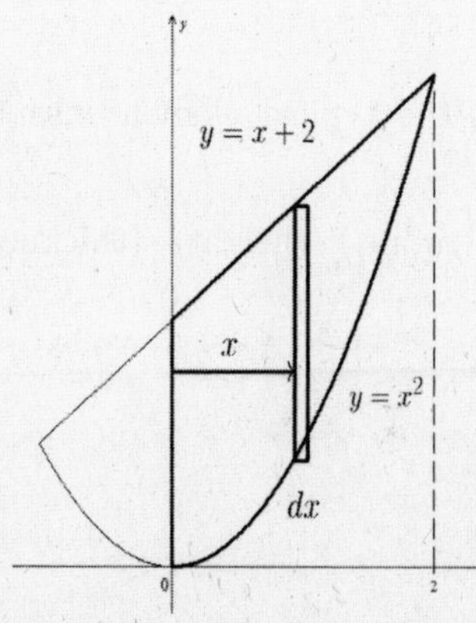

$$\begin{aligned} y &= x^2 \\ y &= x + 2 \\ \hline 0 &= x^2 - x - 2 \qquad \text{(subtracting)} \\ (x+1)(x-2) &= 0 \\ x &= -1, 2 \end{aligned}$$

Volume of shell: $2\pi(\text{radius}) \cdot (\text{height}) \cdot (\text{thickness}) = 2\pi x[(x + 2) - x^2]\, dx$.

$$V = 2\pi \int_0^2 x(x + 2 - x^2)\, dx = 2\pi \left(\frac{1}{3}x^3 + x^2 - \frac{1}{4}x^4\right)\Big|_0^2 = 2\pi\left(\frac{8}{3} + 4 - 4\right) = \frac{16\pi}{3}.$$

9.

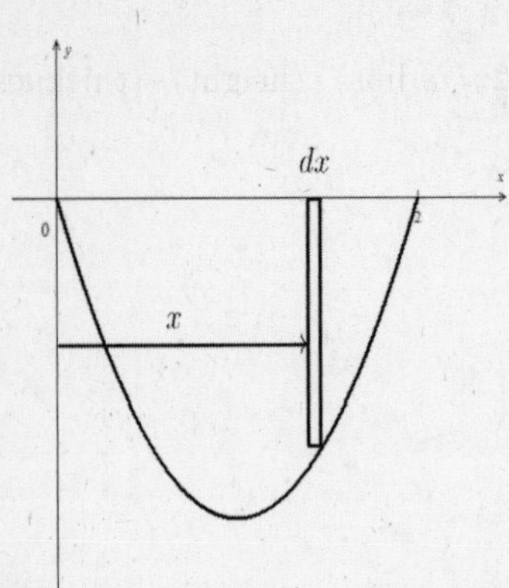

Volume of typical shell:
$2\pi(\text{radius}) \cdot (\text{height}) \cdot (\text{thickness})$
$= 2\pi x[0 - (x^2 - 2x)]\, dx.$

Summing from $x = 0$ to $x = 2$, we get

$$\begin{aligned} V &= \int_0^2 2\pi x[0 - (x^2 - 2x)]\, dx = 2\pi \int_0^2 (-x^3 + 2x^2)\, dx = 2\pi \left(-\frac{1}{4}x^4 + \frac{2}{3}x^3 \right)\Big|_0^2 \\ &= 2\pi \left(-4 + \frac{16}{3} \right) = 2\pi \left(-\frac{12}{3} + \frac{16}{3} \right) = 2\pi \left(\frac{4}{3} \right) = \frac{8\pi}{3}. \end{aligned}$$

11. $y = x^2 - 4x + 3 = (x - 1)(x - 3) = 0$, gives $x = 1, 3$ as the intercepts.

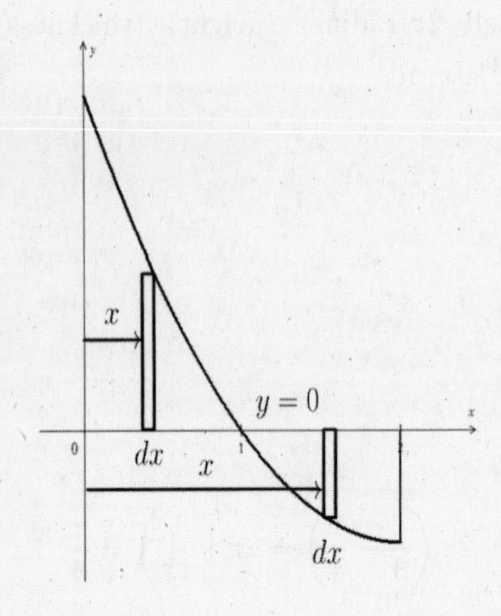

Volume of shell: $2\pi(\text{radius}) \cdot (\text{height}) \cdot (\text{thickness})$.

Left region: $2\pi \displaystyle\int_0^1 x(x^2 - 4x + 3)\, dx = \frac{5\pi}{6}$.

Right region: $2\pi \displaystyle\int_1^2 x[0 - (x^2 - 4x + 3)]\, dx = \frac{13\pi}{6}$.

TOTAL: $\dfrac{18\pi}{6} = 3\pi$.

13.

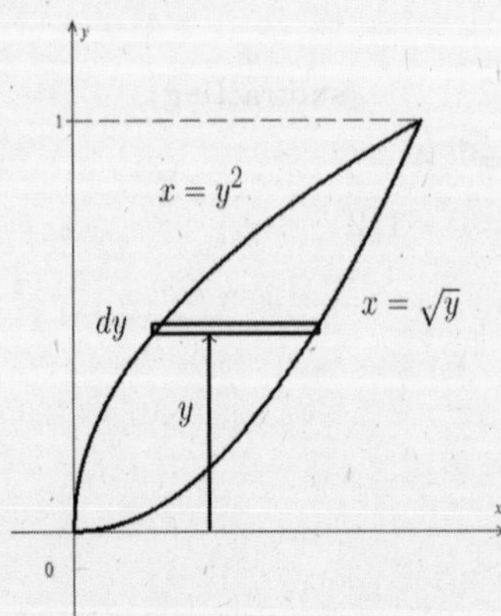

To generate shells, the typical element has to be drawn horizontally.
Volume of shell: $2\pi(\text{radius}) \cdot (\text{height}) \cdot (\text{thickness})$.

$$\begin{aligned} V &= 2\pi \int_0^1 y(\sqrt{y} - y^2)\, dy = 2\pi \int_0^1 (y^{3/2} - y^3)\, dy = 2\pi \left(\frac{2}{5}y^{5/2} - \frac{1}{4}y^4 \right)\Big|_0^1 \\ &= 2\pi \left(\frac{2}{5} - \frac{1}{4} \right) = 2\pi \left(\frac{3}{20} \right) = \frac{3\pi}{10}. \end{aligned}$$

15.

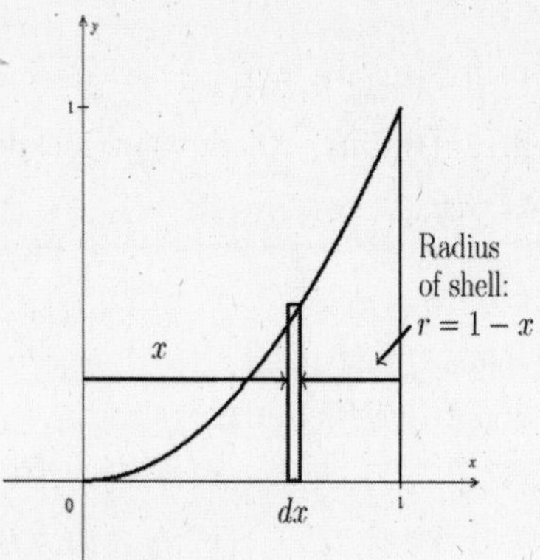

Volume of typical shell: $2\pi(\text{radius}) \cdot (\text{height}) \cdot (\text{thickness}) = 2\pi(1-x) \cdot x^2\,dx$

$$V = \int_0^1 2\pi(1-x)x^2\,dx = 2\pi\int_0^1 (x^2 - x^3)\,dx = 2\pi\left(\frac{1}{3}x^3 - \frac{1}{4}x^4\right)\Big|_0^1 = 2\pi\left(\frac{1}{3} - \frac{1}{4}\right) = \frac{2\pi}{12} = \frac{\pi}{6}$$

17.

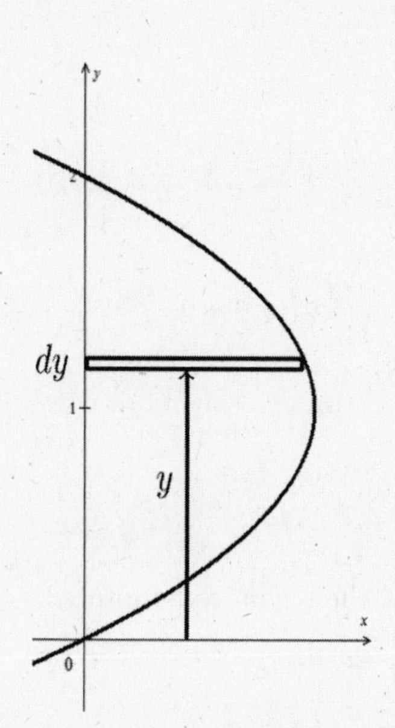

$$x = 2y - y^2 \quad = \quad y(2-y) = 0$$
$$y \quad = \quad 0, 2$$

$$V = 2\pi\int_0^2 y(2y - y^2)\,dy = \frac{8\pi}{3}$$

19.

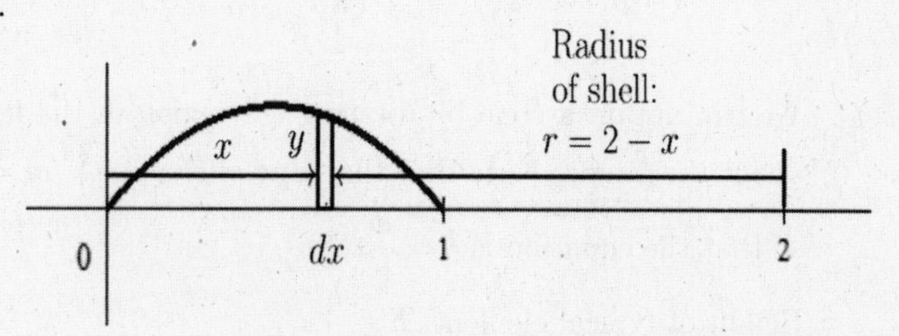

Note that the intercepts of $y = x - x^2 = x(1-x)$ are $x = 0$ and $x = 1$. So the limits of integration are $x = 0$ and $x = 1$, regardless of the position of the (vertical) axis of rotation. (In other words, if we were using Riemann sums instead of the shortcut, we would subdivide the region under the graph.)

Volume of shell: $2\pi(\text{radius}) \cdot (\text{height}) \cdot (\text{thickness}) = 2\pi(2-x) \cdot (x - x^2)\,dx$

$$\begin{aligned} V &= \int_0^1 2\pi(2-x)(x-x^2)\,dx = 2\pi\int_0^1 (2x - 3x^2 + x^3)\,dx \\ &= 2\pi\left(x^2 - x^3 + \frac{1}{4}x^4\right)\Big|_0^1 = 2\pi\left(1 - 1 + \frac{1}{4}\right) = \frac{\pi}{2} \end{aligned}$$

21.

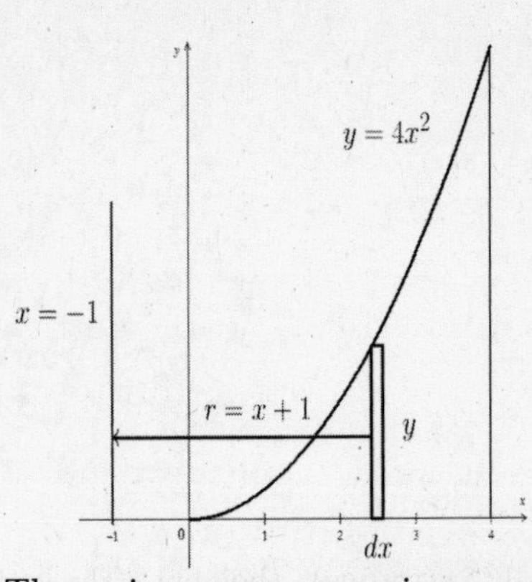

Radius of shell: $x - (-1) = x + 1$.
Volume of shell: $2\pi(\text{radius}) \cdot (\text{height}) \cdot (\text{thickness})$
$= 2\pi(x+1) \cdot 4x^2\,dx$.

The region we are summing over extends from $x = 0$ to $x = 4$:

$$\begin{aligned} V &= \int_0^4 2\pi(x+1)(4x^2)\,dx = 8\pi\int_0^4 (x^3 + x^2)\,dx = 8\pi\left(\frac{1}{4}x^4 + \frac{1}{3}x^3\right)\Big|_0^4 \\ &= 8\pi\left(\frac{1}{4}\cdot 4^4 + \frac{1}{3}\cdot 4^3\right) = 8\pi\left(4^3 + \frac{1}{3}\cdot 4^3\right) = 8\pi(4^3)\left(1 + \frac{1}{3}\right) \\ &= 8\pi(64)\left(\frac{4}{3}\right) = \frac{2048\pi}{3}. \end{aligned}$$

23.

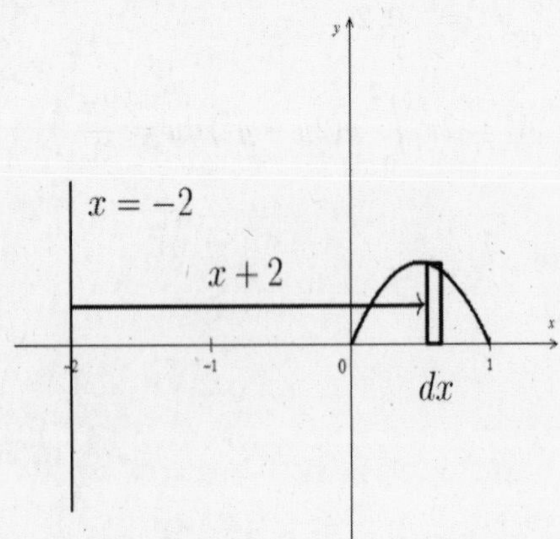

Radius of shell: $x - (-2) = x + 2$.

Summing from $x = 0$ to $x = 1$, the vertical lines that bound the region to be rotated:

$$\begin{aligned} V &= 2\pi\int_0^1 (x+2)(2x - 2x^2)\,dx = 2\pi\int_0^1 (-2x^3 - 2x^2 + 4x)\,dx \\ &= 2\pi\left(-\frac{1}{2}x^4 - \frac{2}{3}x^3 + 2x^2\right)\Big|_0^1 = 2\pi\left(-\frac{1}{2} - \frac{2}{3} + 2\right) = 2\pi\left(\frac{5}{6}\right) = \frac{5\pi}{3}. \end{aligned}$$

25.

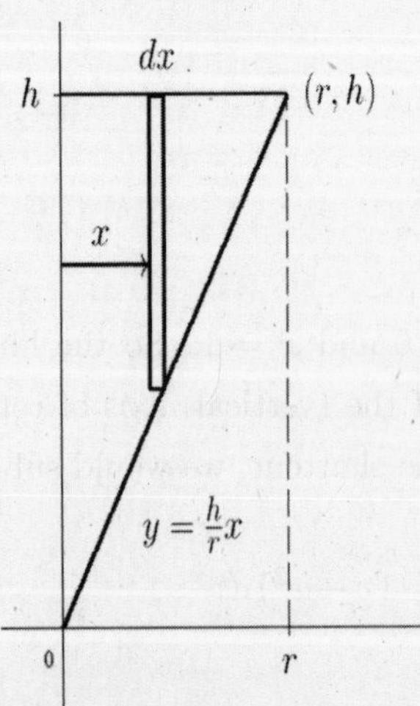

We can obtain a cone by rotating the region in the figure about the y-axis. Note that the slope of the line is $m = \dfrac{h}{r}$, so that the equation is $y = \dfrac{h}{r}x$.

Height of typical element: $h - \dfrac{h}{r}x$.

Volume of shell: $2\pi(\text{radius}) \cdot (\text{height}) \cdot (\text{thickness}) = 2\pi \cdot x \cdot \left(h - \dfrac{h}{r}x\right)\,dx$.

$$\begin{aligned}
V &= \int_0^r 2\pi x\left(h - \frac{h}{r}x\right)\,dx = 2\pi\int_0^r x\cdot h\left(1-\frac{1}{r}x\right)\,dx = 2\pi h\int_0^r\left(x-\frac{1}{r}x^2\right)\,dx \\
&= 2\pi h\left(\frac{1}{2}x^2 - \frac{1}{r}\cdot\frac{1}{3}x^3\right)\Big|_0^r = 2\pi h\left(\frac{1}{2}r^2 - \frac{1}{3}r^2\right) = 2\pi hr^2\left(\frac{1}{2}-\frac{1}{3}\right) \\
&= 2\pi hr^2\frac{1}{6} = \frac{1}{3}\pi r^2 h.
\end{aligned}$$

27.

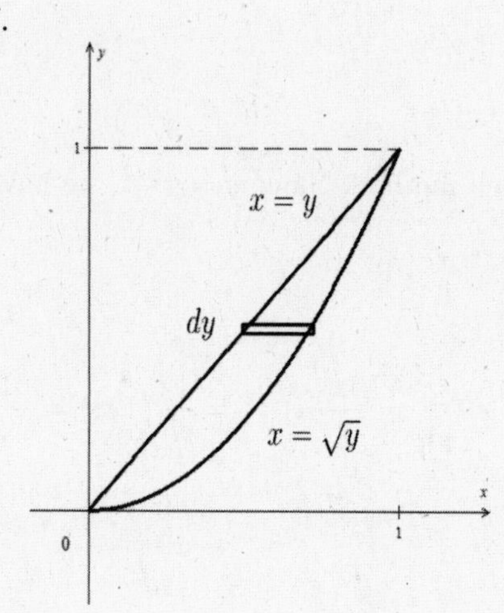

Volume of typical washer: $\pi\left[\left(\sqrt{y}\right)^2 - y^2\right]\,dy$

$$V = \pi\int_0^1 \left(y - y^2\right)\,dy = \frac{\pi}{6}$$

29. (a)

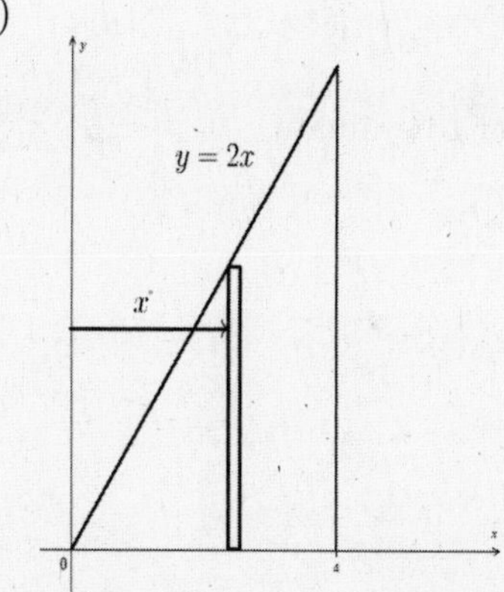

Volume of disk: $\pi(\text{radius})^2\cdot(\text{thickness})$.

$$V = \int_0^4 \pi(2x)^2\,dx = \frac{256\pi}{3}.$$

(b)

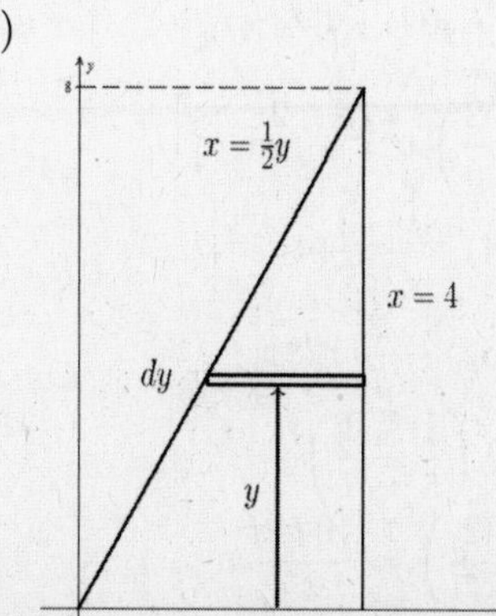

Volume of shell: $2\pi(\text{radius})\cdot(\text{height})\cdot(\text{thickness})$.

$$V = 2\pi\int_0^8 y\left(4 - \frac{1}{2}y\right)\,dy = \frac{256\pi}{3}.$$

31. (a)

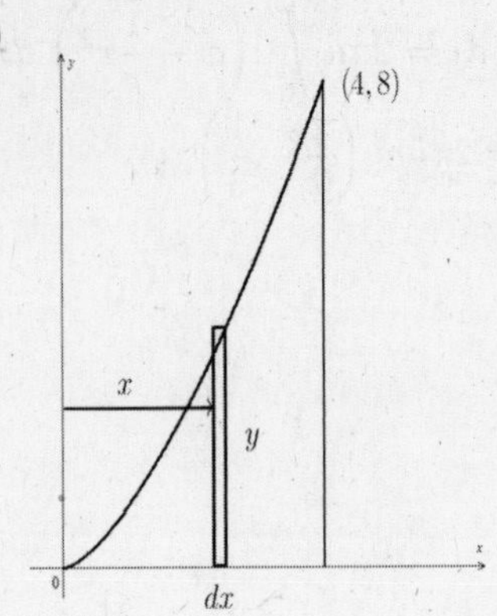

The simplest way to find this volume is by the disk method: $\pi \cdot (\text{radius})^2 \cdot \text{thickness} = \pi y^2\,dx$. Since $y^2 = x^3$, we get

$$V = \pi \int_0^4 x^3\,dx = \pi \frac{x^4}{4}\bigg|_0^4 = 64\pi.$$

(b) The simplest way to find this volume is by the shell method. Since $y = x^{3/2}$, we have $2\pi(\text{radius}) \cdot (\text{height}) \cdot (\text{thickness}) = 2\pi x \cdot x^{3/2}\,dx$.

$$V = 2\pi \int_0^4 x \cdot x^{3/2}\,dx = 2\pi \int_0^4 x^{5/2}\,dx = 2\pi \left(\frac{2}{7}\right) x^{7/2}\bigg|_0^4 = \frac{4\pi}{7}(4)^{7/2} = \frac{4\pi}{7}(128) = \frac{512\pi}{7}.$$

(b) alternate

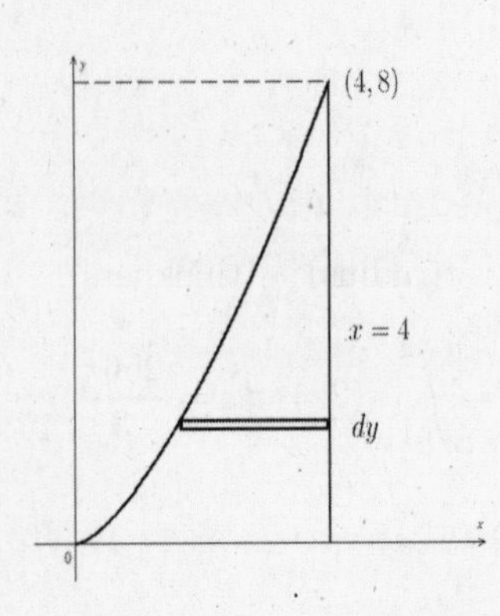

By the washer method, we get, for the typical washer, $\pi(x_2^2 - x_1^2)\,dy$ and $x = y^{2/3}$.

$$V = \pi \int_0^8 [4^2 - (y^{2/3})^2]\,dy,$$

which is harder to evaluate.

(c) Shell: $2\pi(\text{radius}) \cdot (\text{height}) \cdot (\text{thickness})$ with $r = 4 - x$:

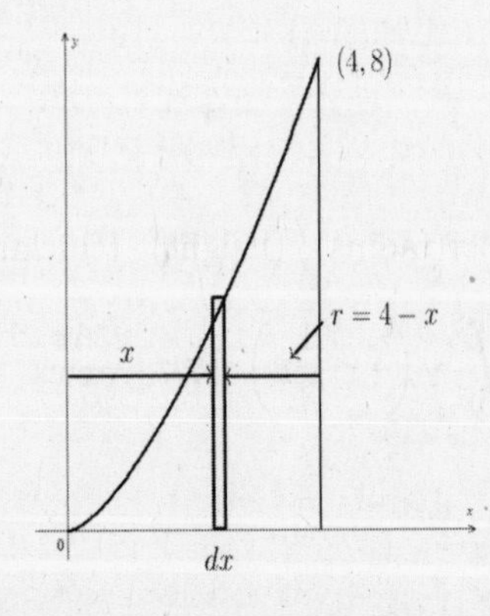

$$\begin{aligned}
V &= 2\pi \int_0^4 (4 - x)x^{3/2}\,dx \\
&= 2\pi \int_0^4 (4x^{3/2} - x^{5/2})\,dx \\
&= 2\pi \left[4\left(\frac{2}{5}\right) x^{5/2} - \frac{2}{7}x^{7/2}\right]\bigg|_0^4 \\
&= 2\pi \left[\frac{8}{5}(4)^{5/2} - \frac{2}{7}(4)^{7/2}\right] \\
&= 2\pi \left[\frac{8}{5}(32) - \frac{2}{7}(128)\right] \\
&= 2\pi(256)\left(\frac{1}{5} - \frac{1}{7}\right) \\
&= 512\pi\left(\frac{2}{35}\right) = \frac{1024\pi}{35}.
\end{aligned}$$

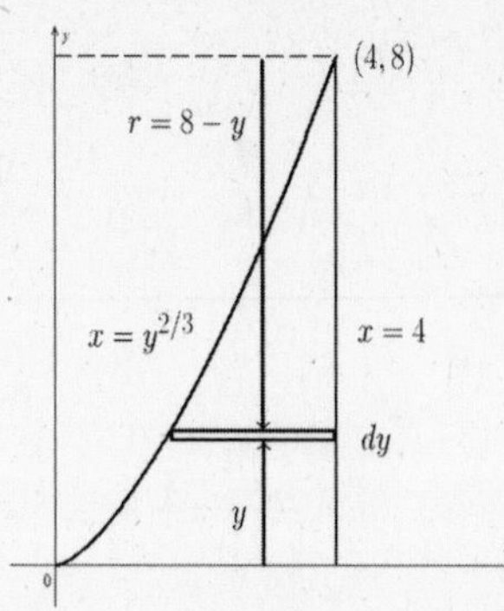

Volume of shell:

$2\pi(\text{radius}) \cdot (\text{height}) \cdot (\text{thickness})$.

Since $r = 8 - y$ and the height of the typical element is $(4 - y^{2/3})\,dy$, we get:

$$\begin{aligned}
V &= 2\pi \int_0^8 (8-y)(4-y^{2/3})\,dy = 2\pi \int_0^8 (32 - 4y - 8y^{2/3} + y^{5/3})\,dy \\
&= 2\pi \left[32y - 2y^2 - 8\left(\frac{3}{5}\right) y^{5/3} + \left(\frac{3}{8}\right) y^{8/3} \right]\Bigg|_0^8 \\
&= 2\pi \left[256 - 128 - \frac{24}{5}(8)^{5/3} + \frac{3}{8}(8)^{8/3} \right] = 2\pi \left[128 - \frac{24}{5}(32) + \frac{3}{8}(256) \right] \\
&= 2\pi \left(128 - \frac{768}{5} + \frac{768}{8} \right) = 2\pi(128)\left(1 - \frac{6}{5} + \frac{6}{8}\right) \\
&= 256\pi \frac{20 - 24 + 15}{20} = 256\pi\left(\frac{11}{20}\right) = \frac{64\pi(11)}{5} = \frac{704\pi}{5}.
\end{aligned}$$

33. Volume of shell: $2\pi(\text{radius}) \cdot (\text{height}) \cdot (\text{thickness})$.

$$\int_0^2 2\pi x \sqrt{1 + \sqrt{x}}\,dx$$

Using the integration capability of a graphing utility, we get 18.317006.

35.

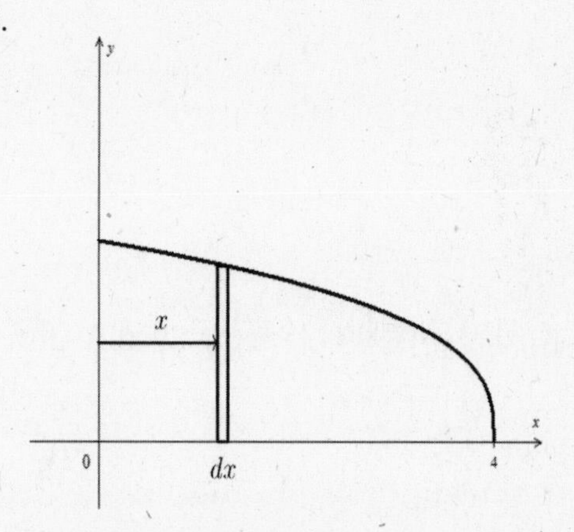

$$V = 2\pi \int_0^4 x \sqrt[3]{4 - x}\,dx = 51.294523.$$

37. Placing the vertex at the origin, the form of the equation is $x^2 = 4py$. One point on the curve is $(3, 6)$. Substituting these values,

$$3^2 = 4p(6) \quad \text{or} \quad 4p = \frac{3}{2}.$$

The equation is $x^2 = \frac{3}{2}y$ or $y = \frac{2}{3}x^2$. The "upper" curve is $y = 6$. By the shell method,

$$\begin{aligned}
V &= 2\pi \int_0^3 x\left(6 - \frac{2}{3}x^2\right)\,dx = 2\pi\left(3x^2 - \frac{2}{3}\frac{x^4}{4}\right)\Bigg|_0^3 \\
&= 2\pi\left(3^3 - \frac{1}{2}\cdot 3^3\right) = 2\pi \cdot 3^3\left(1 - \frac{1}{2}\right) = 27\pi\ \text{m}^3.
\end{aligned}$$

5.4 Centroids

1.

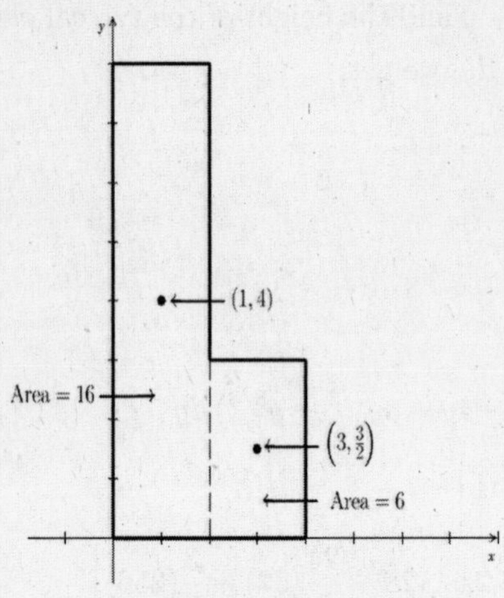

$$\overline{x} = \frac{1 \cdot 16 + 3 \cdot 6}{16 + 6} = \frac{34}{22} = \frac{17}{11}$$
$$\overline{y} = \frac{4 \cdot 16 + \frac{3}{2} \cdot 6}{16 + 6} = \frac{73}{22}$$

3.

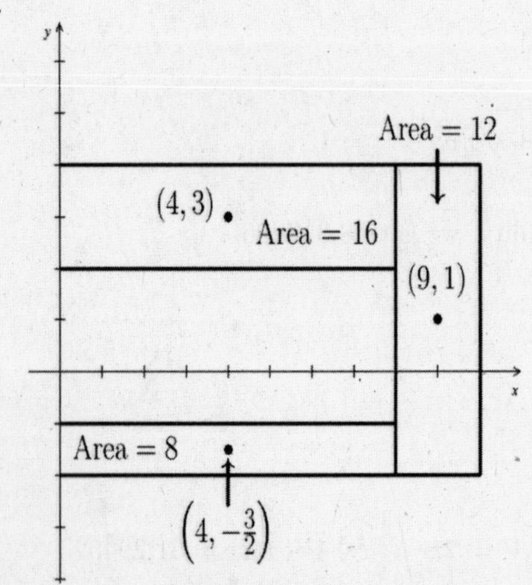

$$\overline{x} = \frac{4 \cdot 16 + 9 \cdot 12 + 4 \cdot 8}{16 + 12 + 8} = \frac{204}{36} = \frac{17}{3}$$
$$\overline{y} = \frac{3 \cdot 16 + 1 \cdot 12 + \left(-\frac{3}{2}\right) \cdot 8}{36} = \frac{48}{36} = \frac{4}{3}$$

5.

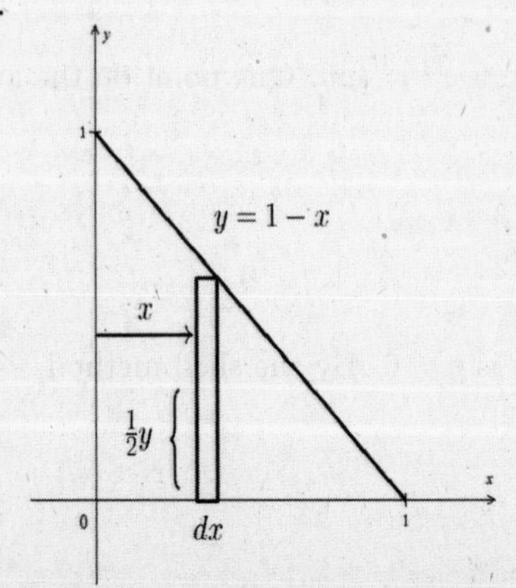

$\overline{x}$: moment of typical element: $xy\,dx = x(1-x)\,dx$.

$$\overline{x} = \frac{M_y}{A} = \frac{\int_0^1 xy\,dx}{\int_0^1 y\,dx} = \frac{\int_0^1 x(1-x)\,dx}{\int_0^1 (1-x)\,dx} = \frac{\frac{1}{2}x^2 - \frac{1}{3}x^3\Big|_0^1}{x - \frac{1}{2}x^2\Big|_0^1} = \frac{\frac{1}{2}-\frac{1}{3}}{1-\frac{1}{2}} = \frac{\frac{1}{6}}{\frac{1}{2}} = \frac{1}{3}.$$

$\overline{y}$: moment of typical element: $\left(\frac{1}{2}y\right)(y\,dx)$.

$$\begin{aligned}\overline{y} &= \frac{M_x}{A} = \frac{\int_0^1 \left(\frac{1}{2}y\right)(y\,dx)}{A} = \frac{\frac{1}{2}\int_0^1 (1-x)(1-x)\,dx}{1/2} = 2\cdot\frac{1}{2}\int_0^1 (1-2x+x^2)\,dx \\ &= x - x^2 + \frac{1}{3}x^3\Big|_0^1 = \frac{1}{3}.\end{aligned}$$

$\overline{y}$ (alternate):

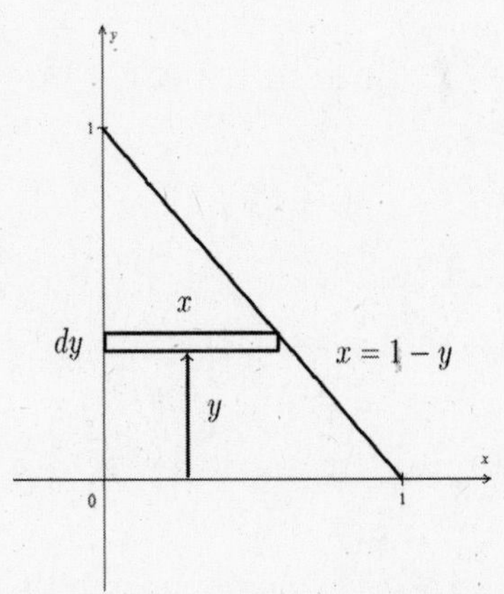

Interchanging the roles of x and y.

Moment of typical element with respect to x-axis: $y \cdot x\,dy = y(1-y)\,dy$.

$$\overline{y} = \frac{M_x}{A} = \frac{\int_0^1 yx\,dy}{A} = \frac{\int_0^1 y(1-y)\,dy}{1/2} = \frac{1}{3}.$$

7.

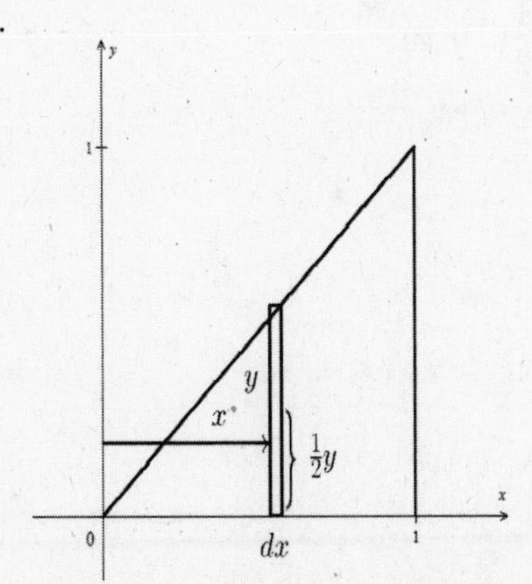

$\overline{x}$: moment of typical element: $xy\,dx$. $\overline{x} = \dfrac{M_y}{A} = \dfrac{\int_0^1 xy\,dx}{\int_0^1 y\,dx} = \dfrac{\int_0^1 x(x)\,dx}{\int_0^1 x\,dx} = \dfrac{\frac{1}{3}x^3\Big|_0^1}{\frac{1}{2}x^2\Big|_0^1} = \dfrac{\frac{1}{3}}{\frac{1}{2}} = \dfrac{2}{3}.$

$\overline{y}$: moment of typical element: $\left(\frac{1}{2}\right)(y\,dx)$.

$$\overline{y} = \frac{M_x}{A} = \frac{\int_0^1 \left(\frac{1}{2}y\right)(y\,dx)}{A} = \frac{\frac{1}{2}\int_0^1 x\cdot x\,dx}{1/2} = \int_0^1 x^2\,dx = \frac{1}{3}.$$

$\overline{y}$ (alternate): moment (with respect to x-axis) of typical element: $y\cdot(\text{height})\cdot dy = y(1-y)\,dy$.

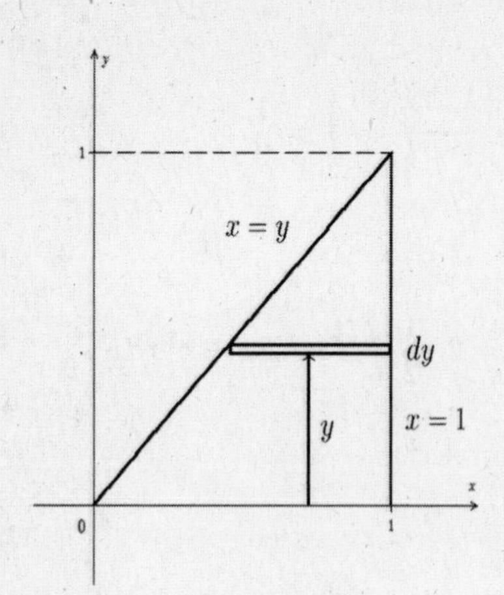

$$\begin{aligned}\overline{y} &= \frac{\int_0^1 y(1-y)\,dy}{1/2}\\ &= 2\int_0^1 (y-y^2)\,dy\\ &= \frac{1}{3}.\end{aligned}$$

9.

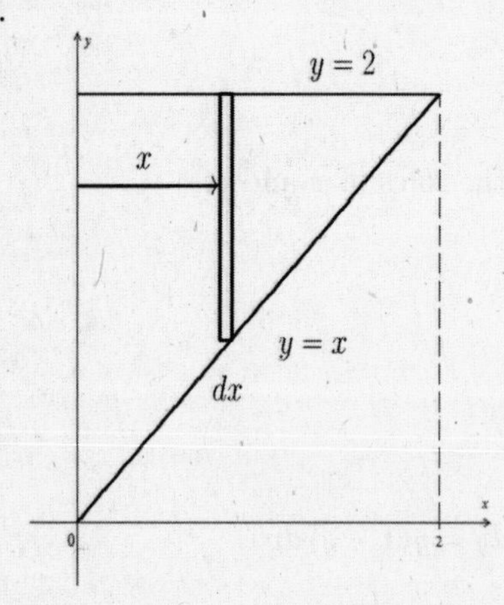

$\overline{x}$: moment of typical element: $x(2-x)\,dx$.

$$\overline{x} = \frac{\int_0^2 x(2-x)\,dx}{\int_0^2 (2-x)\,dx} = \frac{x^2 - \frac{1}{3}x^3\Big|_0^2}{2x - \frac{1}{2}x^2\Big|_0^2} = \frac{4-\frac{8}{3}}{4-2} = \frac{\frac{12}{3}-\frac{8}{3}}{2} = \frac{2}{3}.$$

To obtain $\overline{y}$, observe that

(a) distance from x-axis to center of typical element: $\frac{1}{2}(2+x)$;

(b) height of typical element: $2-x$.

$$\overline{y} = \frac{M_x}{A} = \frac{\int_0^2 \frac{1}{2}(2+x)(2-x)\,dx}{2} = \frac{4}{3}.$$

However, the alternate method is simpler in this case:

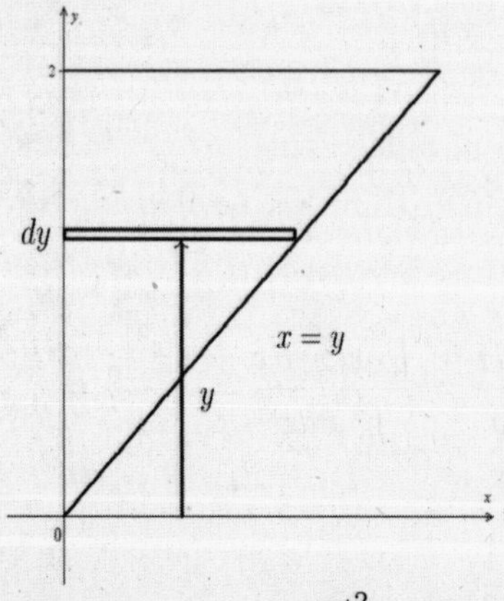

$$\overline{y} = \frac{\int_0^2 y \cdot y\,dy}{2} = \frac{\frac{1}{3}y^3\Big|_0^2}{2} = \frac{4}{3}.$$

11. Intercepts: $4 - x^2 = 0$ when $x = \pm 2$.

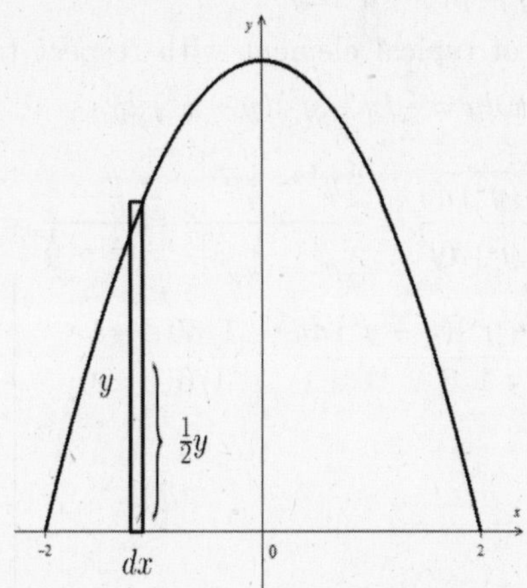

$$A = \int_{-2}^{2} (4 - x^2)\, dx = \frac{32}{3}.$$

$\overline{y}$: moment of typical element: $\left(\frac{1}{2}y\right)(y\, dx)$.

$$\begin{aligned}
\overline{y} &= \frac{M_x}{A} = \frac{\int_{-2}^{2} \left(\frac{1}{2}y\right)(y\, dx)}{A} = \frac{\frac{1}{2}\int_{-2}^{2}(4-x^2)(4-x^2)\, dx}{32/3} = \frac{3}{32}\cdot\frac{1}{2}\int_{-2}^{2}(16 - 8x^2 + x^4)\, dx \\
&= \frac{3}{64}\left(16x - \frac{8x^3}{3} + \frac{x^5}{5}\right)\Bigg|_{-2}^{2} = \frac{3}{64}\left(32 - \frac{64}{3} + \frac{32}{5}\right) - \frac{3}{64}\left(-32 + \frac{64}{3} - \frac{32}{5}\right) \\
&= 2\cdot\frac{3}{64}\left(32 - \frac{64}{3} + \frac{32}{5}\right) = \frac{3}{32}\,\frac{480 - 320 + 96}{15} = \frac{3}{32}\cdot\frac{256}{15} = \frac{8}{5}.
\end{aligned}$$

$\overline{x} = 0$ by symmetry.

13.

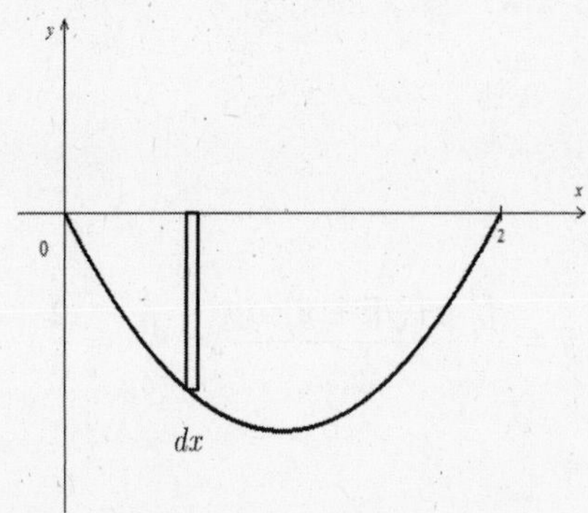

$\overline{x} = 1$ by symmetry.

$\overline{y}$: moment with respect to the x-axis: $\left(\frac{1}{2}y\right) y\, dx = \frac{1}{2}y^2\, dx$.

$$\overline{y} = \frac{\int_0^2 \frac{1}{2}y^2\, dx}{\int_0^2 y\, dx} = \frac{\int_0^2 \frac{1}{2}(x^2 - 2x)^2\, dx}{\int_0^2 (x^2 - 2x)\, dx} = \frac{\frac{8}{15}}{-\frac{4}{3}} = -\frac{2}{5}.$$

That the denominator is negative can be seen from the Riemann sum:

$$\overline{y} = \frac{\int_0^2 \frac{1}{2}y^2\, dx}{\lim_{n\to\infty}\sum_{i=1}^{n} f(x_i)\Delta x_i}.$$ Each $f(x_i) < 0$.

15.

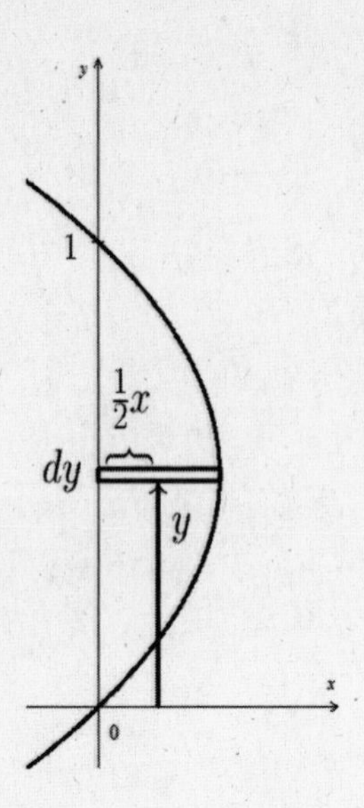

$\overline{y}$: moment of typical element with respect to the x-axis: $yx\,dy = y(y-y^2)\,dy$.

$\overline{x}$: moment of typical element with respect to the y-axis: $\frac{1}{2}x \cdot x\,dy = \frac{1}{2}(y-y^2)(y-y^2)\,dy$.

$$\overline{y} = \frac{\int_0^1 y(y-y^2)\,dy}{\int_0^1 (y-y^2)\,dy} = \frac{\frac{1}{3}y^3 - \frac{1}{4}y^4\Big|_0^1}{\frac{1}{2}y^2 - \frac{1}{3}y^3\Big|_0^1} = \frac{\frac{1}{3}-\frac{1}{4}}{\frac{1}{2}-\frac{1}{3}} = \frac{\frac{1}{12}}{\frac{1}{6}} = \frac{1}{2}.$$

$$\overline{x} = \frac{\int_0^1 \frac{1}{2}(y-y^2)(y-y^2)\,dy}{1/6} = \frac{1/60}{1/6} = \frac{1}{10}.$$

17.

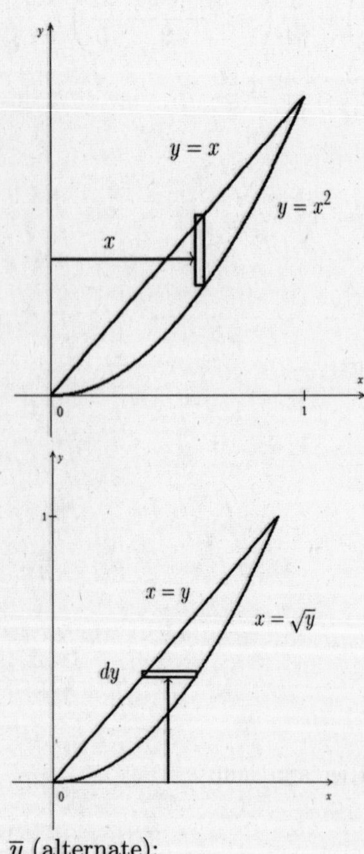

$$\overline{x} = \frac{\int_0^1 x\left(x - x^2\right)\,dx}{\int_0^1 \left(x - x^2\right)\,dx} = \frac{\frac{1}{12}}{\frac{1}{6}} = \frac{1}{2}$$

$$\overline{y} = \frac{\int_0^1 y\left(\sqrt{y} - y\right)\,dy}{\frac{1}{6}} = \frac{\frac{1}{15}}{\frac{1}{6}} = \frac{2}{5}$$

$\overline{y}$ (alternate):

Distance from x-axis to the center of the typical element: $\frac{1}{2}\left(x + x^2\right)$. So

$$\overline{y} = \frac{\int_0^1 \frac{1}{2}\left(x + x^2\right)\left(x - x^2\right)\,dx}{\frac{1}{6}} = \frac{\frac{1}{15}}{\frac{1}{6}} = \frac{2}{5}$$

19.

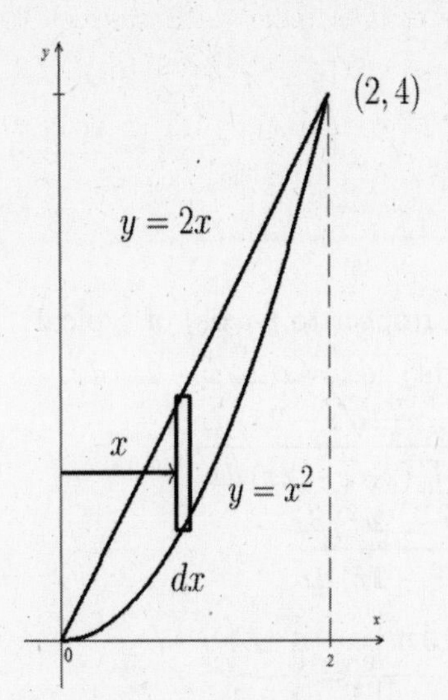

$\overline{x}$: moment with respect to the y-axis: $x(2x - x^2)\,dx$.

$$\overline{x} = \frac{M_y}{A} = \frac{\int_0^2 x(2x - x^2)\,dx}{\int_0^2 (2x - x^2)\,dx} = \frac{\frac{4}{3}}{\frac{4}{3}} = 1.$$

To find $\overline{y}$, observe that

(a) distance from x-axis to center of typical element: $\frac{1}{2}(2x + x^2)$;

(b) height of typical element: $2x - x^2$.

$$\begin{aligned}\overline{y} &= \frac{M_x}{A} = \frac{\int_0^2 \frac{1}{2}(2x + x^2)(2x - x^2)\,dx}{4/3} = \frac{3}{4}\cdot\frac{1}{2}\int_0^2 (4x^2 - x^4)\,dx = \frac{3}{8}\left(\frac{4}{3}x^3 - \frac{1}{5}x^5\right)\Big|_0^2 \\ &= \frac{3}{8}\left(\frac{32}{3} - \frac{32}{5}\right) = \frac{3}{8}\cdot 32\left(\frac{1}{3} - \frac{1}{5}\right) = 12\left(\frac{2}{15}\right) = \frac{8}{5}.\end{aligned}$$

$\overline{y}$ (alternate):

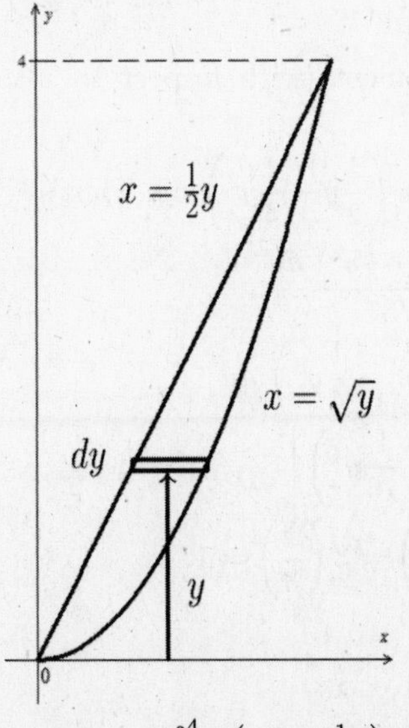

$$\overline{y} = \frac{M_x}{A} = \frac{\int_0^4 y\left(\sqrt{y} - \frac{1}{2}y\right)\,dy}{4/3} = \frac{32/15}{4/3} = \frac{8}{5}.$$

21. Substituting $y = 2x$ in $y^2 = 4x$, we obtain:

$$\begin{aligned} 4x^2 &= 4x \\ x^2 - x &= 0 \\ x(x-1) &= 0 \\ x &= 0,\ 1. \end{aligned}$$

The points of intersection are $(0,0)$ and $(1,2)$.

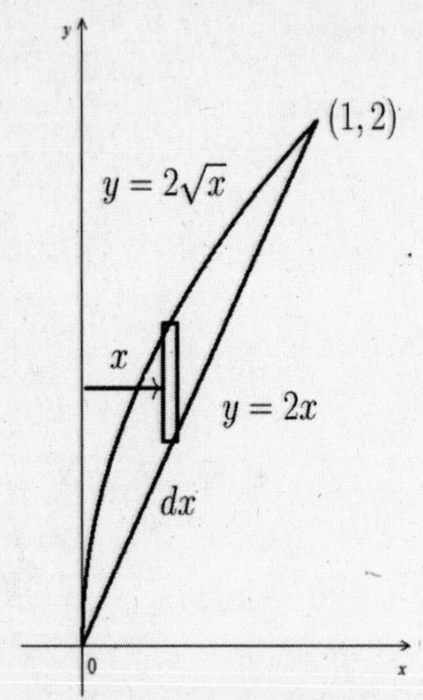

$\overline{x}$: moment (with respect to y-axis) of typical element: $x \cdot (\text{height}) \cdot dx = x(2\sqrt{x} - 2x)\,dx$.

$$\begin{aligned} \overline{x} &= \frac{M_y}{A} = \frac{\int_0^1 x(2\sqrt{x} - 2x)\,dx}{\int_0^1 (2\sqrt{x} - 2x)\,dx} \\ &= \frac{\int_0^1 (2x^{3/2} - 2x^2)\,dx}{\int_0^1 (2x^{1/2} - 2x)\,dx} \\ &= \frac{\frac{4}{5}x^{5/2} - \frac{2}{3}x^3\Big|_0^1}{\frac{4}{3}x^{3/2} - x^2\Big|_0^1} = \frac{\frac{4}{5} - \frac{2}{3}}{\frac{4}{3} - 1} \\ &= \frac{\frac{12-10}{15}}{\frac{1}{3}} = \frac{2}{15} \cdot \frac{3}{1} = \frac{2}{5}. \end{aligned}$$

To obtain $\overline{y}$, note that

(a) distance from x-axis to center of element $= \dfrac{1}{2}(2\sqrt{x} + 2x)$;

(b) height of typical element $= (2\sqrt{x} - 2x)\,dx$.

$$\begin{aligned} \overline{y} &= \frac{M_x}{A} = \frac{\int_0^1 \frac{1}{2}(2\sqrt{x} + 2x)(2\sqrt{x} - 2x)\,dx}{1/3} = \frac{3}{2}\int_0^1 (4x - 4x^2)\,dx \\ &= \frac{3}{2}\left(2x^2 - \frac{4}{3}x^3\right)\Big|_0^1 = \frac{3}{2}\left(2 - \frac{4}{3}\right) = \frac{3}{2} \cdot \frac{2}{3} = 1. \end{aligned}$$

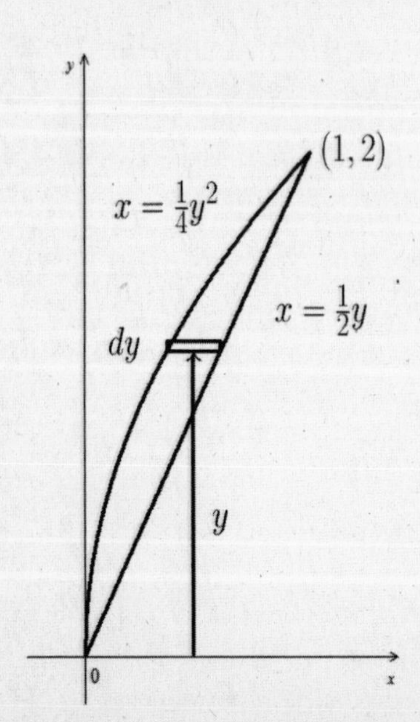

$\overline{y}$ (alternate): moment (with respect to x-axis) of typical element:

$y \cdot (\text{height}) \cdot dy = y\left(\dfrac{1}{2}y - \dfrac{1}{4}y^2\right)dy$. Thus

$$\begin{aligned} \overline{y} &= \frac{\int_0^2 y\left(\frac{1}{2}y - \frac{1}{4}y^2\right)dy}{1/3} \\ &= 3\int_0^2 \left(\frac{1}{2}y^2 - \frac{1}{4}y^3\right)dy \\ &= 3\left(\frac{1}{6}y^3 - \frac{1}{16}y^4\right)\Big|_0^2 \\ &= 3\left(\frac{4}{3} - 1\right) = 3\left(\frac{1}{3}\right) = 1. \end{aligned}$$

23.

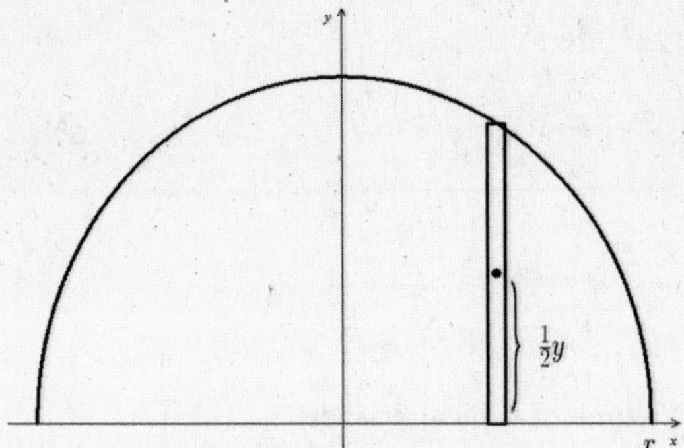

$\overline{x} = 0$ by symmetry.
$\overline{y}$: Area $= \dfrac{1}{2}\pi r^2$.

$$\begin{aligned}\overline{y} &= \frac{\int_{-r}^{r} \frac{1}{2}\sqrt{r^2-x^2}\cdot\sqrt{r^2-x^2}\,dx}{A} = \frac{\int_{-r}^{r}\frac{1}{2}(r^2-x^2)\,dx}{\frac{1}{2}\pi r^2} = \frac{\frac{1}{2}\left(r^2x-\frac{1}{3}x^3\right)\Big|_{-r}^{r}}{\frac{1}{2}\pi r^2} \\ &= \frac{\frac{1}{2}\left(r^3-\frac{1}{3}r^3\right)-\frac{1}{2}\left(-r^3+\frac{1}{3}r^3\right)}{\frac{1}{2}\pi r^2} = \frac{r^3-\frac{1}{3}r^3}{\frac{1}{2}\pi r^2} = \frac{\frac{2}{3}r^3}{\frac{1}{2}\pi r^2} = \frac{4r}{3\pi}.\end{aligned}$$

Answer: along the axis of the semicircle, $\dfrac{4r}{3\pi}$ units from the center.

25.

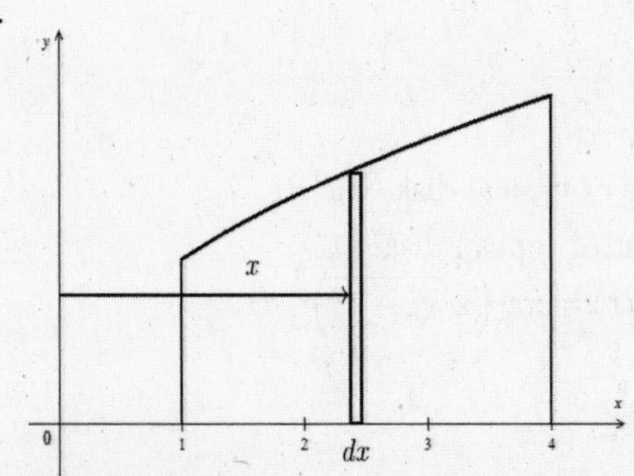

Volume of typical disk: $\pi\,(\text{radius})^2\cdot(\text{thickness})$
$= \pi y^2\,dx = \pi\left(\sqrt{x}\right)^2\,dx$

Moment of typical disk: $x\cdot\pi\left(\sqrt{x}\right)^2\,dx$

$$\begin{aligned}\overline{x} &= \frac{M_y}{V} = \frac{\int_1^4 x\cdot\pi\left(\sqrt{x}\right)^2\,dx}{\int_1^4 \pi\left(\sqrt{x}\right)^2\,dx} = \frac{\pi\frac{x^3}{3}\Big|_1^4}{\pi\frac{x^2}{2}\Big|_1^4} \\ &= \frac{\frac{64}{3}-\frac{1}{3}}{\frac{16}{2}-\frac{1}{2}} = \frac{63}{3}\cdot\frac{2}{15} = \frac{14}{5} \\ \overline{y} &= 0\end{aligned}$$

27.

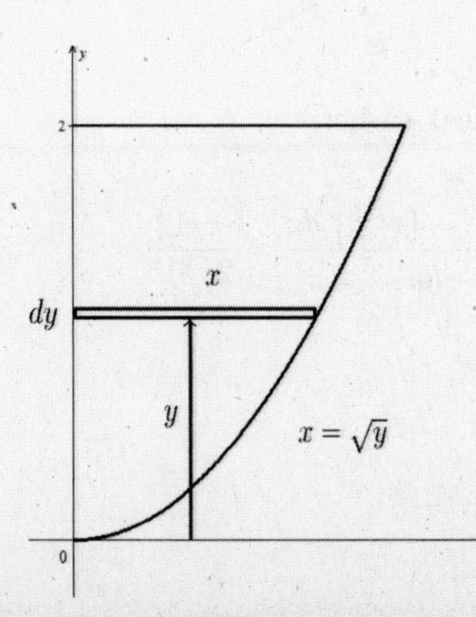

Volume of typical disk: $\pi x^2\,dy = \pi\left(\sqrt{y}\right)^2\,dy$
Moment of typical disk: $y\cdot\pi\left(\sqrt{y}\right)^2\,dy$

$$\begin{aligned}\overline{y} &= \frac{M_x}{V} = \frac{\int_0^2 y\cdot\pi\left(\sqrt{y}\right)^2\,dy}{\int_0^2 \pi\left(\sqrt{y}\right)^2\,dy} = \frac{\pi\frac{y^3}{3}\Big|_0^2}{\pi\frac{y^2}{2}\Big|_0^2} = \frac{4}{3} \\ \overline{x} &= 0\end{aligned}$$

29. Volume of typical disk: $\pi(\text{radius})^2 \cdot (\text{thickness}) = \pi y^2\,dx$.

Moment of typical disk: $x \cdot \pi y^2\,dx = x \cdot \pi(2x^2)^2\,dx$.

$$\overline{x} = \frac{M_y}{V} = \frac{\int_0^1 x \cdot \pi(2x^2)^2\,dx}{\int_0^1 \pi(2x^2)^2\,dx} = \frac{4\pi \int_0^1 x^5\,dx}{4\pi \int_0^1 x^4\,dx} = \frac{1/6}{1/5} = \frac{5}{6}$$

$\overline{y} = 0$

31.

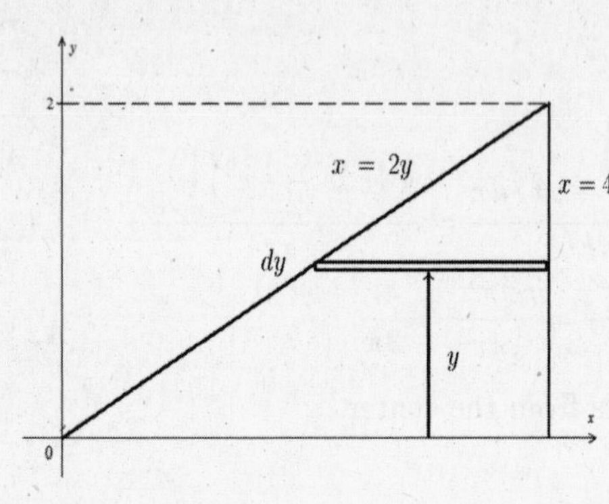

Volume of typical washer:

$\pi\left[4^2 - (2y)^2\right]\,dy = \pi\left(16 - 4y^2\right)\,dy$

$$\overline{y} = \frac{\int_0^2 y \cdot \pi\left(16 - 4y^2\right)\,dy}{\int_0^2 \pi\left(16 - 4y^2\right)\,dy} = \frac{\pi\left(8y^2 - y^4\right)\Big|_0^2}{\pi\left(16y - 4\frac{y^3}{3}\right)\Big|_0^2}$$

$$= \frac{2^5 - 2^4}{2^5 - 2^5/3} = \frac{2^4(2-1)}{2^5(1-1/3)} = \frac{16}{1} \cdot \frac{3}{64} = \frac{3}{4}$$

$\overline{x} = 0$

33.

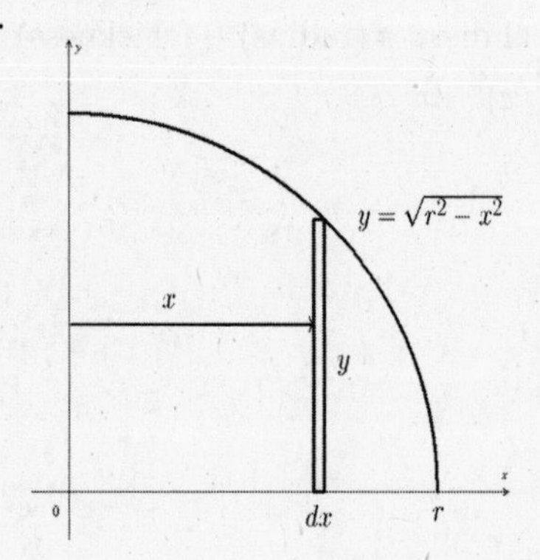

Volume of typical disk: $\pi y^2\,dx$.

Moment of typical disk:

$x(\pi y^2\,dx) = \pi x\left(\sqrt{r^2 - x^2}\right)^2\,dx$.

$$\overline{x} = \frac{M_y}{V} = \frac{\pi \int_0^r x\left(\sqrt{r^2-x^2}\right)^2\,dx}{\pi \int_0^r \left(\sqrt{r^2-x^2}\right)^2\,dx} = \frac{\int_0^r x(r^2 - x^2)\,dx}{\int_0^r (r^2 - x^2)\,dx} = \frac{\int_0^r (r^2 x - x^3)\,dx}{\int_0^r (r^2 - x^2)\,dx}$$

$$= \frac{r^2\frac{x^2}{2} - \frac{x^4}{4}\Big|_0^r}{r^2 x - \frac{x^3}{3}\Big|_0^r} = \frac{\frac{r^4}{2} - \frac{r^4}{4}}{r^3 - \frac{r^3}{3}} = \frac{r^4\left(\frac{1}{2} - \frac{1}{4}\right)}{r^3\left(1 - \frac{1}{3}\right)} = \frac{r\left(\frac{1}{4}\right)}{\frac{2}{3}} = \frac{3}{8}r.$$

$\overline{y} = 0$.

35. (a)

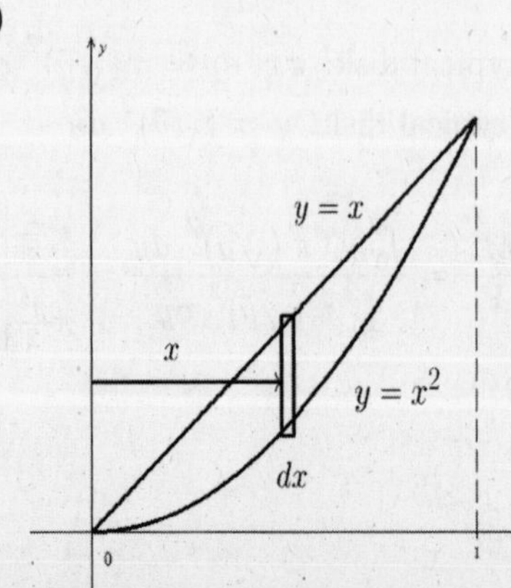

Volume of typical washer:

$\pi\left[x^2 - (x^2)^2\right]\,dx$.

$$\overline{x} = \frac{\int_0^1 x \cdot \pi\left[x^2 - (x^2)^2\right]\,dx}{\int_0^1 \pi\left[x^2 - (x^2)^2\right]\,dx} = \frac{\pi/12}{2\pi/15} = \frac{5}{8}.$$

(b)

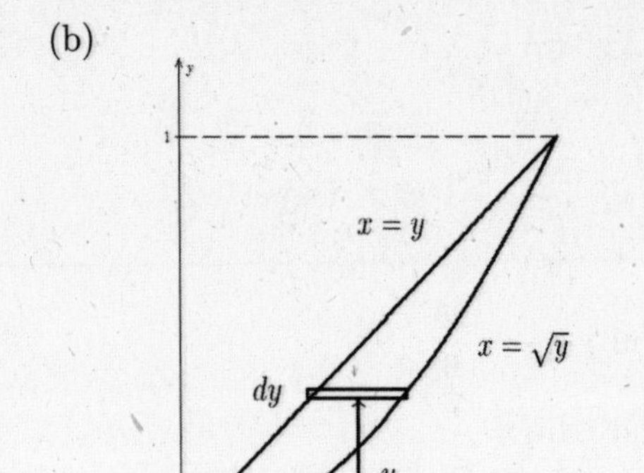

Volume of typical washer:

$\pi\left[(\sqrt{y})^2 - y^2\right]\,dy.$

$$\overline{y} = \frac{\int_0^1 y\cdot\pi\left[(\sqrt{y})^2 - y^2\right]\,dy}{\int_0^1 \pi\left[(\sqrt{y})^2 - y^2\right]\,dy} = \frac{\pi/12}{\pi/6} = \frac{1}{2}.$$

37.

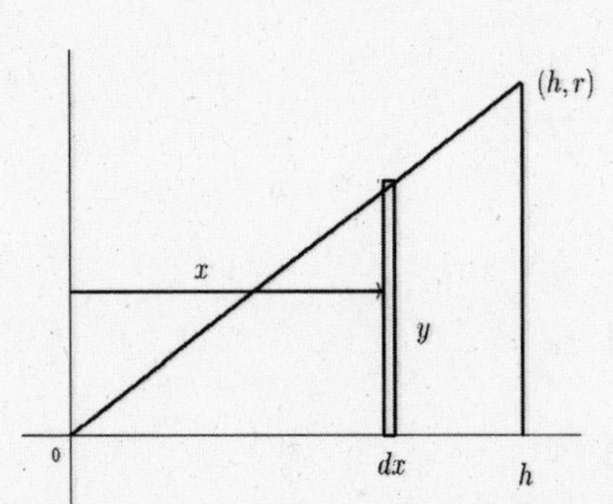

Slope of line: $\dfrac{r}{h}$.

Equation: $y = \dfrac{r}{h}x$.

Moment of typical disk: $x(\pi y^2\,dx) = \pi x\left(\dfrac{r}{h}x\right)^2\,dx.$

$$\overline{x} = \frac{\pi\int_0^h x\left(\frac{r}{h}x\right)^2\,dx}{\pi\int_0^h\left(\frac{r}{h}x\right)^2\,dx} = \frac{(r^2/h^2)\int_0^h x^3\,dx}{(r^2/h^2)\int_0^h x^2\,dx} = \frac{\int_0^h x^3\,dx}{\int_0^h x^2\,dx} = \frac{\left.\frac{x^4}{4}\right|_0^h}{\left.\frac{x^3}{3}\right|_0^h} = \frac{h^4}{4}\cdot\frac{3}{h^3} = \frac{3}{4}h.$$

Since $\overline{y} = 0$, we conclude that the centroid lies on the axis, one-fourth of the way from the base.

39.

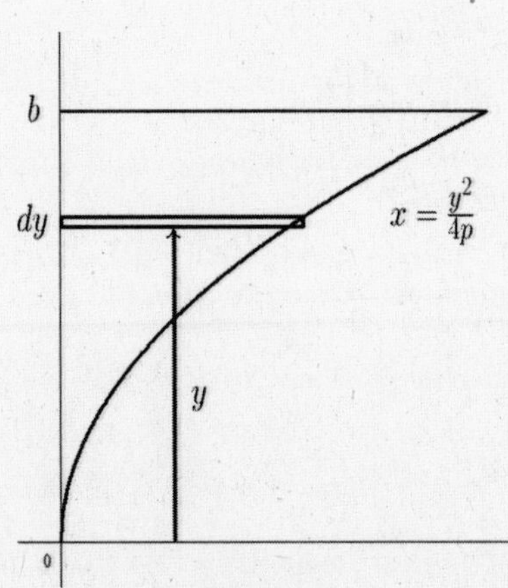

Volume of typical disk:

$\pi\left(\dfrac{y^2}{4p}\right)^2\,dy.$

$$\overline{y} = \frac{\int_0^b y\cdot\pi\left(\frac{y^4}{16p^2}\right)\,dy}{\int_0^b \pi\left(\frac{y^4}{16p^2}\right)\,dy} = \frac{\pi/(16p^2)\int_0^b y^5\,dy}{\pi/(16p^2)\int_0^b y^4\,dy} = \frac{\left.\frac{1}{6}y^6\right|_0^b}{\left.\frac{1}{5}y^5\right|_0^b} = \frac{\frac{1}{6}b^6}{\frac{1}{5}b^5} = \frac{5}{6}b.$$

$\overline{x} = 0.$

41. Form of upper parabola: $x^2 = 4p_1(y-4)$.
Form of lower parabola: $x^2 = 4p_2 y$.
Both parabolas pass through the point $(5, 10)$:

$$25 = 4p_1(10-4) \quad \text{and} \quad 4p_1 = \frac{25}{6},$$
$$25 = 4p_2 \cdot 10 \quad \text{and} \quad 4p_2 = \frac{5}{2}.$$

Equations:

$$x^2 = \frac{25}{6}(y-4) \quad \text{or} \quad y = 0.24(x^2 + 50/3),$$
$$x^2 = \frac{5}{2}y \quad \text{or} \quad y = 0.4x^2.$$

$\overline{y}$: moment of typical element:
$\frac{1}{2}\left[0.24(x^2+50/3)+0.4x^2\right]\left[0.24(x^2+50/3)-0.4x^2\right]\,dx$
$= \frac{1}{2}\left\{\left[0.24(x^2+50/3)\right]^2 - \left[0.4x^2\right]^2\right\}\,dx.$
Using the integration capability of a graphing utility,
$$\overline{y} = \frac{\frac{1}{2}\int_0^5 \left\{\left[0.24(x^2+50/3)\right]^2 - \left[0.4x^2\right]^2\right\}\,dx}{\int_0^5 \left[0.24(x^2+50/3) - 0.4x^2\right]\,dx} = 3.6\,\text{ft}.$$
Finally, $\overline{x} = 0$.

5.5 Moments of Inertia

1.

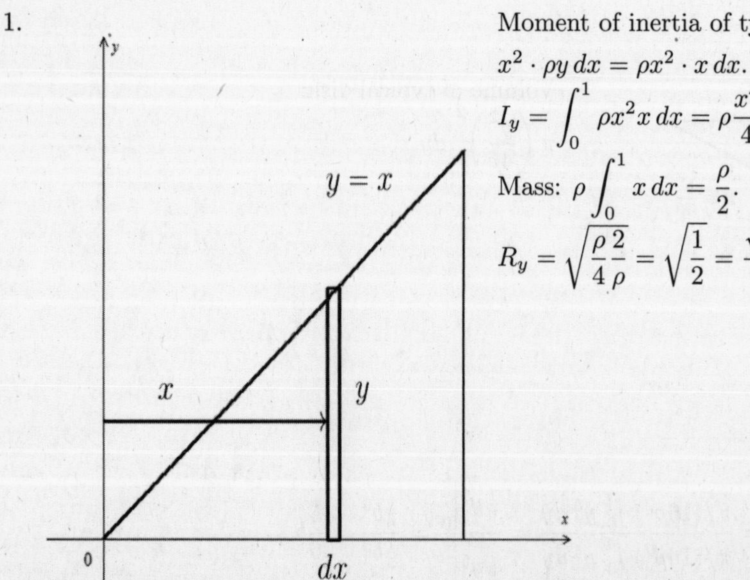

Moment of inertia of typical element:
$x^2 \cdot \rho y\,dx = \rho x^2 \cdot x\,dx.$
$$I_y = \int_0^1 \rho x^2 x\,dx = \rho \frac{x^4}{4}\bigg|_0^1 = \frac{\rho}{4}.$$
$$\text{Mass: } \rho \int_0^1 x\,dx = \frac{\rho}{2}.$$
$$R_y = \sqrt{\frac{\rho}{4}\frac{2}{\rho}} = \sqrt{\frac{1}{2}} = \frac{\sqrt{2}}{2}.$$

3.

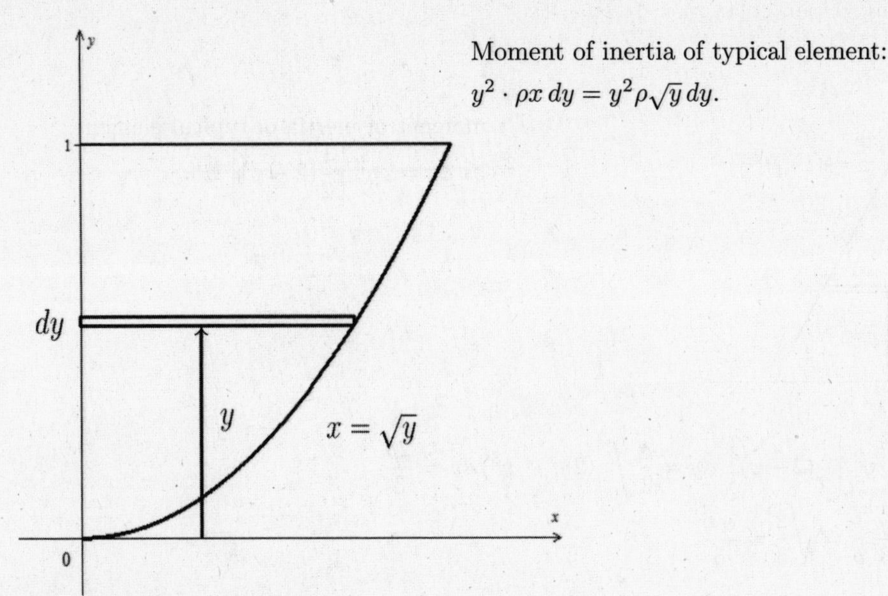

Moment of inertia of typical element:
$y^2 \cdot \rho x\, dy = y^2 \rho \sqrt{y}\, dy$.

Summing from $y = 0$ to $y = 1$, we get

$I_x = \int_0^1 y^2 \cdot \rho \sqrt{y}\, dy = \rho \int_0^1 y^{5/2}\, dy = \rho \frac{y^{7/2}}{7/2}\Big|_0^1 = \frac{2\rho}{7}$.

Mass: $\rho \int_0^1 \sqrt{y}\, dy = \rho \left(\frac{2}{3}\right) y^{3/2}\Big|_0^1 = \frac{2\rho}{3}$.

$R_x = \sqrt{\frac{2\rho}{7} \cdot \frac{3}{2\rho}} = \sqrt{\frac{3}{7}} = \frac{\sqrt{21}}{7}$.

5.

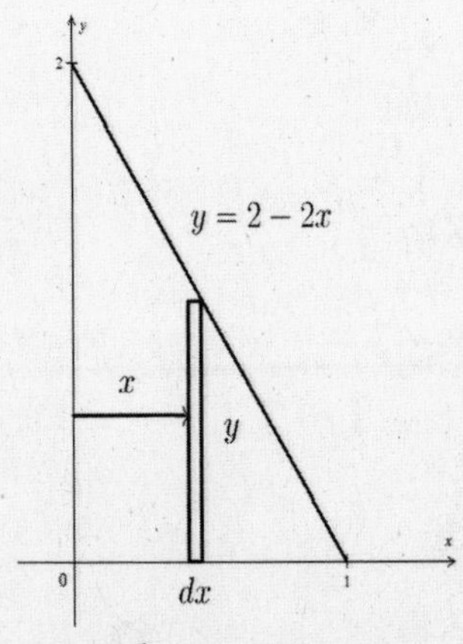

I_y: moment of inertia of typical element:
$x^2 \cdot \rho y\, dx = \rho x^2 \cdot (2 - 2x)\, dx$.

$I_y = \int_0^1 \rho x^2 (2 - 2x)\, dx = \rho \int_0^1 (2x^2 - 2x^3)\, dx = \rho \left(\frac{2}{3}x^3 - \frac{1}{2}x^4\right)\Big|_0^1 = \rho \left(\frac{2}{3} - \frac{1}{2}\right) = \frac{\rho}{6}$.

Since $A = \frac{1}{2}bh = \frac{1}{2}(1)(2) = 1$, the mass $= \rho$. Thus $R_y = \sqrt{\frac{\rho}{6}\frac{1}{\rho}} = \sqrt{\frac{1}{6}} = \frac{\sqrt{6}}{6}$.

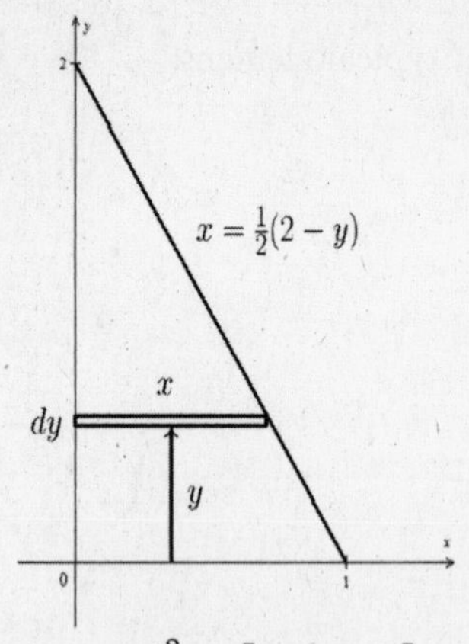

I_x: moment of inertia of typical element:
$y^2 \cdot \rho x\, dy = \rho y^2 \left[\frac{1}{2}(2-y)\right] dy.$

$I_x = \rho \int_0^2 y^2 \left[\frac{1}{2}(2-y)\right] dy = \frac{\rho}{2}\int_0^2 (2y^2 - y^3)\, dy = \frac{2\rho}{3}.$

$R_x = \sqrt{\frac{2\rho}{3}\frac{1}{\rho}} = \sqrt{\frac{2}{3}} = \frac{\sqrt{6}}{3}.$

7.

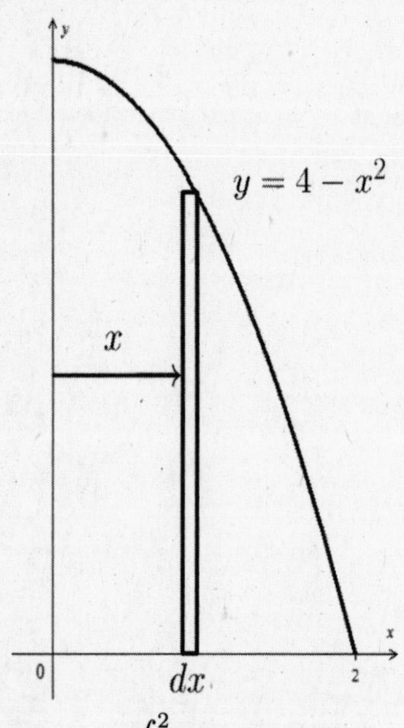

Moment of inertia of typical element:
$x^2 \cdot \rho y\, dx = x^2 \cdot \rho(4 - x^2)\, dx.$

$$\begin{aligned} I_y &= \int_0^2 x^2 \cdot \rho(4-x^2)\, dx = \rho\left(\frac{4}{3}x^3 - \frac{1}{5}x^5\right)\Big|_0^2 \\ &= \rho\left(\frac{32}{3} - \frac{32}{5}\right) = 32\rho\left(\frac{1}{3} - \frac{1}{5}\right) = 32\rho \cdot \frac{5-3}{15} = \frac{64\rho}{15}. \end{aligned}$$

Mass: $\rho \int_0^2 (4 - x^2)\, dx = \rho\left(4x - \frac{1}{3}x^3\right)\Big|_0^2 = \rho\left(8 - \frac{8}{3}\right) = \frac{16\rho}{3}.$

$R_y = \sqrt{\frac{64\rho}{15} \cdot \frac{3}{16\rho}} = \sqrt{\frac{4}{5}} = \frac{2}{\sqrt{5}} = \frac{2\sqrt{5}}{5}.$

9. Substituting $y = \frac{1}{2}x$ in $y^2 = x$, we get

$$\begin{aligned} \frac{1}{4}x^2 &= x \\ x^2 - 4x &= 0 \\ x(x-4) &= 0 \\ x &= 0, 4. \end{aligned}$$

The points of intersection are $(0, 0)$ and $(4, 2)$.

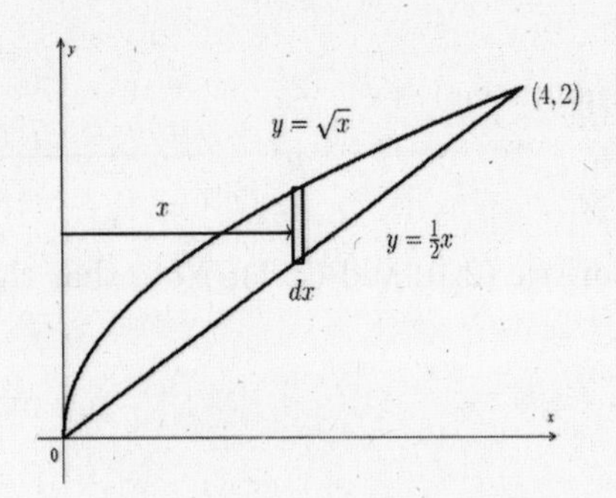

Moment of inertia (with respect to y-axis) of typical element:
$x^2 \cdot \rho \cdot \text{(height)} \cdot dx = \rho x^2 \left(\sqrt{x} - \frac{1}{2}x\right) dx.$

$$\begin{aligned} I_y &= \rho \int_0^4 x^2 \left(\sqrt{x} - \frac{1}{2}x\right) dx = \rho \int_0^4 \left(x^{5/2} - \frac{1}{2}x^3\right) dx = \rho \left(\frac{2}{7}x^{7/2} - \frac{1}{8}x^4\right)\Big|_0^4 \\ &= \rho \left[\frac{2}{7}(4^{7/2}) - \frac{1}{8}(4^4)\right] = \rho \left(\frac{2}{7} \cdot 128 - 32\right) = 32\rho \left(\frac{8}{7} - 1\right) = 32\rho \left(\frac{8-7}{7}\right) = \frac{32\rho}{7}. \end{aligned}$$

Mass: $\rho \int_0^4 \left(\sqrt{x} - \frac{1}{2}x\right) dx = \rho \left(\frac{2}{3}x^{3/2} - \frac{1}{4}x^2\right)\Big|_0^4 = \rho \left(\frac{2}{3} \cdot 8 - 4\right) = \frac{4\rho}{3}.$

$R_y = \sqrt{\frac{32\rho}{7} \frac{3}{4\rho}} = \sqrt{\frac{8 \cdot 3}{7}} = \frac{2\sqrt{2 \cdot 3}}{\sqrt{7}} = \frac{2\sqrt{42}}{7}.$

11.
$$\begin{aligned} y &= 9 - 3x \\ y &= 9 - x^2 \\ \hline 0 &= -3x + x^2 \quad \text{(subtracting)} \end{aligned}$$
$x(x-3) = 0;\ x = 0, 3.$

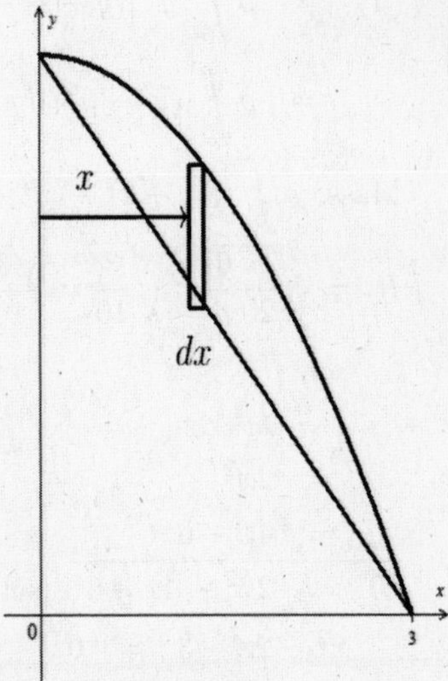

Moment of typical element: $x^2 \cdot \rho(\text{height})\, dx = x^2 \cdot \rho \left[(9 - x^2) - (9 - 3x)\right] dx.$

$$\begin{aligned} I_y &= \rho \int_0^3 x^2(9 - x^2 - 9 + 3x)\, dx = \rho \left(-\frac{1}{5}x^5 + \frac{3}{4}x^4\right)\Big|_0^3 \\ &= \rho \left(-\frac{1}{5} \cdot 3^5 + \frac{3}{4} \cdot 3^4\right) = \rho(3^5)\left(-\frac{1}{5} + \frac{1}{4}\right) = \rho \frac{3^5}{20} = \frac{243\rho}{20}. \end{aligned}$$

Mass: $\rho \int_0^3 \left[(9 - x^2) - (9 - 3x)\right] dx = \frac{9\rho}{2}.$

$R_y = \sqrt{\frac{3^5\rho}{20} \cdot \frac{2}{3^2\rho}} = \sqrt{\frac{3^3}{10}} = \frac{3\sqrt{3}}{\sqrt{10}} = \frac{3\sqrt{30}}{10}.$

13. Solving the equations simultaneously, we get

$$\begin{aligned} x &= y^2+2 \\ x &= y+2 \\ \hline 0 &= y^2-y \qquad \text{(subtracting)} \\ y(y-1) &= 0 \\ y &= 0,1. \end{aligned}$$

The points of intersection are $(2,0)$ and $(3,1)$. Note that the typical element has to be drawn horizontally.

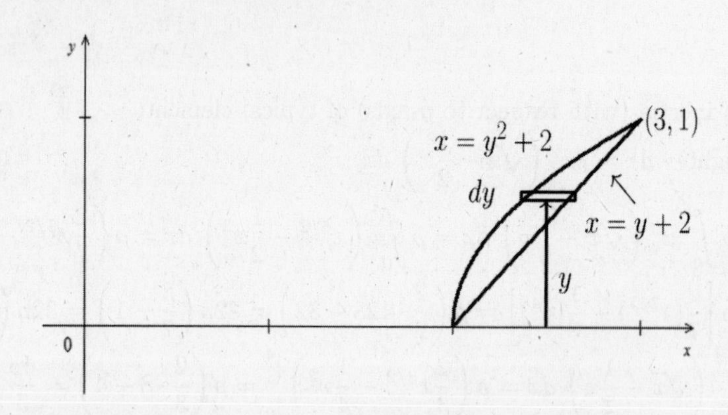

Moment of inertia (with respect to x-axis) of typical element:
$\rho y^2 \cdot (\text{height}) \cdot dy = \rho y^2 \left[(y+2)-(y^2+2)\right] dy.$

$$\begin{aligned} I_x &= \rho \int_0^1 y^2\left[(y+2)-(y^2+2)\right] dy = \rho\int_0^1 y^2(y-y^2)\,dy \\ &= \rho\int_0^1 (y^3-y^4)\,dy = \rho\left(\frac{1}{4}y^4-\frac{1}{5}y^5\right)\Bigg|_0^1 = \rho\left(\frac{1}{4}-\frac{1}{5}\right) = \frac{\rho}{20}. \end{aligned}$$

Mass: $\rho\int_0^1 \left[(y+2)-(y^2+2)\right] dy = \rho\int_0^1 (y-y^2)\,dy = \frac{\rho}{6}.$

$R_x = \sqrt{\frac{\rho}{20}\frac{6}{\rho}} = \sqrt{\frac{3}{10}} = \frac{\sqrt{30}}{10}.$

15. $$\begin{aligned} y &= 2x^2 \\ y &= 4x+6 \\ \hline 0 &= 2x^2-4x-6 \quad \text{(subtracting)} \\ x^2-2x-3 &= 0 \\ (x+1)(x-3) &= 0;\ x=-1,3. \end{aligned}$$

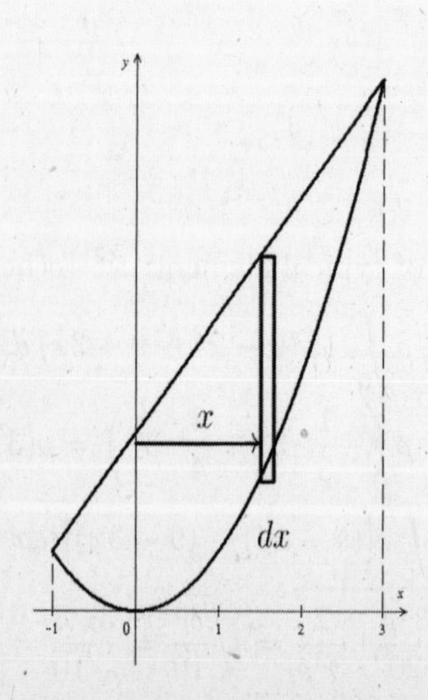

Moment of inertia of typical element:
$x^2\cdot\rho(\text{height})\,dx.$
$I_y = \int_{-1}^3 x^2\cdot\rho(4x+6-2x^2)\,dx = \frac{192\rho}{5}.$

17.

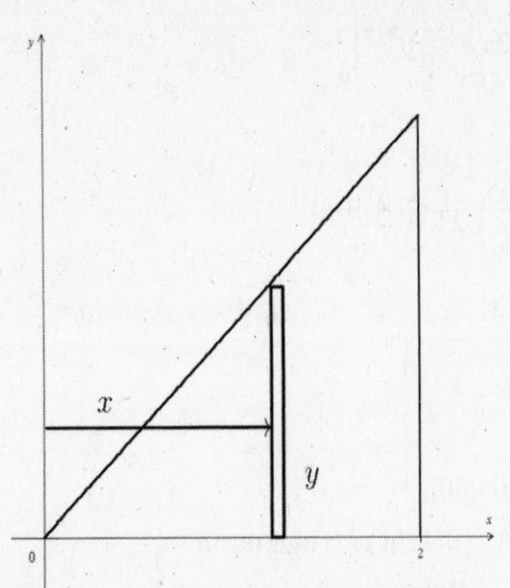

Volume of shell: $2\pi(\text{radius}) \cdot (\text{height}) \cdot (\text{thickness})$
$= 2\pi xy\,dx = 2\pi x \cdot x\,dx$
Mass of shell: $\rho \cdot 2\pi x^2\,dx$

Moment of inertial of typical shell: $x^2 \cdot \rho \cdot 2\pi x^2\,dx$

$$I_y = \int_0^2 x^2 \cdot \rho\left(2\pi x^2\,dx\right) = 2\pi\rho \frac{x^5}{5}\bigg|_0^2 = \frac{64\pi\rho}{5}$$

19.

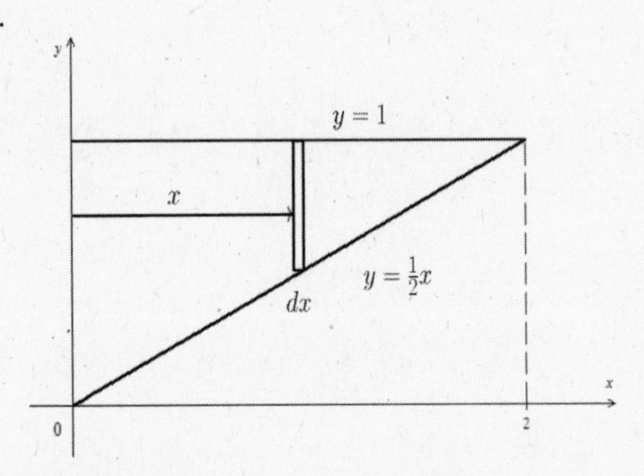

Mass of shell:
$\rho \cdot 2\pi(\text{radius}) \cdot (\text{height}) \cdot (\text{thickness}) = \rho \cdot 2\pi \cdot x \cdot \left(1 - \frac{1}{2}x\right)\,dx = 2\pi\rho x\left(1 - \frac{1}{2}x\right)\,dx.$

Moment of inertia of typical shell: $x^2 \cdot 2\pi\rho x\left(1 - \frac{1}{2}x\right)\,dx.$

$$\begin{aligned} I_y &= \int_0^2 x^2 \cdot 2\pi\rho x\left(1 - \frac{1}{2}x\right)\,dx = 2\pi\rho\int_0^2\left(x^3 - \frac{1}{2}x^4\right)\,dx = 2\pi\rho\left(\frac{x^4}{4} - \frac{x^5}{10}\right)\bigg|_0^2 \\ &= 2\pi\rho\left(4 - \frac{16}{5}\right) = 2\pi\rho\left(\frac{4}{5}\right) = \frac{8\pi\rho}{5} \end{aligned}$$

21.

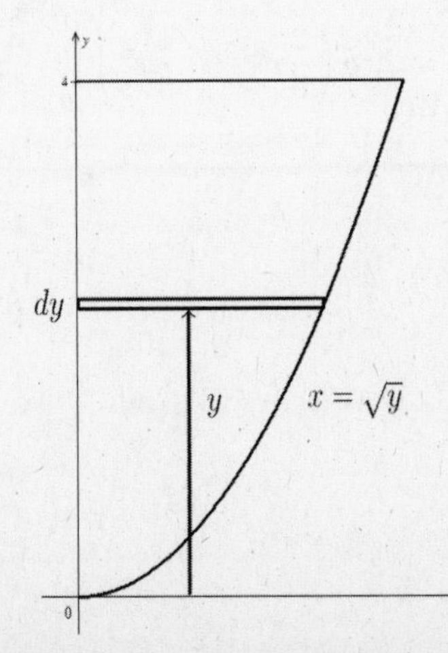

Volume of shell: $2\pi(\text{radius}) \cdot (\text{height}) \cdot (\text{thickness})$
$= 2\pi y \cdot x\,dy = 2\pi y \cdot \sqrt{y}\,dy.$
Mass of shell: $\rho \cdot 2\pi y\sqrt{y}\,dy.$
Moment of inertia of typical shell:
$y^2 \cdot \rho \cdot 2\pi y\sqrt{y}\,dy.$

$$\begin{aligned} I_x &= \int_0^4 y^2 \cdot \rho \cdot 2\pi y\sqrt{y}\,dy = 2\pi\rho \int_0^4 y^{7/2}\,dy = 2\pi\rho \cdot \frac{2}{9} y^{9/2}\Big|_0^4 \\ &= 2\pi\rho \cdot \frac{2}{9}(4)^{9/2} = 2\pi\rho\left(\frac{2}{9}\right)2^9 = \frac{2^{11}\pi\rho}{9}. \end{aligned}$$

Mass: $\int_0^4 \rho \cdot 2\pi y\sqrt{y}\,dy = 2\pi\rho\left(\frac{2}{5}\right)y^{5/2}\Big|_0^4 = 2\pi\rho\left(\frac{2}{5}\right)2^5 = \frac{2^7\pi\rho}{5}.$

$R_x = \sqrt{\frac{2^{11}\pi\rho}{9} \cdot \frac{5}{2^7\pi\rho}} = \frac{4\sqrt{5}}{3}.$

23.

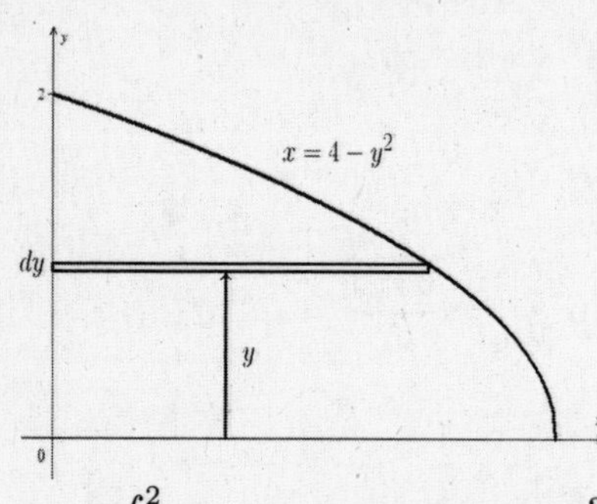

Volume of shell:
$2\pi(\text{radius}) \cdot (\text{height}) \cdot (\text{thickness})$
$= 2\pi y(4 - y^2)\,dy.$
Mass of shell: $\rho \cdot 2\pi y(4 - y^2)\,dy.$

$I_x = \int_0^2 y^2 \cdot \rho \cdot 2\pi y(4 - y^2)\,dy = \frac{32\pi\rho}{3}.$

Mass: $\int_0^2 \rho \cdot 2\pi y(4 - y^2)\,dy = 8\pi\rho.$

$R_x = \sqrt{\frac{32\pi\rho}{3} \cdot \frac{1}{8\pi\rho}} = \sqrt{\frac{4}{3}} = \frac{2}{\sqrt{3}} = \frac{2\sqrt{3}}{3}.$

25.

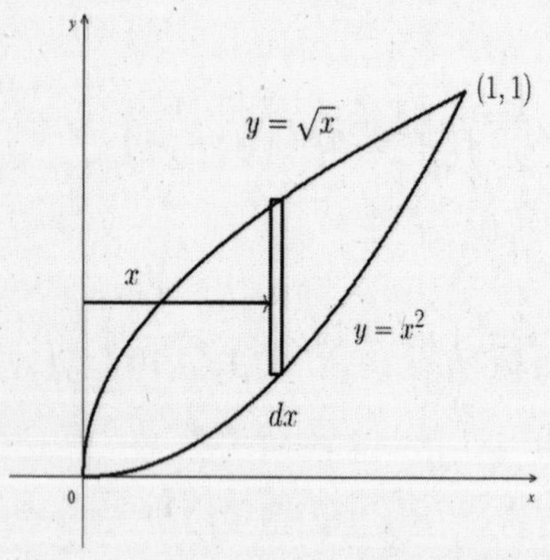

Mass of shell: $\rho \cdot 2\pi x \cdot (\text{height}) \cdot dx$
$= \rho \cdot 2\pi x\left(\sqrt{x} - x^2\right)dx.$
Moment of inertia of shell:
$x^2 \cdot \rho \cdot 2\pi x\left(\sqrt{x} - x^2\right)dx = 2\pi\rho x^3\left(\sqrt{x} - x^2\right)dx.$

$$\begin{aligned} I_y &= 2\pi\rho\int_0^1 x^3\left(\sqrt{x} - x^2\right)dx = 2\pi\rho\int_0^1\left(x^{7/2} - x^5\right)dx = 2\pi\rho\left(\frac{2}{9}x^{9/2} - \frac{1}{6}x^6\right)\Big|_0^1 \\ &= 2\pi\rho\left(\frac{2}{9} - \frac{1}{6}\right) = 2\pi\rho\left(\frac{4}{18} - \frac{3}{18}\right) = \frac{\pi\rho}{9}. \end{aligned}$$

$$\begin{aligned} \text{Mass: } 2\pi\rho\int_0^1 x\left(\sqrt{x} - x^2\right)dx &= 2\pi\rho\int_0^1\left(x^{3/2} - x^3\right)dx \\ &= 2\pi\rho\left(\frac{2}{5}x^{5/2} - \frac{1}{4}x^4\right)\Big|_0^1 \\ &= 2\pi\rho\left(\frac{2}{5} - \frac{1}{4}\right) = \frac{3\pi\rho}{10}. \end{aligned}$$

$R_y = \sqrt{\frac{\pi\rho}{9}\frac{10}{3\pi\rho}} = \sqrt{\frac{10}{9 \cdot 3}} = \frac{\sqrt{10}}{3\sqrt{3}} = \frac{\sqrt{30}}{9}.$

27. Volume of shell: $2\pi(\text{radius}) \cdot (\text{height}) \cdot (\text{thickness}) = 2\pi xy\,dx$.

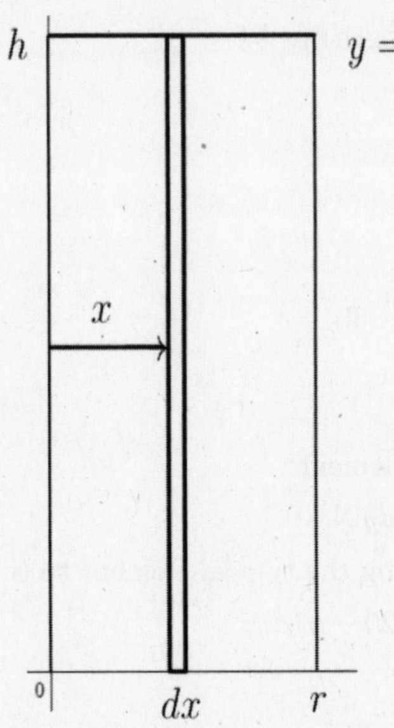

Mass of shell: $\rho(2\pi xy\,dx)$.
Moment of inertia of typical shell (since $y = h$):
$x^2 \cdot \rho(2\pi xy\,dx) = 2\pi\rho x^3 h\,dx$.
$I_y = 2\pi\rho \int_0^r x^3 h\,dx = 2\pi\rho h \frac{x^4}{4}\Big|_0^r = \frac{1}{2}\pi r^4 h\rho$.

Since the mass of the cylinder is $\rho\pi r^2 h$, I_y can also be written as follows:
$I_y = \frac{1}{2}(\rho\pi r^2 h)r^2 = \frac{1}{2}mr^2$.
$R_y = \sqrt{\frac{\pi r^4 h\rho}{2}\frac{1}{\rho\pi r^2 h}} = \sqrt{\frac{r^2}{2}} = \frac{r}{\sqrt{2}} = \frac{r\sqrt{2}}{2}$.

29. By Exercise 27, $I_y = \frac{1}{2}mr^2$.
$\frac{360\text{ rev}}{1\text{ min}} \times \frac{2\pi\text{ rad}}{1\text{ rev}} \times \frac{1\text{ min}}{60\text{ s}} = 12\pi\text{ rad/s}$.
$K = \frac{1}{2}I\omega^2 = \frac{1}{2}\left[\frac{1}{2}(2.0)(0.10)^2\right](12\pi)^2 = 7.1\text{ J}$.
$L = I\omega = \left[\frac{1}{2}(2.0)(0.10)^2\right](12\pi) = 0.38\text{ kg}\cdot\text{m}^2/\text{s}$.

5.6 Work and Fluid Pressure

1. From Hooke's law,

$$\begin{aligned} F &= kx \\ 6 &= k \cdot \frac{1}{8} \quad \text{or} \quad k = 48. \end{aligned}$$

So $F = 48x$. Since the spring is stretched a distance of 0.5 ft,
$W = \int_0^{0.5} 48x\,dx = 24x^2\Big|_0^{0.5} = 6\text{ ft-lb}$.

3. From Hooke's law,

$$\begin{aligned} F &= kx \\ 12 &= k \cdot 2 \quad \text{or} \quad k = 6. \end{aligned}$$

So $F = 6x$.

(a) Since the spring is being compressed from its natural length of 8 ft to 6 ft, a total of 2 ft, we obtain
$W = \int_0^2 6x\,dx = 3x^2\Big|_0^2 = 12$ ft-lb.

(b) The spring is stretched from 2 ft (beyond its natural length) to 5 ft:
$W = \int_2^5 6x\,dx = 3x^2\Big|_2^5 = 3(25-4) = 63$ ft-lb.

5. $F = kx; 2 = k(10)$ or $k = \frac{1}{5}$. So $F = \frac{1}{5}x$.

$$W = \int_0^3 \frac{1}{5}x\,dx = \frac{1}{5}\frac{x^2}{2}\Big|_0^3 = \frac{9}{10} \text{ in.-lb.}$$

7.

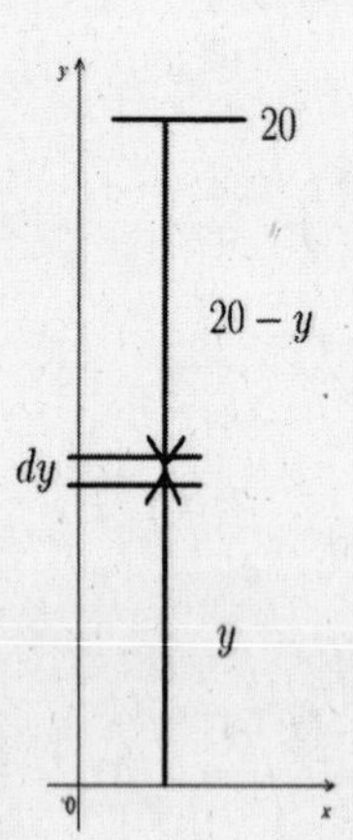

Weight of typical element:
$3\,\text{N/m} \times dy\,\text{m} = 3\,dy\,\text{N}$.
Work done in moving the typical element to the top:
$3\,dy \cdot (20-y) = 3(20-y)\,dy$.

Summing from $y = 0$ to $y = 20$, we get $W = \int_0^{20} 3(20-y)\,dy = 3\left(20y - \frac{1}{2}y^2\right)\Big|_0^{20} = 600$ J.

9.

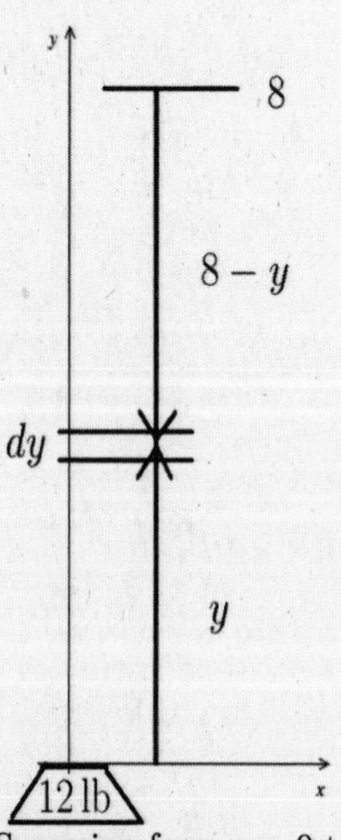

Weight of typical element:
$2\frac{\text{lb}}{\text{ft}} \times dy\,\text{ft} = 2\,dy(\text{lb})$
Work done in moving the typical element to the top:
$(8-y)\cdot 2\,dy$

Summing from $y = 0$ to $y = 8$, we get

$$W = \int_0^8 (8-y)\cdot 2\,dy = 16y - y^2\Big|_0^8 = 64 \text{ ft-lb.}$$

Work done in moving the weight at the end: $12\,\text{lb} \cdot 8\,\text{ft} = 96$ ft-lb.
Total: $64 + 96 = 160$ ft-lb

11.

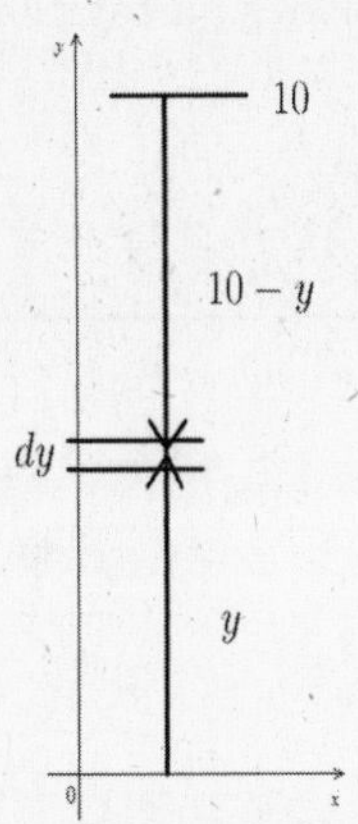

Weight of typical element:
$10\,\text{lb/ft} \times dy\,\text{ft} = 10\,dy\,\text{lb}$.
Work done in moving the typical element to the top:
$10\,dy \cdot (10 - y)$.

Summing from $y = 0$ to $y = 10$, we get
$W = 10\int_0^{10}(10-y)\,dy = 10\left(10y - \frac{1}{2}y^2\right)\Big|_0^{10} = 10(100-50) = 500\,\text{ft-lb}$.
The 20 lb-weight attached to the end is moved 10 ft: $20\,\text{lb} \times 10\,\text{ft} = 200\,\text{ft-lb}$.
Total: $500 + 200 = 700\,\text{ft-lb}$.

13.

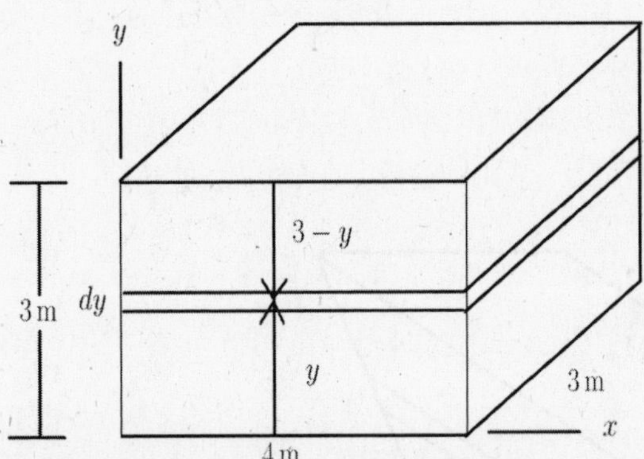

Volume of typical element:
$4 \cdot 3 \cdot dy = 12\,dy$.
Weight of typical element:
$(12\,dy)w = 12w\,dy$.
Work done in moving the typical element to the top: $(3 - y) \cdot 12w\,dy$.

Summing from $y = 0$ to $y = 3$, we get
$W = \int_0^3 (3-y)\cdot 12w\,dy = 12w\left(3y - \frac{1}{2}y^2\right)\Big|_0^3 = 12w\left(9 - \frac{9}{2}\right) = 54w\,\text{J}$.

15.

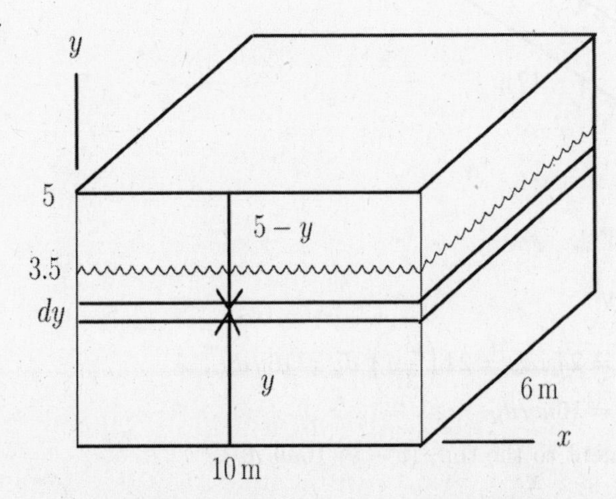

Volume of typical element: $10 \cdot 6 \cdot dy = 60\,dy$.
Weight of typical element: $w \cdot 60\,dy = 60w\,dy$.
Work done in moving the typical element to the top: $(5 - y)(60w\,dy)$.
We sum from $y = 0$ to $y = 3.5$, the level to which the tank is filled:
$W = \int_0^{3.5}(5-y)(60w\,dy) = 60w\left(5y - \frac{1}{2}y^2\right)\Big|_0^{3.5} = 682.5w\,\text{J}$.

17.

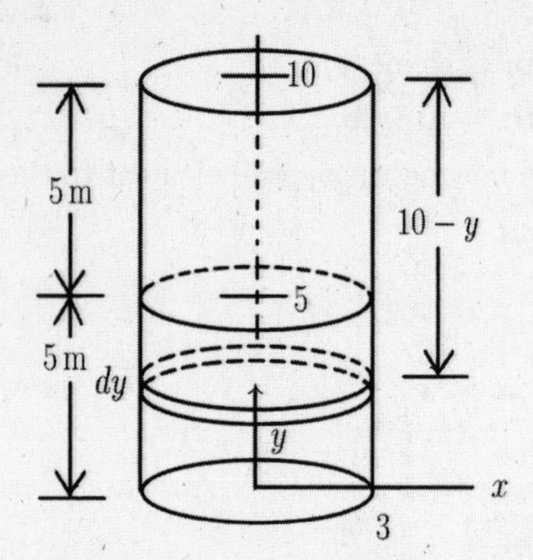

Volume of typical element: $\pi(3^2)\,dy = 9\pi\,dy$.

Weight of typical element: $9\pi\,dy \cdot w = 9\pi w\,dy$.

Work done in moving the typical element to the top: $(10-y)\cdot 9\pi w\,dy$.

Summing from $y=0$ to $y=5$, we get

$$\begin{aligned} W &= \int_0^5 (10-y)\cdot 9\pi w\,dy = 9\pi w\int_0^5 (10-y)\,dy = 9\pi w\left(10y - \frac{1}{2}y^2\right)\Big|_0^5 \\ &= 9\pi w\left(50 - \frac{25}{2}\right) = \frac{675\pi w}{2}\text{ J.} \end{aligned}$$

19.

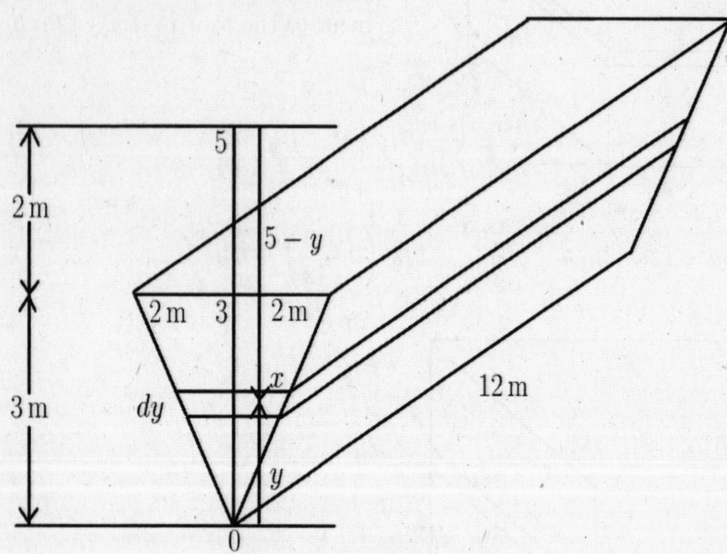

By similar triangles, $\dfrac{x}{y} = \dfrac{2}{3}$ or $x = \dfrac{2}{3}y$.

Volume of typical element: $2x\,dy\cdot 12 = 24x\,dy = 24\left(\dfrac{2}{3}y\right)\,dy = 16y\,dy$.

Weight of typical element: $w\cdot 16y\,dy = 16wy\,dy$.

Work done in moving the typical element to the top: $(5-y)(16wy\,dy)$.

Summing from $y=0$ to $y=3$,

$$W = \int_0^3 (5-y)(16wy\,dy) = 16w\int_0^3 (5y-y^2)\,dy = 16w\left(\frac{5}{2}y^2 - \frac{1}{3}y^3\right)\Big|_0^3 = 216w\text{ J.}$$

21.

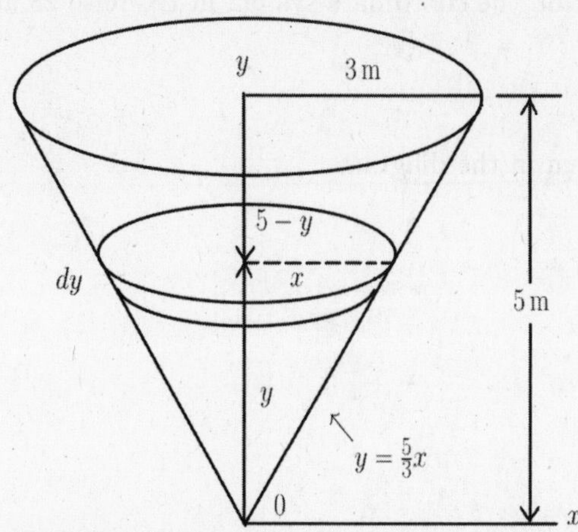

To find the volume and weight of the typical element, we need to determine the radius x. Since the line on the right has slope $m = \frac{5}{3}$, its equation is $y = \frac{5}{3}x$, or $x = \frac{3}{5}y$. Thus:

Weight of typical element: $\pi\left(\frac{3}{5}y\right)^2 dy \cdot w$.

Work done in moving the typical element to the top:

$(5-y)\cdot\pi\left(\frac{3}{5}y\right)^2 dy\cdot w = \frac{9\pi w}{25}(5y^2 - y^3)\,dy$.

Summing from $y = 0$ to $y = 5$, we get

$$\begin{aligned} W &= \int_0^5 \frac{9\pi w}{25}(5y^2 - y^3)\,dy = \frac{9\pi w}{25}\left(\frac{5y^3}{3} - \frac{y^4}{4}\right)\Bigg|_0^5 \\ &= \frac{9\pi w}{25}\left(\frac{5^4}{3} - \frac{5^4}{4}\right) = \frac{9\pi w}{25}\cdot 5^4\left(\frac{1}{3} - \frac{1}{4}\right) = 9\pi w(25)\frac{1}{12} = \frac{75\pi w}{4}\text{ J.} \end{aligned}$$

23.

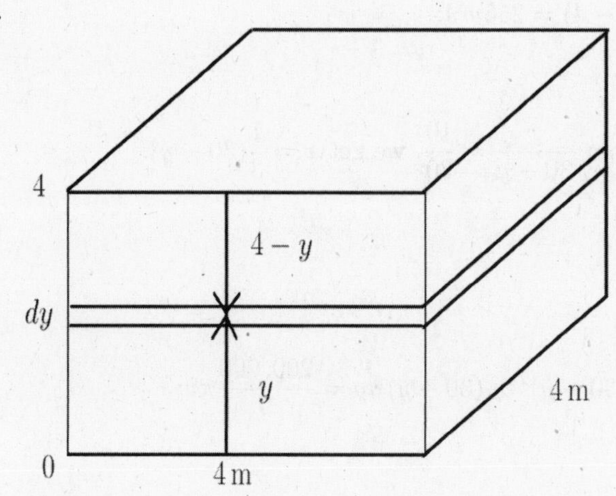

Weight of typical element: $w\cdot 16y\,dy = 16w\,dy$.

Work done in moving the typical element from the bottom: $y\cdot 16w\,dy$.

$W = \int_0^4 y\cdot 16w\,dy = w(8y^2)\Big|_0^4 = 128w$ J.

25.

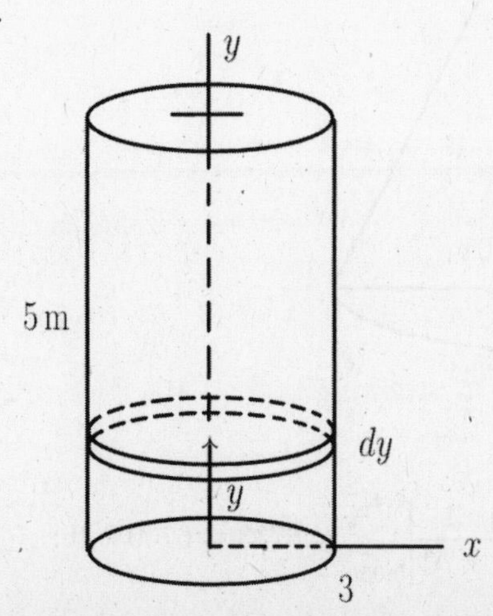

Volume of typical element: $\pi(3^2)\,dy$.

Weight of typical element: $\pi(3^2)\,dy\cdot w = 9\pi w\,dy$.

Work done in moving the typical element from the bottom: $y\cdot 9\pi w\,dy$.

Summing from $y = 0$ to $y = 2$,

$W = \int_0^2 y\cdot 9\pi w\,dy = 9\pi w\frac{y^2}{2}\Big|_0^2 = 18\pi w$ J.

27. The easiest way to solve this problem is to retain the coordinate system in Exercise 23 and to change the y to $y+2$:

$W = \int_0^4 (y+2) \cdot 16w\,dy = 256w$ J.

Alternatively, we can use the coordinate system in the diagram.

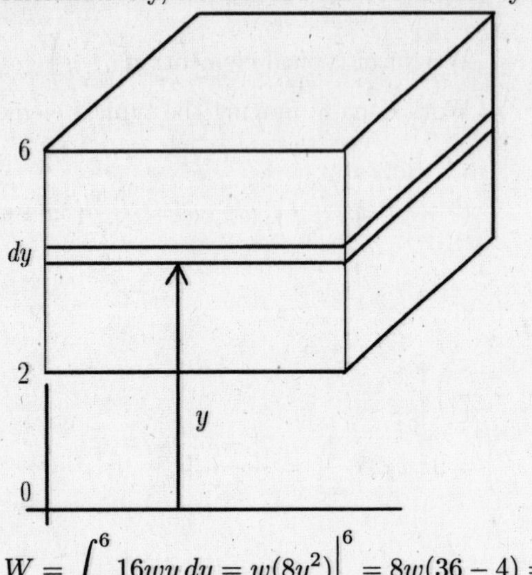

$W = \int_2^6 16wy\,dy = w(8y^2)\Big|_2^6 = 8w(36-4) = 256w$ J.

29. Volume of typical element: $\pi x^2\,dy$. From $\dfrac{x}{30-y} = \dfrac{10}{30}$, we get $x = \dfrac{1}{3}(30-y)$.

Weight: $w\pi\left(\dfrac{1}{3}\right)^2 (30-y)^2\,dy$

Distance to top: $30 - y$

$$W = \int_0^{20} w\pi\left(\frac{1}{3}\right)^2 (30-y)^2 \times (30-y)\,dy = \frac{200{,}000}{9}\pi w \text{ J}$$

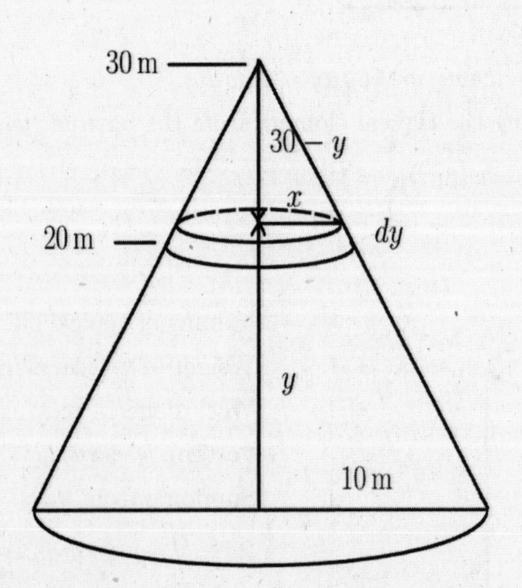

31. Since $k = 1600$, $P = 1600V^{-1.4}$, and

$W = \int_{0.059}^{0.417} 1600V^{-1.4}\,dV = 1600\dfrac{V^{-0.4}}{-0.4}\Big|_{0.059}^{0.417} = -\dfrac{1600}{0.4}\dfrac{1}{V^{0.4}}\Big|_{0.059}^{0.417} = 6733 \approx 6700$ ft-lb.

33.
$$\begin{aligned} F &= \frac{k}{s^2} \\ 20.0 &= \frac{k}{(1.0)^2} \qquad F = 20.0\,\text{dynes},\ s = 1.0\,\text{cm}. \end{aligned}$$

$k = 20.0$ and $F = \dfrac{20.0}{s^2}$.

(a) $W = \displaystyle\int_{1.0}^{10.0} \frac{20.0}{s^2}\,ds = 18\,\text{ergs}.$

(b)
$$\begin{aligned} W &= \int_{1.0}^{\infty} \frac{20.0}{s^2}\,ds = \lim_{b\to\infty} \int_{1.0}^{b} 20.0 s^{-2}\,ds = \lim_{b\to\infty} \left.\frac{-20.0}{s}\right|_{1.0}^{b} \\ &= \lim_{b\to\infty} \left(\frac{-20.0}{b} + \frac{20.0}{1.0}\right) = 20\,\text{ergs}. \end{aligned}$$

35.

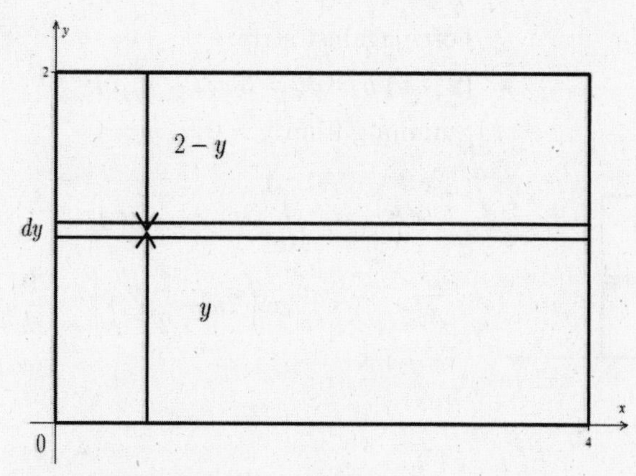

Pressure on strip: $(2-y)w$.

Area of strip: $4\,dy$.

Force against strip: $(2-y)w \cdot 4\,dy = 4w(2-y)\,dy$.

Summing from $y=0$ to $y=2$:

$$F = \int_0^2 4w(2-y)\,dy = 4w\left(2y - \frac{1}{2}y^2\right)\Big|_0^2 = 8w\,\text{N}.$$

37.

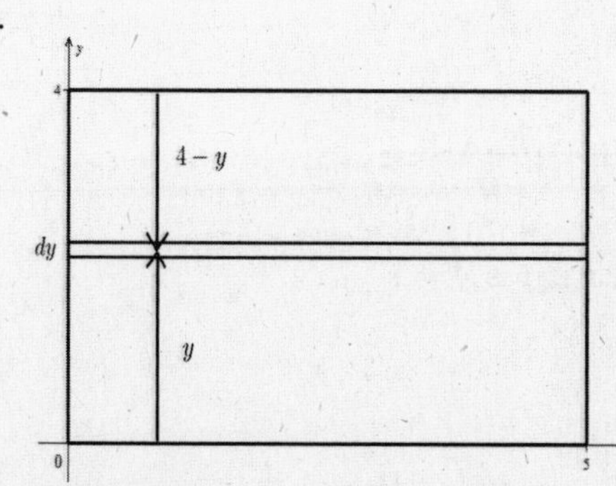

Pressure on strip: $(4-y)w$.

Area of strip: $5\,dy$.

Force against strip: $(4-y)w \cdot 5\,dy$.

Summing from $y=0$ to $y=4$:

$$F = \int_0^4 (4-y)w \cdot 5\,dy = \int_0^4 5w(4-y)\,dy = 5w\left(4y - \frac{1}{2}y^2\right)\Big|_0^4 = 40w\,\text{N}.$$

39.

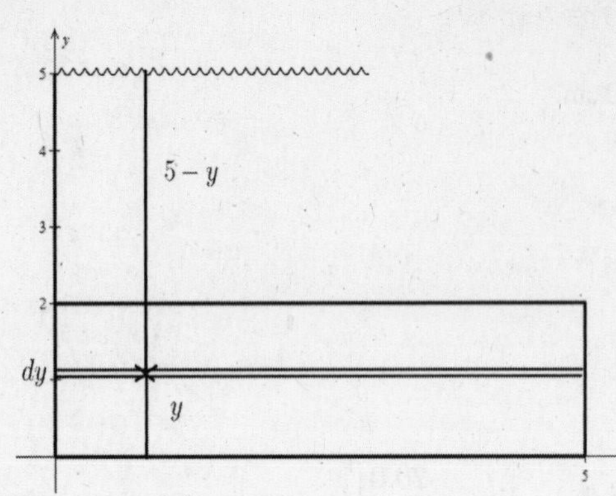

Pressure on strip: $(5 - y)w$.
Area of strip: $5\,dy$.
Force against strip:
$(5 - y)w \cdot 5\,dy = 5w(5 - y)\,dy$.

Summing from $y = 0$ to $y = 2$:

$$F = \int_0^2 5w(5 - y)\,dy = 5w\left(5y - \frac{1}{2}y^2\right)\Big|_0^2 = 40w\,\text{N}.$$

41.

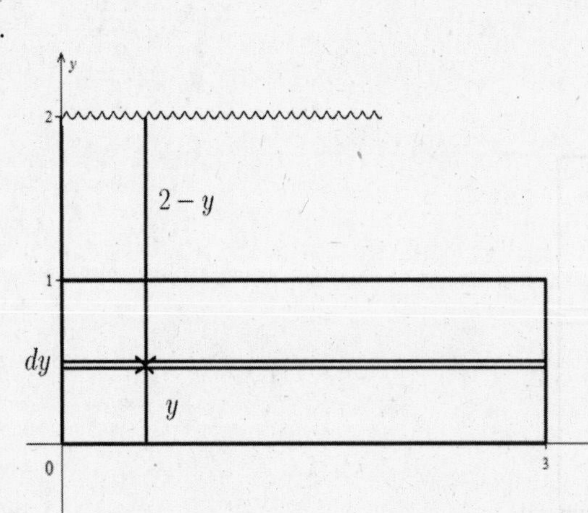

Pressure on strip: $(2 - y)w$.
Area of strip: $3\,dy$.
Force against strip:
$(2 - y)w \cdot 3\,dy = 3w(2 - y)\,dy$.
Summing from $y = 0$ to $y = 1$:

$$\begin{aligned} F &= \int_0^1 3w(2 - y)\,dy \\ &= 3w\left(2y - \frac{1}{2}y^2\right)\Big|_0^1 = \frac{9}{2}w\,\text{N}. \end{aligned}$$

43.

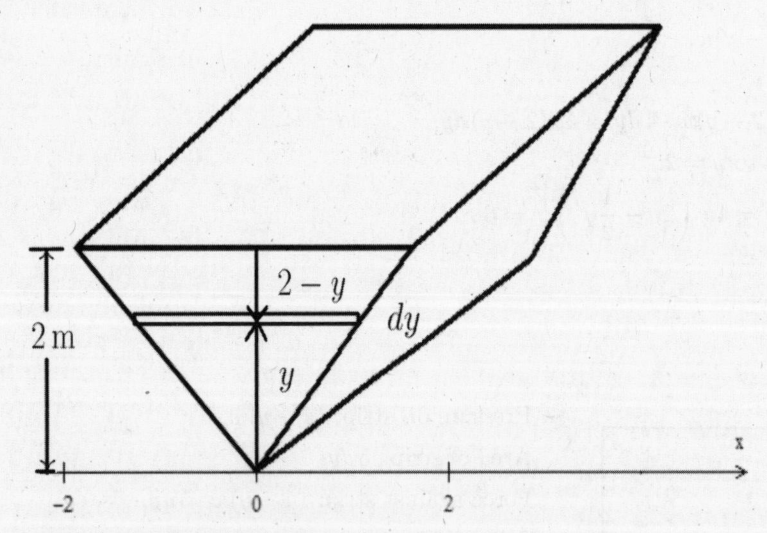

Since the slope of the line on the right is 1, the equation is $y = x$.
Area of strip: $2x\,dy = 2y\,dy$.
Pressure on strip: $(2 - y)w$.
Force against strip: $(2 - y)w \cdot 2y\,dy = 2w(2y - y^2)\,dy$.
Summing from $y = 0$ to $y = 2$, we get

$$F = \int_0^2 2w(2y - y^2)\,dy = 2w\left(y^2 - \frac{1}{3}y^3\right)\Big|_0^2 = 2w\left(4 - \frac{8}{3}\right) = \frac{8w}{3}\,\text{N}.$$

45. We find the force against one side and double the result.

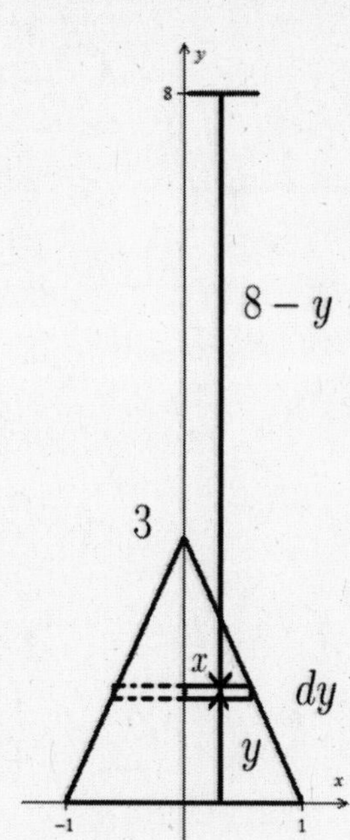

The slope of the line in the first quadrant: $m = -3$.
Thus

$$\begin{aligned} y &= mx + b \\ y &= -3x + 3 \end{aligned}$$

and

$$x = 1 - \frac{1}{3}y.$$

Pressure on strip: $(8 - y)w$.

Area of strip: $2x\,dy = 2\left(1 - \frac{1}{3}y\right)\,dy$.

Force against strip: $(8 - y)w \cdot 2\left(1 - \frac{1}{3}y\right)\,dy$.

Summing from $y = 0$ to $y = 3$,

$F = \int_0^3 (8 - y)w \cdot 2\left(1 - \frac{1}{3}y\right)\,dy = 2w\int_0^3 (8 - y)\left(1 - \frac{1}{3}y\right)\,dy = 21w\,\text{N}.$

So the force against both sides is $42w$ N.

47.

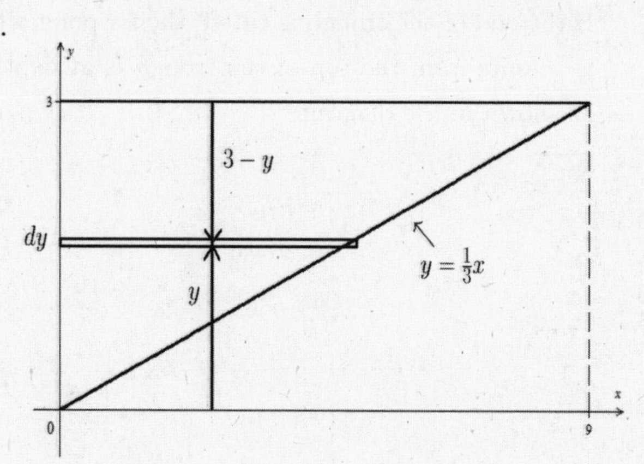

Equation of line: $y = \frac{1}{3}x$, or $x = 3y$.

Area of strip: $x\,dy = 3y\,dy$.

Pressure on strip: $(3 - y)w$.

Force against strip: $(3 - y)w \cdot 3y\,dy$.

Summing from $y = 0$ to $y = 3$:

$$\begin{aligned} F &= \int_0^3 (3 - y)w \cdot 3y\,dy = 3w\int_0^3 (3y - y^2)\,dy = 3w\left(3\frac{y^2}{2} - \frac{y^3}{3}\right)\Big|_0^3 \\ &= 3w\left(\frac{3^3}{2} - \frac{3^3}{3}\right) = 3^4 w\left(\frac{1}{2} - \frac{1}{3}\right) = 81w\left(\frac{1}{6}\right) = \frac{27w}{2}\,\text{N}. \end{aligned}$$

49.

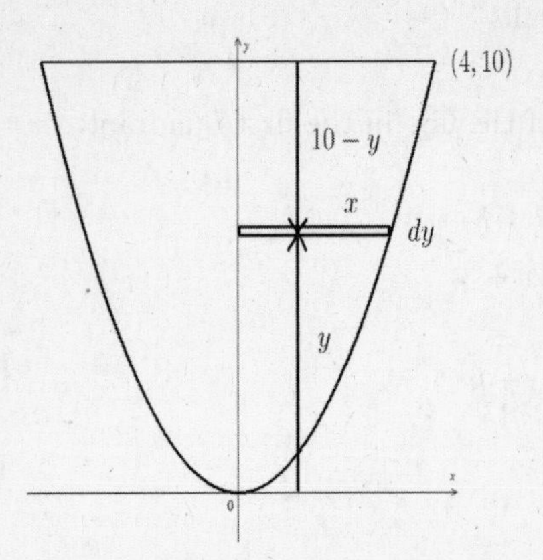

Form of the parabola: $x^2 = 4py$.

From the given information, the point $(4, 10)$ lies on the curve:

$$16 = 4p(10) \text{ or } 4p = \frac{16}{10} = \frac{8}{5}.$$

Equation: $x^2 = \frac{8}{5}y$ or $x = \sqrt{\frac{8}{5}y}$.

Pressure on strip: $(10 - y)w$.

Area of strip: $2x\,dy = 2\sqrt{\frac{8}{5}y}\,dy$.

Force against strip: $(10 - y)w \cdot 2\sqrt{\frac{8}{5}y}\,dy$.

$F = \int_0^{10} (10 - y)w \cdot 2\sqrt{\frac{8}{5}y}\,dy = \frac{640w}{3}$ N.

51.

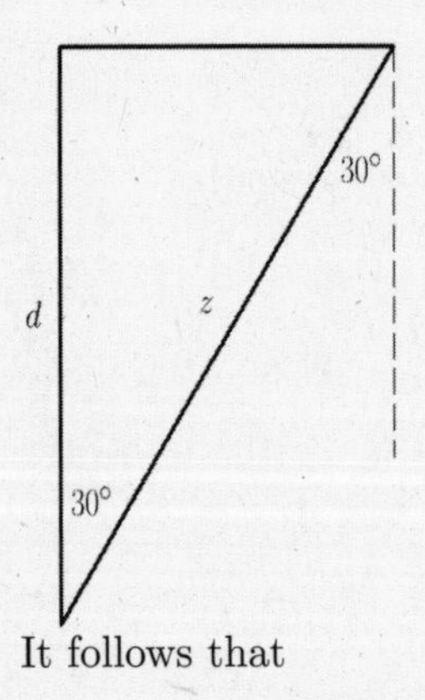

If the end of the trough is tilted, then a point which is z units from the top of the trough is at depth d, as shown in the diagram.

It follows that

$$d = z\cos 30^\circ = \frac{\sqrt{3}}{2}z.$$

The other calculations follow Exercise 43:

Pressure on strip: $\frac{\sqrt{3}}{2}(2 - y)w$.

Force against strip: $\frac{\sqrt{3}}{2}(2 - y)w \cdot 2y\,dy$.

$F = \int_0^2 \frac{\sqrt{3}}{2}(2 - y)w \cdot 2y\,dy = \frac{\sqrt{3}}{2}\int_0^2 2w(2y - y^2)\,dy = \frac{\sqrt{3}}{2}\left(\frac{8w}{3}\right)$ N by Exercise 43.

Chapter 5 Review

1. Taking the upward direction to be positive, $g = -32$. So

$$v = -32t \qquad v = 0 \text{ when } t = 0$$

and

$$s = -16t^2 + k.$$

Since $s = 256$ when $t = 0$, we have

$$256 = 0 + k$$

so that

$$s = -16t^2 + 256.$$

To see how long it takes for the object to reach the ground, we let $s = 0$ and solve for t:

$$0 = -16t^2 + 256 \text{ or } t = 4\,\text{s}.$$

Taking v to be $32t$, we get

$$v_{\text{av}} = \frac{1}{4-0}\int_0^4 32t\,dt = \frac{1}{4}(16t^2)\Big|_0^4 = 64\,\text{ft/s}.$$

3. $Ri_{\text{rms}}^2 = 10i_{\text{rms}}^2 = \dfrac{10}{4.0-0.0}\displaystyle\int_{0.0}^{4.0}(2.1-0.18t^{5/2})^2\,dt = 30.28 \approx 30\,\text{W}.$

5. See Exercise 31, Section 5.3.

7.

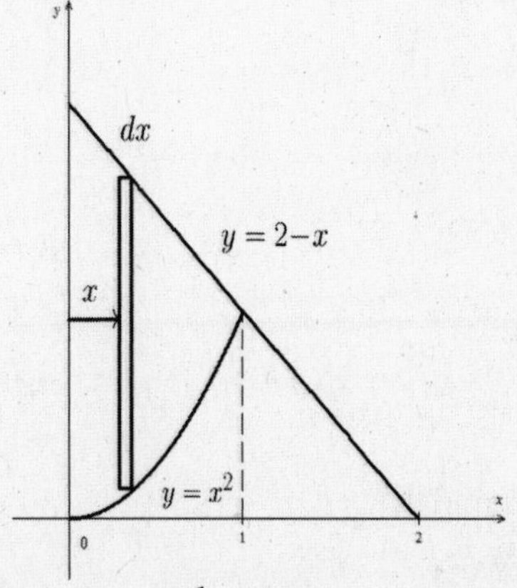

Volume of shell: $2\pi x(\text{height})\,dx$.

$$\begin{aligned} V &= \int_0^1 2\pi x(2-x-x^2)\,dx = 2\pi\int_0^1(2x-x^2-x^3)\,dx \\ &= 2\pi\left(x^2-\frac{1}{3}x^3-\frac{1}{4}x^4\right)\Big|_0^1 = 2\pi\left(1-\frac{1}{3}-\frac{1}{4}\right) = \frac{5\pi}{6}. \end{aligned}$$

9.

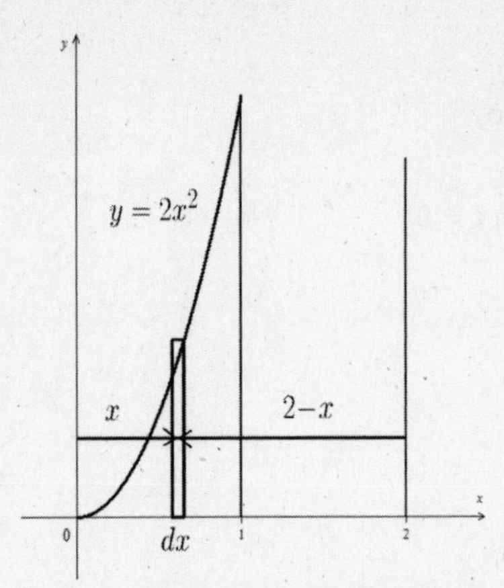

Radius of shell $= 2 - x$.

Volume of typical shell: $2\pi(\text{radius}) \cdot (\text{height}) \cdot (\text{thickness}) = 2\pi(2-x)(2x^2)\,dx$.

$$\begin{aligned} V &= \int_0^1 2\pi(2-x)(2x^2)\,dx = 4\pi\int_0^1 (2-x)(x^2)\,dx = 4\pi\int_0^1 (2x^2 - x^3)\,dx \\ &= 4\pi\left(\frac{2}{3}x^3 - \frac{1}{4}x^4\right)\Big|_0^1 = 4\pi\left(\frac{2}{3} - \frac{1}{4}\right) = 4\pi\frac{8-3}{12} = \frac{5\pi}{3}. \end{aligned}$$

11.

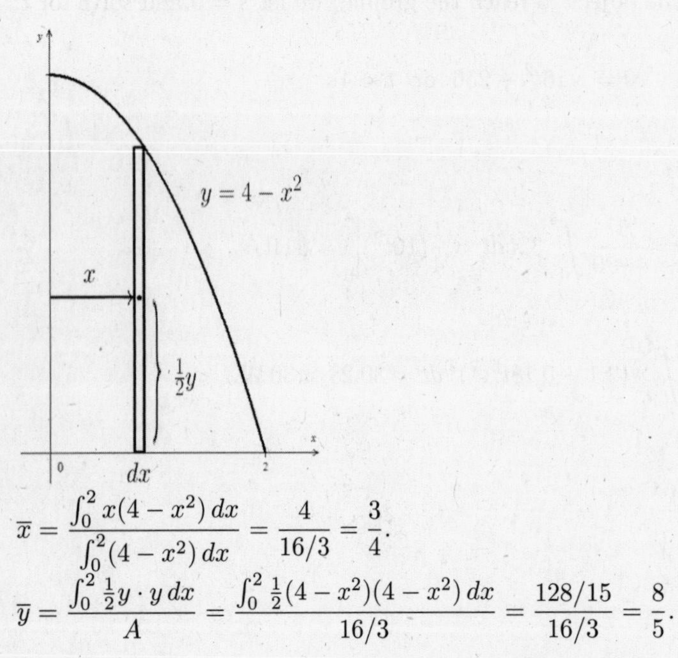

$$\overline{x} = \frac{\int_0^2 x(4-x^2)\,dx}{\int_0^2 (4-x^2)\,dx} = \frac{4}{16/3} = \frac{3}{4}.$$

$$\overline{y} = \frac{\int_0^2 \frac{1}{2}y \cdot y\,dx}{A} = \frac{\int_0^2 \frac{1}{2}(4-x^2)(4-x^2)\,dx}{16/3} = \frac{128/15}{16/3} = \frac{8}{5}.$$

13.

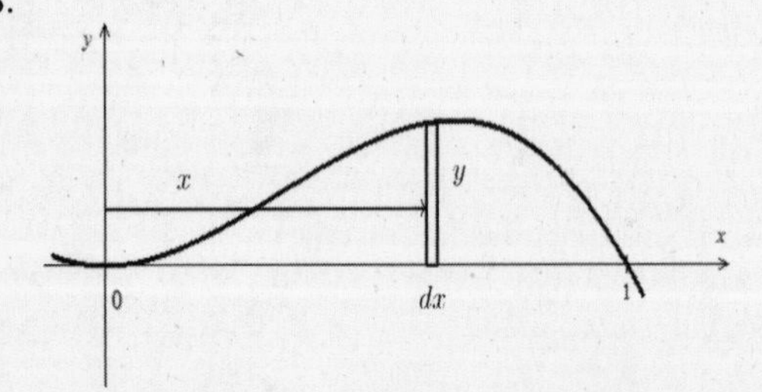

From $y = x^2 - x^3 = x^2(1-x) = 0$, the x-intercepts are 0 and 1.

$\overline{x}$: moment (with respect to y-axis) of typical element: $xy\,dx$.

$$\overline{x} = \frac{M_y}{A} = \frac{\int_0^1 x(x^2 - x^3)\,dx}{\int_0^1 (x^2 - x^3)\,dx} = \frac{\frac{1}{4}x^4 - \frac{1}{5}x^5\Big|_0^1}{\frac{1}{3}x^3 - \frac{1}{4}x^4\Big|_0^1} = \frac{1/20}{1/12} = \frac{3}{5}.$$

$\overline{y}$: moment (with respect to x-axis) of typical element $\left(\frac{1}{2}y\right)(y\,dx)$.

$$\begin{aligned}
\overline{y} &= \frac{M_x}{A} = \frac{\int_0^1 \left(\frac{1}{2}y\right)(y\,dx)}{A} = \frac{\frac{1}{2}\int_0^1 (x^2-x^3)(x^2-x^3)\,dx}{1/12} \\
&= 12\left(\frac{1}{2}\right)\int_0^1 (x^4-2x^5+x^6)\,dx = 6\left(\frac{1}{5}x^5 - \frac{1}{3}x^6 + \frac{1}{7}x^7\right)\Big|_0^1 \\
&= 6\left(\frac{1}{5} - \frac{1}{3} + \frac{1}{7}\right) = 6\left(\frac{21-35+15}{105}\right) = \frac{6}{105} = \frac{2}{35}.
\end{aligned}$$

15.

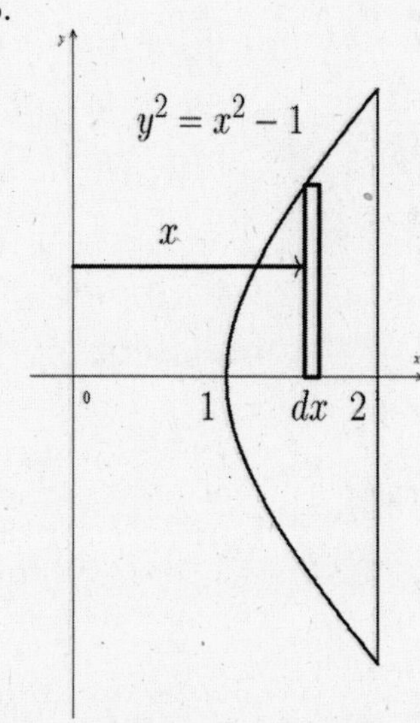

Volume of disk: $\pi(\text{radius})^2 \cdot (\text{thickness})$
$= \pi y^2\,dx = \pi(x^2-1)\,dx$.

$$\overline{x} = \frac{\int_1^2 x\cdot\pi(x^2-1)\,dx}{\int_1^2 \pi(x^2-1)\,dx} = \frac{\frac{9}{4}\pi}{\frac{4}{3}\pi} = \frac{27}{16}.$$

$\overline{y} = 0$.

17.

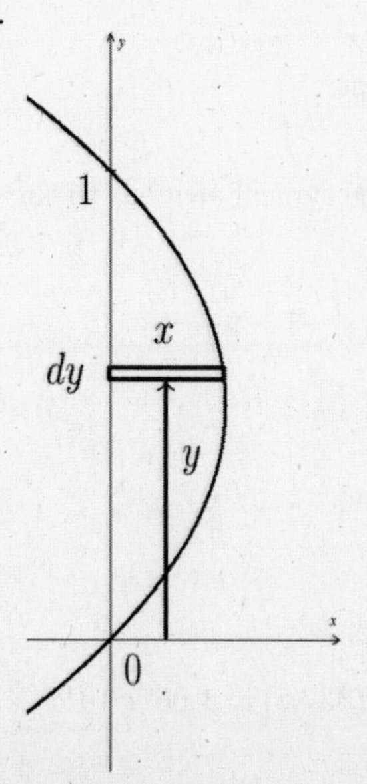

Moment of inertia (with respect to x-axis) of typical element:

$y^2 \cdot \rho x\,dy = y^2 \cdot \rho(y-y^2)\,dy = \rho y^2(y-y^2)\,dy$.

$$\begin{aligned}
I_x &= \rho\int_0^1 y^2(y-y^2)\,dy = \rho\left(\frac{1}{4}y^4 - \frac{1}{5}y^5\right)\Big|_0^1 \\
&= \rho\left(\frac{1}{4} - \frac{1}{5}\right) = \frac{\rho}{20}.
\end{aligned}$$

Mass: $\rho\int_0^1 (y-y^2)\,dy = \frac{1}{2}y^2 - \frac{1}{3}y^3\Big|_0^1 \rho = \frac{\rho}{6}$.

$$R_x = \sqrt{\frac{\rho}{20}\frac{6}{\rho}} = \sqrt{\frac{3}{10}} = \frac{\sqrt{30}}{10}.$$

19.

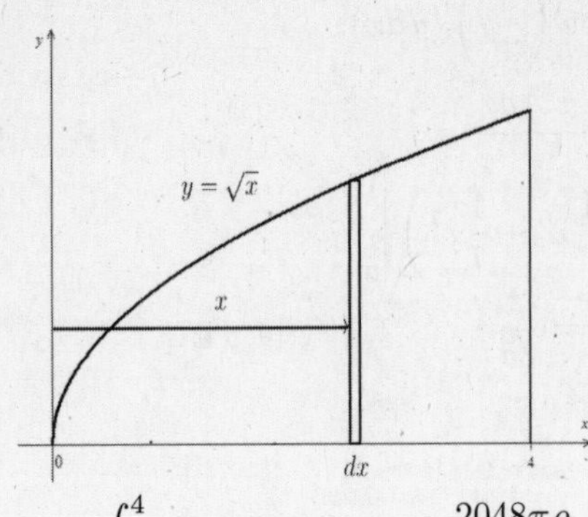

Volume of shell: $2\pi(\text{radius}) \cdot (\text{height}) \cdot (\text{thickness})$
$= 2\pi x\sqrt{x}\,dx$.
Mass of typical shell:
$\rho \cdot 2\pi x\sqrt{x}\,dx$.
Moment of inertia of typical shell:
$x^2 \cdot 2\pi\rho x\sqrt{x}\,dx$.

$I_y = \int_0^4 x^2 \cdot 2\pi\rho x\sqrt{x}\,dx = \frac{2048\pi\rho}{9}$.

Mass: $\int_0^4 2\pi\rho x\sqrt{x}\,dx = \frac{128\pi\rho}{5}$.

$R_y = \sqrt{\frac{2048\pi\rho}{9} \cdot \frac{5}{128\pi\rho}} = \sqrt{\frac{80}{9}} = \sqrt{\frac{16 \cdot 5}{9}} = \frac{4\sqrt{5}}{3}$.

21. By Hooke's law, $F = kx$, $2 = k \cdot \frac{1}{2}$, or $k = 4$. Thus $F = 4x$.

(a) The spring is stretched 2 ft beyond its natural length. Thus
$W = \int_0^2 4x\,dx = 2x^2\Big|_0^2 = 8\text{ ft-lb}$.

(b) $W = \int_1^3 4x\,dx = 2x^2\Big|_1^3 = 16\text{ ft-lb}$.

23.

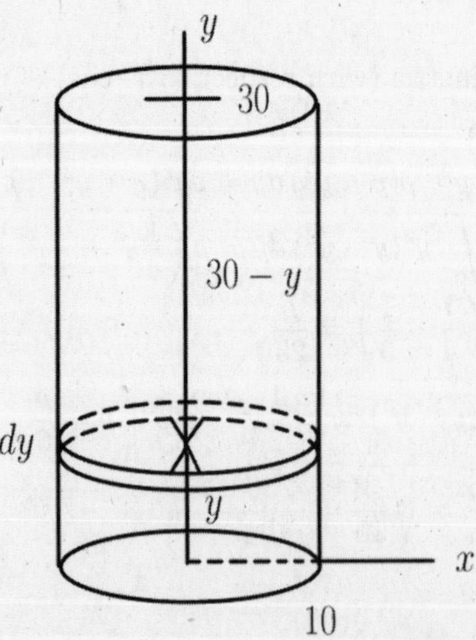

Volume of typical element:
$\pi(10^2)\,dy = 100\pi\,dy$.
Weight of typical element:
$w \cdot 100\pi\,dy$.
Work done in moving the typical element to the top:
$(30 - y)(w \cdot 100\pi\,dy)$.

Since the reservoir is only half-filled, we sum from $y = 0$ to $y = 15$:
$W = \int_0^{15} (30 - y)(w \cdot 100\pi\,dy) = 100\pi w \int_0^{15} (30 - y)\,dy = 100\pi w(337.5) = 1.06 \times 10^9\text{ J}$
($w = 10{,}000\,\text{N/m}^3$).

25.

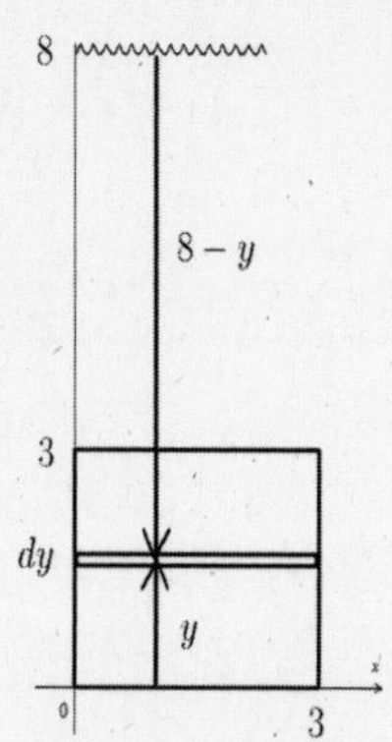

Pressure on strip: $(8 - y)w$.

Area of strip: $3\,dy$.

Force against strip: $(8 - y)w \cdot 3\,dy$.

Summing from $y = 0$ to $y = 3$, we get

$$\begin{aligned} F &= \int_0^3 (8-y)w \cdot 3\,dy = 3w \int_0^3 (8-y)\,dy = 3w\left(8y - \frac{1}{2}y^2\right)\Big|_0^3 \\ &= 3w\left(24 - \frac{9}{2}\right) = \frac{117w}{2}\,\text{N} = \frac{117}{2}(10,000) = 585,000\,\text{N}. \end{aligned}$$

Chapter 6

Derivatives of Transcendental Functions

6.1 Review of Trigonometry

3.

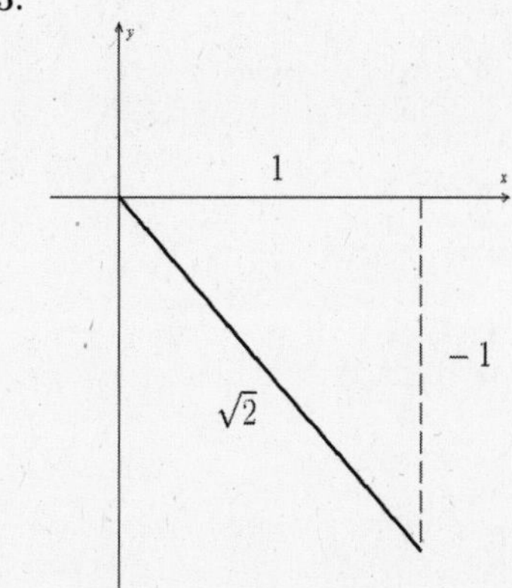

$\tan(-45°) = -1$

5.

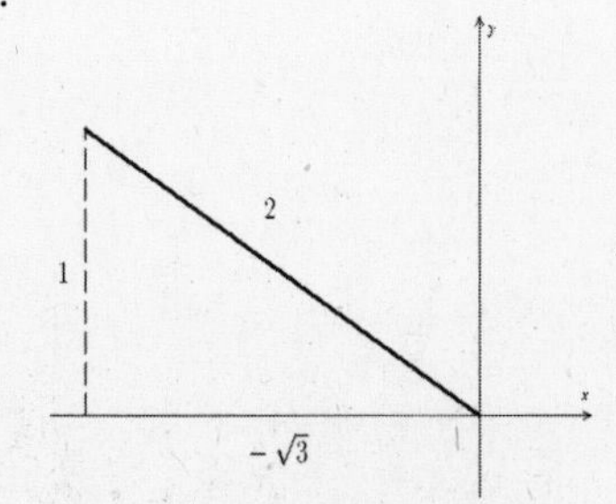

$$\sec 150° = \frac{2}{-\sqrt{3}} = -\frac{2\sqrt{3}}{3}$$

9.

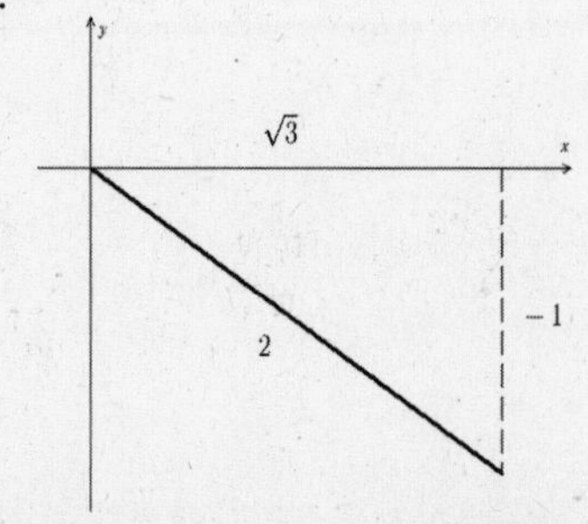

$\csc(-30°) = -2$

13.

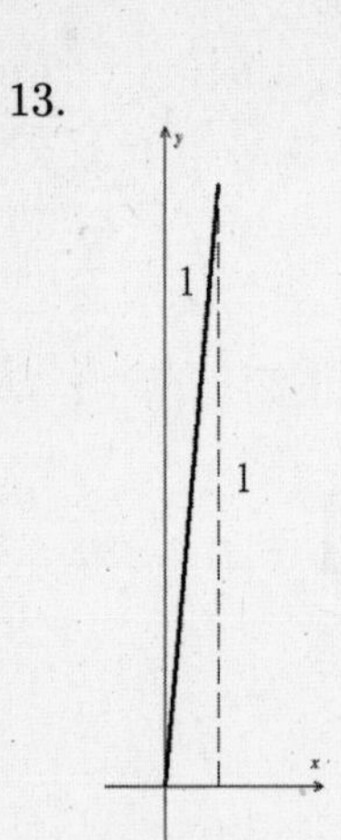

$$\tan 90^\circ = \frac{1}{0} \text{ (undefined)}$$

25. $60^\circ = 60^\circ \cdot \dfrac{\pi}{180^\circ} = \dfrac{\pi}{3}$

29. $135^\circ = 135^\circ \cdot \dfrac{\pi}{180^\circ} = \dfrac{3\pi}{4}$

33. $20^\circ = 20^\circ \cdot \dfrac{\pi}{180^\circ} = \dfrac{\pi}{9}$

37. $\dfrac{\pi}{6} = \dfrac{\pi}{6} \cdot \dfrac{180^\circ}{\pi} = 30^\circ$

41. $\dfrac{11\pi}{10} = \dfrac{11\pi}{10} \cdot \dfrac{180^\circ}{\pi} = 198^\circ$

45. amplitude $= \dfrac{1}{3}$ (coefficient of $\sin 2x$)
period $= \dfrac{2\pi}{2} = \pi$
(see drawing in answer section)

49. amplitude $= \dfrac{1}{2}$ (coefficient of $\cos 3x$)
period $= \dfrac{2\pi}{3}$
(see drawing in answer section)

51. $\tan\theta + \sec\theta = \dfrac{\sin\theta}{\cos\theta} + \dfrac{1}{\cos\theta} = \dfrac{\sin\theta + 1}{\cos\theta}$

53. Since $1 + \tan^2\theta = \sec^2\theta$, we have $\tan^2\theta - \sec^2\theta = -1$. So
$\dfrac{\tan^2\theta - \sec^2\theta}{\sec\theta} = \dfrac{-1}{\sec\theta} = -\cos\theta$.

55. $\csc^2\theta - \cot^2\theta = \dfrac{1}{\sin^2\theta} - \dfrac{\cos^2\theta}{\sin^2\theta} = \dfrac{1 - \cos^2\theta}{\sin^2\theta} = \dfrac{\sin^2\theta}{\sin^2\theta} = 1$
(This also follows from the identity $1 + \cot^2\theta = \csc^2\theta$.)

57. $$\begin{aligned} \frac{1}{\sec^2\theta+\tan^2\theta+\cos^2\theta} &= \frac{1}{(\sin^2\theta+\cos^2\theta)+\tan^2\theta} && \text{rearranging} \\ &= \frac{1}{1+\tan^2\theta} && \sin^2\theta+\cos^2\theta=1 \\ &= \frac{1}{\sec^2\theta} && 1+\tan^2\theta=\sec^2\theta \\ &= \cos^2\theta && \cos\theta=\frac{1}{\sec\theta} \end{aligned}$$

59. $$\begin{aligned} \sec^2\theta+\csc^2\theta &= \frac{1}{\cos^2\theta}+\frac{1}{\sin^2\theta}=\frac{1}{\cos^2\theta}\frac{\sin^2\theta}{\sin^2\theta}+\frac{1}{\sin^2\theta}\frac{\cos^2\theta}{\cos^2\theta} \\ &= \frac{\sin^2\theta+\cos^2\theta}{\sin^2\theta\cos^2\theta}=\frac{1}{\cos^2\theta\sin^2\theta} \end{aligned}$$

61. $$\begin{aligned} \cos\theta\cot\theta+\sin\theta &= \cos\theta\frac{\cos\theta}{\sin\theta}+\sin\theta && \cot\theta=\frac{\cos\theta}{\sin\theta} \\ &= \frac{\cos^2\theta}{\sin\theta}+\frac{\sin^2\theta}{\sin\theta} && \text{common denominator}=\sin\theta \\ &= \frac{\cos^2\theta+\sin^2\theta}{\sin\theta} \\ &= \frac{1}{\sin\theta} && \cos^2\theta+\sin^2\theta=1 \end{aligned}$$

81. $$\begin{aligned} \cos\left(x-\frac{\pi}{2}\right) &= \cos x\cos\frac{\pi}{2}+\sin x\sin\frac{\pi}{2} \\ &= \cos x\cdot 0+\sin x\cdot 1=\sin x \end{aligned}$$

85. $\cos(x-\pi)=\cos x\cos\pi+\sin x\sin\pi=\cos x(-1)+\sin x(0)=-\cos x$

89. Since $\cos 2\theta=\cos^2\theta-\sin^2\theta$, $\cos^2 3x-\sin^2 3x=\cos(2\cdot 3x)=\cos 6x$.

91. $1-2\cos^2 8x=-\left(2\cos^2 8x-1\right)=-\cos 16x$

93. Since $\sin^2\theta=\frac{1}{2}(1-\cos 2\theta)$, $\sin^2 3x=\frac{1}{2}[1-\cos(2\cdot 3x)]=\frac{1}{2}(1-\cos 6x)$.

97. $\sin^2\frac{1}{2}x=\frac{1}{2}\left[1-\cos\left(2\cdot\frac{1}{2}x\right)\right]=\frac{1}{2}(1-\cos x)$

6.2 Derivatives of Sine and Cosine Functions

1. $y=3\sin 4x$
$y'=3\left(\cos 4x\right)\frac{d}{dx}(4x)=3(\cos 4x)4=12\cos 4x$

3. $y=2\cos 4x$
$y'=2\left(-\sin 4x\right)\frac{d}{dx}(4x)=2\left(-\sin 4x\right)(4)=-8\sin 4x.$

5. $y=\sin\sqrt{x}=\sin x^{1/2}$
$y'=\cos x^{1/2}\cdot\frac{d}{dx}\left(x^{1/2}\right)=\cos x^{1/2}\left(\frac{1}{2}x^{-1/2}\right)=\frac{\cos\sqrt{x}}{2\sqrt{x}}$

7. $s = 3\cos t^3$
$$\frac{ds}{dt} = 3\left(-\sin t^3\right)\frac{d}{dt}(t^3) = 3\left(-\sin t^3\right)(3t^2) = -9t^2\sin t^3.$$

9. $y = x\sin x$. By the product rule,
$$y' = x\frac{d}{dx}\sin x + (\sin x)\frac{d}{dx}(x) = x\cos x + \sin x.$$

11. $y = \sin^2 5x = (\sin 5x)^2$. By the power rule,
$$\begin{aligned} y' &= 2(\sin 5x)\tfrac{d}{dx}\sin 5x \\ &= 2\sin 5x \cdot \cos 5x \cdot 5 \\ &= 5(2\sin 5x\cos 5x) \\ &= 5\sin 10x. && \text{double-angle formula} \end{aligned}$$

13. $w = \cos^2 4v = (\cos 4v)^2$. By the power rule,
$$\begin{aligned} \frac{dw}{dv} &= 2(\cos 4v)\frac{d}{dv}\cos 4v \\ &= 2(\cos 4v)(-\sin 4v)(4) \\ &= -8\cos 4v\sin 4v \\ &= -4(2\sin 4v\cos 4v) \\ &= -4\sin(2\cdot 4v) && \text{double-angle formula} \\ &= -4\sin 8v. \end{aligned}$$

15. $w = \cos(v^2+3)$
$$\frac{dw}{dv} = -\sin(v^2+3)\frac{d}{dv}(v^2+3) = -2v\sin(v^2+3).$$

17. $y = x\cos 2x$. By the product rule,
$$\begin{aligned} y' &= x\frac{d}{dx}\cos 2x + \cos 2x\frac{d}{dx}(x) \\ &= x(-\sin 2x)\frac{d}{dx}(2x) + \cos 2x \cdot 1 \\ &= x(-\sin 2x)(2) + \cos 2x \\ &= -2x\sin 2x + \cos 2x \\ &= \cos 2x - 2x\sin 2x. \end{aligned}$$

19. $y = 2x\sin(2x+2)$. By the product rule,
$$\begin{aligned} y' &= 2x\frac{d}{dx}\sin(2x+2) + \sin(2x+2)\frac{d}{dx}(2x) \\ &= 2x\cos(2x+2)\cdot 2 + \sin(2x+2)\cdot 2 \\ &= 4x\cos(2x+2) + 2\sin(2x+2). \end{aligned}$$

21. $y = \sin\dfrac{1}{x}$
$$y' = \left(\cos\frac{1}{x}\right)\frac{d}{dx}\left(\frac{1}{x}\right) = \left(\cos\frac{1}{x}\right)\frac{d}{dx}(x^{-1}) = \left(\cos\frac{1}{x}\right)(-1x^{-2}) = -\frac{\cos(1/x)}{x^2}.$$

23. $y = \dfrac{\sin x}{x}$. By the quotient rule,
$$y' = \frac{x\frac{d}{dx}\sin x - \sin x\frac{d}{dx}(x)}{x^2} = \frac{x\cos x - \sin x}{x^2}.$$

25. $y = \dfrac{x}{\sin 4x}$. By the quotient rule,

$$\begin{aligned} y' &= \frac{(\sin 4x)\left(\frac{d}{dx}\right)x - x\left(\frac{d}{dx}\right)\sin 4x}{(\sin 4x)^2} \\ &= \frac{(\sin 4x)\cdot 1 - x\cos 4x\cdot 4}{\sin^2 4x} \\ &= \frac{\sin 4x - 4x\cos 4x}{\sin^2 4x}. \end{aligned}$$

27. $N = \dfrac{\cos 2\theta}{3\theta}$. By the quotient rule,

$$\begin{aligned} \frac{dN}{d\theta} &= \frac{3\theta\left(\frac{d}{d\theta}\right)\cos 2\theta - \cos 2\theta\left(\frac{d}{d\theta}\right)(3\theta)}{(3\theta)^2} \\ &= \frac{3\theta(-\sin 2\theta)(2) - \cos 2\theta(3)}{9\theta^2} \\ &= -\frac{3(2\theta\sin 2\theta + \cos 2\theta)}{9\theta^2} \\ &= -\frac{2\theta\sin 2\theta + \cos 2\theta}{3\theta^2}. \end{aligned}$$

29. $y = \sqrt{x}\sin x$. By the product rule,

$$\begin{aligned} y' &= x^{1/2}\frac{d}{dx}\sin x + \sin x\frac{d}{dx}(x^{1/2}) = x^{1/2}\cos x + (\sin x)\frac{1}{2}x^{-1/2} \\ &= \sqrt{x}\cos x + \frac{1}{2\sqrt{x}}\sin x. \end{aligned}$$

31. $y = \cos^2 x^3 = (\cos x^3)^2$. By the power rule,
$y' = 2(\cos x^3)\dfrac{d}{dx}(\cos x^3) = 2(\cos x^3)(-\sin x^3)(3x^2) = -6x^2\cos x^3\sin x^3$.

33. $y = \sin x\cos x$. By the product rule,

$$\begin{aligned} y' &= (\sin x)\frac{d}{dx}\cos x + (\cos x)\frac{d}{dx}\sin x = \sin x(-\sin x) + \cos x(\cos x) \\ &= -\sin^2 x + \cos^2 x = \cos^2 x - \sin^2 x = \cos 2x \end{aligned}$$

by the double-angle identity.

35. $y = \dfrac{\sin^3 x}{x} = \dfrac{(\sin x)^3}{x}$. By the quotient rule,

$$\begin{aligned} y' &= \frac{x\left(\frac{d}{dx}\right)(\sin x)^3 - (\sin x)^3\left(\frac{d}{dx}\right)x}{x^2} \\ &= \frac{x\cdot 3(\sin x)^2\cos x - (\sin x)^3\cdot 1}{x^2} \\ &= \frac{\sin^2 x(3x\cos x - \sin x)}{x^2}. \end{aligned}$$

37. $y = x\cos^2 3x = x(\cos 3x)^2$. By the product rule,
$y' = x\dfrac{d}{dx}(\cos 3x)^2 + (\cos 3x)^2\dfrac{d}{dx}(x) = x\dfrac{d}{dx}(\cos 3x)^2 + \cos^2 3x\cdot 1$.
Next, by the generalized power rule,

$$\begin{aligned} y' &= x\cdot 2(\cos 3x)\frac{d}{dx}\cos 3x + \cos^2 3x \\ &= x\cdot 2(\cos 3x)(-\sin 3x\cdot 3) + \cos^2 3x \\ &= -6x\cos 3x\sin 3x + \cos^2 3x. \end{aligned}$$

After factoring $\cos 3x$, we get $y' = \cos 3x(\cos 3x - 6x\sin 3x)$.

39. $y = \sin x,\ y' = \cos x,\ y'' = -\sin x.$

41. $y = \sin x,\ y' = \cos x,\ y'' = -\sin x,\ y^{(3)} = -\cos x,$ and $y^{(4)} = \sin x.$ Thus $\frac{d^4}{dx^4}\sin x = \sin x.$

43. $y = x\sin 2x$. By the product rule,

$$\begin{aligned} y' &= x\cos 2x\cdot 2 + \sin 2x\cdot 1 \\ &= 2x\cos 2x + \sin 2x\Big|_{x=\pi/4} \\ &= 2\left(\frac{\pi}{4}\right)\cos\frac{\pi}{2} + \sin\frac{\pi}{2} = 0 + 1 = 1. \end{aligned}$$

45. $i = 20.0\sin 4.0t.$

$$\begin{aligned} v = L\frac{di}{dt} &= 0.0050(20.0\cos 4.0t)\frac{d}{dt}(4.0t) \\ &= 0.0050(20.0\cos 4.0t)(4.0)\Big|_{t=0.20} \\ &= 0.28\,\mathrm{V}. \end{aligned}$$

47. $s = \frac{1}{8}\sin(20\pi t).$

$$v = \frac{ds}{dt} = \frac{1}{8}\cos(20\pi t)(20\pi)\Big|_{t=0.1} = \frac{20\pi}{8} = \frac{5\pi}{2}\ \mathrm{cm/s}.$$

6.3 Other Trigonometric Functions

1. $y = \sec 5x.$

$$y' = (\sec 5x\tan 5x)\frac{d}{dx}(5x) = (\sec 5x\tan 5x)(5) = 5\sec 5x\tan 5x.$$

3. $y = 2\csc\sqrt{t} = 2\csc t^{1/2}$

$$y' = 2\left(-\csc t^{1/2}\cot t^{1/2}\right)\frac{d}{dt}\left(t^{1/2}\right) = -2\csc t^{1/2}\cot t^{1/2}\cdot\frac{1}{2}t^{-1/2} = -\frac{\csc\sqrt{t}\cot\sqrt{t}}{\sqrt{t}}$$

5. $y = 3\cot 4x.$

$$y' = 3\frac{d}{dx}\cot 4x = 3(-\csc^2 4x)\frac{d}{dx}(4x) = 3(-\csc^2 4x)(4) = -12\csc^2 4x.$$

7. $z = 2\csc w^2.$

$$\frac{dz}{dw} = 2\left(-\csc w^2\cot w^2\right)\frac{d}{dw}w^2 = -4w\csc w^2\cot w^2.$$

9. $s = \tan\sqrt{t} = \tan t^{1/2}$

$$\frac{ds}{dt} = \sec^2 t^{1/2}\frac{d}{dt}t^{1/2} = \sec^2 t^{1/2}\cdot\frac{1}{2}t^{-1/2} = \frac{\sec^2\sqrt{t}}{2\sqrt{t}}$$

11. $y = x\cot 2x$. By the product rule,

$$y' = x\frac{d}{dx}(\cot 2x) + \cot 2x\frac{d}{dx}(x) = x(-\csc^2 2x)(2) + \cot 2x\cdot 1 = \cot 2x - 2x\csc^2 2x.$$

13. $y = x^2\sec 4x$. By the product rule,

$$y' = x^2\frac{d}{dx}\sec 4x + \sec 4x\frac{d}{dx}\left(x^2\right) = x^2\sec 4x\tan 4x\cdot 4 + \sec 4x\cdot 2x = 2x\sec 4x(2x\tan 4x + 1)$$

15. $r = \dfrac{\sec\theta}{\theta}$. By the quotient rule,
$$\frac{dr}{d\theta} = \frac{\theta\left(\frac{d}{d\theta}\right)\sec\theta - \sec\theta\left(\frac{d}{d\theta}\right)\theta}{\theta^2} = \frac{\theta\sec\theta\tan\theta - \sec\theta\cdot 1}{\theta^2} = \frac{\sec\theta(\theta\tan\theta - 1)}{\theta^2}.$$

17. $y = \dfrac{\cot x^2}{x}$. By the quotient rule,
$$y' = \frac{x\frac{d}{dx}\cot x^2 - \cot x^2\cdot 1}{x^2} = \frac{x\left(-\csc^2 x^2\right)(2x) - \cot x^2}{x^2} = -\frac{2x^2\csc^2 x^2 + \cot x^2}{x^2}$$

19. $y = \sqrt{\tan 2x} = (\tan 2x)^{1/2}$. By the power rule,
$$y' = \frac{1}{2}(\tan 2x)^{-1/2}\frac{d}{dx}\tan 2x = \frac{1}{2}(\tan 2x)^{-1/2}(\sec^2 2x)(2) = \frac{\sec^2 2x}{\sqrt{\tan 2x}}.$$

21. $y = 2\tan^4 4x = 2(\tan 4x)^4$. By the power rule,
$$y' = 2\cdot 4(\tan 4x)^3\frac{d}{dx}\tan 4x = 8(\tan 4x)^3\sec^2 4x\cdot 4 = 32\tan^3 4x\sec^2 4x.$$

23. $r = \sqrt{\csc\omega^2} = (\csc\omega^2)^{1/2}$. By the power rule,
$$\frac{dr}{d\omega} = \frac{1}{2}(\csc\omega^2)^{-1/2}(-\csc\omega^2\cot\omega^2)(2\omega) = -\omega(\csc\omega^2)^{1/2}\cot\omega^2 = -\omega\sqrt{\csc\omega^2}\cot\omega^2.$$

25. $T_1 = T_2^2\csc T_2$. By the product rule,
$$\begin{aligned}\frac{dT_1}{dT_2} &= T_2^2\frac{d}{dT_2}\csc T_2 + \csc T_2\frac{d}{dT_2}(T_2^2)\\ &= T_2^2(-\csc T_2\cot T_2) + \csc T_2(2T_2)\\ &= -T_2^2\csc T_2\cot T_2 + 2T_2\csc T_2\\ &= T_2\csc T_2(2 - T_2\cot T_2).\end{aligned}$$

27. $y = \cos^2 x\cot x = (\cos x)^2\cot x$. By the product and power rules,
$$\begin{aligned}y' &= (\cos x)^2\frac{d}{dx}\cot x + \cot x\frac{d}{dx}(\cos x)^2 = \cos^2 x(-\csc^2 x) + \cot x\cdot 2(\cos x)(-\sin x)\\ &= -\frac{\cos^2 x}{\sin^2 x} - 2\left(\frac{\cos x}{\sin x}\right)\cos x\sin x = -\cot^2 x - 2\cos^2 x.\end{aligned}$$

29. $y = \dfrac{x^3}{\tan 3x}$. By the quotient rule,
$$\begin{aligned}y' &= \frac{\tan 3x\left(\frac{d}{dx}\right)x^3 - x^3\left(\frac{d}{dx}\right)\tan 3x}{(\tan 3x)^2} = \frac{\tan 3x(3x^2) - x^3(\sec^2 3x\cdot 3)}{\tan^2 3x}\\ &= \frac{3x^2\tan 3x - 3x^3\sec^2 3x}{\tan^2 3x} = \frac{3x^2(\tan 3x - x\sec^2 3x)}{\tan^2 3x}.\end{aligned}$$

31. $y = \dfrac{x}{(\csc 5x)^2}$.

$$\begin{aligned} y' &= \frac{(\csc 5x)^2 \left(\frac{d}{dx}\right)(x) - x\left(\frac{d}{dx}\right)(\csc 5x)^2}{(\csc 5x)^4} \\ &= \frac{(\csc 5x)^2 \cdot 1 - x \cdot 2(\csc 5x)\left(\frac{d}{dx}\right)\csc 5x}{(\csc 5x)^4} \\ &= \frac{\csc^2 5x - 2x(\csc 5x)(-\csc 5x \cot 5x)(5)}{(\csc 5x)^4} \\ &= \frac{\csc^2 5x(1 + 10x\cot 5x)}{(\csc 5x)^4} \\ &= \frac{1 + 10x \cot 5x}{\csc^2 5x}. \end{aligned}$$

33. $y = \dfrac{\csc 2x^2}{4x}$. By the quotient rule,

$$\begin{aligned} y' &= \frac{4x\left(\frac{d}{dx}\right)\csc 2x^2 - \csc 2x^2\left(\frac{d}{dx}\right)(4x)}{(4x)^2} = \frac{4x(-\csc 2x^2 \cot 2x^2)(4x) - \csc 2x^2 \cdot 4}{4^2x^2} \\ &= -\frac{4x^2 \csc 2x^2 \cot 2x^2 + \csc 2x^2}{4x^2}. \end{aligned}$$

35. $y = \dfrac{\cos 3x}{1 - \cot x^2}$.

$$\begin{aligned} y' &= \frac{(1 - \cot x^2)\left(\frac{d}{dx}\right)(\cos 3x) - \cos 3x\left(\frac{d}{dx}\right)(1 - \cot x^2)}{(1 - \cot x^2)^2} \\ &= \frac{(1 - \cot x^2)(-\sin 3x)(3) - \cos 3x(\csc^2 x^2 \cdot 2x)}{(1 - \cot x^2)^2} \\ &= \frac{-3\sin 3x + 3\sin 3x \cot x^2 - 2x\cos 3x \csc^2 x^2}{(1 - \cot x^2)^2}. \end{aligned}$$

37. By the power rule, $\dfrac{d}{dx}\tan^3 x = (3\tan^2 x)\dfrac{d}{dx}\tan x = 3\tan^2 x \sec^2 x$.
Thus

$$\begin{aligned} \frac{d}{dx}\left(\frac{1}{3}\tan^3 x + \tan x\right) &= \frac{1}{3} \cdot 3\tan^2 x \sec^2 x + \sec^2 x \\ &= \sec^2 x(\tan^2 x + 1) && \text{facoring} \\ &= \sec^2 x \sec^2 x && 1 + \tan^2 x = \sec^2 x \\ &= \sec^4 x. \end{aligned}$$

39. $y = 5\sin 3x; y' = 15\cos 3x; y'' = -45\sin 3x$

41. $y = \tan 2x$; $y' = 2\sec^2 2x = 2(\sec 2x)^2$; $y'' = 4(\sec 2x)(\sec 2x \tan 2x)(2) = 8\sec^2 2x \tan 2x$

43. $y^2 = \tan x$.
$2y\dfrac{dy}{dx} = \sec^2 x$ and $\dfrac{dy}{dx} = \dfrac{1}{2y}\sec^2 x$.

45. Treating y as a function of x, we get $\dfrac{d}{dx}(y) = \dfrac{dy}{dx}$ and $\dfrac{d}{dx}(y^2) = 2y\dfrac{dy}{dx}$:

$$\begin{aligned} y^2 &= x \sec x \\ 2y\frac{dy}{dx} &= x\sec x \tan x + \sec x && \text{product rule} \\ \frac{dy}{dx} &= \frac{x \sec x \tan x + \sec x}{2y} \\ &= \frac{\sec x(x\tan x + 1)}{2y}. \end{aligned}$$

47. $y^2 = \sin(x + y^2)$.

$$\begin{aligned} 2y\frac{dy}{dx} &= \cos(x+y^2)\frac{d}{dx}(x+y^2) \\ &= \cos(x+y^2)\left(1+2y\frac{dy}{dx}\right) \\ &= \cos(x+y^2) + 2y\cos(x+y^2)\frac{dy}{dx} \\ \frac{dy}{dx}\left[2y - 2y\cos(x+y^2)\right] &= \cos(x+y^2) \\ \frac{dy}{dx} &= \frac{\cos(x+y^2)}{2y-2y\cos(x+y^2)}. \end{aligned}$$

49. $y = x\cot y^2$.

$$\begin{aligned} \frac{dy}{dx} &= x\frac{d}{dx}\cot y^2 + \cot y^2 \cdot 1 && \text{product rule} \\ \frac{dy}{dx} &= x(-\csc^2 y^2)\left(2y\frac{dy}{dx}\right) + \cot y^2 && \frac{d}{dx}y^2 = 2y\frac{dy}{dx} \\ \frac{dy}{dx} + (2xy\csc^2 y^2)\frac{dy}{dx} &= \cot y^2 \\ \frac{dy}{dx}(1 + 2xy\csc^2 y^2) &= \cot y^2 && \text{factoring } \frac{dy}{dx} \\ \frac{dy}{dx} &= \frac{\cot y^2}{1+2xy\csc^2 y^2}. \end{aligned}$$

51.

$$\begin{aligned} \cos y &= x^2y - 2x \\ -\sin y\frac{dy}{dx} &= x^2\frac{dy}{dx} + 2xy - 2 \\ 2 - 2xy &= x^2\frac{dy}{dx} + \sin y\frac{dy}{dx} \\ \frac{dy}{dx} &= \frac{2-2xy}{x^2+\sin y}. \end{aligned}$$

53. Slope of tangent line:

$$\frac{dy}{dx} = 2(-\csc^2 2x)\cdot 2 = -4\csc^2 2x\Big|_{x=\pi/8} = -4\left(\csc\frac{\pi}{4}\right)^2 = -4(\sqrt{2})^2 = -8.$$

Slope of normal line: $-\dfrac{1}{-8} = \dfrac{1}{8}$. negative reciprocal

6.4 Inverse Trigonometric Functions

19. Let $\theta = \text{Arctan}\,2 = \text{Arctan}\,\frac{2}{1}$. We place 2 on the side opposite θ and 1 on the side adjacent. Since $\sqrt{2^2+1^2} = \sqrt{5}$, the length of the hypotenuse, we get

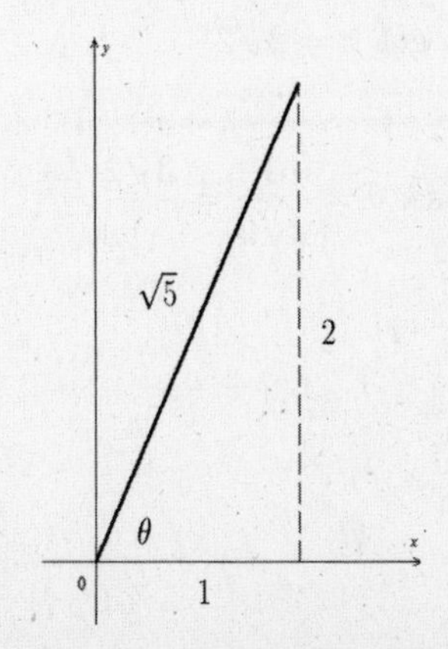

$$\sin(\text{Arctan}\,2) = \sin\theta = \frac{2}{\sqrt{5}} = \frac{2\sqrt{5}}{5}.$$

21. Let $\theta = \text{Arctan}\, 6 = \text{Arctan}\, \frac{6}{1}$. We place 6 on the side opposite θ and 1 on the side adjacent. Since $\sqrt{6^2 + 1^2} = \sqrt{37}$, the length of the hypotenuse, we get

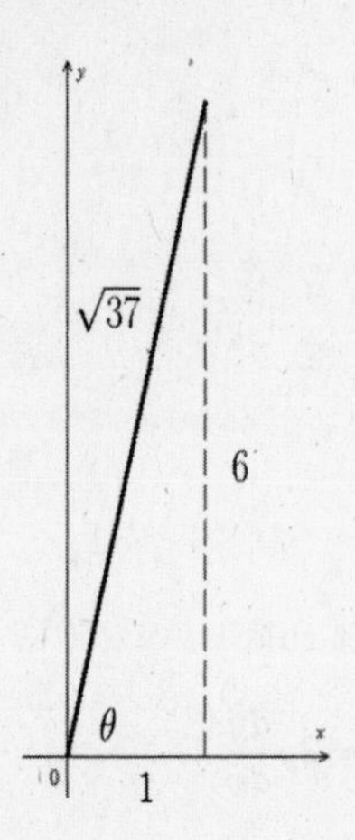

$$\cos(\text{Arctan}\, 6) = \cos\theta = \frac{1}{\sqrt{37}} = \frac{\sqrt{37}}{37}.$$

23. Let $\theta = \text{Arcsin}\, \frac{2}{3}$. So we place 2 on the side opposite θ and 3 on the hypotenuse. From $x^2 + 2^2 = 3^2$, we get $x = \sqrt{5}$.

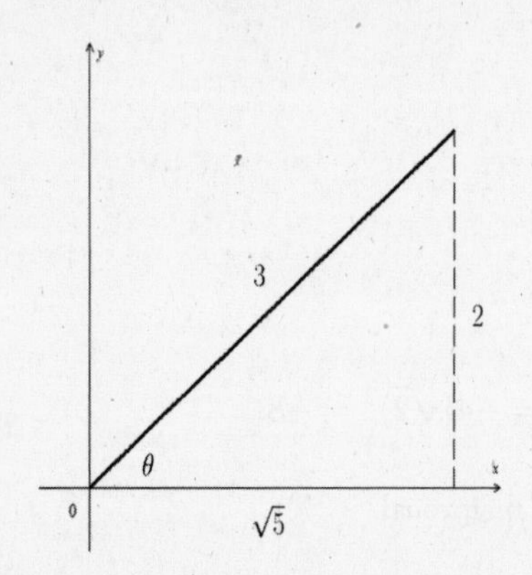

$$\tan\left(\text{Arcsin}\, \frac{2}{3}\right) = \tan\theta = \frac{2}{\sqrt{5}} = \frac{2\sqrt{5}}{5}.$$

25. Let $\theta = \text{Arcsin}\left(-\frac{1}{3}\right)$, where θ is between 0 and $-\pi/2$. So we place -1 on the side opposite (see figure) and 3 on the hypotenuse. From $x^2 + (-1)^2 = 3^2$, we get $x = 2\sqrt{2}$.

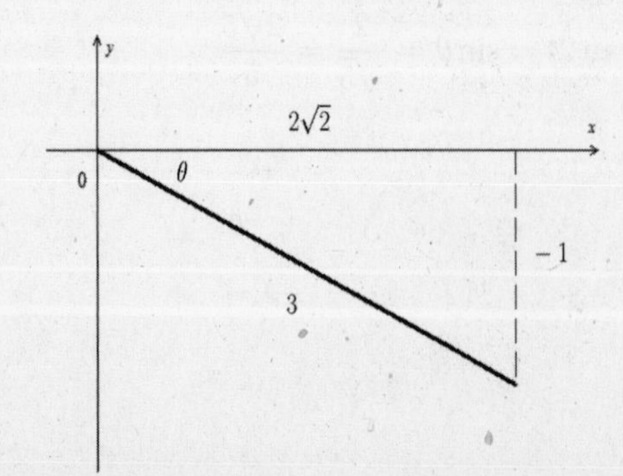

$$\sec\left[\text{Arcsin}\left(-\frac{1}{3}\right)\right] = \sec\theta = \frac{3}{2\sqrt{2}} = \frac{3\sqrt{2}}{4}.$$

27.

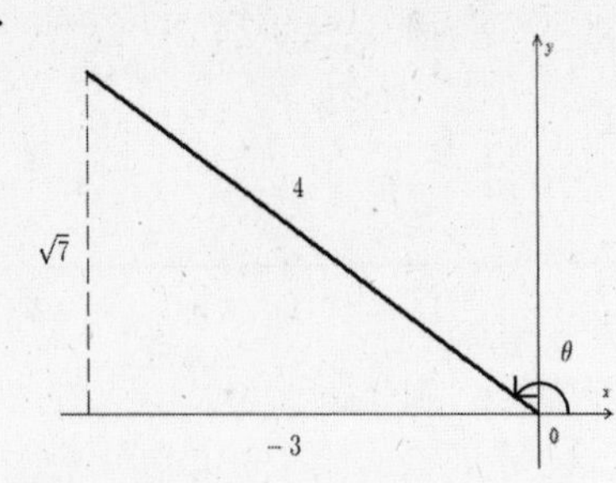

$$\tan\left[\operatorname{Arccos}\left(-\frac{3}{4}\right)\right] = \tan\theta = -\frac{\sqrt{7}}{3}.$$

29.

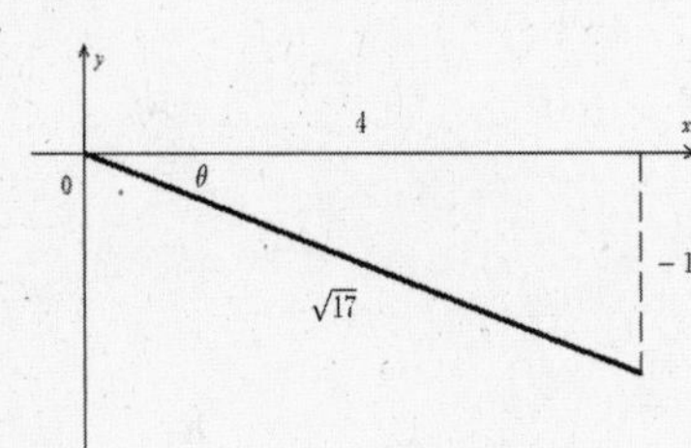

$$\csc\left[\operatorname{Arctan}\left(-\frac{1}{4}\right)\right] = \csc\theta = -\sqrt{17}$$

31.

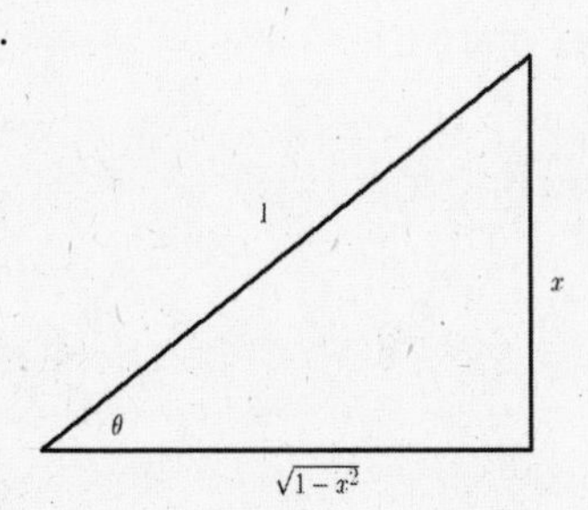

Let $\theta = \operatorname{Arcsin}\dfrac{x}{1}$. By the Pythagorean theorem, the length of the remaining side is $\sqrt{1-x^2}$.
$\csc(\operatorname{Arcsin} x) = \csc\theta = \dfrac{1}{x}$.

33.

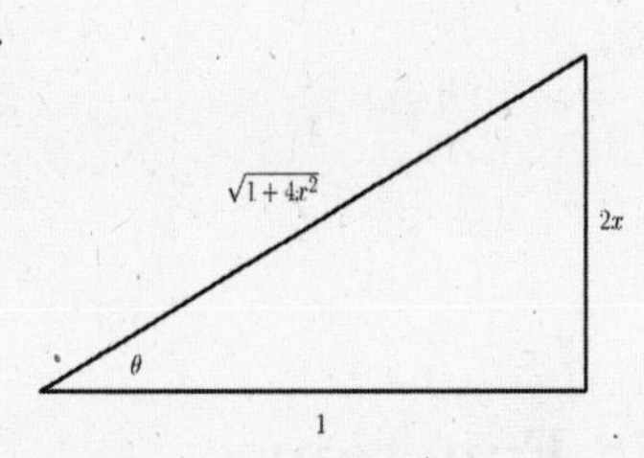

Let $\theta = \operatorname{Arctan}\dfrac{2x}{1}$.
Length of hypotenuse:
$\sqrt{1^2+(2x)^2} = \sqrt{1+4x^2}$.
$\cos(\operatorname{Arctan} 2x) = \cos\theta = \dfrac{1}{\sqrt{1+4x^2}}$.

35.

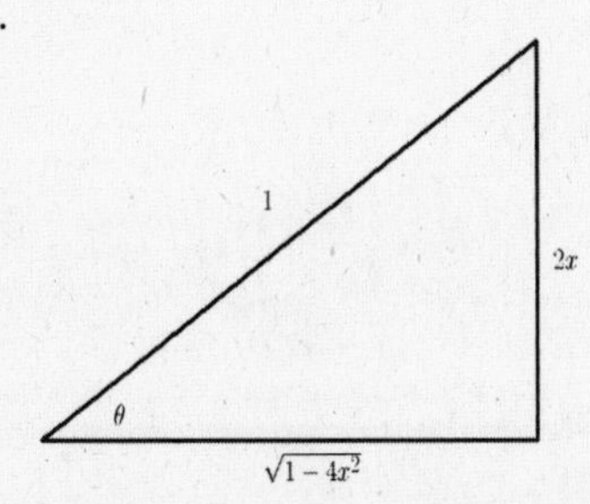

Let $\theta = \operatorname{Arcsin}\dfrac{2x}{1}$. By the Pythagorean theorem, the length of the remaining side is $\sqrt{1-4x^2}$.
$\cot\left(\operatorname{Arcsin}\dfrac{2x}{1}\right) = \cot\theta = \dfrac{\sqrt{1-4x^2}}{2x}$.

37.

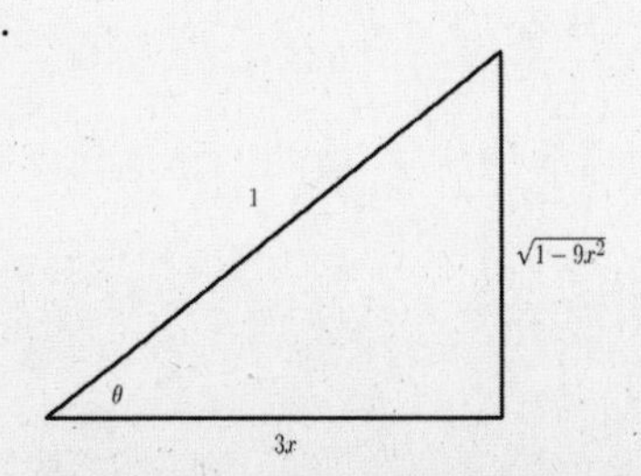

Let $\theta = \operatorname{Arccos}\dfrac{3x}{1}$. By the Pythagorean theorem, the length of the remaining side is $\sqrt{1-9x^2}$. Thus $\sin(\operatorname{Arccos} 3x) = \sin\theta = \dfrac{\sqrt{1-9x^2}}{1}$.

39.
$$\begin{aligned} y &= 2\sin x \\ \sin x &= \frac{y}{2} \\ x &= \text{Arcsin}\,\frac{y}{2} \end{aligned}$$

41.
$$\begin{aligned} y &= 1 - 2\sin 2x \\ \sin 2x &= \frac{1-y}{2} \\ 2x &= \text{Arcsin}\,\frac{1-y}{2} \\ x &= \frac{1}{2}\text{Arcsin}\,\frac{1-y}{2} \end{aligned}$$

43.
$$\begin{aligned} y &= 2\cos 3x - 1 \\ \cos 3x &= \frac{y+1}{2} \\ 3x &= \text{Arccos}\,\frac{y+1}{2} \\ x &= \frac{1}{3}\text{Arccos}\,\frac{y+1}{2} \end{aligned}$$

45.
$$\begin{aligned} y &= 2\tan\frac{1}{2}x + 2 \\ \tan\frac{1}{2}x &= \frac{y-2}{2} \\ \frac{1}{2}x &= \text{Arctan}\,\frac{y-2}{2} \\ x &= 2\text{Arctan}\,\frac{y-2}{2} \end{aligned}$$

47. $\frac{1}{2}\sin y$

6.5 Derivatives of Inverse Trigonometric Functions

1. $y = \text{Arctan}\,3x.$
$$y' = \frac{1}{1+(3x)^2}\frac{d}{dx}(3x) = \frac{3}{1+9x^2}.$$

3. $y = \text{Arccos}\,5x.$
$$y' = -\frac{1}{\sqrt{1-(5x)^2}}\frac{d}{dx}(5x) = \frac{-5}{\sqrt{1-25x^2}}.$$

5. $s = \text{Arctan}\,2t^2.$
$$\frac{ds}{dt} = \frac{1}{1+(2t^2)^2}\frac{d}{dt}(2t^2) = \frac{4t}{1+4t^4}.$$

7. $u = \text{Arcsin}\,3v^2.$
$$\frac{du}{dv} = \frac{1}{\sqrt{1-(3v^2)^2}}\frac{d}{dv}(3v^2) = \frac{6v}{\sqrt{1-9v^4}}.$$

9. $y = \text{Arcsin}\sqrt{2w} = \text{Arcsin}\,(2w)^{1/2}$.
$$\frac{dy}{dw} = \frac{1}{\sqrt{1-2w}} \cdot \frac{1}{2}(2w)^{-1/2} \cdot 2 = \frac{1}{\sqrt{1-2w}\sqrt{2w}} = \frac{1}{\sqrt{2w(1-2w)}}.$$

11. $v_1 = \text{Arccos}\, v_2^2$.
$$\frac{dv_1}{dv_2} = -\frac{1}{\sqrt{1-(v_2^2)^2}}\frac{d}{dv_2}(v_2^2) = -\frac{2v_2}{\sqrt{1-v_2^4}}.$$

13. $y = \text{Arcsin}\, 2x^2$.
$$y' = \frac{1}{\sqrt{1-(2x^2)^2}}\frac{d}{dx}(2x^2) = \frac{4x}{\sqrt{1-4x^4}}.$$

15. $y = \text{Arctan}\, 5x$.
$$y' = \frac{1}{1+(5x)^2} \cdot 5 = \frac{5}{1+25x^2}.$$

17. $y = x\text{Arctan}\, x$. By the product rule,
$$y' = x\frac{d}{dx}\text{Arctan}\, x + \text{Arctan}\, x \cdot 1 = \frac{x}{1+x^2} + \text{Arctan}\, x.$$

19. $y = x\text{Arccos}\, x^2$. By the product rule,
$$\begin{aligned} y' &= x\frac{d}{dx}(\text{Arccos}\, x^2) + (\text{Arccos}\, x^2)\frac{d}{dx}(x) \\ &= x\left(-\frac{2x}{\sqrt{1-(x^2)^2}}\right) + \text{Arccos}\, x^2 \cdot 1 \\ &= -\frac{2x^2}{\sqrt{1-x^4}} + \text{Arccos}\, x^2. \end{aligned}$$

21. $r = \theta\text{Arcsin}\, 3\theta$. By the product rule,
$$\begin{aligned} \frac{dr}{d\theta} &= \theta\frac{d}{d\theta}\text{Arcsin}\, 3\theta + \text{Arcsin}\, 3\theta \cdot 1 \\ &= \frac{\theta}{\sqrt{1-(3\theta)^2}}\frac{d}{d\theta}(3\theta) + \text{Arcsin}\, 3\theta \\ &= \frac{3\theta}{\sqrt{1-9\theta^2}} + \text{Arcsin}\, 3\theta. \end{aligned}$$

23. $R = 2V\text{Arctan}\, 3V$. By the product rule,
$$\begin{aligned} \frac{dR}{dV} &= 2V\frac{d}{dV}\text{Arctan}\, 3V + (\text{Arctan}\, 3V)\frac{d}{dV}(2V) \\ &= 2V\frac{3}{1+(3V)^2} + 2\text{Arctan}\, 3V \\ &= \frac{6V}{1+9V^2} + 2\text{Arctan}\, 3V. \end{aligned}$$

25. $y = \dfrac{\text{Arcsin}\, x}{x}$. By the quotient rule,

$y' = \dfrac{x\frac{d}{dx}\text{Arcsin}\, x - \text{Arcsin}\, x \cdot 1}{x^2} = \dfrac{\dfrac{x}{\sqrt{1-x^2}} - \text{Arcsin}\, x}{x^2}$.

Now simplify the complex fraction by multiplying numerator and denominator by $\sqrt{1-x^2}$:

$\dfrac{\dfrac{x}{\sqrt{1-x^2}} - \text{Arcsin}\, x}{x^2} \cdot \dfrac{\sqrt{1-x^2}}{\sqrt{1-x^2}} = \dfrac{x - \sqrt{1-x^2}\,\text{Arcsin}\, x}{x^2\sqrt{1-x^2}}$.

27. $y = \dfrac{\text{Arctan}\, x}{x^2}$. By the quotient rule,

$y' = \dfrac{x^2\frac{d}{dx}\text{Arctan}\, x - \text{Arctan}\, x(2x)}{x^4} = \dfrac{x^2\dfrac{1}{1+x^2} - 2x\text{Arctan}\, x}{x^4}$

Multiply numerator and denominator by $1 + x^2$:

$y' = \dfrac{x^2 - 2x\left(1+x^2\right)\text{Arctan}\, x}{x^4\left(1+x^2\right)} = \dfrac{x - 2\left(1+x^2\right)\text{Arctan}\, x}{x^3\left(1+x^2\right)}$

29. $y = \dfrac{\text{Arccos}\, x^2}{x}$. By the quotient rule,

$y' = \dfrac{x\frac{d}{dx}\text{Arccos}\, x^2 - \text{Arccos}\, x^2\frac{d}{dx}(x)}{x^2} = \dfrac{x\left(\dfrac{-1}{\sqrt{1-x^4}}\right)(2x) - \text{Arccos}\, x^2}{x^2}$

Multiply numerator and denominator by $\sqrt{1-x^4}$:

$y' = \dfrac{-2x^2 - \sqrt{1-x^4}\,\text{Arccos}\, x^2}{x^2\sqrt{1-x^4}} = -\dfrac{2x^2 + \sqrt{1-x^4}\,\text{Arccos}\, x^2}{x^2\sqrt{1-x^4}}$

31. $y = \text{Arccos}\sqrt{1-x}$.

$$\begin{aligned} y' &= -\frac{1}{\sqrt{1-(\sqrt{1-x})^2}}\frac{d}{dx}(1-x)^{1/2} = -\frac{1}{\sqrt{1-(1-x)}} \cdot \frac{1}{2}(1-x)^{-1/2} \cdot (-1) \\ &= \frac{1}{\sqrt{x}} \cdot \frac{1}{2(1-x)^{1/2}} = \frac{1}{2\sqrt{x(1-x)}} = \frac{1}{2\sqrt{x-x^2}}. \end{aligned}$$

33. $y = (\text{Arcsin}\, x)^2$. By the power rule,

$y' = 2(\text{Arcsin}\, x)\dfrac{1}{\sqrt{1-x^2}}$.

35. $y = (\text{Arccos}\, x)^{1/2}$. By the power rule,

$$\begin{aligned} y' &= \frac{1}{2}(\text{Arccos}\, x)^{-1/2}\left(-\frac{1}{\sqrt{1-x^2}}\right) = -\frac{1}{2}(\text{Arccos}\, x)^{-1/2}(1-x^2)^{-1/2} \\ &= -\frac{1}{2}\left[(1-x^2)\text{Arccos}\, x\right]^{-1/2}. \end{aligned}$$

37.

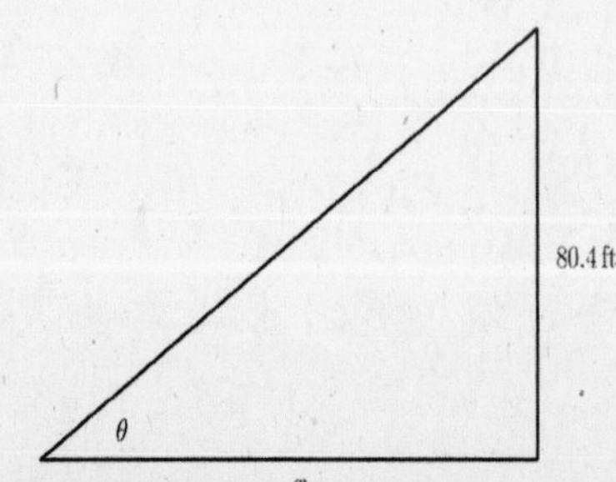

Given: $\dfrac{dx}{dt} = -12.4\,\text{ft/s}$.

Find: $\dfrac{d\theta}{dt}$ when $x = 20.6\,\text{ft}$.

$\tan\theta = \dfrac{80.4}{x}$ and $\theta = \text{Arctan}\,\dfrac{80.4}{x}$.

$$\begin{aligned}\frac{d\theta}{dt} &= \frac{1}{1+(80.4/x)^2}\frac{d}{dt}(80.4x^{-1})\\ &= \frac{1}{1+(80.4/x)^2}(80.4)(-1x^{-2})\frac{dx}{dt}\\ &= \frac{1}{1+(80.4/x)^2}\cdot\frac{-80.4}{x^2}(-12.4).\end{aligned}$$

Finally, let $x = 20.6$: $\dfrac{d\theta}{dt} = 0.145\,\text{rad/s}$.

39. Let $y = \text{Arccot}\, u$, so that $u = \cot y$. Then $\dfrac{du}{dx} = -\csc^2 y\dfrac{dy}{dx}$.
$\dfrac{dy}{dx} = -\dfrac{1}{\csc^2 y}\dfrac{du}{dx} = -\dfrac{1}{1+\cot^2 y}\dfrac{du}{dx} = -\dfrac{1}{1+u^2}\dfrac{du}{dx}$.

6.6 Exponential and Logarithmic Functions

1. $3^3 = 27$; base: 3; exponent: 3. Thus $\log_3 27 = 3$.

5. $(32)^{-1/5} = \dfrac{1}{2}$; base: 32; exponent: $-\dfrac{1}{5}$. Thus $\log_{32}\dfrac{1}{2} = -\dfrac{1}{5}$.

9. $\log_{1/4}\dfrac{1}{16} = 2$; base: $\dfrac{1}{4}$; exponent: 2. Thus $\left(\dfrac{1}{4}\right)^2 = \dfrac{1}{16}$.

13.

x :	-1	0	1	2
y :	$\frac{1}{3}$	1	3	9

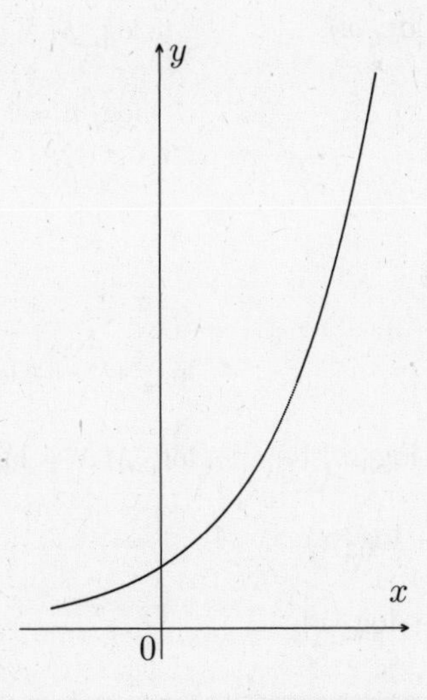

17. $y = \log_3 x$ or $x = 3^y$.

y :	-1	0	1	2
x :	$\frac{1}{3}$	1	3	9

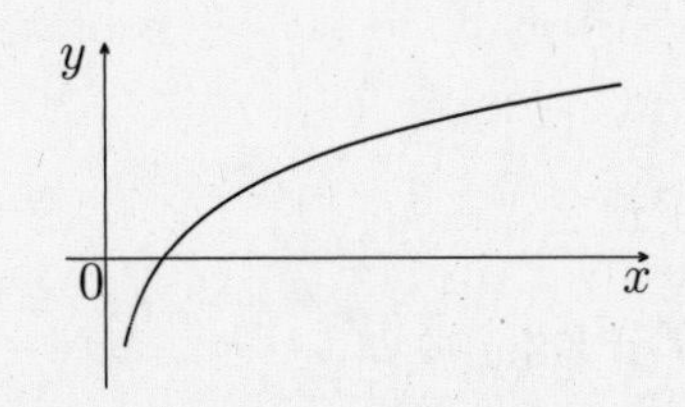

19. $\log_3 4 + \log_3 6 = \log_3 4\cdot 6 = \log_3 24$. $\qquad \log_a M + \log_a N = \log_a MN$

21.
$$\begin{aligned}
5\log_5 2 - 3\log_5 2 &= \log_5 2^5 - \log_5 2^3 && k\log_a M = \log_a M^k \\
&= \log_5 32 - \log_5 8 \\
&= \log_5 \frac{32}{8} && \log_a M - \log_a N = \log_a \frac{M}{N} \\
&= \log_5 4.
\end{aligned}$$

23.
$$\begin{aligned}
\frac{1}{2}\log_b 3 - \frac{1}{2}\log_b 9 &= \log_b 3^{1/2} - \log_b 9^{1/2} && k\log_a M = \log_a M^k \\
&= \log_b \frac{3^{1/2}}{9^{1/2}} = \log_b \frac{\sqrt{3}}{3}. && \log_a M - \log_a N = \log_a \frac{M}{N}
\end{aligned}$$

25.
$$\begin{aligned}
2\log_3 y + \frac{1}{3}\log_3 8 - 2\log_3 5 &= \log_3 y^2 + \log_3 8^{1/3} - \log_3 5^2 && k\log_a M = \log_a M^k \\
&= \log_3 \frac{y^2 \cdot 8^{1/3}}{5^2} && \text{properties (6.16) and (6.17)} \\
&= \log_3 \frac{2y^2}{25}.
\end{aligned}$$

27.
$$\begin{aligned}
\log_3 27 &= \log_3 3^3 \\
&= 3\log_3 3 && \log_a M^k = k\log_a M \\
&= 3 \cdot 1 = 3. && \log_a a = 1
\end{aligned}$$

29.
$$\begin{aligned}
\log_6 \sqrt{6x} &= \log_6 (6x)^{1/2} = \frac{1}{2}\log_6 6x && \log_a M^k = k\log_a M \\
&= \frac{1}{2}(\log_6 6 + \log_6 x) && \log_a MN = \log_a M + \log_a N \\
&= \frac{1}{2}(1 + \log_6 x). && \log_a a = 1
\end{aligned}$$

31.
$$\begin{aligned}
\log_3 \sqrt[3]{9x} &= \log_3 (9x)^{1/3} \\
&= \frac{1}{3}\log_3 9x && \log_a M^k = k\log_a M \\
&= \frac{1}{3}(\log_3 9 + \log_3 x) && \log_a MN = \log_a M + \log_a N \\
&= \frac{1}{3}(\log_3 3^2 + \log_3 x) \\
&= \frac{1}{3}(2\log_3 3 + \log_3 x) \\
&= \frac{1}{3}(2 + \log_3 x) && \log_a a = 1
\end{aligned}$$

33.
$$\begin{aligned}
\log_{10} 1000x^5 &= \log_{10} 10^3 x^5 \\
&= \log_{10} 10^3 + \log_{10} x^5 && \log_a MN = \log_a M + \log_a N \\
&= 3\log_{10} 10 + 5\log_{10} x && \log_a M^k = k\log_a M \\
&= 3 + 5\log_{10} x && \log_a a = 1
\end{aligned}$$

35.
$$\begin{aligned}\log_5 \frac{1}{25x^2} &= \log_5 1 - \log_5 25x^2 && \log_a \frac{M}{N} = \log_a M - \log_a N\\ &= 0 - \log_5 (5x)^2 && \log_a 1 = 0\\ &= -2\log_5(5x) && \log_a M^k = k\log_a M\\ &= -2(\log_5 5 + \log_5 x) && \log_a MN = \log_a M + \log_a N\\ &= -2(1+\log_5 x). && \log_a a = 1\end{aligned}$$

37.
$$\begin{aligned}\log_3 \frac{1}{\sqrt[3]{3x}} &= \log_3 \frac{1}{(3x)^{1/3}}\\ &= \log_3 1 - \log_3(3x)^{1/3} && \log_a \frac{M}{N} = \log_a M - \log_a N\\ &= 0 - \frac{1}{3}\log_3(3x) && \log_a 1 = 0,\ \log_a M^k = k\log_a M\\ &= -\frac{1}{3}(\log_3 3 + \log_3 x) && \log_a MN = \log_a M + \log_a N\\ &= -\frac{1}{3}(1+\log_3 x). && \log_a a = 1\end{aligned}$$

39. $\log_5 \frac{1}{\sqrt{y-2}} = \log_5(y-2)^{-1/2} = -\frac{1}{2}\log_5(y-2).$

41.
$$\begin{aligned}\log_{10} \frac{x}{\sqrt{x+2}} &= \log_{10} \frac{x}{(x+2)^{1/2}}\\ &= \log_{10} x - \log_{10}(x+2)^{1/2} && \log_a \frac{M}{N} = \log_a M - \log_a N\\ &= \log_{10} x - \frac{1}{2}\log_{10}(x+2). && \log_a M^k = k\log_a M\end{aligned}$$

43.
$$\begin{aligned}\log_{10} \frac{\sqrt{x}}{x+1} &= \log_{10} x^{1/2} - \log_{10}(x+1) && \text{by (6.17)}\\ &= \frac{1}{2}\log_{10} x - \log_{10}(x+1). && \text{by (6.18)}\end{aligned}$$

45.
$$\begin{aligned}3.62^x &= 12.4\\ \log_{10}(3.62)^x &= \log_{10} 12.4\\ x\log_{10} 3.62 &= \log_{10} 12.4\\ x &= \frac{\log_{10} 12.4}{\log_{10} 3.62} = 1.96.\end{aligned}$$

47.
$$\begin{aligned}(8.04)^x &= 2.85\\ \log_{10}(8.04)^x &= \log_{10} 2.85\\ x\log_{10} 8.04 &= \log_{10} 2.85\\ x &= \frac{\log_{10} 2.85}{\log_{10} 8.04} = 0.502.\end{aligned}$$

49.
$$\begin{aligned}(17.4)^{x+1} &= 0.935\\ \log_{10}(17.4)^{x+1} &= \log_{10} 0.935\\ (x+1)\log_{10} 17.4 &= \log_{10} 0.935\\ x+1 &= \frac{\log_{10} 0.935}{\log_{10} 17.4} = -0.0235\\ x &= -1.02\end{aligned}$$

51.
$$\begin{aligned}
(5.82)^{3x+1} &= 23.6\\
\log_{10}(5.82)^{3x+1} &= \log_{10} 23.6\\
(3x+1)\log_{10} 5.82 &= \log_{10} 23.6\\
3x+1 &= \frac{\log_{10} 23.6}{\log_{10} 5.82} = 1.7948\\
3x &= 0.7948\\
x &= 0.265
\end{aligned}$$

6.7 Derivative of the Logarithmic Function

1. $y = \ln 2x$.
$y' = \frac{1}{2x}\frac{d}{dx}(2x) = \frac{1}{2x}(2) = \frac{1}{x}$.

3. $y = 4\ln 3x$.
$y' = 4 \cdot \frac{1}{3x}\frac{d}{dx}(3x) = 4 \cdot \frac{1}{3x}(3) = \frac{4}{x}$.

5. $R = \ln s^3$.
$\frac{dR}{ds} = \frac{1}{s^3}\frac{d}{ds}(s^3) = \frac{1}{s^3}(3s^2) = \frac{3}{s}$
or
$R = \ln s^3 = 3\ln s$; thus $\frac{dR}{ds} = \frac{3}{s}$.

7. $y = 2\ln x^3 = 2 \cdot 3\ln x$.
$y' = \frac{6}{x}$.

9. $y = \log_{10} x^3 = 3\log_{10} x$; $y' = \frac{3}{x}\log_{10} e$.

11. $R_1 = \ln \sin R_2$.
$\frac{dR_1}{dR_2} = \frac{1}{\sin R_2}\frac{d}{dR_2}\sin R_2 = \frac{\cos R_2}{\sin R_2} = \cot R_2$.

13. $y = \ln\sqrt{x-2} = \ln(x-2)^{1/2} = \frac{1}{2}\ln(x-2)$.
$y' = \frac{1}{2(x-2)}$.

15. $y = \ln\frac{1}{\sqrt{x+4}} = \ln(x+4)^{-1/2} = -\frac{1}{2}\ln(x+4)$.
$y' = -\frac{1}{2(x+4)}$.

17. $z = 3\ln\sqrt[3]{t^2+1} = 3 \cdot \frac{1}{3}\ln(t^2+1) = \ln(t^2+1)$.
$\frac{dz}{dt} = \frac{1}{t^2+1}\frac{d}{dt}(t^2+1) = \frac{2t}{t^2+1}$.

19. $$\begin{aligned} y &= \ln\frac{x^2}{x+1} = \ln x^2 - \ln(x+1) && \log_a\frac{M}{N} = \log_a M - \log_a N \\ &= 2\ln x - \ln(x+1). && \log_a M^k = k\log_a M \\ y' &= 2\cdot\frac{1}{x} - \frac{1}{x+1} \\ &= \frac{2}{x}\frac{x+1}{x+1} - \frac{1}{x+1}\frac{x}{x} = \frac{2x+2-x}{x(x+1)} && \text{adding fractions} \\ &= \frac{x+2}{x(x+1)}. \end{aligned}$$

21. $$y = \ln\frac{2x}{x^2+1} = \ln 2x - \ln(x^2+1). \qquad \ln\frac{M}{N} = \ln M - \ln N$$
$$y' = \frac{1}{x} - \frac{2x}{x^2+1} = \frac{1}{x}\frac{x^2+1}{x^2+1} - \frac{2x}{x^2+1}\frac{x}{x} = \frac{1-x^2}{x(x^2+1)}.$$

23. $$\begin{aligned} y &= \ln\frac{\sqrt{x}}{2-x^2} = \ln\frac{x^{1/2}}{2-x^2} \\ &= \ln x^{1/2} - \ln(2-x^2) && \log_a\frac{M}{N} = \log_a M - \log_a N \\ &= \frac{1}{2}\ln x - \ln(2-x^2). && \log_a M^k = k\log_a M \\ y' &= \frac{1}{2}\cdot\frac{1}{x} - \frac{1}{2-x^2}\frac{d}{dx}(2-x^2) \\ &= \frac{1}{2x} - \frac{-2x}{2-x^2} = \frac{1}{2x}\cdot\frac{2-x^2}{2-x^2} + \frac{2x}{2-x^2}\cdot\frac{2x}{2x} \\ &= \frac{2-x^2+4x^2}{2x(2-x^2)} = \frac{2+3x^2}{2x(2-x^2)}. \end{aligned}$$

25. $\ln\dfrac{\cos x}{\sqrt{x}} = \ln\cos x - \dfrac{1}{2}\ln x.$
$$y' = \frac{1}{\cos x}(-\sin x) - \frac{1}{2x} = -\left(\tan x + \frac{1}{2x}\right) = -\frac{2x\tan x + 1}{2x}.$$

27. $$\begin{aligned} y &= \ln\frac{\sec^2 x}{\sqrt{x+1}} = \ln\sec^2 x - \ln(x+1)^{1/2} \\ &= 2\ln\sec x - \frac{1}{2}\ln(x+1). \\ y' &= 2\frac{\sec x\tan x}{\sec x} - \frac{1}{2(x+1)} \\ &= 2\tan x - \frac{1}{2(x+1)} = \frac{4(x+1)\tan x - 1}{2(x+1)}. \end{aligned}$$

29. $y = \ln^2 x = (\ln x)^2$. By the power rule,
$$y' = 2(\ln x)\frac{d}{dx}\ln x = \frac{2\ln x}{x}.$$

31. $y = 2x\ln x$. By the product rule,
$$y' = 2x\frac{d}{dx}\ln x + \ln x\frac{d}{dx}(2x) = 2x\left(\frac{1}{x}\right) + \ln x(2) = 2 + 2\ln x = 2(1+\ln x).$$

33. $\dfrac{\ln x}{x}$. By the quotient rule,
$$y' = \frac{x\cdot\frac{1}{x} - \ln x}{x^2} = \frac{1-\ln x}{x^2}.$$

35. $y = (x+1)^{1/2}\ln x$. By the product rule,
$$y' = (x+1)^{1/2}\frac{1}{x} + \frac{1}{2}(x+1)^{-1/2}\ln x = \frac{\sqrt{x+1}}{x} + \frac{\ln x}{2\sqrt{x+1}}.$$

37. $s = (\sec\theta)\ln\theta$. By the product rule,
$\dfrac{ds}{d\theta} = \sec\theta\dfrac{d}{d\theta}\ln\theta + \ln\theta\dfrac{d}{d\theta}\sec\theta = \dfrac{\sec\theta}{\theta} + (\sec\theta\tan\theta)\ln\theta.$

39. $V_1 = \dfrac{\ln V_2}{V_2}$. By the quotient rule,
$$\frac{dV_1}{dV_2} = \frac{V_2 \cdot \dfrac{1}{V_2} - \ln V_2}{V_2^2} = \frac{1 - \ln V_2}{V_2^2}.$$

41. $y = \dfrac{\sin x}{\ln x}$. By the quotient rule,
$$y' = \frac{\ln x(\cos x) - \sin x\left(\dfrac{1}{x}\right)}{(\ln x)^2} = \frac{x(\ln x)\cos x - \sin x}{x\ln^2 x}.$$

43.
$$\begin{aligned} y &= \ln\left[1 + (x^2-1)^{1/2}\right]. \\ y' &= \frac{1}{1+(x^2-1)^{1/2}}\frac{d}{dx}\left[1+(x^2-1)^{1/2}\right] \\ &= \frac{1}{1+\sqrt{x^2-1}}\left[\frac{1}{2}(x^2-1)^{-1/2}(2x)\right] = \frac{x}{(1+\sqrt{x^2-1})\sqrt{x^2-1}} \\ &= \frac{x}{\sqrt{x^2-1}+(\sqrt{x^2-1})^2} = \frac{x}{x^2-1+\sqrt{x^2-1}}. \end{aligned}$$

45.
$$\begin{aligned} y &= \ln(\ln x). \\ y' &= \frac{1}{\ln x}\frac{d}{dx}(\ln x) = \frac{1}{\ln x}\frac{1}{x} = \frac{1}{x\ln x}. \end{aligned}$$

6.8 Derivative of the Exponential Function

1.
$$\begin{aligned} y &= e^{4x}. \\ y' &= e^{4x}\frac{d}{dx}(4x) = 4e^{4x}. \end{aligned}$$

3. $y = e^{x^2}$.
$y' = e^{x^2}\dfrac{d}{dx}x^2 = 2xe^{x^2}.$

5.
$$\begin{aligned} y &= 2e^{-t^2}. \\ \frac{dy}{dt} &= 2e^{-t^2}\frac{d}{dt}(-t^2) = 2e^{-t^2}(-2t) = -4te^{-t^2}. \end{aligned}$$

7. $y = e^{\tan x}$.
$y' = e^{\tan x}\dfrac{d}{dx}\tan x = e^{\tan x}\sec^2 x.$

9.
$$\begin{aligned} y &= 3^{x^3}. \\ y' &= 3^{x^3}(\ln 3)\frac{d}{dx}(x^3) = 3^{x^3}(\ln 3)(3x^2). \end{aligned}$$

11. $y = 4^{x^2}$. By Formula (6.26),
$y' = 4^{x^2}(\ln 4)\dfrac{d}{dx}x^2 = 2x(\ln 4)4^{x^2}.$

13. $C = 2re^r$. By the product rule,
$$\frac{dC}{dr} = 2r\frac{d}{dr}e^r + e^r\frac{d}{dr}(2r) = 2re^r + 2e^r = 2(r+1)e^r.$$

15. $y = e^{\sin x}$.
$$y' = e^{\sin x}\frac{d}{dx}\sin x = e^{\sin x}\cos x.$$

17.
$$\begin{aligned} y &= \sin e^{2x}. \\ y' &= (\cos e^{2x})\frac{d}{dx}e^{2x} = 2e^{2x}\cos e^{2x}. \end{aligned}$$

19. $y = xe^x$. By the product rule,
$$y' = x\frac{d}{dx}e^x + e^x\frac{d}{dx}(x) = xe^x + e^x = (x+1)e^x.$$

21. $y = x^2e^{2x}$. By the product rule,
$$y' = x^2 \cdot e^{2x}(2) + e^{2x}(2x) = \left(2x^2 + 2x\right)e^{2x} = 2x(x+1)e^{2x}.$$

23. $S = e^{2\omega}\sin\omega$. By the product rule,
$$\frac{dS}{d\omega} = e^{2\omega}\frac{d}{d\omega}\sin\omega + \sin\omega\frac{d}{d\omega}e^{2\omega} = e^{2\omega}\cos\omega + (\sin\omega)e^{2\omega}(2) = e^{2\omega}(\cos\omega + 2\sin\omega).$$

25. $y = \dfrac{\tan x}{e^{x^2}}$. By the quotient rule,
$$\begin{aligned} y' &= \frac{e^{x^2}\frac{d}{dx}\tan x - \tan x\frac{d}{dx}e^{x^2}}{(e^{x^2})^2} = \frac{e^{x^2}\sec^2 x - (\tan x)e^{x^2}(2x)}{(e^{x^2})^2} \\ &= \frac{e^{x^2}(\sec^2 x - 2x\tan x)}{(e^{x^2})^2} = \frac{\sec^2 x - 2x\tan x}{e^{x^2}}. \end{aligned}$$

27. $y = (\ln x)e^{\sec x}$. By the product rule,
$$\begin{aligned} y' &= (\ln x)\frac{d}{dx}e^{\sec x} + e^{\sec x}\frac{d}{dx}\ln x = (\ln x)e^{\sec x}\sec x\tan x + e^{\sec x}\cdot\frac{1}{x} \\ &= e^{\sec x}\left(\sec x\tan x\ln x + \frac{1}{x}\right). \end{aligned}$$

29. $y = \ln(1+e^x)\,;\, y' = \dfrac{e^x}{1+e^x}$

31. $y = \dfrac{x}{e^{2x}}$. By the quotient rule,
$$y' = \frac{e^{2x}\cdot 1 - xe^{2x}\cdot 2}{(e^{2x})^2} = \frac{e^{2x}(1-2x)}{(e^{2x})^2} = \frac{1-2x}{e^{2x}}.$$

33. $y = \dfrac{\sin 2x}{e^x+1}$. By the quotient rule,
$$\begin{aligned} y' &= \frac{(e^x+1)\frac{d}{dx}\sin 2x - \sin 2x\frac{d}{dx}(e^x+1)}{(e^x+1)^2} = \frac{(e^x+1)\cos 2x\cdot 2 - \sin 2x\cdot e^x}{(e^x+1)^2} \\ &= \frac{2\cos 2x + 2e^x\cos 2x - e^x\sin 2x}{(e^x+1)^2}. \end{aligned}$$

35.
$$\frac{d}{dx}\sinh x = \frac{d}{dx}\frac{1}{2}(e^x - e^{-x}) = \frac{1}{2}\left[e^x - e^{-x}(-1)\right] = \frac{1}{2}(e^x + e^{-x}) = \cosh x.$$
$$\frac{d}{dx}\cosh x = \frac{d}{dx}\frac{1}{2}(e^x + e^{-x}) = \frac{1}{2}(e^x - e^{-x}) = \sinh x.$$

37.
$$\begin{aligned} y &= \sinh 5x. \\ y' &= (\cosh 5x)\tfrac{d}{dx}(5x) = 5\cosh 5x. \end{aligned}$$

39. $y = \cosh 2x^3$.

$$y' = (\sinh 2x^3)\frac{d}{dx}(2x^3) = 6x^2 \sinh 2x^3.$$

41. $y = x\cosh 3x$. By the product rule,

$$\begin{aligned} y' &= \cosh 3x \cdot \frac{d}{dx}(x) + x\frac{d}{dx}\cosh 3x = \cosh 3x \cdot 1 + x\sinh 3x \cdot \frac{d}{dx}(3x) \\ &= \cosh 3x + 3x\sinh 3x. \end{aligned}$$

43.
$$\begin{aligned} y &= x^{\sin x} \\ \ln y &= \ln x^{\sin x} = \sin x \ln x \\ \frac{1}{y}\frac{dy}{dx} &= \sin x \cdot \frac{1}{x} + \ln x \cos x \\ \frac{dy}{dx} &= y\left(\sin x \cdot \frac{1}{x} + \cos x \ln x\right) \\ &= x^{\sin x}\left(\sin x \cdot \frac{1}{x} + \cos x \ln x\right). \end{aligned}$$

Factoring out x^{-1},

$$\begin{aligned} \frac{dy}{dx} &= x^{\sin x}x^{-1}(\sin x + x\cos x \ln x) \\ &= x^{(\sin x - 1)}(\sin x + x\cos x \ln x). \end{aligned}$$

45.
$$\begin{aligned} y &= (\ln x)^x \\ \ln y &= \ln(\ln x)^x \\ &= x\ln(\ln x). \qquad \text{property (6.18)} \\ \frac{1}{y}\frac{dy}{dx} &= x\frac{d}{dx}\ln(\ln x) + \ln(\ln x)\cdot 1 = x\frac{1}{\ln x}\frac{1}{x} + \ln(\ln x) \\ &= \frac{1}{\ln x} + \ln(\ln x). \\ \frac{dy}{dx} &= y\left[\frac{1}{\ln x} + \ln(\ln x)\right], \text{ where } y = (\ln x)^x. \end{aligned}$$

47.
$$\begin{aligned} y &= (\tan x)^x \\ \ln y &= \ln(\tan x)^x = x\ln\tan x \\ \frac{1}{y}\frac{dy}{dx} &= x \cdot \frac{1}{\tan x}\sec^2 x + \ln\tan x \\ \frac{dy}{dx} &= y\left(\frac{1}{\tan x}\right)(x\sec^2 x + \tan x \ln\tan x) \\ &= (\tan x)^x\left(\frac{1}{\tan x}\right)(x\sec^2 x + \tan x \ln\tan x) \\ &= (\tan x)^{x-1}(x\sec^2 x + \tan x\ln\tan x). \end{aligned}$$

49. $i = \dfrac{dq}{dt} = \dfrac{d}{dt}\left(2.0\mathrm{e}^{-0.60t}\cos 10\pi t\right)$. By the product rule,

$$\begin{aligned} i &= 2.0\left[\mathrm{e}^{-0.60t}\frac{d}{dt}\cos 10\pi t + \cos 10\pi t\frac{d}{dt}\mathrm{e}^{-0.60t}\right] \\ &= 2.0\left[\mathrm{e}^{-0.60t}(-\sin 10\pi t)(10\pi) + (\cos 10\pi t)\mathrm{e}^{-0.60t}(-0.60)\right]. \end{aligned}$$

Substituting $t = 0.20\,\mathrm{s}$, we get

$i = 2.0\left[0 + 1\cdot \mathrm{e}^{(-0.60)(0.20)}(-0.60)\right] = -1.1\,\mathrm{A}.$

51. $p(t) = (1 + ae^{-kt})^{-1}$.
$$\frac{dp}{dt} = -1(1 + ae^{-kt})^{-2}[0 + ae^{-kt}(-k)] = \frac{ake^{-kt}}{(1 + ae^{-kt})^2}.$$

6.9 L'Hospital's Rule

1. $\lim\limits_{x\to-2} \frac{x^2-4}{x+2} = \lim\limits_{x\to-2} \frac{\frac{d}{dx}(x^2-4)}{\frac{d}{dx}(x+2)} = \lim\limits_{x\to-2} \frac{2x}{1} = -4.$

3. $\lim\limits_{x\to\infty} \frac{3x^2-4x}{2x^2+1} = \lim\limits_{x\to\infty} \frac{\frac{d}{dx}(3x^2-4x)}{\frac{d}{dx}(2x^2+1)} = \lim\limits_{x\to\infty} \frac{6x-4}{4x} = \lim\limits_{x\to\infty} \frac{\frac{d}{dx}(6x-4)}{\frac{d}{dx}(4x)} = \frac{6}{4} = \frac{3}{2}.$

5. $\lim\limits_{t\to3} \frac{t^2+t-12}{t-3} = \lim\limits_{t\to3} \frac{\frac{d}{dt}(t^2+t-12)}{\frac{d}{dt}(t-3)} = \lim\limits_{t\to3} \frac{2t+1}{1} = 7.$

7. $\lim\limits_{m\to0} \frac{\sin 6m}{m} = \lim\limits_{m\to0} \frac{\frac{d}{dm}\sin 6m}{\frac{d}{dm}(m)} = \lim\limits_{m\to0} \frac{6\cos 6m}{1} = 6.$

9. $\lim\limits_{x\to0} \frac{\tan 3x}{1-\cos x} = \lim\limits_{x\to0} \frac{3\sec^2 3x}{\sin x} = \infty$ (limit does not exist).

11. $\lim\limits_{x\to0} \frac{e^x - e^{-x}}{\sin x} = \lim\limits_{x\to0} \frac{e^x - e^{-x}(-1)}{\cos x} = \frac{1+1}{1} = 2.$

13. $\lim\limits_{x\to0} \frac{1-e^x}{2x} = \lim\limits_{x\to0} \frac{-e^x}{2} = -\frac{1}{2}.$

15. $\lim\limits_{x\to\infty} \frac{x+\ln x}{x\ln x} = \lim\limits_{x\to\infty} \frac{1+1/x}{x\cdot\frac{1}{x} + \ln x\cdot 1} = \lim\limits_{x\to\infty} \frac{1+1/x}{1+\ln x} = 0.$

17. $\lim\limits_{x\to0} \frac{x+\sin 2x}{x-\sin 2x} = \lim\limits_{x\to0} \frac{1+2\cos 2x}{1-2\cos 2x} = \frac{1+2}{1-2} = -3.$

19. $\lim\limits_{x\to\pi/2} \frac{\cos x}{\pi-2x} = \lim\limits_{x\to\pi/2} \frac{-\sin x}{-2} = \frac{1}{2}.$

21. $\lim\limits_{x\to\infty} e^{-x}\ln x$ has the indeterminate form $0\cdot\infty$. Written as a quotient, $\lim\limits_{x\to\infty} \frac{\ln x}{e^x}$ has the indeterminate form $\frac{\infty}{\infty}$. So by l'Hospital's rule,

$$\lim_{x\to\infty} \frac{\ln x}{e^x} = \lim_{x\to\infty} \frac{1/x}{e^x} = \lim_{x\to\infty} \frac{1}{xe^x} = 0.$$

23.
$$\begin{aligned}\lim_{x\to0+} (\sin x)\ln\sin x &= \lim_{x\to0+} \frac{\ln\sin x}{(\sin x)^{-1}} = \lim_{x\to0+} \frac{\frac{1}{\sin x}\cos x}{-1(\sin x)^{-2}\cos x} \\ &= \lim_{x\to0+} \left(-\frac{\frac{1}{\sin x}}{(\sin x)^{-2}}\right) = \lim_{x\to0+} (-\sin x) = 0.\end{aligned}$$

25. $\lim\limits_{x\to0+} x\ln x = \lim\limits_{x\to0+} \frac{\ln x}{x^{-1}} = \lim\limits_{x\to0+} \frac{1/x}{-1x^{-2}} = \lim\limits_{x\to0+} (-x) = 0.$

27. Form: 1^∞

Let $y = \lim_{x\to\infty}\left(1+\frac{1}{x}\right)^x$. Then

$$\begin{aligned}\ln y &= \lim_{x\to\infty} \ln\left(1+\frac{1}{x}\right)^x \\ &= \lim_{x\to\infty} x\ln\left(1+\frac{1}{x}\right) \qquad \log_a M^k = k\log_a M \\ &= \lim_{x\to\infty}\frac{\ln\left(1+\frac{1}{x}\right)}{x^{-1}}\end{aligned}$$

which tends to the indeterminate form $\frac{0}{0}$. So by L'Hospital's rule,

$$\ln y = \lim_{x\to\infty}\frac{\frac{1}{1+1/x}\frac{d}{dx}\left(1+\frac{1}{x}\right)}{-1x^{-2}} = \lim_{x\to\infty}\frac{\frac{1}{1+1/x}\left(-\frac{1}{x^2}\right)}{-\frac{1}{x^2}} = 1.$$

Since $\ln_e y = 1$, $e^1 = y$, and $y = e$.

6.10 Applications

1. $$\begin{aligned}f(x) &= xe^{-x}\\ f'(x) &= xe^{-x}(-1)+e^{-x} = e^{-x} - xe^{-x}\\ f''(x) &= -e^{-x} - xe^{-x}(-1) - e^{-x} = xe^{-x} - 2e^{-x}\end{aligned}$$

Step 1. Critical points:

$f'(x) = e^{-x} - xe^{-x} = e^{-x}(1-x) = 0$;

thus $x = 1$. From $y = xe^{-x}$, the point $(1, 1/e)$ is the critical point.

Step 2. Test of critical point:

$f''(1) = xe^{-x} - 2e^{-x}\Big|_{x=1} = e^{-1} - 2e^{-1} = -\frac{1}{e} < 0.$

Thus $(1, 1/e)$ is a maximum.

Step 3. Concavity and points of inflection:

$f''(x) = xe^{-x} - 2e^{-x} = 0.$

$e^{-x}(x-2) = 0$; thus $x = 2$. From $y = xe^{-x}$, the point is $(2, 2/e^2)$.

If $x < 2$, $y'' < 0$ (concave down).

If $x > 2$, $y'' > 0$ (concave up).

Step 4. Other:

$\lim_{x\to+\infty} xe^{-x} = \lim_{x\to+\infty}\frac{x}{e^x} = \lim_{x\to+\infty}\frac{1}{e^x} = 0$ by L'Hospital's rule. So the positive x-axis is an asymptote.

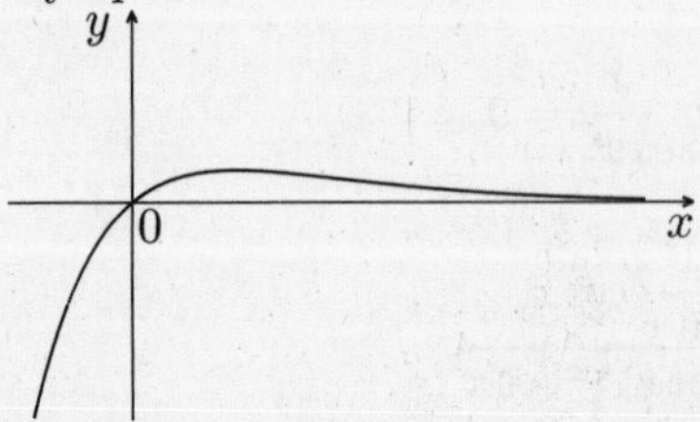

3. $f(x) = x\ln x,\ x > 0$

$f'(x) = x\cdot\frac{1}{x} + \ln x \cdot 1 = 1 + \ln x$

$f''(x) = \frac{1}{x}$

Step 1. Critical points:

$$\begin{aligned} f'(x) = 1 + \ln x &= 0 \\ \ln x &= -1 \\ \log_e x &= -1 && \ln = \log_e \\ x &= e^{-1} = \frac{1}{e}. && \text{definition of log} \end{aligned}$$

From $y = x \ln x$, we have

$y = \frac{1}{e} \ln \frac{1}{e} = \frac{1}{e}(\ln 1 - \ln e) = \frac{1}{e}(0 - 1) = -\frac{1}{e}$.

Critical point: $\left(\frac{1}{e}, -\frac{1}{e}\right)$.

Step 2. Test of critical point:

$f''(e^{-1}) = \frac{1}{e^{-1}} = e > 0$ (minimum).

Step 3. Since $x > 0$, $1/x > 0$. The graph is concave up everywhere.

Other: By L'Hospital's rule, $\lim_{x \to 0+} x \ln x = 0$ (see Exercise 25, Section 6.9, and graph in answer section of book).

5.
$$\begin{aligned} f(x) &= \cosh x = \frac{1}{2}(e^x + e^{-x}) \\ f'(x) &= \sinh x = \frac{1}{2}(e^x - e^{-x}) \\ f''(x) &= \cosh x = \frac{1}{2}(e^x + e^{-x}) \end{aligned}$$

Step 1. Critical points:

$f'(x) = \frac{1}{2}(e^x - e^{-x}) = 0$ when $x = 0$.

Step 2. Test of critical point:

$f''(0) = \frac{1}{2}(1 + 1) = 1 > 0$. The point $(0, 1)$ is a minimum.

Step 3. Concavity:

$f''(x) = \frac{1}{2}(e^x + e^{-x}) > 0$. The graph is concave up. (There are no inflection points.)

7. $i = 2 \cos 10\pi t + 2 \sin 10\pi t$.

$$\begin{aligned} \frac{di}{dt} = 2(-10\pi \sin 10\pi t + 10\pi \cos 10\pi t) &= 0 \\ -\sin 10\pi t + \cos 10\pi t &= 0 \\ \frac{\sin 10\pi t}{\cos 10\pi t} &= 1 \\ \tan 10\pi t &= 1 \\ 10\pi t &= \text{Arctan}\, 1 = \frac{\pi}{4} \\ t &= \frac{1}{40} \text{ s.} \end{aligned}$$

9. $i = 3.0 \cos 80t + 2.0 \sin 80t$

$v = 0.0020(-240 \sin 80t + 160 \cos 80t)\big|_{t=0} = 0.32\,\text{V}$

11. $N = 15e^{-3.0t}$

$\frac{dN}{dt} = 15e^{-3.0t}(-3.0) = -3.0\left(15e^{-3.0t}\right)$ or $\frac{dN}{dt} = -3.0\,\text{N}$

13.
$$\begin{aligned} N &= 6.0\mathrm{e}^{-0.25t} \\ \frac{dN}{dt} &= 6.0\mathrm{e}^{-0.25t}(-0.25)\Big|_{t=8.5} \\ &= 6.0\mathrm{e}^{(-0.25)(8.5)}(-0.25) = -0.18\,\mathrm{g/min}. \end{aligned}$$

15. $N = 200\mathrm{e}^{(1/2)t}$.
$$\frac{dN}{dt} = 200\left(\frac{1}{2}\right)\mathrm{e}^{(1/2)t}\Big|_{t=3} = 448 \text{ bacteria per hour.}$$

17. $T = 160 - 150\mathrm{e}^{-0.025t}$
$$\frac{dT}{dt} = -150\mathrm{e}^{-0.025t}(-0.025)\big|_{t=20} = 2.27°\mathrm{C/min}$$

19. $P = Ax\mathrm{e}^{-x/n}$.
$$\begin{aligned} \frac{dP}{dx} = A\mathrm{e}^{-x/n} + Ax\mathrm{e}^{-x/n}\left(-\frac{1}{n}\right) &= 0 \\ A\mathrm{e}^{-x/n}\left(1 - \frac{x}{n}\right) &= 0 \\ x &= n. \end{aligned}$$

If $x < n$, then $dP/dx > 0$ and if $x > n$, then $dP/dx < 0$. So the critical value corresponds to a maximum by the first-derivative test.

21. $x = 5\cos 2t$.
$v = \dfrac{dx}{dt} = -10\sin 2t$.
Critical value:
$$\begin{aligned} v' = -20\cos 2t &= 0 \\ 2t &= \frac{\pi}{2} \\ t &= \frac{\pi}{4}. \end{aligned}$$
When $t = \pi/4$, then $x = 5\cos 2(\pi/4) = 0$, showing that the particle is at the origin. (At $t = \pi/4$, v is actually a minimum, but its magnitude is a maximum.)

23. In a problem on related rates, we label all quantities that vary by letters, as shown in the figure.

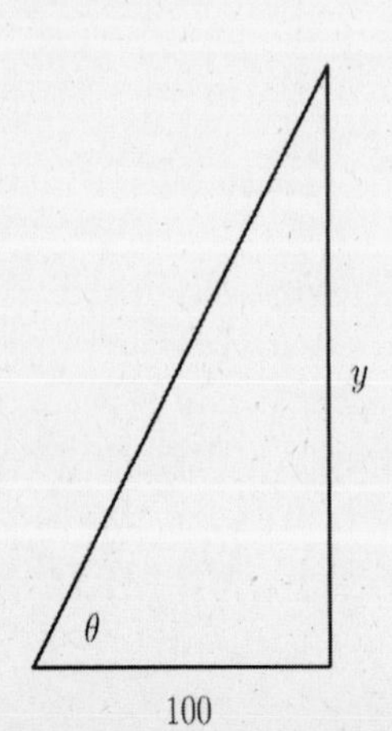

Given $\dfrac{dy}{dt} = 6.0$, find $\dfrac{d\theta}{dt}$ when $y = 150$. From the figure:
$$\begin{aligned} \tan\theta &= \frac{y}{100} \\ \theta &= \mathrm{Arctan}\,\frac{y}{100} \\ \frac{d\theta}{dt} &= \frac{1}{1+(y/100)^2}\left(\frac{1}{100}\right)\frac{dy}{dt} \\ &= \frac{1}{1+(y/100)^2}\left(\frac{1}{100}\right)(6.0)\Big|_{y=150} \\ &= 0.018\,\mathrm{rad/s}. \end{aligned}$$

25.

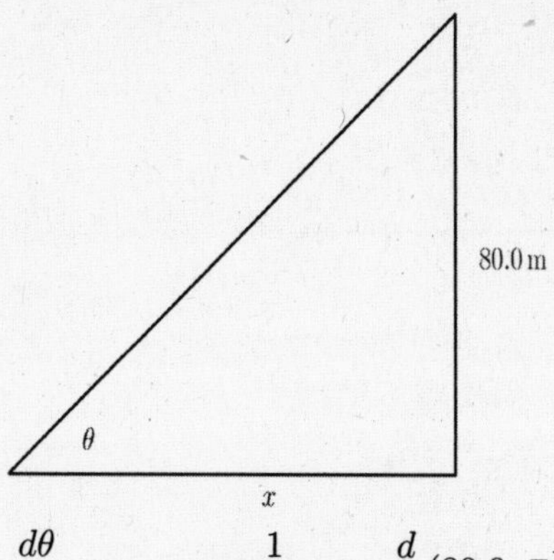

Given: $\dfrac{dx}{dt} = -1.5\,\text{m/s}$.

Find: $\dfrac{d\theta}{dt}$ when $x = 40.0\,\text{m}$.

$$\tan\theta = \frac{80.0}{x}$$
$$\theta = \text{Arctan}\,\frac{80.0}{x}.$$

$$\begin{aligned}\frac{d\theta}{dt} &= \frac{1}{1+(80.0/x)^2}\frac{d}{dt}(80.0x^{-1})\\ &= \frac{1}{1+(80.0/x)^2}(80.0)(-1x^{-2})\frac{dx}{dt}\\ &= \frac{1}{1+(80.0/x)^2}(80.0)\left(-\frac{1}{x^2}\right)(-1.5).\end{aligned}$$

When $x = 40.0\,\text{m}$, $\dfrac{d\theta}{dt} = 0.015\,\text{rad/s}$.

27.

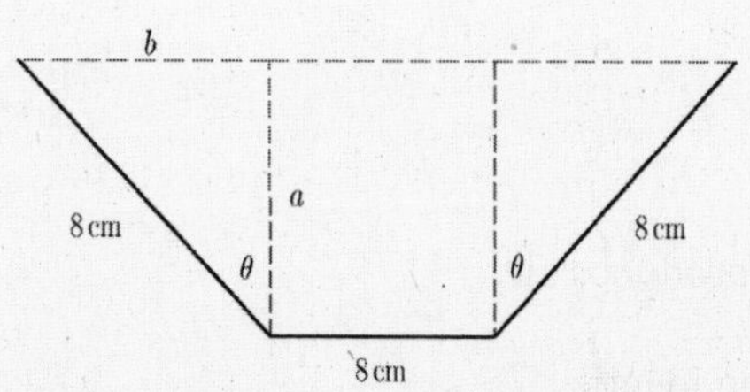

To find an expression for the area, we need to determine a and b in the figure. Observe that $\cos\theta = \dfrac{a}{8}$ and $\sin\theta = \dfrac{b}{8}$.
Hence $a = 8\cos\theta$ and $b = 8\sin\theta$.

Area of each triangle: $\dfrac{1}{2}ab = \dfrac{1}{2}(64)\sin\theta\cos\theta$.

Area of rectangle: $8a = 64\cos\theta$.

Total area: $A = 64\cos\theta + 64\sin\theta\cos\theta = 64(\cos\theta + \sin\theta\cos\theta)$.

$$\begin{aligned}\frac{dA}{d\theta} &= 64[-\sin\theta + \sin\theta(-\sin\theta) + \cos\theta(\cos\theta)]\\ &= 64(-\sin\theta - \sin^2\theta + \cos^2\theta)\\ &= 64(-\sin\theta - \sin^2\theta + 1 - \sin^2\theta) && \cos^2\theta = 1 - \sin^2\theta\\ &= -64(2\sin^2\theta + \sin\theta - 1)\\ &= -64(2\sin\theta - 1)(\sin\theta + 1) = 0. && \text{factoring}\end{aligned}$$

$$\begin{aligned}2\sin\theta &= 1\\ \sin\theta &= \frac{1}{2} \text{ and } \theta = \frac{\pi}{6}.\end{aligned}$$

29.

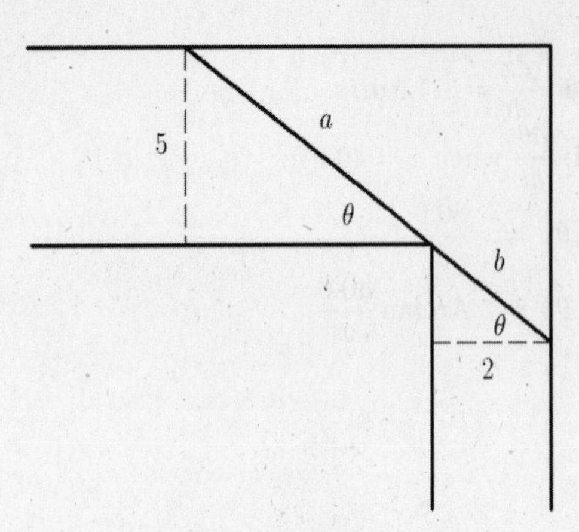

Let $x = a + b$, the length of the pole. Observe that

$$\csc\theta = \frac{a}{5} \qquad \sec\theta = \frac{b}{2}$$
$$a = 5\csc\theta \qquad b = 2\sec\theta.$$

So

$$\begin{aligned} x &= 5\csc\theta + 2\sec\theta \\ \frac{dx}{d\theta} &= -5\csc\theta\cot\theta + 2\sec\theta\tan\theta \\ &= -5\frac{1}{\sin\theta}\frac{\cos\theta}{\sin\theta} + 2\frac{1}{\cos\theta}\frac{\sin\theta}{\cos\theta} = 0. \end{aligned}$$

Multiply by $\sin^2\theta\cos^2\theta$:

$$\begin{aligned} -5\cos^3\theta + 2\sin^3\theta &= 0 \\ \frac{\sin^3\theta}{\cos^3\theta} &= \frac{5}{2} \\ \tan^3\theta &= \frac{5}{2} \\ \tan\theta &= \sqrt[3]{\frac{5}{2}} \\ \theta &= \text{Arctan}\sqrt[3]{\frac{5}{2}} \approx 53.62°. \end{aligned}$$

So the length of the ladder is
$x = 5\csc 53.62° + 2\sec 53.62° = 9.58\,\text{m}$.

31. $R = k\cos\theta\sin(\theta - \alpha)$, where $k = \dfrac{2v_0^2}{g\cos^2\alpha}$. By the product rule,
$\dfrac{dR}{d\theta} = k[\cos\theta\cos(\theta - \alpha) - \sin\theta\sin(\theta - \alpha)]$. From the identity
$\cos(A + B) = \cos A\cos B - \sin A\sin B$,
$\dfrac{dR}{d\theta} = k\cos[\theta + (\theta - \alpha)] = k\cos(2\theta - \alpha) = 0$.
Thus, $2\theta - \alpha = \dfrac{\pi}{2}$ and $\theta = \dfrac{\pi}{4} + \dfrac{\alpha}{2}$.

6.11 Newton's Method

1. $f(x) = x^2 + 2x - 11;\ f'(x) = 2x + 2;\ x_1 = x_0 - \dfrac{x_0^2 + 2x_0 - 11}{2x_0 + 2}$

For the positive root, let $x_0 = 2$. The iteration yields $2.5, 2.4643, 2.464102$.

3. $f(x) = 2x^3 - 4x^2 - x + 4$; $f'(x) = 6x^2 - 8x - 1$.

$$x_1 = x_0 - \frac{2x_0^3 - 4x_0^2 - x_0 + 4}{6x_0^2 - 8x_0 - 1}$$

Start: $x_0 = -1$

$$x_1 = -1 - \frac{2(-1)^3 - 4(-1)^2 - (-1) + 4}{6(-1)^2 - 8(-1) - 1} = -0.923.$$

$$x_2 = -0.923 - \frac{2(-0.923)^3 - 4(-0.923)^2 - (-0.923) + 4}{6(-0.923)^2 - 8(-0.923) - 1} = -0.9180$$

$$x_3 = -0.917988.$$

5. A graphing calculator will quickly locate the roots near -1, between 0 and 1, and near 6. Suppose we find the middle root by letting $x_0 = 1$:

$$f(x) = x^3 - 6x^2 - 4x + 6$$

$$f'(x) = 3x^2 - 12x - 4$$

$$x_1 = x_0 - \frac{x_0^3 - 6x_0^2 - 4x_0 + 6}{3x_0^2 - 12x_0 - 4} = 1 - \frac{1^3 - 6\cdot 1^2 - 4\cdot 1 + 6}{3\cdot 1^2 - 12\cdot 1 - 4} = 0.7692.$$

$$x_2 = 0.7692 - \frac{(0.7692)^3 - 6(0.7692)^2 - 4(0.7692) + 6}{3(0.7692)^2 - 12(0.7692) - 4} = 0.754211$$

$$x_3 = 0.754211 - \frac{(0.754211)^3 - 6(0.754211)^2 - 4(0.754211) + 6}{3(0.754211)^2 - 12(0.754211) - 4} = 0.754138.$$

7.
$$f(x) = \cos x + x$$

$$f'(x) = -\sin x + 1$$

By Newton's method, $x_1 = x_0 - \dfrac{\cos x_0 + x_0}{-\sin x_0 + 1} = x_0 + \dfrac{\cos x_0 + x_0}{\sin x_0 - 1}$.

Since the graph of $y = -x$ intersects the graph of $y = \cos x$ between $-\dfrac{\pi}{2}$ and 0, we choose $x_0 = -1$. Thus

$$x_1 = -1 + \frac{\cos(-1) + (-1)}{\sin(-1) - 1}.$$

Setting the calculator in radian mode, we get $x_1 = -0.75$. It is best to store this value in the memory when computing the next approximation x_2:

$$x_2 = -0.75 + \frac{\cos(-0.75) + (-0.75)}{\sin(-0.75) - 1} = -0.739.$$

$$x_3 = -0.739 + \frac{\cos(-0.739) + (-0.739)}{\sin(-0.739) - 1} = -0.7390851.$$

From this point on, these digits no longer change, so that $x = -0.7390851$ is the root of the equation to 7 decimal places.

9. $f(x) = 4\sin x - x$; $f'(x) = 4\cos x - 1$. The graphs of $y = 4\sin x$ and $y = x$ intersect at the origin and (on the positive side) near $x = 3$. By Newton's method,

$$x_1 = x_0 - \frac{4\sin x_0 - x_0}{4\cos x_0 - 1}$$

$$x_1 = 3 - \frac{4\sin 3 - 3}{4\cos 3 - 1} = 2.509$$

$$x_2 = 2.509 - \frac{4\sin 2.509 - 2.509}{4\cos 2.509 - 1} = 2.4749$$

$$x_3 = 2.4749 - \frac{4\sin 2.4749 - 2.4749}{4\cos 2.4749 - 1} = 2.474577.$$

From this point on these digits no longer change, so that $x = 2.474577$ is the root to 6 decimal places.

11. $f(x) = \sin x - x^2$

$f'(x) = \cos x - 2x$

By Newton's method $x_1 = x_0 - \dfrac{\sin x_0 - x_0^2}{\cos x_0 - 2x_0} = x_0 + \dfrac{\sin x_0 - x_0^2}{2x_0 - \cos x_0}$.

The parabola $y = x^2$ intersects the curve $y = \sin x$ between 0 and $\dfrac{\pi}{2}$. So we choose $x_0 = 1$:

$x_1 = 1 + \dfrac{\sin 1 - 1}{2 - \cos 1} = 0.89.$

Store 0.89 in the memory and evaluate

$x_2 = 0.89 + \dfrac{\sin(0.89) - (0.89)^2}{2(0.89) - \cos(0.89)} = 0.8769.$

Store 0.8769 in the memory and evaluate

$x_3 = 0.8769 + \dfrac{\sin(0.8769) - (0.8769)^2}{2(0.8769) - \cos(0.8769)} = 0.8767262.$

From this point on, these digits no longer change, so that $x = 0.8767262$ is the root of the equation to 7 decimal places.

13. $f(x) = e^x - \dfrac{5}{x}$; $f'(x) = e^x + \dfrac{5}{x^2}$. The graphs of $y = e^x$ and $y = \dfrac{5}{x}$ intersect near $x = 1$. By Newton's method,

$$\begin{aligned} x_1 &= x_0 - \frac{e^{x_0} - 5/x_0}{e^{x_0} + 5/x_0^2} \\ x_1 &= 1 - \frac{e - 5/1}{e + 5/1^2} = 1.2956 \\ x_2 &= 1.2956 - \frac{e^{1.2956} - 5/1.2956}{e^{1.2956} + 5/(1.2956)^2} = 1.32666 \\ x_3 &= 1.326725. \end{aligned}$$

15. $\tan x + x = 2$; since the line $y = 2 - x$ intersects $y = \tan x$ between 0 and $\dfrac{\pi}{2}$, we choose $x_0 = 1$, a convenient value.

Let $f(x) = \tan x + x - 2$; then $f'(x) = \sec^2 x + 1$.

$x_1 = x_0 - \dfrac{\tan x_0 + x_0 - 2}{\sec^2 x_0 + 1} = 1 - \dfrac{\tan 1 + 1 - 2}{\sec^2 1 + 1} = 0.874.$

$x_2 = 0.874 - \dfrac{\tan(0.874) + 0.874 - 2}{\sec^2(0.874) + 1} = 0.85387.$

$x_3 = 0.85387 - \dfrac{\tan(0.85387) + 0.85387 - 2}{\sec^2(0.85387) + 1} = 0.853530.$

Chapter 6 Review

1. In the interval $\left[-\dfrac{\pi}{2}, \dfrac{\pi}{2}\right]$, $\sin\theta = -1$ only for $\theta = -\dfrac{\pi}{2}$.

3. On the interval $[0, \pi]$ the only permissible value is $x = 120° = 2\pi/3$.

5. Let $\theta = \text{Arctan}\,\dfrac{3}{1}$. Then the length of the hypotenuse is $\sqrt{10}$. Thus

$$\begin{aligned}\cos(\text{Arctan}\,3) &= \cos\theta \\ &= \frac{1}{\sqrt{10}} \\ &= \frac{\sqrt{10}}{10}.\end{aligned}$$

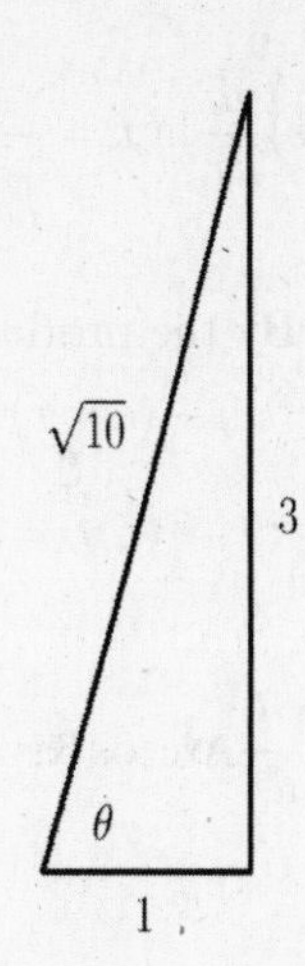

7.

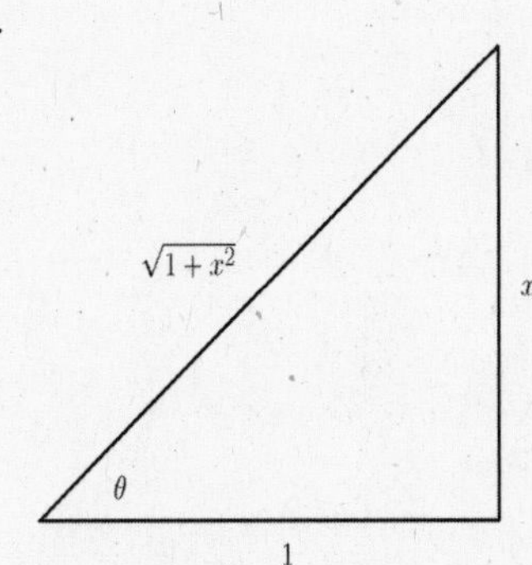

Let $\theta = \text{Arctan}\,\dfrac{x}{1}$;
then $\sin\theta = \dfrac{x}{\sqrt{1+x^2}}$.

9. $y = x^2 \tan 3x$. By the product rule,

$$\begin{aligned}y' &= x^2 \frac{d}{dx}\tan 3x + \tan 3x \frac{d}{dx}(x^2) \\ &= x^2 \sec^2 3x \cdot 3 + \tan 3x(2x) \\ &= 3x^2 \sec^2 3x + 2x \tan 3x.\end{aligned}$$

11. $v = \dfrac{e^{2t}}{t}$. By the quotient rule,

$$\frac{dv}{dt} = \frac{t\frac{d}{dt}e^{2t} - e^{2t}\frac{d}{dt}(t)}{t^2} = \frac{te^{2t}(2) - e^{2t}(1)}{t^2} = \frac{e^{2t}(2t-1)}{t^2}.$$

13.
$$\begin{aligned}y &= \ln\frac{1}{\sqrt{x+3}} = \ln(x+3)^{-1/2} \\ &= -\frac{1}{2}\ln(x+3) \qquad \log_a M^k = k\log_a M \\ y' &= -\frac{1}{2}\frac{1}{x+3} = -\frac{1}{2(x+3)}.\end{aligned}$$

15. $y = \ln\dfrac{\sqrt{2x^2+1}}{x} = \dfrac{1}{2}\ln(2x^2+1) - \ln x$.

$$\begin{aligned}y' &= \frac{1}{2}\frac{1}{2x^2+1}(4x) - \frac{1}{x} = \frac{2x}{2x^2+1}\cdot\frac{x}{x} - \frac{1}{x}\frac{2x^2+1}{2x^2+1} \\ &= \frac{2x^2 - 2x^2 - 1}{x(2x^2+1)} = -\frac{1}{x(2x^2+1)}.\end{aligned}$$

17. $y = \sqrt{\ln 2x} = (\ln 2x)^{1/2}$. By the power rule,

$$y' = \frac{1}{2}(\ln 2x)^{-1/2}\frac{d}{dx}\ln 2x = \frac{1}{2}(\ln 2x)^{-1/2}\frac{1}{2x}(2) = \frac{1}{2x\sqrt{\ln 2x}}.$$

19. $y = \cos(\ln x)$.
$y' = -\sin(\ln x)\dfrac{d}{dx}\ln x = -\dfrac{\sin(\ln x)}{x}$.

21. $y = e^{2x}\cot x$. By the product rule,
$y' = e^{2x}(-\csc^2 x) + (\cot x)e^{2x}(2) = e^{2x}(2\cot x - \csc^2 x)$.

23. $y = e^{\text{Arccos}\,3x}$.
$y' = e^{\text{Arccos}\,3x}\dfrac{d}{dx}\text{Arccos}\,3x = e^{\text{Arccos}\,3x}\left(-\dfrac{1}{\sqrt{1-(3x)^2}}\right)\dfrac{d}{dx}(3x) = -\dfrac{3e^{\text{Arccos}\,3x}}{\sqrt{1-9x^2}}$.

25. $y = (\cot x)^x$.

$$\begin{aligned}
\ln y &= \ln(\cot x)^x \\
&= x\ln\cot x \qquad\qquad \log_a M^k = k\log_a M \\
\frac{1}{y}\frac{dy}{dx} &= x\frac{d}{dx}\ln\cot x + \ln\cot x\cdot 1 \\
&= x\frac{1}{\cot x}(-\csc^2 x) + \ln\cot x.
\end{aligned}$$

Note that $\dfrac{1}{\cot x}\csc^2 x = \dfrac{\sin x}{\cos x}\dfrac{1}{\sin^2 x} = \dfrac{1}{\cos x\sin x} = \sec x\csc x$.
Thus

$$\begin{aligned}
\frac{1}{y}\frac{dy}{dx} &= -x\sec x\csc x + \ln\cot x \\
\frac{dy}{dx} &= y(\ln\cot x - x\csc x\sec x) \\
&= (\cot x)^x(\ln\cot x - x\csc x\sec x).
\end{aligned}$$

27.
$$\begin{aligned}
e^{\sin y} + \csc x &= 1 \\
e^{\sin y}\frac{d}{dx}\sin y - \csc x\cot x &= 0 \\
e^{\sin y}\cos y\frac{dy}{dx} &= \csc x\cot x \\
\frac{dy}{dx} &= \frac{\csc x\cot x}{e^{\sin y}\cos y} \\
&= e^{-\sin y}\sec y\csc x\cot x.
\end{aligned}$$

29.
$$\begin{aligned}
\lim_{x\to 0}\frac{\sin 4x}{x} &= \lim_{x\to 0}\frac{\frac{d}{dx}\sin 4x}{\frac{d}{dx}(x)} \qquad \text{by L'Hospital's rule} \\
&= \lim_{x\to 0}\frac{4\cos 4x}{1} = 4.
\end{aligned}$$

31.
$$\begin{aligned}
\lim_{x\to\pi/4}(1-\tan x)\sec 2x &= \lim_{x\to\pi/4}\frac{1-\tan x}{\cos 2x} \\
&= \lim_{x\to\pi/4}\frac{-\sec^2 x}{-2\sin 2x} = \frac{-(\sqrt{2})^2}{-2\sin\frac{\pi}{2}} = \frac{2}{2\cdot 1} = 1.
\end{aligned}$$

33. $\displaystyle\lim_{x\to\infty} x^3e^{-2x} = \lim_{x\to\infty}\frac{x^3}{e^{2x}} = \lim_{x\to\infty}\frac{3x^2}{2e^{2x}} = \lim_{x\to\infty}\frac{6x}{4e^{2x}} = \lim_{x\to\infty}\frac{6}{8e^{2x}} = 0$

35.

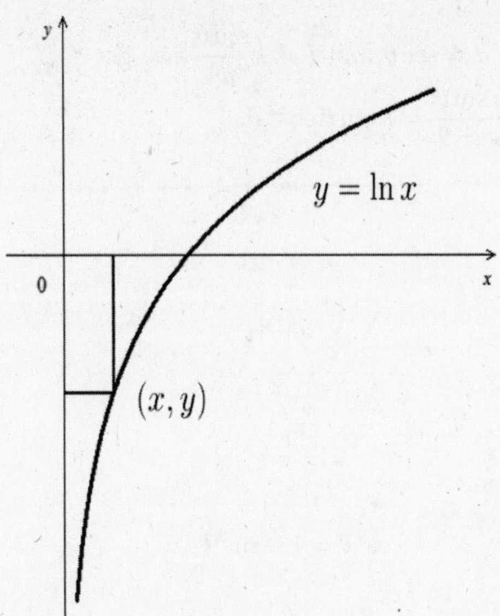

$A = xy = x\ln x$ (since $y = \ln x$).

$$\frac{dA}{dx} = x\frac{d}{dx}\ln x + \ln x\frac{d}{dx}(x)$$

$$\begin{aligned} 1 + \ln x &= 0 \\ \ln x &= -1 \\ \log_e x &= -1 \\ e^{-1} &= x. \end{aligned}$$

So $x = \frac{1}{e}$ and $y = \ln e^{-1} = -1\ln e = -1$. The dimensions are $\frac{1}{e} \times 1$.

37. (a) From $P = Se^{rt}$, we get $P = 5000e^{(0.10)(15)} = \$22,408.45$.

(b) $\frac{dP}{dt} = Se^{rt}\frac{d}{dt}(rt) = Se^{rt}r = Pr = rP$. ($P$ grows at a rate that is proportional to the amount present.)

39. Given: $\frac{dT}{dt} = 2.00\,\text{K/min}$.

Find: $\frac{dp}{dt}$ when $T = 300.0\,\text{K}$.

$$\begin{aligned} \log_{10} p &= -\frac{1706.4}{T} - 7.7760 \\ \frac{1}{p}(\log_{10} e)\frac{dp}{dt} &= \frac{1706.4}{T^2}\frac{dT}{dt} \\ &= \frac{1706.4}{T^2}(2.00) \\ \frac{dp}{dt} &= \frac{p}{\log_{10} e}\frac{1706.4}{T^2}(2.00). \end{aligned}$$

When $T = 300.0$, $\log_{10} p = -\frac{1706.4}{300} - 7.7760 = -13.464$ and $p = 10^{-13.464}$.

Substituting these values, we get

$\frac{dp}{dt} = \frac{10^{-13.464}}{\log_{10} e}\frac{1706.4}{(300.0)^2}(2.00) = 3.00 \times 10^{-15}$ mm of mercury per min.

41.

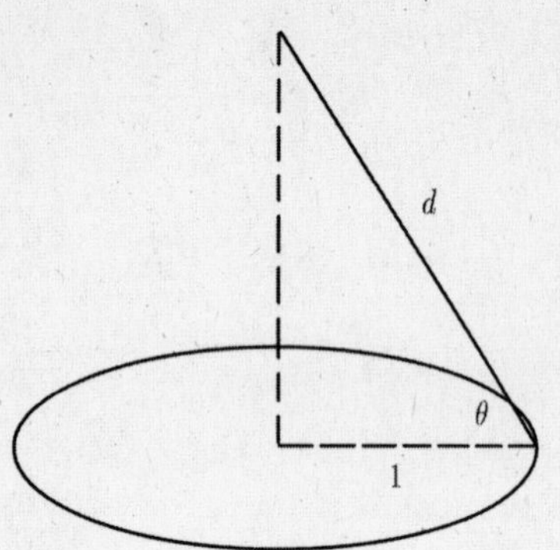

Since $d = \sec\theta$ and $I = k\dfrac{\sin\theta}{d^2}$, we get

$$I = \frac{k\sin\theta}{\sec^2\theta} = k\sin\theta\cos^2\theta.$$

Then

$$\begin{aligned}
\frac{dI}{d\theta} &= k\left[\sin\theta(2\cos\theta)(-\sin\theta) + \cos^2\theta\cos\theta\right] \\
&= k\cos\theta(-2\sin^2\theta + \cos^2\theta) \\
&= k\cos\theta(-2\sin^2\theta + 1 - \sin^2\theta) \qquad \cos^2\theta = 1 - \sin^2\theta \\
&= k\cos\theta(1 - 3\sin^2\theta) = 0.
\end{aligned}$$

$$\begin{aligned}
\sin^2\theta &= \frac{1}{3} \\
\sin\theta &= \frac{1}{\sqrt{3}}.
\end{aligned}$$

Hence $\theta = \text{Arcsin}\,\dfrac{1}{\sqrt{3}} \approx 35.3°$.

43. $f(x) = \cos x - x - \dfrac{1}{2}$; $f'(x) = -\sin x - 1$.

$$x_1 = x_0 - \frac{\cos x_0 - x_0 - 1/2}{-\sin x_0 - 1} = x_0 + \frac{\cos x_0 - x_0 - 0.5}{\sin x_0 + 1}.$$

Let $x_0 = 0.5$:

$$\begin{aligned}
x_1 &= 0.5 + \frac{\cos 0.5 - 0.5 - 0.5}{\sin 0.5 + 1} = 0.41725 \\
x_2 &= 0.415084 \\
x_3 &= 0.415083.
\end{aligned}$$

Chapter 7

Integration Techniques

7.1 The Power Formula Again

1. $\int x\sqrt{x^2+1}\,dx = \int (x^2+1)^{1/2} x\,dx$ $\qquad u = x^2+1;\ du = 2x\,dx.$

$\frac{1}{2}\int (x^2+1)^{1/2}(2x)\,dx = \frac{1}{2}\int u^{1/2}\,du + C = \frac{1}{2}\cdot\frac{2}{3}u^{3/2} + C = \frac{1}{3}(x^2+1)^{3/2} + C.$

3. $\int \frac{x\,dx}{\sqrt{x^2-1}} = \int (x^2-1)^{-1/2} x\,dx$ $\qquad u = x^2-1;\ du = 2x\,dx.$

$= \frac{1}{2}\int (x^2-1)^{-1/2}(2x\,dx) = \frac{1}{2}\int u^{-1/2}\,du$

$= \frac{1}{2}\frac{u^{1/2}}{1/2} + C = \sqrt{u} + C = \sqrt{x^2-1} + C.$

5. $\int \sin^2 x\cos x\,dx$ $\qquad u = \sin x;\ du = \cos x\,dx.$

$\int u^2\,du = \frac{1}{3}u^3 + C = \frac{1}{3}\sin^3 x + C.$

7. $\int \tan^2 2x\sec^2 2x\,dx = \frac{1}{2}\int \tan^2 2x(2\sec^2 2x\,dx)$ $\qquad u = \tan 2x;\ du = 2\sec^2 2x\,dx.$

$= \frac{1}{2}\int u^2\,du = \frac{1}{6}u^3 + C = \frac{1}{6}\tan^3 2x + C.$

9. $\int (1+\tan 3t)^3\sec^2 3t\,dt$ $\qquad u = 1+\tan 3t;\ du = 3\sec^2 3t\,dt.$

$= \frac{1}{3}\int (1+\tan 3t)^3(3\sec^2 3t\,dt) = \frac{1}{3}\int u^3\,du = \frac{1}{3}\frac{u^4}{4} + C = \frac{1}{12}(1+\tan 3t)^4 + C.$

11. $\int (1+4e^r)^3 e^{4r}\,dr = \frac{1}{4}\int (1+e^{4r})^3(4e^{4r}\,dr)$ $\qquad u = 1+e^{4r};\ du = 4e^{4r}\,dr.$

$= \frac{1}{4}\int u^3\,du = \frac{1}{16}u^4 + C = \frac{1}{16}(1+e^{4r})^4 + C.$

13. $\int (1-\cos 5x)^3\sin 5x\,dx$ $\qquad u = 1-\cos 5x;\ du = 5\sin 5x\,dx.$

$\frac{1}{5}\int (1-\cos 5x)^3\cdot 5\sin 5x\,dx = \frac{1}{5}\int u^3\,du = \frac{1}{5}\frac{u^4}{4} + C = \frac{1}{20}(1-\cos 5x)^4 + C.$

15. $\displaystyle\int \frac{\sec^2 3x}{(\tan 3x+1)^3}\,dx = \int (\tan 3x+1)^{-3}\sec^2 3x\,dx \qquad u=\tan 3x+1;\ du = 3\sec^2 3x\,dx.$

$\displaystyle = \frac{1}{3}\int (\tan 3x+1)^{-3}(3\sec^2 3x\,dx)$

$\displaystyle = \frac{1}{3}\int u^{-3}\,du = \frac{1}{3}\frac{u^{-2}}{-2}+C = -\frac{1}{6(\tan 3x+1)^2}+C.$

17. $\displaystyle\int (\ln x)\frac{1}{x}\,dx \qquad u=\ln x;\ du=\frac{1}{x}\,dx.$

$\displaystyle\int u\,du = \frac{1}{2}u^2+C = \frac{1}{2}\ln^2 x + C.$

19. $\displaystyle\int \frac{dx}{x\sqrt{\ln x}} = \int (\ln x)^{-1/2}\frac{1}{x}\,dx \qquad u=\ln x;\ du=\frac{1}{x}\,dx.$

$\displaystyle = \int u^{-1/2}\,du = \frac{u^{1/2}}{1/2}+C = 2\sqrt{\ln x}+C.$

21. $\displaystyle\int \frac{\text{Arccos}\,x\,dx}{\sqrt{1-x^2}} \qquad u=\text{Arccos}\,x;\ du=-\frac{dx}{\sqrt{1-x^2}}.$

$\displaystyle -\int \text{Arccos}\,x\left(-\frac{dx}{\sqrt{1-x^2}}\right) = -\int u\,du = -\frac{1}{2}u^2+C = -\frac{1}{2}(\text{Arccos}\,x)^2+C.$

23. $\displaystyle\int_{\pi/6}^{\pi/2} \cos^2 x\sin x\,dx = -\int_{\pi/6}^{\pi/2}(\cos x)^2(-\sin x\,dx).$

$u=\cos x;\ du=-\sin x\,dx$

Lower limit: if $x=\frac{\pi}{6}$, then $u=\cos\frac{\pi}{6}=\frac{\sqrt{3}}{2}$.

Upper limit: if $x=\frac{\pi}{2}$, then $u=\cos\frac{\pi}{2}=0$.

$\displaystyle -\int_{\sqrt{3}/2}^{0} u^2\,du = -\frac{1}{3}u^3\Big|_{\sqrt{3}/2}^{0} = 0+\frac{1}{3}\left(\frac{\sqrt{3}}{2}\right)^3 = \frac{1}{3}\frac{(\sqrt{3})^2\sqrt{3}}{8} = \frac{\sqrt{3}}{8}.$

25. $\displaystyle\int_1^{e} \frac{\sqrt{\ln x}}{x}\,dx = \int (\ln x)^{1/2}\left(\frac{1}{x}\,dx\right). \qquad u=\ln x;\ du=\frac{1}{x}\,dx.$

$\displaystyle\int_1^{e}(\ln x)^{1/2}\frac{1}{x}\,dx = \frac{2}{3}(\ln x)^{3/2}\Big|_1^{e} = \frac{2}{3}\left[(\ln e)^{3/2}-(\ln 1)^{3/2}\right] = \frac{2}{3}(1-0) = \frac{2}{3}$

(since $\ln e = 1$ and $\ln 1 = 0$).

27. $\displaystyle\int \frac{\cot 2x\,dx}{\sin^2 2x} = \int \cot 2x\csc^2 2x\,dx \qquad u=\cot 2x;\ du=-2\csc^2 2x\,dx.$

$\displaystyle = -\frac{1}{2}\int \cot 2x(-2\csc^2 2x\,dx)$

$\displaystyle = -\frac{1}{2}\int u\,du = -\frac{1}{2}\frac{u^2}{2}+C = -\frac{1}{4}\cot^2 2x + C.$

29. $\displaystyle\int \sqrt{\tan x}\sec^2 dx \qquad u=\tan x;\ du=\sec^2 x\,dx.$

$\displaystyle\int u^{1/2}\,du = \frac{2}{3}u^{3/2}+C = \frac{2}{3}(\tan x)^{3/2}+C.$

31. $\displaystyle\int \text{Arctan}\,4R\left(\frac{dR}{1+16R^2}\right) = \frac{1}{4}\int \text{Arctan}\,4R\frac{4dR}{1+16R^2}$

$\displaystyle u=\text{Arctan}\,4R;\ du=\frac{4\,dR}{1+16R^2}.$

$\displaystyle = \frac{1}{4}\int u\,du = \frac{1}{8}u^2+C = \frac{1}{8}(\text{Arctan}\,4R)^2+C.$

33. $$\begin{aligned}\int_1^e (1+\ln x)^2 \frac{1}{x}\,dx &= \frac{1}{3}(1+\ln x)^3\Big|_1^e \qquad u = 1+\ln x;\, du = \frac{1}{x}\,dx\\ &= \frac{1}{3}(1+\ln e)^3 - \frac{1}{3}(1+\ln 1)\\ &= \frac{1}{3}(1+1)^3 - \frac{1}{3}(1+0) = \frac{1}{3}(8) - \frac{1}{3}(1) = \frac{7}{3}\end{aligned}$$

7.2 The Logarithmic and Exponential Forms

1. $\displaystyle\int \frac{dx}{x-1}$ $\qquad u = x-1;\ du = dx.$

$$\int \frac{dx}{x-1} = \int \frac{du}{u} = \ln|u| + C = \ln|x-1| + C.$$

3. $$\begin{aligned}\int \frac{dx}{5+4x} &= \frac{1}{4}\int \frac{4\,dx}{5+4x} \qquad u = 5+4x;\ du = 4\,dx.\\ &= \frac{1}{4}\int \frac{du}{u} = \frac{1}{4}\ln|u| + C = \frac{1}{4}\ln|5+4x| + C.\end{aligned}$$

5. $\displaystyle\int \frac{ds}{(1-3s)^2}$ $\qquad u = 1-3s;\ du = -3\,ds.$

$$-\frac{1}{3}\int \frac{-3\,ds}{(1-3s)^2} = -\frac{1}{3}\int \frac{du}{u^2} = -\frac{1}{3}\int u^{-2} + C = -\frac{1}{3}\frac{u^{-1}}{-1} + C = \frac{1}{3u} + C = \frac{1}{3(1-3s)} + C.$$

7. $$\begin{aligned}\int_0^1 \frac{x\,dx}{x^2+1} &= \frac{1}{2}\int_0^1 \frac{2x\,dx}{x^2+1} \qquad u = x^2+1;\ du = 2x\,dx.\\ &= \frac{1}{2}\ln(x^2+1)\Big|_0^1 = \frac{1}{2}\ln 2 - \frac{1}{2}\ln 1 = \frac{1}{2}\ln 2 - 0\\ &= \ln 2^{1/2} = \ln\sqrt{2}.\end{aligned}$$

9. $\displaystyle\int e^{-2x}\,dx$ $\qquad u = -2x;\ du = -2dx.$

$$-\frac{1}{2}\int e^{-2x}(-2dx) = -\frac{1}{2}\int e^u\,du = -\frac{1}{2}e^u + C = -\frac{1}{2}e^{-2x} + C.$$

11. $\displaystyle\int_0^2 2e^{3x}\,dx = 2\int_0^2 e^{3x}\,dx = \frac{2}{3}\int_0^2 e^{3x}(3\,dx).$ $\qquad u = 3x;\ du = 3\,dx.$

Lower limit: if $x = 0$, then $u = 0$.

Upper limit: if $x = 2$, then $u = 6$.

$$\frac{2}{3}\int_0^6 e^u\,du = \frac{2}{3}e^u\Big|_0^6 = \frac{2}{3}(e^6 - e^0) = \frac{2}{3}(e^6 - 1).$$

13. $\displaystyle\int e^{4x}\,dx$ $\qquad u = 4x;\ du = 4\,dx.$

$$\frac{1}{4}\int e^{4x}(4\,dx) = \frac{1}{4}\int e^u\,du = \frac{1}{4}e^u + C = \frac{1}{4}e^{4x} + C.$$

15. $$\begin{aligned}\int te^{t^2}\,dt &= \int e^{t^2}t\,dt \qquad u = t^2;\ du = 2t\,dt.\\ &= \frac{1}{2}\int e^{t^2}(2t\,dt) = \frac{1}{2}\int e^u\,du = \frac{1}{2}e^u + C = \frac{1}{2}e^{t^2} + C.\end{aligned}$$

17. $\int e^{\sin R} \cos R\, dR$ $\qquad u = \sin R;\ du = \cos R\, dR.$

$\int e^u\, du = e^u + C = e^{\sin R} + C.$

19. $\int \frac{\sec^2 2x}{1+\tan 2x}\, dx$ $\qquad u = 1 + \tan 2x;\ du = 2\sec^2 2x\, dx.$

$= \frac{1}{2}\int \frac{2\sec^2 2x\, dx}{1+\tan 2x} = \frac{1}{2}\int \frac{du}{u} = \frac{1}{2}\ln|u| + C = \frac{1}{2}\ln|1+\tan 2x| + C.$

21. $\int e^{\text{Arctan}\, z}\frac{dz}{1+z^2}\, dz$ $\qquad u = \text{Arctan}\, z;\ du = \frac{dz}{1+z^2}.$

$= \int e^u\, du = e^u + C = e^{\text{Arctan}\, z} + C.$

23. $\int \frac{dx}{x\ln x} = \int \frac{1}{\ln x}\frac{1}{x}\, dx$ $\qquad u = \ln x;\ du = \frac{1}{x}\, dx.$

$= \int \frac{du}{u} = \ln|u| + C = \ln|\ln x| + C.$

25.
$$\begin{aligned}\int \frac{e^x+1}{e^x}\, dx &= \int\left(\frac{e^x}{e^x}+\frac{1}{e^x}\right) dx\\ &= \int (1+e^{-x})\, dx\\ &= \int 1\, dx + \int e^{-x}\, dx\\ &= x + \int e^{-x}\, dx\\ &= x - \int e^{-x}(-dx)\\ &= x - e^{-x} + C.\end{aligned}$$
$u = -x;\ du = -dx.$

27.
$$\begin{aligned}\int \frac{e^{-3W}}{1-e^{-3W}}\, dW &= \frac{1}{3}\int \frac{3e^{-3W}\, dW}{1-e^{-3W}}\\ &= \frac{1}{3}\int\frac{du}{u} = \frac{1}{3}\ln|u| + C = \frac{1}{3}\ln|1-e^{-3W}| + C.\end{aligned}$$
$u = 1 - e^{-3W};\ du = 3e^{-3W}\, dW.$

29.
$$\begin{aligned}\int (1+e^x)^2\, dx &= \int (1+2e^x+e^{2x})\, dx\\ &= \int dx + 2\int e^x\, dx + \frac{1}{2}\int e^{2x}(2\, dx)\\ &= x + 2e^x + \frac{1}{2}e^{2x} + C.\end{aligned}$$
$u = 2x;\ du = 2\, dx.$

31.
$$\begin{aligned}\int \frac{\cos x}{(\sin x)^2}\, dx &= \int (\sin x)^{-2}\cos x\, dx\\ &= \int u^{-2}\, du = \frac{u^{-1}}{-1} + C = -\frac{1}{u} + C\\ &= -\frac{1}{\sin x} + C = -\csc x + C.\end{aligned}$$
$u = \sin x;\ du = \cos x\, dx.$

33. $\int_0^{\pi/4} \frac{\sec^2 x\, dx}{(1+2\tan x)^2} = \frac{1}{2}\int_0^{\pi/4}\frac{2\sec^2 x\, dx}{(1+2\tan x)^2}.$ $\qquad u = 1 + 2\tan x;\ du = 2\sec^2 x\, dx.$

Lower limit: if $x = 0$, then $u = 1$.

Upper limit: if $x = \frac{\pi}{4}$, then $u = 3$.

$\frac{1}{2}\int_1^3 \frac{du}{u^2} = \frac{1}{2}\int_1^3 u^{-2}\, du = \frac{1}{2}\frac{u^{-1}}{-1}\Big|_1^3 = \frac{-1}{2u}\Big|_1^3 = -\frac{1}{6}+\frac{1}{2} = \frac{1}{3}.$

35. $$\begin{aligned}\int \frac{\cos 2x}{1+\sin 2x}\,dx &= \frac{1}{2}\int \frac{2\cos 2x\,dx}{1+\sin 2x} \qquad u = 1+\sin 2x;\ du = 2\cos 2x\,dx.\\ &= \frac{1}{2}\int \frac{du}{u} = \frac{1}{2}\ln|u| + C\\ &= \frac{1}{2}\ln|1+\sin 2x| + C\\ &= \frac{1}{2}\ln(1+\sin 2x) + C,\end{aligned}$$

since $1+\sin 2x \geq 0$.

37. $$\int \frac{\cos 2x}{(1+\sin 2x)^2}\,dx \qquad u = 1+\sin 2x;\ du = 2\cos 2x\,dx.$$
$$\begin{aligned}\frac{1}{2}\int (1+\sin 2x)^{-2}(2\cos 2x\,dx) &= \frac{1}{2}\int u^{-2}\,du = \frac{1}{2}\frac{u^{-1}}{-1} + C\\ &= -\frac{1}{2u} + C = -\frac{1}{2(1+\sin 2x)} + C.\end{aligned}$$

39. $$\int 5^{3x}\,dx \qquad u = 3x;\ du = 3\,dx.$$
$$\frac{1}{3}\int 5^{3x}(3\,dx) = \frac{1}{3}\int 5^{u}\,du = \frac{1}{3}(5^{u})\log_5 \mathrm{e} = \frac{1}{3}(5^{3x})\log_5 \mathrm{e} + C.$$

41. $$\int 2^{\cot x}\csc^2 x\,dx \qquad u = \cot x;\ du = -\csc^2 x\,dx.$$
$$\begin{aligned}-\int 2^{\cot x}(-\csc^2 x\,dx) &= -\int 2^{u}\,du = -2^{u}\log_2 \mathrm{e}\\ &= -2^{\cot x}\log_2 \mathrm{e} + C \text{ or } -\frac{2^{\cot x}}{\ln 2} + C.\end{aligned}$$

43.

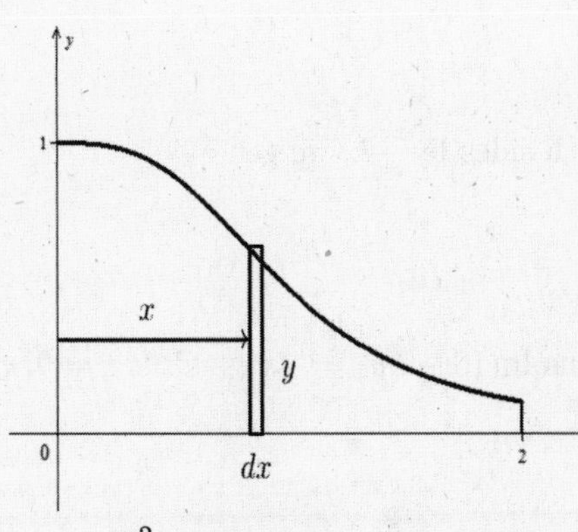

Moment of inertia (with respect to y-axis) of typical element:
$$x^2 \cdot \rho y\,dy = x^2 \cdot \rho \cdot \frac{1}{1+x^3}\,dx.$$

$$I_y = \rho \int_0^2 x^2 \frac{1}{1+x^3}\,dx. \qquad \text{Let } u = 1+x^3;\text{ then } du = 3x^2\,dx.$$

For the indefinite integral $\int \frac{x^2}{1+x^3}\,dx$ we have
$\frac{1}{3}\int \frac{3x^2\,dx}{1+x^3} = \frac{1}{3}\int \frac{du}{u} = \frac{1}{3}\ln|u| + C$. Thus

$$\begin{aligned}I_y &= \frac{\rho}{3}\int_0^2 \frac{3x^2\,dx}{1+x^3} = \frac{\rho}{3}\ln|1+x^3|\Big|_0^2 = \frac{\rho}{3}[\ln 9 - \ln 1]\\ &= \frac{\rho}{3}[\ln 3^2 - 0] = \frac{\rho}{3}(2\ln 3) = \frac{2}{3}\rho\ln 3.\end{aligned}$$

45. $$\begin{aligned} i_{av} &= \frac{1}{2-0}\int_0^2 e^{(-1/3)t}\,dt \qquad u=-\frac{1}{3}t;\ du=-\frac{1}{3}\,dt. \\ &= \frac{1}{2}(-3)\int_0^2 e^{(-1/3)t}\left(-\frac{1}{3}\,dt\right) \\ &= -\frac{3}{2}e^{(-1/3)t}\Big|_0^2 = -\frac{3}{2}\left(e^{-2/3}-e^0\right) \\ &= -\frac{3}{2}\left(e^{-2/3}-1\right) = \frac{3}{2}\left(1-e^{-2/3}\right)\text{ A.} \end{aligned}$$

47.

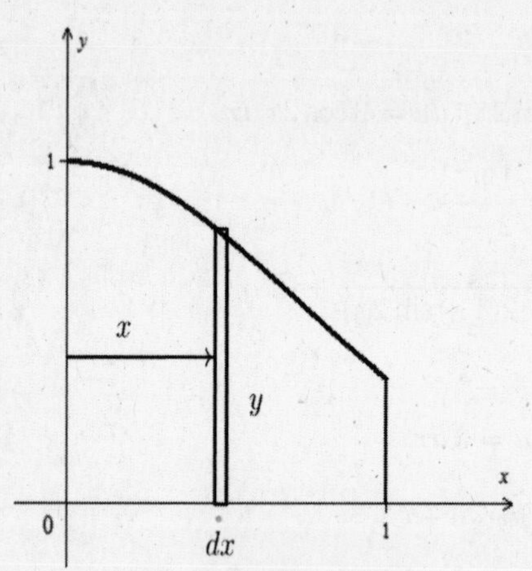

Volume of shell:
$2\pi\cdot(\text{radius})\cdot(\text{height})\cdot(\text{thickness})$
$= 2\pi xy\,dx.$

$V = 2\pi\displaystyle\int_0^1 xe^{-x^2}\,dx.$ Let $u=-x^2$; then $du=-2x\,dx$.

$$V = \frac{2\pi}{-2}\int_0^1 e^{-x^2}(-2x)\,dx = -\pi e^{-x^2}\Big|_0^1 = -\pi e^{-1}+\pi\cdot 1 = \pi(-e^{-1}+1) = \pi\left(1-\frac{1}{e}\right).$$

49. Volume of shell: $2\pi(\text{radius})(\text{height})(\text{thickness}) = 2\pi x\left(\dfrac{4}{x^2}\right)\,dx.$

$$\begin{aligned} V &= 2\pi\int_1^2 x\left(\frac{4}{x^2}\right)\,dx = 8\pi\int_1^2 \frac{1}{x}\,dx \\ &= 8\pi\ln|x|\Big|_1^2 = 8\pi(\ln 2-\ln 1) = 8\pi\ln 2. \end{aligned}$$

51. Let $u = 10-kv$; then $du=-k\,dv$. Multiplying both sides by $-k$, we get

$$\begin{aligned} \frac{-k\,dv}{10-kv} &= -k\,dt \\ \ln|10-kv| &= -kt+C. \qquad \text{integrating} \end{aligned}$$

Since $10-kv>0$, $\ln|10-kv| = \ln(10-kv)$. Thus $\ln(10-kv) = -kt+C$. If $t=0$, $v=0$.
Hence, $\ln 10 = 0 + C$ and

$$\begin{aligned} \ln(10-kv) &= -kt+\ln 10 \\ \ln(10-kv)-\ln 10 &= -kt \\ \ln\frac{10-kv}{10} &= -kt. \qquad \log_a M-\log_a N = \log_a\frac{M}{N} \end{aligned}$$

By definition of logarithm,

$$\begin{aligned} e^{-kt} &= \frac{10-kv}{10} \\ 10e^{-kt} &= 10-kv \\ kv &= 10(1-e^{-kt}) \\ v &= \frac{10}{k}(1-e^{-kt})\text{ m/s.} \end{aligned}$$

53. $$\text{mass} = \int (3.00 + 0.0300e^{0.0100x})\,dx = 3.00x + 0.0300\left(\frac{1}{0.0100}\right)e^{0.0100x} + C$$
$$= 3.00x + 3e^{0.0100x} + C.$$
When $x = 0$, mass $= 0$; so
$0 = 0 + 3 + C$ and $C = -3$.
$$\text{mass} = 3.00x + 3e^{0.0100x} - 3\Big|_{x=6.00} = 18.2\,\text{kg}.$$

7.3 Trigonometric Forms

1. $\int \sec^2 2x\,dx \qquad u = 2x;\ du = 2\,dx.$
$$\frac{1}{2}\int \sec^2 2x(2\,dx) = \frac{1}{2}\int \sec^2 u\,du = \frac{1}{2}\tan u + C = \frac{1}{2}\tan 2x + C.$$

3. $\int \sec 3x \tan 3x\,dx \qquad u = 3x;\ du = 3\,dx.$
$$\frac{1}{3}\int \sec 3x \tan 3x(3\,dx) = \frac{1}{3}\int \sec u \tan u\,du = \frac{1}{3}\sec u + C = \frac{1}{3}\sec 3x + C.$$

5. $\int \csc 4x \cot 4x\,dx \qquad u = 4x;\ du = 4\,dx.$
$$\frac{1}{4}\int \csc 4x \cot 4x(4\,dx) = \frac{1}{4}\int \csc u \cot u\,du = \frac{1}{4}(-\csc u) + C = -\frac{1}{4}\csc 4x + C.$$

7. $\int \tan \frac{1}{2}x\,dx \qquad u = \frac{1}{2}x;\ du = \frac{1}{2}\,dx.$
$$2\int \tan \frac{1}{2}x\left(\frac{1}{2}\,dx\right) = 2\int \tan u\,du = 2\ln|\sec u| + C = 2\ln\left|\sec \frac{1}{2}x\right| + C.$$

9. $\int \cos 2t\,dt \qquad u = 2t;\ du = 2\,dt.$
$$\frac{1}{2}\int \cos 2t(2\,dt) = \frac{1}{2}\int \cos u\,du = \frac{1}{2}\sin u + C = \frac{1}{2}\sin 2t + C.$$

11. $\int y \csc y^2\,dy \qquad u = y^2;\ du = 2y\,dy.$
$$\frac{1}{2}\int \csc y^2(2y\,dy) = \frac{1}{2}\int \csc u\,du = \frac{1}{2}\ln|\csc u - \cot u| + C = \frac{1}{2}\ln|\csc y^2 - \cot y^2| + C.$$

13. $\int \frac{\csc^2 y}{1 + \cot y}\,dy \qquad u = 1 + \cot y;\ du = -\csc^2 y\,dy.$
$$-\int \frac{-\csc^2 y\,dy}{1 + \cot y} = -\int \frac{du}{u} = -\ln|u| + C = -\ln|1 + \cot y| + C.$$

15. $\int \tan^2 2V \sec^2 2V\,dV \qquad u = \tan 2V; du = 2\sec^2 2V\,dV.$
$$\frac{1}{2}\int (\tan 2V)^2(2\sec^2 2V\,dV) = \frac{1}{2}\int u^2\,du = \frac{1}{2}\frac{u^3}{3} + C = \frac{1}{6}\tan^3 2V + C.$$

17. $\int x \sin x^2\,dx \qquad u = x^2;\ du = 2x\,dx.$
$$\frac{1}{2}\int \sin x^2(2x)\,dx = \frac{1}{2}\int \sin u\,du = -\frac{1}{2}\cos u + C = -\frac{1}{2}\cos x^2 + C.$$

19. $\int s \cot \frac{1}{2}s^2\, ds$ $\quad u = \frac{1}{2}s^2; du = s\, ds.$

$\int \cot u\, du = \ln|\sin u| + C = \ln\left|\sin \frac{1}{2}s^2\right| + C.$

21. $\int \frac{\cos\sqrt{x}}{\sqrt{x}}\, dx = \int \frac{\cos x^{1/2}}{x^{1/2}}\, dx.$ $\quad u = x^{1/2};\ du = \frac{1}{2}x^{-1/2}\, dx = \frac{dx}{2x^{1/2}}.$

$2\int \cos x^{1/2} \frac{dx}{2x^{1/2}} = 2\int \cos u\, du = 2\sin u + C = 2\sin\sqrt{x} + C.$

23. $\int \tan^3 4x \sec^2 4x\, dx$ $\quad u = \tan 4x;\ du = 4\sec^2 4x\, dx.$

$\frac{1}{4}\int \tan^3 4x(4\sec^2 4x\, dx) = \frac{1}{4}\int u^3\, du = \frac{1}{4}\frac{u^4}{4} + C = \frac{1}{16}\tan^4 4x + C.$

25. $\int \cos(\ln x)\cdot\frac{1}{x}\, dx$ $\quad u = \ln x; du = \frac{1}{x}\, dx.$

$\int \cos u\, du = \sin u + C = \sin(\ln x) + C.$

27. $\int e^{2x}\csc^2 e^{2x}\, dx$ $\quad u = e^{2x}; du = 2e^{2x}\, dx.$

$\frac{1}{2}\int \csc^2 e^{2x}\,(2e^{2x}\, dx) = \frac{1}{2}\int \csc^2 u\, du = -\frac{1}{2}\cot e^{2x} + C.$

29. $\int (1+\sec x)^2\, dx = \int \left(1 + 2\sec x + \sec^2 x\right)\, dx = x + 2\ln|\sec x + \tan x| + \tan x + C$

31. $u = 1 + 2\cos x; du = -2\sin x\, dx.$

$$\begin{aligned}\int_0^{\pi/2} \frac{\sin x}{1+2\cos x}\, dx &= -\frac{1}{2}\ln|1+2\cos x|\Big|_0^{\pi/2} \\ &= -\frac{1}{2}\left[\ln\left|1+2\cos\frac{\pi}{2}\right| - \ln|1+2\cos 0|\right] \\ &= -\frac{1}{2}(\ln 1 - \ln 3) = -\frac{1}{2}(0 - \ln 3) = \frac{1}{2}\ln 3.\end{aligned}$$

33. $\int_0^{\sqrt{\pi/2}} \omega\cos\omega^2\, d\omega$ $\quad$ Let $u = \omega^2$; then $du = 2\omega\, d\omega$.

Lower limit: if $\omega = 0,\ u = 0.$

Upper limit: if $\omega = \sqrt{\frac{\pi}{2}},\ u = \omega^2 = \frac{\pi}{2}.$

$$\begin{aligned}\int_0^{\sqrt{\pi/2}} \omega\cos\omega^2\, d\omega &= \frac{1}{2}\int_0^{\sqrt{\pi/2}} (\cos\omega^2)(2\omega\, d\omega) = \frac{1}{2}\int_0^{\pi/2} \cos u\, du \\ &= \frac{1}{2}\sin u\Big|_0^{\pi/2} = \frac{1}{2}\sin\frac{\pi}{2} - \frac{1}{2}\sin 0 = \frac{1}{2}\cdot 1 - \frac{1}{2}\cdot 0 = \frac{1}{2}.\end{aligned}$$

35. $\int_0^{\pi/6} \frac{\cos x}{1-\sin x}\, dx$ $\quad u = 1 - \sin x;\ du = -\cos x\, dx.$

$$\begin{aligned}-\int_0^{\pi/6} \frac{-\cos x\, dx}{1-\sin x} &= -\ln|1-\sin x|\Big|_0^{\pi/6} \\ &= -\ln\left(1 - \sin\frac{\pi}{6}\right) + \ln(1 - \sin 0) \\ &= -\ln\left(1 - \frac{1}{2}\right) + \ln 1 = -\ln\frac{1}{2} + 0 \\ &= -(\ln 1 - \ln 2) = \ln 2.\end{aligned}$$

37. Recall that $v = \dfrac{q}{C} = \dfrac{1}{C}\int i\,dt$. Thus $v = \dfrac{1}{100 \times 10^{-6}}\int 2.00\cos 100t\,dt$. Let $u = 100t$; then $du = 100\,dt$.

$$\begin{aligned} v &= \frac{2.00}{10^{-4}}\int \cos 100t\,dt = \frac{2.00}{10^{-4}}\frac{1}{100}\int \cos 100t(100)\,dt \\ &= \frac{2.00}{10^{-2}}\int \cos u\,du = 200\sin u + C = 200\sin 100t + C. \end{aligned}$$

If $t = 0$, $v = 0$. Thus $C = 0$.

$v = 200\sin 100t\Big|_{t=0.200} = 183\text{ V}$, (calculator set in radian mode).

39. Volume of shell: $2\pi(\text{radius})(\text{height})(\text{thickness}) = 2\pi x\cos x^2\,dx$.

$$\begin{aligned} V &= 2\pi\int_0^{\sqrt{\pi/2}} x\cos x^2\,dx \qquad u = x^2;\ du = 2x\,dx. \\ &= \pi\int_0^{\sqrt{\pi/2}} \cos x^2(2x\,dx) \\ &= \pi\sin x^2\Big|_0^{\sqrt{\pi/2}} = \pi\sin\frac{\pi}{2} = \pi\cdot 1 = \pi. \end{aligned}$$

41. Moment (with respect to y-axis) of typical element: $xy\,dx$.

$M_y = \displaystyle\int_0^{\sqrt{\pi}/2} x\tan x^2\,dx$. Let $u = x^2$; then $du = 2x\,dx$.

$$\begin{aligned} M_y &= \frac{1}{2}\int_0^{\sqrt{\pi}/2}\tan x^2(2x)\,dx = \frac{1}{2}\ln|\sec x^2|\Big|_0^{\sqrt{\pi}/2} = \frac{1}{2}\left(\ln\sec\frac{\pi}{4} - \ln\sec 0\right) \\ &= \frac{1}{2}\left(\ln\sqrt{2} - \ln 1\right) = \frac{1}{2}\left(\ln 2^{1/2} - 0\right) = \frac{1}{2}\cdot\frac{1}{2}\ln 2 = \frac{1}{4}\ln 2. \end{aligned}$$

43. Period $= \dfrac{2\pi}{\omega}$.

$$\begin{aligned} v_{\text{av}} &= \frac{1}{\pi/\omega - 0}\int_0^{\pi/\omega} E\sin\omega t\,dt \qquad u = \omega t;\ du = \omega\,dt. \\ &= \frac{1}{\pi/\omega}E\left(\frac{1}{\omega}\right)(-\cos\omega t)\Big|_0^{\pi/\omega} \\ &= \frac{E}{\pi}(-\cos\pi + \cos 0) = \frac{E}{\pi}(1+1) = \frac{2E}{\pi}\text{ V}. \end{aligned}$$

45. $\displaystyle\int_1^2 \frac{\sin x}{x}\,dx$; $h = \dfrac{2-1}{8} = \dfrac{1}{8} = 0.125$.

$x_0 = 1$, $x_1 = 1.125$, $x_2 = 1.125 + 0.125 = 1.25$, $x_3 = 1.25 + 0.125 = 1.375$, $\ldots$, $x_8 = 2$.

The function values are listed next:

$$\begin{aligned}
f(1) &= \frac{\sin 1}{1} = 0.8415. \\
f(1.125) &= \frac{\sin 1.125}{1.125} = 0.8020. \\
f(1.25) &= \frac{\sin 1.25}{1.25} = 0.7592. \\
f(1.375) &= 0.7134. \\
f(1.5) &= 0.6650. \\
f(1.625) &= 0.6145. \\
f(1.75) &= 0.5623. \\
f(1.875) &= 0.5088. \\
f(2) &= 0.4546.
\end{aligned}$$

$$\int_1^2 \frac{\sin x}{x} \approx 0.125\left[\frac{1}{2}(0.8415) + 0.8020 + 0.7592 + \cdots + \frac{1}{2}(0.4546)\right] = 0.66.$$

7.4 Further Trigonometric Forms

1. $\int \sin^3 x \cos^2 x\, dx$

TYPE 1, n odd, identity (7.14). $u = \cos x;\ du = -\sin x\, dx.$

$$\begin{aligned}
\int \sin^2 x \cos^2 x(\sin x\, dx) &= \int \left(1 - \cos^2 x\right)\left(\cos^2 x\right)(\sin x\, dx) \\
&= -\int \left(1 - \cos^2 x\right)\left(\cos^2 x\right)(-\sin x\, dx) \\
&= -\int \left(1 - u^2\right) u^2\, du = -\int \left(u^2 - u^4\right) du \\
&= -\frac{1}{3}u^3 + \frac{1}{5}u^5 + C = \frac{1}{5}\cos^5 x - \frac{1}{3}\cos^3 x + C.
\end{aligned}$$

3. $\int \cos^3 x\, dx$

TYPE 1, m odd, identity (7.14). $u = \sin x;\ du = \cos x\, dx.$

$$\begin{aligned}
\int \cos^2 x \cos x\, dx &= \int \left(1 - \sin^2 x\right)\cos x\, dx \\
&= \int \left(1 - u^2\right) du = u - \frac{1}{3}u^3 + C = \sin x - \frac{1}{3}\sin^3 x + C.
\end{aligned}$$

5. $\int \sin^2 2x \cos^3 2x\, dx$

TYPE 1, m odd, identity (7.14). $u = \sin 2x;\ du = 2\cos 2x\, dx.$

$$\begin{aligned}
\int \sin^2 2x \cos^2 2x(\cos 2x\, dx) &= \int \sin^2 2x(1 - \sin^2 2x)(\cos 2x\, dx) \\
&= \frac{1}{2}\int \sin^2 2x(1 - \sin^2 2x)(2\cos 2x\, dx) \\
&= \frac{1}{2}\int u^2(1 - u^2)\, du = \frac{1}{2}\int (u^2 - u^4)\, du \\
&= \frac{1}{2}\left(\frac{1}{3}u^3 - \frac{1}{5}u^5\right) + C \\
&= \frac{1}{6}\sin^3 2x - \frac{1}{10}\sin^5 2x + C.
\end{aligned}$$

7. $\int \sin^3 3x \cos^4 3x\, dx$

TYPE 1, n odd, identity (7.14). $u = \cos 3x; du = -3\sin 3x\, dx.$

$$\begin{aligned}
\int \sin^2 3x \cos^4 3x(\sin 3x\, dx) &= \int \left(1-\cos^2 3x\right)\left(\cos^4 3x\right)(\sin 3x\, dx) \\
&= -\frac{1}{3}\int \left(1-\cos^2 3x\right)\left(\cos^4 3x\right)(-3\sin 3x\, dx) \\
&= -\frac{1}{3}\int \left(1-u^2\right)u^4\, du = -\frac{1}{3}\int \left(u^4-u^6\right)\, du \\
&= -\frac{1}{3}\left(\frac{1}{5}u^5 - \frac{1}{7}u^7\right) + C \\
&= -\frac{1}{15}\cos^5 3x + \frac{1}{21}\cos^7 3x + C = \frac{1}{21}\cos^7 3x - \frac{1}{15}\cos^5 3x + C.
\end{aligned}$$

9. $\int \cos^2 4x\, dx$

TYPE 1, even powers, identity (7.17).

$$\begin{aligned}
\int \cos^2 4x\, dx &= \frac{1}{2}\int (1+\cos 8x)\, dx \\
&= \frac{1}{2}\int 1\, dx + \frac{1}{2}\int \cos 8x\, dx \qquad u = 8x;\ du = 8\, dx. \\
&= \frac{1}{2}\int dx + \frac{1}{2}\cdot\frac{1}{8}\int \cos 8x(8\, dx) \\
&= \frac{1}{2}\int dx + \frac{1}{16}\int \cos u\, du \\
&= \frac{1}{2}x + \frac{1}{16}\sin 8x + C.
\end{aligned}$$

11. $\int \sin^2 x \cos^2 x\, dx$

TYPE 1, even powers, identities (7.17) and (7.18).

$$\begin{aligned}
\int \frac{1}{2}(1-\cos 2x)\cdot\frac{1}{2}(1+\cos 2x)\, dx &= \frac{1}{4}\int (1-\cos^2 2x)\, dx \\
&= \frac{1}{4}\int dx - \frac{1}{4}\int \cos^2 2x\, dx \\
&= \frac{1}{4}x - \frac{1}{4}\int \frac{1}{2}(1+\cos 4x)\, dx \\
&= \frac{1}{4}x - \frac{1}{8}\left(x + \frac{1}{4}\sin 4x\right) \qquad u = 4x;\ du = 4\, dx. \\
&= \frac{1}{4}x - \frac{1}{8}x - \frac{1}{32}\sin 4x + C \\
&= \frac{1}{8}x - \frac{1}{32}\sin 4x + C.
\end{aligned}$$

13. $\int \sin^3 2t \cos^2 2t\, dt$

TYPE 1, n odd. $u = \cos 2t;\ du = -2\sin 2t\, dt.$

$$\begin{aligned}
\int \sin^2 2t\cos^2 2t(\sin 2t\, dt) &= \int (1-\cos^2 2t)\cos^2 2t(\sin 2t\, dt) \\
&= -\frac{1}{2}\int (1-\cos^2 2t)\cos^2 2t(-2\sin 2t\, dt) \\
&= -\frac{1}{2}\int (1-u^2)u^2\, du = -\frac{1}{2}\int (u^2-u^4)\, du \\
&= -\frac{1}{2}\left(\frac{u^3}{3}-\frac{u^5}{5}\right) + C = \frac{1}{10}u^5 - \frac{1}{6}u^3 + C. \\
&= \frac{1}{10}\cos^5 2t - \frac{1}{6}\cos^3 2t + C.
\end{aligned}$$

15. $\int \sin^4 4x \cos^3 4x\, dx$

TYPE 1, m odd, identity (7.14). $u = \sin 4x$; $du = 4\cos 4x\, dx$.

$$\begin{aligned}\int \sin^4 4x \cos^2 4x \cos 4x\, dx &= \int \sin^4 4x(1-\sin^2 4x)\cos 4x\, dx\\ &= \frac{1}{4}\int \sin^4 4x(1-\sin^2 4x)(4\cos 4x\, dx)\\ &= \frac{1}{4}\int u^4(1-u^2)\, du = \frac{1}{4}\left(\frac{1}{5}u^5 - \frac{1}{7}u^7\right) + C\\ &= \frac{1}{20}\sin^5 4x - \frac{1}{28}\sin^7 4x + C.\end{aligned}$$

17. $\int \tan^2 x \sec^4 x\, dx$

TYPE 2, m even, identity (7.15). $u = \tan x$; $du = \sec^2 x\, dx$.

$$\begin{aligned}\int \tan^2 x \sec^2 x \sec^2 x\, dx &= \int \tan^2 x(1+\tan^2 x)\sec^2 x\, dx\\ &= \int u^2(1+u^2)\, du = \frac{1}{3}u^3 + \frac{1}{5}u^5 + C\\ &= \frac{1}{3}\tan^3 x + \frac{1}{5}\tan^5 x + C.\end{aligned}$$

19. $\int \tan^6 3x \sec^4 3x\, dx$

TYPE 2, m even, identity (7.15). $u = \tan 3x$; $du = 3\sec^2 3x\, dx$.

$$\begin{aligned}\int \tan^6 3x \sec^2 3x \left(\sec^2 3x\, dx\right) &= \int \tan^6 3x\left(1+\tan^2 3x\right)\left(\sec^2 3x\, dx\right)\\ &= \frac{1}{3}\int \tan^6 3x\left(1+\tan^2 3x\right)\left(3\sec^2 3x\, dx\right)\\ &= \frac{1}{3}\int u^6\left(1+u^2\right) du = \frac{1}{3}\int \left(u^6+u^8\right) du\\ &= \frac{1}{3}\left(\frac{1}{7}u^7 + \frac{1}{9}u^9\right) + C = \frac{1}{21}\tan^7 3x + \frac{1}{27}\tan^9 3x + C.\end{aligned}$$

21. $\int \tan y \sec^3 y\, dy$

TYPE 2, n odd.

$\int \tan y \sec^3 y\, dy = \int \sec^2 y(\sec y \tan y)\, dy.$

Let $u = \sec y$; then $du = \sec y \tan y\, dy$.

$\int u^2\, du = \frac{1}{3}u^3 + C = \frac{1}{3}\sec^3 y + C.$

23. $\displaystyle\int \tan^3 x \sec^3 x\, dx$

TYPE 2, n odd, identity (7.15). $u = \sec x;\ du = \sec x \tan x\, dx.$

$$\begin{aligned}\int \tan^2 x \sec^2 x (\sec x \tan x\, dx) &= \int (\sec^2 x - 1)\sec^2 x(\sec x \tan x\, dx)\\ &= \int (u^2-1)u^2\, du = \int (u^4 - u^2)\, du\\ &= \frac{1}{5}u^5 - \frac{1}{3}u^3 + C = \frac{1}{5}\sec^5 x - \frac{1}{3}\sec^3 x + C.\end{aligned}$$

25. $\displaystyle\int \cot^6 2x \csc^4 2x\, dx$

TYPE 3, m even, identity (7.16). The technique is the same as for TYPE 2 with m even: set aside $\csc^2 2x$ for du, and change the remaining cosecants to cotangents.

$\displaystyle\int \cot^6 2x \csc^4 2x\, dx = \int \cot^6 2x \csc^2 2x (\csc^2 2x)\, dx = \int \cot^6 2x(1+\cot^2 2x)(\csc^2 2x\, dx).$

Let $u = \cot 2x$; then $du = -2\csc^2 2x\, dx$.

$$\begin{aligned}-\frac{1}{2}\int \cot^6 2x(1+\cot^2 2x)(-2\csc^2 2x\, dx) &= -\frac{1}{2}\int u^6(1+u^2)\, du\\ &= -\frac{1}{2}\int (u^6+u^8)\, du = -\frac{1}{2}\left(\frac{1}{7}u^7 + \frac{1}{9}u^9\right) + C\\ &= -\frac{1}{14}\cot^7 2x - \frac{1}{18}\cot^9 2x + C.\end{aligned}$$

27. $\displaystyle\int \csc^6 x\, dx$

TYPE 3, m even, identity (7.16). $u = \cot x;\ du = -\csc^2 x\, dx.$

$$\begin{aligned}\int \csc^4 x \csc^2 x\, dx &= \int (1+\cot^2 x)^2 \csc^2 x\, dx\\ &= -\int (1+\cot^2 x)^2(-\csc^2 x\, dx)\\ &= -\int (1+u^2)^2\, du = -\int (1 + 2u^2 + u^4)\, du\\ &= -u - \frac{2}{3}u^3 - \frac{1}{5}u^5 + C\\ &= -\cot x - \frac{2}{3}\cot^3 x - \frac{1}{5}\cot^5 x + C.\end{aligned}$$

29. $\displaystyle\int_0^{\pi/4} (\tan x)^{1/2} \sec^4\, dx$

TYPE 2, m even, identity (7.15).

$\displaystyle\int (\tan x)^{1/2}\sec^2 x(\sec^2 x\, dx) = \int (\tan x)^{1/2}(1+\tan^2 x)(\sec^2 x\, dx).$

Let $u = \tan x$; $du = \sec^2 x\, dx$.

$\displaystyle\int u^{1/2}(1+u^2)\, du = \int (u^{1/2} + u^{5/2})\, du = \frac{2}{3}u^{3/2} + \frac{2}{7}u^{7/2} + C.$

$$\begin{aligned}\int_0^{\pi/4} (\tan x)^{1/2}\sec^4 x\, dx &= \frac{2}{3}(\tan x)^{3/2} + \frac{2}{7}(\tan x)^{7/2}\Big|_0^{\pi/4}\\ &= \frac{2}{3}\left(\tan\frac{\pi}{4}\right)^{3/2} + \frac{2}{7}\left(\tan\frac{\pi}{4}\right)^{7/2} - 0\\ &= \frac{2}{3} + \frac{2}{7} = \frac{20}{21}.\end{aligned}$$

31.
$$\begin{aligned}\int_0^\pi (1+\sin x)^2\,dx &= \int_0^\pi (1+2\sin x+\sin^2 x)\,dx\\ &= \int_0^\pi dx + 2\int_0^\pi \sin x\,dx + \frac{1}{2}\int_0^\pi (1-\cos 2x)\,dx\\ &= x\Big|_0^\pi - 2\cos x\Big|_0^\pi + \frac{1}{2}\left(x-\frac{1}{2}\sin 2x\right)\Big|_0^\pi\\ &= \pi - 2(\cos\pi-\cos 0)+\frac{1}{2}(\pi-0)\\ &= \pi-2(-1-1)+\frac{1}{2}\pi-0 = 4+\frac{3}{2}\pi.\end{aligned}$$

33. $\displaystyle\int_0^{\pi/4} \frac{\sec^2 x}{1+\tan x}\,dx$

Let $u = 1+\tan x$; then $du = \sec^2 x\,dx$.

Lower limit: if $x=0$, $u = 1+\tan 0 = 1$.

Upper limit: if $x = \frac{\pi}{4}$, $u = 1+\tan\frac{\pi}{4} = 1+1 = 2$.

$$\int_0^{\pi/4} \frac{\sec^2 x}{1+\tan x}\,dx = \int_1^2 \frac{du}{u} = \ln|u|\Big|_1^2 = \ln 2 - \ln 1 = \ln 2 - 0 = \ln 2.$$

35. $\displaystyle\int \tan^5 4x\sec^4 4x\,dx$

TYPE 2, m even. (This integral can also be treated as TYPE 2 with n odd.)

$u = \tan 4x$; $du = 4\sec^2 4x\,dx$.

$$\begin{aligned}\int \tan^5 4x\sec^2 4x\sec^2 4x\,dx &= \frac{1}{4}\int \tan^5 4x(1+\tan^2 4x)(4\sec^2 4x\,dx)\\ &= \frac{1}{4}\int u^5(1+u^2)\,du = \frac{1}{4}\left(\frac{1}{6}u^6+\frac{1}{8}u^8\right)+C\\ &= \frac{1}{24}\tan^6 4x + \frac{1}{32}\tan^8 4x + C.\end{aligned}$$

37. Volume of disk: $\pi(\text{radius})^2 \cdot \text{thickness} = \pi y^2\,dx$.

$V = \pi\displaystyle\int_0^\pi \sin^2 x\,dx$

TYPE 1, even powers, identity (7.18).

$$\begin{aligned}V &= \frac{\pi}{2}\int_0^\pi (1-\cos 2x)\,dx\\ &= \frac{\pi}{2}\int_0^\pi dx - \frac{\pi}{2}\int_0^\pi \cos 2x\,dx \qquad u=2x;\ du = 2\,dx.\\ &= \frac{\pi}{2}\int_0^\pi dx - \frac{\pi}{2}\cdot\frac{1}{2}\int_0^\pi \cos 2x(2\,dx)\\ &= \frac{\pi}{2}x\Big|_0^\pi - \frac{\pi}{4}\sin 2x\Big|_0^\pi\\ &= \frac{\pi^2}{2} - 0 = \frac{\pi^2}{2}.\end{aligned}$$

39. $i = 20\cos 100\pi t$. Period: $\dfrac{2\pi}{100\pi} = \dfrac{1}{50}$ second.

$$\begin{aligned} i_{\text{rms}}^2 &= \frac{1}{1/50-0}\int_0^{1/50} (20)^2\cos^2 100\pi t\,dt \\ &= 50(400)\int_0^{1/50}\frac{1}{2}(1+\cos 200\pi t)\,dt \\ &= 10000\left(t+\frac{1}{200\pi}\sin 200\pi t\right)\Big|_0^{1/50} \\ &= 10000\left(\frac{1}{50}\right) = 200. \end{aligned}$$

$i_{\text{rms}} = \sqrt{200} = \sqrt{100\cdot 2} = 10\sqrt{2}\,\text{A}.$

41. Since $\sin\left(2t-\dfrac{\pi}{3}\right) = \sin 2t\cos\dfrac{\pi}{3} - \cos 2t\sin\dfrac{\pi}{3} = \dfrac{1}{2}\sin 2t - \dfrac{\sqrt{3}}{2}\cos 2t$, we get, by Exercise 40 with $T=\pi$:

$$\begin{aligned} P &= \frac{1}{\pi}\int_0^{\pi}(3\sin 2t)(5)\left(\frac{1}{2}\sin 2t - \frac{\sqrt{3}}{2}\cos 2t\right)dt \\ &= \frac{15}{2\pi}\int_0^{\pi}\sin^2 2t\,dt - \frac{15\sqrt{3}}{2\pi}\cdot\frac{1}{2}\int_0^{\pi}\sin 2t\cos 2t(2\,dt). \\ P &= \frac{15}{2\pi}\cdot\frac{1}{2}\int_0^{\pi}(1-\cos 4t)\,dt - \frac{15\sqrt{3}}{2\pi}\cdot\frac{1}{2}\int_0^{\pi}\sin 2t\cos 2t(2\,dt). \end{aligned}$$

$u = 4t;\ du = 4\,dt.$ $\qquad u = \sin 2t;\ du = 2\cos 2t\,dt.$

$$\begin{aligned} P &= \frac{15}{4\pi}\left(t-\frac{1}{4}\sin 4t\right)\Big|_0^{\pi} - \frac{15\sqrt{3}}{4\pi}\left(\frac{1}{2}\sin^2 2t\right)\Big|_0^{\pi} \\ &= \frac{15}{4\pi}(\pi-0) - 0 = \frac{15}{4}\,\text{W}. \end{aligned}$$

43. $\displaystyle\int_0^{\pi/2}\frac{dx}{\sqrt{1+\cos x}}\quad (n=8)$

$h = \dfrac{\pi/2-0}{8} = \dfrac{\pi}{16}$

$x_0 = 0,\ x_1 = \dfrac{\pi}{16},\ x_2 = \dfrac{\pi}{8},\ x_3 = \dfrac{3\pi}{16},\ x_4 = \dfrac{\pi}{4},$

$x_5 = \dfrac{5\pi}{16},\ x_6 = \dfrac{3\pi}{8},\ x_7 = \dfrac{7\pi}{16},\ x_8 = \dfrac{\pi}{2}.$

Some of the function values are

$$\begin{aligned} f(x_0) &= \frac{1}{\sqrt{1+\cos 0}} = 0.7071 \\ f(x_1) &= \frac{1}{\sqrt{1+\cos(\pi/16)}} = 0.7105 \\ f(x_2) &= \frac{1}{\sqrt{1+\cos(\pi/8)}} = 0.7210 \\ f(x_3) &= \frac{1}{\sqrt{1+\cos(3\pi/16)}} = 0.7389 \\ &\cdots \\ f(x_8) &= \frac{1}{\sqrt{1+\cos(\pi/2)}} = 1. \end{aligned}$$

$$\int_0^{\pi/2}\frac{dx}{\sqrt{1+\cos x}} \approx \frac{\pi/16}{3}\left[0.7071 + 4(0.7105) + 2(0.7210) + 4(0.7389) + \cdots + 1\right] = 1.25.$$

45. Since the frequency is 60 cycles per second, the period is $(1/60)\,s$; or:

$$\text{Period} = \frac{120\pi}{2\pi} = \frac{1}{60}\text{ s.}$$

Then

$$\begin{aligned} i^2_{\text{rms}} &= \frac{1}{1/60-0}\int_0^{1/60} 155^2\sin^2 120\pi t\,dt \\ &= 60\cdot 155^2\cdot\frac{1}{2}\int_0^{1/60}(1-\cos 240\pi t)\,dt \\ &= 60\cdot 155^2\cdot\frac{1}{2}\left(t-\frac{1}{240\pi}\sin 240\pi t\right)\Big|_0^{1/60} \qquad u=240\pi t;\ du=240\pi\,dt. \\ &= 60\cdot 155^2\cdot\frac{1}{2}\left(\frac{1}{60}-0\right) = 155^2\cdot\frac{1}{2}. \\ i_{\text{rms}} &= \sqrt{155^2\cdot\frac{1}{2}} = \frac{155}{\sqrt{2}} \approx 110\,\text{V}. \end{aligned}$$

7.5 Inverse Trigonometric Forms

1. $\displaystyle\int\frac{dx}{\sqrt{1-x^2}} = \text{Arcsin}\,x + C$ by (7.19) with $a=1$.

3. $\displaystyle\int\frac{dx}{9+x^2} = \frac{1}{3}\text{Arctan}\,\frac{x}{3} + C$ by (7.20) with $a=3$.

5. $\displaystyle\int\frac{dx}{16+9x^2} = \int\frac{dx}{4^2+(3x)^2}$ $\qquad u=3x;\ du=3\,dx.$

$$\begin{aligned}\frac{1}{3}\int\frac{3\,dx}{4^2+(3x)^2} &= \frac{1}{3}\int\frac{du}{4^2+u^2} \\ &= \frac{1}{3}\cdot\frac{1}{4}\text{Arctan}\,\frac{u}{4}+C \quad \text{(7.20) with } a=4 \\ &= \frac{1}{12}\text{Arctan}\,\frac{3x}{4}+C.\end{aligned}$$

7. $\displaystyle\int\frac{dx}{\sqrt{4-5x^2}} = \int\frac{dx}{\sqrt{2^2-(\sqrt{5}x)^2}}$ $\qquad u=\sqrt{5}x;\ du=\sqrt{5}\,dx.$

$$\begin{aligned}\frac{1}{\sqrt{5}}\int\frac{\sqrt{5}\,dx}{\sqrt{2^2-(\sqrt{5}x)^2}} &= \frac{1}{\sqrt{5}}\int\frac{du}{\sqrt{2^2-u^2}} \\ &= \frac{1}{\sqrt{5}}\text{Arcsin}\,\frac{u}{2}+C = \frac{1}{\sqrt{5}}\text{Arcsin}\,\frac{\sqrt{5}x}{2}+C.\end{aligned}$$

9. $\displaystyle\int\frac{x\,dx}{\sqrt{1-x^2}} = \int(1-x^2)^{-1/2}x\,dx.$ $\qquad u=1-x^2;\ du=-2x\,dx.$

$$-\frac{1}{2}\int(1-x^2)^{-1/2}(-2x\,dx) = -\frac{1}{2}\int u^{-1/2}\,du = -\frac{1}{2}\frac{u^{1/2}}{1/2}+C = -\sqrt{u}+C = -\sqrt{1-x^2}+C.$$

11. $\displaystyle\int\frac{x\,dx}{16+9x^2}$ $\qquad u=16+9x^2;\ du=18x\,dx.$

$$\frac{1}{18}\int\frac{18x\,dx}{16+9x^2} = \frac{1}{18}\ln|u|+C = \frac{1}{18}\ln|16+9x^2|+C = \frac{1}{18}\ln(16+9x^2)+C.$$

13. $\displaystyle\int\frac{t\,dt}{4+t^4} = \int\frac{t\,dt}{4+(t^2)^2}.$ $\qquad u=t^2;\ du=2t\,dt.$

$$\frac{1}{2}\int\frac{2t\,dt}{4+(t^2)^2} = \frac{1}{2}\int\frac{du}{4+u^2} = \frac{1}{2}\cdot\frac{1}{2}\text{Arctan}\,\frac{u}{2}+C \text{ by (7.20) with } a=2.$$

We now get: $\frac{1}{4}\text{Arctan}\,\frac{1}{2}t^2+C.$

15. $\displaystyle\int \frac{\csc^2 x\,dx}{\sqrt{4-\cot^2 x}}$ $\qquad u = \cot x;\ du = -\csc^2 x\,dx.$

$$-\int \frac{-\csc^2 x\,dx}{\sqrt{4-\cot^2 x}} = -\int \frac{du}{\sqrt{4-u^2}} = -\text{Arcsin}\,\frac{u}{2} + C = -\text{Arcsin}\left(\frac{1}{2}\cot x\right) + C.$$

17. $\displaystyle\int \frac{dx}{\sqrt{5-3x^2}} = \int \frac{dx}{\sqrt{5-\left(\sqrt{3}x\right)^2}}.$ $\qquad u = \sqrt{3}x;\ du = \sqrt{3}\,dx.$

$$\frac{1}{\sqrt{3}}\int \frac{\sqrt{3}\,dx}{\sqrt{5-\left(\sqrt{3}x\right)^2}} = \frac{1}{\sqrt{3}}\int \frac{du}{\sqrt{5-u^2}} = \frac{1}{\sqrt{3}}\text{Arcsin}\,\frac{u}{\sqrt{5}} + C \text{ by (7.19) with } a = \sqrt{5}.$$

We get: $\displaystyle\frac{1}{\sqrt{3}}\text{Arcsin}\,\frac{\sqrt{3}x}{\sqrt{5}} + C = \frac{\sqrt{3}}{3}\text{Arcsin}\,\frac{\sqrt{15}x}{5} + C.$

19. $\displaystyle\int \frac{\cos y\,dy}{2+\sin y}$ $\qquad u = 2 + \sin y;\ du = \cos y\,dy.$

$$\int \frac{du}{u} = \ln|u| + C = \ln|2+\sin y| + C = \ln(2+\sin y) + C.$$

21. $$\int_3^{3\sqrt{3}} \frac{3\,dx}{9+x^2} = 3\cdot\frac{1}{3}\text{Arctan}\,\frac{x}{3}\Big|_3^{3\sqrt{3}} = \text{Arctan}\,\frac{3\sqrt{3}}{3} - \text{Arctan}\,\frac{3}{3}$$
$$= \text{Arctan}\,\sqrt{3} - \text{Arctan}\,1 = \frac{\pi}{3} - \frac{\pi}{4} = \frac{\pi}{12}.$$

23. $\displaystyle\int \frac{dx}{\sqrt{4-3x^2}} = \int \frac{dx}{\sqrt{4-(\sqrt{3}x)^2}}.$ $\qquad u = \sqrt{3}x;\ du = \sqrt{3}\,dx.$

$$\frac{1}{\sqrt{3}}\int \frac{\sqrt{3}\,dx}{\sqrt{4-(\sqrt{3}x)^2}} = \frac{1}{\sqrt{3}}\int \frac{du}{\sqrt{4-u^2}} = \frac{1}{\sqrt{3}}\text{Arcsin}\,\frac{u}{2} + C = \frac{1}{\sqrt{3}}\text{Arcsin}\,\frac{\sqrt{3}}{2}x + C.$$

25. $\displaystyle\int \frac{x}{\sqrt{4-3x^2}}\,dx = \int (4-3x^2)^{-1/2}x\,dx.$ $\qquad u = 4-3x^2;\ du = -6x\,dx.$

$$-\frac{1}{6}\int (4-3x^2)^{-1/2}(-6x\,dx) = -\frac{1}{6}\int u^{-1/2}\,du = -\frac{1}{6}\cdot 2u^{1/2} + C = -\frac{1}{3}\sqrt{4-3x^2} + C.$$

27. $\displaystyle\int \frac{dx}{x^2-6x+9} = \int \frac{dx}{(x-3)^2}.$ $\qquad u = x-3;\ du = dx.$

$$\int (x-3)^{-2}\,dx = \int u^{-2}\,du = \frac{u^{-1}}{-1} + C = -\frac{1}{x-3} + C.$$

29. Completing the square, we have

$$x^2 - 6x + 10 = x^2 - 6x + \underline{9} - \underline{9} + 10 = (x^2-6x+9) + 1 = (x-3)^2 + 1.$$

$\displaystyle\int \frac{dx}{x^2-6x+10} = \int \frac{dx}{(x-3)^2+1}.$ $\qquad u = x-3;\ du = dx.$

$$\int \frac{du}{u^2+1} = \text{Arctan}\,u + C = \text{Arctan}\,(x-3) + C \text{ by (7.20) with } a = 1.$$

31. $$1 - x^2 - 4x = -(x^2+4x-1) = -(x^2+4x+4-4-1)$$
$$= -(x^2+4x+4) + 5 = 5 - (x+2)^2.$$

$\displaystyle\int \frac{dx}{\sqrt{1-x^2-4x}} = \int \frac{dx}{\sqrt{5-(x+2)^2}}.$ $\qquad u = x+2;\ du = dx.$

$$\int \frac{du}{\sqrt{5-u^2}} = \text{Arcsin}\,\frac{u}{\sqrt{5}} + C = \text{Arcsin}\,\frac{x+2}{\sqrt{5}} + C.$$

33. Completing the square, we have

$$x^2+3x+3 = x^2+3x+\frac{9}{4}-\frac{9}{4}+3 = \left(x^2+3x+\frac{9}{4}\right)+\frac{3}{4} = \left(x+\frac{3}{2}\right)^2+\frac{3}{4}.$$

$$\int \frac{dx}{x^2+3x+3} = \int \frac{dx}{\left(x+\frac{3}{2}\right)^2+\frac{3}{4}}. \qquad u = x+\frac{3}{2};\ du = dx.$$

$$\int \frac{du}{u^2+\frac{3}{4}} = \int \frac{du}{\left(\frac{\sqrt{3}}{2}\right)^2+u^2} = \frac{2}{\sqrt{3}}\text{Arctan}\,\frac{2}{\sqrt{3}}u + C \text{ by (7.20) with } a = \frac{\sqrt{3}}{2}.$$

We obtain $\frac{2}{\sqrt{3}}\text{Arctan}\,\frac{2}{\sqrt{3}}\left(x+\frac{3}{2}\right)+C = \frac{2}{\sqrt{3}}\text{Arctan}\,\frac{2x+3}{\sqrt{3}}+C.$

35. $$\int \frac{x+4}{x^2+16}\,dx = \int \frac{x\,dx}{x^2+16} + \int \frac{4\,dx}{x^2+16}.$$

For the first integral, let $u = x^2+16$:

$$\frac{1}{2}\int \frac{2x\,dx}{x^2+16} + 4\int \frac{dx}{x^2+16} = \frac{1}{2}\ln(x^2+16)+4\cdot\frac{1}{4}\text{Arctan}\,\frac{x}{4}+C$$
$$= \frac{1}{2}\ln(x^2+16)+\text{Arctan}\,\frac{1}{4}x+C.$$

37. $$\int_0^{\pi/2} \frac{\cos x}{1+\sin^2 x}\,dx \qquad u = \sin x;\ du = \cos x\,dx.$$

$$\int_0^{\pi/2} \frac{\cos x}{1+\sin^2 x}\,dx = \text{Arctan}\,(\sin x)\Big|_0^{\pi/2} = \text{Arctan}\,1 - \text{Arctan}\,0 = \frac{\pi}{4}.$$

39. $$\int \frac{e^{2x}}{1+e^{2x}}\,dx \qquad u = 1+e^{2x};\ du = 2e^{2x}\,dx.$$

$$\frac{1}{2}\int \frac{2e^{2x}\,dx}{1+e^{2x}} = \frac{1}{2}\int \frac{du}{u} = \frac{1}{2}\ln|u|+C = \frac{1}{2}\ln(1+e^{2x})+C.$$

41. $$\int \frac{e^x}{1+e^{2x}}\,dx = \int \frac{e^x\,dx}{1+(e^x)^2} \qquad u = e^x,\ du = e^x\,dx$$
$$= \int \frac{du}{1+u^2} = \text{Arctan}\,e^x + C.$$

43. $$A = \int_0^{\infty} \frac{1}{x^2+1}\,dx = \lim_{b\to\infty}\int_0^b \frac{1}{x^2+1}\,dx = \lim_{b\to\infty} \text{Arctan}\,x\Big|_0^b = \frac{\pi}{2}-0 = \frac{\pi}{2}.$$

45. Volume of shell: $2\pi\cdot(\text{radius})\cdot(\text{height})\cdot(\text{thickness}) = 2\pi xy\,dx$.

$$V = 2\pi\int_0^{\infty} x\frac{1}{4+x^4}\,dx = 2\pi\lim_{b\to\infty}\int_0^b \frac{x\,dx}{4+(x^2)^2} = 2\pi\lim_{b\to\infty}\frac{1}{2}\int_0^b \frac{2x\,dx}{4+(x^2)^2}$$
$$= \pi\lim_{b\to\infty}\frac{1}{2}\text{Arctan}\,\frac{x^2}{2}\Big|_0^b = \frac{\pi}{2}\lim_{b\to\infty}\left(\text{Arctan}\,\frac{b^2}{2}-0\right).$$

Since $\tan\theta$ is undefined for $\theta = \frac{\pi}{2}$, we obtain:

$$V = \frac{\pi}{2}\left(\frac{\pi}{2}\right) = \frac{\pi^2}{4}.$$

7.6 Integration by Trigonometric Substitution

1. $\int \frac{\sqrt{4-x^2}}{x^2}\,dx$. Let $x = 2\sin\theta,\ dx = 2\cos\theta\,d\theta$. Note that
$\sqrt{4-x^2} = \sqrt{4-4\sin^2\theta} = \sqrt{4(1-\sin^2\theta)} = 2\sqrt{\cos^2\theta} = 2\cos\theta$.

$$\begin{aligned}\int \frac{\sqrt{4-x^2}\,dx}{x^2} &= \int \frac{(2\cos\theta)(2\cos\theta\,d\theta)}{4\sin^2\theta} = \int \frac{\cos^2\theta}{\sin^2\theta}\,d\theta \\ &= \int \cot^2\theta\,d\theta = \int (\csc^2\theta - 1)\,d\theta = -\cot\theta - \theta + C.\end{aligned}$$

From the substituted expression $x = 2\sin\theta$, we get $\sin\theta = \frac{x}{2}$ or $\theta = \text{Arcsin}\,\frac{x}{2}$. To change $\cot\theta$, we use a diagram. Note that the opposite side has length x and the hypotenuse length 2. The remaining side is of length $\sqrt{4-x^2}$ by the Pythagorean theorem.

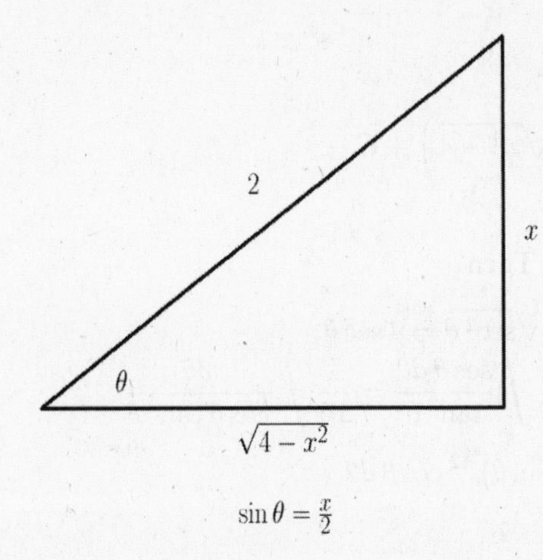

$\sin\theta = \frac{x}{2}$

Since $\cot\theta = \frac{\sqrt{4-x^2}}{x}$, we get

$$\begin{aligned}\int \frac{\sqrt{4-x^2}}{x^2}\,dx &= -\cot\theta - \theta + C \\ &= -\frac{\sqrt{4-x^2}}{x} - \text{Arcsin}\,\frac{x}{2} + C.\end{aligned}$$

3.
$$\begin{aligned}\int x\sqrt{x^2+25}\,dx &= \frac{1}{2}\int \left(x^2+25\right)^{1/2}(2x\,dx) \qquad u = x^2+25;\ du = 2x\,dx. \\ &= \frac{1}{2}\int u^{1/2}\,du = \frac{1}{2}\cdot\frac{2}{3}u^{3/2} + C \\ &= \frac{1}{3}\left(x^2+25\right)^{3/2} + C.\end{aligned}$$

5. $\int \frac{dx}{(x^2+1)^{3/2}}$. Let $x = \tan\theta;\ dx = \sec^2\theta\,d\theta$. Then
$(x^2+1)^{3/2} = (\tan^2\theta+1)^{3/2} = (\sec^2\theta)^{3/2} = \sec^3\theta$
and we get:
$$\int \frac{\sec^2\theta\,d\theta}{\sec^3\theta} = \int \frac{d\theta}{\sec\theta} = \int \cos\theta\,d\theta = \sin\theta + C.$$
The diagram is constructed from the substituted expression $\tan\theta = \frac{x}{1}$.

Thus $\sin\theta + C$
$= \frac{x}{\sqrt{x^2+1}} + C.$

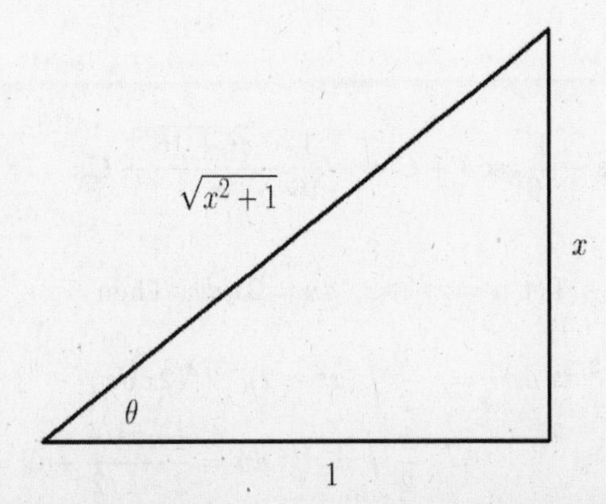

7. $\displaystyle\int \frac{dx}{x\sqrt{x^2-4}}$. Let $x = 2\sec\theta$; $dx = 2\sec\theta\tan\theta\,d\theta$. Then
$\sqrt{x^2-4} = \sqrt{4\sec^2\theta - 4} = \sqrt{4(\sec^2\theta - 1)} = 2\sqrt{\sec^2\theta - 1} = 2\sqrt{\tan^2\theta} = 2\tan\theta$ and
$\displaystyle\int \frac{dx}{x\sqrt{x^2-4}} = \int \frac{2\sec\theta\tan\theta\,d\theta}{2\sec\theta\cdot 2\tan\theta} = \int \frac{1}{2}\,d\theta = \frac{1}{2}\theta + C = \frac{1}{2}\text{Arcsec}\frac{x}{2} + C$
since $\sec\theta = \dfrac{x}{2}$. This relationship also yields the following diagram:

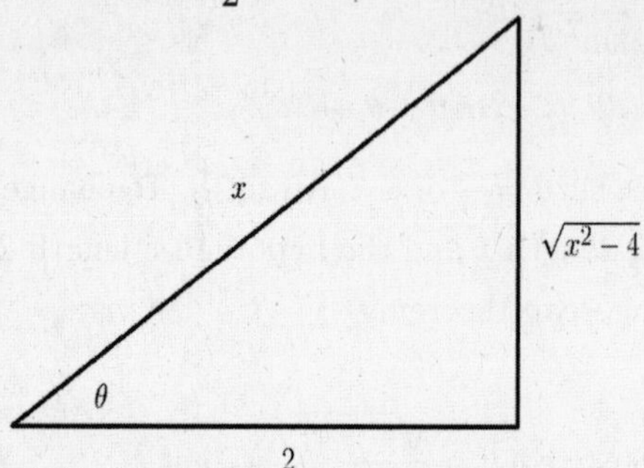

So the answer can also be written: $\dfrac{1}{2}\text{Arctan}\left(\dfrac{1}{2}\sqrt{x^2-4}\right) + C$.

9. $\displaystyle\int \frac{dx}{x^2\sqrt{x^2+16}}$. Let $x = 4\tan\theta$; $dx = 4\sec^2\theta\,d\theta$. Then
$\sqrt{x^2+16} = \sqrt{16\tan^2\theta + 16} = 4\sqrt{\tan^2\theta + 1} = 4\sqrt{\sec^2\theta} = 4\sec\theta$.

$$\begin{aligned}
\int \frac{dx}{x^2\sqrt{x^2+16}} &= \int \frac{4\sec^2\theta\,d\theta}{(16\tan^2\theta)(4\sec\theta)} = \frac{1}{16}\int \frac{\sec\theta\,d\theta}{\tan^2\theta} = \frac{1}{16}\int \frac{d\theta}{\cos\theta\tan^2\theta} \\
&= \frac{1}{16}\int \frac{\cos^2\theta\,d\theta}{\cos\theta\sin^2\theta} = \frac{1}{16}\int (\sin\theta)^{-2}\cos\theta\,d\theta \\
&\quad [u = \sin\theta;\ du = \cos\theta\,d\theta.] \\
&= \frac{1}{16}\int u^{-2}\,du + C = \frac{1}{16}\frac{u^{-1}}{-1} + C \\
&= -\frac{1}{16}\frac{1}{\sin\theta} + C = -\frac{1}{16}\csc\theta + C.
\end{aligned}$$

From $\tan\theta = \dfrac{x}{4}$, we construct the diagram shown at the right.

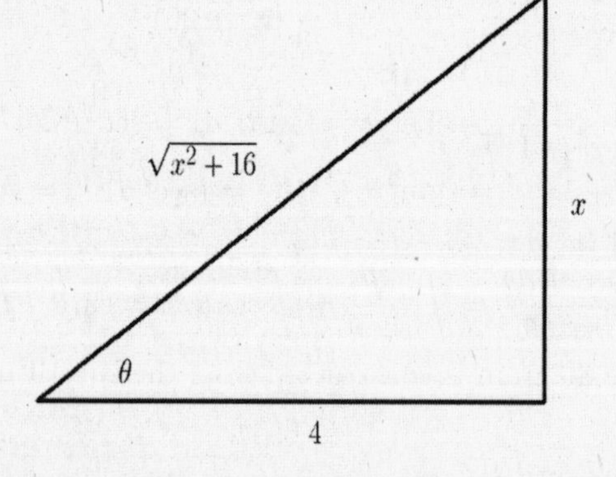

$\tan\theta = \frac{x}{4}$

We now have $-\dfrac{1}{16}\csc\theta + C = -\dfrac{1}{16}\dfrac{\sqrt{x^2+16}}{x} + C$.

11. $\displaystyle\int \frac{x\,dx}{(x^2-2)^{3/2}}$. Let $u = x^2 - 2$; $du = 2x\,dx$. Then

$$\begin{aligned}
\int (x^2-2)^{-3/2}x\,dx &= \frac{1}{2}\int (x^2-2)^{-3/2}(2x\,dx) \\
&= \frac{1}{2}\int u^{-3/2}\,du = \frac{1}{2}\frac{u^{-1/2}}{-1/2} + C = -\frac{1}{\sqrt{x^2-2}} + C.
\end{aligned}$$

13. $\int \frac{x^3\,dx}{\sqrt{x^2-3}}$. Let $x = \sqrt{3}\sec\theta$; $dx = \sqrt{3}\sec\theta\tan\theta\,d\theta$. Then
$\sqrt{x^2-3} = \sqrt{3\sec^2\theta - 3} = \sqrt{3}\sqrt{\sec^2\theta - 1} = \sqrt{3}\sqrt{\tan^2\theta} = \sqrt{3}\tan\theta$.

$$\begin{aligned}
\int \frac{x^3\,dx}{\sqrt{x^2-3}} &= \int \frac{(\sqrt{3})^3\sec^3\theta(\sqrt{3}\sec\theta\tan\theta\,d\theta)}{\sqrt{3}\tan\theta} \\
&= 3\sqrt{3}\int \sec^4\theta\,d\theta = 3\sqrt{3}\int \sec^2\theta\sec^2\theta\,d\theta \\
&= 3\sqrt{3}\int (1+\tan^2\theta)\sec^2\theta\,d\theta \\
&\quad [u = \tan\theta;\ du = \sec^2\theta\,d\theta.] \\
&= 3\sqrt{3}\int (1+u^2)\,du = 3\sqrt{3}\left(u + \frac{1}{3}u^3\right) + C \\
&= 3\sqrt{3}\left(\tan\theta + \frac{1}{3}\tan^3\theta\right) + C.
\end{aligned}$$

To construct the diagram, we use
$\sec\theta = \frac{x}{\sqrt{3}}$.

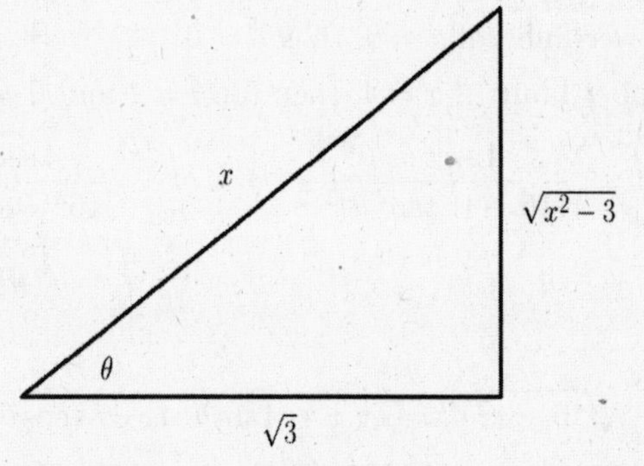

$\sec\theta = \frac{x}{\sqrt{3}}$

$$\begin{aligned}
3\sqrt{3}\left(\tan\theta + \frac{1}{3}\tan^3\theta\right) &= 3\sqrt{3}\left[\frac{\sqrt{x^2-3}}{\sqrt{3}} + \frac{1}{3}\left(\frac{\sqrt{x^2-3}}{\sqrt{3}}\right)^3\right] + C \\
&= 3\sqrt{x^2-3} + \sqrt{3}\frac{(x^2-3)^{3/2}}{3\sqrt{3}} + C \\
&= 3\sqrt{x^2-3} + \frac{1}{3}(x^2-3)^{3/2} + C \\
&= \sqrt{x^2-3}\left[3 + \frac{1}{3}(x^2-3)\right] + C \\
&= \frac{1}{3}(x^2+6)\sqrt{x^2-3} + C.
\end{aligned}$$

15. $\int \frac{\sqrt{x^2+1}}{x^2}\,dx$. Let $x = \tan\theta$; $dx = \sec^2\theta\,d\theta$. Then,
$\sqrt{x^2+1} = \sqrt{\tan^2\theta + 1} = \sqrt{\sec^2\theta} = \sec\theta$.

$$\begin{aligned}
\int \frac{\sqrt{x^2+1}\,dx}{x^2} &= \int \frac{\sec\theta\sec^2\theta\,d\theta}{\tan^2\theta} = \int \frac{\sec\theta(1+\tan^2\theta)\,d\theta}{\tan^2\theta} \\
&= \int \left(\frac{\sec\theta}{\tan^2\theta} + \sec\theta\right)d\theta = \int \frac{\cos^2\theta}{\sin^2\theta}\frac{1}{\cos\theta}\,d\theta + \int \sec\theta\,d\theta \\
&= \int (\sin\theta)^{-2}\cos\theta\,d\theta + \ln|\sec\theta + \tan\theta| + C \\
&= -(\sin\theta)^{-1} + \ln|\sec\theta + \tan\theta| + C.
\end{aligned}$$

From $\tan\theta = \dfrac{x}{1}$, we get the following diagram:

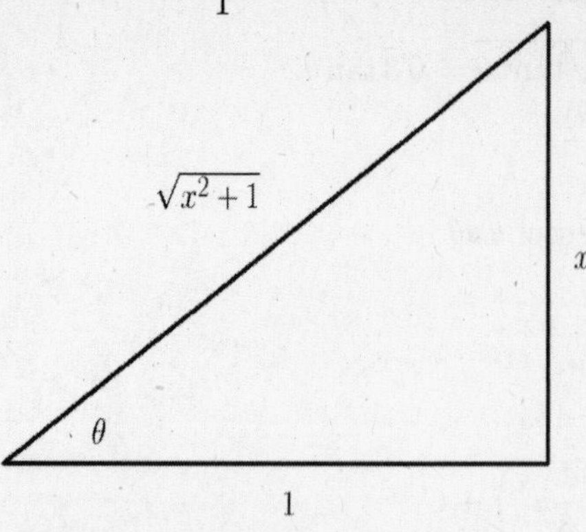

The result is: $-\dfrac{\sqrt{x^2+1}}{x} + \ln|x+\sqrt{x^2+1}| + C.$

17. $\displaystyle\int_0^4 \frac{dx}{(16+x^2)^{3/2}}$. Let $x = 4\tan\theta$; $dx = 4\sec^2\theta\,d\theta$.
Lower limit: if $x = 0$, then $\theta = 0$.
Upper limit: if $x = 4$, then $\tan\theta = 1$ and $\theta = \dfrac{\pi}{4}$.

$$\begin{aligned}\int_0^{\pi/4} \frac{4\sec^2\theta\,d\theta}{(16+16\tan^2\theta)^{3/2}} &= \int_0^{\pi/4} \frac{4\sec^2\theta\,d\theta}{16^{3/2}(\sec^2\theta)^{3/2}} = \frac{4}{16^{3/2}}\int_0^{\pi/4}\frac{d\theta}{\sec\theta} \\ &= \frac{1}{16}\int_0^{\pi/4}\cos\theta\,d\theta = \frac{1}{16}\sin\theta\Big|_0^{\pi/4} = \frac{1}{16}\cdot\frac{\sqrt{2}}{2} = \frac{\sqrt{2}}{32}.\end{aligned}$$

19. $\displaystyle\int_0^4 \sqrt{16-x^2}\,dx$. Let $x = 4\sin\theta; dx = 4\cos\theta\,d\theta$.
$\sqrt{16-x^2} = \sqrt{16-16\sin^2\theta} = 4\sqrt{1-\sin^2\theta} = 4\cos\theta$
Lower limit: if $x = 0$, then $\theta = 0$.
Upper limit: if $x = 4$, then $4 = 4\sin\theta$ and $\theta = \dfrac{\pi}{2}$.

$$\begin{aligned}\int_0^4 \sqrt{16-x^2}\,dx &= \int_0^{\pi/2} 4\cos\theta\cdot 4\cos\theta\,d\theta \\ &= 16\int_0^{\pi/2}\cos^2\theta\,d\theta = 16\int_0^{\pi/2}\frac{1}{2}(1+\cos 2\theta)\,d\theta \\ &= 8\theta + 4\sin 2\theta\Big|_0^{\pi/2} = 8\left(\frac{\pi}{2}\right) + 4\sin\pi = 4\pi.\end{aligned}$$

21.

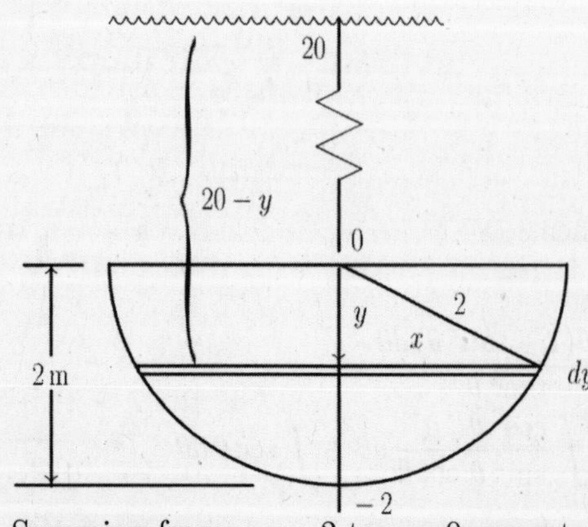

Length of strip: $2x = 2\sqrt{4-y^2}$.
Area of strip: $2\sqrt{4-y^2}\,dy$.
Pressure on strip: $(20-y)w$.
Force against strip:
$(20-y)w(2\sqrt{4-y^2}\,dy)$.

Summing from $y = -2$ to $y = 0$, we get:
$$F = \int_{-2}^0 2w(20-y)\sqrt{4-y^2}\,dy = 2w\int_{-2}^0 20\sqrt{4-y^2}\,dy - 2w\int_{-2}^0 y\sqrt{4-y^2}\,dy.$$

To evaluate $\int \sqrt{4-y^2}\,dy$, we let $y = 2\sin\theta$, so that $dy = 2\cos\theta\,d\theta$. Thus $\sqrt{4-y^2} = \sqrt{4-4\sin^2\theta} = 2\sqrt{1-\sin^2\theta} = 2\cos\theta$.

$$\begin{aligned}
\int \sqrt{4-y^2}\,dy &= \int (2\cos\theta)(2\cos\theta\,d\theta) = 4\int \cos^2\theta\,d\theta \\
&= 4\int \frac{1}{2}(1+\cos 2\theta)\,d\theta = 2\left(\theta + \frac{1}{2}\sin 2\theta\right) \\
&= 2(\theta + \sin\theta\cos\theta) = 2\text{Arcsin}\,\frac{y}{2} + 2\left(\frac{y}{2}\right)\frac{\sqrt{4-y^2}}{2} \\
&= 2\text{Arcsin}\,\frac{y}{2} + \frac{1}{2}y\sqrt{4-y^2}.
\end{aligned}$$

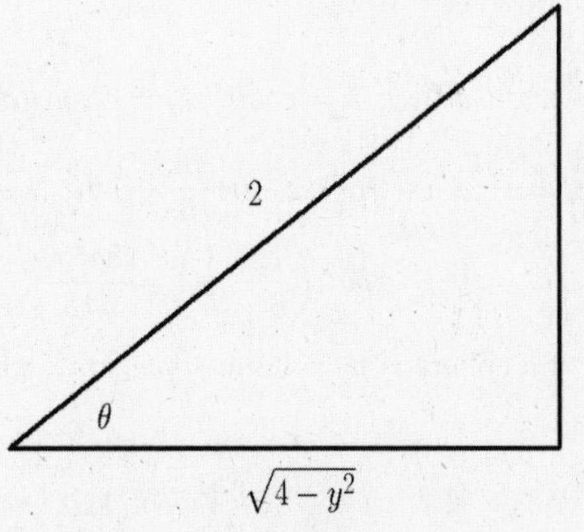

$\sin\theta = \frac{y}{2}$

To evaluate $\int y\sqrt{4-y^2}\,dy$, let $u = 4-y^2$; $du = -2y\,dy$. Then $-\frac{1}{2}\int (4-y^2)^{1/2}(-2y)\,dy = -\frac{1}{2}\frac{2}{3}(4-y^2)^{3/2} = -\frac{1}{3}(4-y^2)^{3/2}$.

We now have

$$\begin{aligned}
F &= 40w\left[2\text{Arcsin}\,\frac{y}{2} + \frac{1}{2}y\sqrt{4-y^2}\right] - 2w\left[-\frac{1}{3}(4-y^2)^{3/2}\right]\Bigg|_{-2}^{0} \\
&= -40w\left[2\cdot\left(-\frac{\pi}{2}\right)\right] - 2w\left(-\frac{1}{3}\cdot 4^{3/2}\right) \qquad \left(\text{since Arcsin}\,(-1) = -\frac{\pi}{2}\right) \\
&= 40w\pi + \left(\frac{16}{3}\right)w = \left(40\pi + \frac{16}{3}\right)w\,\text{N}.
\end{aligned}$$

23.

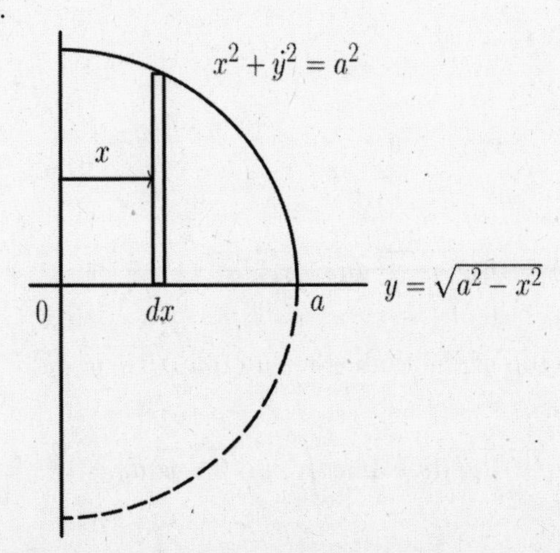

Mass of shell: $\rho\cdot 2\pi(\text{radius})\cdot(\text{height})\cdot(\text{thickness}) = \rho\cdot 2\pi x\sqrt{a^2-x^2}\,dx$.

Moment of inertia of typical shell: $x^2\cdot\rho\cdot 2\pi x\sqrt{a^2-x^2}\,dx$.

$I_y = 2\int_0^a x^2 \cdot \rho \cdot 2\pi x\sqrt{a^2-x^2}\,dx = 4\pi\rho\int_0^a x^3\sqrt{a^2-x^2}\,dx.$

Let $x = a\sin\theta$; then $dx = a\cos\theta\,d\theta$.

Lower limit: if $x=0$, then $\theta=0$.

Upper limit: if $x=a$, $\sin\theta=1$ and $\theta=\dfrac{\pi}{2}$.

$4\pi\rho\int_0^{\pi/2} a^3\sin^3\theta\sqrt{a^2-a^2\sin^2\theta}(a\cos\theta\,d\theta) = 4\pi\rho a^5\int_0^{\pi/2}\sin^3\theta\cos^2\theta\,d\theta.$

The definite integral

$$\int \sin^3\theta\cos^2\theta\,d\theta$$

is an odd power of sine. Thus

$$\int \sin^2\theta\cos^2\theta(\sin\theta\,d\theta) = -\int(1-\cos^2\theta)\cos^2\theta(-\sin^2\theta\,d\theta) \qquad u=\cos\theta;\ du=-\sin\theta\,d\theta.$$

$$= -\int(1-u^2)u^2\,du = -\frac{1}{3}u^3+\frac{1}{5}u^5 = -\frac{1}{3}\cos^3\theta+\frac{1}{5}\cos^5\theta.$$

The result is: $I_y = 4\pi\rho a^5\left(-\frac{1}{3}\cos^3\theta+\frac{1}{5}\cos^5\theta\right)\Big|_0^{\pi/2} = 0+4\pi\rho a^5\left(\frac{1}{3}-\frac{1}{5}\right) = \frac{8a^5\pi\rho}{15}.$

I_y can also be written as follows: since the mass of a sphere is $\rho\cdot\text{volume} = \rho\cdot\frac{4}{3}\pi a^3$, we get

$I_y = \frac{8a^5\pi\rho}{15} = \frac{2}{5}\left(\frac{4}{3}\pi a^3\rho\right)a^2 = \frac{2}{5}ma^2.$

25.

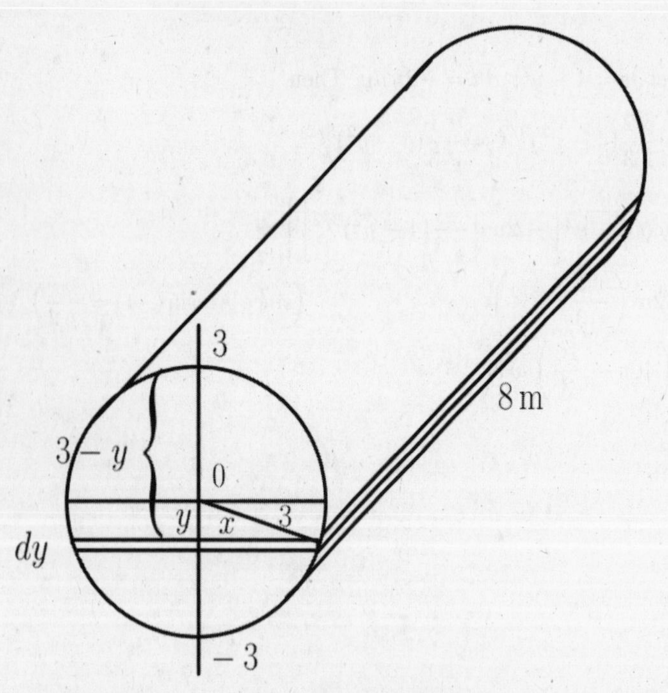

Volume of element: $2x(8\,dy) = 2\sqrt{9-y^2}(8\,dy) = 16\sqrt{9-y^2}\,dy.$

Weight of element: $16w\sqrt{9-y^2}\,dy.$

Work done in moving a typical element to the top of the tank: $(3-y)(16w\sqrt{9-y^2}\,dy).$

Summing from $y=-3$ to $y=0$, we get

$W = 16w\int_{-3}^0(3-y)\sqrt{9-y^2}\,dy = (16w)3\int_{-3}^0\sqrt{9-y^2}\,dy - 16w\int_{-3}^0 y\sqrt{9-y^2}\,dy.$

The first integral is similar to Exercise 19:

$$(16w)3\int_{-3}^0 dy,\ x=3\sin\theta;\ dx=3\cos\theta\,d\theta.$$

Since $\sqrt{9-y^2} = \sqrt{9-9\sin^2\theta} = 3\cos\theta$,

$$3(16w)\int_{-\pi/2}^{0} 3\cos\theta \cdot 3\cos\theta\, d\theta = 108\pi w.$$

In the second integral we let $u = 9 - y^2$ and $du = -2y\,dy$. The result is $144w$ for a total of $(108\pi + 144)w$ J.

7.7 Integration by Parts

1. $\displaystyle\int x\mathrm{e}^x\,dx$

$$\begin{aligned} u &= x & dv &= \mathrm{e}^x\,dx \\ du &= dx & v &= \mathrm{e}^x \end{aligned}$$

$$\int x\mathrm{e}^x\,dx = uv - \int v\,du = x\mathrm{e}^x - \int \mathrm{e}^x\,dx = x\mathrm{e}^x - \mathrm{e}^x + C.$$

3. $\displaystyle\int x\sin 2x\,dx$

$$\begin{aligned} u &= x & dv &= \sin 2x\,dx \\ du &= dx & v &= -\frac{1}{2}\cos 2x \end{aligned}$$

$$\begin{aligned} \int x\sin 2x\,dx &= uv - \int v\,du = x\left(-\frac{1}{2}\cos 2x\right) - \int\left(-\frac{1}{2}\cos 2x\right)dx \\ &= -\frac{1}{2}x\cos 2x + \frac{1}{2}\int \cos 2x\,dx = -\frac{1}{2}x\cos 2x + \frac{1}{2}\cdot\frac{1}{2}\int \cos 2x(2\,dx) \\ &= \frac{1}{4}\sin 2x - \frac{1}{2}x\cos 2x + C. \end{aligned}$$

5. $\displaystyle\int x\sec^2 x\,dx$

$$\begin{aligned} u &= x & dv &= \sec^2 x\,dx \\ du &= dx & v &= \tan x \end{aligned}$$

$$\int x\sec^2 x\,dx = x\tan x - \int \tan x\,dx = x\tan x + \ln|\cos x| + C.$$

7. $\displaystyle\int \frac{1}{x}\ln x\,dx = \int u\,du = \frac{1}{2}\ln^2 x + C \qquad u = \ln x;\ du = \frac{1}{x}\,dx$

9. $\displaystyle\int x^3\ln x\,dx$

$$\begin{aligned} u &= \ln x & dv &= x^3\,dx \\ du &= \frac{1}{x}\,dx & v &= \frac{1}{4}x^4 \end{aligned}$$

$$\begin{aligned} \int x^3\ln x\,dx &= uv - \int v\,du = (\ln x)\left(\frac{1}{4}x^4\right) - \int \frac{1}{4}x^4\cdot\frac{1}{x}\,dx \\ &= \frac{1}{4}x^4\ln x - \frac{1}{4}\int x^3\,dx = \frac{1}{4}x^4\ln x - \frac{1}{16}x^4 + C \\ &= \frac{1}{16}x^4(4\ln x - 1) + C. \end{aligned}$$

11. $$\begin{aligned} \int x\cos x^2\,dx &= \frac{1}{2}\int \cos x^2(2x\,dx) \qquad u = x^2;\ dx = 2x\,dx \\ &= \frac{1}{2}\int \cos u\,du \\ &= \frac{1}{2}\sin u + C = \frac{1}{2}\sin x^2 + C. \end{aligned}$$

13. $\int \text{Arcsin}\, x\, dx$

$$u = \text{Arcsin}\, x \qquad dv = dx$$
$$du = \frac{dx}{\sqrt{1-x^2}} \qquad v = x$$

$$\begin{aligned}
\int \text{Arcsin}\, x\, dx &= x\text{Arcsin}\, x - \int \frac{x}{\sqrt{1-x^2}}\, dx \\
&= x\text{Arcsin}\, x - \int (1-x^2)^{-1/2} x\, dx \\
&= x\text{Arcsin}\, x + \frac{1}{2}\int (1-x^2)^{-1/2}(-2x)\, dx \quad (u = 1-x^2;\ du = -2x\, dx.) \\
&= x\text{Arcsin}\, x + \frac{1}{2}\frac{(1-x^2)^{1/2}}{1/2} + C \\
&= x\text{Arcsin}\, x + \sqrt{1-x^2} + C.
\end{aligned}$$

15. $\int x^2 \sin x\, dx$

$$u = x^2 \qquad dv = \sin x\, dx$$
$$du = 2x\, dx \qquad v = -\cos x$$

$-x^2 \cos x + 2\int x \cos x\, dx$

$$u = x \qquad dv = \cos x\, dx$$
$$du = dx \qquad v = \sin x$$

$$-x^2\cos x + 2\left[x \sin x - \int \sin x\, dx\right] = -x^2 \cos x + 2x \sin x + 2\cos x + C.$$

17. $\int x^2 \cos x\, dx$

$$u = x^2 \qquad dv = \cos x\, dx$$
$$du = 2x\, dx \qquad v = \sin x$$

$x^2 \sin x - 2\int x \sin x\, dx$

$$u = x \qquad dv = \sin x\, dx$$
$$du = dx \qquad v = -\cos x$$

$$x^2 \sin x - 2\left[-x\cos x + \int \cos x\, dx\right] = x^2 \sin x + 2x\cos x - 2\sin x + C.$$

19. $\int e^x \sin x\,dx$

$$\begin{aligned} u &= e^x & dv &= \sin x\,dx \\ du &= e^x\,dx & v &= -\cos x \end{aligned}$$

$$\int e^x \sin x\,dx = -e^x \cos x + \int e^x \cos x\,dx.$$

$$\begin{aligned} u &= e^x & dv &= \cos x\,dx \\ du &= e^x\,dx & v &= \sin x \end{aligned}$$

$$\begin{aligned} \int e^x \sin x\,dx &= -e^x \cos x + e^x \sin x - \int e^x \sin x\,dx \\ 2\int e^x \sin x\,dx &= -e^x \cos x + e^x \sin x \\ \int e^x \sin x\,dx &= \frac{1}{2}e^x(\sin x - \cos x) + C. \end{aligned}$$

21. $\int e^{-x} \cos \pi x\,dx$

(See also Example 5.)

$$\begin{aligned} u &= e^{-x} & dv &= \cos \pi x\,dx \\ du &= -e^{-x}\,dx & v &= \frac{1}{\pi}\sin \pi x \end{aligned}$$

$$\int e^{-x} \cos \pi x\,dx = \frac{1}{\pi}e^{-x}\sin \pi x + \frac{1}{\pi}\left[\int e^{-x}\sin \pi x\,dx\right].$$

$$\begin{aligned} u &= e^{-x} & dv &= \sin \pi x\,dx \\ du &= -e^{-x}\,dx & v &= -\frac{1}{\pi}\cos \pi x \end{aligned}$$

$$\begin{aligned} \int e^{-x} \cos \pi x\,dx &= \frac{1}{\pi}e^{-x}\sin \pi x + \frac{1}{\pi}\left[-\frac{1}{\pi}e^{-x}\cos \pi x - \frac{1}{\pi}\int e^{-x}\cos \pi x\,dx\right] \\ &= \frac{1}{\pi}e^{-x}\sin \pi x - \frac{1}{\pi^2}e^{-x}\cos \pi x - \frac{1}{\pi^2}\int e^{-x}\cos \pi x\,dx. \end{aligned}$$

Now solve for the integral:

$$\begin{aligned} \int e^{-x}\cos \pi x\,dx + \frac{1}{\pi^2}\int e^{-x}\cos \pi x\,dx &= \frac{1}{\pi}e^{-x}\sin \pi x - \frac{1}{\pi^2}e^{-x}\cos \pi x \\ \left(1+\frac{1}{\pi^2}\right)\int e^{-x}\cos \pi x\,dx &= \frac{1}{\pi^2}e^{-x}(\pi \sin \pi x - \cos \pi x) \\ \left(\frac{\pi^2+1}{\pi^2}\right)\int e^{-x}\cos \pi x\,dx &= \frac{1}{\pi^2}e^{-x}(\pi \sin \pi x - \cos \pi x). \end{aligned}$$

We conclude that

$$\int e^{-x}\cos \pi x\,dx = \frac{1}{\pi^2+1}\left[e^{-x}(\pi \sin \pi x - \cos \pi x)\right] + C.$$

23. $\int_0^1 x2^x\,dx$

$$\begin{aligned} u &= x & dv &= 2^x\,dx \\ du &= dx & v &= \frac{2^x}{\ln 2} \end{aligned}$$

$$\begin{aligned} \int_0^1 x2^x\,dx &= \left.\frac{x2^x}{\ln 2}\right|_0^1 - \int_0^1 \frac{2^x\,dx}{\ln 2} = \left.\frac{x2^x}{\ln 2}\right|_0^1 - \left.\frac{2^x}{(\ln 2)^2}\right|_0^1 \\ &= \frac{2}{\ln 2} - \frac{2}{(\ln 2)^2} + \frac{1}{(\ln 2)^2} = \frac{2\ln 2 - 1}{(\ln 2)^2}. \end{aligned}$$

25. $A = \int_0^\pi x \sin x\,dx$

$$\begin{aligned} u &= x & dv &= \sin x\,dx \\ du &= dx & v &= -\cos x \end{aligned}$$

$$\begin{aligned} A &= -x\cos x\Big|_0^\pi + \int_0^\pi \cos x\,dx = -x\cos x\Big|_0^\pi + \sin x\Big|_0^\pi \\ &= -\pi\cos\pi + 0 = -\pi(-1) = \pi. \end{aligned}$$

27. $I_y = \int_0^{\pi/2} x^2 \cdot \rho \cos x\, dx.$

For the indefinite integral $\int x^2 \cos x\, dx$:

$$\begin{aligned} u &= x^2 & dv &= \cos x\, dx \\ du &= 2x\, dx & v &= \sin x \end{aligned}$$

$\int x^2 \cos x\, dx = x^2 \sin x - 2\int x \sin x\, dx.$

$$\begin{aligned} u &= x & dv &= \sin x\, dx \\ du &= dx & v &= -\cos x \end{aligned}$$

$$\int x^2 \cos x\, dx = x^2 \sin x - 2\left[-x\cos x + \int \cos x\, dx\right] = x^2 \sin x + 2x\cos x - 2\sin x + C.$$

$$I_y = \rho(x^2 \sin x + 2x\cos x - 2\sin x)\Big|_0^{\pi/2} = \rho\left(\frac{\pi^2}{4}\cdot 1 + 0 - 2\cdot 1\right) = \left(\frac{\pi^2}{4} - 2\right)\rho.$$

29. Recall that $v = \dfrac{q}{C} = \dfrac{1}{C}\displaystyle\int i\, dt = \frac{1}{10\times 10^{-6}}\int e^{-t}\sin 4t\, dt = 10^5 \int e^{-t}\sin 4t\, dt.$ (1)

$\int e^{-t}\sin 4t\, dt$

$$\begin{aligned} u &= e^{-t} & dv &= \sin 4t\, dt \\ du &= -e^{-t}\, dt & v &= -\frac{1}{4}\cos 4t \end{aligned}$$

$$\int e^{-t}\sin 4t\, dt = -\frac{1}{4}e^{-t}\cos 4t - \frac{1}{4}\left[\int e^{-t}\cos 4t\, dt\right].$$

$$\begin{aligned} u &= e^{-t} & dv &= \cos 4t\, dt \\ du &= -e^{-t}\, dt & v &= \frac{1}{4}\sin 4t \end{aligned}$$

$$\begin{aligned} \int e^{-t}\sin 4t\, dt &= -\frac{1}{4}e^{-t}\cos 4t - \frac{1}{4}\left[\frac{1}{4}e^{-t}\sin 4t + \frac{1}{4}\int e^{-t}\sin 4t\, dt\right] \\ &= -\frac{1}{4}e^{-t}\cos 4t - \frac{1}{16}e^{-t}\sin 4t - \frac{1}{16}\int e^{-t}\sin 4t\, dt. \end{aligned}$$

Solving for the integral, we get:

$$\begin{aligned} \frac{17}{16}\int e^{-t}\sin 4t\, dt &= -\frac{1}{16}e^{-t}(4\cos 4t + \sin 4t) \\ \int e^{-t}\sin 4t\, dt &= -\frac{1}{17}e^{-t}(4\cos 4t + \sin 4t) \end{aligned}$$

By (1), $v = 10^5\left(-\frac{1}{17}e^{-t}\right)(4\cos 4t + \sin 4t) + C.$

If $t = 0$, $v = 0$. Thus

$0 = 10^5\left(-\frac{1}{17}\right)(4) + C$ or $C = \dfrac{4(10^5)}{17}$,

and

$$v = 10^5\left(-\frac{1}{17}e^{-t}\right)(4\cos 4t + \sin 4t) + \frac{4(10)^5}{17}\Big|_{t=0.05} = 482\text{ V}.$$

7.8 Integration of Rational Functions

1. $\displaystyle\int \frac{dx}{x^2-4}$

$\displaystyle\frac{1}{x^2-4} = \frac{1}{(x-2)(x+2)} = \frac{A}{x-2} + \frac{B}{x+2} = \frac{A(x+2)+B(x-2)}{(x-2)(x+2)}$

$A(x+2)+B(x-2) = 1$

$\underline{x=-2}:\ 0 + B(-4) = 1$ or $B = -\frac{1}{4}$.

$\underline{x=2}:\ A(4)+0 = 1$ or $A = \frac{1}{4}$.

$$\begin{aligned}\int \frac{dx}{x^2-4} &= \int \left(\frac{1}{4}\frac{1}{x-2} - \frac{1}{4}\frac{1}{x+2}\right) dx \\ &= \frac{1}{4}\ln|x-2| - \frac{1}{4}\ln|x+2| + C = \frac{1}{4}\ln\left|\frac{x-2}{x+2}\right| + C.\end{aligned}$$

3. $\displaystyle\int \frac{3x+4}{2x^2-3x-2}\,dx$

$\displaystyle\frac{3x+4}{(x-2)(2x+1)} = \frac{A}{x-2} + \frac{B}{2x+1} = \frac{A(2x+1)+B(x-2)}{(x-2)(2x+1)}$

$A(2x+1)+B(x-2) = 3x+4$

$\underline{x=-\frac{1}{2}}:\ 0 - \frac{5}{2}B = \frac{5}{2}$ or $B = -1$.

$\underline{x=2}:\ 5A + 0 = 10$ or $A = 2$.

$$\begin{aligned}\int \frac{3x+4}{2x^2-3x-2}\,dx &= \int \left(\frac{2}{x-2} + \frac{-1}{2x+1}\right) dx \qquad u = 2x+1;\, dx = 2\,dx \\ &= 2\ln|x-2| - \frac{1}{2}\ln|2x+1| + C.\end{aligned}$$

5. $\displaystyle\int \frac{x+4}{x-2}\,dx$. Since the integrand is not a proper fraction, we must perform the long division:

$\displaystyle\frac{x+4}{x-2} = \frac{x-2+6}{x-2} = \frac{x-2}{x-2} + \frac{6}{x-2} = 1 + \frac{6}{x-2}$

$\displaystyle\int \frac{x+4}{x-2}\,dx = \int \left(1 + \frac{6}{x-2}\right) dx = x + 6\ln|x-2| + C.$

7. $\displaystyle\int \frac{x^3\,dx}{x^2-2x-3}$

Since the integrand is not a proper fraction, we must first perform the long division:

$$\begin{array}{r|l} & x \ +2 + \dfrac{7x+6}{x^2-2x-3} \\ x^2-2x-3 & x^3 \\ & \underline{x^3 - 2x^2 - 3x} \\ & \quad 2x^2 + 3x \\ & \quad \underline{2x^2 - 4x - 6} \\ & \qquad 7x + 6 \end{array}$$

So the integrand becomes: $\displaystyle\frac{x^3}{x^2-2x-3} = x + 2 + \frac{7x+6}{x^2-2x-3}$ and we proceed to split up the fraction:

$\displaystyle\frac{7x+6}{x^2-2x-3} = \frac{7x+6}{(x-3)(x+1)} = \frac{A}{x-3} + \frac{B}{x+1} = \frac{A(x+1)+B(x-3)}{(x-3)(x+1)}.$

$A(x+1)+B(x-3) = 7x+6$

$\underline{x=-1}:\ 0 + B(-4) = -1$ or $B = \frac{1}{4}$.

$\underline{x=3}:\ A(4) + 0 = 27$ or $A = \frac{27}{4}$.

$$\int\left(x+2+\frac{27}{4}\frac{1}{x-3}+\frac{1}{4}\frac{1}{x+1}\right)dx=\frac{1}{2}x^2+2x+\frac{27}{4}\ln|x-3|+\frac{1}{4}\ln|x+1|+C.$$

9. $\displaystyle\int\frac{x^2+10x-20}{(x-4)(x-1)(x+2)}\,dx$

$$\frac{x^2+10x-20}{(x-4)(x-1)(x+2)}=\frac{A}{x-4}+\frac{B}{x-1}+\frac{C}{x+2}$$
$$=\frac{A(x-1)(x+2)+B(x-4)(x+2)+C(x-4)(x-1)}{(x-4)(x-1)(x+2)}$$

$A(x-1)(x+2)+B(x-4)(x+2)+C(x-4)(x-1)=x^2+10x-20$

$\underline{x=1}$: $0+B(-3)(3)+0=-9$ or $B=1$.

$\underline{x=-2}$: $0+0+C(-6)(-3)=-36$ or $C=-2$.

$\underline{x=4}$: $A(3)(6)+0+0=36$ or $A=2$.

$$\int\frac{x^2+10x-20}{(x-4)(x-1)(x+2)}\,dx=\int\left(\frac{2}{x-4}+\frac{1}{x-1}+\frac{-2}{x+2}\right)dx$$
$$=2\ln|x-4|+\ln|x-1|-2\ln|x+2|+C$$
$$=\ln|x-4|^2+\ln|x-1|-\ln|x+2|^2+C$$
$$=\ln\left|\frac{(x-4)^2(x-1)}{(x+2)^2}\right|+C.$$

11. $\displaystyle\int\frac{3x+5}{(x+1)^2}\,dx$

$$\frac{3x+5}{(x+1)^2}=\frac{A}{x+1}+\frac{B}{(x+1)^2}=\frac{A}{x+1}\frac{x+1}{x+1}+\frac{B}{(x+1)^2}=\frac{A(x+1)+B}{(x+1)^2}$$

$A(x+1)+B=3x+5$

$\underline{x=-1}$: $0+B=2$ or $B=2$.

Thus $A(x+1)+2=3x+5$. Now let $x=$ any value (such as $x=0$):

$\underline{x=0}$: $A(1)+2=5$ or $A=3$.

$$\int\frac{3x+5}{(x+1)^2}\,dx=\int\left(\frac{3}{x+1}+\frac{2}{(x+1)^2}\right)dx=\int\left(\frac{3}{x+1}+2(x+1)^{-2}\right)dx$$
$$=3\ln|x+1|+2\frac{(x+1)^{-1}}{-1}+C=3\ln|x+1|-\frac{2}{x+1}+C.$$

13. $\displaystyle\int\frac{-2x^2+9x-7}{(x-2)^2(x+1)}\,dx$

$$\frac{-2x^2+9x-7}{(x-2)^2(x+1)}=\frac{A}{x-2}+\frac{B}{(x-2)^2}+\frac{C}{x+1}$$
$$=\frac{A(x-2)(x+1)+B(x+1)+C(x-2)^2}{(x-2)^2(x+1)}$$

$A(x-2)(x+1)+B(x+1)+C(x-2)^2=-2x^2+9x-7$

$\underline{x=2}$: $0+B(3)+0=3$ or $B=1$.

$\underline{x=-1}$: $0+0+C(9)=-18$ or $C=-2$.

Using the values obtained for B and C, we get

$A(x-2)(x+1)+1(x+1)-2(x-2)^2=-2x^2+9x-7$.

Now let $x=$ any value (such as $x=0$):

$\underline{x=0}$: $A(-2)(1)+(1)-2(4)=-7$ or $A=0$.

$$\int\left(\frac{1}{(x-2)^2}+\frac{-2}{x+1}\right)dx=\int\left[(x-2)^{-2}-\frac{2}{x+1}\right]dx \qquad u=x-2;\ du=dx.$$
$$=\frac{(x-2)^{-1}}{-1}-2\ln|x+1|+C$$
$$=-\frac{1}{x-2}-2\ln|x+1|+C.$$

15. $$\frac{2x^2+1}{(x-2)^3} = \frac{A}{x-2} + \frac{B}{(x-2)^2} + \frac{C}{(x-2)^3}$$
$$= \frac{A}{x-2}\frac{(x-2)^2}{(x-2)^2} + \frac{B}{(x-2)^2}\frac{x-2}{x-2} + \frac{C}{(x-2)^3} = \frac{A(x-2)^2 + B(x-2) + C}{(x-2)^3}.$$

$A(x-2)^2 + B(x-2) + C = 2x^2 + 1$

$\underline{x=2}$: $C = 9$.

Thus

$A(x-2)^2 + B(x-2) + 9 = 2x^2 + 1$

$A(x-2)^2 + B(x-2) = 2x^2 - 8$.

Now let $x =$ any two values (such as $x = 0$ and $x = 1$):

$\underline{x=0}$: $A(4) + B(-2) = -8$.

$\underline{x=1}$: $A(1) + B(-1) = -6$.

$2A - B = -4$ (dividing by 2)

$A - B = -6$

$A = 2$ (subtracting)

$B = A + 6 = 8$.

$$\int \frac{2x^2+1}{(x-2)^3}\,dx = \int \left(\frac{2}{x-2} + \frac{8}{(x-2)^2} + \frac{9}{(x-2)^3}\right) dx$$
$$= \int \left[\frac{2}{x-2} + 8(x-2)^{-2} + 9(x-2)^{-3}\right] dx$$
$$= 2\ln|x-2| + 8\frac{(x-2)^{-1}}{-1} + 9\frac{(x-2)^{-2}}{-2} + C$$
$$= 2\ln|x-2| - \frac{8}{x-2} - \frac{9}{2(x-2)^2} + C.$$

17. $$\int \frac{x^2-3x-2}{(x+2)(x^2+4)}\,dx$$

$$\frac{x^2-3x-2}{(x+2)(x^2+4)} = \frac{A}{x+2} + \frac{Bx+C}{x^2+4} = \frac{A(x^2+4) + (Bx+C)(x+2)}{(x+2)(x^2+4)}$$

$A(x^2+4) + (Bx+C)(x+2) = x^2 - 3x - 2$

$\underline{x=-2}$: $A(8) + 0 = 8$ or $A = 1$.

$1(x^2+4) + (Bx+C)(x+2) = x^2 - 3x - 2$ $\qquad A = 1$

Now let $x = 0$ and solve for C:

$(4) + C(2) = -2$ or $C = -3$. $\qquad x = 0$

Since $A = 1$ and $C = -3$, we get

$1(x^2+4) + (Bx-3)(x+2) = x^2 - 3x - 2$.

At this point we may let x be equal to any value (such as $x = 1$):

$5 + (B-3)(3) = -4$ or $B = 0$.

So the integral becomes

$$\int \left(\frac{1}{x+2} + \frac{-3}{x^2+4}\right) dx = \ln|x+2| - 3\cdot\frac{1}{2}\text{Arctan}\,\frac{x}{2} + C = \ln|x+2| - \frac{3}{2}\text{Arctan}\,\frac{x}{2} + C.$$

19. $\dfrac{3x^2+4x+3}{(x+1)(x^2+1)} = \dfrac{A}{x+1} + \dfrac{Bx+C}{x^2+1} = \dfrac{A(x^2+1)+(Bx+C)(x+1)}{(x+1)(x^2+1)}$

$A(x^2+1)+(Bx+C)(x+1) = 3x^2+4x+3$

$\underline{x=-1}:\ A(2) = 2$ or $A = 1$.

Thus $1(x^2+1)+(Bx+C)(x+1) = 3x^2+4x+3$.

Let $x = 0$ and solve for C:

$1 + C(1) = 3$ or $C = 2$.

$1(x^2+1)+(Bx+2)(x+1) = 3x^2+4x+3$.

Now let $x =$ any value (such as $x = 1$):

$\underline{x=1}:\ 2+(B+2)(2) = 10$ or $B = 2$.

$$\begin{aligned}\int \frac{3x^2+4x+3}{(x+1)(x^2+1)}\,dx &= \int\left(\frac{1}{x+1}+\frac{2x+2}{x^2+1}\right)dx\\ &= \int\left(\frac{1}{x+1}+\frac{2x}{x^2+1}+2\cdot\frac{1}{x^2+1}\right)dx\\ &= \ln|x+1| + \ln|x^2+1| + 2\text{Arctan}\,x + C\\ &= \ln|(x+1)(x^2+1)| + 2\text{Arctan}\,x + C.\end{aligned}$$

21. $\displaystyle\int \frac{x^5\,dx}{(x^2+4)^2} = \int \frac{x^5\,dx}{x^4+8x^2+16}$.

Since the fraction is not a proper fraction, we must first perform the long division:

$$\begin{array}{r|l} & x\\ x^4+8x^2+16 & x^5\\ & \underline{x^5+8x^3+16x}\\ & -8x^3-16x\end{array}$$

The integrand can therefore be written as: $x + \dfrac{-8x^3-16x}{(x^2+4)^2}$.

The fractional part has the form

$\dfrac{-8x^3-16x}{(x^2+4)^2} = \dfrac{Ax+B}{x^2+4} + \dfrac{Cx+D}{(x^2+4)^2} = \dfrac{(Ax+B)(x^2+4)+(Cx+D)}{(x^2+4)^2}$.

Equating numerators:

$$\begin{aligned}(Ax+B)(x^2+4)+(Cx+D) &= -8x^3-16x\\ Ax^3+Bx^2+4Ax+4B+Cx+D &= -8x^3-16x\\ Ax^3+Bx^2+(4A+C)x+(4B+D) &= -8x^3-16x.\end{aligned}$$

Comparing coefficients:

$A=-8,\ B=0,\ 4A+C=-16,\ 4B+D=0$, whence $C = 16$ and $D = 0$. Thus

$$\begin{aligned}\int \frac{x^5\,dx}{(x^2+4)^2} &= \int\left(x+\frac{-8x}{x^2+4}+\frac{16x}{(x^2+4)^2}\right)dx\\ &= \int x\,dx - 8\int\frac{x\,dx}{x^2+4} + 16\int (x^2+4)^{-2}x\,dx\\ &= \int x\,dx - \frac{8}{2}\int\frac{2x\,dx}{x^2+4} + \frac{16}{2}\int (x^2+4)^{-2}(2x\,dx)\end{aligned}$$

$u = x^2+4;\ du = 2x\,dx$

$$\begin{aligned}&= \frac{1}{2}x^2 - 4\ln(x^2+4) + 8\frac{(x^2+4)^{-1}}{-1} + C\\ &= \frac{1}{2}x^2 - 4\ln(x^2+4) - \frac{8}{x^2+4} + C.\end{aligned}$$

23. $\dfrac{x^2-3x+5}{x(x^2-2x+5)} = \dfrac{A}{x} + \dfrac{Bx+C}{x^2-2x+5} = \dfrac{A(x^2-2x+5)+(Bx+C)x}{x(x^2-2x+5)}$

$A(x^2-2x+5)+(Bx+C)x = x^2-3x+5$

$\underline{x=0}$: $A(5)=5$ or $A=1$.

$1(x^2-2x+5)+(Bx+C)x = x^2-3x+5$.

Now let $x =$ any two values (such as $x=1$ and $x=-1$):

$\underline{x=1}$: $4+(B+C)(1)=3$

$\underline{x=-1}$: $8+(-B+C)(-1)=9$.

$$\begin{aligned} B+C &= -1 \\ B-C &= 1 \\ \hline 2B \quad &= 0, \quad B=0 \\ C &= -1. \end{aligned}$$

$$\int\left(\frac{1}{x}+\frac{-1}{x^2-2x+5}\right)dx = \int\left(\frac{1}{x}+\frac{-1}{(x-1)^2+4}\right)dx \qquad u=x-1;\ du=dx.$$

$$= \ln|x| - \frac{1}{2}\text{Arctan}\,\frac{1}{2}(x-1)+C.$$

25. $\displaystyle\int\frac{dx}{x(x^2+2x+2)}$

$\dfrac{1}{x(x^2+2x+2)} = \dfrac{A}{x}+\dfrac{Bx+C}{x^2+2x+2} = \dfrac{A(x^2+2x+2)+(Bx+C)x}{x(x^2+2x+2)}$

$$\begin{aligned} A(x^2+2x+2)+(Bx+C)x &= 1 \\ Ax^2+2Ax+2A+Bx^2+Cx &= 1 \\ (A+B)x^2+(2A+C)x+2A &= 1 \end{aligned}$$

Comparing coefficients:

$$\begin{aligned} A+B &= 0 \quad \text{coefficients of } x^2 \\ 2A+C &= 0 \quad \text{coefficients of } x \\ 2A &= 1 \quad \text{constants} \end{aligned}$$

Solution set: $A=\frac{1}{2}$, $B=-\frac{1}{2}$, $C=-1$.

$$\int\frac{dx}{x(x^2+2x+2)} = \int\left(\frac{1}{2}\frac{1}{x}+\frac{-\frac{1}{2}x-1}{x^2+2x+2}\right)dx = \frac{1}{2}\int\frac{1}{x}\,dx - \frac{1}{2}\int\frac{x+2}{x^2+2x+2}\,dx.$$

For the second integral, if $u=x^2+2x+2$, then $du=(2x+2)\,dx$. So the integral has to be split as follows:

$$\begin{aligned} \frac{1}{2}\int\frac{1}{x}\,dx - \frac{1}{2}\cdot\frac{1}{2}\int\frac{2(x+2)}{x^2+2x+2}\,dx &= \frac{1}{2}\int\frac{1}{x}\,dx - \frac{1}{4}\int\frac{2x+4}{x^2+2x+2}\,dx \\ &= \frac{1}{2}\int\frac{1}{x}\,dx - \frac{1}{4}\int\frac{(2x+2)+2}{x^2+2x+2}\,dx \\ &= \frac{1}{2}\int\frac{1}{x}\,dx - \frac{1}{4}\int\frac{2x+2}{x^2+2x+2}\,dx - \frac{1}{4}\int\frac{2\,dx}{x^2+2x+2} \\ &= \frac{1}{2}\int\frac{1}{x}\,dx - \frac{1}{4}\int\frac{2x+2}{x^2+2x+2}\,dx - \frac{1}{2}\int\frac{dx}{(x+1)^2+1} \\ &= \frac{1}{2}\ln|x| - \frac{1}{4}\ln|x^2+2x+2| - \frac{1}{2}\text{Arctan}\,(x+1)+C. \end{aligned}$$

7.9 Integration by Use of Tables

1. Formula 5; $a=2$, $b=1$.

$$\int\frac{dx}{x(2+x)} = \frac{1}{2}\ln\left|\frac{x}{2+x}\right|+C.$$

3. Formula 27; $a = \sqrt{7},\ a^2 = 7$.
$$\int \sqrt{x^2 - 7}\,dx = \frac{1}{2}\left(x\sqrt{x^2-7} - 7\ln|x + \sqrt{x^2-7}|\right) + C.$$

5. Formula 16; $a = \sqrt{5}$.
$$\int \frac{dx}{5 - x^2} = \frac{1}{2\sqrt{5}} \ln\left|\frac{\sqrt{5}+x}{\sqrt{5}-x}\right| + C.$$

7. $$\int \frac{dx}{x^2\sqrt{5x^2+4}} = \int \frac{dx}{x\sqrt{(\sqrt{5}x)^2+4}} \qquad u = \sqrt{5}x;\ du = \sqrt{5}\,dx.$$
$$= \sqrt{5}\int \frac{\sqrt{5}\,dx}{(\sqrt{5}x)(\sqrt{5}x)\sqrt{(\sqrt{5}x)^2+4}} = \sqrt{5}\int \frac{du}{u^2\sqrt{u^2+4}}.$$
By Formula 35 with $a = 2$, we get
$$\sqrt{5}\left(-\frac{\sqrt{u^2+4}}{4u}\right) + C = \sqrt{5}\left(-\frac{\sqrt{5x^2+4}}{4(\sqrt{5}x)}\right) + C = -\frac{\sqrt{5x^2+4}}{4x} + C.$$

9. Formula 61; $m = 2,\ n = 1$.
$$\int \sin 2x \sin x\,dx = -\frac{\sin(2+1)x}{2(2+1)} + \frac{\sin(2-1)x}{2(2-1)} + C$$
$$= -\frac{1}{6}\sin 3x + \frac{1}{2}\sin x + C.$$

11. $$\int \frac{dx}{\sqrt{3x^2+5}} = \int \frac{dx}{\sqrt{(\sqrt{3}x)^2+5}} \qquad u = \sqrt{3}x;\ du = \sqrt{3}\,dx.$$
$$= \frac{1}{\sqrt{3}}\int \frac{\sqrt{3}\,dx}{\sqrt{(\sqrt{3}x)^2+5}} = \frac{1}{\sqrt{3}}\int \frac{du}{\sqrt{u^2+5}}.$$
By Formula 32 with $a = \sqrt{5}$, we get
$$\frac{1}{\sqrt{3}}\ln\left|u + \sqrt{u^2+5}\right| + C = \frac{1}{\sqrt{3}}\ln\left|\sqrt{3}x + \sqrt{3x^2+5}\right| + C.$$

13. Formula 42; $n = 2,\ a = 2$.
$$\int x^2 e^{2x}\,dx = \frac{x^2 e^{2x}}{2} - \frac{2}{2}\int x e^{2x}\,dx = \frac{1}{2}x^2 e^{2x} - \int x e^{2x}\,dx.$$
Now by formula 41 with $a = 2$, we get:
$$\frac{1}{2}x^2 e^{2x} - \left[\frac{e^{2x}}{4}(2x-1)\right] + C = \frac{1}{2}x^2 e^{2x} - \frac{1}{2}x e^{2x} + \frac{1}{4}e^{2x} + C.$$

15. Formula 69
$$\int \tan^6 x\,dx = \frac{\tan^5 x}{5} - \int \tan^4 x\,dx$$
$$= \frac{1}{5}\tan^5 x - \left[\frac{\tan^3 x}{3} - \int \tan^2 x\,dx\right]$$
$$= \frac{1}{5}\tan^5 x - \frac{1}{3}\tan^3 x + \int (\sec^2 x - 1)\,dx$$
$$= \frac{1}{5}\tan^5 x - \frac{1}{3}\tan^3 x + \tan x - x + C.$$

17. $$\int \frac{dx}{4x^2 - 9} = \int \frac{dx}{(2x)^2 - 9} = \frac{1}{2}\int \frac{du}{u^2 - 9}. \qquad u = 2x;\ du = 2\,dx.$$
By Formula 17 with $a = 3$, we get:
$$\frac{1}{2}\cdot\frac{1}{6}\ln\left|\frac{u-3}{u+3}\right| + C = \frac{1}{12}\ln\left|\frac{2x-3}{2x+3}\right| + C.$$

19. Formula 12; $a = 3 > 0,\ b = 1$.
$$\int \frac{dx}{x\sqrt{3+x}} = \frac{1}{\sqrt{3}} \ln \left| \frac{\sqrt{3+x} - \sqrt{3}}{\sqrt{3+x} + \sqrt{3}} \right| + C.$$

21. $$\int \frac{dx}{3x^2 - 5} = \int \frac{dx}{(\sqrt{3}x)^2 - 5} = \frac{1}{\sqrt{3}} \int \frac{\sqrt{3}\,dx}{(\sqrt{3}x)^2 - 5} = \frac{1}{\sqrt{3}} \int \frac{du}{u^2 - 5}. \qquad u = \sqrt{3}x;\ du = \sqrt{3}\,dx.$$
By formula 17 with $a = \sqrt{5}$, we get:
$$\frac{1}{\sqrt{3}} \frac{1}{2\sqrt{5}} \ln \left| \frac{u - \sqrt{5}}{u + \sqrt{5}} \right| + C = \frac{1}{2\sqrt{15}} \ln \left| \frac{\sqrt{3}x - \sqrt{5}}{\sqrt{3}x + \sqrt{5}} \right| + C.$$

23. Formula 29; $a = \sqrt{10}$.
$$\int \frac{\sqrt{x^2 - 10}}{x}\,dx = \sqrt{x^2 - 10} - \sqrt{10}\text{Arccos}\,\frac{\sqrt{10}}{x} + C.$$

25. Formula 63, $m = 3,\ n = 2$.
$$\int \sin 3x \cos 2x\,dx = -\frac{\cos(3+2)x}{2(3+2)} - \frac{\cos(3-2)x}{2(3-2)} = -\frac{1}{10}\cos 5x - \frac{1}{2}\cos x + C.$$

27. $$\int \frac{dx}{(4x^2+5)^{3/2}} = \int \frac{dx}{[(2x)^2+5]^{3/2}} = \frac{1}{2} \int \frac{2\,dx}{[(2x)^2+5]^{3/2}} = \frac{1}{2} \int \frac{du}{(u^2+5)^{3/2}}. \qquad u = 2x;\ du = 2\,dx.$$
By Formula 37 with $a = \sqrt{5}$, we get
$$\frac{1}{2} \frac{u}{5\sqrt{u^2+5}} + C = \frac{1}{2} \frac{2x}{5\sqrt{4x^2+5}} + C = \frac{x}{5\sqrt{4x^2+5}} + C.$$

29. $\displaystyle\int \frac{\sin\sqrt{x}}{\sqrt{x}}$. Formula 47 with $u = \sqrt{x}$; $du = \frac{1}{2}x^{-1/2}\,dx$:
$$2 \int \sin x^{1/2} \left(\frac{1}{2} x^{-1/2}\,dx \right) = 2 \int \sin u\,du = -2\cos u + C = -2\cos\sqrt{x} + C.$$

31. $\displaystyle\int e^u\,du$ with $u = \text{Arctan}\,x^2$; $du = \dfrac{2x\,dx}{1+x^4}$.

33. $\displaystyle\int \sec^3 5x\,dx$. Reduction Formula 71 with $n = 3$: let $u = 5x;\ du = 5\,dx$.

$$\begin{aligned} \frac{1}{5} \int \sec^3 u\,du &= \frac{1}{5} \left[\frac{\sec u \tan u}{2} + \frac{1}{2} \int \sec u\,du \right] \\ &= \frac{1}{10} \sec u \tan u + \frac{1}{10} \ln|\sec u + \tan u| + C \\ &= \frac{1}{10} \sec 5x \tan 5x + \frac{1}{10} \ln|\sec 5x + \tan 5x| + C. \end{aligned}$$

Chapter 7 Review

1. $\displaystyle\int \frac{x\,dx}{x^2+1} = \frac{1}{2}\int \frac{2x\,dx}{x^2+1}$ $u = x^2+1$; $du = 2x\,dx$.

$$= \frac{1}{2}\ln|u| + C$$
$$= \frac{1}{2}\ln(x^2+1) + C.$$

3. $\displaystyle\int \frac{2\,dx}{x^2+1} = 2\int \frac{dx}{x^2+1} = 2\text{Arctan}\,x + C.$

5. $\displaystyle\int \frac{t\,dt}{\sqrt{9-t^2}} = \int (9-t^2)^{-1/2} t\,dt$ $u = 9-t^2$; $du = -2t\,dt$.

$$= -\frac{1}{2}\int (9-t^2)^{-1/2}(-2t)\,dt$$
$$= -\frac{1}{2}\frac{(9-t^2)^{1/2}}{1/2} + C$$
$$= -\sqrt{9-t^2} + C.$$

7. $\displaystyle\int x\cos 2x^2\,dx = \int \cos 2x^2 \cdot x\,dx$ $u = 2x^2$; $du = 4x\,dx$.

$$= \frac{1}{4}\int \cos 2x^2 (4x\,dx) = \frac{1}{4}\int \cos u\,du$$
$$= \frac{1}{4}\sin u + C = \frac{1}{4}\sin 2x^2 + C.$$

9. $\displaystyle\int \frac{e^x\,dx}{4+e^{2x}} = \int \frac{e^x\,dx}{4+(e^x)^2}.$ $u = e^x$; $du = e^x\,dx$.

$$\int \frac{du}{4+u^2} = \frac{1}{2}\text{Arctan}\,\frac{u}{2} + C = \frac{1}{2}\text{Arctan}\,\frac{1}{2}e^x + C.$$

11. $\displaystyle\int \frac{e^x\,dx}{(4+e^x)^2} = \int (4+e^x)^{-2}e^x\,dx$ $u = 4+e^x$; $du = e^x\,dx$.

$$= \int u^{-2}\,du = \frac{u^{-1}}{-1} + C = \frac{-1}{u} + C = \frac{-1}{4+e^x} + C.$$

13. $\displaystyle\int \frac{\ln x}{x}\,dx$ $u = \ln x$; $du = \frac{1}{x}\,dx$.

$$\int (\ln x)\frac{1}{x}\,dx = \int u\,du = \frac{1}{2}u^2 + C = \frac{1}{2}\ln^2 x + C.$$

15. $\displaystyle\int \frac{x+2}{x^2+4x+5}\,dx$

Let $u = x^2+4x+5$; then $du = (2x+4)\,dx = 2(x+2)\,dx$.

$$\frac{1}{2}\int \frac{2(x+2)\,dx}{x^2+4x+5} = \frac{1}{2}\int \frac{du}{u} = \frac{1}{2}\ln|u| + C = \frac{1}{2}\ln|x^2+4x+5| + C.$$

17. $\displaystyle\int \frac{dx}{x^2+4x+4} = \int \frac{dx}{(x+2)^2}$ $u = x+2$; $du = dx$.

$$= \int (x+2)^{-2}\,dx = \frac{(x+2)^{-1}}{-1} + C$$
$$= -\frac{1}{x+2} + C.$$

19. $\int \ln^2 x\,dx$

$$u = (\ln x)^2 \qquad dv = dx$$
$$du = 2(\ln x)\frac{1}{x}\,dx \qquad v = x$$

$$\int (\ln x)^2\,dx = uv - \int v\,du = x(\ln x)^2 - \int x \cdot 2(\ln x)\frac{1}{x}\,dx = x\ln^2 x - 2\int \ln x\,dx.$$

$$u = \ln x \qquad dv = dx$$
$$du = \frac{1}{x}\,dx \qquad v = x$$

$$\begin{aligned}\int \ln^2 x\,dx &= x\ln^2 x - 2\left(x\ln x - \int x \cdot \frac{1}{x}\,dx\right)\\ &= x\ln^2 x - 2x\ln x + 2\int dx = x\ln^2 x - 2x\ln x + 2x + C.\end{aligned}$$

21. $\int \frac{dx}{x\sqrt{4-x^2}}$. Let $x = 2\sin\theta$; then $dx = 2\cos\theta\,d\theta$.

$\sqrt{4-x^2} = \sqrt{4 - 4\sin^2\theta} = 2\sqrt{1-\sin^2\theta} = 2\sqrt{\cos^2\theta} = 2\cos\theta.$

$\int \frac{2\cos\theta\,d\theta}{(2\sin\theta)(2\cos\theta)} = \frac{1}{2}\int \csc\theta\,d\theta = \frac{1}{2}\ln|\csc\theta - \cot\theta| + C$

2

x

θ

$\sqrt{4-x^2}$

$\sin\theta = \frac{x}{2}$

$$\begin{aligned}&= \frac{1}{2}\ln\left|\frac{2}{x} - \frac{\sqrt{4-x^2}}{x}\right| + C\\ &= \frac{1}{2}\ln\left|\frac{2-\sqrt{4-x^2}}{x}\right| + C.\end{aligned}$$

23.
$$\begin{aligned}\int \sin^3 2x\,dx &= \int \sin^2 2x\,(\sin 2x\,dx)\\ &= \int \left(1-\cos^2 2x\right)(\sin 2x\,dx) \qquad u = \cos 2x;\ du = -2\sin 2x\,dx\\ &= -\frac{1}{2}\int \left(1-\cos^2 2x\right)(-2\sin 2x\,dx)\\ &= -\frac{1}{2}\int \left(1-u^2\right)\,du = -\frac{1}{2}\left(u - \frac{1}{3}u^3\right) + C\\ &= -\frac{1}{2}\cos 2x + \frac{1}{6}\cos^3 2x + C = \frac{1}{6}\cos^3 2x - \frac{1}{2}\cos 2x + C.\end{aligned}$$

25.
$$\begin{aligned}\int \sin^2 2x\,dx &= \frac{1}{2}\int (1-\cos 4x)\,dx = \frac{1}{2}\int dx - \frac{1}{2}\int \cos 4x\,dx\\ &= \frac{1}{2}x - \frac{1}{2}\left(\frac{1}{4}\sin 4x\right) + C \qquad u = 4x;\ du = 4\,dx\\ &= \frac{1}{2}x - \frac{1}{8}\sin 4x + C.\end{aligned}$$

27. $\int xe^{2x}\,dx$

$$\begin{aligned} u &= x & dv &= e^{2x}\,dx \\ du &= dx & v &= \frac{1}{2}e^{2x} \end{aligned}$$

$$\begin{aligned} uv - \int v\,du &= x\left(\frac{1}{2}e^{2x}\right) - \int \frac{1}{2}e^{2x}\,dx \\ &= \frac{1}{2}xe^{2x} - \frac{1}{2}\cdot\frac{1}{2}\int e^{2x}(2\,dx) \\ &= \frac{1}{2}xe^{2x} - \frac{1}{4}e^{2x} + C = \frac{1}{4}e^{2x}(2x-1) + C. \end{aligned}$$

29. $\int \text{Arctan}\,2x\,dx$

$$\begin{aligned} u &= \text{Arctan}\,2x & dv &= dx \\ du &= \frac{2\,dx}{1+4x^2} & v &= x \end{aligned}$$

(Integration by parts.)

$$\begin{aligned} x\text{Arctan}\,2x - \int \frac{2x\,dx}{1+4x^2} &= x\text{Arctan}\,2x - \frac{1}{4}\int \frac{8x\,dx}{1+4x^2} \qquad (u = 1+4x^2;\ du = 8x\,dx.) \\ &= x\text{Arctan}\,2x - \frac{1}{4}\ln(1+4x^2) + C. \end{aligned}$$

31.
$$\begin{aligned} \int \frac{e^{\tan x}}{\cos^2 x}\,dx &= \int e^{\tan x}\sec^2 x\,dx \qquad u = \tan x;\ du = \sec^2 x\,dx. \\ &= \int e^u\,du = e^u + C = e^{\tan x} + C. \end{aligned}$$

33. $(x^2-1) \div (x^2+3) = 1 - \dfrac{4}{x^2+3}$.

$$\int \frac{x^2-1}{x^2+3}\,dx = \int \left[1 - \frac{4}{x^2+3}\right]dx = x - \frac{4}{\sqrt{3}}\text{Arctan}\,\frac{x}{\sqrt{3}} + C.$$

35.
$$\begin{aligned} \int \frac{1-\cos\omega}{\sin\omega}\,d\omega &= \int \left(\frac{1}{\sin\omega} - \frac{\cos\omega}{\sin\omega}\right)d\omega \\ &= \int \csc\omega\,d\omega - \int \frac{\cos\omega\,d\omega}{\sin\omega} \qquad u = \sin\omega;\ du = \cos\omega\,d\omega. \\ &= \ln|\csc\omega - \cot\omega| - \int \frac{du}{u} = \ln|\csc\omega - \cot\omega| - \ln|\sin\omega| + C. \end{aligned}$$

This result can be rewritten as follows:

$$\begin{aligned} \ln|\csc\omega - \cot\omega| - \ln\frac{1}{|\csc\omega|} + C &= \ln|\csc\omega - \cot\omega| - \ln 1 + \ln|\csc\omega| + C \\ &= \ln|\csc\omega(\csc\omega - \cot\omega)| + C. \end{aligned}$$

Another form comes from

$$\begin{aligned} \csc^2\omega - \csc\omega\cot\omega &= \frac{1}{\sin^2\omega} - \frac{1}{\sin\omega}\frac{\cos\omega}{\sin\omega} = \frac{1-\cos\omega}{\sin^2\omega} = \frac{1-\cos\omega}{1-\cos^2\omega} \\ &= \frac{1-\cos\omega}{(1-\cos\omega)(1+\cos\omega)} = \frac{1}{1+\cos\omega} = (1+\cos\omega)^{-1}. \end{aligned}$$

The result is: $\ln|1+\cos\omega|^{-1} + C = -\ln|1+\cos\omega| + C$.

37. $$\begin{aligned}\int \cot^4 x \csc^4 x\,dx &= \int \cot^4 x \csc^2 x(\csc^2 x\,dx)\\ &= \int \cot^4 x(1+\cot^2 x)(\csc^2 x\,dx)\\ &= -\int \cot^4 x(1+\cot^2 x)(-\csc^2 x\,dx) \qquad u=\cot x;\ du=-\csc^2 x\,dx.\\ &= -\int u^4(1+u^2)\,du = -\frac{1}{5}u^5 - \frac{1}{7}u^7 + C\\ &= -\frac{1}{5}\cot^5 x - \frac{1}{7}\cot^7 x + C.\end{aligned}$$

39. $$\begin{aligned}\int \sec^3 2x \tan^3 2x\,dx &= \int \sec^2 2x \tan^2 2x(\sec 2x \tan 2x\,dx) \qquad \text{(odd power of tangent)}\\ &= \frac{1}{2}\int \sec^2 2x(\sec^2 2x - 1)(2\sec 2x \tan 2x\,dx) \qquad u=\sec 2x;\\ &= \frac{1}{2}\int u^2(u^2-1)\,du = \frac{1}{2}\left(\frac{1}{5}u^5 - \frac{1}{3}u^3\right) + C \qquad du = 2\sec 2x \tan 2x\,dx.\\ &= \frac{1}{10}\sec^5 2x - \frac{1}{6}\sec^3 2x + C.\end{aligned}$$

41. $\int \tan^2 x \sec^2 x\,dx \qquad u=\tan x;\ du=\sec^2 x\,dx.$

$$\int u^2\,du = \frac{1}{3}u^3 + C = \frac{1}{3}\tan^3 x + C.$$

43. $$\begin{aligned}\int \sec^4 3x\,dx &= \int \sec^2 3x \sec^2 3x\,dx \qquad \text{(even power of secant)}\\ &= \int (1+\tan^2 3x)\sec^2 3x\,dx \qquad u=\tan 3x;\ du=3\sec^2 3x\,dx.\\ &= \frac{1}{3}\int (1+\tan^2 3x)(3\sec^2 3x\,dx)\\ &= \frac{1}{3}\int (1+u^2)\,du = \frac{1}{3}\left(u+\frac{1}{3}u^3\right) + C\\ &= \frac{1}{3}\tan 3x + \frac{1}{9}\tan^3 3x + C.\end{aligned}$$

45. $\int e^x \cos 4x\,dx \qquad u=e^x \quad dv=\cos 4x\,dx; \quad du=e^x\,dx \quad v=\frac{1}{4}\sin 4x$

$$\int e^x \cos 4x\,dx = \frac{1}{4}e^x \sin 4x - \frac{1}{4}\int e^x \sin 4x\,dx.$$

$u=e^x \quad dv=\sin 4x\,dx; \quad du=e^x\,dx \quad v=-\frac{1}{4}\cos 4x$

$$\begin{aligned}\int e^x \cos 4x\,dx &= \frac{1}{4}e^x \sin 4x - \frac{1}{4}\left[-\frac{1}{4}e^x \cos 4x + \frac{1}{4}\int e^x \cos 4x\,dx\right]\\ &= \frac{1}{4}e^x \sin 4x + \frac{1}{16}e^x \cos 4x - \frac{1}{16}\int e^x \cos 4x\,dx.\end{aligned}$$

Solving for the integral:

$$\begin{aligned}\left(1+\frac{1}{16}\right)\int e^x \cos 4x\,dx &= \frac{1}{4}e^x \sin 4x + \frac{1}{16}e^x \cos 4x\\ \frac{17}{16}\int e^x \cos 4x\,dx &= \frac{1}{16}e^x(4\sin 4x + \cos 4x).\end{aligned}$$

We conclude that

$$\int e^x \cos 4x\,dx = \frac{1}{17}e^x(4\sin 4x + \cos 4x) + C.$$

47. $\displaystyle\int \frac{dx}{5x^2+4} = \frac{dx}{(\sqrt{5}x)^2+4}$ $\qquad u = \sqrt{5}x;\ du = \sqrt{5}\,dx.$

$$= \frac{1}{\sqrt{5}}\int \frac{\sqrt{5}\,dx}{(\sqrt{5}x)^2+4} = \frac{1}{\sqrt{5}}\int \frac{du}{u^2+4}$$

$$= \frac{1}{\sqrt{5}}\cdot\frac{1}{2}\text{Arctan}\,\frac{u}{2} + C = \frac{1}{2\sqrt{5}}\text{Arctan}\,\frac{\sqrt{5}}{2}x + C.$$

49. $\displaystyle\int \frac{x\,dx}{(x^2+9)^{3/2}}$. Let $u = x^2+9;\ du = 2x\,dx.$

$$\frac{1}{2}\int \frac{2x\,dx}{(x^2+9)^{3/2}} = \frac{1}{2}\int u^{-3/2}\,du = \frac{1}{2}\frac{u^{-1/2}}{-1/2} + C = -\frac{1}{\sqrt{x^2+9}} + C.$$

51. $\displaystyle\int \frac{x+1}{x^2+2x-8}\,dx = \frac{1}{2}\int \frac{2(x+1)}{x^2+2x-8}\,dx$ $\qquad u = x^2+2x-8;\ du = (2x+2)\,dx.$

$$= \frac{1}{2}\ln|x^2+2x-8| + C.$$

53. $\displaystyle\frac{3x^2-4x+9}{(x-2)(x^2+9)} = \frac{A}{x-2} + \frac{Bx+C}{x^2+9} = \frac{A(x^2+9)+(Bx+C)(x-2)}{(x-2)(x^2+9)}$

$A(x^2+9)+(Bx+C)(x-2) = 3x^2-4x+9$

$\underline{x=2}$: $A(13) = 13$ or $A = 1$.

$1(x^2+9)+(Bx+C)(x-2) = 3x^2-4x+9$. Now let $x = 0$ and find C:

$9 + C(-2) = 9$ or $C = 0$.

$1(x^2+9)+(Bx+0)(x-2) = 3x^2-4x+9$

Finally, let $x =$ any value (such as $x = 1$):

$\underline{x=1}$: $10 + B(-1) = 8$ or $B = 2$.

$$\int \frac{3x^2-4x+9}{(x-2)(x^2+9)}\,dx = \int\left(\frac{1}{x-2} + \frac{2x}{x^2+9}\right)dx$$

$$= \ln|x-2| + \ln|x^2+9| + C = \ln|(x-2)(x^2+9)| + C.$$

55. $\displaystyle\int \frac{9x}{(x-1)^2(x+2)}\,dx$

$$\frac{9x}{(x-1)^2(x+2)} = \frac{A}{x-1} + \frac{B}{(x-1)^2} + \frac{C}{x+2} = \frac{A(x-1)(x+2)+B(x+2)+C(x-1)^2}{(x-1)^2(x+2)}$$

$A(x-1)(x+2)+B(x+2)+C(x-1)^2 = 9x$

$\underline{x=1}$: $3B = 9$ or $B = 3$.

$\underline{x=-2}$: $9C = -18$ or $C = -2$

Using the values of B and C, we have $A(x-1)(x+2)+3(x+2)-2(x-1)^2 = 9x$.

Now let $x =$ any value (such as $x = 0$): $A(-2)+6-2 = 0$ and $A = 2$.

$$\int\left(\frac{2}{x-1} + \frac{3}{(x-1)^2} + \frac{-2}{x+2}\right)dx = 2\ln|x-1| - \frac{3}{x-1} - 2\ln|x+2| + C.$$

Chapter 8

Parametric Equations, Vectors, and Polar Coordinates

8.1 Vectors and Parametric Equations

1. $x = 3t,\ y = t + 1$. From the second equation, $t = y - 1$. Substituting in the first equation we get $x = 3(y - 1)$, and $x - 3y + 3 = 0$.

3.
$$\begin{array}{rcl} x &=& t^2 + 1 \\ y &=& t \\ \hline x &=& y^2 + 1 \quad \text{(substitution)} \\ y^2 &=& x - 1 \end{array}$$

5.
$$\begin{array}{rcl} x &=& \cos\theta + 1 \\ y &=& \sin\theta \\ \hline x - 1 &=& \cos\theta \\ y &=& \sin\theta \\ \hline (x-1)^2 &=& \cos^2\theta \\ y^2 &=& \sin^2\theta \\ \hline (x-1)^2 + y^2 &=& \cos^2\theta \ + \ \sin^2\theta \quad \text{(adding)} \\ (x-1)^2 + y^2 &=& 1 \end{array}$$

7.
$$\begin{array}{rcl} x &=& -3 - \sin t \\ y &=& 1 + \cos t \\ \hline x + 3 &=& -\sin t \\ y - 1 &=& \cos t \\ \hline (x+3)^2 &=& \sin^2 t \\ (y-1)^2 &=& \cos^2 t \\ \hline (x+3)^2 + (y-1)^2 &=& \sin^2 t + \cos^2 t \\ (x+3)^2 + (y-1)^2 &=& 1 \end{array}$$

9.
$$\begin{array}{rcl} x &=& \tan\theta \\ y &=& \sec^2\theta \\ \hline x^2 &=& \tan^2\theta \\ y &=& \sec^2\theta \\ \hline 1+x^2 &=& 1+\tan^2\theta \\ y &=& \sec\theta \\ \hline y &=& x^2+1 \qquad \text{since } 1+\tan^2\theta = \sec^2\theta \end{array}$$

11. From $x = \ln t$, we get $e^x = t$. Substituting in $y = t+2$, we have $y = e^x + 2$.

13. $x = 2t^2 + 3;\ v_x = 4t\Big|_{t=1} = 4.$
$y = 3t + 1;\ v_y = 3.$
Thus $\vec{v} = 4\vec{i} + 3\vec{j}$.

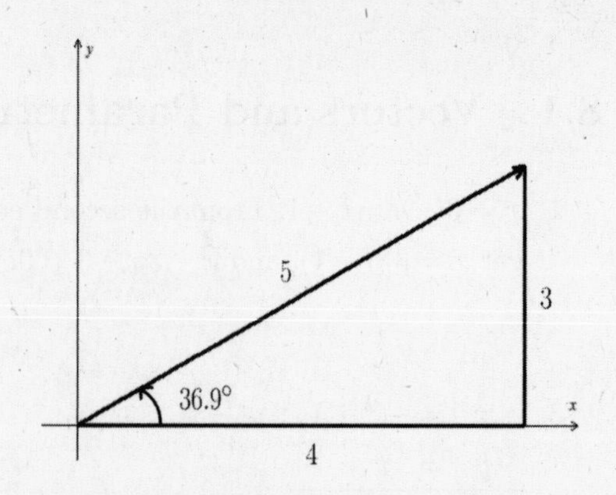

15. $x = t^2;\ v_x = 2t\Big|_{t=2} = 4.$
$y = 2 - t;\ v_y = -1.$
Thus $\vec{v} = 4\vec{i} - \vec{j}$.

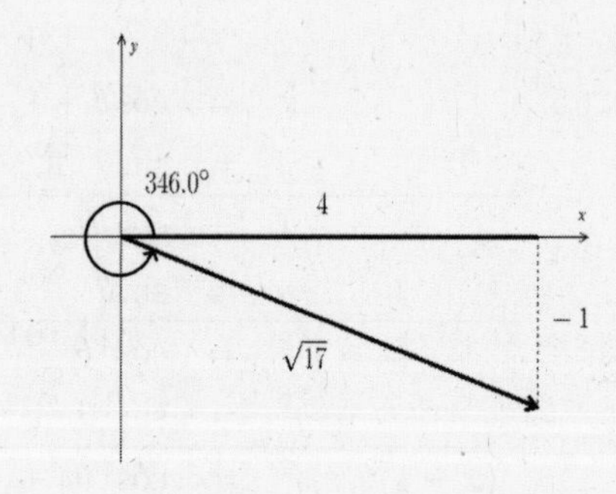

17. $x = t^2 - 2;\ v_x = 2t\Big|_{t=-1} = -2.$
$y = \frac{1}{3}t^3 + 2t + 1;\ v_y = t^2 + 2\Big|_{t=-1} = 3.$
Thus $\vec{v} = -2\vec{i} + 3\vec{j}$.

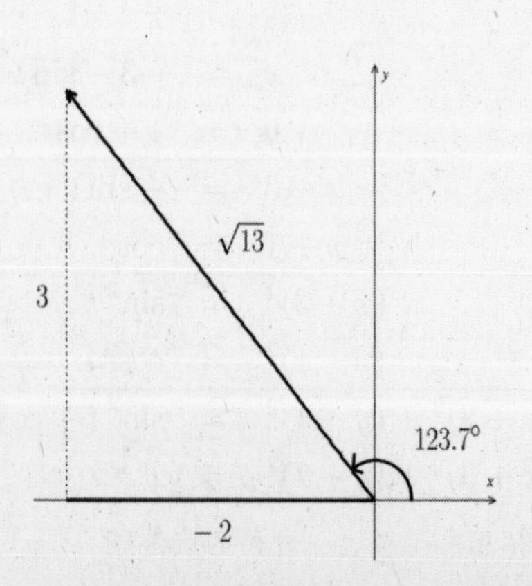

19. $x = (t-3)^2;\ v_x = 2(t-3)\Big|_{t=2} = -2.$
$y = 2t;\ v_y = 2.$
Thus $\vec{v} = -2\vec{i} + 2\vec{j}$.

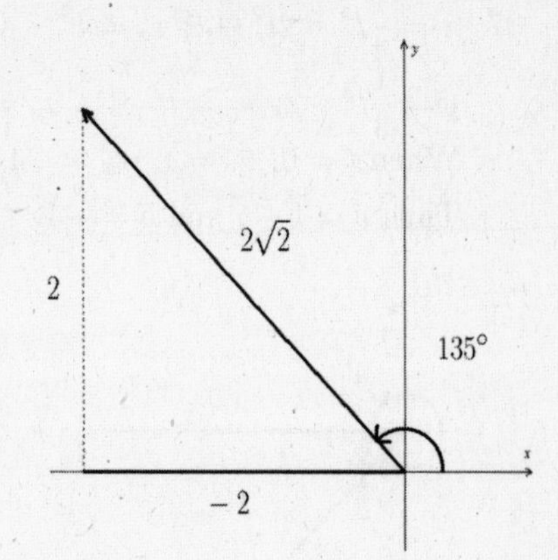

21. $x = e^t;\ v_x = e^t\Big|_{t=0} = 1.$
$y = e^{-t};\ v_y = -e^{-t}\Big|_{t=0} = -1.$
Thus $\vec{v} = \vec{i} - \vec{j}$.

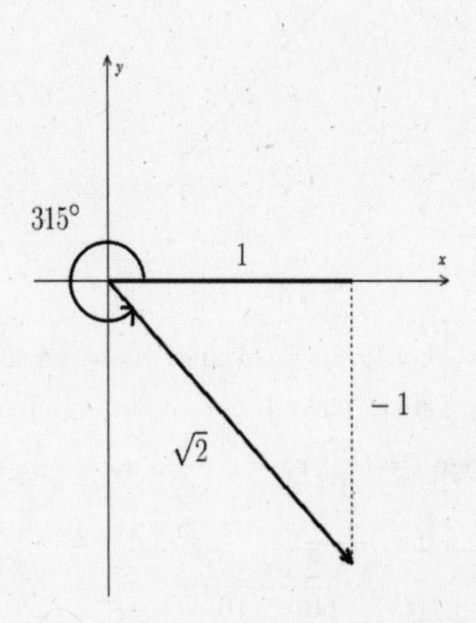

23. $x = \sec t;\ v_x = \sec t \tan t\Big|_{t=\pi/6} = \dfrac{2}{\sqrt{3}} \cdot \dfrac{1}{\sqrt{3}} = \dfrac{2}{3}.$
$y = 2\tan t;\ v_y = 2\sec^2 t\Big|_{t=\pi/6} = 2\left(\dfrac{2}{\sqrt{3}}\right)^2 = \dfrac{8}{3}.$
Thus $\vec{v} = \dfrac{2}{3}\vec{i} + \dfrac{8}{3}\vec{j}$.
$|\vec{v}| = \sqrt{\left(\dfrac{2}{3}\right)^2 + \left(\dfrac{8}{3}\right)^2} = \sqrt{\dfrac{4+64}{9}} = \dfrac{\sqrt{4\cdot 17}}{3} = \dfrac{2}{3}\sqrt{17};\ \theta = 76.0°.$

25. $x = 2t^2;\ v_x = 4t,\ a_x = 4.$
$y = \dfrac{1}{3}t^3;\ v_y = t^2,\ a_y = 2t.$
When $t = -1,\ v_x = -4,\ a_x = 4$ and $v_y = 1,\ a_y = -2.$
Thus $\vec{v} = -4\vec{i} + \vec{j}$ and $\vec{a} = 4\vec{i} - 2\vec{j}$.

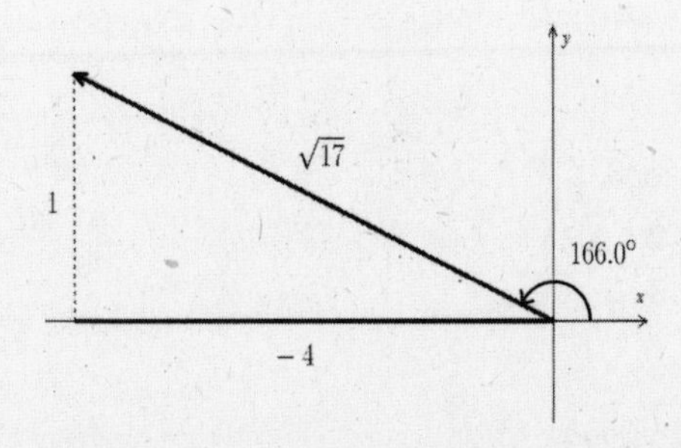

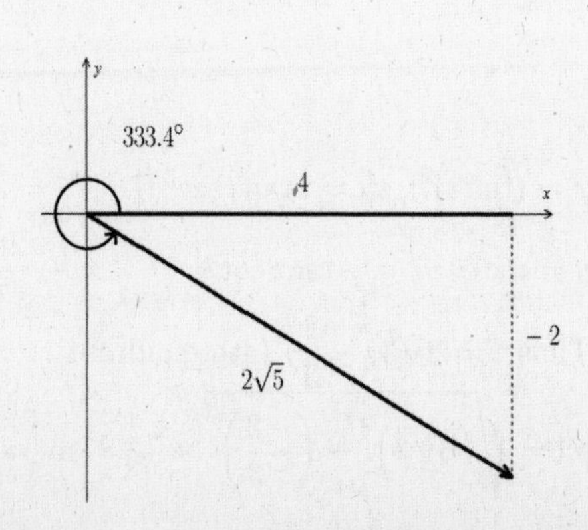

27. $x = \frac{1}{4}t^4 - 2t^2 + t;\ v_x = t^3 - 4t + 1,\ a_x = 3t^2 - 4.$

$y = \frac{1}{3}t^3 - t;\ v_y = t^2 - 1,\ a_y = 2t.$

When $t = 0$, $v_x = 1$, $a_x = -4$ and $v_y = -1$, $a_y = 0$.

Thus $\vec{v} = \vec{\imath} - \vec{\jmath}$ and $\vec{a} = -4\vec{\imath}$.

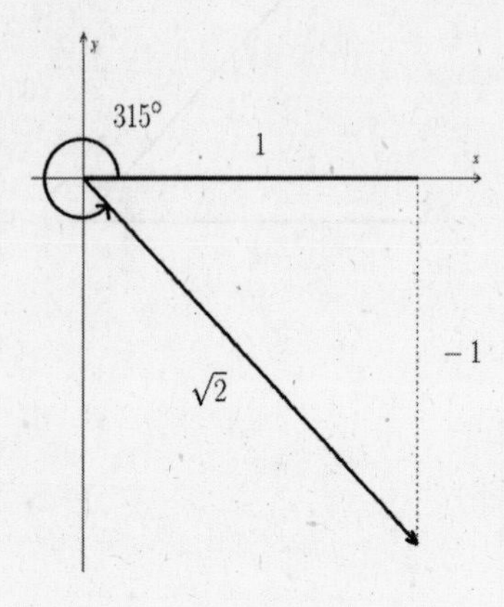

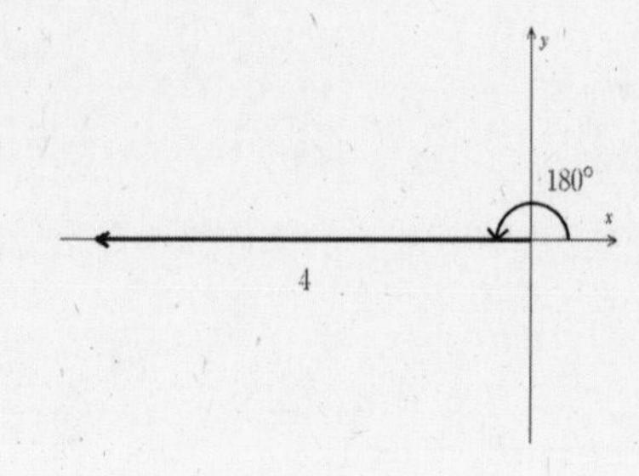

29. $x = 4\cos t;\ v_x = -4\sin t,\ a_x = -4\cos t.$

$y = 4\sin t;\ v_y = 4\cos t,\ a_y = -4\sin t.$

When $t = \frac{3\pi}{4}$, $v_x = -\frac{4}{\sqrt{2}}$, $a_x = \frac{4}{\sqrt{2}}$ and $v_y = -\frac{4}{\sqrt{2}}$, $a_y = -\frac{4}{\sqrt{2}}$.

Thus $\vec{v} = -\frac{4}{\sqrt{2}}\vec{\imath} - \frac{4}{\sqrt{2}}\vec{\jmath}$ and $\vec{a} = \frac{4}{\sqrt{2}}\vec{\imath} - \frac{4}{\sqrt{2}}\vec{\jmath}$.

$|\vec{v}| = |\vec{a}| = \sqrt{\frac{16}{2} + \frac{16}{2}} = \sqrt{16} = 4.$

The direction of $\vec{v}$ is 225° and the direction of $\vec{a}$ is 315°. The curve is a circle of radius 4 centered at the origin.

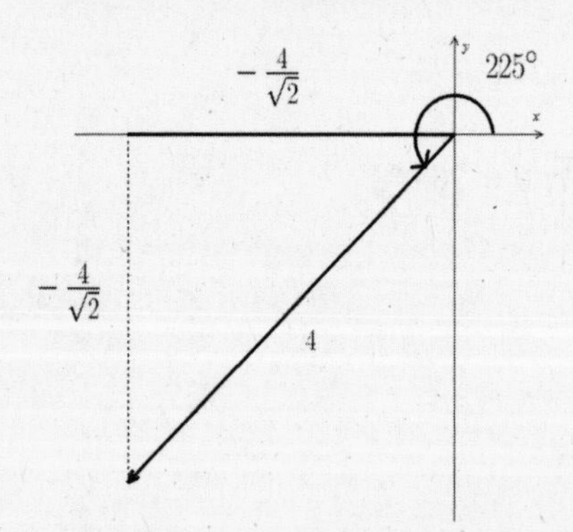

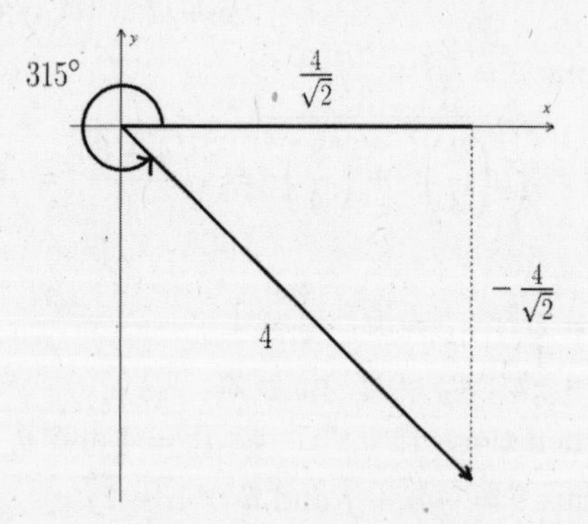

31. $x = (\tan t)^2;\ v_x = 2\tan t \sec^2 t\Big|_{t=\pi/3} = 2\sqrt{3}(2^2) = 8\sqrt{3}.$

$y = \csc t;\ v_y = -\csc t \cot t\Big|_{t=\pi/3} = -\frac{2}{\sqrt{3}} \cdot \frac{1}{\sqrt{3}} = -\frac{2}{3}.$

Thus $\vec{v} = 8\sqrt{3}\,\vec{\imath} - \frac{2}{3}\vec{\jmath}$ (4th quadrant).

$|\vec{v}| = \sqrt{\left(8\sqrt{3}\right)^2 + \left(-\frac{2}{3}\right)^2} \approx 13.87\,\text{m/s};\ \theta = \text{Arctan}\,\frac{-2/3}{8\sqrt{3}} = -2.75°.$

33. By Galileo's principle, the horizontal and vertical components of the velocity can be treated separately.

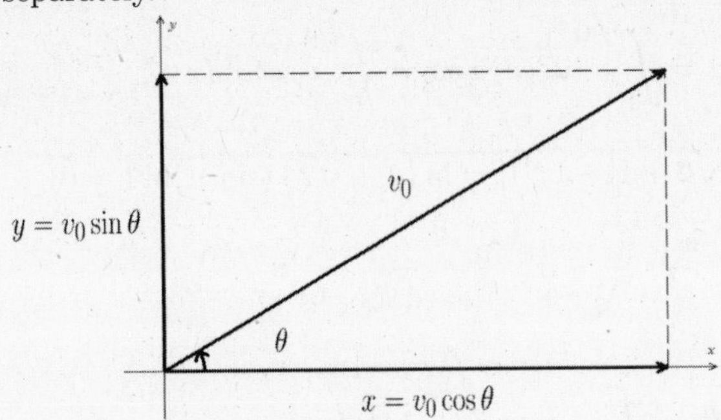

(a) In the horizontal direction, we need only the relationship distance = rate × time to obtain $x = (v_0 \cos\theta)t = v_0 t \cos\theta$.
In the vertical direction, we have $g = -10\,\text{m/s}^2$ by the assumption that the upward direction is positive. So if v denotes the velocity, then $v = -10t + C$. If $t = 0$, $v = v_0 \sin\theta$, so that $v_0 \sin\theta = 0 + C$. Thus $v = -10t + v_0 \sin\theta$.
Integrating, we get $y = -5t^2 + (v_0 \sin\theta)t + k$.
If $t = 0$, then $0 = k$, so that $y = v_0 t \sin\theta - 5t^2$.

(b) If $v_0 = 40\,\text{m/s}$ and $\theta = 30°$, then
$x = 40t\cos 30°,\ y = 40t \sin 30° - 5t^2$
$x = 40t\left(\frac{\sqrt{3}}{2}\right),\ y = 40t\left(\frac{1}{2}\right) - 5t^2$
$x = 20\sqrt{3}t,\ y = 20t - 5t^2$
$v_x = 20\sqrt{3},\ v_y = 20 - 10t\Big|_{t=1} = 10.$
Magnitude: $\sqrt{(20\sqrt{3})^2 + (10)^2} = \sqrt{1200 + 100} = 10\sqrt{13}\,\text{m/s}.$
Direction: Arctan $\left[\frac{10}{20\sqrt{3}}\right] = 16.1°.$
Also, $a_x = 0,\ a_y = -10.$
Magnitude: $\sqrt{0 + 100} = 10\,\text{m/s}^2$. Direction: $-90°$.

(c) The range is the value of x when $y = 0$:
$0 = v_0 t \sin\theta - 5t^2$
$t(v_0 \sin\theta - 5t) = 0$
$t = 0,\ t = \frac{1}{5}v_0 \sin\theta.$
Substituting the nonzero value in the equation $x = v_0 t \cos\theta$, we get:
$R = x = v_0\left(\frac{1}{5}v_0 \sin\theta\right)\cos\theta$
$R = \frac{1}{5}v_0^2 \sin\theta\cos\theta.$

8.2 Arc Length

1. $y = \frac{2}{3}x^{3/2};\ y' = \frac{2}{3}\cdot\frac{3}{2}x^{1/2} = x^{1/2};\ (y')^2 = x.$
$s = \int_0^3 \sqrt{1+x}\,dx = \frac{2}{3}(1+x)^{3/2}\Big|_0^3 = \frac{2}{3}\left[4^{3/2} - 1\right] = \frac{14}{3}.$

3. $y = \ln \cos x; y' = \dfrac{1}{\cos x}(-\sin x) = -\tan x.$

$$\begin{aligned} s &= \int_0^{\pi/4} \sqrt{1+(-\tan x)^2}\,dx = \int_0^{\pi/4} \sqrt{\sec^2 x}\,dx = \int_0^{\pi/4} \sec x\,dx \\ &= \ln|\sec x \tan x|\Big|_0^{\pi/4} = \ln\left|\sqrt{2}+1\right| - \ln|1| = \ln\left(1+\sqrt{2}\right) \text{ (since } \ln 1 = 0). \end{aligned}$$

5. $y = \ln \sin x; \ y' = \dfrac{\cos x}{\sin x} = \cot x.$

$$\begin{aligned} s &= \int_{\pi/4}^{\pi/3} \sqrt{1+\cot^2 x}\,dx = \int_{\pi/4}^{\pi/3} \csc x\,dx = \ln|\csc x - \cot x|\Big|_{\pi/4}^{\pi/3} \\ &= \ln\left(\frac{2}{\sqrt{3}} - \frac{1}{\sqrt{3}}\right) - \ln\left(\sqrt{2}-1\right) = \ln\frac{1}{\sqrt{3}} - \ln\left(\sqrt{2}-1\right) \\ &= \ln\frac{1}{\sqrt{3}\,(\sqrt{2}-1)} = \ln\frac{1}{\sqrt{6}-\sqrt{3}} \end{aligned}$$

7. $y = \dfrac{1}{2}(e^x + e^{-x}); \ y' = \dfrac{1}{2}(e^x - e^{-x}).$

$1 + (y')^2 = 1 + \dfrac{1}{4}(e^{2x} - 2 + e^{-2x}) = \dfrac{1}{4}(e^{2x} + 2 + e^{-2x}) = \left[\dfrac{1}{2}(e^x + e^{-x})\right]^2.$

$$\begin{aligned} s &= \int_0^1 \sqrt{\left[\frac{1}{2}(e^x + e^{-x})\right]^2}\,dx = \int_0^1 \frac{1}{2}(e^x + e^{-x})\,dx \\ &= \frac{1}{2}(e^x - e^{-x})\Big|_0^1 = \frac{1}{2}(e - e^{-1}) \approx 1.18. \end{aligned}$$

9. $y = \dfrac{1}{6}x^3 + \dfrac{1}{2x} = \dfrac{1}{6}x^3 + \dfrac{1}{2}x^{-1}; \ y' = \dfrac{1}{2}x^2 - \dfrac{1}{2}x^{-2}; \ (y')^2 = \dfrac{1}{4}x^4 - \dfrac{1}{2} + \dfrac{1}{4}x^{-4};$

$1 + (y')^2 = 1 + \dfrac{1}{4}x^4 - \dfrac{1}{2} + \dfrac{1}{4}x^{-4} = \dfrac{1}{4}x^4 + \dfrac{1}{2} + \dfrac{1}{4}x^{-4} = \left(\dfrac{1}{2}x^2 + \dfrac{1}{2}x^{-2}\right)^2.$

$$\begin{aligned} s &= \int_1^3 \sqrt{\left(\frac{1}{2}x^2 + \frac{1}{2}x^{-2}\right)^2}\,dx = \int_1^3 \left(\frac{1}{2}x^2 + \frac{1}{2}x^{-2}\right)dx \\ &= \frac{1}{2}\int_1^3 (x^2 + x^{-2})\,dx = \frac{1}{2}\left(\frac{1}{3}x^3 + \frac{x^{-1}}{-1}\right)\Big|_1^3 = \frac{1}{2}\left(\frac{1}{3}x^3 - \frac{1}{x}\right)\Big|_1^3 \\ &= \frac{1}{2}\left[\left(9 - \frac{1}{3}\right) - \left(\frac{1}{3} - 1\right)\right] = \frac{1}{2}\left(10 - \frac{2}{3}\right) = \frac{14}{3}. \end{aligned}$$

11. $x = 3t^2, \ y = t^3; \ \dfrac{dx}{dt} = 6t, \ \dfrac{dy}{dt} = 3t^2.$

$$\begin{aligned} s &= \int_0^{\sqrt{5}} \sqrt{(6t)^2 + (3t^2)^2}\,dt = \int_0^{\sqrt{5}} \sqrt{36t^2 + 9t^4}\,dt \\ &= \int_0^{\sqrt{5}} \sqrt{9t^2(4+t^2)}\,dt = 3\int_0^{\sqrt{5}} \sqrt{4+t^2}\,t\,dt \\ &= \frac{3}{2}\int_0^{\sqrt{5}} (4+t^2)^{1/2}(2t\,dt) \qquad u = 4 + t^2; \ du = 2t\,dt. \\ &= \frac{3}{2}\cdot\frac{2}{3}(4+t^2)^{3/2}\Big|_0^{\sqrt{5}} = (4+5)^{3/2} - 4^{3/2} = 27 - 8 = 19. \end{aligned}$$

13. $x = \cos^3\theta, \ y = \sin^3\theta; \ \dfrac{dx}{d\theta} = 3\cos^2\theta(-\sin\theta), \ \dfrac{dy}{d\theta} = 3\sin^2\theta\cos\theta;$

$\left(\dfrac{dx}{d\theta}\right)^2 = 9\cos^4\theta\sin^2\theta; \ \left(\dfrac{dy}{d\theta}\right)^2 = 9\sin^4\theta\cos^2\theta.$

$$\begin{aligned} s &= \int_0^{\pi/2} \sqrt{9\cos^4\theta\sin^2\theta + 9\sin^4\theta\cos^2\theta}\,d\theta \\ &= \int_0^{\pi/2} \sqrt{9\cos^2\theta\sin^2\theta(\cos^2\theta+\sin^2\theta)}\,d\theta \qquad \sin^2\theta+\cos^2\theta = 1 \\ &= \int_0^{\pi/2} 3\cos\theta\sin\theta\,d\theta = 3\int_0^{\pi/2}\sin\theta\cos\theta\,d\theta \qquad u=\sin\theta;\ du=\cos\theta. \\ &= \frac{3}{2}\sin^2\theta\Big|_0^{\pi/2} = \frac{3}{2}. \end{aligned}$$

15. $x = e^t\sin t;\ \dfrac{dx}{dt} = e^t\cos t + e^t\sin t.$
$y = e^t\cos t;\ \dfrac{dy}{dt} = -e^t\sin t + e^t\cos t.$

$$\begin{aligned} \left(\frac{dx}{dt}\right)^2 &= e^{2t}\cos^2 t + 2e^{2t}\sin t\cos t + e^{2t}\sin^2 t \\ &= e^{2t}(\cos^2 t+\sin^2 t) + 2e^{2t}\sin t\cos t = e^{2t} + 2e^{2t}\sin t\cos t. \\ \left(\frac{dy}{dt}\right)^2 &= e^{2t}\sin^2 t - 2e^{2t}\sin t\cos t + e^{2t}\cos^2 t \\ &= e^{2t}(\sin^2 t+\cos^2 t) - 2e^{2t}\sin t\cos t = e^{2t} - 2e^{2t}\sin t\cos t. \end{aligned}$$

By Formula (8.5),

$$s = \int_0^1 \sqrt{2e^{2t}}\,dt = \sqrt{2}\int_0^1 e^t\,dt = \sqrt{2}e^t\Big|_0^1 = \sqrt{2}(e-1).$$

17. $y = \left(a^{2/3}-x^{2/3}\right)^{3/2};\ y' = \dfrac{3}{2}\left(a^{2/3}-x^{2/3}\right)^{1/2}\left(-\dfrac{2}{3}x^{-1/3}\right) = -x^{-1/3}\left(a^{2/3}-x^{2/3}\right)^{1/2};$

$(y')^2 = \left(-x^{-1/3}\right)^2\left[\left(a^{2/3}-x^{2/3}\right)^{1/2}\right]^2 = x^{-2/3}\left(a^{2/3}-x^{2/3}\right) = x^{-2/3}a^{2/3} - 1.$

$1+(y')^2 = x^{-2/3}a^{2/3};\ \sqrt{1+(y')^2} = x^{-1/3}a^{1/3}.$

$$\begin{aligned} \text{Arc length} &= 4\int_0^a x^{-1/3}a^{1/3}\,dx = 4a^{1/3}\int_0^a x^{-1/3}\,dx \\ &= 4a^{1/3}\cdot\frac{3}{2}x^{2/3}\Big|_0^a = 6a^{1/3}a^{2/3} = 6a. \end{aligned}$$

19. $y = 2x^2,\ y' = 4x.$

$$\text{Arclength} = \int_0^3 \sqrt{1+(4x)^2}\,dx = 18.4599.$$

21. $y = \ln x,\ y' = \dfrac{1}{x}.$

$$\text{Arc length} = \int_1^4 \sqrt{1+(1/x)^2}\,dx = 3.3428.$$

23. $y = e^{-x},\ y' = -e^{-x}.$

$$s = \int_1^5 \sqrt{1+(-e^{-x})^2}\,dx = \int_1^5 \sqrt{1+e^{-2x}}\,dx = 4.0333.$$

25. The corresponding parametric equations are $x = \cos\theta, y = \sqrt{2}\sin\theta$, so that

$$\frac{dx}{d\theta} = -\sin\theta \text{ and } \frac{dy}{d\theta} = \sqrt{2}\cos\theta.$$

$$s = \int_0^{2\pi} \sqrt{(-\sin\theta)^2 + \left(\sqrt{2}\cos\theta\right)^2}\,d\theta \approx 7.64$$

8.3 Polar Coordinates

1.

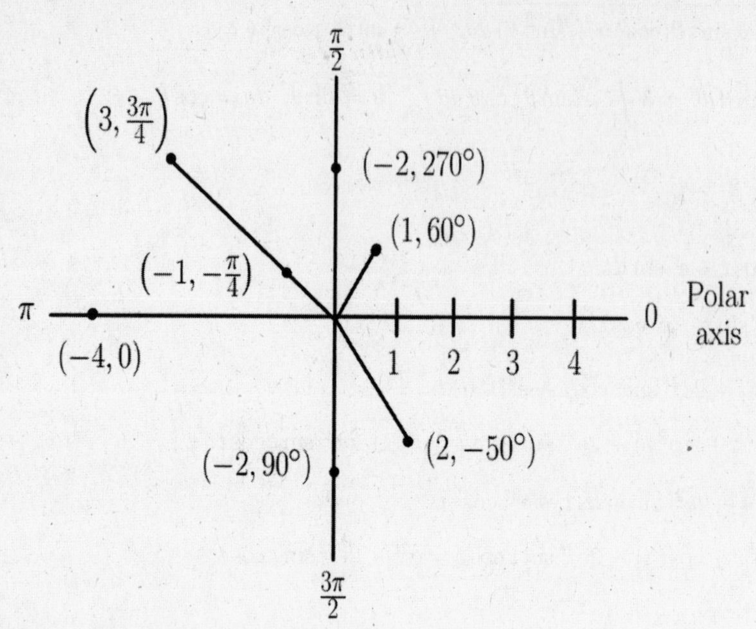

3. $x = r\cos\theta = 2\cos 120° = 2\left(-\frac{1}{2}\right) = -1.$

   $y = r\sin\theta = 2\sin 120° = 2\left(\frac{\sqrt{3}}{2}\right) = \sqrt{3}.$

5. $x = r\cos\theta = -6\cos\left(\frac{\pi}{3}\right) = -6\left(\frac{1}{2}\right) = -3.$

   $y = r\sin\theta = -6\sin\left(\frac{\pi}{3}\right) = -6\left(\frac{\sqrt{3}}{2}\right) = -3\sqrt{3}.$

7. $x = r\cos\theta = -2\cos 170° = 1.97.$

   $y = r\sin\theta = -2\sin 170° = -0.35.$

9. $r^2 = x^2 + y^2 = 1^2 + (-1)^2 = 2,\ r = \sqrt{2},\ \tan\theta = \frac{-1}{1} = -1.$

   Since θ is in the fourth quadrant, $\theta = 315° = \frac{7\pi}{4}$.

11. $r = \sqrt{3^2 + 4^2} = 5,\ \tan\theta = \frac{4}{3}$; since θ is in the first quadrant, $\theta = \text{Arctan}\,\frac{4}{3} = 0.93.$

13.
$$\begin{aligned} x &= 2 \\ r\cos\theta &= 2 \\ r &= \frac{2}{\cos\theta} \\ r &= 2\sec\theta. \end{aligned}$$

15. $x^2 + y^2 = r^2 = 2$, or $r = \pm\sqrt{2}$. To generate the circle, it is sufficient to take the positive square root: $r = \sqrt{2}$.

17. $x^2+y^2+3x-4y=0$; we have: $x^2+y^2=r^2, x=r\cos\theta$, and $y=r\sin\theta$. Substituting we get

$$r^2+3r\cos\theta-4r\sin\theta=0 \quad \text{and} \quad r=4\sin\theta-3\cos\theta.$$

19. Since $x=r\cos\theta$ and $y=r\sin\theta$, $(r\cos\theta)(r\sin\theta)=6$ or $r^2\cos\theta\sin\theta=6$.

21. $x=2y^2$; $r\cos\theta=2r^2\sin^2\theta$ or $\cos\theta=2r\sin^2\theta$. Solving for r we get

$$\begin{aligned} r &= \frac{1}{2}\frac{\cos\theta}{\sin^2\theta}=\frac{1}{2}\frac{\cos\theta}{\sin\theta}\cdot\frac{1}{\sin\theta} \\ r &= \frac{1}{2}\cot\theta\csc\theta. \end{aligned}$$

23. $2x^2+4y^2=3$; $2r^2\cos^2\theta+4r^2\sin^2\theta=3$

$$\begin{aligned} r^2 &= \frac{3}{2\cos^2\theta+4\sin^2\theta} \\ r^2 &= \frac{3}{2\left(1-\sin^2\theta\right)+4\sin^2\theta} && \sin^2\theta+\cos^2\theta=1 \\ r^2 &= \frac{3}{2\left(1+\sin^2\theta\right)} \end{aligned}$$

25. $r=2\csc\theta=\dfrac{2}{\sin\theta}$ or $r\sin\theta=2$, which yields $y=2$.

27.
$$\begin{aligned} \theta &= \frac{\pi}{4} \\ \tan\theta &= \tan\frac{\pi}{4} \\ \frac{y}{x} &= 1 \text{ and } y=x. \end{aligned}$$

29.
$$\begin{aligned} r &= \cos\theta \\ r^2 &= r\cos\theta && \text{multiplying by } r \\ x^2+y^2 &= x. \end{aligned}$$

Alternatively, $r=\cos\theta=\dfrac{x}{r}$.

From $r=\dfrac{x}{r}$ we have $r^2=x$ or $x^2+y^2=x$.

31.
$$\begin{aligned} r &= 1+\cos\theta \\ r^2 &= r+r\cos\theta && \text{multiplying by } r \\ (1)\ \ x^2+y^2 &= \pm\sqrt{x^2+y^2}+x \\ x^2+y^2-x &= \pm\sqrt{x^2+y^2} \\ (x^2+y^2-x)^2 &= x^2+y^2. && \text{squaring both sides} \end{aligned}$$

Alternatively,

$$\begin{aligned} r &= 1+\cos\theta \\ r &= 1+\frac{x}{r} && \text{since } \cos\theta=\frac{x}{r} \\ \pm\sqrt{x^2+y^2} &= 1+\frac{x}{\pm\sqrt{x^2+y^2}} \\ x^2+y^2 &= \pm\sqrt{x^2+y^2}+x, \text{ which agrees with (1).} \end{aligned}$$

33.
$$\begin{aligned} r &= 1-\cos\theta \\ r^2 &= r - r\cos\theta && \text{multiplying by } r \\ x^2+y^2 &= \pm\sqrt{x^2+y^2} - x \\ x^2+y^2+x &= \pm\sqrt{x^2+y^2} \\ (x^2+y^2+x)^2 &= x^2+y^2. && \text{squaring both sides} \end{aligned}$$

Alternatively,

$$\begin{aligned} r &= 1-\cos\theta \\ r &= 1-\frac{x}{r} && \cos\theta = \frac{x}{r}. \\ \pm\sqrt{x^2+y^2} &= 1-\frac{x}{\pm\sqrt{x^2+y^2}} \\ x^2+y^2 &= \pm\sqrt{x^2+y^2} - x && \text{multiplying by } \pm\sqrt{x^2+y^2} \\ x^2+y^2+x &= \pm\sqrt{x^2+y^2} \\ (x^2+y^2+x)^2 &= x^2+y^2. \end{aligned}$$

35.
$$\begin{aligned} r &= 1-2\sin\theta \\ r^2 &= r-2r\sin\theta && \text{multiplying by } r \\ x^2+y^2 &= \pm\sqrt{x^2+y^2} - 2y \\ x^2+y^2+2y &= \pm\sqrt{x^2+y^2} \\ (x^2+y^2+2y)^2 &= x^2+y^2. && \text{squaring both sides} \end{aligned}$$

37.
$$\begin{aligned} r &= -2+4\sin\theta \\ r^2 &= -2r+4r\sin\theta && \text{multiplying by } r \\ x^2+y^2 &= -2\left(\pm\sqrt{x^2+y^2}\right)+4y \\ (x^2+y^2-4y)^2 &= 4(x^2+y^2). \end{aligned}$$

39. By the double-angle formula,

$$\begin{aligned} r^2 &= \sin 2\theta = 2\sin\theta\cos\theta \\ r^4 &= 2(r\sin\theta)(r\cos\theta) && \text{multiplying by } r^2 \\ (x^2+y^2)^2 &= 2xy. \end{aligned}$$

41. $r = \dfrac{2}{1-\cos\theta}$. Dividing both sides by r, we get

$$\begin{aligned} 1 &= \frac{2}{r(1-\cos\theta)} = \frac{2}{r-r\cos\theta} \\ 1 &= \frac{2}{\pm\sqrt{x^2+y^2}-x} \\ \pm\sqrt{x^2+y^2}-x &= 2 \\ \pm\sqrt{x^2+y^2} &= x+2 \\ x^2+y^2 &= x^2+4x+4 && \text{squaring both sides} \\ y^2 &= 4(x+1). \end{aligned}$$

43.
$$\begin{aligned} r &= \frac{4}{1-\sin\theta} \\ r-r\sin\theta &= 4 && \text{multipling by } (1-\sin\theta) \\ \pm\sqrt{x^2+y^2}-y &= 4 \\ \pm\sqrt{x^2+y^2} &= y+4 \\ x^2+y^2 &= y^2+8y+16 \\ x^2 &= 8(y+2). \end{aligned}$$

45.
$$\begin{aligned} r^2 &= 3\cos 2\theta = 3(\cos^2\theta - \sin^2\theta) && \text{double-angle formula} \\ r^4 &= 3(r^2\cos^2\theta - r^2\sin^2\theta) && \text{multiplying by } r^2 \\ (x^2+y^2)^2 &= 3(x^2-y^2). \end{aligned}$$

47. $r = a\sin 3\theta = a\sin(2\theta+\theta) = a(\sin 2\theta\cos\theta + \cos 2\theta\sin\theta)$.
$$\begin{aligned} r &= a\left[(2\sin\theta\cos\theta)\cos\theta + (\cos^2\theta - \sin^2\theta)\sin\theta\right] \\ &= a(2\sin\theta\cos^2\theta + \cos^2\theta\sin\theta - \sin^3\theta) \\ &= a(3\sin\theta\cos^2\theta - \sin^3\theta) \\ r^4 &= a\left[3(r\sin\theta)(r^2\cos^2\theta) - r^3\sin^3\theta\right] && \text{multiplying by } r^3 \\ (x^2+y^2)^2 &= a(3x^2y - y^3). \end{aligned}$$

49.
$$\begin{aligned} r^2 &= \sec 2\theta = \frac{1}{\cos 2\theta} \\ &= \frac{1}{\cos^2\theta - \sin^2\theta} && \text{double-angle identity} \\ r^2\cos^2\theta - r^2\sin^2\theta &= 1 \\ x^2 - y^2 &= 1 \end{aligned}$$

8.4 Curves in Polar Coordinates

1. $r = 2$ is a circle of radius 2.

3. $r = 2\cos\theta$ is a circle but may be viewed as a special case of a rose with one leaf.

5. $r\cos\theta = 4$ is the line $x = 4$.

7. $r = 2(1+\cos\theta)$.
   Cardioid.
   If $\theta = 0°$, $r = 4$.
   If $\theta = 90°$ or $270°$, $r = 2$.
   If $\theta = 180°$, $r = 0$.

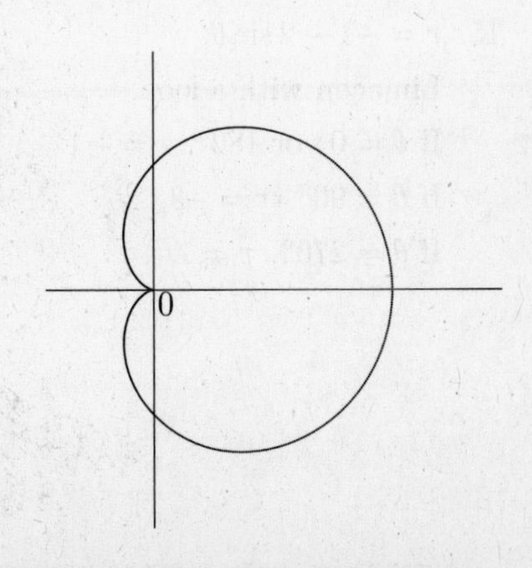

9. Limaçon.
   If $\theta = 90°$ or $270°$, $r = 9$.
   If $\theta = 0°$, $r = 14$.
   If $\theta = 180°$, $r = 4$.

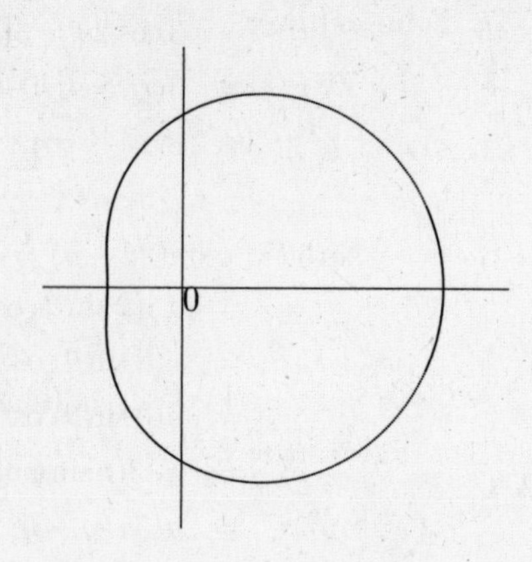

11. $r = 3 - 2\cos\theta$.
    Limaçon.
    If $\theta = 0°$, $r = 1$.
    8 If $\theta = 90°$ or $270°$, $r = 3$.
    If $\theta = 180°$, $r = 5$.

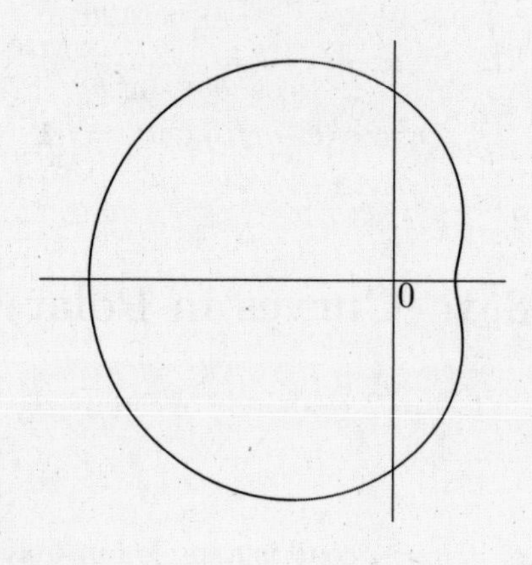

13. Limaçon with a loop.
    If $\theta = 0°$ or $180°$, $r = 1$.
    If $\theta = 90°$, $r = -1$.
    If $\theta = 270°$, $r = 3$.
    Also, $r = 0$ when

$$\begin{aligned} 1 - 2\sin\theta &= 0 \\ \sin\theta &= \frac{1}{2} \\ \theta &= 30°,\ 150°. \end{aligned}$$

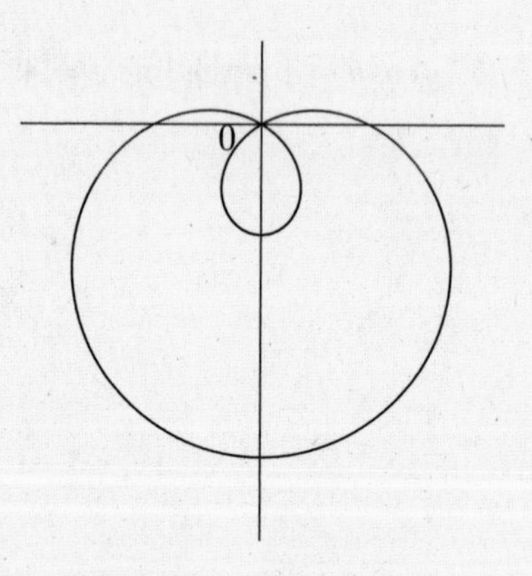

15. $r = -1 - 2\sin\theta$.
    Limaçon with a loop.
    If $\theta = 0°$ or $180°$, $r = -1$.
    If $\theta = 90°$, $r = -3$.
    If $\theta = 270°$, $r = 1$.

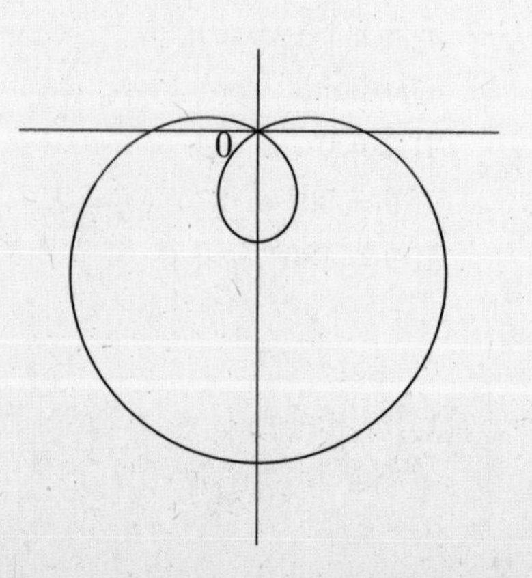

17. Four-leaf rose.

$r = 0$ when $\theta = 0°,\ 90°,\ 180°,\ 270°$.

$|r| = 4$ when $\theta = 45°,\ 135°,\ 225°,\ 315°$.

19. $r = 3\cos 3\theta$, three-leaf rose. When $\theta = 0°,\ r = 1$; when $\theta = \pm 30°,\ r = \cos(\pm 90°) = 0$. So one leaf is traced out in the interval $\theta = -30°$ to $\theta = +30°$. The location of the other two leaves follows from the symmetry.

21. $r^2 = 4\sin 2\theta$, lemniscate. When $\theta = 0°$ or $90°$, we get $r = 0$. When $\theta = 45°,\ r^2 = 4\sin 90° = 4$ and $r = \pm 2$. So the upper portion of the figure is traced out in the interval $\theta = 0°$ to $\theta = 90°$ with $r = 2$ at $\theta = 45°$.

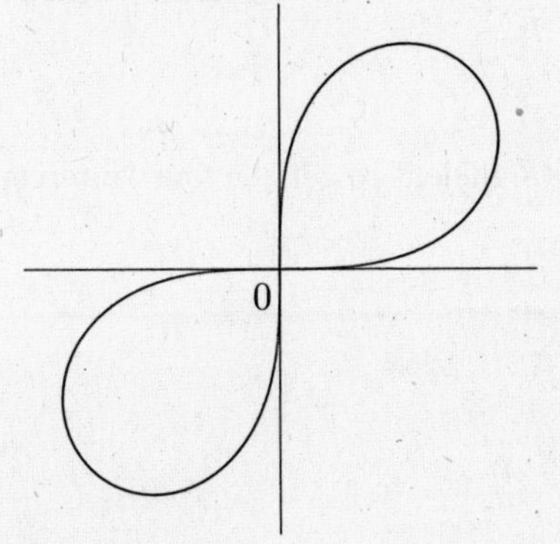

23. See Example 6.

25. $r = 3\sqrt{\sin 2\theta}$, $0 \leq \theta \leq 90°$; $r = 0$ when $\theta = 0°$ or $90°$; r attains its largest value when $\theta = 45°$: $r = 3$.

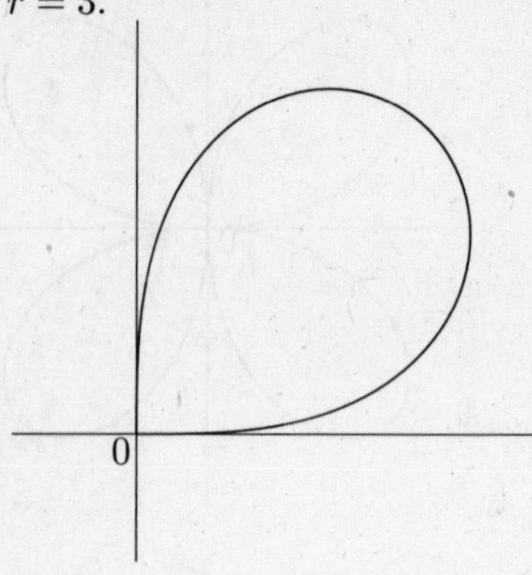

27. $r = \sin 3\theta$, three-leaf rose. When $\theta = 0°$ or $60°$, we get $r = 0$. When $\theta = 30°$, $r = \sin 3 \cdot 30° = \sin 90° = 1$. So one leaf is traced out in the interval $\theta = 0°$ to $\theta = 60°$, attaining a maximum value of $r = 1$ at $\theta = 30°$. The location of the other two leaves follows from the symmetry.

29. $r = \dfrac{2}{\theta}$; as $\theta \to 0$, $r \to \infty$; as $\theta \to \infty$, $r \to 0$.
(In other words, as the curve circles the pole, r gets closer to 0; see curve in answer section.)

31. Multiply both sides of the equation by $2 - \cos\theta$:

$$\begin{aligned} 2r - r\cos\theta &= 3 \\ \pm 2\sqrt{x^2+y^2} - x &= 3 \\ \pm 2\sqrt{x^2+y^2} &= x+3 \\ 4(x^2+y^2) &= (x+3)^2 = x^2+6x+9 \\ 3x^2+4y^2-6x-9 &= 0 \end{aligned}$$

which is the equation of an ellipse. To plot the curve, it is sufficient to get the four "intercepts":

$\theta:$	$0°$	$90°$	$180°$	$270°$
$r:$	3	$\frac{3}{2}$	1	$\frac{3}{2}$

33. If $\theta = 0°$ or $180°$, $r = 3$.
If $\theta = 90°$, $r = \frac{6}{5}$.
If $\theta = 270°$, $r = -6$.

Some other points are given in the following table:

$\theta:$	$-10°$	$190°$	$250°$	$260°$	$280°$	$290°$
$r:$	4.1	4.1	-7.3	-6.3	-6.3	-7.3

The resulting curve is a hyperbola. (See the sketch in the answer section.)

8.5 Areas in Polar Coordinates

1. $A = \frac{1}{2}\int_0^{\pi/6} 2^2\,d\theta = 2\theta\Big|_0^{\pi/6} = \frac{\pi}{3}.$

3. $A = \frac{1}{2}\int_0^{\pi/4} \sec^2\theta\,d\theta = \frac{1}{2}\tan\theta\Big|_0^{\pi/4} = \frac{1}{2}.$

5. $A = \frac{1}{2}\int_0^{2} \theta^2\,d\theta = \frac{1}{2}\frac{1}{3}\theta^3\Big|_0^{2} = \frac{4}{3}.$

7. 

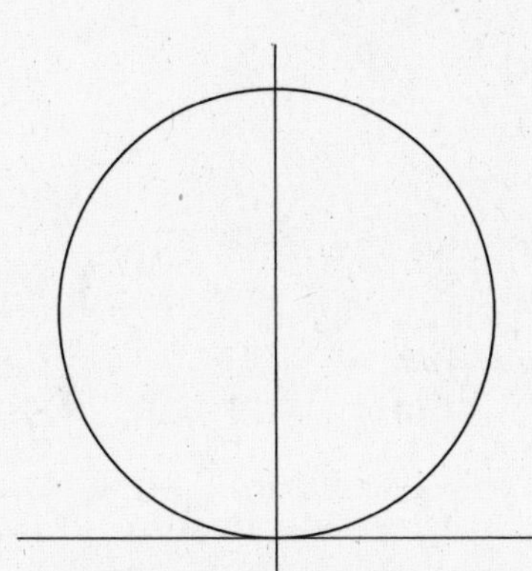

$r = 2\sin\theta$ is a circle (one-leaf rose), similar to Figure 8.19. The curve is traced out in the interval $\theta = 0$ to $\theta = \pi$:

$A = \frac{1}{2}\int_0^{\pi}(2\sin\theta)^2\,d\theta = 2\int_0^{\pi}\frac{1}{2}(1-\cos 2\theta)\,d\theta = \int_0^{\pi} 1\,d\theta - \int_0^{\pi}\cos 2\theta\,d\theta.$

For the second integral, let $u = 2\theta$; $du = 2\,d\theta$. So we get

$\int_0^{\pi} d\theta - \frac{1}{2}\int_0^{\pi}\cos 2\theta(2\,d\theta) = \theta - \frac{1}{2}\sin 2\theta\Big|_0^{\pi} = \pi - 0 = \pi.$

9. 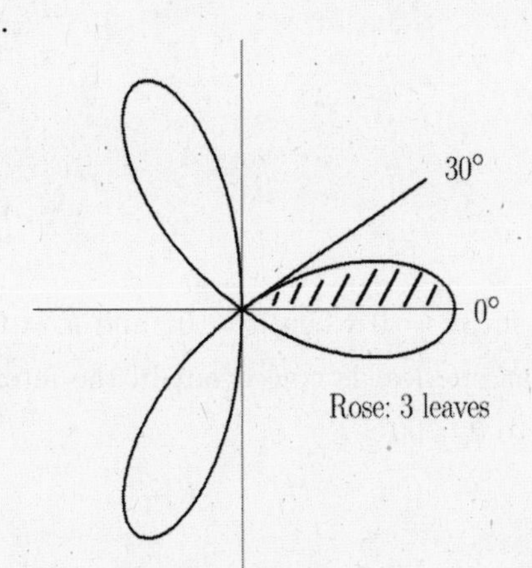

Rose: 3 leaves

Since $r = 4\cos 3\theta = 0$ when $\theta = 30°$, it follows that half a leaf is traced out in the interval $\theta = 0°$ to $\theta = 30°$.

So integrating from $\theta = 0$ to $\theta = \frac{\pi}{6}$ yields one-sixth of the area:

$$\begin{aligned}\frac{1}{6}A &= \frac{1}{2}\int_0^{\pi/6}(4\cos 3\theta)^2\,d\theta = 8\int_0^{\pi/6}\cos^2 3\theta\,d\theta \\ &= 8\int_0^{\pi/6}\frac{1}{2}(1+\cos 6\theta)\,d\theta = 4\int_0^{\pi/6}(1+\cos 6\theta)\,d\theta.\end{aligned}$$

To integrate $\int \cos 6\theta\,d\theta$, let $u = 6\theta$; $du = 6\,d\theta$.

$\frac{1}{6}\int\cos 6\theta(6\,d\theta) = \frac{1}{6}\sin 6\theta + C.$

It follows that $\frac{1}{6}A = 4\left(\theta + \frac{1}{6}\sin 6\theta\right)\Big|_0^{\pi/6} = 4\left(\frac{\pi}{6}\right) = \frac{2\pi}{3}.$

So $A = 6\left(\frac{2\pi}{3}\right) = 4\pi.$

11. $r = 3\sin 2\theta$, four-leaf rose.

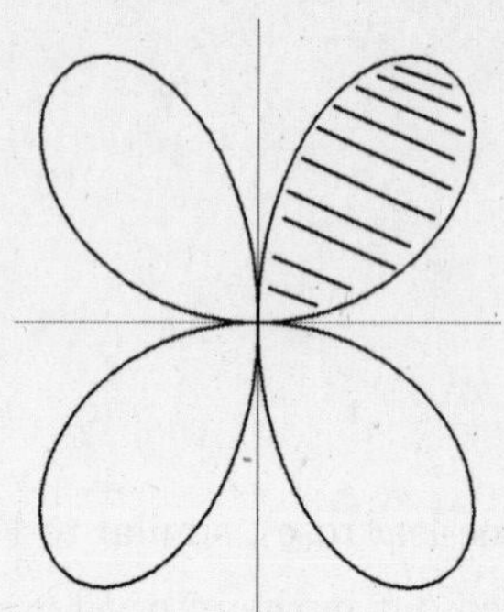

When $\theta = 0°$, $r = 0$ and when $\theta = 90°$, $r = 3\sin 2(90°) = 3\sin 180° = 0$. So one leaf is traced out in the interval $\theta = 0$ to $\theta = \frac{\pi}{2}$. The entire area is equal to four times the area of one leaf.

$$\begin{aligned} A &= 4 \cdot \frac{1}{2}\int_0^{\pi/2} (3\sin 2\theta)^2\, d\theta = 18\int_0^{\pi/2} \sin^2 2\theta\, d\theta \\ &= 18\int_0^{\pi/2} \frac{1}{2}(1-\cos 4\theta)\, d\theta \\ &= 9\int_0^{\pi/2} 1\, d\theta - 9\int_0^{\pi/2} \cos 4\theta\, d\theta \qquad u = 4\theta;\ du = 4\, d\theta. \\ &= 9\theta\Big|_0^{\pi/2} - \frac{9}{4}\sin 4\theta\Big|_0^{\pi/2} = \frac{9\pi}{2}. \end{aligned}$$

13.

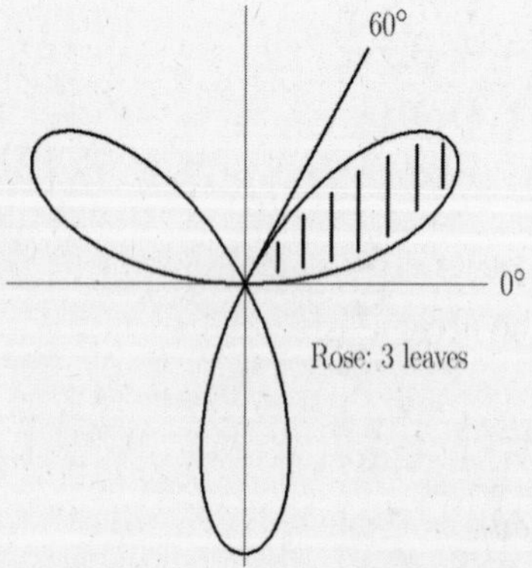

Since $r = 6\sin 3\theta = 0$ when $\theta = 0°$ and $\theta = 60°$, it follows that one leaf is traced out in the interval from $\theta = 0°$ to $\theta = 60°$.

So integrating from $\theta = 0$ to $\theta = \frac{\pi}{3}$ yields one-third of the area:

$$\begin{aligned} \frac{1}{3}A &= \frac{1}{2}\int_0^{\pi/3} (6\sin 3\theta)^2\, d\theta = \frac{1}{2}\int_0^{\pi/3} 36\sin^2 3\theta\, d\theta \\ &= 18\int_0^{\pi/3} \frac{1}{2}(1-\cos 6\theta)\, d\theta = 9\int_0^{\pi/3} (1-\cos 6\theta)\, d\theta \\ &= 9\left(\theta - \frac{1}{6}\sin 6\theta\right)\Big|_0^{\pi/3} = 9\left(\frac{\pi}{3}\right) = 3\pi. \qquad u = 6\theta;\ du = 6\, d\theta. \end{aligned}$$

So $A = 9\pi$.

15. $r = \sqrt{\sin\theta}$.

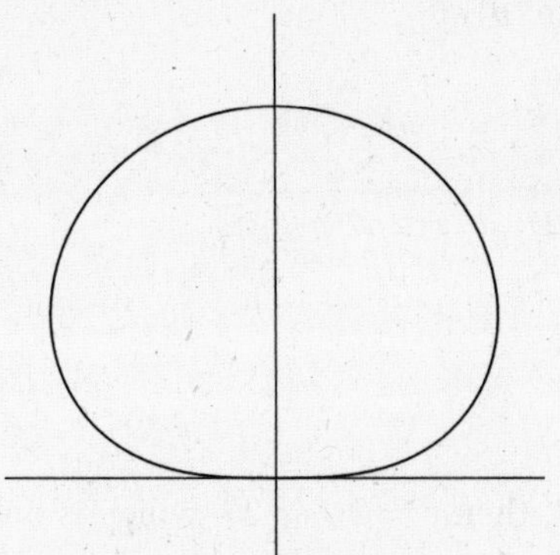

This curve is traced out in the interval $\theta = 0$ to $\theta = \pi$.

$$\begin{aligned} A &= \frac{1}{2}\int_0^{\pi}\left(\sqrt{\sin\theta}\right)^2\,d\theta = \frac{1}{2}\int_0^{\pi}\sin\theta\,d\theta \\ &= -\frac{1}{2}\cos\theta\Big|_0^{\pi} = -\frac{1}{2}(\cos\pi - \cos 0) = -\frac{1}{2}(-1-1) = 1. \end{aligned}$$

17. For a limaçon without a loop, the limits of integration are 0 and 2π:

$$A = \frac{1}{2}\int_0^{2\pi}(2+\cos\theta)^2\,d\theta = \frac{1}{2}\int_0^{2\pi}(4+4\cos\theta+\cos^2\theta)\,d\theta.$$

Since $\cos^2\theta = \frac{1}{2}(1+\cos 2\theta)$, we get

$$\begin{aligned} A &= \frac{1}{2}\int_0^{2\pi}\left[4+4\cos\theta+\frac{1}{2}(1+\cos 2\theta)\right]d\theta \\ &= \frac{1}{2}\left[4\theta+4\sin\theta+\frac{1}{2}\left(\theta+\frac{1}{2}\sin 2\theta\right)\right]\Bigg|_0^{2\pi} \qquad u = 2\theta;\ du = 2\,d\theta. \\ &= \frac{1}{2}\left[4(2\pi)+\frac{1}{2}(2\pi)\right] = \frac{9\pi}{2}. \end{aligned}$$

19. For any cardioid, the limits of integration are 0 and 2π:

$$\begin{aligned} A &= \frac{1}{2}\int_0^{2\pi}[2(1-\cos\theta)]^2\,d\theta = 2\int_0^{2\pi}(1-2\cos\theta+\cos^2\theta)\,d\theta \\ &= 2\int_0^{2\pi}\left[1-2\cos\theta+\frac{1}{2}(1+\cos 2\theta)\right]d\theta = 2\left[\theta-2\sin\theta+\frac{1}{2}\left(\theta+\frac{1}{2}\sin 2\theta\right)\right]\Bigg|_0^{2\pi} \\ &= 2\left[2\pi+\frac{1}{2}(2\pi)\right] = 6\pi. \end{aligned}$$

21. For a limaçon without a loop, the limits of integration are 0 and 2π:

$$\begin{aligned} A &= \frac{1}{2}\int_0^{2\pi}(2-\cos\theta)^2\,d\theta = \frac{1}{2}\int_0^{2\pi}\left(4-4\cos\theta+\cos^2\theta\right)\,d\theta \\ &= \frac{1}{2}\int_0^{2\pi}\left[4-4\cos\theta+\frac{1}{2}\left(1+\cos 2\theta\right)\right]d\theta \qquad u = 2\theta;\ du = 2\,d\theta \\ &= \frac{1}{2}\left[4\theta-4\sin\theta+\frac{1}{2}\left(\theta+\frac{1}{2}\sin 2\theta\right)\right]\Bigg|_0^{2\pi} \\ &= \frac{1}{2}[8\pi-0+\pi+0] = \frac{9\pi}{2} \end{aligned}$$

23. For any cardioid, the limits of integration are 0 and 2π:

$$\begin{aligned}
A &= \frac{1}{2}\int_0^{2\pi} (3+3\sin\theta)^2\,d\theta = \frac{9}{2}\int_0^{2\pi} (1+2\sin\theta+\sin^2\theta)\,d\theta \\
&= \frac{9}{2}\int_0^{2\pi} \left[1+2\sin\theta+\frac{1}{2}(1-\cos 2\theta)\right]\,d\theta \\
&= \frac{9}{2}\left[\theta-2\cos\theta+\frac{1}{2}\left(\theta-\frac{1}{2}\sin 2\theta\right)\right]\Bigg|_0^{2\pi} \qquad u=2\theta;\ du=2\,d\theta \\
&= \frac{9}{2}[2\pi-2+\pi-0]-\frac{9}{2}[-2]=\frac{9}{2}(3\pi)=\frac{27\pi}{2}
\end{aligned}$$

25.

If $r^2 = 4\sin 2\theta$, then $r = 2\sqrt{\sin 2\theta}$, which is only the upper half of the lemniscate. Note that $r = 0$ when $\theta = 0°$ and $\theta = 90°$.

The total area is therefore given by:

$$\begin{aligned}
A &= 2\cdot\frac{1}{2}\int_0^{\pi/2}\left(2\sqrt{\sin 2\theta}\right)^2\,d\theta = \int_0^{\pi/2} 4\sin 2\theta\,d\theta \qquad u=2\theta;\ du=2\,d\theta. \\
&= 4\left(-\frac{1}{2}\cos 2\theta\right)\Bigg|_0^{\pi/2} = -2\cos 2\theta\Big|_0^{\pi/2} = -2(-1-1)=4.
\end{aligned}$$

27. $r^2 = 2\cos 2\theta$, lemniscate.

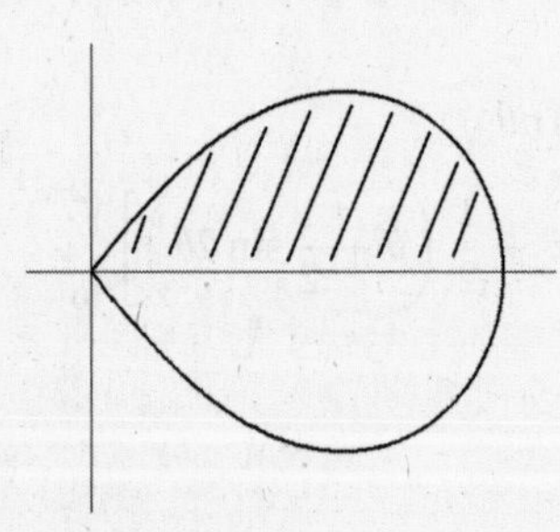

The function $r = \sqrt{2\cos 2\theta}$ is one-half of the lemniscate, shown in the figure. When $\theta = 0$, $r = \sqrt{2}$, the largest possible value. When $\theta = 45°$, $r = \sqrt{2\cos 90°} = 0$.

So integrating from $\theta = 0$ to $\theta = \frac{\pi}{4}$ yields the area of the shaded region. By symmetry, the area enclosed by the entire lemniscate is four times this value.

$$\begin{aligned}
A &= 4\cdot\frac{1}{2}\int_0^{\pi/4}\left(\sqrt{2\cos 2\theta}\right)^2\,d\theta = 2\int_0^{\pi/4} 2\cos 2\theta\,d\theta \\
&= 2\int_0^{\pi/4}\cos 2\theta(2\,d\theta) = 2\sin 2\theta\Big|_0^{\pi/4} = 2\sin 2\left(\frac{\pi}{4}\right) = 2.
\end{aligned}$$

29. Similar to Exercise 25. Since $r = \sqrt{6}\sqrt{\sin 2\theta}$,

$$\begin{aligned}
A &= 2\cdot\frac{1}{2}\int_0^{\pi/2}\left(\sqrt{6}\sqrt{\sin 2\theta}\right)^2\,d\theta = \int_0^{\pi/2} 6\sin 2\theta\,d\theta \qquad u=2\theta;\ du=2\,d\theta. \\
&= -3\cos 2\theta\Big|_0^{\pi/2} = -3(-1-1)=6.
\end{aligned}$$

31. $r = 1 - 2\sin\theta$.

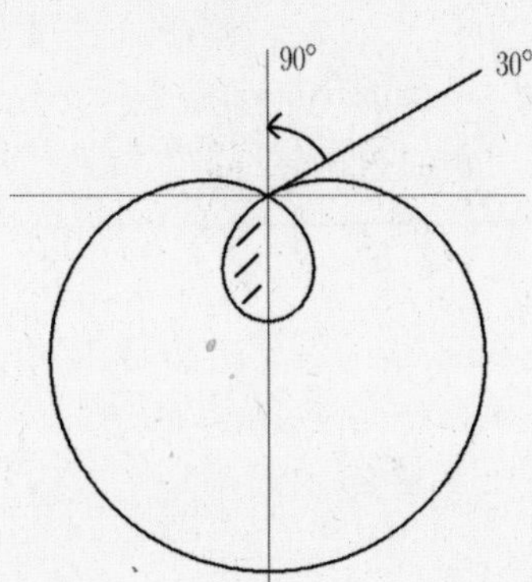

$$\begin{aligned} r = 0 \text{ when } \quad 1 - 2\sin\theta &= 0 \\ \sin\theta &= \frac{1}{2} \\ \theta &= 30°,\ 150°. \end{aligned}$$

Half the loop is traced out in the interval $\frac{\pi}{6}$ to $\frac{\pi}{2}$. So

$$\begin{aligned} A &= 2 \cdot \frac{1}{2} \int_{\pi/6}^{\pi/2} (1 - 2\sin\theta)^2 \, d\theta = \int_{\pi/6}^{\pi/2} (1 - 4\sin\theta + 4\sin^2\theta) \, d\theta \\ &= \int_{\pi/6}^{\pi/2} [1 - 4\sin\theta + 2(1 - \cos 2\theta)] \, d\theta = \theta + 4\cos\theta + 2\left(\theta - \frac{1}{2}\sin 2\theta\right)\Bigg|_{\pi/6}^{\pi/2} \\ &= \left(\frac{\pi}{2} + \pi\right) - \left(\frac{\pi}{6} + 4\cos\frac{\pi}{6} + \frac{\pi}{3} - \sin\frac{\pi}{3}\right) \\ &= \frac{3\pi}{2} - \frac{\pi}{6} - 4\left(\frac{\sqrt{3}}{2}\right) - \frac{\pi}{3} + \frac{\sqrt{3}}{2} = \pi - 3\left(\frac{\sqrt{3}}{2}\right) = \frac{1}{2}(2\pi - 3\sqrt{3}). \end{aligned}$$

33.

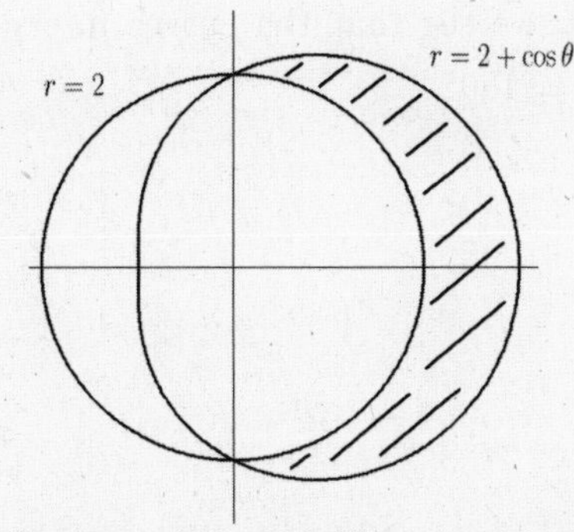

To determine where the curves intersect, we substitute $r = 2$ in $r = 2 + \cos\theta$ and solve for θ:

$$\begin{aligned} 2 &= 2 + \cos\theta \\ \cos\theta &= 0. \end{aligned}$$

So $\theta = -90°,\ 90°$. We can therefore obtain the upper half of the shaded region by integrating from $\theta = 0$ to $\theta = \frac{\pi}{2}$:

$$\begin{aligned} \frac{1}{2}A &= \frac{1}{2}\int_0^{\pi/2} (2 + \cos\theta)^2 \, d\theta - \frac{1}{2}\int_0^{\pi/2} 2^2 \, d\theta \\ A &= \int_0^{\pi/2} (2 + \cos\theta)^2 \, d\theta - \int_0^{\pi/2} 4 \, d\theta \\ &= \int_0^{\pi/2} (4 + 4\cos\theta + \cos^2\theta) \, d\theta - 4\theta\Big|_0^{\pi/2} \\ &= \int_0^{\pi/2} \left[4 + 4\cos\theta + \frac{1}{2}(1 + \cos 2\theta)\right] d\theta - 4\left(\frac{\pi}{2}\right) \\ &= 4\theta + 4\sin\theta + \frac{1}{2}\left(\theta + \frac{1}{2}\sin 2\theta\right)\Bigg|_0^{\pi/2} - 2\pi \\ &= \left[2\pi + 4 + \frac{1}{2}\left(\frac{\pi}{2}\right)\right] - 0 - 2\pi = 4 + \frac{\pi}{4} = \frac{16 + \pi}{4}. \end{aligned}$$

35.

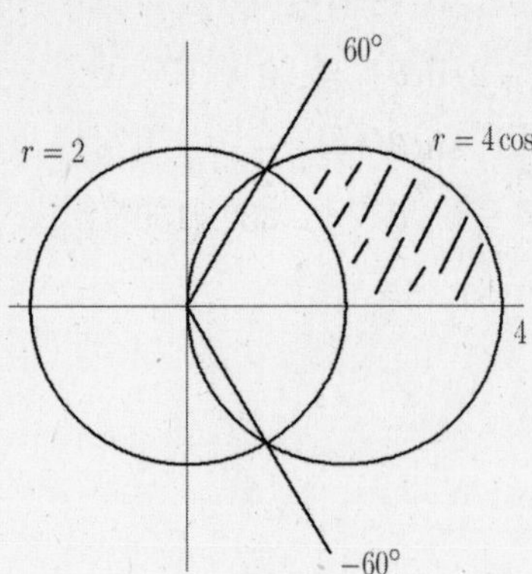

$$\begin{aligned} r &= 4\cos\theta \\ r &= 2 \\ \hline 2 &= 4\cos\theta \qquad \text{(substitution)} \\ \cos\theta &= \frac{1}{2}, \qquad \theta = \pm 60^\circ = \pm\frac{\pi}{3}. \end{aligned}$$

$$\begin{aligned} A &= 2\cdot\frac{1}{2}\int_0^{\pi/3}\left[(4\cos\theta)^2 - 2^2\right]d\theta = \int_0^{\pi/3}(16\cos^2\theta - 4)\,d\theta \\ &= \int_0^{\pi/3}[8(1+\cos 2\theta) - 4]\,d\theta = \int_0^{\pi/3}(4 + 8\cos 2\theta)\,d\theta \qquad u = 2\theta;\ du = 2\,d\theta. \\ &= 4\theta + 4\sin 2\theta\Big|_0^{\pi/3} = \frac{4\pi}{3} + 4\sin\frac{2\pi}{3} \\ &= \frac{4\pi}{3} + 4\left(\frac{\sqrt{3}}{2}\right) = 4\left(\frac{\pi}{3} + \frac{\sqrt{3}}{2}\right). \end{aligned}$$

37.

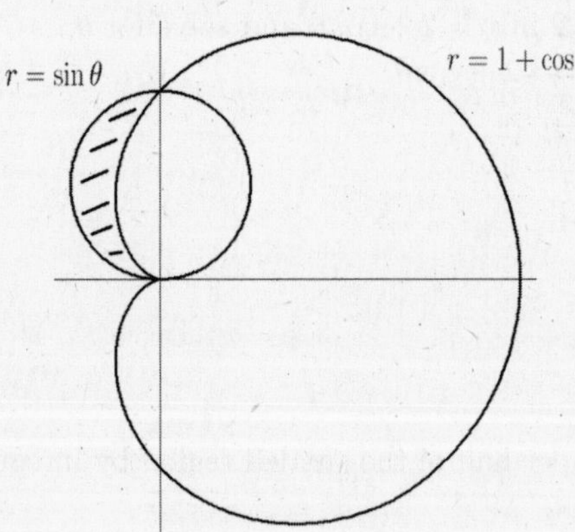

By inspection, we see that the curves intersect at $\theta = 90^\circ$ and $\theta = 180^\circ$.

$$\begin{aligned} A &= \frac{1}{2}\int_{\pi/2}^{\pi}\sin^2\theta\,d\theta - \frac{1}{2}\int_{\pi/2}^{\pi}(1+\cos\theta)^2\,d\theta \\ &= \frac{1}{2}\int_{\pi/2}^{\pi}\sin^2\theta\,d\theta - \frac{1}{2}\int_{\pi/2}^{\pi}(1 + 2\cos\theta + \cos^2\theta)\,d\theta \\ &= \frac{1}{4}\int_{\pi/2}^{\pi}(1-\cos 2\theta)\,d\theta - \frac{1}{2}\int_{\pi/2}^{\pi}\left[1 + 2\cos\theta + \frac{1}{2}(1+\cos 2\theta)\right]d\theta \\ &= \frac{1}{4}\left(\theta - \frac{1}{2}\sin 2\theta\right)\Big|_{\pi/2}^{\pi} - \frac{1}{2}\left[\theta + 2\sin\theta + \frac{1}{2}\left(\theta + \frac{1}{2}\sin 2\theta\right)\right]\Big|_{\pi/2}^{\pi} \\ &= \frac{1}{4}\left(\pi - \frac{\pi}{2}\right) - \frac{1}{2}\left[\left(\pi + \frac{\pi}{2}\right) - \left(\frac{\pi}{2} + 2 + \frac{\pi}{4}\right)\right] \\ &= \frac{1}{4}\left(\frac{\pi}{2}\right) - \frac{1}{2}\left(\frac{3\pi}{4} - 2\right) = 1 - \frac{\pi}{4}. \end{aligned}$$

39. (Partial solution)

Shaded area:

$$\frac{1}{2}\int_0^{\pi/2} 16(1+\sin\theta)^2\,d\theta - \frac{1}{2}\int_0^{\pi/2} 64\sin^2\theta\,d\theta.$$

Area below polar axis (right side):

$$\frac{1}{2}\int_{-\pi/2}^{0} 16(1+\sin\theta)^2\,d\theta.$$

Total area $= (32-4\pi)+(12\pi-32)=8\pi$.

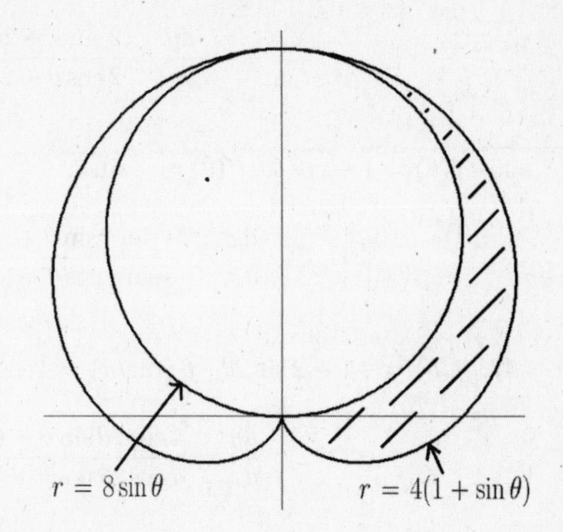

41.

Substituting $r = 3\cos\theta$ in $r = 1+\cos\theta$, we get:

$$3\cos\theta = 1+\cos\theta$$

$$\cos\theta = \frac{1}{2}$$

$$\theta = \frac{\pi}{3},\ -\frac{\pi}{3}.$$

Area 1:

$$\frac{1}{2}\int_{\pi/3}^{\pi/2}(3\cos\theta)^2\,d\theta$$

$$= \frac{9}{2}\int_{\pi/3}^{\pi/2}\cos^2\theta\,d\theta$$

$$= \frac{9}{4}\int_{\pi/3}^{\pi/2}(1+\cos 2\theta)\,d\theta$$

$$= \frac{9}{4}\left(\theta+\frac{1}{2}\sin 2\theta\right)\Big|_{\pi/3}^{\pi/2}$$

$$= \frac{9}{4}\left[\left(\frac{\pi}{2}\right)-\left(\frac{\pi}{3}+\frac{1}{2}\frac{\sqrt{3}}{2}\right)\right] = \frac{9}{4}\left(\frac{\pi}{2}-\frac{\pi}{3}-\frac{\sqrt{3}}{4}\right) = \frac{9}{4}\left(\frac{\pi}{6}-\frac{\sqrt{3}}{4}\right).$$

Area 2:

$$\begin{aligned}
\frac{1}{2}\int_0^{\pi/3}(1+\cos\theta)^2\,d\theta &= \frac{1}{2}\int_0^{\pi/3}(1+2\cos\theta+\cos^2\theta)\,d\theta\\
&= \frac{1}{2}\int_0^{\pi/3}\left[1+2\cos\theta+\frac{1}{2}(1+\cos 2\theta)\right]d\theta\\
&= \frac{1}{2}\left[\theta+2\sin\theta+\frac{1}{2}\left(\theta+\frac{1}{2}\sin 2\theta\right)\right]\Big|_0^{\pi/3}\\
&= \frac{1}{2}\left(\frac{\pi}{3}+\frac{2\sqrt{3}}{2}+\frac{1}{2}\frac{\pi}{3}+\frac{1}{4}\frac{\sqrt{3}}{2}\right)\\
&= \frac{1}{2}\left(\frac{\pi}{2}+\frac{9\sqrt{3}}{8}\right).
\end{aligned}$$

$$\begin{aligned}
\text{Total area} &= 2(\text{Area } 1+\text{Area } 2)\\
&= \frac{9}{2}\left(\frac{\pi}{6}-\frac{\sqrt{3}}{4}\right)+\left(\frac{\pi}{2}+\frac{9\sqrt{3}}{8}\right)\\
&= \frac{9\pi}{12}-\frac{9\sqrt{3}}{8}+\frac{\pi}{2}+\frac{9\sqrt{3}}{8}=\frac{3\pi}{4}+\frac{2\pi}{4}=\frac{5\pi}{4}.
\end{aligned}$$

43. $f(\theta) = 2\theta,\ f'(\theta) = 2.$

$$\frac{dy}{dx} = \left.\frac{2\sin\theta + 2\theta\cos\theta}{2\cos\theta - 2\theta\sin\theta}\right|_{\theta=\pi/2} = \frac{2+0}{0-2(\pi/2)} = -\frac{2}{\pi}$$

45. $f(\theta) = 1+\cos\theta,\ f'(\theta) = -\sin\theta.$

$$\frac{dy}{dx} = \left.\frac{-\sin\theta\sin\theta + (1+\cos\theta)\cos\theta}{-\sin\theta\cos\theta - (1+\cos\theta)\sin\theta)}\right|_{\theta=3\pi/2} = \frac{-1+0}{0+1} = -1$$

47. $f(\theta) = -1+2\sin 2\theta,\ f'(theta) = 4\cos 2\theta.$

$$\frac{dy}{dx} = \left.\frac{4\cos 2\theta\sin\theta + (-1+2\sin 2\theta)\cos\theta}{4\cos 2\theta\cos\theta - (-1+2\sin 2\theta)\sin\theta}\right|_{\theta=0} = \frac{0-1}{4-0} = -\frac{1}{4}$$

49. $f(\theta) = \sqrt{4\sin 2\theta},\ f'(\theta) = \dfrac{4\cos 2\theta}{\sqrt{4\sin 2\theta}}$; at $\theta = \dfrac{\pi}{4},\ f'(\theta) = 0.$. So

$$\frac{dy}{dx} = \left.\frac{0(\sin\theta) + \sqrt{4\sin 2\theta}\cos\theta}{0(\cos\theta) - \sqrt{4\sin 2\theta}\sin\theta}\right|_{\theta=\pi/4} = \frac{2\left(\frac{1}{\sqrt{2}}\right)}{-2\left(\frac{1}{\sqrt{2}}\right)} = -1$$

Chapter 8 Review

1.
$$\begin{aligned} x &= 4\cos^2\theta \\ y &= 4\sin\theta \\ \hline \frac{x}{4} &= \cos^2\theta \\ \frac{y^2}{16} &= \sin^2\theta \\ \hline \frac{x}{4} + \frac{y^2}{16} &= \cos^2\theta + \sin^2\theta = 1 \\ 4x + y^2 &= 16. \end{aligned}$$

3. (a) $x = 5\cos t;\ v_x = -5\sin t,\ a_x = -5\cos t.$
$y = 5\sin t;\ v_y = 5\cos t,\ a_y = -5\sin t.$
When $t = \dfrac{\pi}{4}$, we have
$\vec{v} = -\dfrac{5}{\sqrt{2}}\vec{i} + \dfrac{5}{\sqrt{2}}\vec{j},\ |\vec{v}| = \sqrt{\left(-\dfrac{5}{\sqrt{2}}\right)^2 + \left(\dfrac{5}{\sqrt{2}}\right)^2} = \sqrt{25} = 5\,\text{m/s};\ \theta = 135°;$
$\vec{a} = -\dfrac{5}{\sqrt{2}}\vec{i} - \dfrac{5}{\sqrt{2}}\vec{j},\ |\vec{a}| = 5\,\text{m/s}^2;\ \theta = 225°.$

(b) $x = 5\sin t;\ v_x = 5\cos t,\ a_x = -5\sin t.$
$y = 5\cos t;\ v_y = -5\sin t,\ a_y = -5\cos t.$
When $t = \dfrac{\pi}{4}$, we get
$\vec{v} = \dfrac{5}{\sqrt{2}}\vec{i} - \dfrac{5}{\sqrt{2}}\vec{j},\ |\vec{v}| = 5\,\text{m/s};\ \theta = 315°;$
$\vec{a} = -\dfrac{5}{\sqrt{2}}\vec{i} - \dfrac{5}{\sqrt{2}}\vec{j}$ (same as in part (a)).

5. $y = \ln \sec x;\ y' = \dfrac{\sec x \tan x}{\sec x} = \tan x;\ (y')^2 = \tan^2 x.$
$1 + (y')^2 = 1 + \tan^2 x = \sec^2 x.$

$$\begin{aligned} s &= \int_0^{\pi/4} \sqrt{\sec^2 x}\, dx = \int_0^{\pi/4} \sec x\, dx = \ln|\sec x + \tan x|\Big|_0^{\pi/4} \\ &= \ln(\sqrt{2}+1) - \ln 1 = \ln(1+\sqrt{2}). \qquad \text{(since } \ln 1 = 0\text{)} \end{aligned}$$

7. Since $y = r\sin\theta$ and $x = r\cos\theta$, $y = 3x$ becomes

$$\begin{aligned} (r\sin\theta)^2 &= 3(r\cos\theta) \\ r^2\sin^2\theta &= 3r\cos\theta, \end{aligned}$$

which can be rewritten as follows:

$$r = \frac{3\cos\theta}{\sin^2\theta} = 3\frac{\cos\theta}{\sin\theta}\frac{1}{\sin\theta} = 3\cot\theta\csc\theta.$$

9.
$$\begin{aligned} r^2 &= 4\tan\theta = 4\frac{\sin\theta}{\cos\theta} = 4\frac{r\sin\theta}{r\cos\theta} \\ x^2+y^2 &= 4\frac{y}{x} \\ x(x^2+y^2) &= 4y. \end{aligned}$$

11. $r = \sin\theta$ is a circle (one-leaf rose).

13. Four-leaf rose.

$r = \pm 2$ when $\theta = 0°,\ 90°,\ 180°,\ 270°.$

$r = 0$ when $\theta = 45°,\ 135°,\ 225°,\ 315°.$

15. $r^2 = 9\sin 2\theta$ is a lemniscate: $r = 0$ when $\theta = 0°$ and $\theta = 90°$, $r = \pm\sqrt{9} = \pm 3$ when $\theta = 45°$.

17.
$$\begin{aligned} A &= \frac{1}{2}\int_0^{\pi/8} \tan^2 2\theta\, d\theta = \frac{1}{2}\int_0^{\pi/8} (\sec^2 2\theta - 1)\, d\theta \\ &= \frac{1}{2}\left(\frac{1}{2}\tan 2\theta - \theta\right)\Big|_0^{\pi/8} = \frac{1}{2}\left(\frac{1}{2} - \frac{\pi}{8}\right) = \frac{1}{4} - \frac{\pi}{16}. \end{aligned}$$

19. The function $r = \sqrt{a^2\cos 2\theta}$ is the right half of the lemniscate, shown in the figure. When $\theta = 0$, $r = a$ and when $\theta = 45°$, $r = 0$. So the shaded region is traced out in the interval $\theta = 0$ to $\theta = \dfrac{\pi}{4}$. The area enclosed by the entire lemniscate is four times the shaded area. Thus

$$\begin{aligned} A &= 4\cdot\frac{1}{2}\int_0^{\pi/4}\left(\sqrt{a^2\cos 2\theta}\right)^2 d\theta = 2\int_0^{\pi/4} a^2\cos 2\theta\, d\theta \\ &= a^2\int_0^{\pi/4}\cos 2\theta(2\, d\theta) = a^2\sin 2\theta\Big|_0^{\pi/4} = a^2. \end{aligned}$$

21.

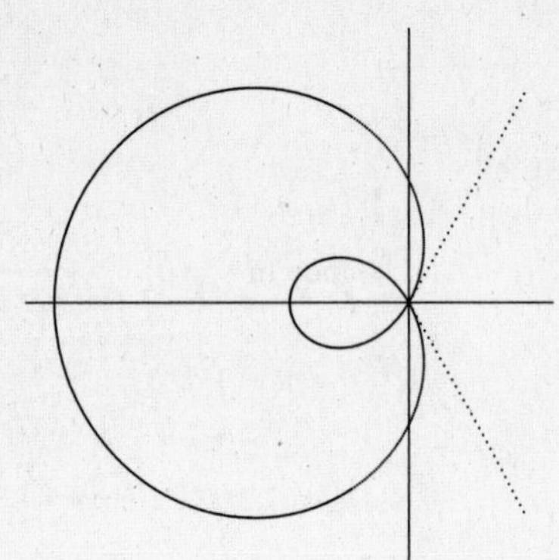

$$\begin{aligned} r = 0 \text{ when } 1 - 2\cos\theta &= 0 \\ \cos\theta &= \frac{1}{2} \\ \theta &= -\frac{\pi}{3}, \frac{\pi}{3}. \end{aligned}$$

$$\begin{aligned} A &= 2 \cdot \frac{1}{2} \int_0^{\pi/3} (1 - 2\cos\theta)^2 \, d\theta = \int_0^{\pi/3} (1 - 4\cos\theta + 4\cos^2\theta) \, d\theta \\ &= \int_0^{\pi/3} [1 - 4\cos\theta + 2(1 + \cos 2\theta)] \, d\theta = \theta - 4\sin\theta + 2\theta + \sin 2\theta \Big|_0^{\pi/3} \\ &= \frac{\pi}{3} - 4\frac{\sqrt{3}}{2} + \frac{2\pi}{3} + \frac{\sqrt{3}}{2} = \pi - \frac{3}{2}\sqrt{3} = \frac{1}{2}\left(2\pi - 3\sqrt{3}\right). \end{aligned}$$

23.

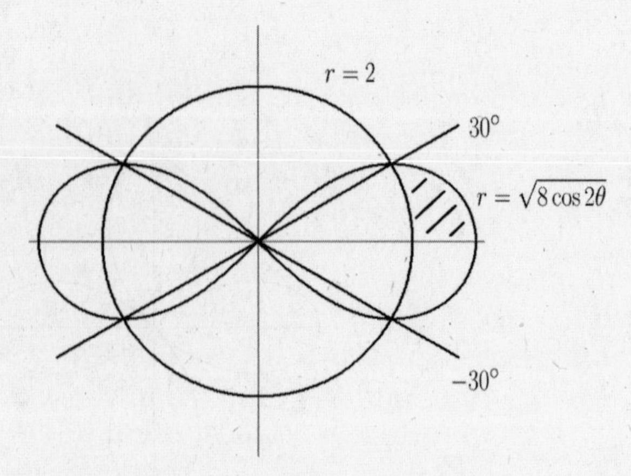

Substituting $r = 2$ in $r^2 = 8\cos 2\theta$, we get

$$\begin{aligned} 4 &= 8\cos 2\theta \\ \cos 2\theta &= \frac{1}{2} \\ 2\theta &= \pm 60° \\ \theta &= \pm 30° = \pm\frac{\pi}{6}. \end{aligned}$$

The desired area is equal to four times the shaded area:

$$\begin{aligned} A &= 4 \cdot \frac{1}{2} \int_0^{\pi/6} \left[\left(\sqrt{8\cos 2\theta}\right)^2 - 2^2 \right] d\theta = 2\int_0^{\pi/6} (8\cos 2\theta - 4) \, d\theta \\ &= 8\sin 2\theta - 8\theta \Big|_0^{\pi/6} = 8\sin\frac{\pi}{3} - 8\left(\frac{\pi}{6}\right) \\ &= 8\left(\frac{\sqrt{3}}{2}\right) - \frac{4\pi}{3} = 4\left(\sqrt{3} - \frac{\pi}{3}\right). \end{aligned}$$

25. $f(\theta) = 1 - \sin\theta$; $f'(\theta) = -\cos\theta$.

$$\frac{dy}{dx} = \frac{-\cos\theta\sin\theta + (1 - \sin\theta)\cos\theta}{-\cos\theta\cos\theta - (1 - \sin\theta)\sin\theta} = \frac{0 + 1}{-1 + 0} = -1$$

Chapter 9

Three Dimensional Space; Partial Derivatives; Multiple Integrals

9.1 Surfaces in Three Dimensions

1. The trace in the xy-plane is the circle $x^2 + y^2 = 4$.

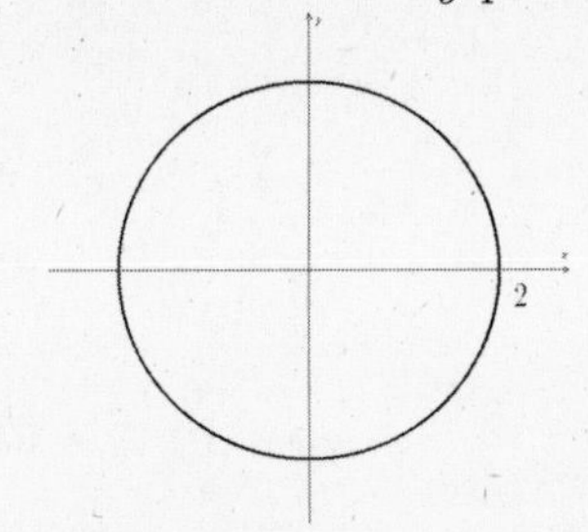

In three-space, $x^2 + y^2 = 4$ represents a cylinder (note the missing z-variable) whose axis is the z-axis.

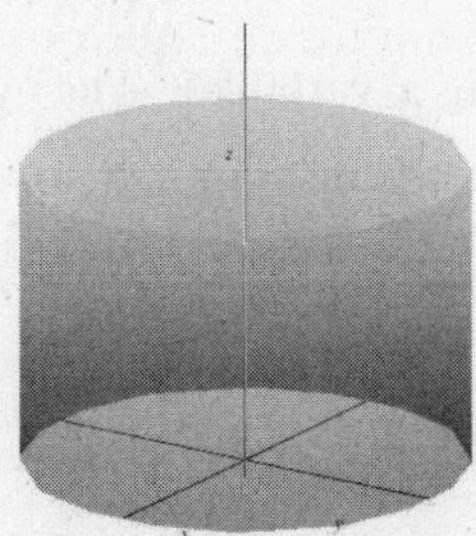

3. The trace in the xy-plane is the parabola $y^2 = 9x$.

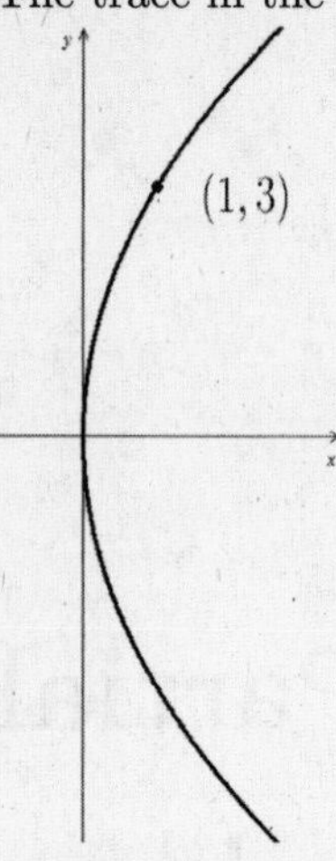

In three-space, $y^2 = 9x$ represents a cylinder extending along the z-axis: every cross-section parallel to the xy-plane is the parabola $y^2 = 9x$.

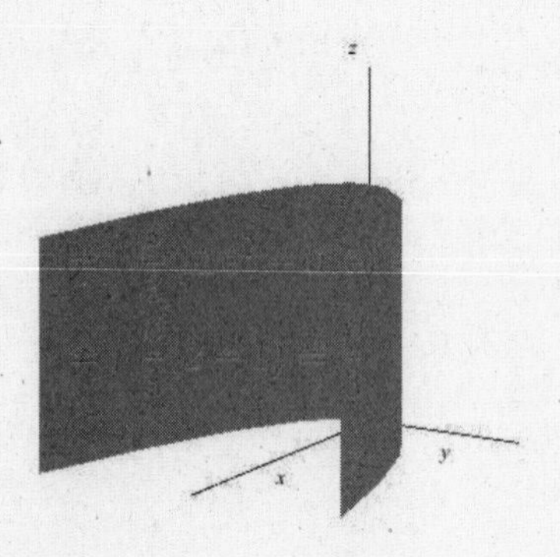

5. The trace in the yz-plane is the ellipse $y^2 + 4z^2 = 4$.

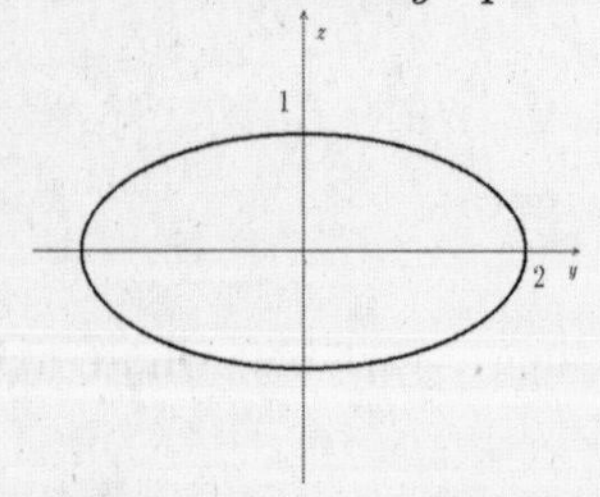

In three-space, $y^2 + 4z^2 = 4$ represents a cylinder (note the missing x-variable). The cylinder extends along the x-axis: every cross-section parallel to the yz-plane is the ellipse $y^2 + 4z^2 = 4$.

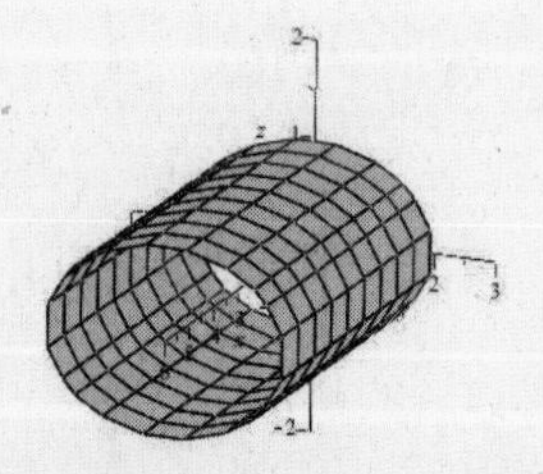

7. The trace in the xz-plane is the parabola $z = 3x - 3x^2$.

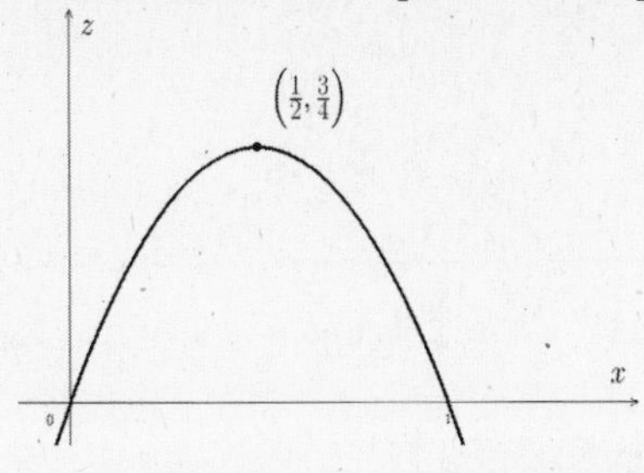

In three-space, $z = 3x - 3x^2$ represents a cylinder extending along the y-axis: every cross-section parallel to the xz-plane is the parabola $z = 3x - 3x^2$.

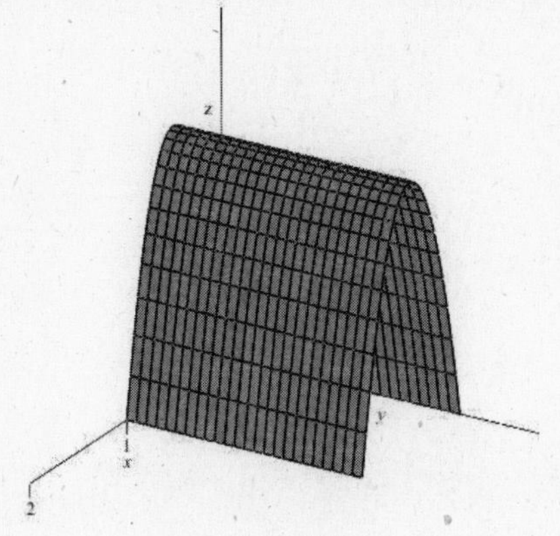

9. Trace in the xy-plane: $y = e^x$.

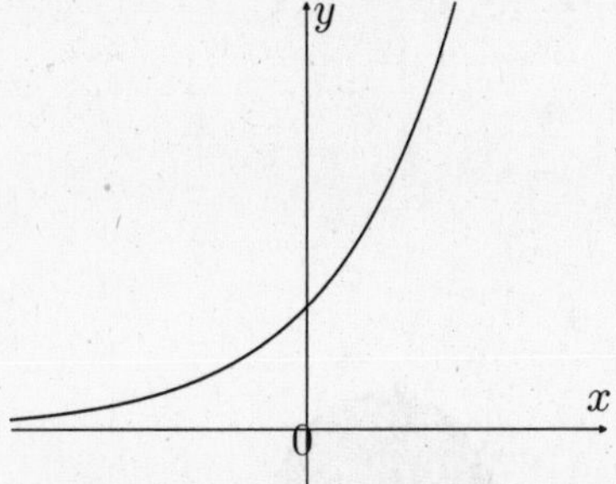

In three-space, $y = e^x$ represents a cylinder extending along the z-axis. (See graph in Answer Section.)

11. The plane $2x - 3y + z = 6$ can be sketched from the intercepts, since 3 noncollinear points determine a plane. (See graph in Answer Section.)

13. $z = x$ is a line in the xz-plane. Because of the missing y-variable, the plane $z = x$ extends along the y-axis.

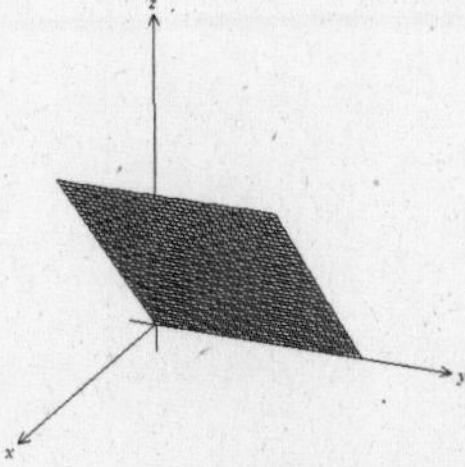

15. $x^2 + y^2 + z^2 = 9$ is a sphere of radius 3.

17. $9x^2 + 4y^2 + z^2 = 36$ (ellipsoid)

1. Trace in xy-plane ($z = 0$): $9x^2 + 4y^2 = 36$.

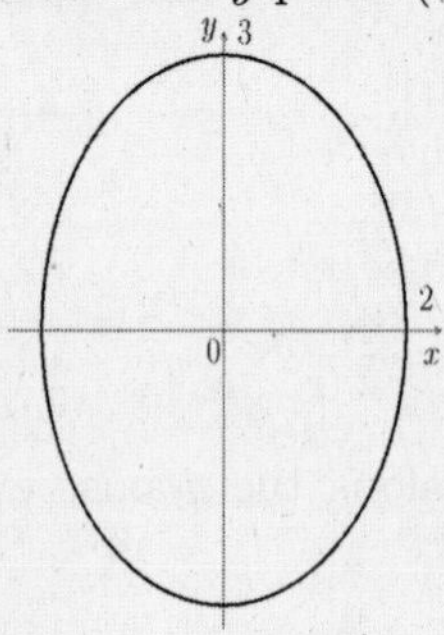

2. Trace in xz-plane ($y = 0$): $9x^2 + z^2 = 36$.

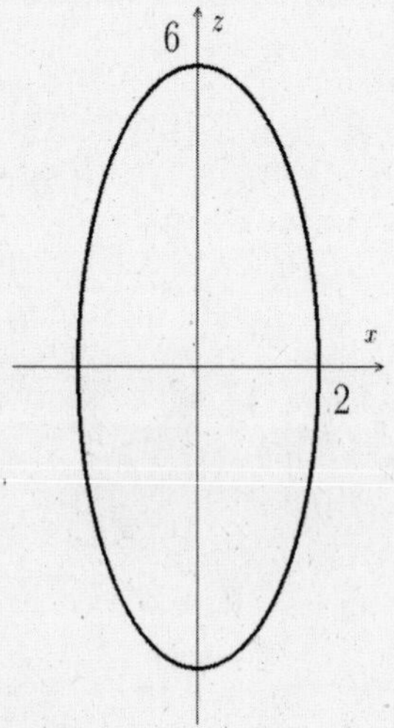

3. Trace in yz-plane ($x = 0$): $4y^2 + z^2 = 36$.

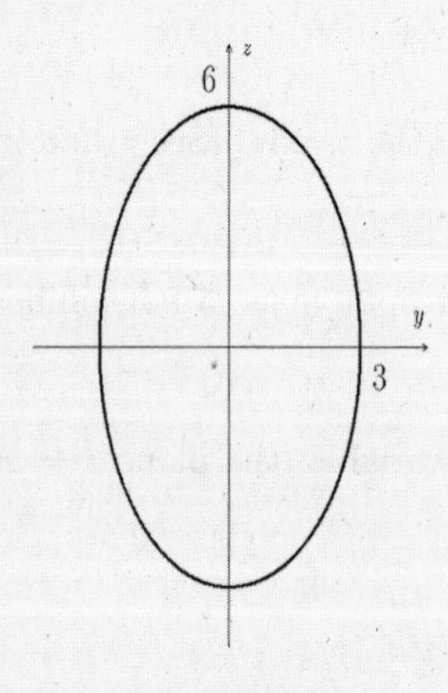

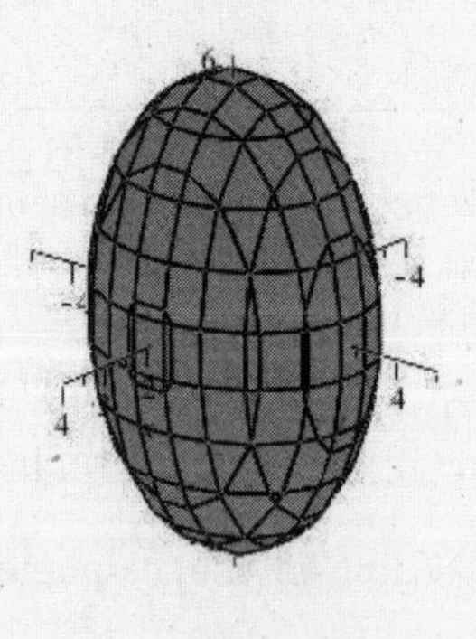

19. $z = x^2 + y^2$ is a paraboloid similar to Example 1.

21. $z = 1 + 2x^2 + 4y^2$ (paraboloid)

1. Trace in xy-plane ($z = 0$): $0 = 1 + 2x^2 + 4y^2$ or $2x^2 + 4y^2 = -1$ (imaginary locus).

2. Trace in xz-plane ($y = 0$): $z = 1 + 2x^2$.

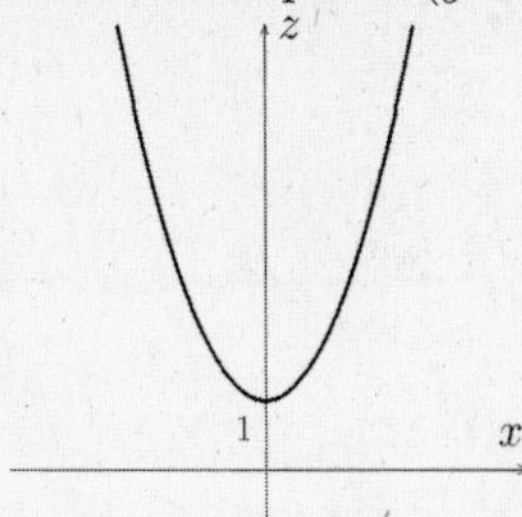

3. Trace in yz-plane ($x = 0$): $z = 1 + 4y^2$.

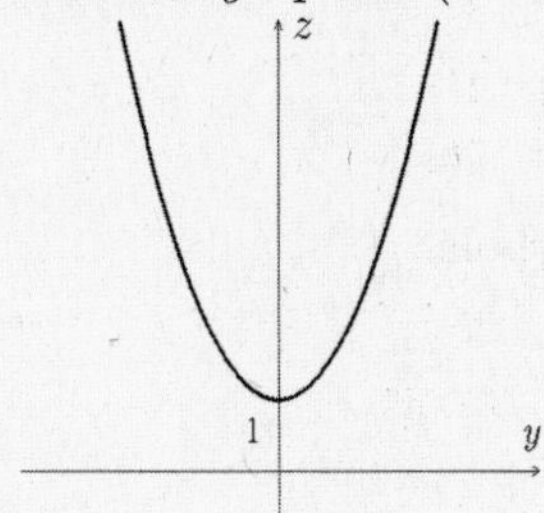

Cross-section: let $z = 2$ in the given equation. The resulting ellipse, $2x^2 + 4y^2 = 1$, is parallel to the xy-plane and two units above.

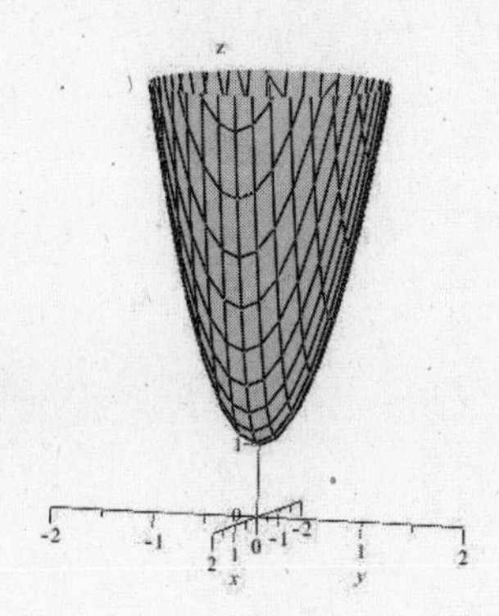

23. $4x^2 + y^2 - z^2 = 4$ (hyperboloid of one sheet)

1. Trace in xy-plane ($z = 0$): $4x^2 + y^2 = 4$.

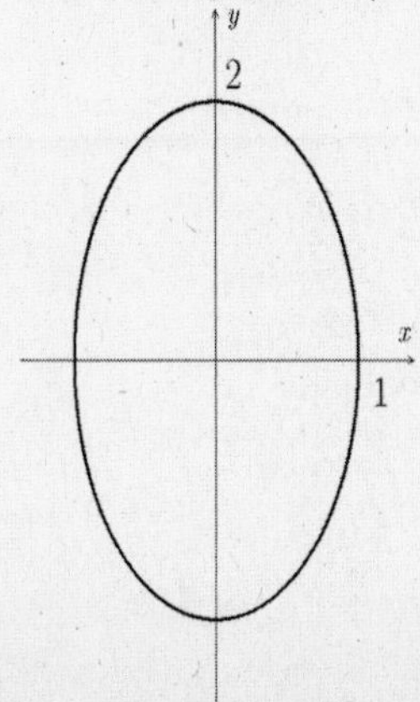

2. Trace in yz-plane ($x = 0$): $y^2 - z^2 = 4$.

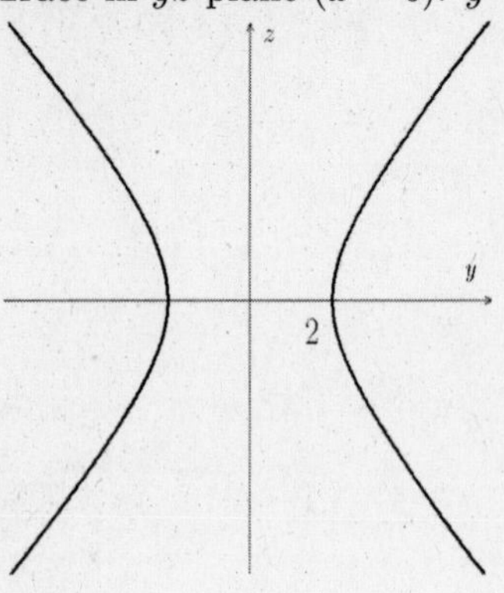

3. Trace in xz-plane ($y = 0$): $4x^2 - z^2 = 4$, another hyperbola.

Cross-section: let $z = \pm 2 : 4x^2 + y^2 = 8$ (ellipse).

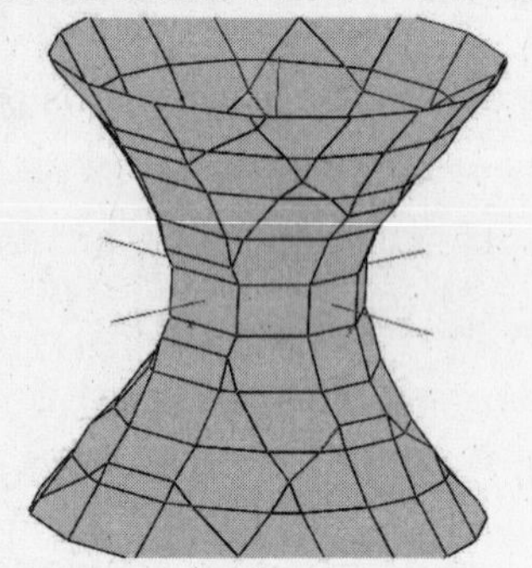

25. $-2x^2 - y^2 + z^2 = 6$ (hyperboloid of two sheets)

1. Trace in xy-plane ($z = 0$): $-2x^2 - y^2 = 6$ (imaginary locus).

2. Trace in xz-plane ($y = 0$): $-2x^2 + z^2 = 6$.

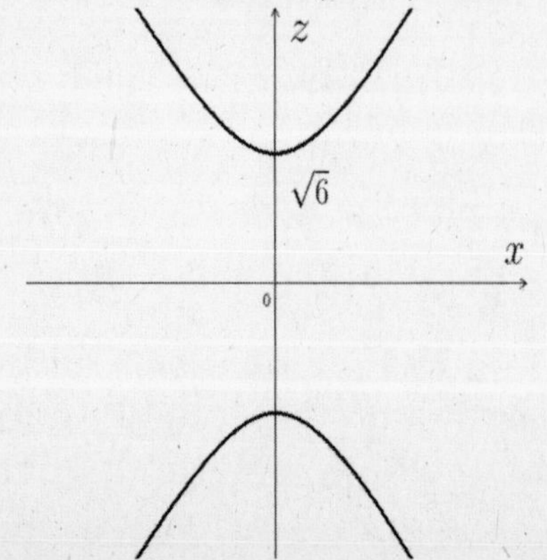

3. Trace in yz-plane ($x = 0$): $-y^2 + z^2 = 6$.

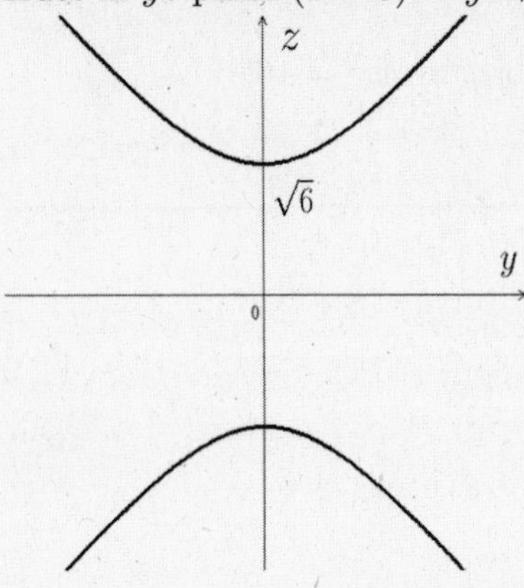

Cross-section: let $z = 3$ and $z = -3$ in the given equation. In each case, the result is an ellipse:

$$-2x^2 - y^2 + 9 = 6 \text{ or } 2x^2 + y^2 = 3.$$

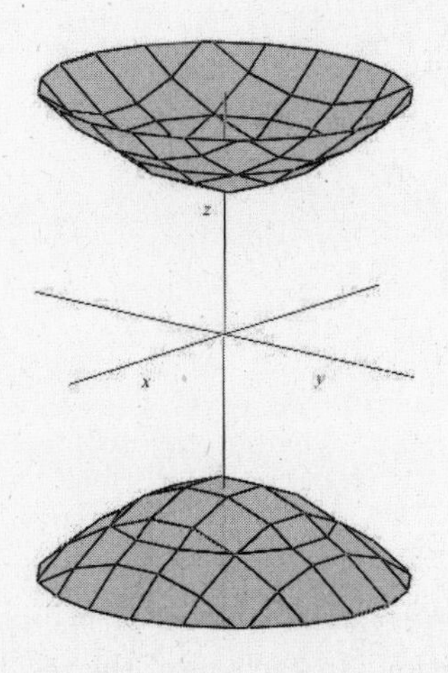

27. $x^2 - 4y^2 + z^2 = 4$ (hyperboloid of one sheet)

1. Trace in xz-plane ($y = 0$): $x^2 + z^2 = 4$ (circle).
2. Trace in yz-plane ($x = 0$): $-4y^2 + z^2 = 4$.

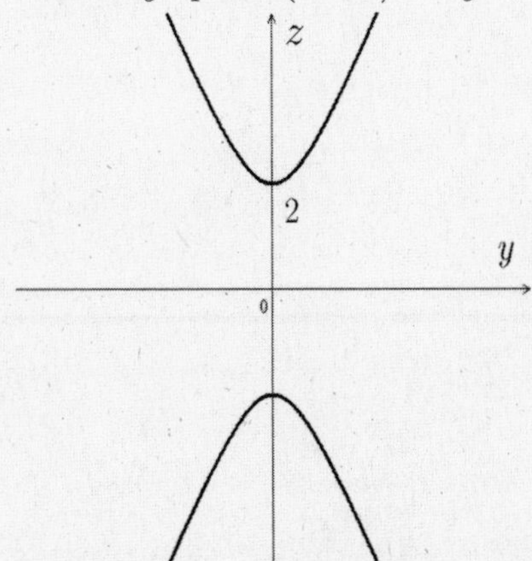

3. Trace in xy-plane ($z = 0$): $x^2 - 4y^2 = 4$, another hyperbola.

Cross-section: $y = \pm 1 : x^2 + z^2 = 8$ (circle). (See graph in Answer Section.)

29. $9y^2 - 36x^2 - 16z^2 = 144$ (hyperboloid of two sheets)

1. Trace in xy-plane ($z = 0$): $9y^2 - 36x^2 = 144$ or $y^2 - 4x^2 = 16$.

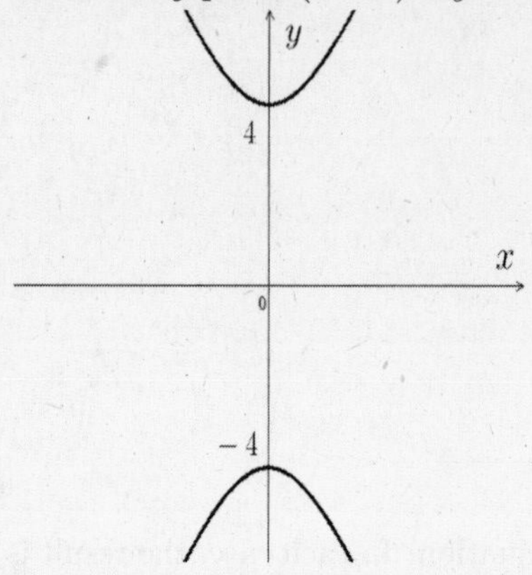

2. Trace in xz-plane ($y = 0$): $-36x^2 - 16z^2 = 144$ (imaginary locus).
3. Trace in yz-plane ($x = 0$): $9y^2 - 16z^2 = 144$.

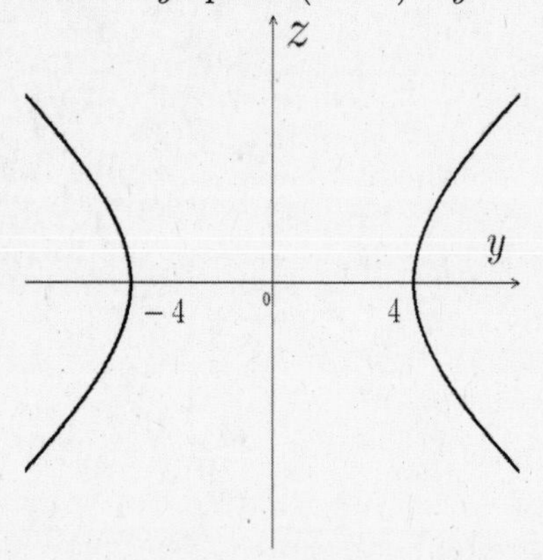

Cross-section: let $y = 8$ and $y = -8$ in the given equation. In each case, the resulting ellipse,

$$9(8)^2 - 36x^2 - 16z^2 = 144 \text{ or } 9x^2 + 4z^2 = 108$$

is parallel to the xz-plane and 8 units to the right (respectively left).

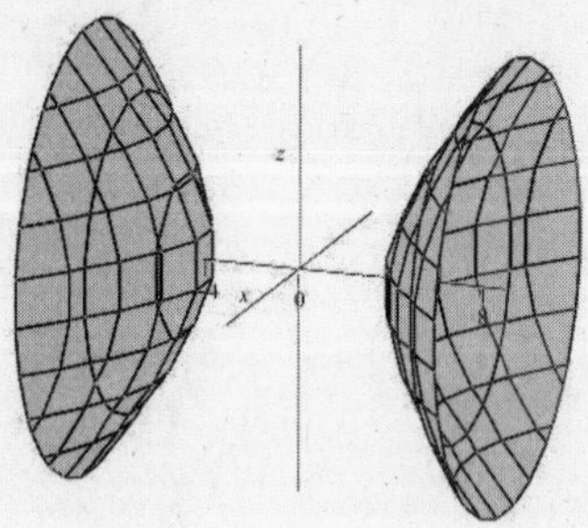

31. $z^2 = 2x^2 + y^2$ (cone)

1. Trace in yz-plane: $z = \pm y$ (2 intersecting lines).
2. Trace in xz-plane: $z = \pm\sqrt{2}x$ (2 intersecting lines).
3. Cross-section: let $z = 1$ to get the ellipse $2x^2 + y^2 = 1$. (See graph in Answer Section.)

33. Saddle shape; see graph in Answer Section.

9.2 Partial Derivatives

1. $f(x, y) = 2x^2 + 5y^2 + 1$

   (a) $\dfrac{\partial f}{\partial x} = 4x + 0 = 4x$, since y is a constant.

   (b) $\dfrac{\partial f}{\partial y} = 0 + 10y = 10y$, since x is a constant.

3. $f(x, y) = x + \sin 2y$

   (a) $\dfrac{\partial f}{\partial x} = 1 + 0 = 1$, since $\sin 2y$ is a constant.

   (b) $\dfrac{\partial f}{\partial y} = 0 + \cos 2y \cdot 2 = 2\cos 2y$.

5. $f(x, y) = 2x^2 + 3\tan y$

   (a) $\dfrac{\partial f}{\partial y} = 4x + 0 = 4x$, since y is a constant.

   (b) $\dfrac{\partial f}{\partial y} = 0 + 3\sec^2 y = 3\sec^2 y$, since x is a constant.

7. $f(x, y) = 2\ln x - \operatorname{Arctan} y$

   (a) Since y is treated as a constant, $\dfrac{\partial}{\partial x}(-\operatorname{Arctan} y) = 0$; so $\dfrac{\partial f}{\partial x} = \dfrac{2}{x}$.

   (b) Similarly, if x is a constant, then $\dfrac{\partial}{\partial y}(2\ln x) = 0$; so $\dfrac{\partial f}{\partial y} = -\dfrac{1}{1 + y^2}$.

9. $f(x, y) = x\mathrm{e}^{2y}$

   (a) Since e^{2y} is a constant coefficient, $\dfrac{\partial f}{\partial x} = \mathrm{e}^{2y} \cdot 1 = \mathrm{e}^{2y}$.

   (b) Now x is a constant coefficient: $\dfrac{\partial f}{\partial y} = x\dfrac{\partial}{\partial y}\mathrm{e}^{2y} = x\mathrm{e}^{2y}(2) = 2x\mathrm{e}^{2y}$.

11. $f(x, y) = 2y^2 \ln x$

    (a) $\dfrac{\partial f}{\partial x} = \left(2y^2\right)\dfrac{\partial}{\partial x}\ln x = 2y^2\left(\dfrac{1}{x}\right) = \dfrac{2y^2}{x}$.

    (b) $\dfrac{\partial f}{\partial y} = \ln x \dfrac{\partial}{\partial y}\left(2y^2\right) = 4y\ln x$.

13. $f(x, y) = x^2 \cos xy$

    (a) For this case we need the product rule:

    $$\frac{\partial f}{\partial x} = x^2\left(-\sin xy\right)(y) + 2x\cos xy = -x^2 y \sin xy + 2x\cos xy$$

    (b) Since x^2 is a constant coefficient,

    $$\frac{\partial f}{\partial y} = x^2\frac{\partial}{\partial y}\cos xy = x^2(-\sin xy)x = -x^3 \sin xy$$

15. $f(x, y) = y^2 \tan xy$

    (a) Since y^2 is a constant coefficient,
    $\dfrac{\partial f}{\partial x} = y^2\dfrac{\partial}{\partial x}\tan xy = y^2\sec^2 xy \cdot y = y^3\sec^2 xy$.

(b) For this case, we need the product rule:

$$\begin{aligned}\frac{\partial f}{\partial y} &= y^2\frac{\partial}{\partial y}\tan xy + \tan xy\frac{\partial}{\partial y}y^2 \\ &= y^2\sec^2 xy\cdot x + \tan xy\cdot 2y \\ &= xy^2\sec^2 xy + 2y\tan xy.\end{aligned}$$

17. $f(x,y) = \dfrac{\cos x^2 y}{y}$

(a) $f(x,y) = \dfrac{1}{y}\cos x^2 y$. Since y is a constant coefficient, we get

$$\begin{aligned}\frac{\partial f}{\partial x} &= \frac{1}{y}\frac{\partial}{\partial x}\cos x^2 y = \frac{1}{y}(-\sin x^2 y)\frac{\partial}{\partial x}(x^2 y) \\ &= \frac{1}{y}(-\sin x^2 y)(2xy) = -2x\sin x^2 y.\end{aligned}$$

(b) Since y and $\cos x^2 y$ are both functions of y, we need the quotient rule:

$$\begin{aligned}\frac{\partial f}{\partial y} &= \frac{y\frac{\partial}{\partial y}\cos x^2 y - \cos x^2 y\frac{\partial}{\partial y}(y)}{y^2} \\ &= \frac{y(-\sin x^2 y)\frac{\partial}{\partial y}(x^2 y) - \cos x^2 y\cdot 1}{y^2} \\ &= \frac{y(-\sin x^2 y)(x^2) - \cos x^2 y}{y^2} \\ &= \frac{-x^2 y\sin x^2 y - \cos x^2 y}{y^2}.\end{aligned}$$

19. $f(x,y) = \dfrac{(x+y^2)^{1/2}}{x}$

(a) By the quotient rule,

$$\frac{\partial f}{\partial x} = \frac{x\frac{\partial}{\partial x}(x+y^2)^{1/2} - (x+y^2)^{1/2}\frac{\partial}{\partial x}(x)}{x^2} = \frac{x\cdot\frac{1}{2}(x+y^2)^{-1/2}(1) - (x+y^2)^{1/2}\cdot 1}{x^2}.$$

To clear fractions, multiply numerator and denominator by $2(x+y^2)^{1/2}$:

$$\frac{\partial f}{\partial x} = \frac{x - 2(x+y^2)}{2x^2(x+y^2)^{1/2}} = \frac{-x-2y^2}{2x^2\sqrt{x+y^2}}.$$

(b) Writing $f(x) = \dfrac{1}{x}(x+y^2)^{1/2}$, $\dfrac{1}{x}$ is now a constant coefficient.

$$\frac{\partial f}{\partial y} = \frac{1}{x}\cdot\frac{1}{2}(x+y^2)^{-1/2}\frac{\partial}{\partial y}(x+y^2) = \frac{1}{2x(x+y^2)^{1/2}}(0+2y) = \frac{y}{x\sqrt{x+y^2}}.$$

21. $f(x,y) = 5x - 2y - 3$

(a) $\dfrac{\partial f}{\partial x} = 5$ (b) $\dfrac{\partial f}{\partial y} = -2$ (c) $\dfrac{\partial^2 f}{\partial x^2} = 0$

(d) $\dfrac{\partial^2 f}{\partial y^2} = 0$ (e) $\dfrac{\partial^2 f}{\partial x\partial y} = 0$

23. $f(x,y) = 2x^2 + 3y^2 + 3x^2y^2 + 5xy$

(a) $\dfrac{\partial f}{\partial x} = 4x + 6xy^2 + 5y$ (b) $\dfrac{\partial f}{\partial y} = 6y + 6x^2y + 5x$

(c) $\dfrac{\partial^2 f}{\partial x^2} = 4 + 6y^2$ (d) $\dfrac{\partial^2 f}{\partial y^2} = 6 + 6x^2$

(e) $\dfrac{\partial^2 f}{\partial x\partial y} = \dfrac{\partial}{\partial x}\left(\dfrac{\partial f}{\partial y}\right) = \dfrac{\partial}{\partial x}(6y + 6x^2y + 5x) = 12xy + 5$

or

$$\frac{\partial^2 f}{\partial x\partial y} = \frac{\partial^2 f}{\partial y\partial x} = \frac{\partial}{\partial y}\left(\frac{\partial f}{\partial x}\right) = \frac{\partial}{\partial y}(4x + 6xy^2 + 5y) = 12xy + 5$$

25. $f(x,y) = \sin(2x+y)$

(a) $\dfrac{\partial f}{\partial x} = 2\cos(2x+y)$ (b) $\dfrac{\partial f}{\partial y} = \cos(2x+y)$

(c) $\dfrac{\partial^2 f}{\partial x^2} = \dfrac{\partial}{\partial x}[2\cos(2x+y)] = -4\sin(2x+y)$

(d) $\dfrac{\partial^2 f}{\partial y^2} = \dfrac{\partial}{\partial y}[\cos(2x+y)] = -\sin(2x+y)$

(e) $\dfrac{\partial^2 f}{\partial x \partial y} = \dfrac{\partial}{\partial x}\left(\dfrac{\partial f}{\partial y}\right) = \dfrac{\partial}{\partial x}[\cos(2x+y)] = -2\sin(2x+y)$

or

$$\frac{\partial^2 f}{\partial x \partial y} = \frac{\partial^2 f}{\partial y \partial x} = \frac{\partial}{\partial y}\left(\frac{\partial f}{\partial x}\right) = \frac{\partial}{\partial y}[2\cos(2x+y)] = -2\sin(2x+y)$$

27. $f(x,y) = \ln(x^2 y) + \tan y$

(a) $\dfrac{\partial f}{\partial x} = \dfrac{1}{x^2 y}\dfrac{\partial}{\partial x}(x^2 y) + 0 = \dfrac{2xy}{x^2 y} = \dfrac{2}{x}$

(b) $\dfrac{\partial f}{\partial y} = \dfrac{1}{x^2 y}\dfrac{\partial}{\partial y}(x^2 y) + \sec^2 y = \dfrac{1}{y} + \sec^2 y$

(c) $\dfrac{\partial^2 f}{\partial x^2} = -\dfrac{2}{x^2}$

(d) $\dfrac{\partial^2 f}{\partial y^2} = -\dfrac{1}{y^2} + 2(\sec y)\dfrac{\partial}{\partial y}\sec y = -\dfrac{1}{y^2} + 2\sec^2 y \tan y$

(e) $\dfrac{\partial^2 f}{\partial x \partial y} = \dfrac{\partial}{\partial x}\left(\dfrac{\partial f}{\partial y}\right) = \dfrac{\partial}{\partial x}\left(\dfrac{1}{y} + \sec^2 y\right) = 0$

or

$\dfrac{\partial^2 f}{\partial y \partial x} = \dfrac{\partial}{\partial y}\left(\dfrac{\partial f}{\partial x}\right) = \dfrac{\partial}{\partial y}\left(\dfrac{2}{x}\right) = 0$

29. $f(x,y) = \operatorname{Arctan}\dfrac{y}{x}$

(a) $\dfrac{\partial f}{\partial x} = \dfrac{1}{1+\left(\frac{y}{x}\right)^2}\left(-\dfrac{y}{x^2}\right) = -\dfrac{y}{x^2+y^2}$

(b) $\dfrac{\partial f}{\partial y} = \dfrac{1}{1+\left(\frac{y}{x}\right)^2}\left(\dfrac{1}{x}\right) = \dfrac{1}{1+\frac{y^2}{x^2}} \cdot \dfrac{1}{x} \cdot \dfrac{x}{x} = \dfrac{x}{x^2+y^2}$

(c) $\dfrac{\partial^2 f}{\partial x^2} = \dfrac{\partial}{\partial x}\left[-y(x^2+y^2)^{-1}\right] = y(x^2+y^2)^{-2}(2x) = \dfrac{2xy}{(x^2+y^2)^2}$

(d) $\dfrac{\partial^2 f}{\partial y^2} = \dfrac{\partial}{\partial y}\left[x(x^2+y^2)^{-1}\right] = -x(x^2+y^2)^{-2}(2y) = -\dfrac{2xy}{(x^2+y^2)^2}$

(e)
$$\begin{aligned}
\frac{\partial^2 f}{\partial x \partial y} &= \frac{\partial}{\partial x}\left(\frac{\partial f}{\partial y}\right) = \frac{\partial}{\partial x}\frac{x}{x^2+y^2} \\
&= \frac{(x^2+y^2)\cdot 1 - x \cdot 2x}{(x^2+y^2)^2} && \text{quotient rule} \\
&= \frac{-x^2+y^2}{(x^2+y^2)^2} \quad \text{or}
\end{aligned}$$

$$\begin{aligned}
\frac{\partial^2 f}{\partial x \partial y} &= \frac{\partial^2 f}{\partial y \partial x} = \frac{\partial}{\partial y}\left(\frac{\partial f}{\partial x}\right) = \frac{\partial}{\partial y}\left(-\frac{y}{x^2+y^2}\right) \\
&= -\frac{(x^2+y^2)\cdot 1 - y \cdot 2y}{(x^2+y^2)^2} && \text{quotient rule} \\
&= -\frac{x^2-y^2}{(x^2+y^2)^2} = \frac{-x^2+y^2}{(x^2+y^2)^2}
\end{aligned}$$

31. $f(x,y,z) = \sqrt{xyz} = x^{1/2}y^{1/2}z^{1/2}$

(a) $\dfrac{\partial f}{\partial x} = \dfrac{1}{2}x^{-1/2}y^{1/2}z^{1/2} = \dfrac{\sqrt{yz}}{2\sqrt{x}}\cdot\dfrac{\sqrt{x}}{\sqrt{x}} = \dfrac{\sqrt{xyz}}{2x}$

(b) $\dfrac{\partial f}{\partial y} = x^{1/2}\cdot\dfrac{1}{2}y^{-1/2}z^{1/2} = \dfrac{\sqrt{xz}}{2\sqrt{y}}\cdot\dfrac{\sqrt{y}}{\sqrt{y}} = \dfrac{\sqrt{xyz}}{2y}$

(c) $\dfrac{\partial^2 f}{\partial x^2} = \dfrac{\partial}{\partial x}\left(\dfrac{1}{2}x^{-1/2}y^{1/2}z^{1/2}\right) = \dfrac{1}{2}\left(-\dfrac{1}{2}\right)x^{-3/2}y^{1/2}z^{1/2} = -\dfrac{\sqrt{yz}}{4x^{3/2}}\cdot\dfrac{\sqrt{x}}{\sqrt{x}} = -\dfrac{\sqrt{xyz}}{4x^2}$

(d) $\dfrac{\partial^2 f}{\partial y^2} = \dfrac{\partial}{\partial y}\left(x^{1/2}\cdot\dfrac{1}{2}y^{-1/2}z^{1/2}\right) = -\dfrac{1}{4}y^{-3/2}x^{1/2}z^{1/2} = -\dfrac{\sqrt{xz}}{4y^{3/2}}\cdot\dfrac{\sqrt{y}}{\sqrt{y}} = -\dfrac{\sqrt{xyz}}{4y^2}$

(e) $\dfrac{\partial^2 f}{\partial x\partial y} = \dfrac{\partial}{\partial x}\left(x^{1/2}\cdot\dfrac{1}{2}y^{-1/2}z^{1/2}\right) = \dfrac{1}{2}x^{-1/2}\cdot\dfrac{1}{2}y^{-1/2}z^{1/2} = \dfrac{\sqrt{z}}{4\sqrt{xy}}\cdot\dfrac{\sqrt{xy}}{\sqrt{xy}} = \dfrac{\sqrt{xyz}}{4xy}$

33. (a) $\dfrac{\partial}{\partial x}(ye^x) = ye^x\Big|_{(1,-1,-e)} = (-1)e = -e$

(b) $\dfrac{\partial}{\partial y}(ye^x) = e^x\Big|_{(1,-1,-e)} = e$

35. $z = \ln(x^2+y)$ at $\left(\dfrac{1}{\sqrt{2}},\dfrac{1}{2},0\right)$

(a) $\dfrac{\partial}{\partial x}\ln(x^2+y) = \dfrac{1}{x^2+y}(2x) = \dfrac{2\left(1/\sqrt{2}\right)}{\left(1/\sqrt{2}\right)^2+1/2} = \dfrac{2}{\sqrt{2}}\cdot\dfrac{\sqrt{2}}{\sqrt{2}} = \sqrt{2}$

(b) $\dfrac{\partial}{\partial y}\ln(x^2+y) = \dfrac{1}{x^2+y} = \dfrac{1}{\left(1/\sqrt{2}\right)^2+1/2} = 1$

37. $Z = \dfrac{RX}{R+X}$. By the quotient rule,

$\dfrac{\partial Z}{\partial R} = \dfrac{(R+X)(X)-RX(1)}{(R+X)^2} = \dfrac{RX+X^2-RX}{(R+X)^2} = \dfrac{X^2}{(R+X)^2}$.

If $R = 8.0$ and $X = 5.0$, $\dfrac{\partial Z}{\partial R} = \dfrac{(5.0)^2}{(8.0+5.0)^2} = 0.15$.

39. $\dfrac{1}{R} = \dfrac{R_1+R_2}{R_1R_2}$ and $R = \dfrac{R_1R_2}{R_1+R_2}$. By the quotient rule,

$\dfrac{\partial R}{\partial R_1} = \dfrac{(R_1+R_2)(R_2)-R_1R_2(1)}{(R_1+R_2)^2} = \dfrac{R_2^2}{(R_1+R_2)^2}$.

When $R_1 = 10.0\,\Omega$ and $R_2 = 20.0\,\Omega$,

$\dfrac{\partial R}{\partial R_1} = \dfrac{(20.0)^2}{(10.0+20.0)^2} = 0.444$.

41. $g_m = \dfrac{\partial}{\partial v_g}\left[0.50(v_g+0.1v_p)^{4/3}\right] = 0.50\left(\dfrac{4}{3}\right)(v_g+0.1v_p)^{1/3}(1) = 0.67(v_g+0.1v_p)^{1/3}\Omega^{-1}$.

43. $\dfrac{\partial T}{\partial x} = \dfrac{\partial}{\partial x}(1+x^3y) = 3x^2y\Big|_{(2,1)} = 3(4)(1) = 12°\,\text{C/m}$.

45. $y = A\cos\omega\left(t-\dfrac{x}{v}\right)$.

$$\begin{aligned}\frac{\partial y}{\partial x} &= -A\sin\omega\left(t-\frac{x}{v}\right)\frac{\partial}{\partial x}\omega\left(t-\frac{x}{v}\right) = \left[-A\sin\omega\left(t-\frac{x}{v}\right)\right]\left(-\frac{\omega}{v}\right)\\ &= \left(\frac{\omega}{v}\right)A\sin\omega\left(t-\frac{x}{v}\right).\\ \frac{\partial^2 y}{\partial x^2} &= \left[\left(\frac{\omega}{v}\right)A\cos\omega\left(t-\frac{x}{v}\right)\right]\left(-\frac{\omega}{v}\right) = \left(-\frac{\omega^2}{v^2}\right)A\cos\omega\left(t-\frac{x}{v}\right)\\ &= \frac{1}{v^2}\frac{\partial^2 y}{\partial t^2}\text{ since }\frac{\partial^2 y}{\partial t^2} = -\omega^2\left[A\cos\omega\left(t-\frac{x}{v}\right)\right].\end{aligned}$$

9.3 Applications of Partial Derivatives

1. $P = 2LV^2$. Pattern: $dP = \frac{\partial}{\partial L}(\ \ )dL + \frac{\partial}{\partial V}(\ \ )dV$.

$$dP = \frac{\partial}{\partial L}\left(2LV^2\right) dL + \frac{\partial}{\partial V}\left(2LV^2\right) dV = 2V^2\, dL + 4LV\, dV$$

3. $L = \frac{y}{\sqrt{z}} = yz^{-1/2}$. Pattern: $dL = \frac{\partial}{\partial y}(\ \ )dy + \frac{\partial}{\partial z}(\ \ )dz$.

$$dL = \frac{\partial}{\partial y}\left(yz^{-1/2}\right) dy + \frac{\partial}{\partial z}\left(yz^{-1/2}\right) dz = z^{-1/2}\, dy + y\left(-\frac{1}{2}\right) z^{-3/2}\, dz = \frac{dy}{\sqrt{z}} - \frac{y\, dz}{2z^{3/2}}$$

5. $M = \frac{\sin\theta}{\sin\phi} = \sin\theta\csc\phi$. Pattern: $dM = \frac{\partial}{\partial\theta}(\ \ )d\theta + \frac{\partial}{\partial\phi}(\ \ )d\phi$.

$$\begin{aligned} dM &= \frac{\partial}{\partial\theta}(\sin\theta\csc\phi)d\theta + \frac{\partial}{\partial\phi}(\sin\theta\csc\phi)d\phi \\ &= \cos\theta\csc\phi\, d\theta + (\sin\theta)(-\csc\phi\cot\phi)d\phi \\ &= \frac{\cos\theta}{\sin\phi}\, d\theta - \frac{\sin\theta}{\sin\phi\tan\phi}\, d\phi. \end{aligned}$$

7. $R = \mathrm{e}^{-2t/L} = \mathrm{e}^{-2tL^{-1}}$

$$\begin{aligned} dR &= \frac{\partial}{\partial L}\left(\mathrm{e}^{-2tL^{-1}}\right) dL + \frac{\partial}{\partial t}\left(\mathrm{e}^{-2tL^{-1}}\right) dt \\ &= \mathrm{e}^{-2tL^{-1}}(-2t)\left(-L^{-2}\right) dL + \mathrm{e}^{-2tL^{-1}}\left(-2L^{-1}\right) dt \\ &= \mathrm{e}^{-2t/L}\left(\frac{2t\, dL}{L^2} - \frac{2\, dt}{L}\right) \end{aligned}$$

9. $V = \frac{L}{P} = LP^{-1}$.

Differential:

$$\begin{aligned} dV &= \frac{\partial}{\partial L}(LP^{-1})dL + \frac{\partial}{\partial P}(LP^{-1})dP \\ dV &= P^{-1}dL + L(-P^{-2})dP \\ dV &= \frac{1}{P}dL - \frac{L}{P^2}dP. \end{aligned}$$

$L = 6.0,\ dL = \pm 0.05$

$P = 4.0,\ dP = \pm 0.02$

Approximate maximum error: either

$dV = \frac{1}{(4.0)}(0.05) - \frac{(6.0)}{(4.0)^2}(-0.02)$

or

$dV = \frac{1}{(4.0)}(-0.05) - \frac{(6.0)}{(4.0)^2}(0.02)$.

Thus $dV = \pm 0.02$. Approximate maximum percentage error:

$$\frac{dV}{V} \times 100 = \frac{0.02}{6.0/4.0} \times 100 = 1.3\%.$$

11. $f = \frac{\sqrt{r}}{s^2} = r^{1/2}s^{-2}$.

$$\begin{aligned} df &= \frac{\partial}{\partial r}\left(r^{1/2}s^{-2}\right) dr + \frac{\partial}{\partial s}\left(r^{1/2}s^{-2}\right) ds \\ &= \frac{1}{2}r^{-1/2}s^{-2}\, dr + r^{1/2}\left(-2s^{-3}\right) ds \\ &= \frac{1}{2\sqrt{r}s^2}\, dr - \frac{2\sqrt{r}}{s^3}\, ds \end{aligned}$$

$r = 2.50,\ dr = \pm 0.04;\ s = 3.75,\ ds = \pm 0.01$

Approximate maximum possible error: either

$$df = \frac{1}{2\sqrt{2.50}(3.75)^2}(0.04) - \frac{2\sqrt{2.50}}{(3.75)^3}(-0.01)$$

or

$$df = \frac{1}{2\sqrt{2.50}(3.75)^2}(-0.04) - \frac{2\sqrt{2.50}}{(3.75)^3}(0.01).$$

$df = \pm 0.0015$

Approximate maximum percentage error:

$$\frac{df}{f} \times 100 = \frac{0.0015}{\sqrt{2.50}/(3.75)^2} \times 100 = 1.3\%.$$

13. $V = \pi r^2 h.$

$$dV = \frac{\partial}{\partial r}(\pi r^2 h)\,dr + \frac{\partial}{\partial h}(\pi r^2 h)\,dh = 2\pi rh\,dr + \pi r^2\,dh;$$

$h = 10.0\,\text{cm},\ dh = 0.1\,\text{cm}$

$r = 5.00\,\text{cm},\ dr = 0.08\,\text{cm}$

Approximate maximum error: $\Delta V \approx dV = 2\pi(5.00)(10.0)(0.08) + \pi(5.00)^2(0.1) = 33\,\text{cm}^3.$

15. $i = 1 - e^{-R/L}.$

Differential:

$$\begin{aligned} di &= \frac{\partial}{\partial L}\left(1 - e^{-R/L}\right)dL + \frac{\partial}{\partial R}\left(1 - e^{-R/L}\right)dR \\ &= -e^{-R/L}\left(\frac{R}{L^2}\right)dL + \left(-e^{-R/L}\right)\left(-\frac{1}{L}\right)dR \\ di &= e^{-R/L}\left[-\frac{R}{L^2}\,dL + \frac{1}{L}dR\right]. \end{aligned}$$

$R = 1.2,\ dR = \pm 0.05$

$L = 0.70,\ dL = \pm 0.01$

Approximate maximum error: either

$$di = e^{-1.2/0.70}\left[-\frac{1.2}{(0.70)^2}(-0.01) + \frac{1}{0.70}(0.05)\right]$$

or

$$di = e^{-1.2/0.70}\left[-\frac{1.2}{(0.70)^2}(0.01) + \frac{1}{0.70}(-0.05)\right].$$

Thus $di = \pm 0.017\,\text{A}.$

Approximate maximum percentage error: $\dfrac{di}{i} \times 100 = \dfrac{0.017}{1 - e^{-1.2/0.70}} \times 100 = 2.1\%.$

17. $T = 2\pi\sqrt{L/g} = 2\pi L^{1/2} g^{-1/2}.$

$$\begin{aligned} dT &= \frac{\partial}{\partial L}\left(2\pi L^{1/2} g^{-1/2}\right)dL + \frac{\partial}{\partial g}\left(2\pi L^{1/2} g^{-1/2}\right)dg \\ &= \pi L^{-1/2} g^{-1/2}\,dL - \pi L^{1/2} g^{-3/2}\,dg \\ &= \pi\left(\frac{1}{\sqrt{Lg}}\,dL - \frac{\sqrt{L}}{g^{3/2}}\,dg\right). \end{aligned}$$

$L = 15.0\,\text{cm},\ dL = \pm 0.2\,\text{cm}$

$g = 980\,\text{cm/s}^2,\ dg = \pm 6, \text{cm/s}^2$

Approximate maximum error: either

$$dT = \pi\left[\frac{1}{\sqrt{(15.0)(980)}}(+0.2) - \frac{\sqrt{15.0}}{980^{3/2}}(-6)\right]$$

or

$$dT = \pi\left[\frac{1}{\sqrt{(15.0)(980)}}(-0.2) - \frac{\sqrt{15.0}}{980^{3/2}}(+6)\right].$$

Thus, $dT = \pm 0.0076\,\text{s}.$

Approximate maximum percentage error: $\dfrac{dT}{T} \times 100 = \dfrac{0.0076}{2\pi\sqrt{15.0/980}} \times 100 = 0.98\%.$

19. $\frac{1}{R} = \frac{1}{R_1} + \frac{1}{R_2} = \frac{R_1 + R_2}{R_1 R_2}$ and $R = \frac{R_1 R_2}{R_1 + R_2}$.
Given: $\frac{dR_1}{dt} = 5.0$ and $\frac{dR_2}{dt} = 10.0$, find $\frac{dR}{dt}$ when $R_1 = 75$ and $R_2 = 100$.

$$\begin{aligned}\frac{dR}{dt} &= \frac{\partial}{\partial R_1}\left(\frac{R_1 R_2}{R_1 + R_2}\right)\frac{dR_1}{dt} + \frac{\partial}{\partial R_2}\left(\frac{R_1 R_2}{R_1 + R_2}\right)\frac{dR_2}{dt} \\ &= \frac{R_2^2}{(R_1 + R_2)^2}\frac{dR_1}{dt} + \frac{R_1^2}{(R_1 + R_2)^2}\frac{dR_2}{dt} \\ &= \frac{100^2}{(75 + 100)^2}(5.0) + \frac{75^2}{(75 + 100)^2}(10.0) = 3.47\frac{\Omega}{\text{min}}\end{aligned}$$

21. $V = \frac{1}{3}\pi r^2 h$.
Given: $\frac{dr}{dt} = 1.0\,\frac{\text{cm}}{\text{min}}$, $\frac{dh}{dt} = 1.0\,\frac{\text{cm}}{\text{min}}$.
Find: $\frac{dV}{dt}$ when $r = 10\,\text{cm}$ and $h = 20\,\text{cm}$.

$$\begin{aligned}\frac{dV}{dt} &= \frac{\partial V}{\partial r}\frac{dr}{dt} + \frac{\partial V}{\partial h}\frac{dh}{dt} \\ &= \left(\frac{2}{3}\pi r h\right)\frac{dr}{dt} + \left(\frac{1}{3}\pi r^2\right)\frac{dh}{dt} \\ &= \left(\frac{2}{3}\pi r h\right)(1.0) + \left(\frac{1}{3}\pi r^2\right)(1.0)\Big|_{r=10,\,h=20} \\ &= \frac{500\pi}{3} = 520\,\text{cm}^3/\text{min}.\end{aligned}$$

23. $P = i^2 R$.
Given: $\frac{di}{dt} = 2.0\,\text{A/s}$ and $\frac{dR}{dt} = 3.0\,\Omega/\text{s}$.
Find: $\frac{dP}{dt}$ when $i = 10\,\text{A}$ and $R = 50\,\Omega$.
$\frac{dP}{dt} = \frac{\partial}{\partial i}(i^2 R)\frac{di}{dt} + \frac{\partial}{\partial R}(i^2 R)\frac{dR}{dt} = 2iR\frac{di}{dt} + i^2\frac{dR}{dt} = 2iR(2.0) + i^2(3.0)$.
When $i = 10$ and $R = 50$, we get
$\frac{dP}{dt} = 2(10)(50)(2.0) + (10)^2(3.0) = 2300\,\text{W/s}$.

25. $z = 3x^2 + 2y^2 + 4x - 4y - 1$
Critical points:
$\frac{\partial z}{\partial x} = 6x + 4 = 0$ or $x = -\frac{2}{3}$
$\frac{\partial z}{\partial y} = 4y - 4 = 0$ or $y = 1$.
The critical point is at $\left(-\frac{2}{3},\ 1\right)$.
$\frac{\partial^2 z}{\partial x^2} = 6$, $\frac{\partial^2 z}{\partial y^2} = 4$, $\frac{\partial^2 z}{\partial x \partial y} = 0$.
Thus $A = 6 \cdot 4 - 0^2 = 24 > 0$.
Since $A > 0$ and $\frac{\partial^2 z}{\partial x^2} > 0$, $f(x, y)$ has a minimum at $\left(-\frac{2}{3},\ 1\right)$.

$$\begin{aligned}\text{Also,}\quad z &= 3\left(-\frac{2}{3}\right)^2 + 2(1)^2 + 4\left(-\frac{2}{3}\right) - 4(1) - 1 \\ &= \frac{4}{3} + 2 - \frac{8}{3} - 4 - 1 = -3 - \frac{4}{3} = -\frac{13}{3}.\end{aligned}$$

27. $z = 2xy - x^2 - 2y^2 + 3x + 5$

Critical points:

$$\begin{aligned}
\frac{\partial z}{\partial x} &= 2y - 2x + 3 = 0 \\
\frac{\partial z}{\partial y} &= 2x - 4y = 0 \\
\hline
-2y + 3 &= 0 \qquad \text{(adding)} \\
y &= \frac{3}{2},\ x = 2y = 3.
\end{aligned}$$

The critical point is at $\left(3, \frac{3}{2}\right)$.

$\frac{\partial^2 z}{\partial x^2} = -2,\ \frac{\partial^2 z}{\partial y^2} = -4,\ \frac{\partial^2 z}{\partial x \partial y} = \frac{\partial}{\partial x}\left(\frac{\partial z}{\partial y}\right) = \frac{\partial}{\partial x}(2x - 4y) = 2.$

Thus $A = (-2)(-4) - (2)^2 = 4 > 0$.

Since $A > 0$ and $\frac{\partial^2 z}{\partial x^2} < 0$, $f(x, y)$ has a maximum at $\left(3, \frac{3}{2}\right)$.

Finally, $z = 2(3)\left(\frac{3}{2}\right) - (3)^2 - 2\left(\frac{3}{2}\right)^2 + 3(3) + 5 = \frac{19}{2}$,

so that the maximum point is $\left(3, \frac{3}{2}, \frac{19}{2}\right)$.

29. $z = y^2 - x^2 - 2xy - 4y$

Critical points:

$$\begin{aligned}
\frac{\partial z}{\partial x} &= -2x - 2y = 0 \\
\frac{\partial z}{\partial y} &= 2y - 2x - 4 = 0 \\
\hline
-4x - 4 &= 0 \qquad \text{adding} \\
x &= -1 \\
y &= 1.
\end{aligned}$$

The critical point is at $(-1, 1)$; $z = 1^2 - (-1)^2 - 2(-1)(1) - 4(1) = -2$.

$\frac{\partial^2 z}{\partial x^2} = -2,\ \frac{\partial^2 z}{\partial y^2} = 2,\ \frac{\partial^2 z}{\partial x \partial y} = -2.$

Thus $A = (-2)(2) - (-2)^2 = -8 < 0$.

Since $A < 0$, the point $(-1, 1, -2)$ is a saddle point.

31. $z = x^2 + 2y^3 - x - 12y - 4$

Critical points:

$\frac{\partial z}{\partial x} = 2x - 1 = 0$ or $x = \frac{1}{2}$.

$\frac{\partial z}{\partial y} = 6y^2 - 12 = 0$ or $y = \pm\sqrt{2}$.

Substituting in the given equation, we find that the critical points are $\left(\frac{1}{2}, \sqrt{2}, -15.6\right)$ and $\left(\frac{1}{2}, -\sqrt{2}, 7.1\right)$.

$\frac{\partial^2 z}{\partial x^2} = 2,\ \frac{\partial^2 z}{\partial y^2} = 12y,\ \frac{\partial^2 z}{\partial x \partial y} = 0.\ A = 2(12y) - 0^2 = 24y.$

For the first critical point, $A > 0$ and $\frac{\partial^2 z}{\partial x^2} > 0$. So $\left(\frac{1}{2}, \sqrt{2}, -15.6\right)$ is a minimum.

For the other critical point, $A < 0$. So $\left(\frac{1}{2}, -\sqrt{2}, 7.1\right)$ is a saddle point.

33. $z = 3x^3 - xy^2 + x$

Critical points:

$\frac{\partial z}{\partial x} = 9x^2 - y^2 + 1 = 0$, $\frac{\partial z}{\partial y} = -2xy = 0$, which implies that $x = 0$ or $y = 0$.

Substituting $y = 0$ in the first equation, we get $9x^2 + 1 = 0$ which has no real roots. If we substitute $x = 0$ in the first equation, we get

$-y^2 + 1 = 0$, whence $y = \pm 1$;

so the critical points are at $(0, 1)$ and $(0, -1)$.

$\frac{\partial^2 z}{\partial x^2} = 18x$, $\frac{\partial^2 z}{\partial y^2} = -2x$, $\frac{\partial^2 z}{\partial x \partial y} = -2y$.

So $A = (18x)(-2x) - (-2y)^2$ or $A = -36x^2 - 4y^2$.

For both critical points, $A = -4 < 0$.

We conclude that $f(x, y)$ has saddle points at $(0, 1)$ and $(0, -1)$.

35. $z = 9xy - x^3 - y^3$

$\frac{\partial z}{\partial x} = 9y - 3x^2 = 0$ or $3y - x^2 = 0$.

$\frac{\partial z}{\partial y} = 9x - 3y^2 = 0$ or $3x - y^2 = 0$.

Substituting $x = \frac{1}{3}y^2$ in the first equation, we get

$$\begin{aligned} 3y - \left(\frac{1}{3}y^2\right)^2 &= 0 \\ 3y - \frac{1}{9}y^4 &= 0 \\ 27y - y^4 &= 0 \\ y(27 - y^3) &= 0 \\ y &= 0,\ 3. \end{aligned}$$

When $y = 0$, $x = 0$ and when $y = 3$, $x = 3$. So the critical points are $(0, 0, 0)$ and $(3, 3, 27)$.

$\frac{\partial^2 z}{\partial x^2} = -6x$, $\frac{\partial^2 z}{\partial y^2} = -6y$, $\frac{\partial^2 z}{\partial x \partial y} = 9$.

$A = (-6x)(-6y) - 9^2 = 36xy - 81$.

We conclude that $(0, 0, 0)$ is a saddle point since $A < 0$. For the other critical point, $A = 36(3 \cdot 3) - 81 > 0$ and $\frac{\partial^2 z}{\partial x^2} < 0$. So $(3, 3, 27)$ is a maximum.

37. Let x and y be the first two numbers; then $60 - x - y$ is the third. The quantity to be maximized is the product P:

$$\begin{aligned} P &= xy(60 - x - y) \\ P &= 60xy - x^2y - xy^2 \\ \frac{\partial P}{\partial x} &= 60y - 2xy - y^2 = 0 \\ \frac{\partial P}{\partial y} &= 60x - x^2 - 2xy = 0 \end{aligned}$$

$$\begin{aligned} -2xy &= x^2 - 60x && \text{second equation} \\ y &= -\frac{1}{2}x + 30. \end{aligned}$$

Substituting in first equation:

$$\begin{aligned}
60\left(-\frac{1}{2}x+30\right)-2x\left(-\frac{1}{2}x+30\right)-\left(-\frac{1}{2}x+30\right)^2 &= 0\\
-30x+1800+x^2-60x-\frac{1}{4}x^2+30x-900 &= 0\\
\frac{3}{4}x^2-60x+900 &= 0\\
x^2-80x+1200 &= 0\\
(x-20)(x-60) &= 0\\
x &= 20\\
y &= 20\\
\text{third number} &= 60-20-20=20.
\end{aligned}$$

39. x units of W_1 and y units of W_2 selling at $\$P_1$ and $\$P_2$, respectively, produce a total revenue of xP_1+yP_2 (dollars). So the profit P is given by

$$\begin{aligned}
P(x,y) &= xP_1+yP_2-C(x,y)\\
\frac{\partial P}{\partial x} &= P_1-\frac{\partial C(x,y)}{\partial x}=0\\
\frac{\partial P}{\partial y} &= P_2-\frac{\partial C(x,y)}{\partial y}=0.
\end{aligned}$$

So if the profit is indeed a maximum at (a,b), the critical point, then $\dfrac{\partial C(a,b)}{\partial x}=P_1$ and $\dfrac{\partial C(a,b)}{\partial y}=P_2$.

41.
$$\begin{aligned}
f(x,y) &= 0\\
\frac{\partial f}{\partial x}dx+\frac{\partial f}{\partial y}dy &= 0\\
\frac{\partial f}{\partial x}+\frac{\partial f}{\partial y}\frac{dy}{dx} &= 0\\
\frac{dy}{dx} &= -\frac{\frac{\partial f}{\partial x}}{\frac{\partial f}{\partial y}}
\end{aligned}$$

43. Let $f=x^6+2x^5y^2-6x^3y^3-4y^4+10$;
$\dfrac{\partial f}{\partial x}=6x^5+10x^4y^2-18x^2y^3$;
$\dfrac{\partial f}{\partial y}=4x^5y-18x^3y^2-16y^3$.
By Exercise 41,
$$\frac{dy}{dx}=-\frac{\partial f/\partial x}{\partial f/\partial y}=-\frac{6x^5+10x^4y^2-18x^2y^3}{4x^5y-18x^3y^2-16y^3}.$$

45. Let $f=2x^5+3x^4y-4x^3y^2+7x^2y^2+1$;
$\dfrac{\partial f}{\partial x}=10x^4+12x^3y-12x^2y^2+14xy^2$; $\dfrac{\partial f}{\partial y}=3x^4-8x^3y+14x^2y$.
By Exercise 41,
$$\frac{dy}{dx}=-\frac{\partial f/\partial x}{\partial f/\partial y}=-\frac{10x^4+12x^3y-12x^2y^2+14xy^2}{3x^4-8x^3y+14x^2y}.$$

47. Let $f = 7x^4y^8 + 16x^3y^5 + 25x^2 - 7y^2 + 2$;
$\frac{\partial f}{\partial x} = 28x^3y^8 + 48x^2y^5 + 50x$;
$\frac{\partial f}{\partial y} = 56x^4y^7 + 80x^3y^4 - 14y$.
$\frac{dy}{dx} = -\frac{\partial f/\partial x}{\partial f/\partial y} = -\frac{28x^3y^8 + 48x^2y^5 + 50x}{56x^4y^7 + 80x^3y^4 - 14y}$.

49. Let $f = y^2 + y\cos x - 2$;
$\frac{\partial f}{\partial x} = -y\sin x$; $\frac{\partial f}{\partial y} = 2y + \cos x$.
By Exercise 41,
$\frac{dy}{dx} = -\frac{\partial f/\partial x}{\partial f/\partial y} = -\frac{-y\sin x}{2y + \cos x} = \frac{y\sin x}{2y + \cos x}$.

51. Let $f = \cot(x^2 - y^2) - 3xy + 7$;
$\frac{\partial f}{\partial x} = -\csc^2(x^2 - y^2)(2x) - 3y$;
$\frac{\partial f}{\partial y} = -\csc^2(x^2 - y^2)(-2y) - 3x$.
$\frac{dy}{dx} = -\frac{-2x\csc^2(x^2 - y^2) - 3y}{2y\csc^2(x^2 - y^2) - 3x} = \frac{2x\csc^2(x^2 - y^2) + 3y}{2y\csc^2(x^2 - y^2) - 3x}$.

53. Let $f = \ln y - x\sin y + 3y^2$;
$\frac{\partial f}{\partial x} = -\sin y$; $\frac{\partial f}{\partial y} = \frac{1}{y} - x\cos y + 6y$.
By Exercise 41,
$\frac{dy}{dx} = -\frac{\partial f/\partial x}{\partial f/\partial y} = -\frac{-\sin y}{1/y - x\cos y + 6y} = \frac{\sin y}{1/y - x\cos y + 6y} = \frac{y\sin y}{1 - xy\cos x + 6y^2}$.

9.4 Iterated Integrals

1. $$\begin{aligned}\int_0^1\left(\int_0^x x^2y^2\,dy\right)dx &= \int_0^1 x^2\frac{y^3}{3}\bigg|_0^x dx = \int_0^1 x^2\left(\frac{x^3}{3} - 0\right)dx \\ &= \int_0^1 \frac{1}{3}x^5\,dx = \frac{1}{3}\frac{x^6}{6}\bigg|_0^1 = \frac{1}{18}.\end{aligned}$$

3. $$\begin{aligned}\int_0^1\left(\int_{\sqrt{x}}^1 y\,dy\right)dx &= \int_0^1 \frac{1}{2}y^2\bigg|_{\sqrt{x}}^1 dx = \int_0^1 \frac{1}{2}\left[1^2 - (\sqrt{x})^2\right]dx \\ &= \frac{1}{2}\int_0^1(1 - x)\,dx = \frac{1}{2}\left(x - \frac{1}{2}x^2\right)\bigg|_0^1 = \frac{1}{2}\left(1 - \frac{1}{2}\right) = \frac{1}{4}.\end{aligned}$$

5. $$\begin{aligned}\int_0^2\int_0^{\sqrt{y-1}} xy\,dx\,dy &= \int_0^2 y\cdot\frac{1}{2}x^2\bigg|_0^{\sqrt{y-1}} dy = \int_0^2 y\cdot\frac{1}{2}\left(\sqrt{y-1}\right)^2 dy \\ &= \frac{1}{2}\int_0^2 y(y-1)\,dy = \frac{1}{2}\int_0^2(y^2 - y)\,dy \\ &= \frac{1}{2}\left(\frac{1}{3}y^3 - \frac{1}{2}y^2\right)\bigg|_0^2 = \frac{1}{2}\left(\frac{8}{3} - 2\right) = \frac{1}{3}.\end{aligned}$$

7. $$\begin{aligned}\int_1^3\left(\int_0^{\sqrt{9-y^2}} y\,dx\right)dy &= \int_1^3 yx\Big|_0^{\sqrt{9-y^2}} dy = \int_1^3 y\left(\sqrt{9-y^2}-0\right)dy \\ &= \int_1^3 (9-y^2)^{1/2}y\,dy \qquad\qquad u=9-y^2;\ du=-2y\,dy. \\ &= -\frac{1}{2}\int_1^3 (9-y^2)^{1/2}(-2y)\,dy = -\frac{1}{2}\frac{(9-y^2)^{3/2}}{3/2}\Big|_1^3 \\ &= -\frac{1}{3}(9-y^2)^{3/2}\Big|_1^3 = 0+\frac{1}{3}(8)^{3/2} = \frac{1}{3}\left(\sqrt{8}\right)^3 \\ &= \frac{1}{3}\left(\sqrt{8}\right)^2\left(\sqrt{8}\right) = \frac{1}{3}(8)\cdot 2\sqrt{2} = \frac{16\sqrt{2}}{3}.\end{aligned}$$

9. $$\begin{aligned}\int_0^{\sqrt{\pi/6}}\left(\int_0^x \cos x^2\,dy\right)dx &= \int_0^{\sqrt{\pi/6}}(\cos x^2)y\Big|_0^x dx = \int_0^{\sqrt{\pi/6}}(\cos x^2)x\,dx \\ &= \frac{1}{2}\int_0^{\sqrt{\pi/6}}\cos x^2(2x)\,dx \qquad\qquad u=x^2;\ du=2x\,dx. \\ &= \frac{1}{2}\sin x^2\Big|_0^{\sqrt{\pi/6}} = \frac{1}{2}\sin\frac{\pi}{6} = \frac{1}{4}.\end{aligned}$$

11. $$\int_0^{\pi/4}\left(\int_0^{\sec y} 2x\,dx\right)dy = \int_0^{\pi/4} x^2\Big|_0^{\sec y} dy = \int_0^{\pi/4}\sec^2 y\,dy = \tan y\Big|_0^{\pi/4} = 1.$$

13. $$\begin{aligned}\int_2^3\int_1^x\left(\int_0^{6y} xy\,dz\right)dy\,dx &= \int_2^3\int_1^x xyz\Big|_0^{6y} dy\,dx \\ &= \int_2^3\int_1^x xy(6y)\,dy\,dx = \int_2^3 x\cdot 2y^3\Big|_1^x dx \\ &= \int_2^3 2x(x^3-1)\,dx = \int_2^3(2x^4-2x)\,dx = \frac{397}{5}.\end{aligned}$$

15. $$\begin{aligned}\int_0^1\int_0^{y^2}\int_0^{e^x} 2y\,dz\,dx\,dy &= \int_0^1\int_0^{y^2} 2yz\Big|_0^{e^x} dx\,dy = \int_0^1\int_0^{y^2} 2ye^x\,dx\,dy \\ &= \int_0^1 2ye^x\Big|_0^{y^2} dy = \int_0^1 2y\left(e^{y^2}-1\right)dy = e^{y^2}-y^2\Big|_0^1 = e-2\end{aligned}$$

17.

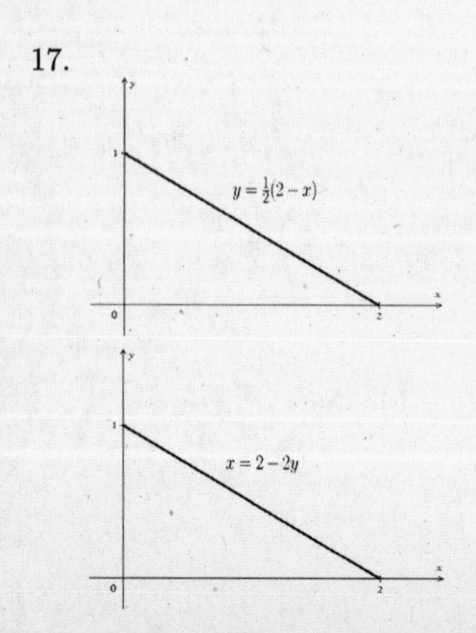

$$\begin{aligned}A &= \int_0^2\int_0^{(2-x)/2} dy\,dx = \int_0^2 y\Big|_0^{(2-x)/2} dx \\ &= \int_0^2 \frac{1}{2}(2-x)\,dx = x-\frac{1}{4}x^2\Big|_0^2 = 1.\end{aligned}$$

$$\begin{aligned}A &= \int_0^1\int_0^{2-2y} dx\,dy = \int_0^1 x\Big|_0^{2-2y} dy \\ &= \int_0^1 (2-2y)\,dy = 2y-y^2\Big|_0^1 = 2-1 = 1.\end{aligned}$$

19.

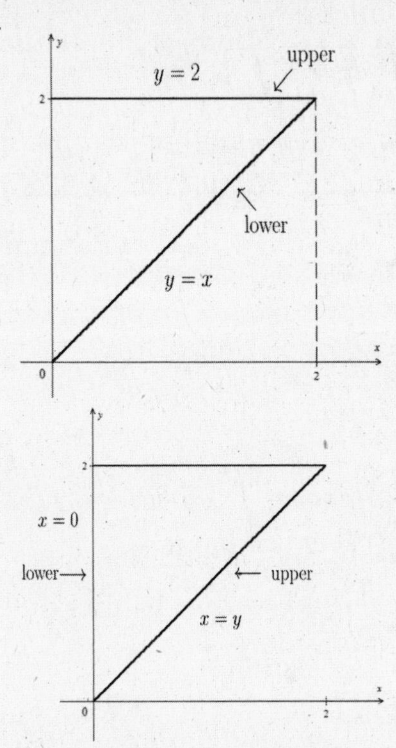

$$
\begin{aligned}
A &= \int_0^2 \int_x^2 dy\,dx = \int_0^2 y\Big|_x^2 dx \\
&= \int_0^2 (2-x)\,dx = 2x - \frac{1}{2}x^2\Big|_0^2 = 4 - 2 = 2.
\end{aligned}
$$

$$
\begin{aligned}
A &= \int_0^2 \int_0^y dx\,dy = \int_0^2 x\Big|_0^y dy \\
&= \int_0^2 y\,dy = \frac{1}{2}y^2\Big|_0^2 = 2.
\end{aligned}
$$

21.

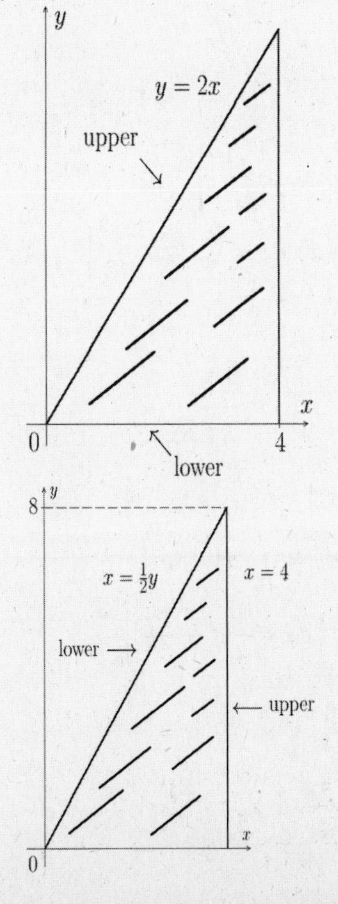

$$
\begin{aligned}
A &= \int_0^4 \int_0^{2x} dy\,dx = \int_0^4 y\Big|_0^{2x} dx \\
&= \int_0^4 2x\,dx = x^2\Big|_0^4 = 16.
\end{aligned}
$$

$$
\begin{aligned}
A &= \int_0^8 \int_{(1/2)y}^4 dx\,dy = \int_0^8 x\Big|_{(1/2)y}^4 dy \\
&= \int_0^8 \left(4 - \frac{1}{2}y\right) dy = 4y - \frac{1}{4}y^2\Big|_0^8 \\
&= 32 - 16 = 16.
\end{aligned}
$$

23.

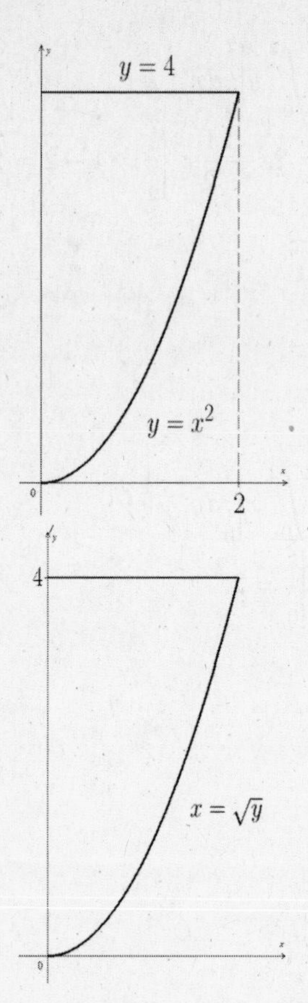

$$\begin{aligned} A &= \int_0^2\int_{x^2}^4 dy\,dx = \int_0^2 y\Big|_{x^2}^4 dx \\ &= \int_0^2 (4-x^2)\,dx = 4x - \frac{1}{3}x^3\Big|_0^2 \\ &= 8 - \frac{8}{3} = \frac{24}{3} - \frac{8}{3} = \frac{16}{3}. \end{aligned}$$

$$\begin{aligned} A &= \int_0^4\int_0^{\sqrt{y}} dx\,dy = \int_0^4 \sqrt{y}\,dy \\ &= \frac{2}{3}y^{3/2}\Big|_0^4 = \frac{2}{3}(8-0) = \frac{16}{3}. \end{aligned}$$

25.

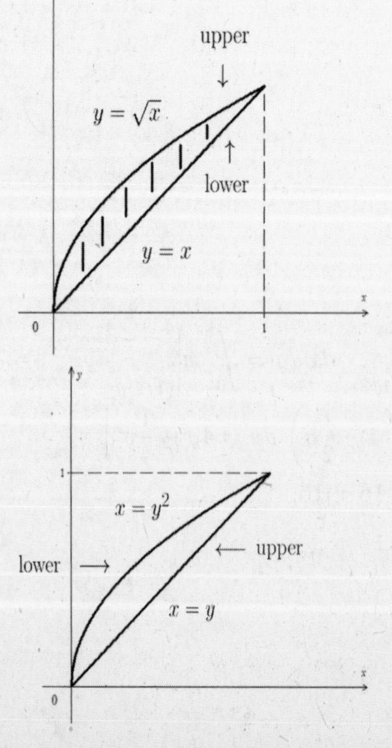

$$\begin{aligned} A &= \int_0^1\int_x^{\sqrt{x}} dy\,dx = \int_0^1 y\Big|_x^{\sqrt{x}} dx \\ &= \int_0^1 \left(\sqrt{x} - x\right)\,dx = \frac{2}{3}x^{3/2} - \frac{x^2}{2}\Big|_0^1 \\ &= \frac{2}{3} - \frac{1}{2} = \frac{1}{6}. \end{aligned}$$

$$\begin{aligned} A &= \int_0^1\int_{y^2}^{y} dx\,dy = \int_0^1 x\Big|_{y^2}^{y} dy \\ &= \int_0^1 (y - y^2)\,dy = \frac{1}{2}y^2 - \frac{1}{3}y^3\Big|_0^1 \\ &= \frac{1}{2} - \frac{1}{3} = \frac{1}{6}. \end{aligned}$$

27.

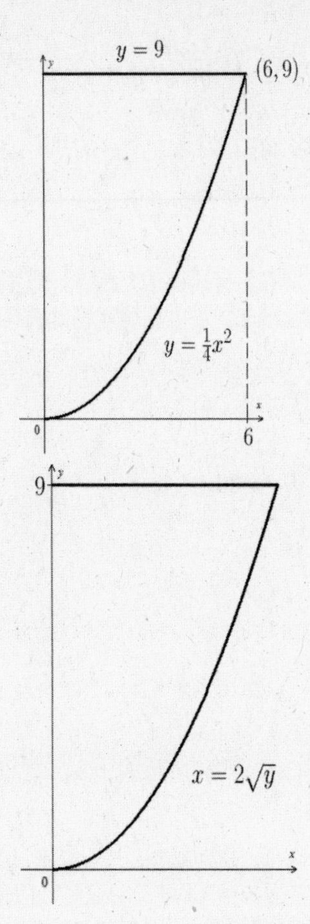

$$\begin{aligned} A &= \int_0^6 \int_{x^2/4}^9 dy\,dx = \int_0^6 y\Big|_{x^2/4}^9 dx \\ &= \int_0^6 \left(9 - \frac{1}{4}x^2\right) dx = 36. \end{aligned}$$

$$\begin{aligned} A &= \int_0^9 \int_0^{2\sqrt{y}} dx\,dy = \int_0^9 x\Big|_0^{2\sqrt{y}} dy \\ &= \int_0^9 2\sqrt{y}\,dy = 2 \cdot \frac{2}{3}x^{3/2}\Big|_0^9 = 36. \end{aligned}$$

29.

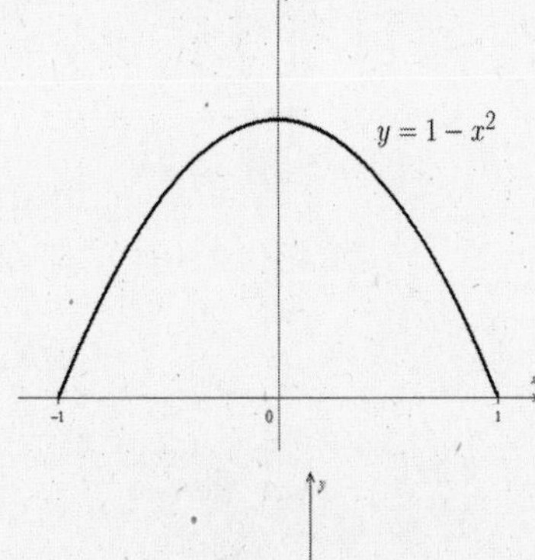

$$\begin{aligned} A &= \int_{-1}^1 \int_0^{1-x^2} dy\,dx = \int_{-1}^1 y\Big|_0^{1-x^2} dx \\ &= \int_{-1}^1 (1 - x^2)\,dx = x - \frac{1}{3}x^3\Big|_{-1}^1 \\ &= \left(1 - \frac{1}{3}\right) - \left(-1 + \frac{1}{3}\right) = \frac{4}{3}. \end{aligned}$$

$$\begin{aligned} A &= \int_0^1 \int_{-\sqrt{1-y}}^{\sqrt{1-y}} dx\,dy = \int_0^1 x\Big|_{-\sqrt{1-y}}^{\sqrt{1-y}} dy \\ &= \int_0^1 \left(\sqrt{1-y} + \sqrt{1-y}\right) dy \\ &= 2\int_0^1 \sqrt{1-y}\,dy \qquad u = 1 - y;\; du = -dy. \\ &= -2\int_0^1 (1-y)^{1/2}(-dy) \\ &= -2 \cdot \frac{2}{3}(1-y)^{3/2}\Big|_0^1 \\ &= 0 + \frac{4}{3} \cdot 1^{3/2} = \frac{4}{3}. \end{aligned}$$

$x = -\sqrt{1-y}$ $x = \sqrt{1-y}$

lower ⟶ ⟵ upper

0

31.

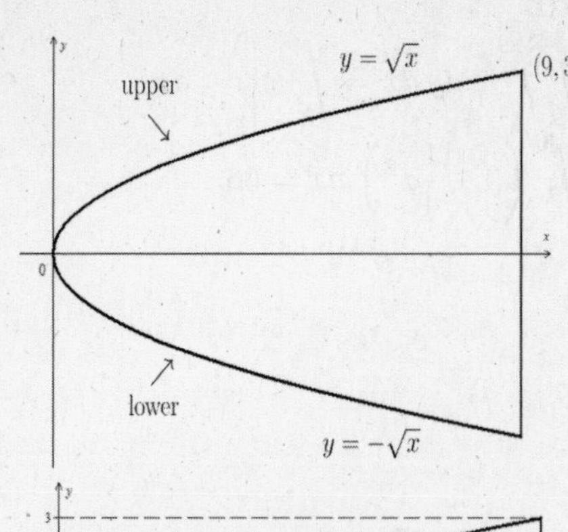

$$A = \int_0^9 \int_{-\sqrt{x}}^{\sqrt{x}} dy\,dx = \int_{0.}^{9} y\Big|_{-\sqrt{x}}^{\sqrt{x}} dx$$
$$= \int_0^9 2\sqrt{x}\,dx = 36.$$

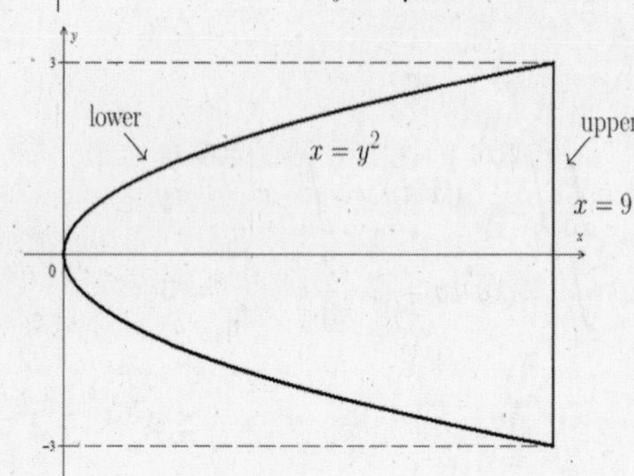

$$A = \int_{-3}^{3} \int_{y^2}^{9} dx\,dy = \int_{-3}^{3} x\Big|_{y^2}^{9} dy$$
$$= \int_{-3}^{3} (9 - y^2)\,dy = 36.$$

33.

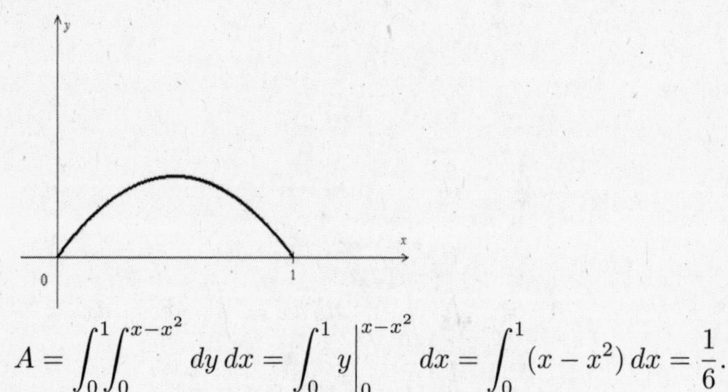

$$A = \int_0^1 \int_0^{x-x^2} dy\,dx = \int_0^1 y\Big|_0^{x-x^2} dx = \int_0^1 (x - x^2)\,dx = \frac{1}{6}.$$

35.

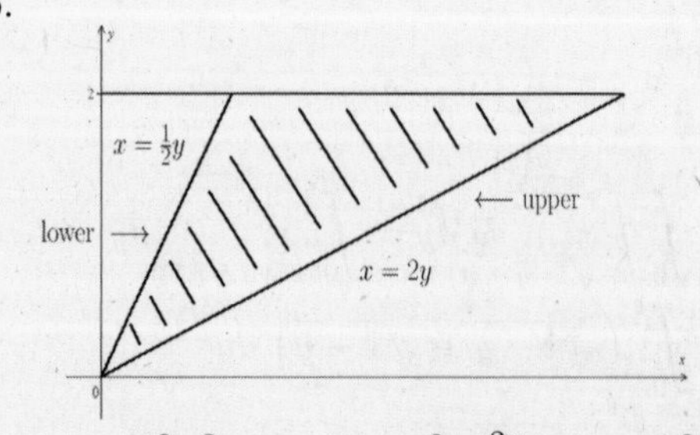

$$A = \int_0^2 \int_{y/2}^{2y} dx\,dy = \int_0^2 x\Big|_{y/2}^{2y} dy = \int_0^2 \left(2y - \frac{1}{2}y\right) dy = \int_0^2 \frac{3}{2}y\,dy = \frac{3}{2} \cdot \frac{y^2}{2}\Big|_0^2 = 3.$$

37. $A = \int_0^1 \int_{\sqrt{y}}^{2-y} dx\,dy = \int_0^1 (2 - y - \sqrt{y})\,dy = \frac{5}{6}$

9.5 Volumes by Double Integration

1.

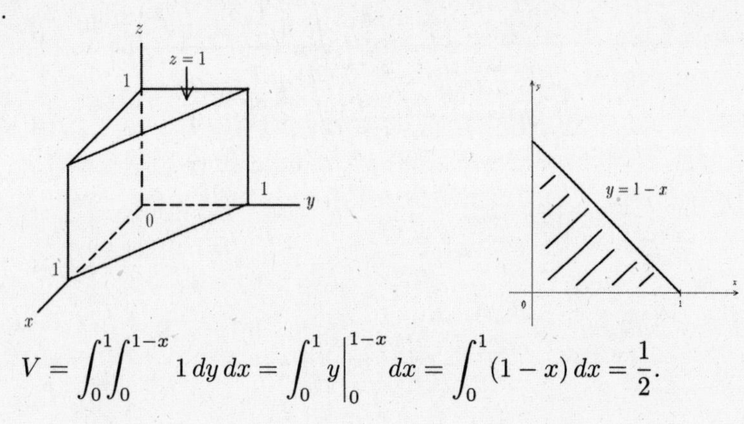

$$V = \int_0^1 \int_0^{1-x} 1\, dy\, dx = \int_0^1 y\Big|_0^{1-x} dx = \int_0^1 (1-x)\, dx = \frac{1}{2}.$$

3.

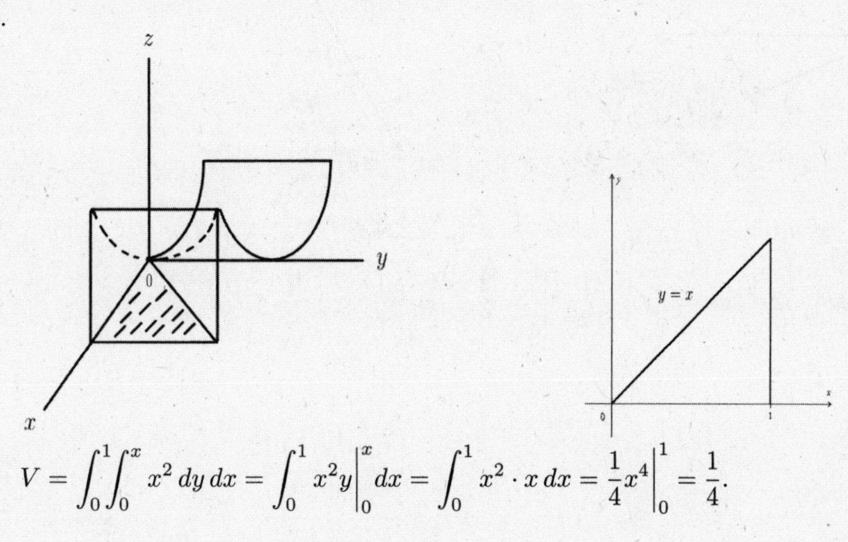

$$V = \int_0^1 \int_0^x x^2\, dy\, dx = \int_0^1 x^2 y\Big|_0^x dx = \int_0^1 x^2 \cdot x\, dx = \frac{1}{4}x^4\Big|_0^1 = \frac{1}{4}.$$

5.

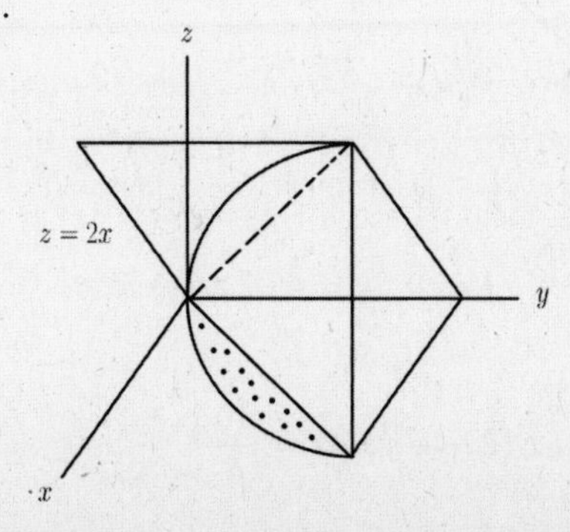

$$\begin{aligned} V &= \int_0^1 \int_{x^2}^x 2x\, dy\, dx \\ &= \int_0^1 2xy\Big|_{x^2}^x dx \\ &= \int_0^1 2x\left(x - x^2\right)\, dx = \frac{1}{6} \end{aligned}$$

7.

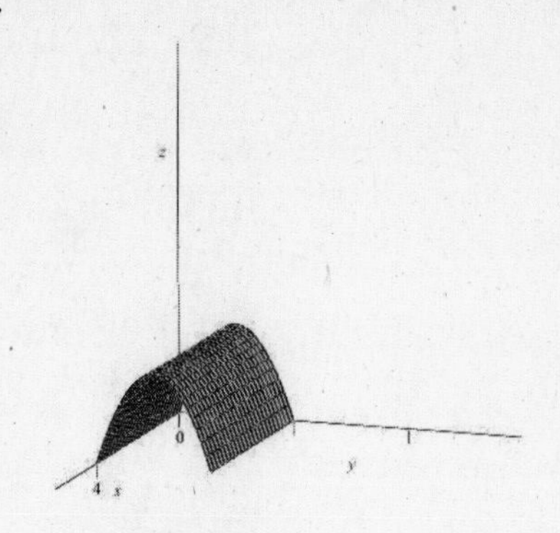

$$\begin{aligned} V &= \int_0^4\int_0^1 y(1-y)\,dy\,dx \\ &= \int_0^4 \left(\frac{1}{2}y^2 - \frac{1}{3}y^3\right)\Big|_0^1\,dx \\ &= \int_0^4 \frac{1}{6}\,dx = \frac{2}{3} \end{aligned}$$

9.

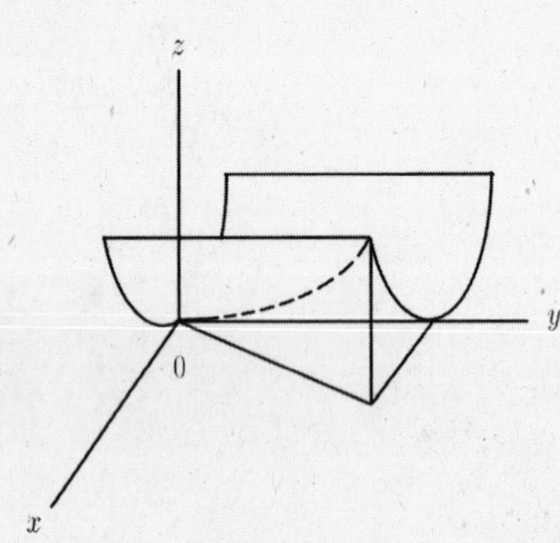

y = 2

y = 2x

0

1

$$\begin{aligned} V &= \int_0^1\int_{2x}^2 x^2\,dy\,dx = \int_0^1 x^2y\Big|_{2x}^2\,dx = \int_0^1 x^2(2-2x)\,dx \\ &= \int_0^1 (2x^2-2x^3)\,dx = \frac{2}{3}x^3 - \frac{1}{2}x^4\Big|_0^1 = \frac{2}{3}-\frac{1}{2} = \frac{1}{6}. \end{aligned}$$

11.

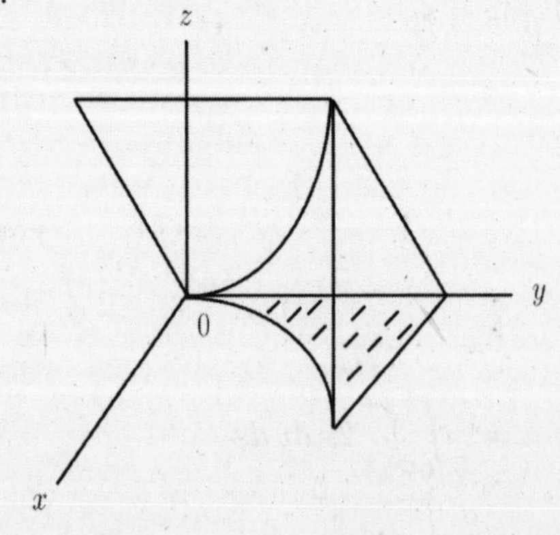

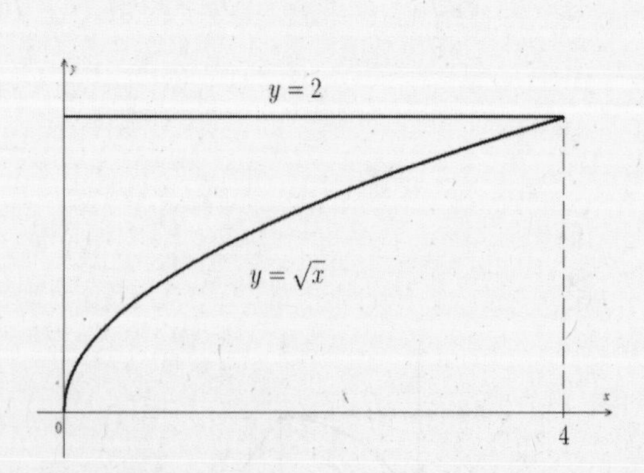

$$V = \int_0^4\int_{\sqrt{x}}^2 x\,dy\,dx = \int_0^4 xy\Big|_{\sqrt{x}}^2\,dx = \int_0^4 x\left(2-\sqrt{x}\right)\,dx = \frac{16}{5}$$

13.

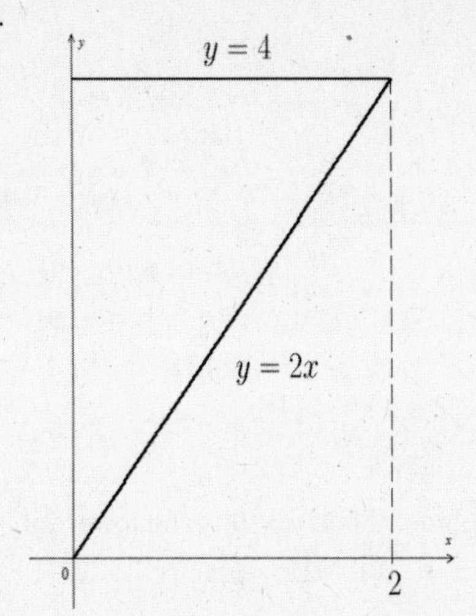

$$\begin{aligned} V &= \int_0^2\int_{2x}^4 xy\,dy\,dx \\ &= \int_0^2 x\left(\frac{1}{2}y^2\right)\Big|_{2x}^4\,dx \\ &= \int_0^2 x\left(8-2x^2\right)\,dx = 8 \end{aligned}$$

15.

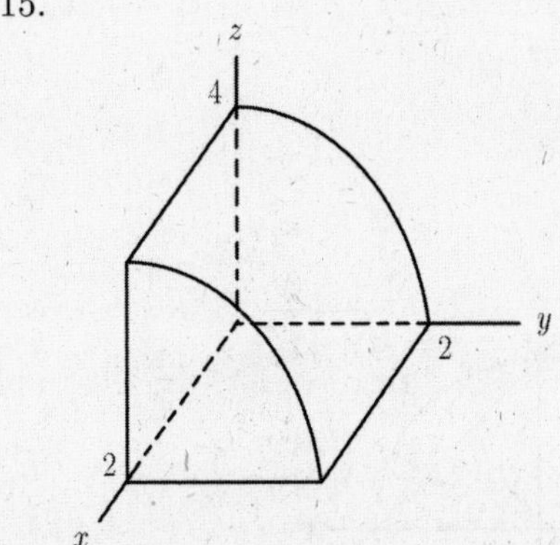

$$\begin{aligned} V &= \int_0^2\int_0^2 (4-y^2)\,dy\,dx \\ &= \int_0^2\left(4y-\frac{1}{3}y^3\right)\Big|_0^2\,dx \\ &= \int_0^2\left(8-\frac{8}{3}\right)\,dx = \frac{16}{3}\int_0^2 dx \\ &= \frac{16}{3}(2) = \frac{32}{3}. \end{aligned}$$

17. Consider the figure for the first-octant volume:

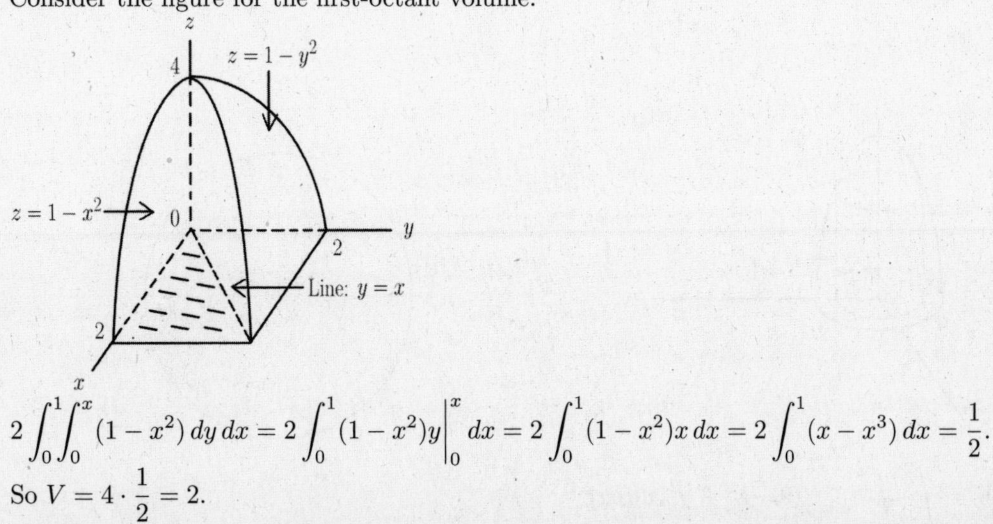

$$2\int_0^1\int_0^x (1-x^2)\,dy\,dx = 2\int_0^1 (1-x^2)y\Big|_0^x\,dx = 2\int_0^1 (1-x^2)x\,dx = 2\int_0^1 (x-x^3)\,dx = \frac{1}{2}.$$

So $V = 4\cdot\frac{1}{2} = 2$.

19.

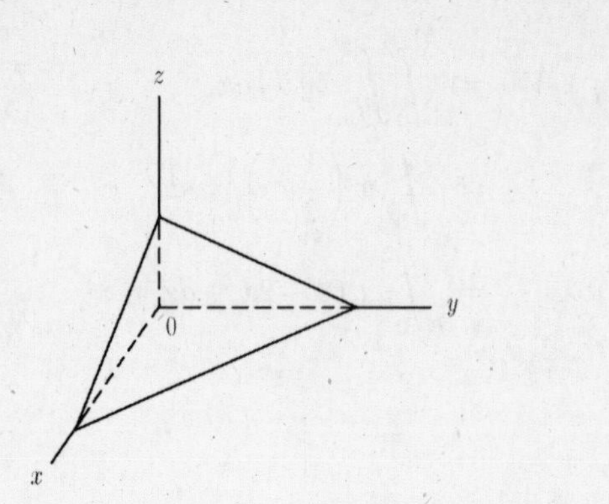

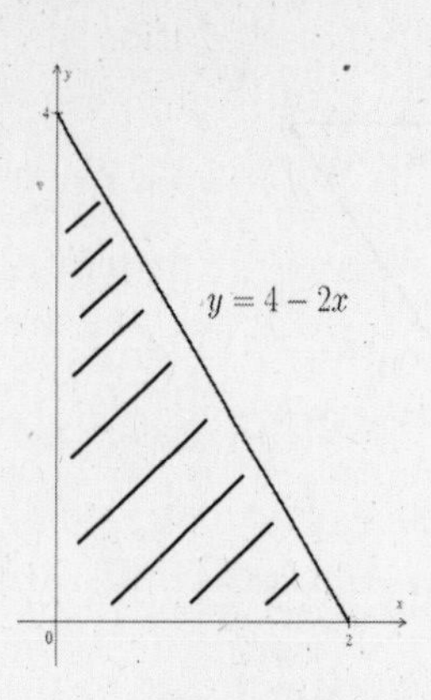

If $z = 0$, we get $y = 4 - 2x$, which is the trace in the xy-plane. Thus $y = 0$ is the lower function and $y = 4 - 2x$ is the upper function. The integrand is $z = \frac{1}{2}(4 - 2x - y)$.

$$V = \int_0^2 \int_0^{4-2x} \left(\frac{1}{2}\right)(4 - 2x - y)\, dy\, dx$$

21.

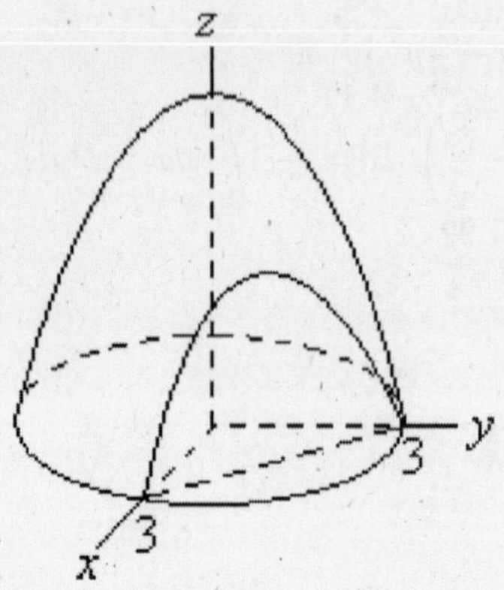

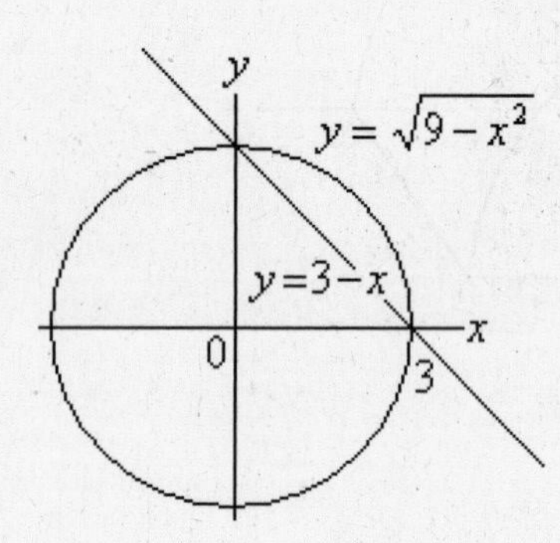

$$V = \int_0^3 \int_{3-x}^{\sqrt{9-x^2}} (9 - x^2 - y^2)\, dy\, dx$$

23.

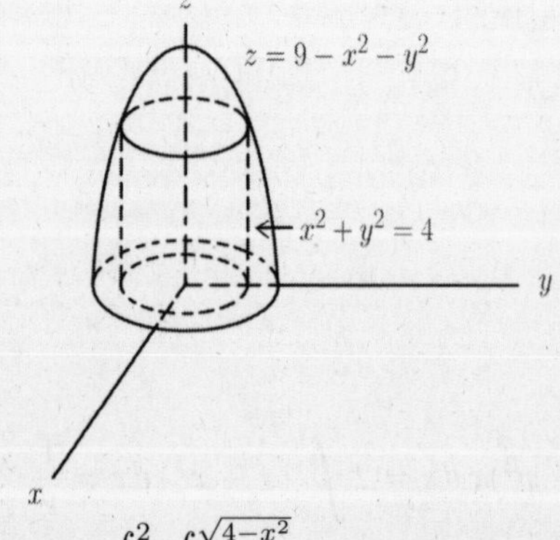

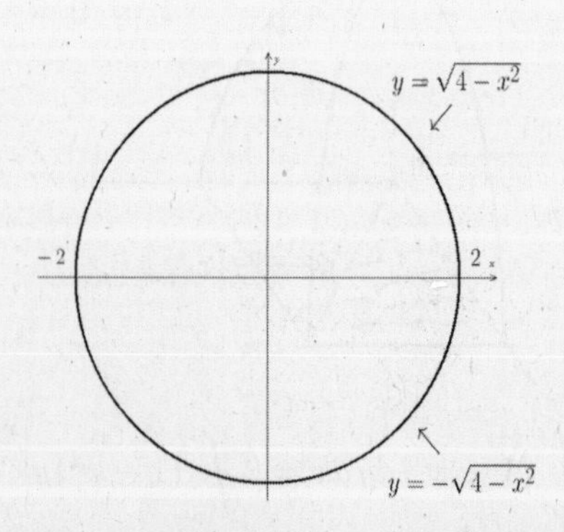

$$V = \int_{-2}^2 \int_{-\sqrt{4-x^2}}^{\sqrt{4-x^2}} (9 - x^2 - y^2)\, dy\, dx$$

25.

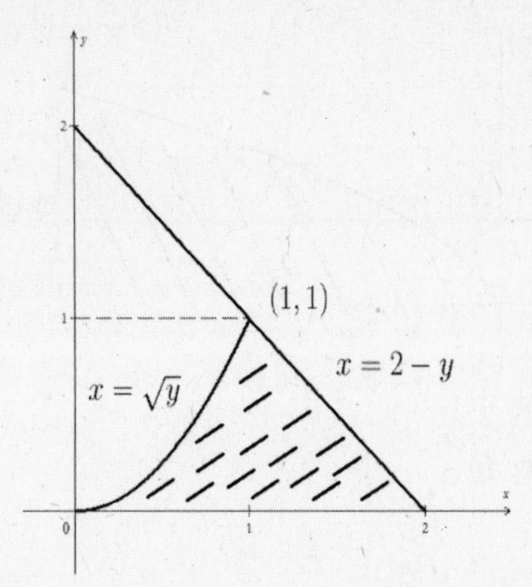

By (9.12) with $h_1(y) = \sqrt{y}$ and $h_2(y) = 2 - y$, we get:
$V = \int_0^1 \int_{\sqrt{y}}^{2-y} \sqrt{8 - 2x^2 - y^2}\, dx\, dy.$

27.

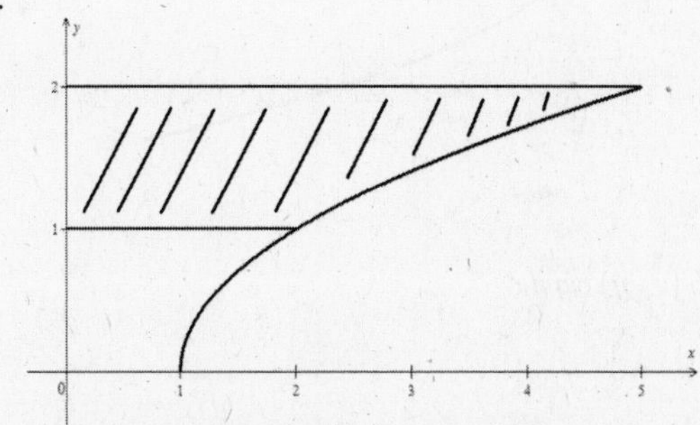

By (9.12) with $h_1(y) = 0$ and $h_2(y) = y^2 + 1$, we get:
$V = \int_1^2 \int_0^{y^2+1} xy\, dx\, dy.$

29.

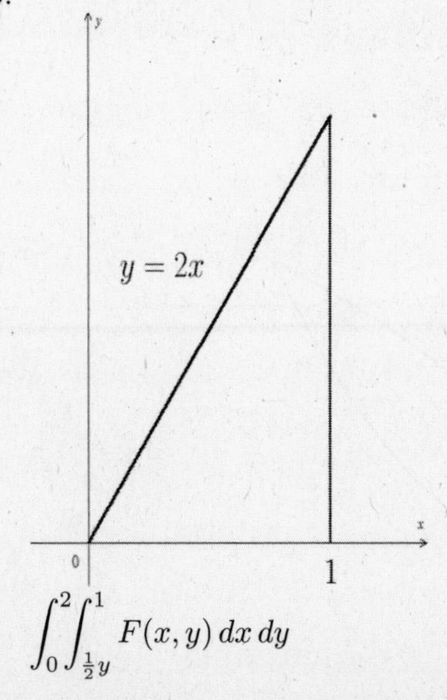

y
2
lower $\longrightarrow$
$\longleftarrow$ upper
$x = 1$
$x = \frac{1}{2}y$
0
x

$\int_0^2 \int_{\frac{1}{2}y}^1 F(x, y)\, dx\, dy$

31.

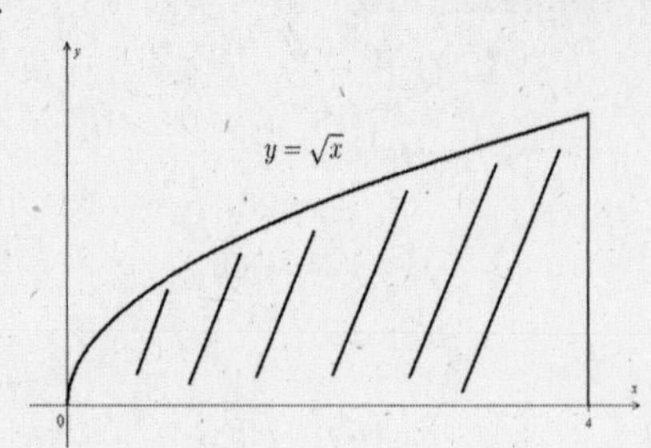

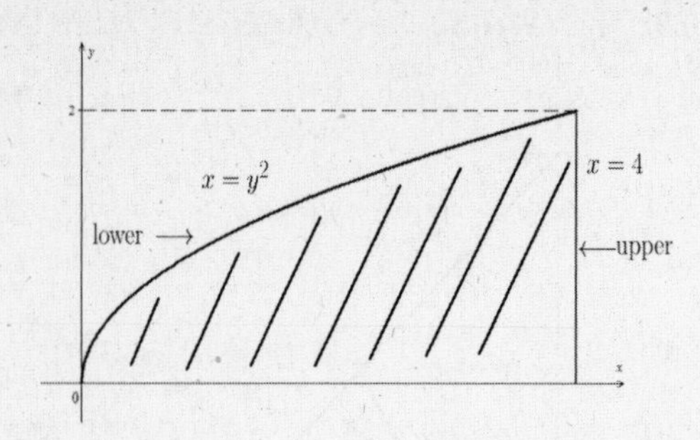

$$\int_0^2 \int_{y^2}^4 F(x, y)\, dx\, dy$$

33.

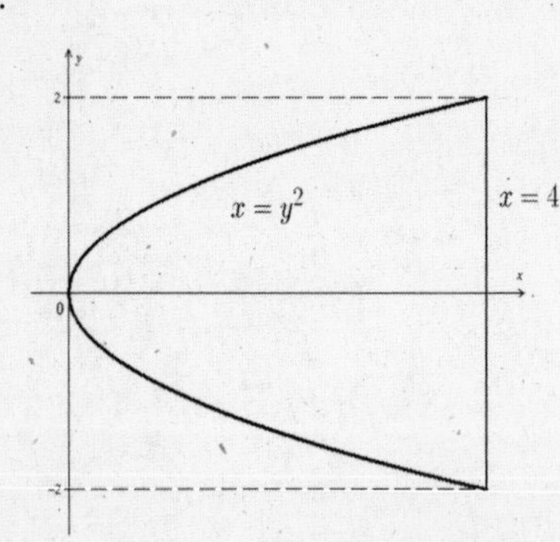

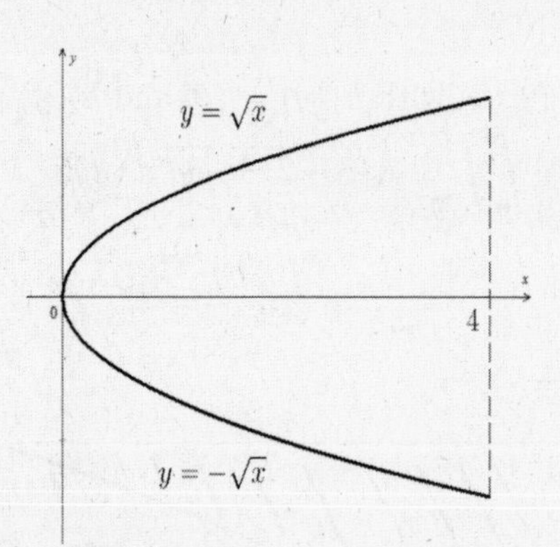

$$\int_0^4 \int_{-\sqrt{x}}^{\sqrt{x}} F(x, y)\, dy\, dx$$

35.

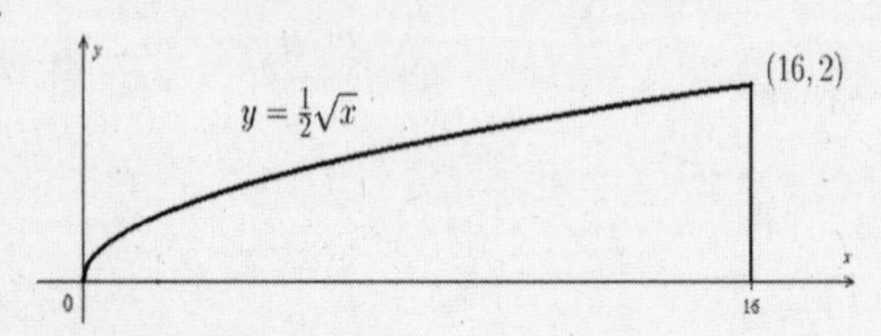

From $x = 4y^2$, we get $y = \dfrac{1}{2}\sqrt{x}$. So by (9.11) with $g_1(x) = 0$ and $g_2(x) = \dfrac{1}{2}\sqrt{x}$, we get:
$V = \displaystyle\int_0^{16} \int_0^{\sqrt{x}/2} F(x, y) dy\, dx.$

37.

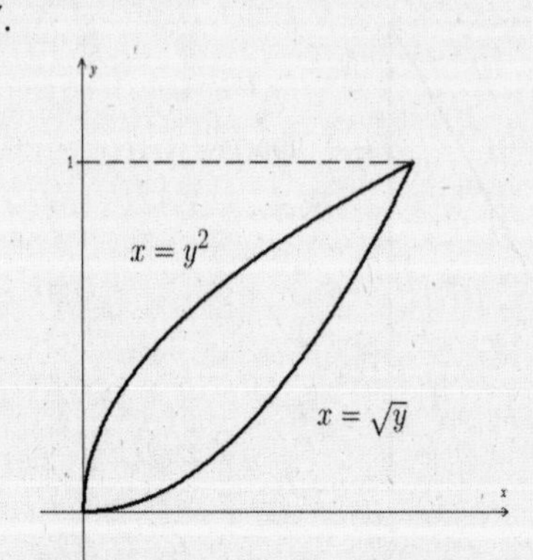

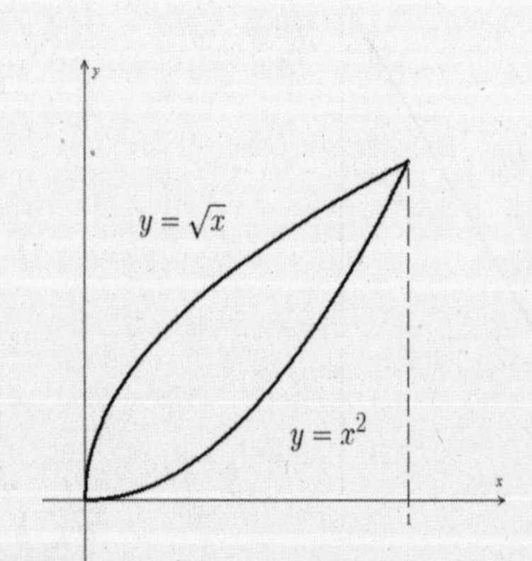

$$\int_0^1 \int_{x^2}^{\sqrt{x}} F(x, y)\, dy\, dx$$

9.6 Mass, Centroids, and Moments of Inertia

1. Area of typical element: $dy\,dx$.

$$A = \int_0^4 \int_0^{\frac{1}{3}x} dy\,dx = \int_0^4 y\Big|_0^{\frac{1}{3}x} dx = \int_0^4 \frac{1}{3}x\,dx = \frac{8}{3}.$$

Mass of typical element: $\frac{1}{10}x\,dy\,dx$.

$$\begin{aligned} m &= \int_0^4 \int_0^{\frac{1}{3}x} \frac{1}{10}x\,dy\,dx \\ &= \int_0^4 \frac{1}{10}xy\Big|_0^{x/3} dx \\ &= \int_0^4 \frac{1}{10}x\left(\frac{1}{3}x\right) dx = \frac{32}{45}. \end{aligned}$$

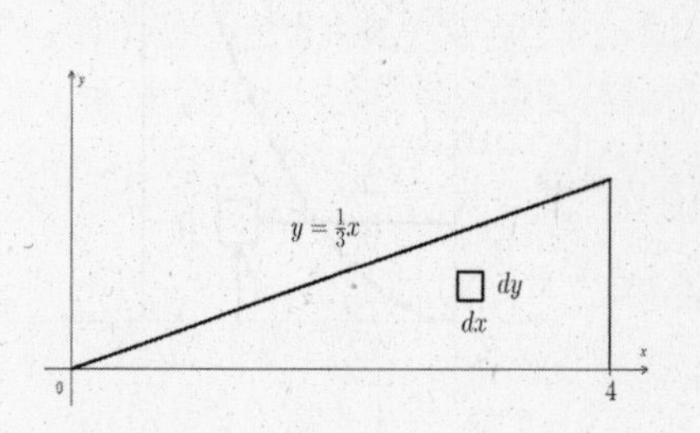

3. Area of typical element: $dx\,dy$.

$$A = \int_0^2 \int_0^{y^2} dx\,dy = \int_0^2 y^2\,dy = \frac{8}{3}$$

Mass of typical element: $(x+1)\,dx\,dy$.

$$\begin{aligned} m &= \int_0^2 \int_0^{y^2} (x+1)\,dx\,dy \\ &= \int_0^2 \left(\frac{1}{2}x^2 + x\right)\Big|_0^{y^2} dx \\ &= \int_0^2 \left(\frac{1}{2}y^4 + y^2\right) dy = \frac{88}{15} \end{aligned}$$

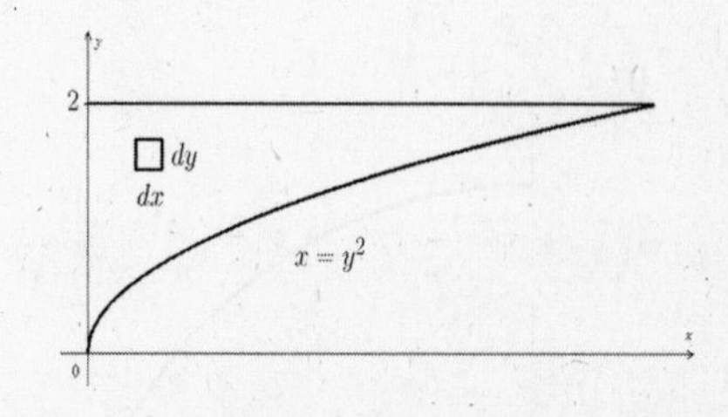

5. Area of typical element: $dy\,dx$.

$$A = \int_{-2}^2 \int_0^{4-x^2} dy\,dx = \int_{-2}^2 (4-x^2)\,dx = \frac{32}{3}.$$

Mass of typical element: $x^2\,dy\,dx$.

$$\begin{aligned} m &= \int_{-2}^2 \int_0^{4-x^2} x^2\,dy\,dx = \int_{-2}^2 x^2 y\Big|_0^{4-x^2} dx \\ &= \int_{-2}^2 x^2(4-x^2)\,dx = \frac{4}{3}x^3 - \frac{1}{5}x^5\Big|_{-2}^2 \\ &= \left(\frac{32}{3} - \frac{32}{5}\right) - \left(-\frac{32}{3} + \frac{32}{5}\right) \\ &= 2\left(\frac{32}{3} - \frac{32}{5}\right) = 64\left(\frac{1}{3} - \frac{1}{5}\right) \\ &= 64\left(\frac{2}{15}\right) = \frac{128}{15}. \end{aligned}$$

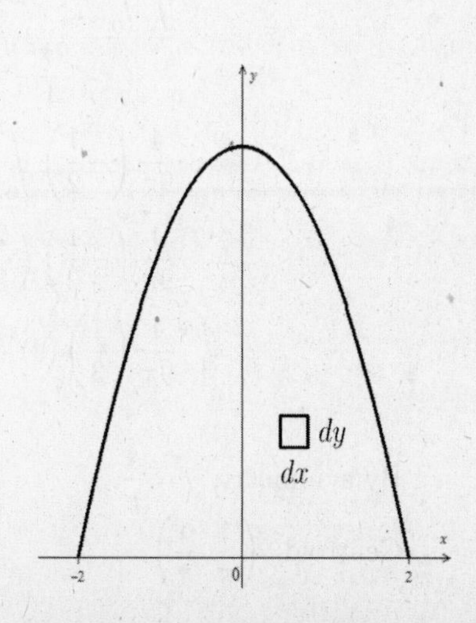

7.

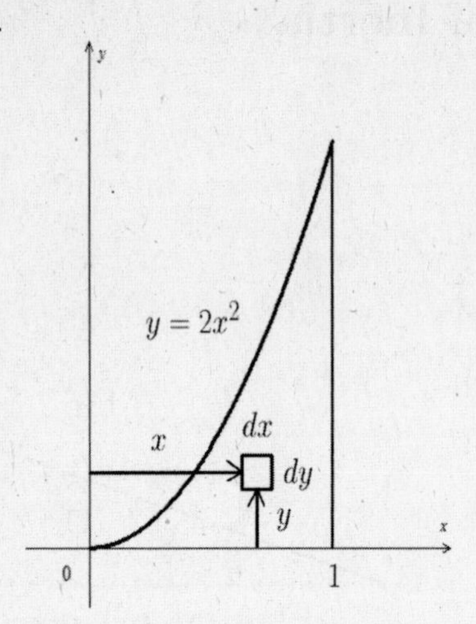

Moment of typical element with respect to the y-axis: $x\,dy\,dx$.

$$\overline{x} = \frac{M_y}{A} = \frac{\int_0^1 \int_0^{2x^2} x\,dy\,dx}{\int_0^1 \int_0^{2x^2} dy\,dx} = \frac{\int_0^1 xy\Big|_0^{2x^2} dx}{\int_0^1 y\Big|_0^{2x^2} dx} = \frac{\int_0^1 x \cdot 2x^2\,dx}{\int_0^1 2x^2\,dx} = \frac{1/2}{2/3} = \frac{3}{4}.$$

Moment of typical element with respect to the x-axis: $y\,dy\,dx$.

$$\overline{y} = \frac{M_x}{A} = \frac{\int_0^1 \int_0^{2x^2} y\,dy\,dx}{2/3} = \frac{3}{2}\int_0^1 \frac{1}{2}y^2\Big|_0^{2x^2} dx = \frac{3}{2}\int_0^1 \frac{1}{2}\left(4x^4\right)\,dx = \frac{3}{5}.$$

9.

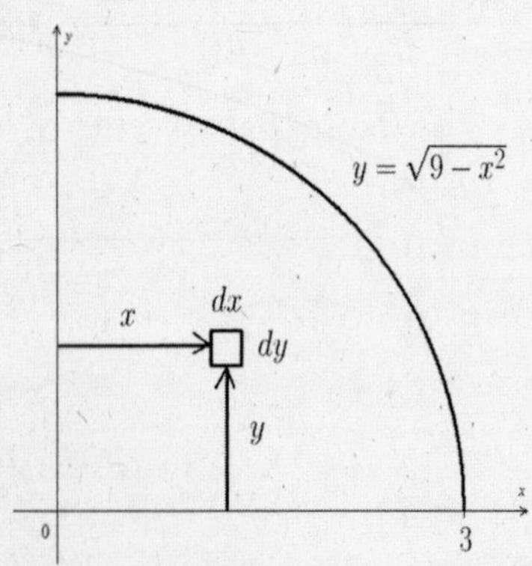

Moments: $x\,dy\,dx$ and $y\,dy\,dx$. Area of region: $\frac{1}{4}\pi \cdot r^2 = \frac{1}{4}\pi \cdot 3^2 = \frac{9\pi}{4}$.

$$\begin{aligned}
\overline{x} &= \frac{\int_0^3 \int_0^{\sqrt{9-x^2}} x\,dy\,dx}{\frac{9\pi}{4}} = \frac{4}{9\pi}\int_0^3 xy\Big|_0^{\sqrt{9-x^2}} dx \\
&= \frac{4}{9\pi}\left(-\frac{1}{2}\right)\int_0^3 (9-x^2)^{1/2}(-2x\,dx) \qquad u = 9 - x^2,\ du = -2x\,dx \\
&= \frac{4}{9\pi}\left(-\frac{1}{2}\right)\left(\frac{2}{3}\right)(9-x^2)^{3/2}\Big|_0^3 \\
&= \frac{4}{9\pi}\left(\frac{1}{3}\right)(9)^{3/2} = \frac{4}{\pi}
\end{aligned}$$

By symmetry, $\overline{y} = \frac{4}{\pi}$.

Centroid: $\left(\frac{4}{\pi}, \frac{4}{\pi}\right)$.

11.

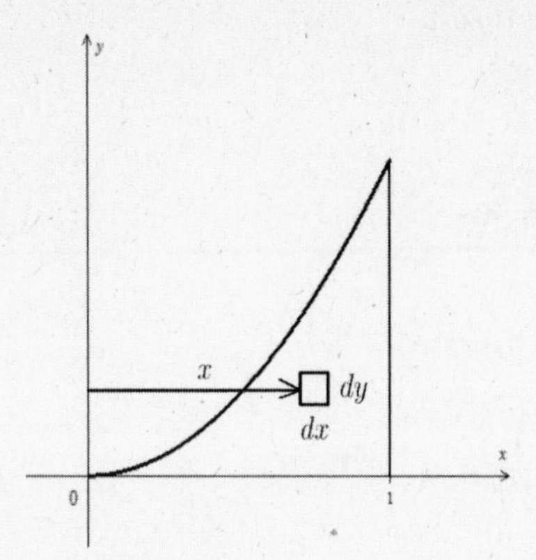

Moment of inertia of typical element: $x^2 \cdot \rho\, dy\, dx = \rho x^2\, dy\, dx$.

$I_y = \rho \int_0^1 \int_0^{x^3} x^2\, dy\, dx = \rho \int_0^1 x^2 y \Big|_0^{x^3} dx = \rho \int_0^1 x^5\, dx = \frac{\rho}{6}$.

Mass: $\rho \int_0^1 \int_0^{x^3} dy\, dx = \rho \int_0^1 x^3\, dx = \frac{\rho}{4}$. $R_y = \sqrt{\frac{\rho}{6}\frac{4}{\rho}} = \sqrt{\frac{2}{3}} = \frac{\sqrt{6}}{3}$.

13.

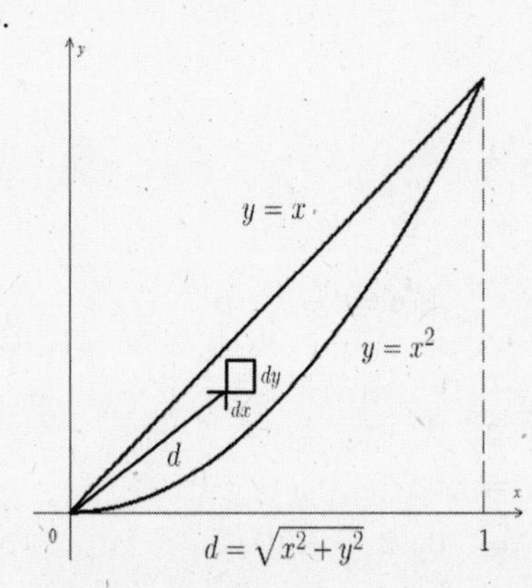

Moment of inertia of typical element: $\rho\, d^2\, dy\, dx = \rho(x^2 + y^2)\, dy\, dx$. [See Equation (9.22).]

$$\begin{aligned} I_o &= \int_0^1 \int_{x^2}^{x} \rho(x^2 + y^2)\, dy\, dx = \rho \int_0^1 \left(x^2 y + \frac{1}{3} y^3 \right) \Big|_{x^2}^{x} dx \\ &= \rho \int_0^1 \left[\left(x^3 + \frac{1}{3} x^3 \right) - \left(x^4 + \frac{1}{3} x^6 \right) \right] dx = \rho \int_0^1 \left(\frac{4}{3} x^3 - x^4 - \frac{1}{3} x^6 \right) dx = \frac{3\rho}{35}. \end{aligned}$$

15.

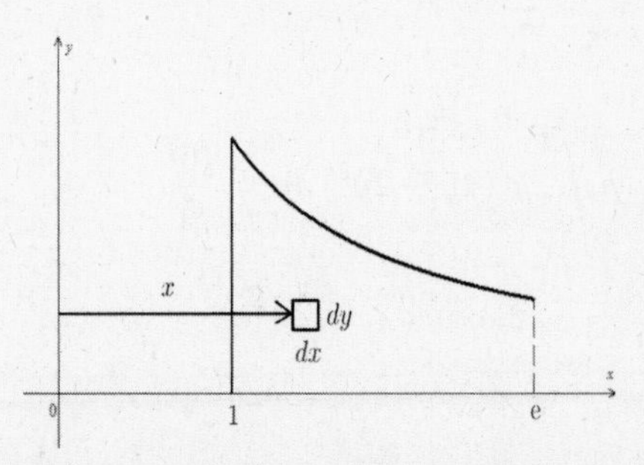

Moment of inertia of typical element: $x^2 \cdot \rho y\, dx = \rho x^2\, dy\, dx$.

$I_y = \rho \int_1^e \int_0^{1/x} x^2\, dy\, dx = \rho \int_1^e x^2 y \Big|_0^{1/x} dx = \rho \int_1^e x\, dx = \frac{\rho}{2} x^2 \Big|_1^e = \frac{\rho}{2}(e^2 - 1)$.

Mass: $\rho \int_1^e \int_0^{1/x} dy\, dx = \rho \int_1^e \frac{1}{x}\, dx = \rho \ln|x| \Big|_1^e = \rho(\ln e - \ln 1) = \rho(1 - 0) = \rho$.

$R_y = \sqrt{\frac{\rho(e^2 - 1)}{2} \frac{1}{\rho}} = \sqrt{\frac{e^2 - 1}{2}}$.

17. Moment of typical element with respect to the y-axis: $x\,dy\,dx$.

$$\overline{x} = \frac{M_y}{A} = \frac{\int_1^{e}\int_0^{1/x} x\,dy\,dx}{\int_1^{e}\int_0^{1/x} dy\,dx}.$$

By Exercise 15, $A = 1$. So $\overline{x} = \int_1^{e} xy\Big|_0^{1/x}\,dx = \int_1^{e} 1\,dx = e - 1.$

19.

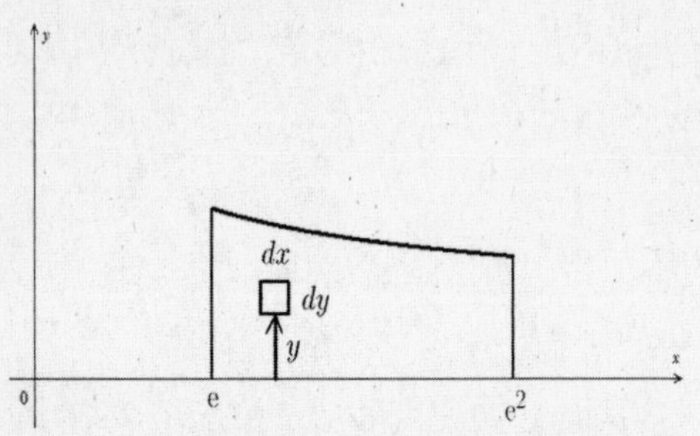

$$y = \frac{1}{\sqrt[3]{x}} = x^{-1/3}$$

Moment of inertia of typical element: $y^2 \cdot \rho\,dy\,dx = \rho y^2\,dy\,dx$.

$$\begin{aligned}
I_x &= \rho\int_e^{e^2}\int_0^{x^{-1/3}} y^2\,dy\,dx = \rho\int_e^{e^2} \frac{1}{3}y^3\Big|_0^{x^{-1/3}}\,dx \\
&= \rho\int_e^{e^2} \frac{1}{3}(x^{-1/3})^3\,dx = \frac{\rho}{3}\int_e^{e^2} x^{-1}\,dx = \frac{\rho}{3}\ln|x|\Big|_e^{e^2} \\
&= \frac{\rho}{3}(\ln e^2 - \ln e) = \frac{\rho}{3}(2\ln e - \ln e) \\
&= \frac{\rho}{3}(2-1) = \frac{\rho}{3}. \qquad \ln e = 1
\end{aligned}$$

21.

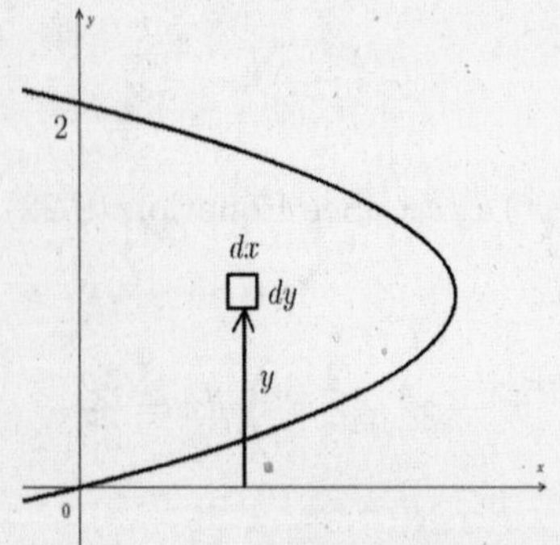

$$\begin{aligned}
x = 4y - 2y^2 &= 0 \\
2y(2-y) &= 0 \\
y &= 0,\ 2
\end{aligned}$$

Moment of typical element: $y^2 \cdot \rho\,dx\,dy$.

$$I_x = \int_0^2\int_0^{4y-2y^2} y^2 \cdot \rho\,dx\,dy = \rho\int_0^2 y^2 x\Big|_0^{4y-2y^2}\,dy = \rho\int_0^2 y^2(4y - 2y^2)\,dy = \frac{16\rho}{5}.$$

23.

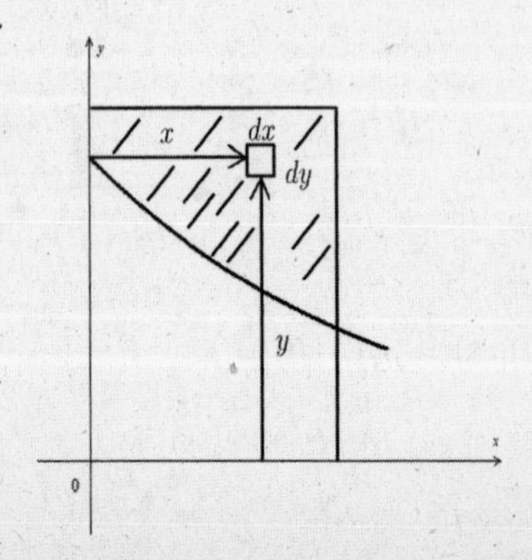

Moment (with respect to y-axis) of typical element: $x\,dy\,dx$.

$$M_y = \int_0^1\int_{e^{-x}}^1 x\,dy\,dx = \int_0^1 xy\Big|_{e^{-x}}^1 dx = \int_0^1 x(1-e^{-x})\,dx = \int_0^1 x\,dx - \int_0^1 xe^{-x}\,dx.$$

[Integration by parts] $\quad u = x \quad dv = e^{-x}\,dx$

$\quad du = dx \quad v = -e^{-x}$

$$\frac{1}{2}x^2\Big|_0^1 - \left[-xe^{-x}\Big|_0^1 + \int_0^1 e^{-x}\,dx\right] = \frac{1}{2}x^2\Big|_0^1 - \left[-xe^{-x}\Big|_0^1 - e^{-x}\Big|_0^1\right]$$
$$= \frac{1}{2}x^2\Big|_0^1 + xe^{-x}\Big|_0^1 + e^{-x}\Big|_0^1 = \frac{1}{2} + e^{-1} + e^{-1} - 1 = \frac{2}{e} - \frac{1}{2}.$$

$$A = \int_0^1\int_{e^{-x}}^1 dy\,dx = \int_0^1 (1-e^{-x})\,dx = x + e^{-x}\Big|_0^1 = 1 + e^{-1} - 1 = \frac{1}{e}.$$

$$\overline{x} = \frac{2/e - 1/2}{1/e} = 2 - \frac{1}{2}e = \frac{4-e}{2}.$$

$$M_x = \int_0^1\int_{e^{-x}}^1 y\,dy\,dx = \int_0^1 \frac{1}{2}y^2\Big|_{e^{-x}}^1 dx = \frac{1}{2}\int_0^1 (1-e^{-2x})\,dx \qquad u = -2x;\ du = -2\,dx.$$
$$= \frac{1}{2}\left(x + \frac{1}{2}e^{-2x}\right)\Big|_0^1 = \frac{1}{2}\left(1 + \frac{1}{2}e^{-2}\right) - \frac{1}{2}\left(\frac{1}{2}\right)$$
$$= \frac{1}{2} + \frac{1}{4}e^{-2} - \frac{1}{4} = \frac{1}{4e^2} + \frac{1}{4}.$$

$$\overline{y} = \frac{1/(4e^2) + 1/4}{1/e} = \frac{1}{4e} + \frac{e}{4} = \frac{e^2+1}{4e}.$$

25.

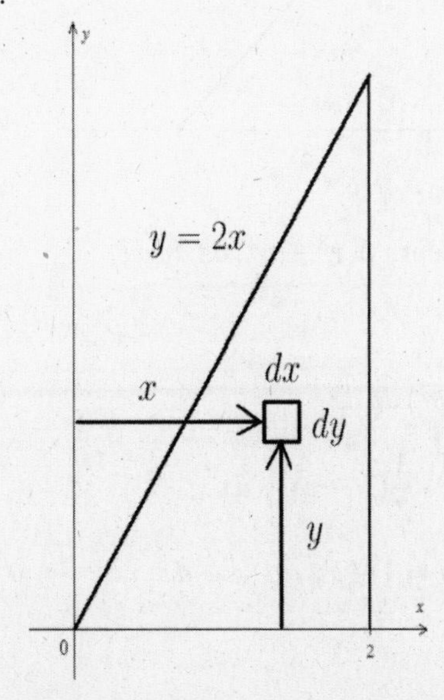

$$\overline{x} = \frac{\int_0^2\int_0^{2x} x\cdot\frac{1}{5}x\,dy\,dx}{\int_0^2\int_9^{2x}\frac{1}{5}x\,dy\,dx} = \frac{\int_0^2 \frac{1}{5}x^2y\Big|_0^{2x} dx}{\int_0^2 \frac{1}{5}xy\Big|_0^{2x} dx}$$
$$= \frac{\int_0^2 \frac{1}{5}x^2(2x)\,dx}{\int_0^2 \frac{1}{5}x(2x)\,dx} = \frac{\frac{8}{5}}{\frac{16}{15}} = \frac{3}{2}$$

$$\overline{y} = \frac{\int_0^2\int_0^{2x} y\cdot\frac{1}{5}x\,dy\,dx}{\frac{16}{15}} = \frac{15}{16}\int_0^2 \frac{1}{5}x\cdot\frac{y^2}{2}\Big|_0^{2x} dx$$
$$= \frac{15}{16}\int_0^2 \frac{1}{5}x\frac{4x^2}{2}\,dx = \frac{15}{16}\cdot\frac{8}{5} = \frac{3}{2}$$

27.

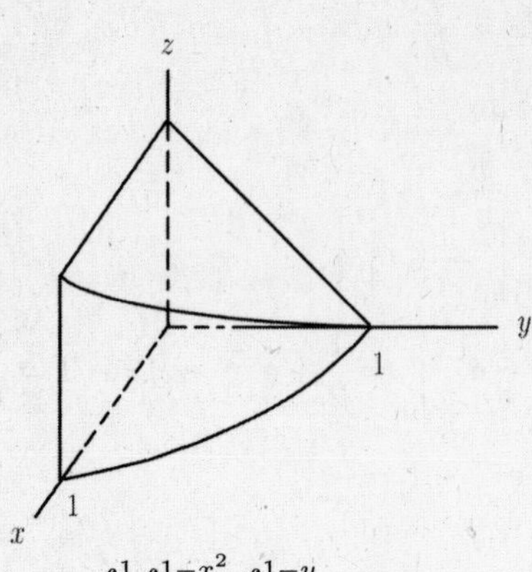

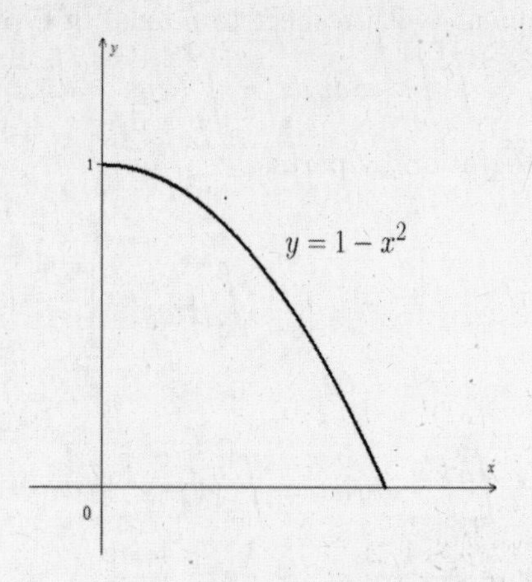

$$\overline{x} = \frac{\displaystyle\int_0^1\int_0^{1-x^2}\int_0^{1-y} x\,dz\,dy\,dx}{\displaystyle\int_0^1\int_0^{1-x^2}\int_0^{1-y} dz\,dy\,dx}.$$

Partial evaluation:

$$\begin{aligned}
\text{Numerator} &= \int_0^1\int_0^{1-x^2} xz\Big|_0^{1-y} dy\,dx = \int_0^1\int_0^{1-x^2} x(1-y)\,dy\,dx \\
&= \int_0^1 x\left(y-\frac{1}{2}y^2\right)\Big|_0^{1-x^2} dx = \int_0^1 x\left[(1-x^2)-\frac{1}{2}(1-x^2)^2\right] dx \\
&= \frac{1}{2}\int_0^1 (x-x^5)\,dx = \frac{1}{6}.
\end{aligned}$$

29.

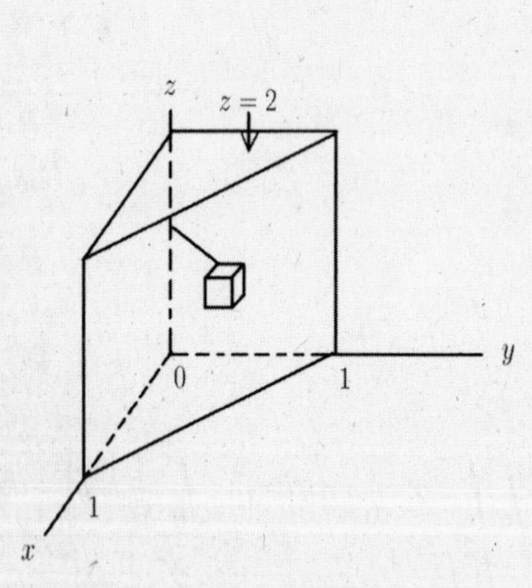

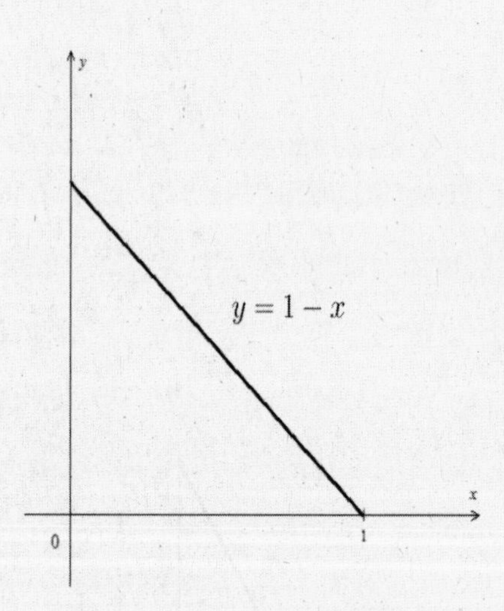

Moment of inertia (with respect to z-axis) of typical element: $\rho(x^2+y^2)\,dz\,dy\,dx$.

$$\begin{aligned}
I_z &= \rho\int_0^1\int_0^{1-x}\int_0^2 (x^2+y^2)\,dz\,dy\,dx \\
&= \rho\int_0^1\int_0^{1-x} (x^2+y^2)z\Big|_0^2 dy\,dx = 2\rho\int_0^1\int_0^{1-x}(x^2+y^2)\,dy\,dx \\
&= 2\rho\int_0^1\left(x^2y+\frac{1}{3}y^3\right)\Big|_0^{1-x} dx = 2\rho\int_0^1\left[x^2(1-x)+\frac{1}{3}(1-x)^3\right] dx \\
&= 2\rho\int_0^1 x^2(1-x)\,dx + \frac{2\rho}{3}\int_0^1 (1-x)^3\,dx \qquad u = 1-x;\ du = -\,dx. \\
&= 2\rho\left(\frac{1}{3}x^3-\frac{1}{4}x^4\right)\Big|_0^1 - \frac{2\rho}{3}\frac{(1-x)^4}{4}\Big|_0^1 \\
&= 2\rho\left(\frac{1}{3}-\frac{1}{4}\right)+\frac{2\rho}{12} = 2\rho\left(\frac{1}{12}\right)+\frac{2\rho}{12} = \frac{1}{3}\rho.
\end{aligned}$$

31. $I_y = \rho \int_0^1 \int_{x^2}^{x} \int_0^{xy} (x^2 + z^2)\, dz\, dy\, dx.$

33. Moment of inertia of typical element with respect to the x-axis: $(y^2 + z^2)\rho\, dz\, dy\, dx$.

$$\begin{aligned} I_x &= \int_0^1 \int_0^2 \int_0^{x^2} (y^2 + z^2)\rho\, dz\, dy\, dx = \rho \int_0^1 \int_0^2 \left(y^2 z + \frac{1}{3}z^3 \right)\Big|_0^{x^2} dy\, dx \\ &= \rho \int_0^1 \int_0^2 \left(y^2 x^2 + \frac{1}{3}x^6 \right) dy\, dx = \rho \int_0^1 \left(x^2 \cdot \frac{1}{3}y^3 + \frac{1}{3}x^6 y \right)\Big|_0^2 dx \\ &= \rho \int_0^1 \left(\frac{8}{3}x^2 + \frac{2}{3}x^6 \right) dx = \frac{62\rho}{63}. \end{aligned}$$

35.
$$\begin{aligned} V &= \int_0^1 \int_0^{1-x} \int_0^2 dz\, dy\, dx = \int_0^1 \int_0^{1-x} z\Big|_0^2 dy\, dx = 2 \int_0^1 \int_0^{1-x} dy\, dx \\ &= 2 \int_0^1 y\Big|_0^{1-x} dx = 2 \int_0^1 (1 - x)\, dx = 1. \end{aligned}$$

$$\begin{aligned} M_{yz} &= \int_0^1 \int_0^{1-x} \int_0^2 x\, dz\, dy\, dx = \int_0^1 \int_0^{1-x} xz\Big|_0^2 dy\, dx \\ &= 2 \int_0^1 \int_0^{1-x} x\, dy\, dx = 2 \int_0^1 xy\Big|_0^{1-x} dx = 2 \int_0^1 x(1 - x)\, dx = \frac{1}{3}. \end{aligned}$$

$\overline{x} = \dfrac{M_{yz}}{V} = \dfrac{1}{3}.$

$$\begin{aligned} M_{xy} &= \int_0^1 \int_0^{1-x} \int_0^2 z\, dz\, dy\, dx = \int_0^1 \int_0^{1-x} \frac{1}{2}z^2\Big|_0^2 dy\, dx \\ &= 2 \int_0^1 \int_0^{1-x} dy\, dx = 2 \int_0^1 (1 - x)\, dx = 1. \end{aligned}$$

$\overline{z} = \dfrac{M_{xy}}{V} = 1.$

37.

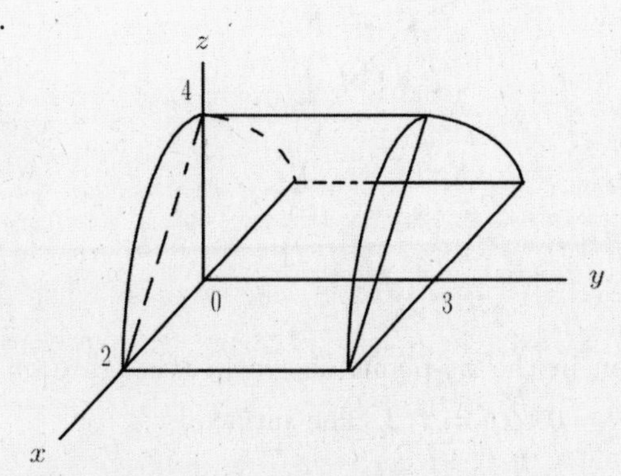

Lower surface: $z = 4 - 2x$.

Upper surface: $z = 4 - x^2$.

$V = \int_0^2 \int_0^3 \int_{4-2x}^{4-x^2} (1 + x)\, dz\, dy\, dx.$

9.7 Volumes in Cylindrical Coordinates

1.

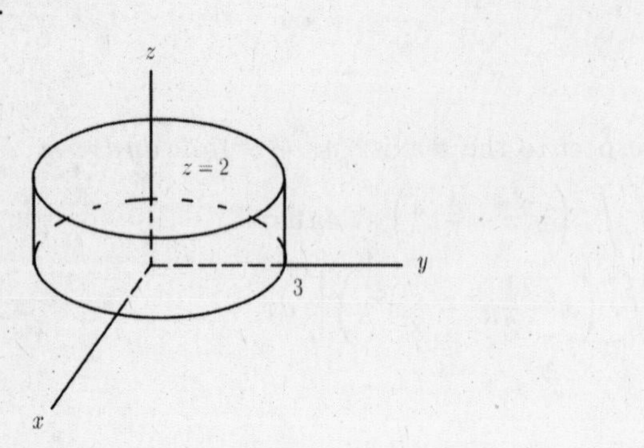

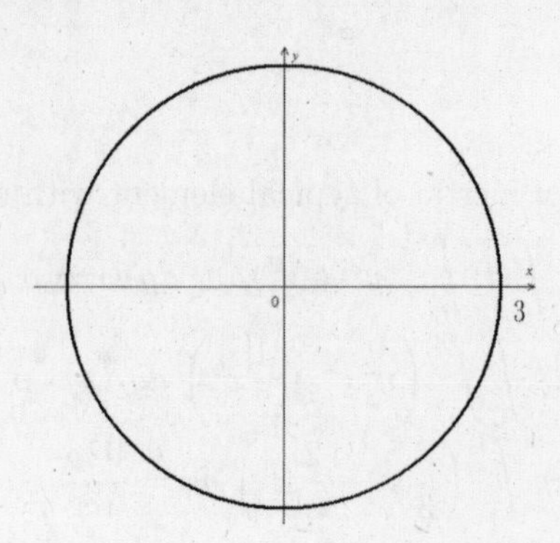

The limits of integration depend on the region in the xy-plane. The circle $x^2+y^2=9$ becomes $r=3$, while θ ranges from $\theta=0$ to $\theta=2\pi$. (See figure on the right.) The equation of the surface is $z=2$.

$$V=\int_0^{2\pi}\int_0^3 2r\,dr\,d\theta=\int_0^{2\pi} r^2\Big|_0^3\,d\theta=\int_0^{2\pi}9\,d\theta=9\theta\Big|_0^{2\pi}=18\pi.$$

3. The paraboloid $z=9-x^2-y^2$ is shown in Exercise 23, Section 9.5. The region in the xy-plane is the trace $0=9-x^2-y^2$, which is a circle of radius 3, that is, $r=3$ in cylindrical coordinates. The equation of the surface becomes $z=9-(x^2+y^2)=9-r^2$.

$$\begin{aligned}V&=\int_0^{2\pi}\int_0^3(9-r^2)r\,dr\,d\theta=\int_0^{2\pi}\int_0^3(9r-r^3)\,dr\,d\theta\\&=\int_0^{2\pi}\left(\frac{9}{2}r^2-\frac{1}{4}r^4\right)\Big|_0^3\,d\theta=\int_0^{2\pi}\left(\frac{81}{2}-\frac{81}{4}\right)d\theta=\frac{81}{4}\theta\Big|_0^{2\pi}=\frac{81\pi}{2}.\end{aligned}$$

5.

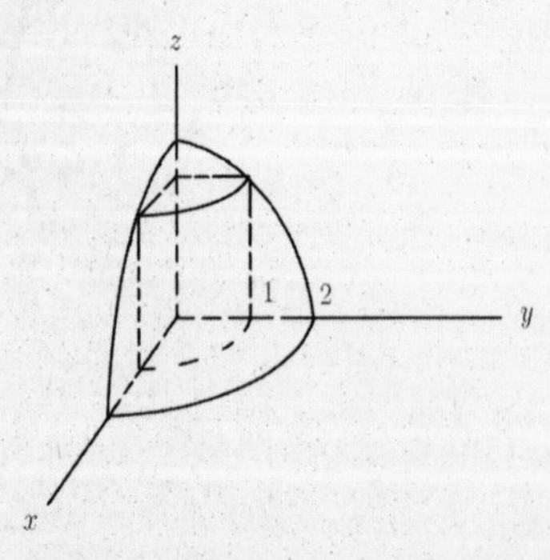

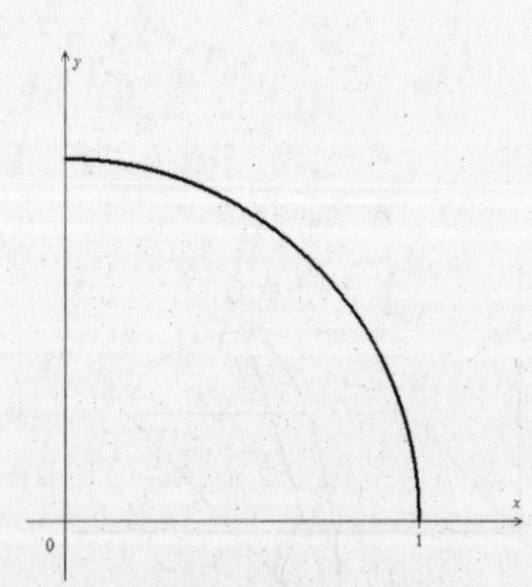

The limits of integration depend on the region in the xy-plane, shown in the figure on the right: the circle is $r=1$, while θ ranges from $\theta=0$ to $\theta=\pi/2$. The surface is $z=4-x^2-y^2=4-(x^2+y^2)=4-r^2$. So

$$\begin{aligned}V&=\int_0^{\pi/2}\int_0^1(4-r^2)r\,dr\,d\theta=\int_0^{\pi/2}\int_0^1(4r-r^3)\,dr\,d\theta\\&=\int_0^{\pi/2}\left(2r-\frac{1}{4}r^4\right)\Big|_0^1\,dr=\int_0^{\pi/2}\frac{7}{4}\,d\theta=\frac{7\pi}{8}.\end{aligned}$$

7.

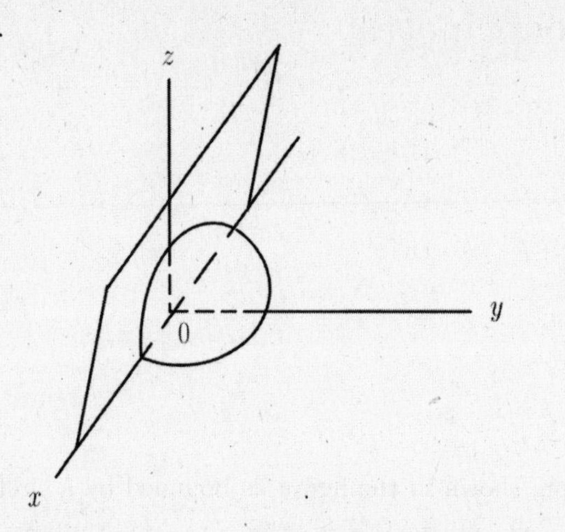

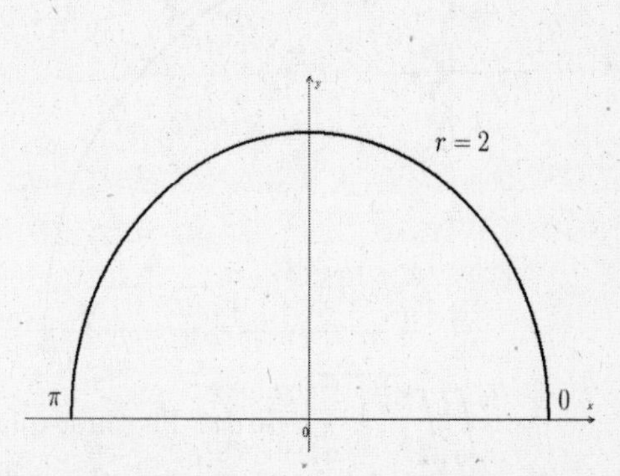

The limits of integration depend on the region in the xy-plane, shown in the figure on the right. The circle is $r = 2$ and θ ranges from 0 to π. The equation of the surface is $z = 3y = 3r\sin\theta$.

$$\begin{aligned} V &= \int_0^{\pi}\int_0^2 (3r\sin\theta) r\,dr\,d\theta = 3\int_0^{\pi} (\sin\theta)\frac{1}{3}r^3\Big|_0^2\,d\theta = 8\int_0^{\pi}\sin\theta\,d\theta \\ &= -8\cos\theta\Big|_0^{\pi} = -8\cos\pi + 8\cos 0 = 8 + 8 = 16. \end{aligned}$$

9.

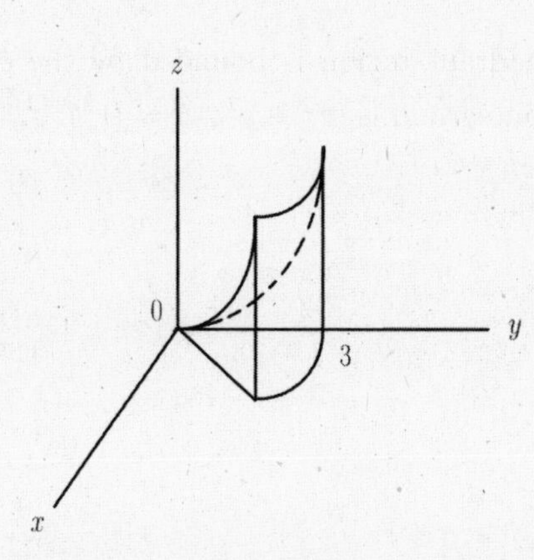

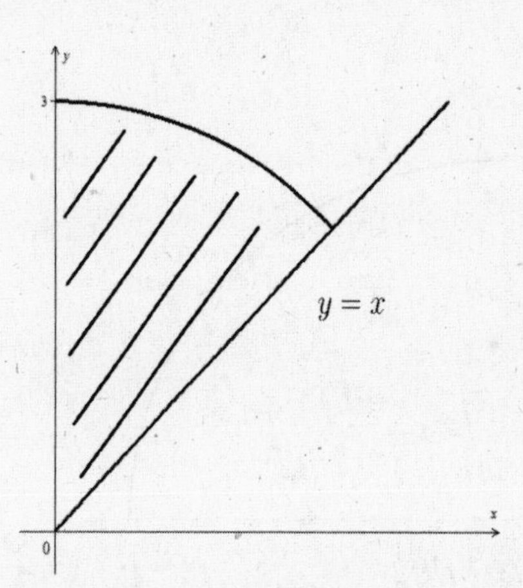

The region bounded by $y = x$ and $x^2 + y^2 = 9$ is shown in the figure on the right: the circle is $r = 3$, while θ ranges from $\theta = \pi/4$ to $\theta = \pi/2$. The surface is $z = x^2 + y^2 = r^2$. So

$$\begin{aligned} V &= \int_{\pi/4}^{\pi/2}\int_0^3 r^2\,r\,dr\,d\theta = \int_{\pi/4}^{\pi/2} \frac{1}{4}r^4\Big|_0^3\,d\theta = \frac{1}{4}(81)\int_{\pi/4}^{\pi/2} d\theta = \frac{1}{4}(81)\theta\Big|_{\pi/4}^{\pi/2} \\ &= \frac{1}{4}(81)\left(\frac{\pi}{2} - \frac{\pi}{4}\right) = \frac{1}{4}(81)\frac{\pi}{4} = \frac{81\pi}{16}. \end{aligned}$$

11. Equation of the circle: $x^2 + y^2 = 4 : r = 2$.
$z = xy = (r\cos\theta)(r\sin\theta) = r^2\sin\theta\cos\theta$

$$\begin{aligned} V &= \int_0^{\pi/2}\int_0^2 (r^2\sin\theta\cos\theta) r\,dr\,d\theta = \int_0^{\pi/2} (\sin\theta\cos\theta)\frac{r^4}{4}\Big|_0^2\,d\theta \\ &= 4\int_0^{\pi/2}\sin\theta\cos\theta\,d\theta \qquad u = \sin\theta;\ du = \cos\theta\,d\theta \\ &= 4\frac{\sin^2\theta}{2}\Big|_0^{\pi/2} = 2 \end{aligned}$$

13.

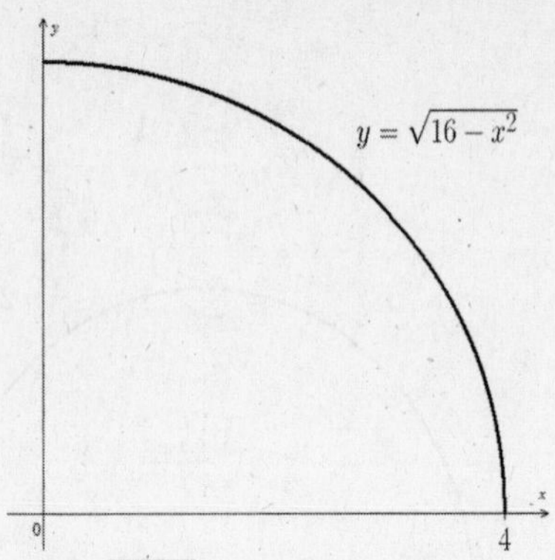

$\int_0^4 \int_0^{\sqrt{16-x^2}} x\,dy\,dx$. The first-quadrant region, shown in the figure, is bounded by a circle of radius 4 ($r = 4$); θ ranges from $\theta = 0$ to $\theta = \pi/2$. The integrand x is replaced by $r\cos\theta$.

$$\begin{aligned}
\int_0^{\pi/2} \int_0^4 (r\cos\theta) r\,dr\,d\theta &= \int_0^{\pi/2} \int_0^4 (\cos\theta) r^2\,dr\,d\theta \\
&= \int_0^{\pi/2} (\cos\theta) \frac{1}{3} r^3 \Big|_0^4 \,d\theta = \frac{4^3}{3} \int_0^{\pi/2} \cos\theta\,d\theta \\
&= \frac{64}{3} \sin\theta \Big|_0^{\pi/2} = \frac{64}{3}.
\end{aligned}$$

15.

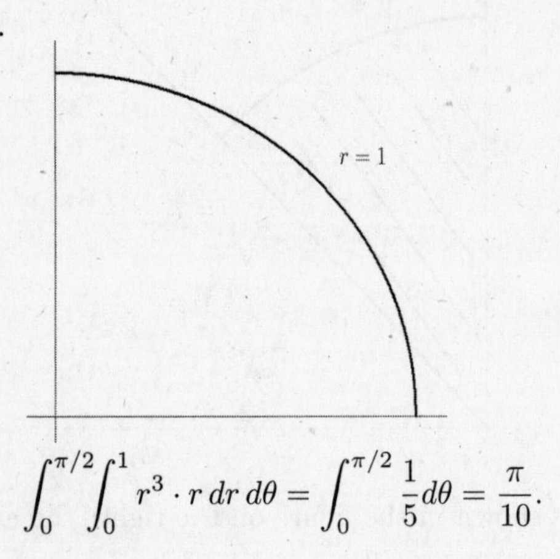

The first-quadrant region is bounded by the circle $r = 1$. The integrand is $(x^2 + y^2)^{3/2} = (r^2)^{3/2} = r^3$.

$$\int_0^{\pi/2} \int_0^1 r^3 \cdot r\,dr\,d\theta = \int_0^{\pi/2} \frac{1}{5}\,d\theta = \frac{\pi}{10}.$$

17.

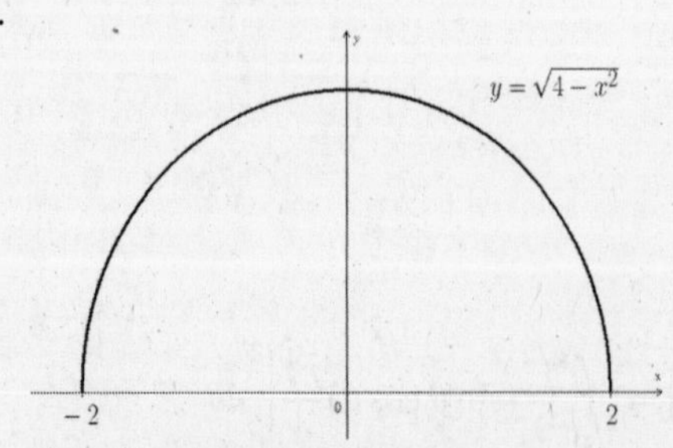

The region, which is bounded by a semicircle of radius 2, is shown in the figure: $r = 2$ and $\theta = 0$ to $\theta = \pi$. Since $x = r\cos\theta$, we have

$$\int_0^{\pi} \int_0^2 (r\cos\theta) r\,dr\,d\theta = \int_0^{\pi} (\cos\theta) \frac{1}{3} r^3 \Big|_0^2 \,d\theta = \frac{8}{3} \sin\theta \Big|_0^{\pi} = 0.$$

19.

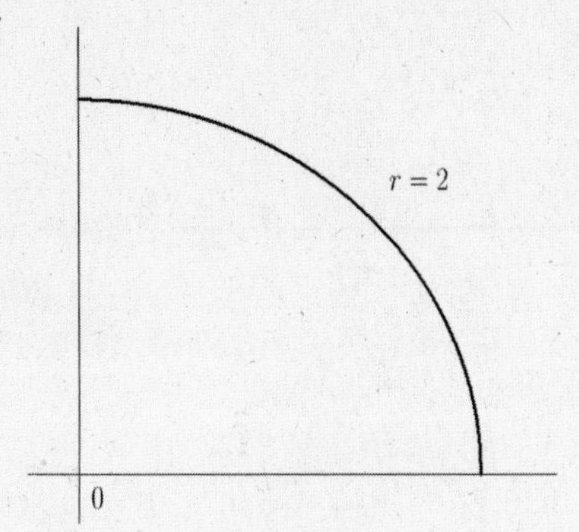

$$\begin{aligned}\int_0^2\int_0^{\sqrt{4-y^2}} x^2y\,dx\,dy &= \int_0^{\pi/2}\int_0^2 (r^2\cos^2\theta)(r\sin\theta)r\,dr\,d\theta \\ &= \int_0^{\pi/2}(\cos^2\theta\sin\theta)\left.\frac{r^5}{5}\right|_0^2 d\theta \qquad u=\cos\theta;\, du=-\sin\theta \\ &= \frac{32}{5}\left.\left(-\frac{\cos^3\theta}{3}\right)\right|_0^{\pi/2} = \frac{32}{15}\end{aligned}$$

Chapter 9 Review

1. Plane; the intercepts are $x = 2$, $y = 2$, and $z = 1$. (See graph in Answer Section.)

3. Cylinder; $2y^2 + z^2 = 4$ is an ellipse in the yz-plane. (See graph in Answer Section.)

5. Trace in xy-plane: $y = 3x^2$.

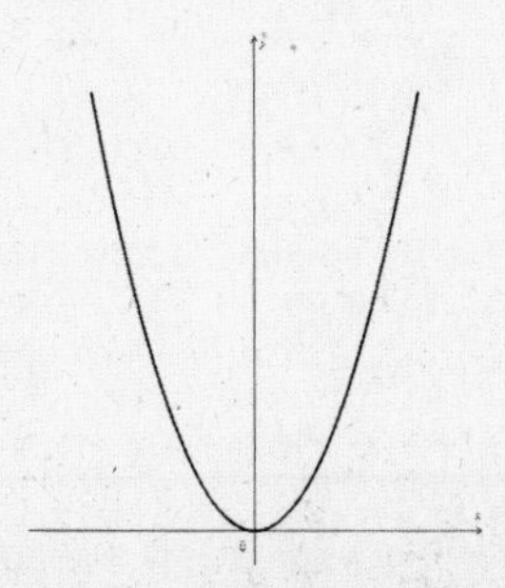

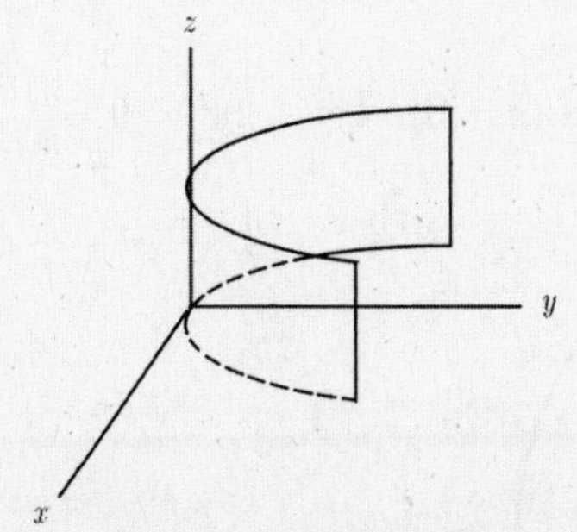

The cylinder $y = 3x^2$ extends along the z-axis (note the missing z-variable).

7. $z^2 - 4x^2 - y^2 = 4$ (hyperboloid of two sheets)

   1. Trace in xy-plane ($z = 0$): $-4x^2 - y^2 = 4$ (no trace; imaginary locus).

2. Trace in yz-plane ($x = 0$): $z^2 - y^2 = 4$.

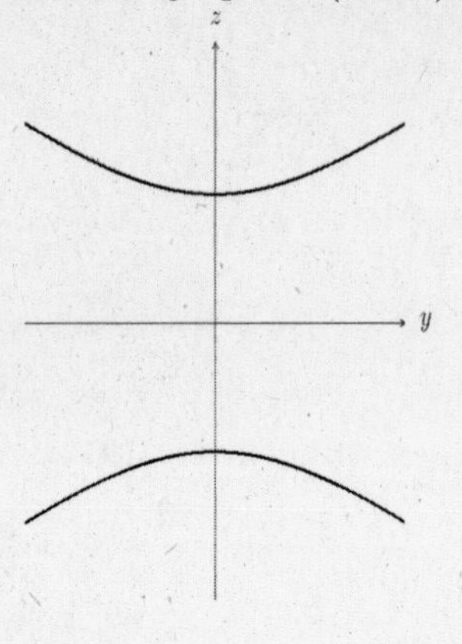

3. Trace in xz-plane ($y = 0$): $z^2 - 4x^2 = 4$ (another hyperbola).

Cross-section: let $z = 3$ in the original equation:

$$9 - 4x^2 - y^2 = 4 \text{ or } 4x^2 + y^2 = 5.$$

The resulting ellipse is in a plane parallel to the xy-plane and 3 units above it. (See graph in Answer Section.)

9. $4x^2 + 2y^2 - z^2 = 9$ (hyperboloid of one sheet)

1. Trace in xy-plane ($z = 0$): $4x^2 + 2y^2 = 9$.

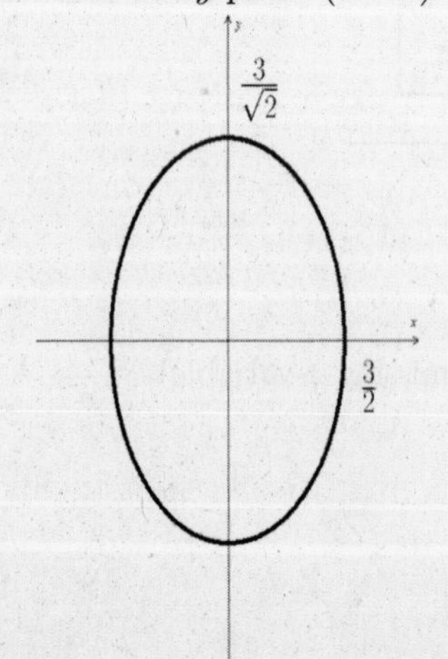

2. Trace in xz-plane ($y = 0$): $4x^2 - z^2 = 9$.

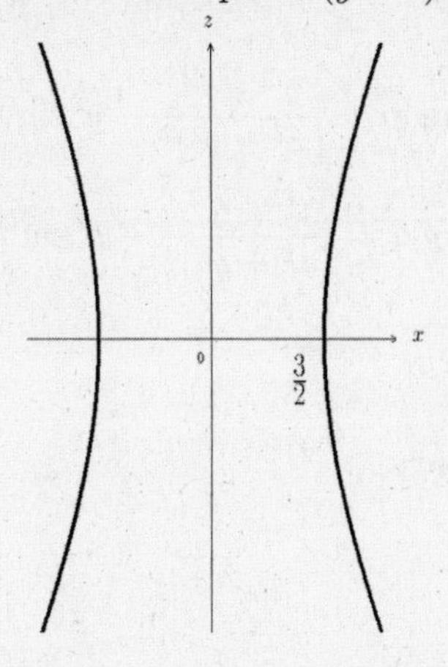

3. Trace in yz-plane ($x = 0$): $2y^2 - z^2 = 9$.

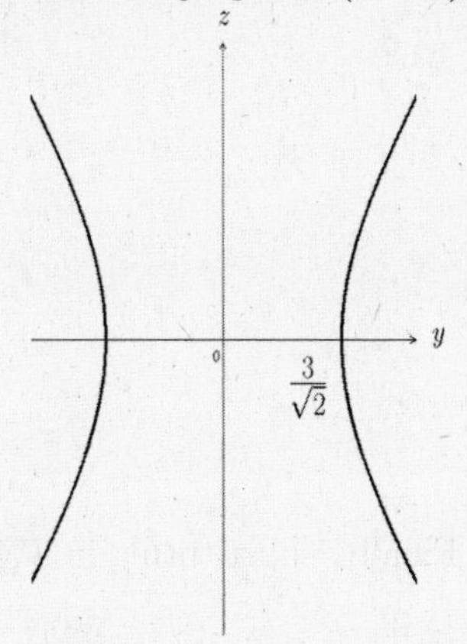

Cross-section: let $z = 4$ in the original equation:

$$4x^2 + 2y^2 - 4^2 = 9 \text{ or } 4x^2 + 2y^2 = 25.$$

The resulting ellipse is parallel to the xy-plane and 4 units above.

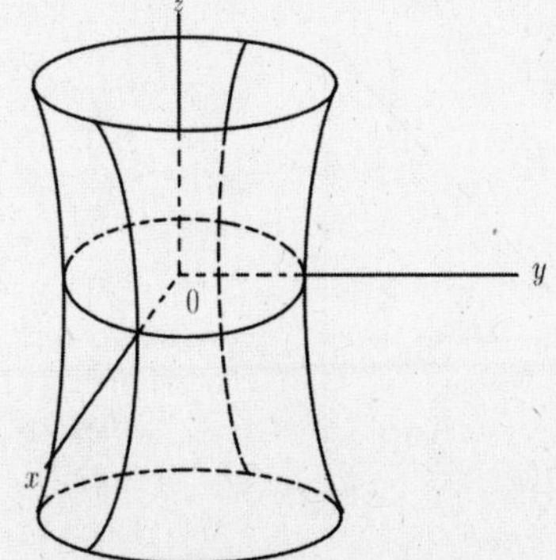

11. $z = \ln(x^2 + y^2)^{1/2} + \sin xy = \dfrac{1}{2}\ln(x^2 + y^2) + \sin xy$

(a) $\dfrac{\partial z}{\partial x} = \dfrac{1}{2}\dfrac{1}{x^2 + y^2}(2x) + (\cos xy)y = \dfrac{x}{x^2 + y^2} + y\cos xy$

(b) $\dfrac{\partial z}{\partial y} = \dfrac{1}{2}\dfrac{1}{x^2+y^2}(2y) + (\cos xy)x = \dfrac{y}{x^2+y^2} + x\cos xy$

(c) $\dfrac{\partial^2 z}{\partial x^2} = \dfrac{\partial}{\partial x}\left(\dfrac{\partial z}{\partial x}\right) = \dfrac{(x^2+y^2)(1) - x(2x)}{(x^2+y^2)^2} + y(-\sin xy)y = \dfrac{y^2-x^2}{(x^2+y^2)^2} - y^2\sin xy$

(d) $\dfrac{\partial^2 z}{\partial y^2} = \dfrac{\partial}{\partial y}\left(\dfrac{\partial z}{\partial y}\right) = \dfrac{(x^2+y^2)(1) - y(2y)}{(x^2+y^2)^2} + x(-\sin xy)x = \dfrac{x^2-y^2}{(x^2+y^2)^2} - x^2\sin xy$

(e)
$$\begin{aligned}\frac{\partial^2 z}{\partial x\partial y} &= \frac{\partial}{\partial x}\left(\frac{\partial z}{\partial y}\right) = \frac{\partial}{\partial x}\left[y(x^2+y^2)^{-1} + x\cos xy\right]\\ &= -y(x^2+y^2)^{-2}(2x) + x(-\sin xy)y + \cos xy\\ &= -\frac{2xy}{(x^2+y^2)^2} - xy\sin xy + \cos xy\end{aligned}$$

13. $T = 2xy - x^2 - 2y^2 + 3x + 5$

Critical point:

$$\begin{aligned}\frac{\partial T}{\partial x} &= 2y - 2x + 3 = 0\\ \frac{\partial T}{\partial y} &= 2x - 4y = 0\\ \hline -2y + 3 &= 0 \quad \text{adding}\\ y &= \frac{3}{2}, \quad x = 3.\end{aligned}$$

$\dfrac{\partial^2 T}{\partial x^2} = -2,\ \dfrac{\partial^2 T}{\partial y^2} = -4,\ \dfrac{\partial^2 T}{\partial x\partial y} = 2$

$A = (-2)(-4) - (2)^2 = 4 > 0$

Since $A > 0$ and $\dfrac{\partial^2 T}{\partial x^2} < 0$, T has a maximum at $\left(3, \dfrac{3}{2}\right)$. Finally, substituting in T, we get

$T\Big|_{(3,3/2)} = \dfrac{19}{2} = 9.5°$ C (warmest point).

15. $T = 2xy - x^2 - 2y^2 + 3x + 5$; $\dfrac{\partial T}{\partial x} = 2y - 2x + 3\Big|_{(-3,1)} = 11°$ C/cm.

17.

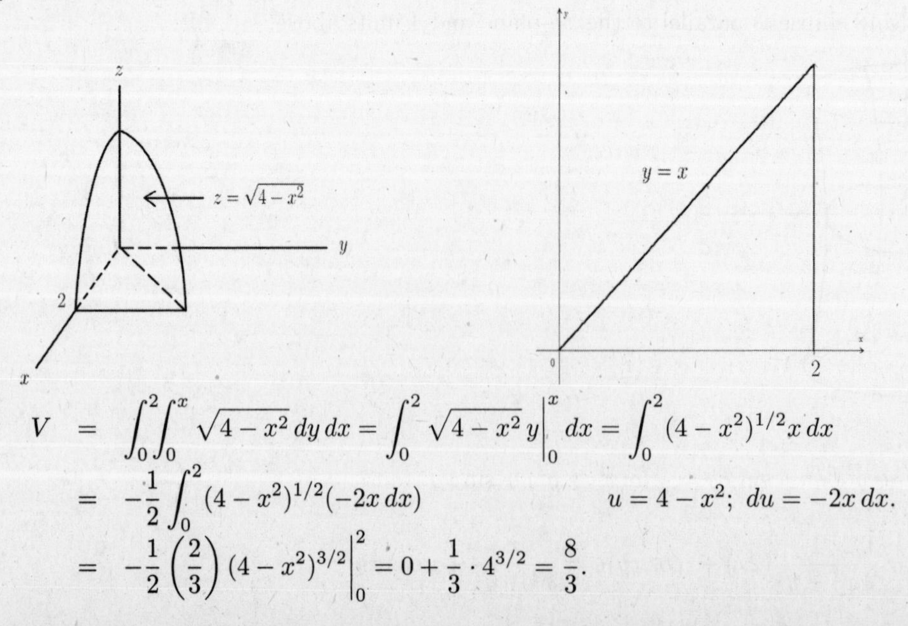

$$\begin{aligned}V &= \int_0^2\int_0^x \sqrt{4-x^2}\,dy\,dx = \int_0^2 \sqrt{4-x^2}\,y\Big|_0^x dx = \int_0^2 (4-x^2)^{1/2}x\,dx\\ &= -\frac{1}{2}\int_0^2 (4-x^2)^{1/2}(-2x\,dx) \qquad u = 4-x^2;\ du = -2x\,dx.\\ &= -\frac{1}{2}\left(\frac{2}{3}\right)(4-x^2)^{3/2}\Big|_0^2 = 0 + \frac{1}{3}\cdot 4^{3/2} = \frac{8}{3}.\end{aligned}$$

19.

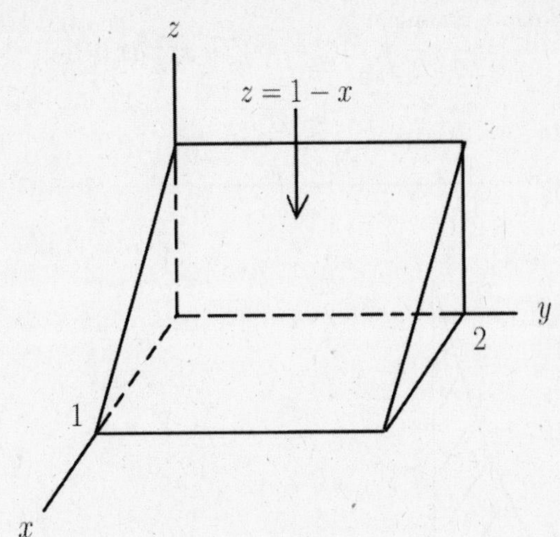

$$\begin{aligned} V &= \int_0^1\int_0^2 (1-x)\,dy\,dx = \int_0^1 (1-x)y\Big|_0^2 dx = \int_0^1 (1-x)(2)\,dx \\ &= 2\left(x - \frac{1}{2}x^2\right)\Big|_0^1 = 2\left(\frac{1}{2}\right) = 1. \end{aligned}$$

21.

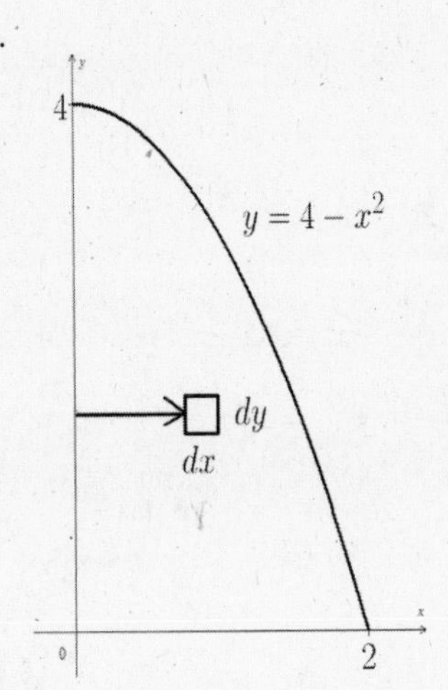

Moment of inertia of typical element: $x^2 \cdot \rho\,dy\,dx$.

$$I_y = \rho \int_0^2\int_0^{4-x^2} x^2\,dy\,dx = \frac{64\rho}{15}.$$

23.

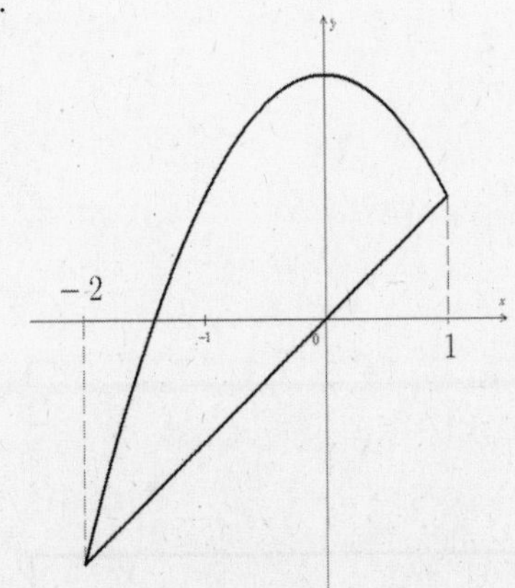

$$\begin{aligned} y &= x \\ y &= 2 - x^2 \\ \hline 0 &= x - 2 + x^2 \quad \text{(subtracting)} \end{aligned}$$

$$\begin{aligned} x^2 + x - 2 &= 0 \\ (x+2)(x-1) &= 0 \\ x &= -2,\ 1 \end{aligned}$$

Moment of typical element with respect to the y-axis: $x\,dy\,dx$.

$$\overline{x} = \frac{M_y}{A} = \frac{\int_{-2}^1\int_x^{2-x^2} x\,dy\,dx}{\int_{-2}^1\int_x^{2-x^2} dy\,dx} = \frac{\int_{-2}^1 xy\Big|_x^{2-x^2} dx}{\int_{-2}^1 y\Big|_x^{2-x^2} dx} = \frac{\int_{-2}^1 x(2-x^2-x)\,dx}{\int_{-2}^1 (2-x^2-x)\,dx} = \frac{-9/4}{9/2} = -\frac{1}{2}.$$

Moment of typical element with respect to the x-axis: $y\,dy\,dx$.

$$\overline{y} = \frac{M_x}{A} = \frac{\int_{-2}^{1}\int_{x}^{2-x^2} y\,dy\,dx}{\int_{-2}^{1}\int_{x}^{2-x^2} dy\,dx} = \frac{\int_{-2}^{1} \frac{1}{2}y^2\Big|_{x}^{2-x^2} dx}{9/2} = \frac{2}{9}\cdot\frac{1}{2}\int_{-2}^{1}\left[(2-x^2)^2 - x^2\right] dx = \frac{2}{5}.$$

25.

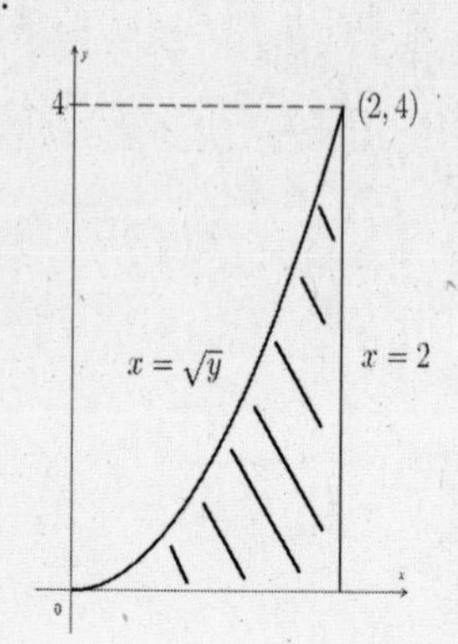

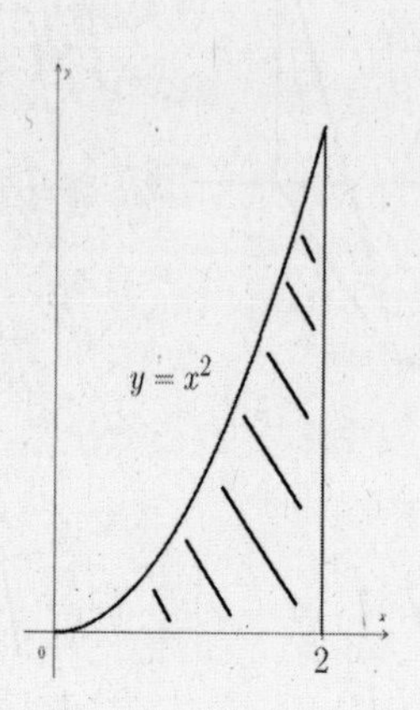

$$\int_0^2\int_0^{x^2} F(x,y)\,dy\,dx$$

27. $V = \pi r^2 h$

$r = 10.00\,\text{cm},\ dr = \pm 0.1\,\text{mm} = \pm 0.01\,\text{cm}$

$h = 15.000\,\text{cm},\ dh = \pm 0.05\,\text{mm} = \pm 0.005\,\text{cm}.$

$$\begin{aligned} dV &= \frac{\partial}{\partial r}(\pi r^2 h)\,dr + \frac{\partial}{\partial h}(\pi r^2 h)\,dh \\ &= 2\pi r h\,dr + \pi r^2\,dh \\ &= 2\pi(10.00)(15.000)(\pm 0.01) + \pi(10.00)^2(\pm 0.005) \\ &= \pm 11.0\,\text{cm}^3. \end{aligned}$$

$$\frac{dV}{V}\times 100 = \frac{11.0}{\pi(10.00)^2(15.000)}\times 100 = 0.23\%.$$

29.

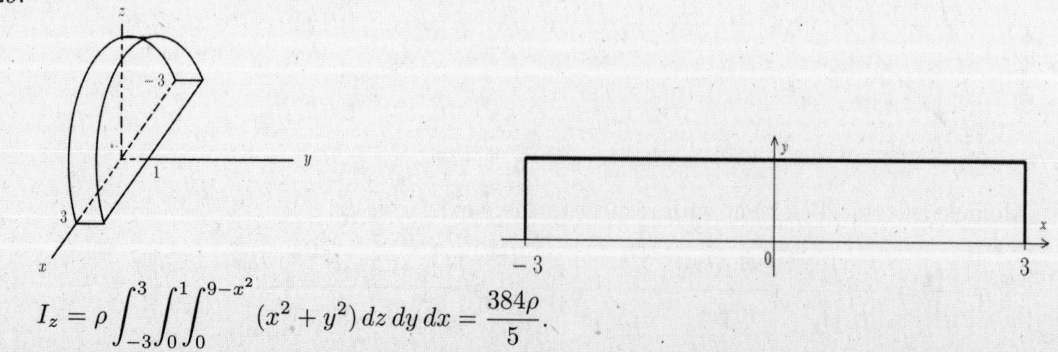

$$I_z = \rho\int_{-3}^{3}\int_0^1\int_0^{9-x^2}(x^2+y^2)\,dz\,dy\,dx = \frac{384\rho}{5}.$$

31.

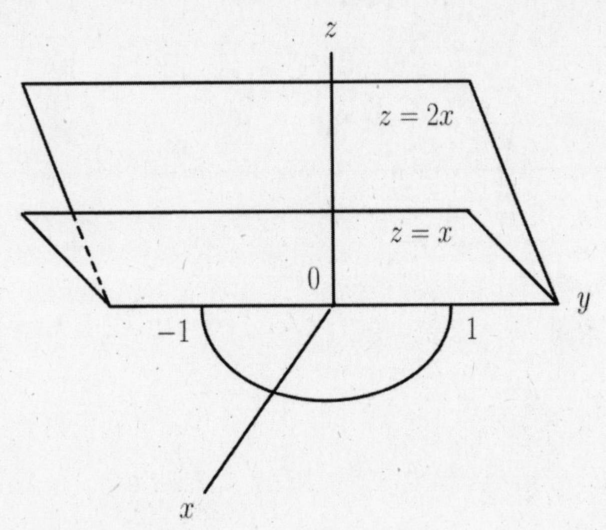

Lower surface: $z = x$.

Upper surface: $z = 2x$.

$$I_x = \int_{-1}^{1}\int_{0}^{\sqrt{1-y^2}}\int_{x}^{2x} \rho(y^2 + z^2)\, dz\, dx\, dy.$$

33.

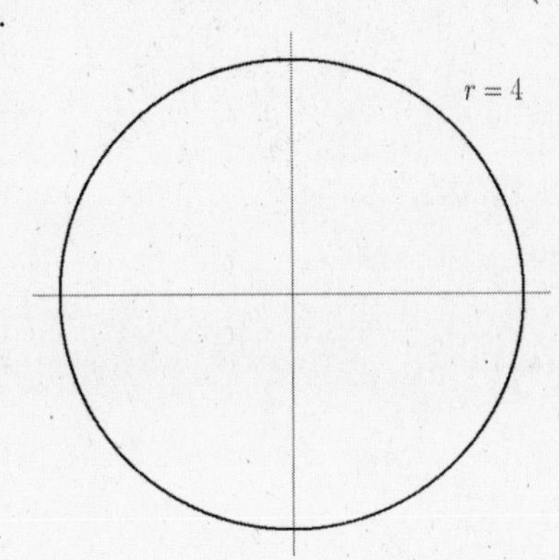

The trace in the xy-plane ($z = 0$) is the circle $x^2 + y^2 = 16$. So the region is bounded by $r = 4$, while θ ranges from 0 to 2π. The integrand is $z = 16 - (x^2 + y^2) = 16 - r^2$. So

$$V = \int_0^{2\pi}\int_0^4 (16 - r^2)r\, dr\, d\theta = \int_0^{2\pi} \left(8r^2 - \frac{1}{4}r^4\right)\Big|_0^4 d\theta = \int_0^{2\pi} 64\, d\theta = 128\pi.$$

Chapter 10

Infinite Series

10.1 Introduction to Infinite Series

1. $a = 1,\ r = \frac{1}{3};\ S = \frac{a}{1-r} = \frac{1}{1-(1/3)} = \frac{3}{2}.$

3. $a = \frac{2}{3},\ ar = \frac{2}{3^2},\ \frac{2}{3}r = \frac{2}{3^2}$ and $r = \frac{1}{3};$
$$S = \frac{a}{1-r} = \frac{2/3}{1-1/3} = \frac{2/3}{2/3} = 1.$$

5. $a = 1,\ r = \frac{3}{4};\ S = \frac{1}{1-(3/4)} = \frac{1}{(1/4)} = 4.$

7. $a = \frac{4}{9},\ ar = \frac{4}{9^2},\ \frac{4}{9}a = \frac{4}{9^2}$ and $r = \frac{1}{9};$
$$S = \frac{a}{1-r} = \frac{4/9}{1-1/9} = \frac{4/9}{8/9} = \frac{1}{2}.$$

9. $\frac{2}{3} - \frac{4}{9} + \frac{8}{27} - \dots$
$$\begin{aligned} a = \frac{2}{3},\ ar &= -\frac{4}{9} \\ \frac{2}{3}r &= -\frac{4}{9} \\ r &= -\frac{2}{3}. \end{aligned}$$
$$S = \frac{a}{1-r} = \frac{2/3}{1-(-2/3)} = \frac{2/3}{1+2/3} = \frac{2/3}{5/3} = \frac{2}{3}\cdot\frac{3}{5} = \frac{2}{5}.$$
Alternatively, the series can be written $S = \frac{2}{3} + \frac{2}{3}\left(-\frac{2}{3}\right) + \frac{2}{3}\left(-\frac{2}{3}\right)^2 - \cdots$, showing that $a = \frac{2}{3},\ r = -\frac{2}{3}.$

11. $a = 1$ and $r = -\frac{3}{4};\ S = \frac{1}{1-r} = \frac{1}{1-(-3/4)} = \frac{1}{7/4} = \frac{4}{7}.$

13. $$\begin{aligned} a = \frac{2}{7},\ ar &= -\left(\frac{2}{7}\right)^2 \\ \frac{2}{7}r &= -\left(\frac{2}{7}\right)^2 \\ r &= -\frac{2}{7} \end{aligned}$$

$$S = \frac{a}{1-r} = \frac{2/7}{1-(-2/7)} = \frac{2}{7}\cdot\frac{7}{9} = \frac{2}{9}$$

15. Since $a = 1$ and $r = -\frac{2}{7}$, $S = \frac{a}{1-r} = \frac{1}{1-(-2/7)} = \frac{1}{1+2/7} = \frac{7}{9}$.

17. $2 - 1 + \frac{1}{2} - \frac{1}{4} + \cdots$

$$\begin{aligned} a = 2,\ ar &= -1 \\ 2r &= -1 \\ r &= -\frac{1}{2} \end{aligned}$$

$$S = \frac{a}{1-r} = \frac{2}{1-(-1/2)} = \frac{2}{3/2} = \frac{4}{3}$$

Alternatively, the series can be written:

$$S = 2 + 2\left(-\frac{1}{2}\right) + 2\left(-\frac{1}{2}\right)^2 + 2\left(-\frac{1}{2}\right)^3 + \cdots,$$

showing that $a = 2$ and $r = -\frac{1}{2}$.

19. $$\begin{aligned} a = 4,\ ar &= -2 \\ 4r &= -2 \\ r &= -\frac{1}{2} \end{aligned}$$

$$S = \frac{a}{1-r} = \frac{4}{1-(-1/2)} = \frac{4}{3/2} = \frac{8}{3}$$

Alternatively, the series can be written:

$$S = 4 + 4\left(-\frac{1}{2}\right) + 4\left(-\frac{1}{2}\right)^2 + 4\left(-\frac{1}{2}\right)^3 + \cdots,$$

showing that $a = 4$ and $r = -\frac{1}{2}$.

21. $0.212121\ldots = \frac{21}{10^2} + \frac{21}{10^4} + \frac{21}{10^6} + \cdots$

$a = \frac{21}{10^2}$, $ar = \frac{21}{10^4}$,

$\frac{21}{10^2}r = \frac{21}{10^4}$ and $r = \frac{1}{10^2}$;

$S = \frac{a}{1-r} = \frac{21/10^2}{1-1/10^2} = \frac{21}{10^2-1} = \frac{21}{99} = \frac{7}{33}$.

23.
$$\begin{aligned}
0.757575\ldots &= 0.75 + 0.0075 + 0.000075 + \cdots \\
&= \frac{75}{100} + \frac{75}{10000} + \frac{75}{1000000} + \cdots \\
&= \frac{75}{10^2} + \frac{75}{10^4} + \frac{75}{10^6} + \cdots \\
a = \frac{75}{10^2},\ ar &= \frac{75}{10^4} \\
\frac{75}{10^2} r &= \frac{75}{10^4} \\
r &= \frac{1}{10^2}.
\end{aligned}$$
$$S = \frac{a}{1-r} = \frac{75/10^2}{1-1/10^2} = \frac{75}{10^2-1} = \frac{75}{99} = \frac{25}{33}.$$

25. $0.001001001\ldots = 0.001 + 0.000001 + 0.000000001 + \cdots = \dfrac{1}{10^3} + \dfrac{1}{10^6} + \dfrac{1}{10^9} + \cdots$

$a = \dfrac{1}{10^3},\ ar = \dfrac{1}{10^6},$

$\dfrac{1}{10^3} r = \dfrac{1}{10^6}$ and $r = \dfrac{1}{10^3};$

$S = \dfrac{a}{1-r} = \dfrac{1/10^3}{1-1/10^3} = \dfrac{1}{10^3-1} = \dfrac{1}{999}.$

27.
$$\begin{aligned}
0.50707\ldots &= 0.5 + 0.007 + 0.00007 + 0.0000007 + \ldots \\
&= \frac{1}{2} + \frac{7}{10^3} + \frac{7}{10^5} + \frac{7}{10^7} + \cdots \\
&= \frac{1}{2} + \left(\frac{7}{10^3} + \frac{7}{10^3}\frac{1}{10^2} + \frac{7}{10^3}\frac{1}{10^4} + \cdots\right)
\end{aligned}$$
$a = \dfrac{7}{10^3},\ r = \dfrac{1}{10^2}.$

$S = \dfrac{1}{2} + \dfrac{7/10^3}{1-1/10^2} = \dfrac{1}{2} + \dfrac{7}{10^3-10} = \dfrac{1}{2} + \dfrac{7}{990} = \dfrac{495+7}{990} = \dfrac{502}{990} = \dfrac{251}{495}.$

29.
$$\begin{aligned}
0.3777\ldots &= 0.3 + 0.07 + 0.007 + 0.0007 + \cdots \\
&= \frac{3}{10} + \frac{7}{10^2} + \frac{7}{10^3} + \frac{7}{10^4} + \cdots \\
&= \frac{3}{10} + \left(\frac{7}{10^2} + \frac{7}{10^2}\frac{1}{10} + \frac{7}{10^3}\frac{1}{10} + \cdots\right)
\end{aligned}$$
Since $a = \dfrac{7}{10^2}$ and $r = \dfrac{1}{10}$, we get $\dfrac{3}{10} + \dfrac{7/10^2}{1-1/10} = \dfrac{17}{45}.$

10.2 Tests for Convergence

1. $\displaystyle\lim_{n\to\infty} \frac{n}{2n+2} = \lim_{n\to\infty} \frac{1}{2+2/n} = \frac{1}{2}$
(n^{th} term does not go to 0).

3. Dividing numerator and denominator by n^2, we get for the limit
$\displaystyle\lim_{n\to\infty} \frac{5n^2}{2n^2-2} = \lim_{n\to\infty} \frac{5}{2-2/n^2} = \frac{5}{2}$
(n^{th} term does not go to 0).

5. $$\int_1^\infty \frac{dx}{(x+1)^2}\,dx = \lim_{b\to\infty}\int_1^b (x+1)^{-2}\,dx \qquad u=x+1;\ du=dx.$$
$$= \lim_{b\to\infty}\left.\frac{(x+1)^{-1}}{-1}\right|_1^b = \lim_{b\to\infty}\left.\left(-\frac{1}{x+1}\right)\right|_1^b$$
$$= \lim_{b\to\infty}\left[-\frac{1}{b+1}+\frac{1}{2}\right] = \frac{1}{2}. \qquad \text{(convergent)}$$

7. $$\int_1^\infty \frac{x}{x^2+1}\,dx = \lim_{b\to\infty}\frac{1}{2}\int_1^b \frac{2x\,dx}{x^2+1} \qquad u=x^2+1;\ du=2x\,dx.$$
$$= \lim_{b\to\infty}\left.\frac{1}{2}\ln(x^2+1)\right|_1^b = \lim_{b\to\infty}\frac{1}{2}\left[\ln(b^2+1)-\ln 2\right].$$
This limit does not exist; so the series diverges.

9. $$\int_0^\infty \frac{dx}{(2x+2)^2} = \lim_{b\to\infty}\int_0^b (2x+2)^{-2}\,dx \qquad u=2x+2;\ du=2\,dx.$$
$$= \lim_{b\to\infty}\frac{1}{2}\int_0^b (2x+2)^{-2}(2\,dx) = \lim_{b\to\infty}\left.\frac{1}{2}\frac{(2x+2)^{-1}}{-1}\right|_0^b$$
$$= \lim_{b\to\infty}\left.\left(-\frac{1}{2(2x+2)}\right)\right|_0^b$$
$$= \lim_{b\to\infty}\left(-\frac{1}{2(2b+2)}+\frac{1}{4}\right) = \frac{1}{4}. \qquad \text{(convergent)}$$

11. $$\int_3^\infty \frac{x^2}{x^3-8}\,dx = \lim_{b\to\infty}\frac{1}{3}\int_2^b \frac{3x^2\,dx}{x^3-8} \qquad u=x^3-8;\ du=3x^2\,dx$$
$$= \lim_{b\to\infty}\left.\frac{1}{3}\ln(x^3-8)\right|_3^b = \lim_{b\to\infty}\frac{1}{3}\left[\ln(b^3-8)-\ln 19\right].$$
Since this limit does not exist, the series diverges.

13. $$\int_1^\infty \frac{x}{e^x}\,dx = \int_1^\infty xe^{-x}\,dx$$

Integration by parts:
$$u = x \qquad dv = e^{-x}dx$$
$$du = dx \qquad v = -e^{-x}$$

$$\lim_{b\to\infty}\left[\left.-xe^{-x}\right|_1^b + \int_1^b e^{-x}dx\right] = \lim_{b\to\infty}\left[\left.-xe^{-x}\right|_1^b - \left.e^{-x}\right|_1^b\right]$$
$$= \lim_{b\to\infty}\left[(-be^{-b}+e^{-1})-(e^{-b}-e^{-1})\right]$$
$$= \lim_{b\to\infty}\left[\left(-\frac{b}{e^b}+\frac{1}{e}\right)-\left(\frac{1}{e^b}-\frac{1}{e}\right)\right]$$
$$= \frac{2}{e}. \qquad \text{(convergent)}$$

15. $$\int_3^\infty \frac{x\,dx}{(x^2-4)^{3/2}} = \lim_{b\to\infty}\frac{1}{2}\int_3^b \frac{2x\,dx}{(x^2-4)^{3/2}} \qquad u=x^2-4;\ du=2x\,dx.$$
$$= \lim_{b\to\infty}\frac{1}{2}\int_3^b (x^2-4)^{-3/2}(2x\,dx) = \lim_{b\to\infty}\left.\frac{1}{2}\frac{(x^2-4)^{-1/2}}{-1/2}\right|_3^b$$
$$= \lim_{b\to\infty}\left.\frac{1}{2}\left(\frac{-2}{\sqrt{x^2-4}}\right)\right|_3^b = \lim_{b\to\infty}\left(-\frac{1}{\sqrt{b^2-4}}+\frac{1}{\sqrt{5}}\right)$$
$$= \frac{1}{\sqrt{5}}. \qquad \text{(convergent)}$$

17. $$\int_0^\infty \frac{x\,dx}{(x^2+2)^2} = \frac{1}{4}. \qquad \text{(convergent)}$$

19.
$$\begin{aligned}\int_2^\infty \frac{x}{\sqrt{x^2+2}}\,dx &= \lim_{b\to\infty}\int_2^b (x^2+2)^{-1/2}\,dx \\ &= \lim_{b\to\infty}\frac{1}{2}\int_2^b (x^2+2)^{-1/2}(2x)\,dx \qquad u = x^2+2;\ du = 2x\,dx. \\ &= \lim_{b\to\infty}\frac{1}{2}\frac{(x^2+2)^{1/2}}{1/2}\Big|_2^b = \lim_{b\to\infty}\left(\sqrt{b^2+2}-\sqrt{6}\right).\end{aligned}$$

Since this limit does not exist, the integral diverges.

21. $\dfrac{1}{n^2+2} < \dfrac{1}{n^2}$ (converges by comparison to the p-series with $p = 2$).

23. $\dfrac{1}{n-6} > \dfrac{1}{n}$ (diverges by comparison to the harmonic series).

25. $\dfrac{1}{n^3+2} < \dfrac{1}{n^3}$ (converges by comparison to the p-series with $p = 3$).

27. $\dfrac{1}{n^2+n} < \dfrac{1}{n^2}$ (converges by comparison to the p-series with $p = 2$).

29. $\dfrac{1}{3^n+1} < \dfrac{1}{3^n}$ (converges by comparison to the geometric series).

31. $\dfrac{1}{3^n-1} < \dfrac{1}{2^n}$ for $n \geq 2$ (converges by comparison to the geometric series with $r = \dfrac{1}{2}$).

33. $\dfrac{1}{\sqrt{n}-1} = \dfrac{1}{n^{1/2}-1} > \dfrac{1}{n^{1/2}}$ (diverges by comparison to the p-series with $p = 1/2$).

35. $\dfrac{1}{n^3-1} < \dfrac{1}{n^2}$ for $n \geq 2$ (converges by comparison to the p-series with $p = 2$).

37. $\dfrac{1}{\ln n} > \dfrac{1}{n}$ (diverges by comparison to the harmonic series).

39. $\displaystyle\lim_{n\to\infty}\frac{a_{n+1}}{a_n} = \lim_{n\to\infty}\frac{1/3^{n+1}}{1/3^n} = \lim_{n\to\infty}\frac{1}{3^{n+1}}\frac{3^n}{1} = \lim_{n\to\infty}\frac{1}{3\cdot 3^n}\frac{3^n}{1} = \lim_{n\to\infty}\frac{1}{3} = \frac{1}{3} < 1.$ (convergent)

41. $\displaystyle\lim_{n\to\infty}\frac{a_{n+1}}{a_n} = \lim_{n\to\infty}\frac{\frac{1}{(n+1)!}}{\frac{1}{n!}}$

Now observe that $(n+1)! = (n+1)n!$ (for example, $5! = 5\cdot 4!$).

$\displaystyle\lim_{n\to\infty}\frac{1}{(n+1)n!}\frac{n!}{1} = \lim_{n\to\infty}\frac{1}{n+1} = 0 < 1.$ (convergent)

43. $\displaystyle\lim_{n\to\infty}\frac{a_{n+1}}{a_n} = \lim_{n\to\infty}\frac{\frac{4^{n+1}}{(n+1)!}}{\frac{4^n}{n!}} = \lim_{n\to\infty}\frac{4^{n+1}}{(n+1)!}\frac{n!}{4^n}$

Now observe that $(n+1)! = (n+1)\cdot n!$. (For example $6! = 6\cdot 5!$.) Also, $4^{n+1} = 4^n\cdot 4$.

$\displaystyle\lim_{n\to\infty}\frac{4^n\cdot 4}{(n+1)n!}\frac{n!}{4^n} = \lim_{n\to\infty}\frac{4}{n+1} = 0 < 1.$ (convergent)

45. $$\begin{aligned}\lim_{n\to\infty}\frac{a_{n+1}}{a_n} &= \lim_{n\to\infty}\frac{\dfrac{(n+1)^2}{2^{n+1}}}{\dfrac{n^2}{2^n}} = \lim_{n\to\infty}\frac{(n+1)^2}{2^{n+1}}\cdot\frac{2^n}{n^2}\\ &= \lim_{n\to\infty}\frac{(n+1)^2\,2^n}{2^n\cdot 2\ \ n^2} = \lim_{n\to\infty}\frac{n^2+2n+1}{2n^2}\\ &= \lim_{n\to\infty}\frac{1+2/n+1/n^2}{2} = \frac{1}{2} < 1. \quad \text{(convergent)}\end{aligned}$$

47. $$\begin{aligned}\lim_{n\to\infty}\frac{a_{n+1}}{a_n} &= \lim_{n\to\infty}\frac{(n+1)(2/3)^{n+1}}{n(2/3)^n} = \lim_{n\to\infty}\frac{(n+1)(2/3)^n\cdot(2/3)}{n(2/3)^n}\\ &= \lim_{n\to\infty}\frac{2}{3}\cdot\frac{n+1}{n} = \lim_{n\to\infty}\frac{2}{3}\cdot\frac{1+1/n}{1} \quad \text{dividing numerator and denominator by } n\\ &= \frac{2}{3} < 1. \quad \text{(convergent)}\end{aligned}$$

49. $$\begin{aligned}\lim_{n\to\infty}\frac{a_{n+1}}{a_n} &= \lim_{n\to\infty}\frac{(n+1)\left(\frac{3}{2}\right)^{n+1}}{n\left(\frac{3}{2}\right)^n} = \lim_{n\to\infty}\frac{(n+1)\left(\frac{3}{2}\right)^n\left(\frac{3}{2}\right)}{n\left(\frac{3}{2}\right)^n}\\ &= \lim_{n\to\infty}\frac{3(n+1)}{2n} = \lim_{n\to\infty}\frac{3+3/n}{2} = \frac{3}{2} > 1. \quad \text{(divergent)}\end{aligned}$$

51. $$\lim_{n\to\infty}\frac{\dfrac{2^{n+1}}{(n+2)!}}{\dfrac{2^n}{(n+1)!}} = \lim_{n\to\infty}\frac{2^n\cdot 2}{(n+2)(n+1)!}\,\frac{(n+1)!}{2^n} = \lim_{n\to\infty}\frac{2}{n+2} = 0 < 1 \quad \text{(convergent)}$$

53. $$a_{n+1} = \frac{(n+1)!}{1\cdot 3\cdot 5\cdots(2n-1)(2n+1)}$$

$$\begin{aligned}\lim_{n\to\infty}\frac{a_{n+1}}{a_n} &= \lim_{n\to\infty}\frac{(n+1)n!}{1\cdot 3\cdot 5\cdots(2n-1)(2n+1)}\cdot\frac{1\cdot 3\cdot 5\cdots(2n-1)}{n!}\\ &= \lim_{n\to\infty}\frac{n+1}{2n+1} = \lim_{n\to\infty}\frac{1+1/n}{2+1/n} = \frac{1}{2} < 1. \quad \text{(convergent)}\end{aligned}$$

55. $$a_{n+1} = \frac{1\cdot 4\cdot 7\cdots(3n-2)(3n+1)}{2\cdot 4\cdot 6\cdots(2n)(2n+2)}$$

$$\frac{a_{n+1}}{a_n} = \frac{1\cdot 4\cdot 7\cdots(3n-2)(3n+1)}{2\cdot 4\cdot 6\cdots(2n)(2n+2)}\cdot\frac{2\cdot 4\cdot 6\cdots(2n)}{1\cdot 4\cdot 7\cdots(3n-2)}$$

$$\lim_{n\to\infty}\frac{3n+1}{2n+2} = \lim_{n\to\infty}\frac{3+1/n}{2+2/n} = \frac{3}{2} > 1. \quad \text{(divergent)}$$

57.
$$\begin{aligned}\lim_{n\to\infty}\frac{a_{n+1}}{a_n} &= \frac{\dfrac{(n+1)^2-1}{(n+1)^3}}{\dfrac{n^2-1}{n^3}}\\ &= \lim_{n\to\infty}\frac{n^2+2n}{n^3+3n^2+3n+1}\cdot\frac{n^3}{n^2-1}\\ &= \lim_{n\to\infty}\frac{n^5+2n^4}{n^5+3n^4+2n^3-2n^2-3n-1}=1. \qquad \text{(test fails)}\end{aligned}$$

By the integral test,

$$\int_2^\infty \frac{x^2-1}{x^3}\,dx = \int_2^\infty\left(\frac{1}{x}-\frac{1}{x^3}\right)dx = \int_2^\infty \frac{1}{x}\,dx - \int_2^\infty \frac{1}{x^3}\,dx.$$

Consider the first integral:

$$\lim_{b\to\infty}\int_2^b \frac{1}{x}\,dx = \lim_{b\to\infty}\ln x\Big|_2^b = \lim_{b\to\infty}(\ln b - \ln 2).$$

Since this limit does not exist, the series diverges.

10.3 Maclaurin Series

1.
$$\begin{aligned}f(x) &= \sin x & f(0) &= 0\\ f'(x) &= \cos x & f'(0) &= 1\\ f''(x) &= -\sin x & f''(0) &= 0\\ f'''(x) &= -\cos x & f'''(0) &= -1\\ f^{(4)}(x) &= \sin x & f^{(4)}(0) &= 0\\ f^{(5)}(x) &= \cos x & f^{(5)}(0) &= 1\end{aligned}$$

$$\begin{aligned}\sin x &= 0+1x+\frac{0}{2!}x^2+\frac{-1}{3!}x^3+\frac{0}{4!}x^4+\frac{1}{5!}x^5+\cdots\\ &= x-\frac{x^3}{3!}+\frac{x^5}{5!}-\cdots\end{aligned}$$

3.
$$\begin{aligned}f(x) &= \sin 3x & f(0) &= 0\\ f'(x) &= 3\cos 3x & f'(0) &= 3\\ f''(x) &= -3^2\sin 3x & f''(0) &= 0\\ f'''(x) &= -3^3\cos 3x & f'''(0) &= -3^3\\ f^{(4)}(x) &= 3^4\sin 3x & f^{(4)}(0) &= 0\\ f^{(5)}(x) &= 3^5\cos 3x & f^{(5)}(0) &= 3^5\end{aligned}$$

$$\begin{aligned}\sin 3x &= 0+3x+\frac{0}{2!}x^2+\frac{-3^3}{3!}x^3+\frac{0}{4!}x^4+\frac{3^5}{5!}x^5+\cdots\\ &= 3x-\frac{3^3}{3!}x^3+\frac{3^5}{5!}x^5-\cdots\end{aligned}$$

5.
$$\begin{aligned}f(x) &= e^{-x} & f(0) &= 1\\ f'(x) &= -e^{-x} & f'(0) &= -1\\ f''(x) &= e^{-x} & f''(0) &= 1\\ f'''(x) &= -e^{-x} & f'''(0) &= -1\\ f^{(4)}(x) &= e^{-x} & f^{(4)}(0) &= 1\end{aligned}$$

$$\begin{aligned}e^{-x} &= 1-1x+\frac{1}{2!}x^2+\frac{-1}{3!}x^3+\frac{1}{4!}x^4-\cdots\\ &= 1-x+\frac{x^2}{2!}-\frac{x^3}{3!}+\frac{x^4}{4!}-\cdots\end{aligned}$$

7.
$$\begin{aligned}
f(x) &= \ln(1+x) & f(0) &= 0\\
f'(x) &= \frac{1}{x+1} = (x+1)^{-1} & f'(0) &= 1\\
f''(x) &= -1(x+1)^{-2} & f''(0) &= -1\\
f'''(x) &= 1\cdot 2(x+1)^{-3} & f'''(0) &= 2!\\
f^{(4)}(x) &= -1\cdot 2\cdot 3(x+1)^{-4} & f^{(4)}(0) &= -3!\\
f^{(5)}(x) &= 1\cdot 2\cdot 3\cdot 4(x+1)^{-5} & f^{(5)}(0) &= 4!\\
f^{(6)}(x) &= -1\cdot 2\cdot 3\cdot 4\cdot 5(x+1)^{-6} & f^{(6)}(0) &= -5!
\end{aligned}$$
$$\begin{aligned}
\ln(x+1) &= 0 + 1x + \frac{-1}{2!}x^2 + \frac{2!}{3!}x^3 + \frac{-3!}{4!}x^4 + \frac{4!}{5!}x^5 + \cdots\\
&= x - \frac{x^2}{2} + \frac{x^3}{3} - \frac{x^4}{4} + \frac{x^5}{5} - \cdots
\end{aligned}$$

9.
$$\begin{aligned}
f(x) &= \sinh x = \frac{1}{2}(e^x - e^{-x}) & f(0) &= 0\\
f'(x) &= \frac{1}{2}(e^x + e^{-x}) & f'(0) &= 1\\
f''(x) &= \frac{1}{2}(e^x - e^{-x}) & f''(0) &= 0\\
f'''(x) &= \frac{1}{2}(e^x + e^{-x}) & f'''(0) &= 1\\
f^{(4)}(x) &= \frac{1}{2}(e^x - e^{-x}) & f^{(4)}(0) &= 0\\
f^{(5)}(x) &= \frac{1}{2}(e^x + e^{-x}) & f^{(5)}(0) &= 1
\end{aligned}$$
$$\begin{aligned}
\sinh x &= 0 + 1x + \frac{0}{2!}x^2 + \frac{1}{3!}x^3 + \frac{0}{4!}x^4 + \frac{1}{5!}x^5 + \cdots\\
&= x + \frac{x^3}{3!} + \frac{x^5}{5!} + \cdots
\end{aligned}$$

11.
$$\begin{aligned}
f(x) &= \frac{1}{1-x} = (1-x)^{-1} & f(0) &= 1\\
f'(x) &= (1-x)^{-2} & f'(0) &= 1\\
f''(x) &= 2(1-x)^{-3} & f''(0) &= 2!\\
f'''(x) &= 3!(1-x)^{-4} & f'''(0) &= 3!\\
f^{(4)}(x) &= 4!(1-x)^{-5} & f^{(4)}(0) &= 4!
\end{aligned}$$
$$\begin{aligned}
\frac{1}{1-x} &= 1 + x + \frac{2!}{2!}x^2 + \frac{3!}{3!}x^3 + \frac{4!}{4!}x^4 + \cdots\\
&= 1 + x + x^2 + x^3 + x^4 + \cdots
\end{aligned}$$

13. Verify: $(1-x)^{-2} = 1 + 2x + 3x^2 + 4x^3 + 5x^4 + 6x^5 + \cdots$

(a)
$$\begin{aligned}
(1-x)^{-2} &= 1 + (-2)(-x) + \tfrac{(-2)(-3)}{2!}(-x)^2 + \tfrac{(-2)(-3)(-4)}{3!}(-x)^3 + \cdots\\
&= 1 + 2x + 3x^2 + 4x^3 + 5x^4 + 6x^5 + \cdots
\end{aligned}$$

(b)
$$\begin{aligned} f(x) &= (1-x)^{-2} & f(0) &= 1 \\ f'(x) &= 2(1-x)^{-3} & f'(0) &= 2 \\ f''(x) &= 3!(1-x)^{-4} & f''(0) &= 3! \\ f'''(x) &= 4!(1-x)^{-5} & f'''(0) &= 4! \end{aligned}$$

$$\begin{aligned} (1-x)^{-2} &= 1+2x+\frac{3!}{2!}x^2+\frac{4!}{3!}x^3+\frac{5!}{4!}x^4+\cdots \\ &= 1+2x+3x^2+4x^3+5x^4+6x^5+\cdots \end{aligned}$$

(c)
$$\begin{array}{r l} & 1+2x+3x^2+4x^3+\cdots \\ 1-2x+x^2 \big) & 1 \\ & \underline{1-2x+x^2} \\ & 2x-x^2 \\ & \underline{2x-4x^2+2x^3} \\ & 3x^2-2x^3 \\ & \underline{3x^2-6x^3+3x^4} \\ & 4x^3-3x^4 \end{array}$$

10.4 Operations with Series

1. $\sin x = x - \dfrac{x^3}{3!} + \dfrac{x^5}{5!} - \cdots$. Replacing x by $4x$,
$\sin 4x = 4x - \dfrac{4^3}{3!}x^3 + \dfrac{4^5}{5!}x^5 - \cdots$

3. $e^x = 1 + x + \dfrac{x^2}{2!} + \dfrac{x^3}{3!} + \dfrac{x^4}{4!} + \cdots$. Replacing x by $-2x$,
$e^{-2x} = 1 - 2x + \dfrac{2^2}{2!}x^2 - \dfrac{2^3}{3!}x^3 + \dfrac{2^4}{4!}x^4 - \cdots$.

5. $\sin x = x - \dfrac{x^3}{3!} + \dfrac{x^5}{5!} - \cdots$. Replacing x by x^3,
$\sin x^3 = x^3 - \dfrac{x^9}{3!} + \dfrac{x^{15}}{5!} - \cdots$

7. $\cos x = 1 - \dfrac{x^2}{2!} + \dfrac{x^4}{4!} - \dfrac{x^6}{6!} + \cdots$. Multiplying the series by x, we get
$x\cos x = x - \dfrac{x^3}{2!} + \dfrac{x^5}{4!} - \dfrac{x^7}{6!} + \cdots$.

9.
$$\begin{aligned} x\ln(1+x) &= x\left(x - \frac{x^2}{2} + \frac{x^3}{3} - \frac{x^4}{4} + \cdots\right) \\ &= x^2 - \frac{x^3}{2} + \frac{x^4}{3} - \cdots \end{aligned}$$

11. The expansion for $\operatorname{Arcsin} x$ is given in Example 4. Dividing each term by x yields the expansion of $\dfrac{\operatorname{Arcsin} x}{x}$.

13. Replace x by x^2 in the expansion of $\ln(1+x^2)$:

$$\ln\left(1+x^2\right) = x^2 - \frac{x^4}{2} + \frac{x^6}{3} - \frac{x^8}{4} + \cdots$$

15.
$$\begin{aligned}
\frac{d}{dx}\sin x &= \frac{d}{dx}\left(x - \frac{x^3}{3!} + \frac{x^5}{5!} - \frac{x^7}{7!} + \cdots\right)\\
&= 1 - \frac{3x^2}{3\cdot 2!} + \frac{5x^4}{5\cdot 4!} - \frac{7x^6}{7\cdot 6!} + \cdots\\
&= 1 - \frac{x^2}{2!} + \frac{x^4}{4!} - \frac{x^6}{6!} + \cdots = \cos x
\end{aligned}$$

17.
$$\begin{aligned}
\frac{d}{dx}\ln(1+x) &= \frac{1}{1+x}\\
&= 1 + (-x) + (-x)^2 + (-x)^3 + (-x)^4 + \cdots\\
&= 1 - x + x^2 - x^3 + x^4 - \cdots\\
&\quad \text{(geometric series with } r = -x\text{)}\\
\ln(1+x) &= \int_0^x \frac{dx}{1+x} = \int_0^x \left(1 - x + x^2 - x^3 + x^4 - \cdots\right) dx\\
&= x - \frac{x^2}{2} + \frac{x^3}{3} - \frac{x^4}{4} + \frac{x^5}{5} - \cdots
\end{aligned}$$

19. Carrying only powers up to the fourth power and using the expansion for e^{-x}, we get

$$\begin{aligned}
e^{-x} &= 1 - x + \frac{x^2}{2} - \frac{x^3}{6} + \frac{x^4}{24} - \cdots\\
\cos x &= 1 - \frac{x^2}{2} + \frac{x^4}{24} - \cdots
\end{aligned}$$

$$\begin{array}{l}
1 - x + \dfrac{x^2}{2} - \dfrac{x^3}{6} + \dfrac{x^4}{24} - \cdots\\
\qquad\; - \dfrac{x^2}{2} + \dfrac{x^3}{2} - \dfrac{x^4}{4} + \cdots\\
\qquad\qquad\qquad\qquad + \dfrac{x^4}{24} - \cdots\\
\hline
1 - x \qquad\quad + \dfrac{1}{3}x^3 - \dfrac{1}{6}x^4 + \cdots
\end{array}$$

21.
$$\begin{aligned}
\frac{\sin x - x}{x^2} &= \frac{1}{x^2}(\sin x - x) = \frac{1}{x^2}\left(x - \frac{x^3}{3!} + \frac{x^5}{5!} - \frac{x^7}{7!} + \cdots - x\right)\\
&= \frac{1}{x^2}\left(-\frac{x^3}{3!} + \frac{x^5}{5!} - \frac{x^7}{7!} + \cdots\right) = -\frac{x}{3!} + \frac{x^3}{5!} - \frac{x^5}{7!} + \cdots
\end{aligned}$$

23.
$$\begin{array}{r|l}
 & x + \dfrac{x^3}{3} + \dfrac{2x^5}{15} + \cdots\\
\hline
1 - \dfrac{x^2}{2} + \dfrac{x^4}{24} - \cdots & x - \dfrac{x^3}{6} + \dfrac{x^5}{120} - \cdots\\
 & x - \dfrac{x^3}{2} + \dfrac{x^5}{24} - \cdots\\
\cline{2-2}
 & \dfrac{x^3}{3} - \dfrac{x^5}{30} + \cdots\\
 & \dfrac{x^3}{3} - \dfrac{x^5}{6} + \cdots\\
\cline{2-2}
 & \dfrac{2x^5}{15} - \cdots
\end{array}$$

25. $-\sqrt{3}+j:\ r=2,\ \theta=150^\circ=\dfrac{5\pi}{6}$. Thus $-\sqrt{3}+j=2e^{5\pi j/6}$.

27. $3j:\ r=3,\ \theta=90^\circ=\dfrac{\pi}{2}$. Thus $3j=3e^{\pi j/2}$.

29. $-2+2j:\ r=\sqrt{8}=2\sqrt{2},\ \theta=135^\circ=\dfrac{3\pi}{4}$. Thus $-2+2j=2\sqrt{2}e^{3\pi j/4}$.

31. $-1-j: r=\sqrt{2},\ \theta=225^\circ=\dfrac{5\pi}{4}$. Thus $-1-j=\sqrt{2}e^{5\pi j/4}$.

10.5 Computations with Series; Applications

1.
$$\begin{aligned}\sin x &= x-\frac{x^3}{3!}+\frac{x^5}{5!}-\frac{x^7}{7!}+\cdots\\ \sin(0.7) &= 0.7-\frac{(0.7)^3}{3!}+\frac{(0.7)^5}{5!}=0.644234 \qquad \text{(three terms)}\end{aligned}$$

max. error (fourth term): $-\dfrac{(0.7)^7}{7!}=-0.000016$

(a) 0.644234 sum of first three terms

-0.000016 error

(b) 0.644218

The values of (a) and (b) agree to four decimal places: 0.6442.

3. $10^\circ=\dfrac{10^\circ\pi}{180^\circ}=\dfrac{\pi}{18}$

$$\begin{aligned}\cos x &= 1-\frac{x^2}{2!}+\frac{x^4}{4!}-\frac{x^6}{6!}+\cdots\\ \cos\frac{\pi}{18} &= 1-\frac{1}{2!}\left(\frac{\pi}{18}\right)^2=0.98477 \qquad \text{(two terms)}\end{aligned}$$

max. error (third term): $\dfrac{1}{4!}\left(\dfrac{\pi}{18}\right)^4=0.00004$

(a) 0.98477 sum of first two terms

0.00004 error

(b) 0.98481

The values of (a) and (b) agree to four decimal places (after rounding off): 0.9848.

5.
$$\begin{aligned}e^x &= 1+x+\frac{x^2}{2!}+\frac{x^3}{3!}+\frac{x^4}{4!}+\cdots\\ e^{-0.2} &= 1+(-0.2)+\frac{(-0.2)^2}{2!}+\frac{(-0.2)^3}{3!} \qquad \text{(four terms)}\\ &= 0.818667\end{aligned}$$

max. error (fifth term): $\dfrac{(-0.2)^4}{4!}=0.000067$

(a) 0.818667 sum of first four terms

0.000067 error

(b) 0.818734

The values of (a) and (b) agree to four decimal places (after rounding off): 0.8187.

7. $$\cos x = 1 - \frac{x^2}{2!} + \frac{x^4}{4!} - \frac{x^6}{6!} + \frac{x^8}{8!} - \cdots$$

$$\cos 1.2 = 1 - \frac{(1.2)^2}{2!} + \frac{(1.2)^4}{4!} - \frac{(1.2)^6}{6!} = 0.36225 \quad \text{(four terms)}$$

max. error (fifth term): $\frac{(1.2)^8}{8!} = 0.0001$

(a) 0.36225 sum of first four terms
0.0001 error
(b) 0.36235

The values of (a) and (b) agree to three decimal places: 0.362.

9. $$\ln(1+x) = x - \frac{x^2}{2} + \frac{x^3}{3} - \frac{x^4}{4} + \cdots$$

$$\ln 1.1 = \ln(1+0.1) = 0.1 - \frac{(0.1)^2}{2} + \frac{(0.1)^3}{3} \quad \text{(three terms)}$$

$$= 0.095333$$

max. error (fourth term): $-\frac{(0.1)^4}{4} = -0.000025$

(a) 0.095333 sum of first three terms
−0.000025 error
(b) 0.095308

The values of (a) and (b) agree to four decimal places: 0.0953.

11. $25° = \frac{25°\pi}{180°} = \frac{5\pi}{36}$

$$\sin x = x - \frac{x^3}{3!} + \frac{x^5}{5!} - \frac{x^7}{7!} + \cdots$$

$$\sin\frac{5\pi}{36} = \frac{5\pi}{36} - \frac{1}{3!}\left(\frac{5\pi}{36}\right)^3 + \frac{1}{5!}\left(\frac{5\pi}{36}\right)^5$$

$$= 0.4226189 \quad \text{(3 terms)}$$

max. error (fourth term): $-\frac{1}{7!}\left(\frac{5\pi}{36}\right)^7 = -0.0000006$

(a) 0.4226189 sum of first three terms
−0.0000006 error
(b) 0.4226183

The values of (a) and (b) agree to five decimals (after rounding off): 0.42262.

13. Arcsin $\frac{1}{2} = \frac{\pi}{6}$; so

$$\frac{\pi}{6} = \frac{1}{2} + \frac{1 \cdot (1/2)^3}{2 \cdot 3} + \frac{1 \cdot 3 \cdot (1/2)^5}{2 \cdot 4 \cdot 5} = 0.5232 \text{ and } \pi \approx 6(0.5232) = 3.14.$$

15. $$\frac{\sin x}{x} = \frac{x - \frac{x^3}{3!} + \frac{x^5}{5!} - \frac{x^7}{7!} + \cdots}{x}$$

$$= 1 - \frac{x^2}{3!} + \frac{x^4}{5!} - \frac{x^6}{7!} + \cdots$$

$$\int_0^{0.7} \frac{\sin x}{x}\, dx = x - \frac{x^3}{3 \cdot 3!} + \frac{x^5}{5 \cdot 5!} - \frac{x^7}{7 \cdot 7!} + \cdots \Big|_0^{0.7}$$

$$= 0.6812246 \quad \text{(three terms)}$$

max. error (fourth term): -0.0000023

(a) 0.6812246 sum of first three terms

$\underline{-0.0000023}$

(b) 0.6812223

Result: 0.68122 to five decimal places.

17.
$$\begin{aligned}
\frac{1-\cos x}{x} &= \left[1-\left(1-\frac{x^2}{2!}+\frac{x^4}{4!}-\frac{x^6}{6!}+\frac{x^8}{8!}-\cdots\right)\right]/x \\
&= \frac{x}{2!}-\frac{x^3}{4!}+\frac{x^5}{6!}-\frac{x^7}{8!}+\cdots \\
\int_0^{1/2}\frac{1-\cos x}{x}\,dx &= \int_0^{1/2}\left(\frac{x}{2!}-\frac{x^3}{4!}+\frac{x^5}{6!}-\frac{x^7}{8!}+\cdots\right)dx \\
&= \frac{x^2}{2\cdot 2!}-\frac{x^4}{4\cdot 4!}+\frac{x^6}{6\cdot 6!}-\frac{x^8}{8\cdot 8!}+\cdots\bigg|_0^{1/2} \\
&= 0.06185 \quad \text{using three terms}
\end{aligned}$$

error: $-\dfrac{(0.5)^8}{8\cdot 8!} = -1.2\times 10^{-8}$ (fourth term)

19.
$$\begin{aligned}
e^x &= 1+x+\frac{x^2}{2!}+\frac{x^3}{3!}+\frac{x^4}{4!}+\cdots \\
e^{-x^2} &= 1-x^2+\frac{x^4}{2!}-\frac{x^6}{3!}+\frac{x^8}{4!}-\cdots \\
\int_0^{0.3} e^{-x^2}\,dx &= x-\frac{x^3}{3}+\frac{x^5}{5\cdot 2!}-\frac{x^7}{7\cdot 3!}\bigg|_0^{0.3} = 0.29124 \quad \text{using four terms}
\end{aligned}$$

error: $\dfrac{(0.3)^9}{9\cdot 4!} = 9.1\times 10^{-8}$ (fifth term)

21. $\dfrac{1}{1+x^5}$ can be expanded as a geometric series:

$$\begin{aligned}
\frac{1}{1+x^5} &= \frac{1}{1-(-x^5)} = 1+(-x^5)+(-x^5)^2+(-x^5)^3+(-x^5)^4+\cdots \\
&= 1-x^5+x^{10}-x^{15}+x^{20}-\cdots \\
\int_0^{0.5}\frac{dx}{1+x^5} &= x-\frac{x^6}{6}+\frac{x^{11}}{11}-\frac{x^{16}}{16}+\frac{x^{21}}{21}-\cdots\bigg|_0^{0.5} \\
&= 0.49744 \quad \text{using four terms.}
\end{aligned}$$

error: 0.00000002

23.
$$\begin{aligned}
f(x) &= \sin x & f\left(\frac{\pi}{6}\right) &= \frac{1}{2} \\
f'(x) &= \cos x & f'\left(\frac{\pi}{6}\right) &= \frac{\sqrt{3}}{2} \\
f''(x) &= -\sin x & f''\left(\frac{\pi}{6}\right) &= -\frac{1}{2}
\end{aligned}$$

$\sin x = \dfrac{1}{2}+\dfrac{\sqrt{3}}{2}\left(x-\dfrac{\pi}{6}\right)-\dfrac{1}{2}\dfrac{1}{2!}\left(x-\dfrac{\pi}{6}\right)^2+\cdots$

$29^\circ = 30^\circ - 1^\circ = \dfrac{\pi}{6}-\dfrac{\pi}{180}$

Thus $x-\dfrac{\pi}{6} = \left(\dfrac{\pi}{6}-\dfrac{\pi}{180}\right)-\dfrac{\pi}{6} = -\dfrac{\pi}{180}$.

$\sin 29^\circ = \dfrac{1}{2}+\dfrac{\sqrt{3}}{2}\left(-\dfrac{\pi}{180}\right)-\dfrac{1}{2}\dfrac{1}{2!}\left(-\dfrac{\pi}{180}\right)^2 = 0.4848.$

25.
$$\begin{aligned} f(x) &= \cos x & f\left(\frac{\pi}{6}\right) &= \frac{\sqrt{3}}{2} \\ f'(x) &= -\sin x & f'\left(\frac{\pi}{6}\right) &= -\frac{1}{2} \\ f''(x) &= -\cos x & f''\left(\frac{\pi}{6}\right) &= -\frac{\sqrt{3}}{2} \\ f'''(x) &= \sin x & f'''\left(\frac{\pi}{6}\right) &= \frac{1}{2} \end{aligned}$$

$$\cos x = \frac{\sqrt{3}}{2} + \left(-\frac{1}{2}\right)\left(x - \frac{\pi}{6}\right) + \frac{-\sqrt{3}/2}{2!}\left(x - \frac{\pi}{6}\right)^2 + \frac{1/2}{3!}\left(x - \frac{\pi}{6}\right)^3 + \cdots$$

$$31^\circ = 30^\circ + 1^\circ = \frac{\pi}{6} + \frac{\pi}{180} = x$$

Thus $x - \dfrac{\pi}{6} = \left(\dfrac{\pi}{6} + \dfrac{\pi}{180}\right) - \dfrac{\pi}{6} = \dfrac{\pi}{180}$.

$$\cos 31^\circ = \frac{\sqrt{3}}{2} - \frac{1}{2}\left(\frac{\pi}{180}\right) - \frac{\sqrt{3}}{4}\left(\frac{\pi}{180}\right)^2 + \frac{1}{12}\left(\frac{\pi}{180}\right)^3 = 0.85717.$$

27.
$$\begin{aligned} f(x) &= \cos x & f\left(\frac{\pi}{3}\right) &= \frac{1}{2} \\ f'(x) &= -\sin x & f'\left(\frac{\pi}{3}\right) &= -\frac{\sqrt{3}}{2} \\ f''(x) &= -\cos x & f''\left(\frac{\pi}{3}\right) &= -\frac{1}{2} \end{aligned}$$

$$\cos x = \frac{1}{2} - \frac{\sqrt{3}}{2}\left(x - \frac{\pi}{3}\right) - \frac{1}{2}\frac{1}{2!}\left(x - \frac{\pi}{3}\right)^2 + \cdots$$

$$58^\circ = 60^\circ - 2^\circ = \frac{\pi}{3} - \frac{2^\circ\pi}{180^\circ} = \frac{\pi}{3} - \frac{\pi}{90}$$

Thus $x - \dfrac{\pi}{3} = \left(\dfrac{\pi}{3} - \dfrac{\pi}{90}\right) - \dfrac{\pi}{3} = -\dfrac{\pi}{90}$.

$$\cos 58^\circ = \frac{1}{2} - \frac{\sqrt{3}}{2}\left(-\frac{\pi}{90}\right) - \frac{1}{2}\frac{1}{2!}\left(-\frac{\pi}{90}\right)^2 = 0.5299.$$

29. Observe that $c = 1$.

$$\begin{aligned} f(x) &= \ln x & f(1) &= 0 \\ f'(x) &= \frac{1}{x} = x^{-1} & f'(1) &= 1 \\ f''(x) &= -1x^{-2} & f''(1) &= -1 \\ f'''(x) &= 2 \cdot 1x^{-3} & f'''(1) &= 2! \\ f^{(4)}(x) &= -3 \cdot 2 \cdot 1x^{-4} & f^{(4)}(1) &= -3! \\ f^{(5)}(x) &= 4 \cdot 3 \cdot 2 \cdot 1x^{-5} & f^{(5)}(x) &= 4! \end{aligned}$$

$$\begin{aligned} \ln x &= 0 + 1(x-1) + \frac{-1}{2!}(x-1)^2 + \frac{2!}{3!}(x-1)^3 + \frac{-3!}{4!}(x-1)^4 + \frac{4!}{5!}(x-1)^5 + \cdots \\ &= (x-1) - \frac{1}{2}(x-1)^2 + \frac{1}{3}(x-1)^3 - \frac{1}{4}(x-1)^4 + \cdots \end{aligned}$$

31. Expand the given function about $x = 1$:

$$\begin{aligned} f(x) &= 5x^5 + 10x^4 - 2x^3 + x^2 + 5 & f(1) &= 19 \\ f'(x) &= 25x^4 + 40x^3 - 6x^2 + 2x & f'(1) &= 61 \\ f''(x) &= 100x^3 + 120x^2 - 12x + 2 & f''(1) &= 210 \\ f'''(x) &= 300x^2 + 240x - 12 & f'''(1) &= 528 \\ f^{(4)}(x) &= 600x + 240 & f^{(4)}(1) &= 840 \\ f^{(5)}(x) &= 600 & f^{(5)}(1) &= 600 \end{aligned}$$

$$\begin{aligned} f(x) &= 19+61(x-1)+\frac{210}{2!}(x-1)^2+\frac{528}{3!}(x-1)^3+\frac{840}{4!}(x-1)^4+\frac{600}{5!}(x-1)^5 \\ &= 5(x-1)^5+35(x-1)^4+88(x-1)^3+105(x-1)^2+61(x-1)+19 \end{aligned}$$

33. (a) $\sin\theta = \theta - \frac{\theta^3}{3!} + \frac{\theta^5}{5!} - \frac{\theta^7}{7!} + \cdots$
If $\theta \approx 0$, then the higher powers of θ become negligible, so that $\sin\theta \approx \theta$.

(b) If $\sin\theta$ is replaced by θ, then
$$\frac{d^2\theta}{dt^2} = -\frac{g}{L}\theta,$$
that is, the acceleration is directly proportional to the displacement and oppositely directed. By Example 5, Section 6.10, θ has the form
$$\theta = a\cos\sqrt{\frac{g}{L}}\,t.$$

(c) The period of the cosine function is
$$\frac{2\pi}{\sqrt{\frac{g}{L}}} = 2\pi\sqrt{\frac{L}{g}}.$$

35.
$$\begin{aligned} q &= \int_{0.0}^{0.8} \sin t^2\,dt \\ &= \int_{0.0}^{0.8}\left(t^2 - \frac{t^6}{3!} + \frac{t^{10}}{5!} - \frac{t^{14}}{7!} + \cdots\right)dt \\ &= \left.\frac{t^3}{3} - \frac{t^7}{7\cdot 3!} + \frac{t^{11}}{11\cdot 5!} - \frac{t^{15}}{15\cdot 7!} + \cdots\right|_{0.0}^{0.8} \\ &= 0.16575\,\mathrm{C} \qquad \text{(using three terms; error: } -5\times 10^{-7}) \end{aligned}$$

37.
$$\begin{aligned} e^x &= 1 + x + \frac{x^2}{2!} + \frac{x^3}{3!} + \frac{x^4}{4!} + \frac{x^5}{5!} + \frac{x^6}{6!} + \cdots \\ e^{-x^2/2} &= 1 - \frac{1}{2}x^2 + \frac{1}{4}\frac{x^4}{2!} - \frac{1}{8}\frac{x^6}{3!} + \frac{1}{16}\frac{x^8}{4!} - \frac{1}{32}\frac{x^{10}}{5!} + \frac{1}{64}\frac{x^{12}}{6!} - \cdots \\ \frac{1}{\sqrt{2\pi}}\int_0^1 e^{-x^2/2}dx &= \frac{1}{\sqrt{2\pi}}\left.\left(x - \frac{1}{2}\frac{x^3}{3} + \frac{1}{4}\frac{x^5}{5\cdot 2!} - \frac{1}{8}\frac{x^7}{7\cdot 3!} + \frac{1}{16}\frac{x^9}{9\cdot 4!} - \frac{1}{32}\frac{x^{11}}{11\cdot 5!}\right)\right|_0^1 \\ &= \frac{1}{\sqrt{2\pi}}\left(1 - \frac{1}{2}\frac{1}{3} + \frac{1}{4}\frac{1}{5\cdot 2!} - \frac{1}{8}\frac{1}{7\cdot 3!} + \frac{1}{16}\frac{1}{9\cdot 4!} - \frac{1}{32}\frac{1}{11\cdot 5!}\right) \\ &= 0.3413. \end{aligned}$$

error: $\frac{1}{\sqrt{2\pi}}\frac{1}{64}\frac{1}{13\cdot 6!} = 6.7\times 10^{-7}$

39.
$$\begin{aligned} q(t) = \frac{\ln(1+t^2)}{t^2} &= \frac{t^2 - \frac{t^4}{2} + \frac{t^6}{3} - \frac{t^8}{4} + \frac{t^{10}}{5} - \frac{t^{12}}{6} + \cdots}{t^2} \\ &= 1 - \frac{t^2}{2} + \frac{t^4}{3} - \frac{t^6}{4} + \frac{t^8}{5} - \frac{t^{10}}{6} + \cdots \\ i = \frac{dq}{dt} &= 0 - \frac{2t}{2} + \frac{4t^3}{3} - \frac{6t^5}{4} + \frac{8t^7}{5} - \frac{10t^9}{6} + \cdots \\ &= -t + \frac{4}{3}t^3 - \frac{3}{2}t^5 + \frac{8}{5}t^7 - \frac{5}{3}t^9 + \cdots \end{aligned}$$

Let $t = 0.40$:

$$\begin{aligned} i &= -0.40 + \frac{4}{3}(0.40)^3 - \frac{3}{2}(0.40)^5 + \frac{8}{5}(0.40)^7 \\ &= -0.327405. \quad \text{(using 4 terms)} \end{aligned}$$

error: $-\frac{5}{3}(0.40)^9 = -0.0004$ (fifth term)

(a) -0.327405 sum of first four terms
-0.0004 error

(b) -0.327805

The values of (a) and (b) agree to two decimal places: -0.33 A.

41. $\dfrac{P_\theta}{P} = 1 + \dfrac{1}{4}\left(\sin\dfrac{\theta}{2}\right)^2 + \dfrac{9}{64}\left(\sin\dfrac{\theta}{2}\right)^4 + \cdots$

When $\theta = 5°$, $\dfrac{P_\theta}{P} = 1.0005$.

10.6 Fourier Series

1.

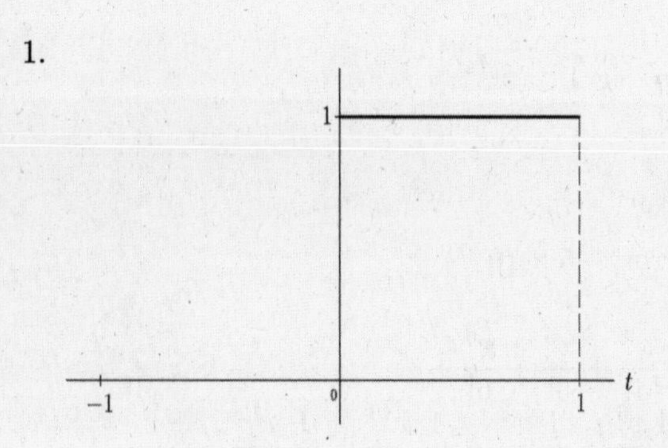

(a) Period: $2p = 2$, so that $p = 1$.

$$\begin{aligned}
a_0 &= \frac{1}{1}\int_{-1}^{1} f(t)\,dt = \int_{-1}^{0} 0\,dt + \int_{0}^{1} 1\,dt = 0 + t\Big|_0^1 = 1 \\
\frac{a_0}{2} &= \frac{1}{2}. \\
a_1 &= \frac{1}{1}\int_{-1}^{1} f(t)\cos\frac{1\pi t}{1}\,dt = \int_{-1}^{0} 0\,dt + \int_{0}^{1} 1\cdot\cos\pi t\,dt \\
&= \frac{1}{\pi}\sin\pi t\Big|_0^1 = 0. \qquad u = \pi t;\ du = \pi\,dt. \\
a_2 &= \frac{1}{1}\int_{-1}^{1} f(t)\cos\frac{2\pi t}{1}\,dt = \int_{-1}^{0} 0\,dt + \int_{0}^{1} 1\cdot\cos 2\pi t\,dt \\
&= \frac{1}{2\pi}\sin 2\pi t\Big|_0^1 = 0. \qquad u = 2\pi t;\ du = 2\pi\,dt. \\
a_n &= \frac{1}{1}\int_{-1}^{1} f(t)\cos\frac{n\pi t}{p}\,dt = \int_{-1}^{0} 0\,dt + \int_{0}^{1} 1\cdot\cos\frac{n\pi t}{1}\,dt \\
&= \frac{1}{n\pi}\sin n\pi t\Big|_0^1 = \frac{1}{n\pi}\sin n\pi = 0. \qquad u = n\pi t;\ du = n\pi\,dt.
\end{aligned}$$

(b)
$$\begin{aligned}
b_1 &= \frac{1}{1}\int_{-1}^{1} f(t)\sin\frac{1\pi t}{1}\,dt = \int_{-1}^{0} 0\,dt + \int_{0}^{1} 1\cdot\sin\pi t\,dt \\
&= -\frac{1}{\pi}\cos\pi t\Big|_0^1 = -\frac{1}{\pi}(\cos\pi - \cos 0) \qquad u = \pi t;\ du = \pi\,dt. \\
&= -\frac{1}{\pi}(-1-1) = \frac{2}{\pi}. \\
b_2 &= \frac{1}{1}\int_{-1}^{1} f(t)\sin\frac{2\pi t}{1}\,dt = \int_{-1}^{0} 0\,dt + \int_{0}^{1} 1\cdot\sin 2\pi t\,dt \\
&= -\frac{1}{2\pi}\cos 2\pi t\Big|_0^1 = -\frac{1}{2\pi}(\cos 2\pi - \cos 0) \qquad u = 2\pi t;\ du = 2\pi\,dt. \\
&= -\frac{1}{2\pi}(1-1) = 0. \\
b_3 &= \frac{1}{1}\int_{-1}^{1} f(t)\sin\frac{3\pi t}{1}\,dt = \int_{-1}^{0} 0\,dt + \int_{0}^{1} 1\cdot\sin 3\pi t\,dt \\
&= -\frac{1}{3\pi}\cos 3\pi t\Big|_0^1 \qquad u = 3\pi t;\ du = 3\pi\,dt. \\
&= -\frac{1}{3\pi}(\cos 3\pi - \cos 0) = -\frac{1}{3\pi}(-1-1) = \frac{2}{3\pi}. \\
b_n &= \frac{1}{1}\int_{-1}^{1} f(t)\sin\frac{n\pi t}{p}\,dt = \int_{-1}^{0} 0\,dt + \int_{0}^{1} 1\cdot\sin\frac{n\pi t}{1}\,dt \qquad u = n\pi t;\ du = n\pi\,dt. \\
&= -\frac{1}{n\pi}\cos n\pi t\Big|_0^1 = -\frac{1}{n\pi}(\cos n\pi - 1) = \frac{1}{n\pi}(1-\cos n\pi).
\end{aligned}$$

Now recall that

$\cos 2\pi = \cos 4\pi = \cos 6\pi = \ldots = 1,$

$\cos\pi = \cos 3\pi = \cos 5\pi = \ldots = -1.$

Hence $1-\cos n\pi = 0$ whenever n is even and $1-\cos n\pi = 1-(-1) = 2$ whenever n is odd.

Thus

$$b_n = \begin{cases} 0, & n \text{ even} \\ \dfrac{2}{n\pi}, & n \text{ odd} \end{cases}$$

It follows that (since $p = 1$)

$$\begin{aligned}
f(t) &= \frac{1}{2} + \frac{2}{\pi}\sin\frac{1\pi t}{1} + 0 + \frac{2}{3\pi}\sin\frac{3\pi t}{1} + 0 + \frac{2}{5\pi}\sin\frac{5\pi t}{1} + \cdots \\
&= \frac{1}{2} + \frac{2}{\pi}\left(\sin\pi t + \frac{1}{3}\sin 3\pi t + \frac{1}{5}\sin 5\pi t + \cdots\right).
\end{aligned}$$

3.

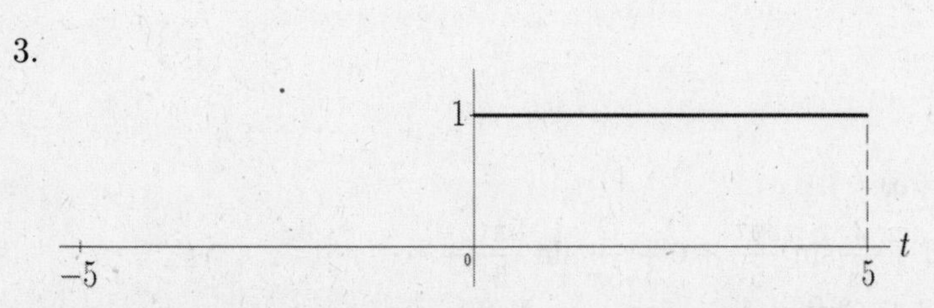

Period: $2p = 10$, so that $p = 5$.

$$
\begin{aligned}
a_0 &= \frac{1}{5}\int_{-5}^{5} f(t)\,dt = \frac{1}{5}\int_{-5}^{0} 0\,dt + \frac{1}{5}\int_{0}^{5} 1\,dt = \frac{1}{5}t\Big|_0^5 = 1 \\
\frac{a_0}{2} &= \frac{1}{2}. \\
a_1 &= \frac{1}{5}\int_{-5}^{5} f(t)\cos\frac{1\pi t}{5}\,dt = \frac{1}{5}\int_{-5}^{0} 0\cos\frac{\pi t}{5}\,dt + \frac{1}{5}\int_{0}^{5} 1\cdot\cos\frac{\pi t}{5}\,dt \\
&= \frac{1}{5}\frac{5}{\pi}\int_{0}^{5}\cos\frac{\pi t}{5}\left(\frac{\pi}{5}\,dt\right) = \frac{1}{\pi}\sin\frac{\pi t}{5}\Big|_0^5 = 0. \qquad u = \frac{\pi t}{5};\ du = \frac{\pi}{5}\,dt. \\
a_2 &= \int_{-5}^{5} f(t)\cos\frac{2\pi t}{5}\,dt = \frac{1}{5}\int_{-5}^{0} 0\cos\frac{2\pi t}{5}\,dt + \frac{1}{5}\int_{0}^{5} 1\cdot\cos\frac{2\pi t}{5}\,dt \\
&= \frac{1}{5}\frac{5}{2\pi}\int_{0}^{5}\cos\frac{2\pi}{5}\left(\frac{2\pi}{5}\,dt\right) = \frac{1}{2\pi}\sin\frac{2\pi t}{5}\Big|_0^5 = 0. \qquad u = \frac{2\pi t}{5};\ du = \frac{2\pi}{5}\,dt. \\
a_n &= \frac{1}{5}\int_{-5}^{5} f(t)\cos\frac{n\pi t}{5}\,dt = \frac{1}{5}\int_{-5}^{0} 0\cos\frac{n\pi t}{5}\,dt + \frac{1}{5}\int_{0}^{5} 1\cdot\cos\frac{n\pi t}{5}\,dt \\
&= \frac{1}{5}\frac{5}{n\pi}\int_{0}^{5}\cos\frac{n\pi t}{5}\left(\frac{n\pi}{5}\,dt\right) = \frac{1}{n\pi}\sin\frac{n\pi t}{5}\Big|_0^5 = 0. \qquad u = \frac{n\pi t}{5};\ du = \frac{n\pi}{5}\,dt. \\
b_1 &= \frac{1}{5}\int_{-5}^{5} f(t)\sin\frac{1\pi t}{5}\,dt = \frac{1}{5}\int_{-5}^{0} 0\sin\frac{\pi t}{5}\,dt + \frac{1}{5}\int_{0}^{5} 1\cdot\sin\frac{\pi t}{5}\,dt \\
&= \frac{1}{5}\frac{5}{\pi}\int_{0}^{5}\sin\frac{\pi t}{5}\left(\frac{\pi}{5}\,dt\right) = -\frac{1}{\pi}\cos\frac{\pi t}{5}\Big|_0^5 \qquad u = \frac{\pi t}{5};\ du = \frac{\pi}{5}\,dt. \\
&= -\frac{1}{\pi}(\cos\pi - \cos 0) = -\frac{1}{\pi}(-1-1) = \frac{2}{\pi}. \\
b_2 &= \frac{1}{5}\int_{-5}^{5} f(t)\sin\frac{2\pi t}{5}\,dt = \frac{1}{5}\int_{-5}^{0} 0\sin\frac{2\pi t}{5}\,dt + \frac{1}{5}\int_{0}^{5} 1\cdot\sin\frac{2\pi t}{5}\,dt \\
&= \frac{1}{5}\frac{5}{2\pi}\int_{0}^{5}\sin\frac{2\pi t}{5}\left(\frac{2\pi}{5}\,dt\right) = -\frac{1}{2\pi}\cos\frac{2\pi t}{5}\Big|_0^5 \qquad u = \frac{2\pi t}{5};\ du = \frac{2\pi}{5}\,dt. \\
&= -\frac{1}{2\pi}(\cos 2\pi - \cos 0) = 0. \\
b_3 &= \frac{1}{5}\int_{-5}^{5} f(t)\sin\frac{3\pi t}{5}\,dt = \frac{1}{5}\int_{-5}^{0} 0\sin\frac{3\pi t}{5}\,dt + \frac{1}{5}\int_{0}^{5} 1\cdot\sin\frac{3\pi t}{5}\,dt \\
&= \frac{1}{5}\frac{5}{3\pi}\int_{0}^{5}\sin\frac{3\pi t}{5}\left(\frac{3\pi}{5}\,dt\right) = -\frac{1}{3\pi}\cos\frac{3\pi t}{5}\Big|_0^5 \qquad u = \frac{3\pi t}{5};\ du = \frac{3\pi}{5}\,dt. \\
&= -\frac{1}{3\pi}(\cos 3\pi - \cos 0) = -\frac{1}{3\pi}(-1-1) = \frac{2}{3\pi}. \\
b_n &= \frac{1}{5}\int_{-5}^{5} f(t)\sin\frac{n\pi t}{5}\,dt = \frac{1}{5}\int_{-5}^{0} 0\sin\frac{n\pi t}{5}\,dt + \frac{1}{5}\int_{0}^{5} 1\cdot\sin\frac{n\pi t}{5}\,dt \\
&= -\frac{1}{5}\frac{5}{n\pi}\int_{0}^{5}\sin\frac{n\pi t}{5}\left(\frac{n\pi}{5}\,dt\right) = -\frac{1}{n\pi}\left(\cos\frac{n\pi t}{5}\right)\Big|_0^5 \\
&= -\frac{1}{n\pi}(\cos n\pi - \cos 0) = -\frac{1}{n\pi}(\cos n\pi - 1) = \frac{1}{n\pi}(1 - \cos n\pi).
\end{aligned}
$$

Now recall that

$\cos 2\pi = \cos 4\pi = \cos 6\pi = \cdots = 1,$

$\cos\pi = \cos 3\pi = \cos 5\pi = \cdots = -1.$

So $1 - \cos n\pi = 0$ whenever n is even and $1 - \cos n\pi = 1 - (-1) = 2$ whenever n is odd. Thus

$$
b_n = \begin{cases} 0, & n \text{ even} \\ \dfrac{2}{n\pi}, & n \text{ odd} \end{cases}
$$

It now follows that (since $p = 5$)

$$
\begin{aligned}
f(t) &= \frac{1}{2} + \frac{2}{\pi}\sin\frac{1\pi t}{5} + 0 + \frac{2}{3\pi}\sin\frac{3\pi t}{5} + 0 + \frac{2}{5\pi}\sin\frac{5\pi t}{5} + \cdots \\
&= \frac{1}{2} + \frac{2}{\pi}\left(\sin\frac{\pi t}{5} + \frac{1}{3}\sin\frac{3\pi t}{5} + \frac{1}{5}\sin\frac{5\pi t}{5} + \cdots\right).
\end{aligned}
$$

5.

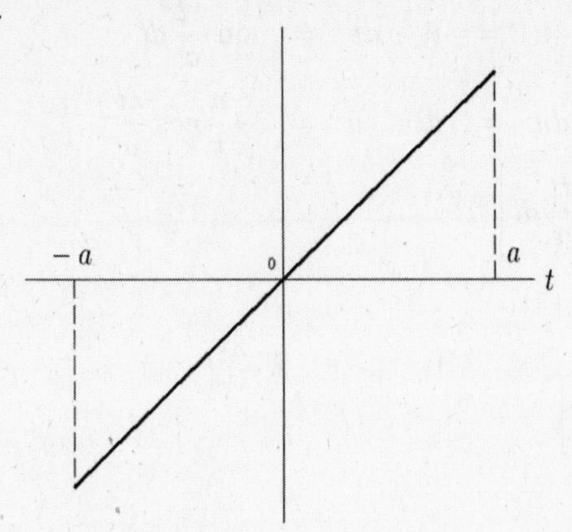

(a) Since the period $2p = 2a,\ p = a$.

$$a_0 = \frac{1}{a}\int_{-a}^{a} t\,dt = \frac{1}{a}\cdot\frac{1}{2}t^2\Big|_{-a}^{a} = 0.$$

$$a_1 = \frac{1}{a}\int_{-a}^{a} t\cos\frac{1\pi t}{a}\,dt = \frac{1}{a}\int_{-a}^{a} t\cos\frac{\pi t}{a}\,dt.$$

$$u = t \quad dv = \cos\frac{\pi t}{a}\,dt$$
$$du = dt \quad v = \frac{a}{\pi}\sin\frac{\pi t}{a}$$

$$\begin{aligned} a_1 &= \frac{1}{a}\left[t\left(\frac{a}{\pi}\right)\sin\frac{\pi t}{a}\Big|_{-a}^{a} - \frac{a}{\pi}\int_{-a}^{a}\sin\frac{\pi t}{a}\,dt\right] \\ &= \frac{1}{a}\left[0 - \frac{a}{\pi}\left(-\frac{a}{\pi}\right)\cos\frac{\pi t}{a}\Big|_{-a}^{a}\right] \\ &= \frac{a}{\pi^2}(\cos\pi - \cos(-\pi)) = 0. \qquad \cos(-\theta) = \cos\theta \end{aligned}$$

$$a_2 = \frac{1}{a}\int_{-a}^{a} t\cos\frac{2\pi t}{a}\,dt$$

$$u = t \quad dv = \cos\frac{2\pi t}{a}\,dt$$
$$du = dt \quad v = \frac{a}{2\pi}\sin\frac{2\pi t}{a}$$

$$\begin{aligned} &= \frac{1}{a}\left[t\left(\frac{a}{2\pi}\right)\sin\frac{2\pi t}{a}\Big|_{-a}^{a} - \frac{a}{2\pi}\int_{-a}^{a}\sin\frac{2\pi t}{a}\,dt\right] \\ &= \frac{1}{a}\left[0 - \frac{a}{2\pi}\left(-\frac{a}{2\pi}\right)\cos\frac{2\pi t}{a}\Big|_{-a}^{a}\right] \\ &= \frac{a}{2^2\pi^2}(\cos 2\pi - \cos(-2\pi)) = 0. \end{aligned}$$

$$a_n = \frac{1}{a}\int_{-a}^{a} t\cos\frac{n\pi t}{a}\,dt$$

$$u = t \quad dv = \cos\frac{n\pi t}{a}\,dt$$
$$du = dt \quad v = \frac{a}{n\pi}\sin\frac{n\pi t}{a}$$

$$\begin{aligned} a_n &= \frac{1}{a}\left[t\left(\frac{a}{n\pi}\right)\sin\frac{n\pi t}{a}\Big|_{-a}^{a} - \frac{a}{n\pi}\int_{-a}^{a}\sin\frac{n\pi t}{a}\,dt\right] \\ &= \frac{1}{a}\left[0 - \frac{a}{n\pi}\left(-\frac{a}{n\pi}\right)\cos\frac{n\pi}{a}\Big|_{-a}^{a}\right] = \frac{a}{n^2\pi^2}\left[\cos\frac{n\pi a}{a} - \cos\left(\frac{-n\pi a}{a}\right)\right] \\ &= \frac{a}{n^2\pi^2}(\cos n\pi - \cos n\pi) = 0, \text{ since } \cos(-\theta) = \cos\theta. \end{aligned}$$

(b)
$$b_1 = \frac{1}{a}\int_{-a}^{a} t\sin\frac{1\pi t}{a}\,dt \qquad \begin{aligned} u &= t & dv &= \sin\frac{\pi t}{a}\,dt \\ du &= dt & v &= -\frac{a}{\pi}\cos\frac{\pi t}{a} \end{aligned}$$

$$\begin{aligned}
&= \frac{1}{a}\left[t\left(-\frac{a}{\pi}\right)\cos\frac{\pi t}{a}\Big|_{-a}^{a} + \frac{a}{\pi}\int_{-a}^{a}\cos\frac{\pi t}{a}\,dt\right] \\
&= \frac{1}{a}\left[-\frac{at}{\pi}\cos\frac{\pi t}{a} + \frac{a^2}{\pi^2}\sin\frac{\pi t}{a}\right]\Big|_{-a}^{a} \\
&= \frac{1}{a}\left[-\frac{a^2}{\pi}\cos\pi + \frac{a(-a)}{\pi}\cos(-\pi) + 0\right] \\
&= \frac{2a}{\pi} \quad \text{since } \cos(-\theta) = \cos\theta.
\end{aligned}$$

$$b_2 = \frac{1}{a}\int_{-a}^{a} t\sin\frac{2\pi t}{a}\,dt \qquad \begin{aligned} u &= t & dv &= \sin\frac{2\pi t}{a}\,dt \\ du &= dt & v &= -\frac{a}{2\pi}\cos\frac{2\pi t}{a} \end{aligned}$$

$$\begin{aligned}
&= \frac{1}{a}\left[t\left(-\frac{a}{2\pi}\right)\cos\frac{2\pi t}{a}\Big|_{-a}^{a} + \frac{a}{2\pi}\int_{-a}^{a}\cos\frac{2\pi t}{a}\,dt\right] \\
&= \frac{1}{a}\left[-\frac{at}{2\pi}\cos\frac{2\pi t}{a} + \frac{a^2}{2^2\pi^2}\sin\frac{2\pi t}{a}\right]\Big|_{-a}^{a} \\
&= \frac{1}{a}\left[-\frac{a^2}{2\pi}\cos 2\pi + \frac{a(-a)}{2\pi}\cos(-2\pi) + 0\right] \\
&= \frac{1}{a}\left[-\frac{a^2}{2\pi} - \frac{a^2}{2\pi}\right] = -\frac{2a}{2\pi}.
\end{aligned}$$

$$b_3 = \frac{1}{a}\int_{-a}^{a} t\sin\frac{3\pi t}{a}\,dt \qquad \begin{aligned} u &= t & dv &= \sin\frac{3\pi t}{a}\,dt \\ du &= dt & v &= -\frac{a}{3\pi}\cos\frac{3\pi t}{a} \end{aligned}$$

$$\begin{aligned}
&= \frac{1}{a}\left[t\left(-\frac{a}{3\pi}\right)\cos\frac{3\pi t}{a}\Big|_{-a}^{a} + \frac{a}{3\pi}\int_{-a}^{a}\cos\frac{3\pi t}{a}\,dt\right] \\
&= \frac{1}{a}\left[-\frac{at}{3\pi}\cos\frac{3\pi t}{a} + \frac{a^2}{3^2\pi^2}\sin\frac{3\pi t}{a}\right]\Big|_{-a}^{a} \\
&= \frac{1}{a}\left[-\frac{a^2}{3\pi}\cos 3\pi + \frac{a(-a)}{3\pi}\cos(-3\pi) + 0\right] \\
&= -\frac{a}{3\pi}\cos 3\pi - \frac{a}{3\pi}\cos 3\pi \qquad \cos(-\theta) = \cos\theta \\
&= -\frac{a}{3\pi}(-1) - \frac{a}{3\pi}(-1) = \frac{2a}{3\pi}.
\end{aligned}$$

$$u = t \quad dv = \sin\frac{n\pi t}{a}\,dt$$

$$du = dt \quad v = -\frac{a}{n\pi}\cos\frac{n\pi t}{a}$$

$$\begin{aligned}
b_n &= \frac{1}{a}\int_{-a}^{a} t\sin\frac{n\pi t}{a}\,dt \\
&= \frac{1}{a}\left[-\frac{at}{n\pi}\cos\frac{n\pi t}{a}\Big|_{-a}^{a} + \frac{a}{n\pi}\int_{-a}^{a}\cos\frac{n\pi t}{a}\,dt\right] \\
&= \frac{1}{a}\left[-\frac{at}{n\pi}\cos\frac{n\pi t}{a}\Big|_{-a}^{a} + \frac{a^2}{n^2\pi^2}\sin\frac{n\pi t}{a}\Big|_{-a}^{a}\right] \\
&= \frac{1}{a}\left[-\frac{a^2}{n\pi}\cos\frac{n\pi a}{a} - \frac{a^2}{n\pi}\cos\left(\frac{-n\pi a}{a}\right) + 0\right] \\
&= -\frac{a}{n\pi}(\cos n\pi + \cos n\pi) = -\frac{2a}{n\pi}\cos n\pi, \text{ since } \cos(-\theta) = \cos\theta.
\end{aligned}$$

Thus

$$b_n = \begin{cases} -\dfrac{2a}{n\pi}, & n \text{ even} \\ \dfrac{2a}{n\pi}, & n \text{ odd} \end{cases}$$

$$\begin{aligned}
f(t) &= \frac{2a}{1\pi}\sin\frac{1\pi t}{a} - \frac{2a}{2\pi}\sin\frac{2\pi t}{a} + \frac{2a}{3\pi}\sin\frac{3\pi t}{a} - \cdots \\
&= \frac{2a}{\pi}\left(\sin\frac{\pi t}{a} - \frac{1}{2}\sin\frac{2\pi t}{a} + \frac{1}{3}\sin\frac{3\pi t}{a} - \cdots\right).
\end{aligned}$$

7.

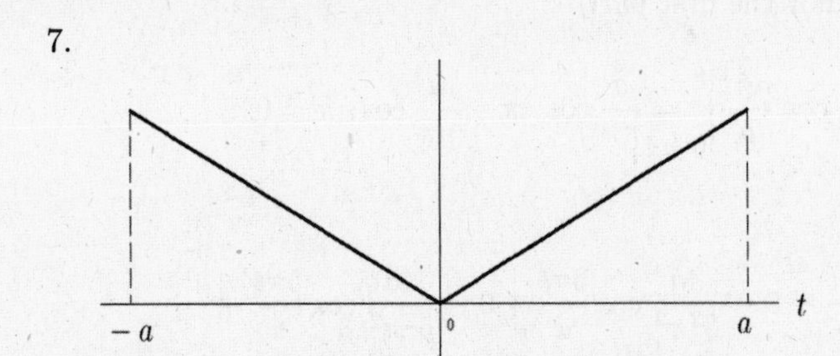

Since one period is $[-a, a]$, we have $p = a$.

$$\begin{aligned}
a_0 &= \frac{1}{a}\int_{-a}^{a} f(t)\,dt = \frac{1}{a}\int_{-a}^{0}(-t)\,dt + \frac{1}{a}\int_{0}^{a} t\,dt \\
&= \frac{1}{a}\left(-\frac{t^2}{2}\right)\Big|_{-a}^{0} + \frac{1}{a}\frac{t^2}{2}\Big|_{0}^{a} \\
&= 0 + \frac{a}{2} + \frac{a}{2} - 0 = a
\end{aligned}$$

$\dfrac{a_0}{2} = \dfrac{a}{2}$ (first term in the series); $a_n = \dfrac{1}{a}\displaystyle\int_{-a}^{a} f(t)\cos\frac{n\pi t}{a}\,dt.$

Since $f(t)\cos\frac{n\pi t}{a}$ is symmetric with respect to the vertical axis, we can integrate from 0 to a and double the result:

$$\begin{aligned} a_n &= 2\cdot\frac{1}{a}\int_0^a t\cos\frac{n\pi t}{a}\,dt \qquad u = t \quad dv = \cos\frac{n\pi t}{a}\,dt \\ & \qquad\qquad\qquad\qquad du = dt \quad v = \frac{a}{n\pi}\sin\frac{n\pi t}{a} \\ &= \frac{2}{a}\left[\frac{at}{n\pi}\sin\frac{n\pi t}{a}\Big|_0^a - \frac{a}{n\pi}\int_0^a \sin\frac{n\pi t}{a}\,dt\right] \\ &= \frac{2}{a}\left(0+\frac{a}{n\pi}\frac{a}{n\pi}\cos\frac{n\pi t}{a}\Big|_0^a\right) \\ &= \frac{2}{a}\frac{a^2}{n^2\pi^2}(\cos n\pi - 1) = \begin{cases} 0, & n \text{ even} \\ -\frac{4a}{n^2\pi^2}, & n \text{ odd}\end{cases} \end{aligned}$$

$$\begin{aligned} b_n &= \frac{1}{a}\int_{-a}^a f(t)\sin\frac{n\pi t}{a}\,dt \qquad u = t \quad dv = \sin\frac{n\pi t}{a}\,dt \\ & \qquad\qquad\qquad\qquad du = dt \quad v = -\frac{a}{n\pi}\cos\frac{n\pi t}{a} \\ &= \frac{1}{a}(-1)\int_{-a}^0 t\sin\frac{n\pi t}{a}\,dt + \frac{1}{a}\int_0^a t\sin\frac{n\pi t}{a}\,dt \\ &= -\frac{1}{a}\left[-\frac{at}{n\pi}\cos\frac{n\pi t}{a}\Big|_{-a}^0 + \frac{a}{n\pi}\int_{-a}^0 \cos\frac{n\pi t}{a}\,dt\right] + \frac{1}{a}\left[-\frac{at}{n\pi}\cos\frac{n\pi t}{a}\Big|_0^a + \frac{a}{n\pi}\int_0^a \cos\frac{n\pi t}{a}\,dt\right] \end{aligned}$$

Now observe that

$$\int \cos\frac{n\pi t}{a}\,dt = \frac{a}{n\pi}\sin\frac{n\pi t}{a} = 0$$

with the above limits of integration. For the first part,

$$\frac{t}{n\pi}\cos\frac{n\pi t}{a}\Big|_{-a}^0 - \frac{t}{n\pi}\cos\frac{n\pi t}{a}\Big|_0^a = \frac{a}{n\pi}\cos n\pi - \frac{a}{n\pi}\cos n\pi = 0$$

So $b_n = 0$. It follows that

$$\begin{aligned} f(t) &= \frac{a}{2} - \frac{4a}{1^2\pi^2}\cos\frac{1\pi t}{a} + 0 - \frac{4a}{3^2\pi^2}\cos\frac{3\pi t}{a} + 0 - \frac{4a}{5^2\pi^2}\cos\frac{5\pi t}{a} + \cdots \\ &= \frac{a}{2} - \frac{4a}{\pi^2}\left(\cos\frac{\pi t}{2} + \frac{1}{3^2}\cos\frac{3\pi t}{a} + \frac{1}{5^2}\cos\frac{5\pi t}{a} + \cdots\right) \end{aligned}$$

9.

Period: $2p = 2a$, $p = a$.

$$\begin{aligned}
a_0 &= \frac{1}{a}\int_{-a}^{a} f(t)\,dt = \frac{1}{a}\int_{-a}^{0} 0\,dt + \frac{1}{a}\int_{0}^{a} t\,dt = \frac{1}{a}\frac{t^2}{2}\bigg|_0^a = \frac{a}{2} \\
\frac{a_0}{2} &= \frac{a}{4}. \\
a_n &= \frac{1}{a}\int_{-a}^{a} \cos\frac{n\pi t}{a}\,dt = \frac{1}{a}\int_{-a}^{0} 0\cos\frac{n\pi t}{a}\,dt + \frac{1}{a}\int_{0}^{a} t\cos\frac{n\pi t}{a}\,dt \\
&= \frac{1}{a}\int_{0}^{a} t\cos\frac{n\pi t}{a}\,dt
\end{aligned}$$

$$\begin{aligned}
u &= t & dv &= \cos\frac{n\pi t}{a}\,dt \\
du &= dt & v &= \frac{a}{n\pi}\sin\frac{n\pi t}{a}
\end{aligned}$$

$$\begin{aligned}
a_n &= \frac{1}{a}\left[\frac{at}{n\pi}\sin\frac{n\pi t}{a}\bigg|_0^a - \frac{a}{n\pi}\int_0^a \sin\frac{n\pi t}{a}\,dt\right] = \frac{1}{a}\left[0 - \frac{a}{n\pi}\left(-\frac{a}{n\pi}\right)\cos\frac{n\pi t}{a}\bigg|_0^a\right] \\
&= \frac{a}{n^2\pi^2}(\cos n\pi - 1) = \begin{cases} 0, & n \text{ even} \\ -\dfrac{2a}{n^2\pi^2}, & n \text{ odd} \end{cases} \\
b_n &= \frac{1}{a}\int_{-a}^{a} \sin\frac{n\pi t}{a}\,dt = \frac{1}{a}\int_{-a}^{0} 0\sin\frac{n\pi t}{a}\,dt + \frac{1}{a}\int_{0}^{a} t\sin\frac{n\pi t}{a}\,dt \\
&= \frac{1}{a}\int_{0}^{a} t\sin\frac{n\pi t}{a}\,dt
\end{aligned}$$

$$\begin{aligned}
u &= t & dv &= \sin\frac{n\pi t}{a}\,dt \\
du &= dt & v &= -\frac{a}{n\pi}\cos\frac{n\pi t}{a}
\end{aligned}$$

$$= \frac{1}{a}\left[-\frac{at}{n\pi}\cos\frac{n\pi t}{a}\bigg|_0^a + \frac{a}{n\pi}\int_0^a \cos\frac{n\pi t}{a}\,dt\right].$$

Now observe that $\int \cos\frac{n\pi t}{a}\,dt = 0$ with the given limits of integration. The first part yields

$$b_n = -\frac{a}{n\pi}\cos n\pi - 0 = \begin{cases} -\dfrac{a}{n\pi}, & n \text{ even} \\ \dfrac{a}{n\pi}, & n \text{ odd} \end{cases}$$

$$\begin{aligned}
f(t) &= \frac{a}{4} - \frac{2a}{1^2\pi^2}\cos\frac{1\pi t}{a} + 0 - \frac{2a}{3^2\pi^2}\cos\frac{3\pi t}{a} + 0 - \frac{2a}{5^2\pi^2}\cos\frac{5\pi t}{a} - \cdots \\
&\quad + \frac{a}{1\pi}\sin\frac{1\pi t}{a} - \frac{a}{2\pi}\sin\frac{2\pi t}{a} + \frac{a}{3\pi}\sin\frac{3\pi t}{a} - \frac{a}{4\pi}\sin\frac{4\pi t}{a} + \cdots \\
f(t) &= \frac{a}{4} - \frac{2a}{\pi^2}\left(\cos\frac{\pi t}{a} + \frac{1}{3^2}\cos\frac{3\pi t}{a} + \frac{1}{5^2}\cos\frac{5\pi t}{a} + \cdots\right) \\
&\quad + \frac{a}{\pi}\left(\sin\frac{\pi t}{a} - \frac{1}{2}\sin\frac{2\pi t}{a} + \frac{1}{3}\sin\frac{3\pi t}{a} - \frac{1}{4}\sin\frac{4\pi t}{a} + \cdots\right).
\end{aligned}$$

11. Since one period is $[-\pi, \pi]$, we have $p = \pi$. Thus

$$\begin{aligned}
a_0 &= \tfrac{1}{\pi}\int_{-\pi}^{\pi} f(t)\,dt = \tfrac{1}{\pi}\int_{-\pi}^{0} 0\cdot dt + \tfrac{1}{\pi}\int_0^{\pi} \sin t\,dt \\
&= \frac{1}{\pi}(-\cos t)\bigg|_0^{\pi} = \frac{1}{\pi}(1+1) = \frac{2}{\pi}
\end{aligned}$$

and $\dfrac{a_0}{2} = \dfrac{1}{\pi}$.

Next,

$$a_n = \frac{1}{\pi}\int_{-\pi}^{\pi} f(t)\cos\frac{n\pi t}{\pi}\,dt = \frac{1}{\pi}\int_0^{\pi} \sin t\cos nt\,dt.$$

Now, by formula 63, Table 2, with $m = 1$, we get

$$\begin{aligned}
a_n &= \frac{1}{\pi}\left[-\frac{\cos(1+n)t}{2(1+n)} - \frac{\cos(1-n)t}{2(1-n)}\right]\bigg|_0^{\pi} \\
&= \frac{1}{2\pi}\left[-\frac{\cos(1+n)\pi}{1+n} - \frac{\cos(1-n)\pi}{1-n} + \frac{1}{1+n} + \frac{1}{1-n}\right] \quad (n \neq 1).
\end{aligned}$$

Observe that this expression is valid for all n except $n = 1$. So if $n \neq 1$, note that

$$\cos(1+n)\pi = \cos(1-n)\pi = \begin{cases} 1, & \text{for } n \text{ odd} \\ -1, & \text{for } n \text{ even} \end{cases}$$

For n odd, we have

$$\frac{1}{2\pi}\left[-\frac{1}{1+n} - \frac{1}{1-n} + \frac{1}{1+n} + \frac{1}{1-n}\right] = 0.$$

For n even, we have

$$\frac{1}{2\pi}\left[\frac{1}{1+n} + \frac{1}{1-n} + \frac{1}{1+n} + \frac{1}{1-n}\right] = \frac{1}{2\pi}\left[\frac{2}{1+n} + \frac{2}{1-n}\right]$$

$$= \frac{1}{\pi}\left[\frac{1}{1+n} + \frac{1}{1-n}\right] = \frac{1}{\pi}\frac{1-n+1+n}{(1+n)(1-n)} = \frac{1}{\pi}\frac{2}{1-n^2} = -\frac{2}{\pi(n^2-1)}.$$ Since these forms are not valid for $n = 1$, the coefficient a_1 must be evaluated separately:

$$a_1 = \frac{1}{\pi}\int_0^{\pi} \sin t\cos t\,dt = \frac{\sin^2 t}{2\pi}\Big|_0^{\pi} = 0. \qquad u = \sin t;\ du = \cos t\,dt.$$

Thus

$$a_n = \begin{cases} -\dfrac{2}{\pi(n^2-1)}, & n \text{ even} \\ 0, & n \text{ odd} \end{cases}$$

Next, $b_n = \dfrac{1}{\pi}\displaystyle\int_{-\pi}^{\pi} f(t)\sin\frac{n\pi t}{\pi}\,dt = \frac{1}{\pi}\int_0^{\pi} \sin t\sin nt\,dt.$

By formula 61, Table 2, with $m = 1$, we get

$$\begin{aligned} b_n &= \frac{1}{\pi}\left[-\frac{\sin(1+n)t}{2(1+n)} + \frac{\sin(1-n)t}{2(1-n)}\right]\Bigg|_0^{\pi} \quad (n \neq 1) \\ &= 0, \text{ provided that } n \neq 1. \end{aligned}$$

If $n = 1$, we have

$$\begin{aligned} b_1 &= \frac{1}{\pi}\int_0^{\pi} \sin t\sin t\,dt = \frac{1}{\pi}\int_0^{\pi} \sin^2 t\,dt \\ &= \frac{1}{2\pi}\int_0^{\pi} (1-\cos 2t)\,dt = \frac{1}{2\pi}\left(t - \frac{1}{2}\sin 2t\right)\Bigg|_0^{\pi} = \frac{1}{2}. \end{aligned}$$

Substituting the first few values in series (10.33), we obtain

$$\begin{aligned} f(t) &= \frac{1}{\pi} + \frac{1}{2}\sin\frac{1\pi t}{\pi} \qquad \frac{a_0}{2} = \frac{1}{\pi},\ b_1 = \frac{1}{2} \\ &\quad +0 - \frac{2}{\pi\cdot 3}\cos\frac{2\pi t}{\pi} + 0 - \frac{2}{\pi\cdot 15}\cos\frac{4\pi t}{\pi} \\ &\quad +0 - \frac{2}{\pi\cdot 35}\cos\frac{6\pi t}{\pi} + 0 - \frac{2}{\pi\cdot 63}\cos\frac{8\pi t}{\pi} + \cdots \end{aligned}$$

or

$$f(t) = \frac{1}{\pi} + \frac{1}{2}\sin t - \frac{2}{\pi}\left(\frac{1}{3}\cos 2t + \frac{1}{15}\cos 4t + \frac{1}{35}\cos 6t + \frac{1}{63}\cos 8t + \cdots\right).$$

Chapter 10 Review

1. $$\begin{aligned} a = 6,\ ar &= -2 \\ 6r &= -2 \\ r &= -\frac{1}{3} \end{aligned}$$

$$S = \frac{a}{1-r} = \frac{6}{1-(-1/3)} = \frac{6}{4/3} = \frac{6}{1}\cdot\frac{3}{4} = \frac{9}{2}$$

3. $\lim_{n\to\infty} \dfrac{2n}{4n+3} = \lim_{n\to\infty} \dfrac{2}{4+3/n} = \dfrac{1}{2}$

Since the nth term does not approach 0, the series diverges.

5. $$\int_1^\infty \frac{x}{x^2+4}\,dx = \lim_{b\to\infty} \frac{1}{2}\int_1^b \frac{2x\,dx}{x^2+4} = \lim_{b\to\infty} \frac{1}{2}\ln\left(x^2+4\right)\Big|_1^b$$
$$= \lim_{b\to\infty} \left(\frac{1}{2}\ln\left(b^2+4\right) - \frac{1}{2}\ln 5\right)$$

This limit does not exist; so the series diverges.

7. $\dfrac{1}{\ln n} > \dfrac{1}{n}$ (diverges by comparison to the harmonic series).

9. $\lim_{n\to\infty} \dfrac{a_{n+1}}{a_n} = \lim_{n\to\infty} \dfrac{6^{n+1}}{(n+1)!}\dfrac{n!}{6^n} = \lim_{n\to\infty} \dfrac{6\cdot 6^n}{(n+1)n!}\dfrac{n!}{6^n} = \lim_{n\to\infty} \dfrac{6}{n+1} = 0 < 1$

The series is therefore convergent by the ratio test.

11.
$$\begin{aligned} f(x) &= e^{-x} & f(0) &= 1 \\ f'(x) &= -e^{-x} & f'(0) &= -1 \\ f''(x) &= e^{-x} & f''(0) &= 1 \\ f'''(x) &= -e^{-x} & f'''(0) &= -1 \\ f^{(4)}(x) &= e^{-x} & f^{(4)}(0) &= 1 \end{aligned}$$
$$\begin{aligned} e^{-x} &= 1 - 1x + \frac{1}{2!}x^2 + \frac{-1}{3!}x^3 + \frac{1}{4!}x^4 + \cdots \\ &= 1 - x + \frac{x^2}{2!} - \frac{x^3}{3!} + \frac{x^4}{4!} - \cdots. \end{aligned}$$

13. $\sin x = x - \dfrac{x^3}{3!} + \dfrac{x^5}{5!} - \cdots$. Replacing x by x^2,

$$\sin x^2 = x^2 - \frac{(x^2)^3}{3!} + \frac{(x^2)^5}{5!} - \cdots = x^2 - \frac{x^6}{3!} + \frac{x^{10}}{5!} - \cdots.$$

15. By (10.14)
$$\begin{aligned} e^x &= 1 + x + \frac{x^2}{2!} + \frac{x^3}{3!} + \frac{x^4}{4!} + \cdots \\ f(x) &= \frac{1-e^x}{x} = \frac{1}{x}\left[1 - \left(1 + x + \frac{x^2}{2!} + \frac{x^3}{3!} + \frac{x^4}{4!} + \cdots\right)\right] \\ &= \frac{1}{x}\left(-x - \frac{x^2}{2!} - \frac{x^3}{3!} - \frac{x^4}{4!} - \cdots\right) = -1 - \frac{x}{2!} - \frac{x^2}{3!} - \frac{x^3}{4!} - \cdots. \end{aligned}$$

17. $\cos x = 1 - \dfrac{x^2}{2!} + \dfrac{x^4}{4!} - \dfrac{x^6}{6!} + \cdots$

$$\cos(0.5) = 1 - \frac{(0.5)^2}{2!} + \frac{(0.5)^4}{4!} \qquad \text{(three terms)}$$
$$= 0.877604.$$

max. error (fourth term): $-\dfrac{(0.5)^6}{6!} = -0.000022.$

(a) 0.877604 sum of first three terms

-0.000022 error

(b) 0.877582

The values of (a) and (b) agree to four decimal places (after rounding off): 0.8776.

19.
$$\begin{aligned}
\cos x &= 1-\frac{x^2}{2!}+\frac{x^4}{4!}-\frac{x^6}{6!}+\frac{x^8}{8!}-\cdots\\
\cos x^3 &= 1-\frac{x^6}{2!}+\frac{x^{12}}{4!}-\frac{x^{18}}{6!}+\frac{x^{24}}{8!}+\cdots\\
\int_0^{0.9}\cos x^3\,dx &= \int_0^{0.9}\left(1-\frac{x^6}{2!}+\frac{x^{12}}{4!}-\frac{x^{18}}{6!}+\frac{x^{24}}{8!}-\cdots\right)dx\\
&= x-\frac{x^7}{7\cdot 2!}+\frac{x^{13}}{13\cdot 4!}-\frac{x^{19}}{19\cdot 6!}+\frac{x^{25}}{25\cdot 8!}+\cdots\Big|_0^{0.9}\\
&= 0.9-\frac{(0.9)^7}{7\cdot 2!}+\frac{(0.9)^{13}}{13\cdot 4!}-\frac{(0.9)^{19}}{19\cdot 6!}\\
&= 0.8666. \qquad \text{(four terms)}
\end{aligned}$$

error (fifth term): $\dfrac{(0.9)^{25}}{25\cdot 8!}=7.1\times 10^{-8}$.

21.
$$\begin{aligned}
e^x &= 1+x+\frac{x^2}{2!}+\frac{x^3}{3!}+\cdots\\
e^{-0.20t^2} &= 1-0.20t^2+\frac{(-0.20t^2)^2}{2!}+\frac{(-0.20t^2)^3}{3!}+\cdots\\
&= 1-0.20t^2+\frac{(0.20)^2}{2!}t^4-\frac{(0.20)^3}{3!}t^6+\cdots
\end{aligned}$$

$$\int_0^{0.10}\left(1-0.20t^2+\frac{(0.20)^2}{2!}t^4-\frac{(0.20)^3}{3!}t^6+\cdots\right)dt$$

$$=t-0.20\frac{t^3}{3}+\frac{(0.20)^2}{2!}\frac{t^5}{5}-\frac{(0.20)^3}{3!}\frac{t^7}{7}\Big|_0^{0.10}$$

$$=0.10-0.20\frac{(0.10)^3}{3}+\frac{(0.20)^2}{2!}\frac{(0.10)^5}{5} \qquad \text{(three terms)}$$

$= 0.0999 \approx 0.10$ coulombs (two significant digits).

error: $-\dfrac{(0.20)^3}{3!}\dfrac{(0.10)^7}{7}=-1.9\times 10^{-11}$.

23.
$$\begin{aligned}
f(x) &= \sin x & f\left(\frac{\pi}{4}\right) &= \frac{1}{\sqrt{2}}\\
f'(x) &= \cos x & f'\left(\frac{\pi}{4}\right) &= \frac{1}{\sqrt{2}}\\
f''(x) &= -\sin x & f''\left(\frac{\pi}{4}\right) &= -\frac{1}{\sqrt{2}}\\
f'''(x) &= -\cos x & f'''\left(\frac{\pi}{4}\right) &= -\frac{1}{\sqrt{2}}
\end{aligned}$$

$$\sin x=\frac{1}{\sqrt{2}}+\frac{1}{\sqrt{2}}\left(x-\frac{\pi}{4}\right)-\frac{1}{\sqrt{2}}\frac{1}{2!}\left(x-\frac{\pi}{4}\right)^2-\frac{1}{\sqrt{2}}\frac{1}{3!}\left(x-\frac{\pi}{4}\right)^3+\cdots$$

$$44^\circ=45^\circ-1^\circ=\frac{\pi}{4}-\frac{\pi}{180}$$

Thus $x-\dfrac{\pi}{4}=\left(\dfrac{\pi}{4}-\dfrac{\pi}{180}\right)-\dfrac{\pi}{4}=-\dfrac{\pi}{180}$.

$$\sin 44^\circ=\frac{1}{\sqrt{2}}+\frac{1}{\sqrt{2}}\left(-\frac{\pi}{180}\right)-\frac{1}{\sqrt{2}}\frac{1}{2!}\left(-\frac{\pi}{180}\right)^2-\frac{1}{\sqrt{2}}\frac{1}{3!}\left(-\frac{\pi}{180}\right)^3=0.69466.$$

25.
$$\lim_{x\to 0}\frac{1-\cos x}{x}=\lim_{x\to 0}\frac{1-\left(1-\frac{x^2}{2!}+\frac{x^4}{4!}-\frac{x^6}{6!}+\cdots\right)}{x}=\lim_{x\to 0}\left(\frac{x}{2!}-\frac{x^3}{4!}+\frac{x^5}{6!}-\cdots\right)=0.$$

Chapter 11

First-Order Differential Equations

11.1 What is a Differential Equation?

1. $y = 2e^{3x}$; $\dfrac{dy}{dx} = 6e^{3x}$. Substituting in $\dfrac{dy}{dx} - 3y = 0$, we get $6e^{3x} - 3(2e^{3x}) = 0$ so that the solution checks.

3. $y = 2e^{-2x}$; $\dfrac{dy}{dx} = -4e^{-2x}$. Substituting in $\dfrac{dy}{dx} + 2y = 0$, we get $-4e^{-2x} + 2\left(2e^{-2x}\right) = 0$. The equation is satisfied.

5. $y = x^2 - 4$, $\dfrac{dy}{dx} = 2x$.
Substituting in the left side of the given equation, we get
$x\dfrac{dy}{dx} - y = x(2x) - (x^2 - 4) = 2x^2 - x^2 + 4 = x^2 + 4$, which is equal to the right side.

7.
$$\begin{aligned} y &= x + x\ln x - 1 \\ \frac{dy}{dx} &= 1 + x \cdot \frac{1}{x} + \ln x \cdot 1 \qquad \text{product rule} \\ &= 2 + \ln x. \end{aligned}$$
Substituting in the left side of the given equation, we have
$x^2(2 + \ln x) - x(x + \ln x - 1) = 2x^2 + x^2\ln x - x^2 - x^2\ln x + x = x^2 + x$, the right side.

9.
$$\begin{aligned} y &= \cos 3x + 3\sin 3x \\ y' &= -3\sin 3x + 9\cos 3x \\ y'' &= -9\cos 3x - 27\sin 3x \end{aligned}$$
Substituting in $y'' + 9y = 0$, we get
$$-9\cos 3x - 27\sin 3x + 9\left(\cos 3x + 3\sin 3x\right) = 0.$$
The solution checks.

11.
$$\begin{aligned} y &= c_1e^x + c_2e^{2x} \\ y' &= c_1e^x + 2c_2e^{2x} \\ y'' &= c_1e^x + 4c_2e^{2x} \end{aligned}$$
Substituting in $y'' - 3y' + 2y = 0$, we get
$c_1e^x + 4c_2e^{2x} - 3\left(c_1e^x + 2c_2e^{2x}\right) + 2\left(c_1e^x + c_2e^{2x}\right)$
$= c_1e^x + 4c_2e^{2x} - 3c_1e^x - 6c_2e^{2x} + 2c_1e^x + 2c_2e^{2x} = 0.$
The solution checks.

13. $y = xe^x + 2e^x$; by the product rule, $\dfrac{dy}{dx} = xe^x + e^x \cdot 1 + 2e^x = xe^x + 3e^x$.
Substituting in the given equation, we get

$$\begin{aligned} \frac{dy}{dx} - y &= e^x && \text{given equation} \\ (xe^x + 3e^x) - (xe^x + 2e^x) &= e^x. && \text{collecting terms} \end{aligned}$$

The solution checks.

15. $y = c\cos 3x - x\cos 3x$. By the product rule,

$$\begin{aligned} y' &= -3c\sin 3x - x(-3\sin 3x) - \cos 3x \cdot 1 \\ &= -3c\sin 3x + 3x\sin 3x - \cos 3x \\ y'' &= -9c\cos 3x + 3x(3\cos 3x) + 3\sin 3x + 3\sin 3x \\ &= -9c\cos 3x + 9x\cos 3x + 6\sin 3x. \end{aligned}$$

Substituting in the left side of the given equation, we get

$$\begin{aligned} y'' + 9y &= (-9c\cos 3x + 9x\cos 3x + 6\sin 3x) + 9(c\cos 3x - x\cos 3x) \\ &= 6\sin 3x, \text{ the right side.} \end{aligned}$$

17.
$$\begin{aligned} y &= c_1e^x + c_2e^{-6x} + xe^x \\ \frac{dy}{dx} &= c_1e^x - 6c_2e^{-6x} + xe^x + e^x \\ \frac{d^2y}{dx^2} &= c_1e^x + 36c_2e^{-6x} + xe^x + 2e^x \end{aligned}$$

Then
$\dfrac{d^2y}{dx^2} + 5\dfrac{dy}{dx} - 6y = c_1e^x + 36c_2e^{-6x} + xe^x + 2e^x + 5c_1e^x - 30c_2e^{-6x} + 5xe^x$
$+ 5e^x - 6c_1e^x - 6c_2e^{-x} - 6xe^x = 7e^x$, the right side.

19.
$$\begin{aligned} \frac{dy}{dx} &= 3x^2 \\ y &= x^3 + c && \text{integrating} \end{aligned}$$

Now let $x = 2$ and $y = 5$:

$5 = 2^3 + c$ or $c = -3$.

So $y = x^3 - 3$.

21.
$$\begin{aligned} \frac{dy}{dx} &= \sec^2 x \\ y &= \tan x + c && \text{integrating} \end{aligned}$$

Now let $y = 1$ and $x = \dfrac{\pi}{4}$:

$$\begin{aligned} 1 &= \tan\frac{\pi}{4} + c \\ 1 &= 1 + c \text{ and } c = 0. \end{aligned}$$

So $y = \tan x$.

23.
$$\begin{aligned} \frac{d^2y}{dx^2} &= e^x \\ \frac{dy}{dx} &= e^x + c_1 && \text{integrating} \\ y &= e^x + c_1x + c_2 && \text{integrating again} \end{aligned}$$

Substituting $(0, 0)$ and $(1, 1)$, respectively, we obtain the system of equations

$$\begin{aligned} 0 &= 1 + c_2 \\ 1 &= e + c_1 + c_2 \end{aligned}$$

Thus $c_2 = -1$. From the second equation, we get

$1 = e + c_1 - 1$ or $c_1 = 2 - e$.

The solution is therefore given by $y = e^x + (2 - e)x - 1$.

11.2 Separation of Variables

1.
$$\begin{aligned} x^2\,dx + y\,dy &= 0 \\ \int x^2\,dx + \int y\,dy &= c_1 \\ \frac{x^3}{3} + \frac{y^2}{2} &= c_1 \\ 2x^3 + 3y^2 &= 6c_1 && \text{multiplying by 6} \\ 2x^3 + 3y^2 &= c \ \ (\text{let } c = 6c_1) \end{aligned}$$

3.
$$\begin{aligned} y^2\left(1+x^2\right)\,dx + 3\,dy &= 0 \\ \left(1+x^2\right)\,dx + \frac{3\,dy}{y^2} &= 0 && \text{dividing by } y^2 \\ \int (1+x^2)\,dx + \int 3y^{-2}\,dy &= 0 && \frac{1}{y^2} = y^{-2} \\ x + \frac{1}{3}x^3 + 3\frac{y^{-1}}{-1} &= c_1 \\ 3x + x^3 - \frac{9}{y} &= 9c_1 && \text{multiplying by 3} \\ 3x + x^3 - \frac{9}{y} &= c && \text{letting } c = 9c_1 \end{aligned}$$

or

$$3xy + x^3y - 9 = cy$$

5.
$$\begin{aligned} 2x\,dx + (1+x^2)\,dy &= 0 \\ \frac{2x\,dx}{1+x^2} + dy &= 0 && \text{dividing by } 1+x^2 \\ \int \frac{2x\,dx}{1+x^2} + \int dy &= c \\ \int \frac{du}{u} + \int dy &= c && u = 1+x^2;\ du = 2x\,dx. \\ \ln(1+x^2) + y &= c \end{aligned}$$

7.
$$\begin{aligned} 2 - y\csc x\frac{dy}{dx} &= 0 \\ 2\,dx - y\csc x\,dy &= 0 && \text{differential form} \\ \frac{2\,dx}{\csc x} - y\,dy &= 0 && \text{dividing by } \csc x \\ \int 2\sin x\,dx - \int y\,dy &= c_1 && \frac{1}{\csc x} = \sin x \\ -2\cos x - \frac{1}{2}y^2 &= c_1 \\ 4\cos x + y^2 &= -2c_1 && \text{multiplying by } -2 \\ 4\cos x + y^2 &= c && c = -2c_1 \end{aligned}$$

9.
$$
\begin{aligned}
1+(x^2y-x^2)\frac{dy}{dx} &= 0 \\
dx+(x^2y-x^2)\,dy &= 0 && \text{differential form} \\
dx+x^2(y-1)\,dy &= 0 && \text{factoring } x^2 \\
\frac{dx}{x^2}+(y-1)\,dy &= 0 && \text{dividing by } x^2 \\
\int x^{-2}\,dx+\int (y-1)\,dy &= c_1 \\
\frac{x^{-1}}{-1}+\frac{1}{2}y^2-y &= c_1 \\
-\frac{1}{x}+\frac{1}{2}y^2-y &= c_1 \\
-2+xy^2-2xy &= 2xc_1 && \text{multiplying by } 2x \\
xy^2-2xy-2 &= cx && \text{letting } c=2c_1
\end{aligned}
$$

11.
$$
\begin{aligned}
dx-y\,dx+x\,dy &= 0 \\
(1-y)\,dx+x\,dy &= 0 && \text{factoring } dx \\
\frac{dx}{x}+\frac{dy}{1-y} &= 0 && \text{dividing by } x(1-y) \\
\int \frac{dx}{x}+\int \frac{dy}{1-y} &= c_1 \\
\int \frac{dx}{x}-\int \frac{-dy}{1-y} &= c_1 && u=1-y;\ du=-dy. \\
\ln|x|-\ln|1-y| &= \ln c_2,\ c_2>0 && \text{letting } \ln c_2=c_1 \\
\ln\left|\frac{x}{1-y}\right| &= \ln c_2 && \ln A-\ln B=\ln\frac{A}{B} \\
\left|\frac{x}{1-y}\right| &= c_2 \\
\frac{x}{1-y} &= \pm c_2 \\
\frac{x}{1-y} &= c,\ c\neq 0 && \text{letting } c=\pm c_2 \\
x &= c(1-y)
\end{aligned}
$$

13.

$$\begin{aligned}
2y\,dx + 3x\,dy &= 0 \\
\frac{2\,dx}{x} + \frac{3\,dy}{y} &= 0 && \text{dividing by } xy \\
\int \frac{2\,dx}{x} + \int \frac{3\,dy}{y} &= c_2 && \text{Form: } \int \frac{du}{u} \\
2\ln|x| + 3\ln|y| &= c_2 && \text{Recall: } n\ln A = \ln A^n \\
\ln x^2 + \ln|y|^3 &= \ln c_1 && (\text{let } \ln c_1 = c_2,\ c_1 > 0) \\
\ln|x^2y^3| &= \ln c_1 && \ln A + \ln B = \ln AB \\
|x^2y^3| &= c_1,\ c_1 > 0 \\
x^2y^3 &= \pm c_1 \\
x^2y^3 &= c,\ c \neq 0 && (\text{let } c = \pm c_1)
\end{aligned}$$

15.

$$\begin{aligned}
\frac{dV}{dP} &= -\frac{V}{P} \\
\frac{dV}{V} &= -\frac{dP}{P} && \text{separating variables} \\
\int \frac{dV}{V} &= -\int \frac{dP}{P} \\
\ln|V| &= -\ln|P| + c_2 && \text{form: } \int \frac{du}{u} = \ln|u| + c \\
\ln|P| + \ln|V| &= \ln c_1 && \text{letting } \ln c_1 = c_2 \\
\ln|PV| &= \ln c_1 && \ln A + \ln B = \ln AB \\
|PV| &= c_1 \\
PV &= \pm c_1 \\
PV &= c && \text{letting } c = \pm c_1
\end{aligned}$$

17.

$$\begin{aligned}
dx + (2\cos^2 x - y\cos^2 x)\,dy &= 0 \\
dx + \cos^2 x(2-y)\,dy &= 0 && \text{factoring } \cos^2 x \\
\frac{dx}{\cos^2 x} + (2-y)\,dy &= 0 && \text{dividing by } \cos^2 x \\
\int \frac{dx}{\cos^2 x} + \int (2-y)\,dy &= c_1 \\
\int \sec^2 x\,dx + \int (2-y)\,dy &= c_1 && \frac{1}{\cos x} = \sec x \\
\tan x + 2y - \frac{1}{2}y^2 &= c_1 \\
2\tan x + 4y - y^2 &= 2c_1 && \text{multiplying by 2} \\
2\tan x + 4y - y^2 &= c && \text{letting } c = 2c_1
\end{aligned}$$

19.

$$\begin{aligned}
\cos^2 t + y\csc t\frac{dy}{dt} &= 0 \\
\cos^2 t\,dt + y\csc t\,dy &= 0 \\
\frac{\cos^2 t\,dt}{\csc t} + y\,dy &= 0 && \text{dividing by } \csc t \\
\int \cos^2 t\sin t\,dt + \int y\,dy &= c_1 && \frac{1}{\csc t} = \sin t \\
-\int (\cos t)^2(-\sin t\,dt) + \int y\,dy &= c_1 && u = \cos t;\ du = -\sin t\,dt. \\
-\int u^2\,du + \int y\,dy &= c_1 \\
-\frac{1}{3}u^3 + \frac{1}{2}y^2 &= c_1 \\
-\frac{1}{3}\cos^3 t + \frac{1}{2}y^2 &= c_1 \\
3y^2 - 2\cos^3 t &= 6c_1 && \text{multiplying by 6} \\
3y^2 - 2\cos^3 t &= c && \text{letting } c = 6c_1
\end{aligned}$$

21.

$$\begin{aligned}
\sqrt{v^2+1}\,dt + vt^2\,dv &= 0 \\
\frac{dt}{t^2} + \frac{v\,dv}{\sqrt{v^2+1}} &= 0 && \text{dividing by } t^2\sqrt{v^2+1} \\
\int \frac{dt}{t^2} + \int \frac{v\,dv}{(v^2+1)^{1/2}} &= c \\
\int t^{-2}\,dt + \int (v^2+1)^{-1/2}v\,dv &= c && u = v^2+1;\ du = 2v\,dv. \\
\int t^{-2}\,dt + \frac{1}{2}\int (v^2+1)^{-1/2}(2v\,dv) &= c \\
\frac{t^{-1}}{-1} + \frac{1}{2}\frac{(v^2+1)^{1/2}}{1/2} &= c \\
-\frac{1}{t} + \sqrt{v^2+1} &= c \\
-1 + t\sqrt{v^2+1} &= ct && \text{multiplying by } t
\end{aligned}$$

23.

$$\begin{aligned}
T_1\,dT_1 + (\csc T_1 + T_2\csc T_1)\,dT_2 &= 0 \\
T_1\,dT_1 + \csc T_1(1+T_2)\,dT_2 &= 0 \\
\frac{T_1\,dT_1}{\csc T_1} + (1+T_2)\,dT_2 &= 0 && \text{dividing by } \csc T_1 \\
\int T_1\sin T_1\,dT_1 + \int (1+T_2)\,dT_2 &= c_1 && \frac{1}{\csc T_1} = \sin T_1
\end{aligned}$$

$$u = T_1 \qquad dV = \sin T_1\,dT_1 \qquad \text{integration by parts}$$
$$du = dT_1 \qquad v = -\cos T_1$$

$$\begin{aligned}
-T_1\cos T_1 + \int \cos T_1\,dT_1 + T_2 + \frac{1}{2}T_2^2 &= c_1 \\
-T_1\cos T_1 + \sin T_1 + T_2 + \frac{1}{2}T_2^2 &= c_1 \\
2\sin T_1 - 2T_1\cos T_1 + 2T_2 + T_2^2 &= c
\end{aligned}$$

25.
$$\begin{aligned}
(y^2-1)\cos x\,dx + 2y\sin x\,dy &= 0 \\
\frac{\cos x}{\sin x}\,dx + \frac{2y}{y^2-1}\,dy &= 0 && \text{separating variables} \\
\ln|\sin x| + \ln|y^2-1| &= \ln c_1,\ c_1>0 && \text{integrating} \\
\ln|(\sin x)(y^2-1)| &= \ln c_1 && \ln A + \ln B = \ln AB \\
|(y^2-1)\sin x| &= c_1,\ c_1>0 \\
(y^2-1)\sin x &= \pm c_1 \\
(y^2-1)\sin x &= c,\ c\neq 0
\end{aligned}$$

27.
$$\begin{aligned}
xe^y\,dx + e^{-x}\,dy &= 0 && \text{multiply by } e^{-y}e^x \\
\int xe^x\,dx + \int e^{-y}\,dy &= c
\end{aligned}$$

$$\begin{aligned}
u &= x & dv &= e^x\,dx \\
du &= dx & v &= e^x
\end{aligned} \qquad \text{integration by parts}$$

$$\begin{aligned}
xe^x - \int e^x\,dx - e^{-y} &= c \\
xe^x - e^x - e^{-y} &= c
\end{aligned}$$

29.
$$\begin{aligned}
(e^x\tan y + \tan y)\frac{dy}{dx} + e^x &= 0 \\
(e^x+1)\tan y\,dy + e^x\,dx &= 0 \\
\int \tan y\,dy + \int \frac{e^x}{e^x+1}\,dx &= \ln c_1 && u = e^x+1;\ du = e^x\,dx. \\
\ln|\sec y| + \ln(e^x+1) &= \ln c_1,\ c_1>0 && \text{integrating} \\
\ln|(e^x+1)\sec y| &= \ln c_1 && \ln A + \ln B = \ln AB \\
|(e^x+1)\sec y| &= c_1,\ c_1>0 \\
(e^x+1)\sec y &= \pm c_1 \\
(e^x+1)\sec y &= c,\ c\neq 0
\end{aligned}$$

31.
$$\begin{aligned}
y\frac{dy}{dx} + 2x\sec y &= 0 \\
y\,dy + 2x\sec y\,dx &= 0 \\
\frac{y\,dy}{\sec y} + 2x\,dx &= 0 && \text{dividing by } \sec y \\
\int y\cos y\,dy + \int 2x\,dx &= c && \frac{1}{\sec y} = \cos y
\end{aligned}$$

$$\begin{aligned}
u &= y & dv &= \cos y\,dy \\
du &= dy & v &= \sin y
\end{aligned} \qquad \text{integration by parts}$$

$$\begin{aligned}
y\sin y - \int \sin y\,dy + x^2 &= c \\
y\sin y + \cos y + x^2 &= c
\end{aligned}$$

33.
$$\begin{aligned}
x\,dy - y\,dx &= 0,\ y=2 \text{ when } x=1 \\
\frac{dy}{y} - \frac{dx}{x} &= 0 \\
\ln y - \ln x &= \ln c && \text{integrating} \\
\ln\left(\frac{y}{x}\right) &= \ln c \\
\frac{y}{x} &= c
\end{aligned}$$

Thus $y = cx$. Substituting the given values, we get $2 = c\cdot 1$ or $y = 2x$.

35.
$$\begin{aligned}
(y+2)\,dx + (x-3)\,dy &= 0 \\
\frac{dx}{x-3} + \frac{dy}{y+2} &= 0 \\
\ln|x-3| + \ln|y+2| &= \ln c_1 && \text{form: } \int \frac{du}{u} \\
\ln|(x-3)(y+2)| &= \ln c_1 && \ln A + \ln B = \ln AB \\
|(x-3)(y+2)| &= c_1 \\
(x-3)(y+2) &= \pm c_1 = c \\
(2-3)(5+2) &= c \text{ or } c = -7 && x = 2,\ y = 5 \\
(y+2)(x-3) &= -7
\end{aligned}$$

37.
$$\begin{aligned}
dx + x\tan y\,dy &= 0,\ y = 0 \text{ when } x = 1 \\
\frac{dx}{x} + \tan y\,dy &= 0 \\
\ln x + \ln\sec y &= \ln c && \text{integrating} \\
\ln x\sec y &= \ln c && \ln A + \ln B = \ln AB \\
x\sec y &= c
\end{aligned}$$

If $x = 1$ and $y = 0$, we get $1 \cdot \sec 0 = 1 = c$, so that

$$\begin{aligned}
x\sec y &= 1 \\
x &= \frac{1}{\sec y} \\
x &= \cos y
\end{aligned}$$

11.3 First-Order Linear Differential Equations

1. Step 1. $\frac{dy}{dx} + 1y = 1$ — already in standard form

Step 2. $I.F. = e^{\int 1\,dx} = e^x$

Step 3. $e^x\left(\frac{dy}{dx} + y\right) = e^x \cdot 1$ — multiplying by e^x

Step 4. $\frac{d}{dx}(ye^x) = e^x$ — by (11.9)

Step 5.
$$\begin{aligned}
ye^x &= \int e^x\,dx = e^x + c \\
y &= 1 + ce^{-x}
\end{aligned}$$

3. $2\frac{dy}{dx} + 3y = 4$

Step 1. $\frac{dy}{dx} + \frac{3}{2}y = 2$

Step 2. $I.F. = e^{(3/2)x}$

Step 3. $e^{(3/2)x}\left(\frac{dy}{dx} + \frac{3}{2}y\right) = 2e^{(3/2)x}$

Step 4. $\frac{d}{dx}\left(ye^{(3/2)x}\right) = 2e^{(3/2)x}$

Step 5.
$$\begin{aligned}
ye^{(3/2)x} &= 2\int e^{(3/2)x}\,dx && u = \frac{3}{2}x;\ du = \frac{3}{2}\,dx \\
&= 2\left(\frac{2}{3}\right)\int e^{(3/2)x}\left(\frac{3}{2}\,dx\right) \\
&= \frac{4}{3}e^{(3/2)x} + c \\
y &= \frac{4}{3} + ce^{-3x/2}
\end{aligned}$$

5. Step 1. $\dfrac{dy}{dx} - 2y = e^{3x}$ already in standard form

Step 2. $I.F. = e^{-2x}$

Step 3. $e^{-2x}\left(\dfrac{dy}{dx} - 2y\right) = e^{x}$

Step 4. $\dfrac{d}{dx}\left(ye^{-2x}\right) = e^{x}$

Step 5.
$$\begin{aligned} e^{-2x}y &= e^{x} + c \\ y &= e^{3x} + ce^{2x} \end{aligned}$$

7. $2\dfrac{dy}{dx} - 8xy = e^{2x^2}$

Step 1. $\dfrac{dy}{dx} - 4xy = \dfrac{1}{2}e^{2x^2}$

Step 2. $I.F. = e^{\int(-4x)\,dx} = e^{-2x^2}$

Step 3. $e^{-2x^2}\left(\dfrac{dy}{dx} - 4xy\right) = e^{-2x^2}\cdot\dfrac{1}{2}e^{2x^2} = \dfrac{1}{2}$

Step 4. $\dfrac{d}{dx}(ye^{-2x^2}) = \dfrac{1}{2}$

Step 5.
$$\begin{aligned} ye^{-2x^2} &= \frac{1}{2}x + c \\ y &= \left(\frac{1}{2}x + c\right)e^{2x^2} \end{aligned}$$

9. $\dfrac{1}{2}\dfrac{dy}{dx} + y\cos x = \cos x$

Step 1. $\dfrac{dy}{dx} + 2y\cos x = 2\cos x$

Step 2. $I.F. = e^{\int 2\cos x\,dx} = e^{2\sin x}$

Step 3. $e^{2\sin x}\left(\dfrac{dy}{dx} + 2y\cos x\right) = e^{2\sin x}\cdot 2\cos x$ multiplying by $I.F.$

Step 4. The left becomes the derivative of the product of

y and $I.F.$

$\dfrac{d}{dx}\left(ye^{2\sin x}\right) = e^{2\sin x}\cdot 2\cos x$

Step 5. $ye^{2\sin x} = \displaystyle\int e^{2\sin x}\cdot 2\cos x\,dx = e^{2\sin x} + c$ $u = 2\sin x;\ du = 2\cos x\,dx.$

$y = 1 + ce^{-2\sin x}$ multiplying by $e^{-2\sin x}$

11. $x\,dy + (y - x)\,dx = 0$

Step 1. $$x\frac{dy}{dx} + y - x = 0$$
$$\frac{dy}{dx} + \left(\frac{1}{x}\right)y = 1$$

Step 2. $I.F. = e^{\int (1/x)\,dx} = e^{\ln x} = x$ by(11.12)

Step 3. $x\left(\frac{dy}{dx} + \frac{y}{x}\right) = x$ multiplying by $I.F.$

Step 4. $\frac{d}{dx}(xy) = x$ derivative of the product of y and $I.F.$

Step 5. $xy = \frac{1}{2}x^2 + c$ integrating
$$y = \frac{1}{2}x + \frac{c}{x}$$

13. $\frac{dy}{dx} = \frac{2e^x}{x} - \frac{y}{x}$

Step 1. $\frac{dy}{dx} + \frac{1}{x}y = \frac{2e^x}{x}$

Step 2. $I.F. = e^{\int \frac{1}{x}\,dx} = e^{\ln x} = x$

Step 3. $x\left(\frac{dy}{dx} + \frac{1}{x}y\right) = 2e^x$

Step 4. $\frac{d}{dx}(xy) = 2e^x$

Step 5. $xy = 2e^x + c$
$$y = \frac{2e^x}{x} + \frac{c}{x}$$

15. $\frac{dy}{dx} - \frac{2y}{x} - x^2\sec^2 x = 0$

Step 1. $\frac{dy}{dx} - \frac{2}{x}y = x^2\sec^2 x$

Step 2. $$I.F. = e^{\int (-2/x)\,dx} = e^{-2\ln x}$$
$$= e^{\ln x^{-2}} = x^{-2} = \frac{1}{x^2}$$

Step 3. $\frac{1}{x^2}\left(\frac{dy}{dx} - \frac{2}{x}y\right) = \frac{1}{x^2}(x^2\sec^2 x)$ multiplying by $\frac{1}{x^2}$

Step 4. $\frac{d}{dx}\left(\frac{1}{x^2}y\right) = \sec^2 x$ derivative of the product of y and $I.F.$

Step 5. $\frac{1}{x^2}y = \tan x + c$ integrating
$$y = x^2\tan x + cx^2$$

17. $xy' = 3y + x^5 \sin x$

Step 1. $y' - \dfrac{3}{x}y = x^4 \sin x$

Step 2. $I.F. = e^{\int(-3/x)\,dx} = e^{-3\ln x} = e^{\ln x^{-3}} = x^{-3} = \dfrac{1}{x^3}$

Step 3. $\dfrac{1}{x^3}\left(y' - \dfrac{3}{x}y\right) = \dfrac{1}{x^3} \cdot x^4 \sin x$ multiplying by $I.F.$

Step 4. $\dfrac{d}{dx}\left(y \cdot \dfrac{1}{x^3}\right) = x \sin x$ $\dfrac{d}{dx}(y \cdot I.F.)$

Step 5. $y\dfrac{1}{x^3} = \displaystyle\int x \sin x\,dx$

Integration by parts: $u = x \quad dv = \sin x\,dx$

$du = dx \quad v = -\cos x$

$$\begin{aligned} \frac{y}{x^3} &= -x\cos x + \int \cos x\,dx \\ &= -x\cos x + \sin x + c \\ y &= x^3 \sin x - x^4 \cos x + cx^3 \end{aligned}$$

19. $xy' - 2y = x^3 e^x$

Step 1. $\dfrac{dy}{dx} - \dfrac{2}{x}y = x^2 e^x$

Step 2. $I.F. = e^{\int(-2/x)\,dx} = e^{-2\ln x} = e^{\ln x^{-2}} = x^{-2} = \dfrac{1}{x^2}$

Step 3. $\dfrac{1}{x^2}\left(\dfrac{dy}{dx} - \dfrac{2}{x}y\right) = \dfrac{1}{x^2}(x^2 e^x)$ multiplying by $I.F.$

Step 4. $\dfrac{d}{dx}\left(\dfrac{1}{x^2}y\right) = e^x$

Step 5. $\dfrac{1}{x^2}y = e^x + c$ integrating

$$y = x^2 e^x + cx^2$$

21. $(y-1)\sin x\,dx + dy = 0$

Step 1. $\dfrac{dy}{dx} + y\sin x = \sin x$

Step 2. $I.F. = e^{\int \sin x\,dx} = e^{-\cos x}$

Step 3. $e^{-\cos x}\left(\dfrac{dy}{dx} + y\sin x\right) = e^{-\cos x}\sin x$ multiplying by $I.F.$

Step 4. $\dfrac{d}{dx}(ye^{-\cos x}) = e^{-\cos x}\sin x$ $\dfrac{d}{dx}(y \cdot I.F.)$

Step 5. $ye^{-\cos x} = \displaystyle\int e^{-\cos x}\sin x\,dx = e^{-\cos x} + c$ $u = -\cos x;\ du = \sin x\,dx.$

$y = 1 + ce^{\cos x}$

23. Step 1. $y' - y\tan x = x$ already in standard form

Step 2. $I.F. = \int e^{\int(-\tan x)\,dx} = e^{-\ln\sec x} = e^{\ln(\sec x)^{-1}} = (\sec x)^{-1} = \cos x$

Step 3. $\cos x\,(y' - y\tan x) = x\cos x$

Step 4. $\frac{d}{dx}(y\cos x) = x\cos x$

Step 5. $y\cos x = \int x\cos x\,dx$

Integrating by parts: $u = x \quad dv = \cos x\,dx$

$du = dx \quad v = \sin x$

$$\begin{aligned} y\cos x &= x\sin x - \int \sin x\,dx \\ y\cos x &= x\sin x + \cos x + c \end{aligned}$$

25. $y' - y\tan x - \cos x = 0$

Step 1. $\frac{dy}{dx} - (\tan x)y = \cos x$

Step 2. $I.F. = e^{-\int\tan x\,dx} = e^{\ln\cos x} = \cos x$

Step 3. $\cos x\left(\frac{dy}{dx} - y\tan x\right) = \cos^2 x$ multiplying by $I.F.$

Step 4. $\frac{d}{dx}(y\cos x) = \frac{1}{2}(1+\cos 2x)$ half-angle formula

Step 5.
$$\begin{aligned} y\cos x &= \frac{1}{2}\int(1+\cos 2x)\,dx \qquad u = 2x;\ du = 2\,dx. \\ y\cos x &= \frac{1}{2}x + \frac{1}{4}\sin 2x + c \\ 4y\cos x &= 2x + \sin 2x + c \end{aligned}$$

27. $y' - \frac{1}{x}y = x^2\sin x^2$ already in standard form (Step 1).

Step 2. $I.F. = e^{\int(-1/x)\,dx} = e^{-\ln x} = e^{\ln x^{-1}} = x^{-1} = \frac{1}{x}$

Step 3. $\frac{1}{x}\left(\frac{dy}{dx} - \frac{1}{x}y\right) = x\sin x^2$ multiplying by $\frac{1}{x}$

Step 4. $\frac{d}{dx}\left(\frac{y}{x}\right) = x\sin x^2$

Step 5.
$$\begin{aligned} \frac{y}{x} &= \int x\sin x^2\,dx \\ &= \frac{1}{2}\int \sin x^2(2x)\,dx \qquad u = x^2;\ du = 2x\,dx. \\ \frac{y}{x} &= -\frac{1}{2}\cos x^2 + c \\ y &= -\frac{1}{2}x\cos x^2 + cx \end{aligned}$$

29. $y' + y\tan t = \sec t$

Step 1. $y' + y\tan t = \sec t$ already in standard form

Step 2. $I.F. = e^{\int \tan t\,dt} = e^{\ln\sec t} = \sec t$

Step 3. $\sec t(y' + y\tan t) = \sec^2 t$

Step 4. $\dfrac{d}{dt}(y\sec t) = \sec^2 t$

Step 5. $y\sec t = \tan t + c$

31. $t\dfrac{dr}{dt} + r = t\ln t$

Step 1. $\dfrac{dr}{dt} + \dfrac{1}{t}r = \ln t$ dividing by t

Step 2. $I.F. = e^{\int(1/t)\,dt} = e^{\ln t} = t$

Step 3. $t\left(\dfrac{dr}{dt} + \dfrac{1}{t}r\right) = t\ln t$ multiplying by $I.F.$

Step 4. $\dfrac{d}{dt}(tr) = t\ln t$

Step 5. Integrating by parts: $u = \ln t \quad dv = t\,dt$

$$du = \frac{1}{t}\,dt \quad v = \frac{1}{2}t^2$$

$$\begin{aligned} tr &= \frac{1}{2}t^2\ln t - \int \frac{1}{t}\cdot\frac{1}{2}t^2\,dt \\ tr &= \frac{1}{2}t^2\ln t - \int \frac{1}{2}t\,dt \\ tr &= \frac{1}{2}t^2\ln t - \frac{1}{4}t^2 + c \\ r &= \frac{1}{2}t\ln t - \frac{1}{4}t + \frac{c}{t} \quad \text{dividing by } t \end{aligned}$$

33. $s\dfrac{dr}{ds} - r = s^3e^{3s}$

Step 1. $\dfrac{dr}{ds} - \dfrac{1}{s}r = s^2e^{3s}$

Step 2. $I.F. = e^{\int(-1/s)\,ds} = e^{-\ln s} = e^{\ln s^{-1}} = s^{-1} = \dfrac{1}{s}$

Step 3. $\dfrac{1}{s}\left(\dfrac{dr}{ds} - \dfrac{1}{s}r\right) = se^{3s}$

Step 4. $\dfrac{d}{ds}\left(r\cdot\dfrac{1}{s}\right) = se^{3s}$

Step 5. $r\cdot\dfrac{1}{s} = \displaystyle\int se^{3s}\,ds$

Integration by parts: $u = s \quad dv = e^{3s}\,ds$

$$du = ds \quad v = \frac{1}{3}e^{3s}$$

$$\begin{aligned} r\cdot\frac{1}{s} &= \frac{1}{3}se^{3s} - \frac{1}{3}\int e^{3s}\,ds \\ &= \frac{1}{3}se^{3s} - \frac{1}{9}e^{3s} + c \\ r &= \frac{1}{3}s^2e^{3s} - \frac{1}{9}se^{3s} + cs \end{aligned}$$

35. Step 1. $\dfrac{dy}{dx} + y = 2e^{-2x}$ already in standard form

Step 2. $I.F. = e^{x}$

Step 3. $e^{x}\left(\dfrac{dy}{dx} + y\right) = 2e^{x}e^{-2x} = 2e^{-x}$

Step 4. $\dfrac{d}{dx}(ye^{x}) = 2e^{-x}$

Step 5. $ye^{x} = -2e^{-x} + c$ $u = -x;\ du = -dx$

$y = -2e^{-2x} + ce^{-x}$

Now let $x = 0$ and $y = 2$ and solve for c:

$$2 = -2(1) + c(1) \text{ and } c = 4.$$

The solution is $y = -2e^{-2x} + 4e^{-x}$.

37. $dy + x^2y\,dx = 2x^2\,dx$

Step 1. $\dfrac{dy}{dx} + x^2y = 2x^2$

Step 2. $I.F. = e^{\int x^2\,dx} = e^{(1/3)x^3}$

Step 3. $e^{(1/3)x^3}\left(\dfrac{dy}{dx} + x^2y\right) = e^{(1/3)x^3}\cdot 2x^2$

Step 4. $\dfrac{d}{dx}\left(ye^{(1/3)x^3}\right) = e^{(1/3)x^3}\cdot 2x^2$

Step 5. $ye^{(1/3)x^3} = 2\displaystyle\int e^{(1/3)x^3}x^2\,dx$ $u = \dfrac{1}{3}x^3;\ du = x^2\,dx.$

$ye^{(1/3)x^3} = 2e^{(1/3)x^3} + c$

$y = 2 + ce^{(-1/3)x^3}$

Now let $y = 3$ and $x = 0$:

$3 = 2 + c$ and $c = 1$; so $y = 2 + e^{(-1/3)x^3}$.

39.

$$\begin{aligned}
L\frac{di}{dt} + Ri &= E \\
\frac{di}{dt} + \frac{R}{L}i &= \frac{E}{L} && \text{dividing by } L \\
e^{(R/L)t}\left(\frac{di}{dt} + \frac{R}{L}i\right) &= \frac{E}{L}e^{(R/L)t} && I.F. = e^{\int (R/L)\,dt} \\
\frac{d}{dt}\left(ie^{(R/L)t}\right) &= \frac{E}{L}e^{(R/L)t} \\
ie^{Rt/L} &= \frac{E}{L}\cdot\frac{L}{R}e^{(R/L)t} + c && u = \frac{R}{L}t;\ du = \frac{R}{L}\,dt. \\
i &= \frac{E}{R} + ce^{-Rt/L}
\end{aligned}$$

From the condition $i = 0$ when $t = 0$, $c = -E/R$. So $i = \dfrac{E}{R}\left(1 - e^{-Rt/L}\right)$.

11.4 Applications of First-Order Differential Equations

1.
$$\begin{aligned}
L\frac{di}{dt} + Ri &= \mathrm{e}(t) \\
0.2\frac{di}{dt} + 5i &= 5 \\
\frac{di}{dt} + \frac{5}{0.2}i &= \frac{5}{0.2} \\
\frac{di}{dt} + 25i &= 25 \qquad\qquad I.F. = \mathrm{e}^{25t} \\
\mathrm{e}^{25t}\left(\frac{di}{dt} + 25i\right) &= 25\mathrm{e}^{25t} \\
\frac{d}{dt}(i\mathrm{e}^{25t}) &= 25\mathrm{e}^{25t} \\
i\mathrm{e}^{25t} &= \int 25\mathrm{e}^{25t}\,dt \qquad u = 25t;\ du = 25\,dt. \\
i\mathrm{e}^{25t} &= \mathrm{e}^{25t} + c \\
i &= 1 + c\mathrm{e}^{-25t}
\end{aligned}$$

Initial condition: if $t = 0,\ i = 0$. Thus $0 = 1 + c$ or $c = -1$, and $i = 1 - \mathrm{e}^{-25t}$.

3. (a) $\frac{dN}{dt} = kN$ by the law of exponential decay.

Step 1. $\frac{dN}{dt} - kN = 0$

Step 2. $I.F. = \mathrm{e}^{-kt}$

Step 3. $\mathrm{e}^{-kt}\left(\frac{dN}{dt} - kN\right) = 0$

Step 4. $\frac{d}{dt}\left(N\mathrm{e}^{-kt}\right) = 0$

Step 5.
$$\begin{aligned}
N\mathrm{e}^{-kt} &= c \\
N &= c\mathrm{e}^{kt}
\end{aligned}$$

Given: when $t = 0,\ N = 100\,\mathrm{g}$; $100 = c\mathrm{e}^0 = c$. So $N = 100\mathrm{e}^{kt}$. To evaluate k, we use the second pair of values: when $t = 10,\ N = 80\,\mathrm{g}$: $80 = 100\mathrm{e}^{10k}$ or $\mathrm{e}^{10k} = 0.8$. Taking natural logarithms,

$$\begin{aligned}
\ln \mathrm{e}^{10k} &= \ln(0.8) \\
10k \ln \mathrm{e} &= \ln(0.8) \qquad\qquad \ln \mathrm{e} = 1
\end{aligned}$$

and

$$k = \frac{\ln(0.8)}{10} = -0.0223.$$

The solution is $N = 100\mathrm{e}^{-0.0223t}$.

(b) Starting with $N = c\mathrm{e}^{kt}$, if $N = N_0$ when $t = 0$, we get $N_0 = c\mathrm{e}^0 = c$ and $N = N_0\mathrm{e}^{kt}$. For the second pair of values we now have: when $t = 10,\ N = 0.80\,N_0$ (80% of the original amount). So $0.80N_0 = N_0\mathrm{e}^{10k}$, and $\ln(0.8) = \ln \mathrm{e}^{10k}$, which leads to $k = -0.0223$ from part (a). Thus $N = N_0\mathrm{e}^{-0.0223t}$.

5. By Example 1, $N = N_0 e^{kt}$. Since the half-life is 5.27 years, we have $N = \frac{N_0}{2}$ when $t = 5.27$:

$$\begin{aligned} \frac{1}{2}N_0 &= N_0 e^{5.27k} \\ \frac{1}{2} &= e^{5.27k} \\ \ln\frac{1}{2} &= \ln e^{5.27k} = 5.27k \ln e = 5.27k \text{ (since } \ln e = 1). \end{aligned}$$

So $k = \frac{\ln(1/2)}{5.27} = -0.1315$. Solution: $N = N_0 e^{-0.1315t}$.

If 80% of the initial amount has decayed, then 20% is left. So we need to find t such that $N = 0.20N_0$:

$$\begin{aligned} 0.20N_0 &= N_0 e^{-0.1315t} \\ 0.20 &= e^{-0.1315t} \\ \ln(0.20) &= \ln e^{-0.1315t} = -0.1315t \ln e = -0.1315t \\ t &= \frac{\ln(0.20)}{-0.1315} = 12.2 \text{ years.} \end{aligned}$$

7. $\frac{dN}{dt} = kN$; by Exercise 3, $N = ce^{kt}$. If N_0 is the initial quantity, then $N_0 = ce^0 = c$ and $N = N_0 e^{kt}$.

If 1% of the given quantity has decayed, then 99% is left: $N = 0.99N_0$ when $t = 20$ years. So

$$\begin{aligned} 0.99N_0 &= N_0 e^{20k} \\ 0.99 &= e^{20k} \\ \ln 0.99 &= \ln e^{2k} = 20k \ln e \\ k &= \frac{\ln 0.99}{20} = -0.0005025 \qquad \ln e = 1 \end{aligned}$$

and

$$N = N_0 e^{-0.0005025t}.$$

The half-life is the time t corresponding to $N = \frac{1}{2}N_0$:

$$\begin{aligned} \frac{1}{2}N_0 &= N_0 e^{-0.0005025t} \\ \frac{1}{2} &= e^{-0.0005025t} \\ \ln\frac{1}{2} &= -0.0005025t \ln e \\ t &= \frac{\ln(1/2)}{-0.0005025} = 1379.4 \approx 1380 \text{ years.} \end{aligned}$$

9. By Example 1, $N = N_0 e^{kt}$. We are given that $N = 3N_0$ when $t = 3h$:

$3N_0 = N_0 e^{3k}$ or $3 = e^{3k}$

$\ln 3 = \ln e^{3k} = 3k \ln e = 3k$. So

$$k = \frac{\ln 3}{3} = 0.3662.$$

Solution: $N = N_0 e^{0.3662t}$. Now find t such that $N = 10N_0$:

$$\begin{aligned} 10N_0 &= N_0 e^{0.3662t} \text{or} 10 = e^{0.3662t} \\ \ln 10 &= \ln e^{0.3662t} = 0.3662t \\ t &= \frac{\ln 10}{0.3662} = 6.3 \text{ h} \end{aligned}$$

11. By Example 1, $N = N_0 e^{kt}$. Since the half-life is 5600 years, we have $N = \frac{1}{2}N_0$ when $t = 5600$ years:

$$\begin{aligned} \frac{1}{2}N_0 &= N_0 e^{5600k} \\ \frac{1}{2} &= e^{5600k} \\ \ln\frac{1}{2} &= \ln e^{5600k} = 5600k \text{ (since } \ln e = 1) \\ k &= \frac{\ln(1/2)}{5600} = -0.000124. \end{aligned}$$

Solution: $N = N_0 e^{-0.000124t}$. Now find t such that $N = 0.25N_0$:

$$\begin{aligned} 0.25N_0 &= N_0 e^{-0.000124t} \\ 0.25 &= e^{-0.000124t} && \text{dividing by } N_0 \\ \ln(0.25) &= -0.000124t \\ t &= \frac{\ln(0.25)}{-0.000124} \approx 11,200 \text{ years.} \end{aligned}$$

13. Since $T_m = 0°$, the equation is $\frac{dT}{dt} = -k(T-0)$ or $\frac{dT}{dt} + kT = 0$. So

$$\begin{aligned} T &= ce^{-kt} \\ 25 &= c \cdot 1 && T = 25 \text{ when } t = 0 \\ T &= 25e^{-kt} \\ 22 &= 25e^{-k \cdot 1} && T = 22 \text{ when } t = 1 \\ \ln\frac{22}{25} &= \ln e^{-k} \text{ or } k = 0.1278 \end{aligned}$$

Solution: $T = 25e^{-0.1278t}\Big|_{t=10} = 7.0°$ C.

15. Since $T_w = -10°$ C, the equation is $\frac{dT}{dt} = -k(T+10)$.

Step 1. $\frac{dT}{dt} + kT = -10k$

Step 2. $I.F. = e^{kt}$

Step 3. $e^{kt}\left(\frac{dT}{dt} + kT\right) = -10ke^{kt}$

Step 4. $\frac{d}{dt}\left(Te^{kt}\right) = -10ke^{kt}$

Step 5.
$$\begin{aligned} Te^{kt} &= -10\int e^{kt}k\,dt && u = kt;\ du = k\,dt \\ &= -10e^{kt} + c \\ T &= -10 + ce^{-kt} \end{aligned}$$

Initial condition: when $t = 0$, $T = 20$.

$$\begin{aligned} 20 &= -10 + c \text{ or } c = 30 \\ T &= -10 + 30e^{-kt} \end{aligned}$$

Second condition: when $t = 1$, $T = 16$.

$$\begin{aligned} 16 &= -10 + 30e^{-k} \text{ or } k = 0.1431 \\ T &= -10 + 30e^{-0.1431t} \end{aligned}$$

Finally, we find t such that $T = 0$:

$$\begin{aligned} 0 &= -10 + 30e^{-0.1431t} \\ \ln\frac{10}{30} &= -0.1431t \\ t &= \frac{\ln(10/30)}{-0.1431} = 7.7 \text{ min.} \end{aligned}$$

17. For the first part of the problem, $T_m = 10°\text{ C} : \dfrac{dT}{dt} = -k(T-10)$.
Solution: $T = 10 + ce^{-kt}$.
Initial condition: when $t = 0$, $T = 21$.

$$\begin{aligned} 21 &= 10 + c \text{ or } c = 11 \\ T &= 10 + 11e^{-kt} \end{aligned}$$

Second condition: when $t = 1$, $T = 15$.

$$\begin{aligned} 15 &= 10 + 11e^{-k} \text{ and } k = 0.78846 \\ T &= 10 + 11e^{-0.78846t} \end{aligned}$$

For the second part of the problem, $T_m = 21°$ C with $k = 0.78846$ already known:

$$\begin{aligned} \frac{dT}{dt} &= -k(T-21) \\ T &= 21 + ce^{-kt} \end{aligned}$$

If 5:01 P.M. is taken as $t = 0$ (when $T = 15$), then 5:02 P.M. corresponds to $t = 1$:

$$\begin{aligned} 15 &= 21 + c \text{ or } c = -6 \\ T &= 21 - 6e^{-0.78846t}\Big|_{t=1} = 18.3°\, C. \end{aligned}$$

19.
$$\begin{aligned} m\frac{dv}{dt} &= mg - kv \\ 2\frac{dv}{dt} &= 64 - 0.3v \\ \frac{dv}{dt} + 0.15v &= 32 \\ \frac{d}{dt}\left(ve^{0.15t}\right) &= 32e^{0.15t} \qquad I.F. = e^{0.15t} \\ ve^{0.15t} &= 32\left(\frac{1}{0.15}\right)e^{0.15t} + c \\ v &= 213 + ce^{-0.15t} \end{aligned}$$

From $v = 0$ when $t = 0$, we get $c = -213$:

$$v = 213\left(1 - e^{-0.15t}\right);\ \text{as } t \to \infty, v \to 213\,\text{ft/s}.$$

21. From the equation $m\dfrac{dv}{dt} = mg - kv$, we get

$$\begin{aligned} 5\frac{dv}{dt} &= 50 - 1v \\ \frac{dv}{dt} + \frac{1}{5}v &= 10 \\ \frac{d}{dt}\left(ve^{(1/5)t}\right) &= 10e^{(1/5)t} \\ ve^{t/5} &= 10 \cdot 5e^{t/5} + c \\ v &= 50 + ce^{-t/5} \\ 1 &= 50 + c \qquad v = 1 \text{ when } t = 0 \end{aligned}$$

Since $c = -49$, $v = 50 - 49e^{-t/5}$; as $t \to \infty$, $v \to 50\,\text{m/s}$.

23.
$$\begin{aligned} \frac{dx}{dt} &= kx \\ \frac{dx}{dt} - kx &= 0 \qquad I.F. = e^{-kt} \\ \frac{d}{dt}(xe^{-kt}) &= 0 \\ xe^{-kt} &= c \text{ and } x = ce^{kt}. \end{aligned}$$

If $x = x_0$ when $t = 0$, then $c = x_0$. The solution is
$x = x_0 e^{kt}$.
From the given condition, $x = \frac{3}{4}x_0$ when $t = 10$, we have

$$\begin{aligned} \frac{3}{4}x_0 &= x_0 e^{10k} \\ \ln\left(\frac{3}{4}\right) &= 10k \text{ or } k = \frac{\ln(3/4)}{10} = -0.02877. \end{aligned}$$

Thus, $x = x_0 e^{-0.02877t}$. If $x = \frac{1}{10}x_0$, then   one-tenth unconverted

$$\begin{aligned} \frac{1}{10}x_0 &= x_0 e^{-0.02877t} \\ \ln\left(\frac{1}{10}\right) &= -0.02877t \\ t &= \frac{\ln(1/10)}{-0.02877} = 80\,\text{s}. \end{aligned}$$

25. $x^2 + y^2 = c^2$; differentiating implicitly,

$$\begin{aligned} 2x + 2y\frac{dy}{dx} &= 0 \\ \frac{dy}{dx} &= -\frac{x}{y}. \end{aligned}$$

The orthogonal trajectories must therefore satisfy the following condition:

$$\begin{aligned} \frac{dy}{dx} &= \frac{y}{x} && \text{negative reciprocal} \\ \frac{dy}{y} &= \frac{dx}{x} \\ \ln|y| &= \ln|x| + \ln k_1 \\ \ln\left|\frac{y}{x}\right| &= \ln k_1 \text{ or } \left|\frac{y}{x}\right| = k_1. \end{aligned}$$

It follows that $\frac{y}{x} = \pm k_1$ and $y = kx$.

27. From $x^2 = ay^3$, we get $2x = 3ay^2\frac{dy}{dx}$. From the give equation, $a = \frac{x^2}{y^3}$. Substituting in the second equation, $2x = 3\frac{x^2}{y^3}y^2\frac{dy}{dx}$ and $\frac{dy}{dx} = \frac{2y}{3x}$. So the orthogonal trajectories satisfy the equation

$$\begin{aligned} \frac{dy}{dx} &= -\frac{3x}{2y} && \text{negative reciprocals} \\ 2y\,dy &= -3x\,dx && \text{separating variables} \\ y^2 &= -3\frac{x^2}{2} + k \text{ or } 3x^2 + 2y^2 = k. \end{aligned}$$

29. $y^2 = 4px$. Differentiating implicitly, we get $2y\frac{dy}{dx} = 4p$.
From the first equation, $4p = \frac{y^2}{x}$. Substituting in the second equation, we get
$2y\frac{dy}{dx} = \frac{y^2}{x}$ or $\frac{dy}{dx} = \frac{y}{2x}$.
So the orthogonal trajectories satisfy the condition

$$\begin{aligned} \frac{dy}{dx} &= -\frac{2x}{y} && \text{negative reciprocal} \\ y\,dy &= -2x\,dx \\ \frac{y^2}{2} &= -x^2 + k_1. \end{aligned}$$

Thus $2x^2 + y^2 = k$.

31. $y = ce^{-2x}$; differentiating, $\dfrac{dy}{dx} = c(-2e^{-2x})$.
To eliminate c_1 observe that from the first equation, $c = ye^{2x}$. Substituting in the second equation, we get $\dfrac{dy}{dx} = ye^{2x}(-2e^{-2x}) = -2y$.
For the orthogonal trajectories, dy/dx is equal to the negative reciprocal:
$\dfrac{dy}{dx} = \dfrac{1}{2y}$ or $2y\,dy = dx$ and $y^2 = x + k$.

33.
$$\begin{aligned} xy &= c && \text{given family} \\ y &= \frac{c}{x} \\ \frac{dy}{dx} &= -\frac{c}{x^2} \end{aligned}$$

Substituting $c = xy$, we get $\dfrac{dy}{dx} = -\dfrac{xy}{x^2} = -\dfrac{y}{x}$. The orthogonal family therefore satisfies the condition

$$\begin{aligned} \frac{dy}{dx} &= \frac{x}{y} && \text{negative reciprocal} \\ y\,dy &= x\,dx \\ \frac{1}{2}y^2 &= \frac{1}{2}x^2 + k_1 \\ y^2 &= x^2 + 2k_1 \\ y^2 - x^2 &= k. \end{aligned}$$

35. Since $M = 10000$, the equation is

$$\begin{aligned} \frac{dN}{dt} &= kN(10000 - N) \\ \frac{dN}{N(10000-N)} &= k\,dt. \end{aligned}$$

To integrate, we need to split the left side into partial fractions:

$$\begin{aligned} \frac{1}{N(10000-N)} &= \frac{A}{N} + \frac{B}{10000-N} \\ &= \frac{A(10000-N) + BN}{N(10000-N)} \\ A(10000-N) + BN &= 1. \end{aligned}$$

$\underline{N = 0}$: $A(10000) + 0 = 1$ and $A = \dfrac{1}{10000}$.

$\underline{N = 10000}$: $0 + B(10000) = 1$ and $B = \dfrac{1}{10000}$.

Integrating,

$$\begin{aligned} \frac{1}{10000}\int\left(\frac{1}{N} + \frac{1}{10000-N}\right) dN &= \int k\,dt \\ \frac{1}{10000}[\ln N - \ln(10000-N)] &= kt + c_1 \\ \ln\frac{N}{10000-N} &= 10000kt + \ln c. \end{aligned}$$

(Here $10000c_1$ was replaced by $\ln c$.)
Initial condition: when $t = 0, N = 1000$.
$\ln\dfrac{1000}{10000-1000} = 0 + \ln c$ or $\ln c = \ln\dfrac{1}{9}$.
This yields $\ln\dfrac{N}{10000-N} = 10000kt + \ln\dfrac{1}{9}$, or $\ln\dfrac{9N}{10000-N} = 10000kt$.
From the second condition (when $t = 1$, $N = 2000$), we get
$\ln\dfrac{9(2000)}{10000-2000} = 10000k$ or $k = \dfrac{1}{10000}\ln\dfrac{9}{4}$.
The solution is therefore
$\ln\dfrac{9N}{10000-N} = 10000\dfrac{1}{10000}\left(\ln\dfrac{9}{4}\right)t$ or $\ln\dfrac{9N}{10000-N} = \left(\ln\dfrac{9}{4}\right)t$.

This solution can be put into an explicit form by solving for N:

$$\begin{aligned}
e^{[\ln(9/4)]t} &= \frac{9N}{10000-N} \\
10000e^{[\ln(9/4)]t} - Ne^{[\ln(9/4)]t} &= 9N \\
N &= \frac{10000e^{[\ln(9/4)]t}}{9+e^{[\ln(9/4)]t}} \cdot \frac{e^{-[\ln(9/4)]t}}{e^{-[\ln(9/4)]t}} \\
N &= \frac{10000}{1+9e^{-[\ln(9/4)]t}}.
\end{aligned}$$

37. $\frac{dx}{dt}$ = rate of gain−rate of loss.

rate of gain $= \left(0.2\frac{\text{lb}}{\text{gal}}\right)\left(4\frac{\text{gal}}{\text{min}}\right) = 0.8\frac{\text{lb}}{\text{min}}$

rate of loss $= \left(\frac{x\,\text{lb}}{200\,\text{gal}}\right)\left(4\frac{\text{gal}}{\text{min}}\right) = \frac{x}{50}\frac{\text{lb}}{\text{min}}$

Thus $\frac{dx}{dt} = 0.8 - \frac{x}{50} = \frac{40-x}{50}$. Separating variables

$$\begin{aligned}
\frac{dx}{40-x} &= \frac{dt}{50} \\
\frac{-dx}{40-x} &= -\frac{dt}{50} \qquad u = 40-x;\ du = -dx. \\
\ln(40-x) &= -\frac{1}{50}t + \ln c \\
\ln\frac{40-x}{c} &= -\frac{1}{50}t \\
e^{-t/50} &= \frac{40-x}{c} \\
ce^{-t/50} &= 40-x \\
x &= 40 - ce^{-t/50}.
\end{aligned}$$

When $t=0,\ x=20:\ 20 = 40-c$, or $c=20$: $x(t) = 40-20e^{-t/50}$.

When $t = 30\,\text{min}$, $x = 29\,\text{lb}$.

39. $\frac{dx}{dt}$ = rate of gain − rate of loss.

rate of gain = 0

rate of loss $= \left(\frac{x\,\text{lb}}{200\,\text{gal}}\right)\left(4\frac{\text{gal}}{\text{min}}\right) = \frac{x}{50}\frac{\text{lb}}{\text{min}}$

The equation is

$$\begin{aligned}
\frac{dx}{dt} = -\frac{x}{50} \quad &\text{or} \quad \frac{dx}{x} = -\frac{1}{50}dt \\
\ln x &= -\frac{1}{50}t + \ln c \\
\ln\frac{x}{c} &= -\frac{1}{50}t \\
\frac{x}{c} &= e^{-t/50} \\
x &= ce^{-t/50}.
\end{aligned}$$

When $t=0,\ x=20:\ 20 = ce^0 = c$.

$x(t) = 20e^{-t/50}$ and $x(30) = 11.0\,\text{lb}$.

11.5 Numerical Solutions

1. (a) $dy = (1 - y)\,dx$, initial conditions: $x = 0,\ y = 2,\ dx = 0.05$.

x	y
0	2
0.05	$(1 - y_0)\,dx + y_0 = (1 - 2)(0.05) + 2 = 1.95$
0.10	$(1 - y_1)\,dx + y_1 = (1 - 1.95)(0.05) + 1.95 = 1.9025$
0.15	$(1 - y_2)\,dx + y_2 = (1 - 1.9025)(0.05) + 1.9025 = 1.857375$
etc.	

To obtain the solution using a spreadsheet, enter `0` in cell $A1$ and `2` in cell $B1$. To generate the x-values, enter `=A1+0.05` in cell $A2$. Now copy the contents of this cell, mark the block $A3 - A41$, and paste, or drag the fill handle in the lower right-hand corner of $A2$ to cell $A41$.

For the corresponding y-values, enter the following expression in cell $B2$:

```
=(1-B1)*0.05+B1
```

Now copy the contents of this cell, mark the block $B3 - B41$, and paste, or use the fill handle.

3. (a) $dy = (x - y)\,dx$, initial conditions: $x = 0,\ y = 3,\ dx = 0.01$.

x	y
0	3
0.01	$(x_0 - y_0)\,dx + y_0 = (0 - 3)(0.01) + 3 = 2.97$
0.02	$(x_1 - y_1)\,dx + y_1 = (0.01 - 2.97)(0.01) + 2.97 = 2.9404$
0.03	$(x_2 - y_2)\,dx + y_2 = (0.02 - 2.9404)(0.01) + 2.9404 = 2.911196$

5. $dy = (y^2 - x^2)\,dx$, initial conditions: $x = 0,\ y = 1,\ dx = 0.01$.

x	y
0	1
0.01	$(y_0^2 - x_0^2)\,dx + y_0 = (1^2 - 0^2)(0.01) + 1 = 1.01$
0.02	$(1.01^2 - 0.01^2)(0.01) + 1.01 = 1.0202$
0.03	$(1.0202^2 - 0.02^2)(0.01) + 1.0202 = 1.0306041$
etc.	

To obtain the solution using a spreadsheet, enter `0` in cell $A1$ and `1` in cell $B1$. To generate the x-values, enter `=A1+0.01` in cell $A2$. Now copy the contents of this cell, mark the block $A3 - A51$, and paste, or use the shortcut.

For the corresponding y-values, enter the following expression in cell $B2$:

```
=(B1^2-A1^2)*0.01+B1
```

Now copy the contents of this cell, mark the block $B3 - B51$, and paste, or use the shortcut.

7. (b) $dy = \mathrm{e}^{-xy}\,dx$, initial conditions: $x = 0,\ y = 0,\ dx = 0.01$.

x	y
0	0
0.01	$\mathrm{e}^{-x_0 y_0}\,dx + y_0 = 1(0.01) + 0 = 0.01$
0.02	$\mathrm{e}^{-x_1 y_1}\,dx + y_1 = \mathrm{e}^{-(0.01)(0.01)}(0.01) + 0.01 = 0.019999$
0.03	$\mathrm{e}^{-x_2 y_2}\,dx + y_2 = \mathrm{e}^{-(0.02)(0.019999)}(0.01) + 0.019999 = 0.029995$

9. We start with $(x_0, y_0) = (0, 2)$ and calculate y_1:

$$\begin{aligned}
\frac{dy}{dx} &= 1 - y \\
x_0 &= 0 \\
y_0 &= 2 \\
h &= 0.05 \\
K_1 &= f(x_0, y_0) = 1 - 2 = -1 \\
K_2 &= f\left(x_0 + \frac{1}{2}h, y_0 + \frac{1}{2}hK_1\right) = 1 - \left[2 + \frac{1}{2}(0.05)(-1)\right] = -0.975 \\
K_3 &= f\left(x_0 + \frac{1}{2}h, y_0 + \frac{1}{2}hK_2\right) = 1 - \left[2 + \frac{1}{2}(0.05)(-0.975)\right] = -0.975625 \\
K_4 &= f\left(x_0 + h, y_0 + hK_3\right) = 1 - [2 + 0.05(-0.975625)] = -0.951219 \\
y_1 &= y_0 + \frac{1}{6}(0.05)(K_1 + 2K_2 + 2K_3 + K_4) = 1.951229
\end{aligned}$$

11. To illustrate the procedure, we calculate y_1 and y_2, starting with $(x_0, y_0) = (0, 3)$.

$$\begin{aligned}
\frac{dy}{dx} &= x - y \\
x_0 &= 0 \\
y_0 &= 3 \\
h &= 0.01 \\
K_1 &= 0 - 3 = -3 \\
K_2 &= 0 + \frac{1}{2}(0.01) - \left[3 + \frac{1}{2}(0.01)(-3)\right] = -2.98 \\
K_3 &= 0 + \frac{1}{2}(0.01) - \left[3 + \frac{1}{2}(0.01)(-2.98)\right] = -2.9801 \\
K_4 &= 0 + 0.01 - [3 + 0.01(-2.9801)] = -2.960199 \\
y_1 &= y_0 + \frac{1}{6}(0.01)(K_1 + 2K_2 + 2K_3 + K_4) = 2.970199 \\
x_1 &= 0.01 \\
y_1 &= 2.970199 \\
h &= 0.01 \\
K_1 &= 0.01 - 2.970199 = -2.960199 \\
K_2 &= 0.01 + \frac{1}{2}(0.01) - \left[2.970199 + \frac{1}{2}(0.01)(-2.960199)\right] = -2.940398 \\
K_3 &= 0.01 + \frac{1}{2}(0.01) - \left[2.970199 + \frac{1}{2}(0.01)(-2.940398)\right] = -2.940497 \\
K_4 &= 0.01 + 0.01 - [2.970199 + 0.01(-2.940497)] = -2.920794 \\
y_2 &= y_1 + \frac{1}{6}(0.01)(K_1 + 2K_2 + 2K_3 + K_4) = 2.940794
\end{aligned}$$

13. To illustrate the procedure, we calculate y_1 and y_2, starting with $(x_0, y_0) = (0, 1)$.

$$\begin{aligned}
\frac{dy}{dx} &= y^2 - x^2,\ dx = 0.05 \\
x_0 &= 0 \\
y_0 &= 1 \\
h &= 0.05 \\
K_1 &= 1^2 - 0^2 = 1 \\
K_2 &= \left[1 + \frac{1}{2}(0.05)(1)\right]^2 - \left[0 + \frac{1}{2}(0.05)\right]^2 = 1.05 \\
K_3 &= \left[1 + \frac{1}{2}(0.05)(1.05)\right]^2 - \left[0 + \frac{1}{2}(0.05)\right]^2 = 1.05256406 \\
K_4 &= [1 + 0.05(1.05256406)]^2 - (0 + 0.05)^2 = 1.10552613
\end{aligned}$$

$$y_1 = y_0 + \frac{1}{6}(0.05)(K_1 + 2K_2 + 2K_3 + K_4) = 1.05258879$$

$$\begin{aligned}
x_1 &= 0.05 \\
y_1 &= 1.05258879 \\
h &= 0.05 \\
K_1 &= 1.05258879^2 - 0.05^2 = 1.10544316 \\
K_2 &= \left[1.05258879 + \frac{1}{2}(0.05)(1.10544316)\right]^2 - \left[0.05 + \frac{1}{2}(0.05)\right]^2 = 1.16126077 \\
K_3 &= \left[1.05258879 + \frac{1}{2}(0.05)1.16126077)\right]^2 - \left[0.05 + \frac{1}{2}(0.05)\right]^2 = 1.16427749 \\
K_4 &= [1.05258879 + 0.05(1.16427749)]^2 - (0.05 + 0.05)^2 = 1.22388256
\end{aligned}$$

$$y_2 = y_1 + \frac{1}{6}(0.05)(K_1 + 2K_2 + 2K_3 + K_4) = 1.110759$$

Chapter 11 Review

1. $y' = x - y$

Step 1. $\frac{dy}{dx} + y = x$

Step 2. $I.F. = e^x$

Step 3. $e^x\left(\frac{dy}{dx} + y\right) = xe^x$

Step 4. $\frac{d}{dx}(ye^x) = xe^x$

Step 5. $ye^x = \int xe^x\, dx$

(Integration by parts) $\begin{aligned} u &= x & dv &= e^x\,dx \\ du &= dx & v &= e^x \end{aligned}$

$$\begin{aligned}
ye^x &= xe^x - \int e^x\, dx = xe^x - e^x + c \\
y &= x - 1 + ce^{-x}
\end{aligned}$$

3. $y' = x - 2xy$. As a linear equation:

Step 1. $\dfrac{dy}{dx} + 2xy = x$

Step 2. $I.F. = e^{\int 2x\,dx} = e^{x^2}$

Step 3. $e^{x^2}\left(\dfrac{dy}{dx} + 2xy\right) = xe^{x^2}$

Step 4. $\dfrac{d}{dx}\left(ye^{x^2}\right) = xe^{x^2}$

Step 5.

$$\begin{aligned} ye^{x^2} &= \int xe^{x^2}\,dx \\ &= \frac{1}{2}\int e^{x^2}(2x\,dx) \qquad u = x^2;\ du = 2x\,dx. \\ &= \frac{1}{2}e^{x^2} + c \\ y &= \frac{1}{2} + ce^{-x^2} \end{aligned}$$

By separation of variables:

$$\begin{aligned} \frac{dy}{dx} &= x(1-2y) \\ \frac{dy}{1-2y} &= x\,dx \qquad u = 1-2y;\ du = -2y\,dy. \\ \frac{-2\,dy}{1-2y} &= -2x\,dx \qquad \text{multiplying both sides by } -2 \\ \ln|1-2y| &= -x^2 + c \qquad \text{form: } \int \frac{du}{u} \end{aligned}$$

To show that the two solutions agree, we need to solve the latter for y in terms of x. First replace c by $\ln c_1$ and observe that $|1-2y| = |2y-1|$.

$$\begin{aligned} \ln|2y-1| &= -x^2 + \ln c_1 \\ \ln\frac{|2y-1|}{c_1} &= -x^2 \\ \frac{|2y-1|}{c_1} &= e^{-x^2} \\ |2y-1| &= c_1e^{-x^2} \\ 2y-1 &= \pm c_1e^{-x^2} \\ y &= \frac{1}{2} \pm \frac{1}{2}c_1e^{-x^2} \\ y &= \frac{1}{2} + ce^{-x^2} \qquad \text{replacing } \pm\frac{1}{2}c_1 \text{ by } c \end{aligned}$$

5.

$$\begin{aligned} (1+y^2)\,dx + (x^2y+y)\,dy &= 0 \\ (1+y^2)\,dx + y(x^2+1)\,dy &= 0 \\ \frac{dx}{1+x^2} + \frac{y\,dy}{1+y^2} &= 0 \qquad u = 1+y^2;\ du = 2y\,dy. \\ \frac{2\,dx}{1+x^2} + \frac{2y\,dy}{1+y^2} &= 0 \qquad \text{multiplying by 2} \\ \int\frac{2\,dx}{1+x^2} + \int\frac{2y\,dy}{1+y^2} &= c \\ 2\text{Arctan}\,x + \ln(1+y^2) &= c \end{aligned}$$

7.
$$\begin{aligned}
2y\,dx + x\,dy &= 0 \\
\frac{2\,dx}{x} + \frac{dy}{y} &= 0 \\
2\ln|x| + \ln|y| &= \ln c_1 \\
\ln|x^2| + \ln|y| &= \ln c_1 \\
\ln|x^2 y| &= \ln c_1 \\
|x^2 y| &= c_1 \\
x^2 y &= \pm c_1 \text{ and } x^2 y = c
\end{aligned}$$

9. $(x^4 + 2y)\,dx - x\,dy = 0$. This equation is linear:

Step 1. $$-x\,dy + 2y\,dx = -x^4\,dx$$
$$\frac{dy}{dx} - \left(\frac{2}{x}\right)y = x^3$$

Step 2. $I.F. = e^{-\int(2/x)\,dx} = e^{-2\ln x} = e^{\ln x^{-2}} = x^{-2}$

Step 3. $x^{-2}\left[\frac{dy}{dx} - \left(\frac{2}{x}\right)y\right] = x^{-2}x^3$

Step 4. $\frac{d}{dx}(x^{-2}y) = x^{-2}x^3 = x$

Step 5. $$x^{-2}y = \frac{1}{2}x^2 + c$$
$$y = \frac{1}{2}x^4 + cx^2$$

11.
$$\begin{aligned}
x\sin^2 y\,dx - \cot y\,dy &= 0 \\
x\,dx - \frac{\cot y}{\sin^2 y}\,dy &= 0 \\
x\,dx - \frac{\cos y}{\sin y}\frac{1}{\sin^2 y} &= 0 \\
\int x\,dx - \int(\sin y)^{-3}\cos y\,dy &= c_1 \qquad u = \sin y;\; du = \cos y\,dy. \\
\int x\,dx - \int u^{-3}\,du &= c_1 \\
\frac{1}{2}x^2 - \frac{u^{-2}}{-2} &= c_1 \\
x^2 + \frac{1}{u^2} &= 2c_1 \\
x^2 + \frac{1}{\sin^2 y} &= 2c_1 \\
x^2 + \csc^2 y &= c
\end{aligned}$$

13. As a linear equation:
$$\begin{aligned}
\frac{dy}{dx} + (\sec x)y &= 0 \qquad I.F. = e^{\int \sec x\,dx} = e^{\ln(\sec x + \tan x)} = \sec x + \tan x \\
\frac{d}{dx}[y(\sec x + \tan x)] &= 0 \\
y(\sec x + \tan x) &= c
\end{aligned}$$

Separation of variables:
$$\begin{aligned}
dy + y\sec x\,dx &= 0 \\
\frac{dy}{y} + \sec x\,dx &= 0 \\
\ln y + \ln(\sec x + \tan x) &= \ln c \qquad \text{integrating} \\
\ln y(\sec x + \tan x) &= \ln c \\
y(\sec x + \tan x) &= c.
\end{aligned}$$

Since $y = 2$ when $x = \frac{\pi}{4}$, we get

$$\begin{aligned} 2\left[\sec\left(\frac{\pi}{4}\right) + \tan\left(\frac{\pi}{4}\right)\right] &= c \\ 2\left(\sqrt{2}+1\right) &= c. \end{aligned}$$

Hence, $y(\sec x + \tan x) = 2\left(\sqrt{2}+1\right)$.

15. $\frac{dN}{dt} = kN$. By Exercise 9, Section 11.4, $N = N_0 e^{kt}$.
Given: when $t = 2\,\text{h}$, $N = 2N_0$.

$$\begin{aligned} 2N_0 &= N_0 e^{2k} \\ 2 &= e^{2k} \\ \ln 2 &= \ln e^{2k} = 2k \ln e = 2k \cdot 1 = 2k \\ k &= \frac{1}{2}\ln 2 \end{aligned}$$

Solution: $N = N_0 e^{[(\ln 2)/2]t}$.
Now find t such that $N = 3N_0$:

$$\begin{aligned} 3N_0 &= N_0 e^{[(\ln 2)/2]t} \\ 3 &= e^{[(\ln 2)/2]t} \\ \ln 3 &= \left(\frac{1}{2}\ln 2\right) t \end{aligned}$$

and

$$t = \frac{\ln 3}{\frac{1}{2}\ln 2} \approx 3.2\,\text{h}.$$

17. Since $T_m = 65°$ F, the equation is

$$\begin{aligned} \frac{dT}{dt} &= -k(T-65) \\ \frac{dT}{dt} + kT &= 65k \qquad I.F. = e^{kt} \\ \frac{d}{dt}(Te^{kt}) &= 65ke^{kt} \\ Te^{kt} &= 65e^{kt} + c \\ T &= 65 + ce^{-kt}. \end{aligned}$$

If $t = 0$, $T = T_0$ (unknown initial temperature). Thus $T_0 = 65 + c$ or $c = T_0 - 65$.
$T = 65 + (T_0 - 65)e^{-kt}$. If $t = 15$, then $T = 0$, and if $t = 30$, then $T = 20$:

$$\begin{aligned} 0 &= 65 + (T_0 - 65)e^{-15k} \\ 20 &= 65 + (T_0 - 65)e^{-30k} \end{aligned}$$

$$\begin{aligned} -65e^{15k} &= T_0 - 65 \qquad (1) \\ -45e^{30k} &= T_0 - 65 \end{aligned}$$

$$\begin{aligned} -65e^{15k} + 45e^{30k} &= 0 \qquad \text{(subtracting)} \\ -65 + 45e^{15k} &= 0 \qquad \text{(dividing by } e^{15k}\text{)} \\ e^{15k} &= \frac{65}{45} \\ 15k &= \ln\left(\frac{65}{45}\right) \\ k &= \frac{1}{15}\ln\left(\frac{65}{45}\right) = 0.0245. \end{aligned}$$

So from Equation (1), $T_0 = 65 - 65e^{15k} = -29°$ F.

19. $m\dfrac{dv}{dt} = w - kv$. Given: $m = 6\,\text{kg}$, $w = 60\,\text{N}$, and $k = 1$.

$6\dfrac{dv}{dt} = 60 - v$

Step 1. $\dfrac{dv}{dt} + \dfrac{1}{6}v = 10$

Step 2. $I.F. = e^{(1/6)t}$

Step 3. $e^{(1/6)t}\left(\dfrac{dv}{dt} + \dfrac{1}{6}v\right) = 10e^{(1/6)t}$

Step 4. $\dfrac{d}{dt}\left[ve^{(1/6)t}\right] = 10e^{(1/6)t}$

Step 5.
$$\begin{aligned} ve^{(1/6)t} &= 10\int e^{(1/6)t}\,dt \qquad u = \frac{1}{6}t;\ du = \frac{1}{6}\,dt. \\ ve^{(1/6)t} &= 10\cdot 6\int e^{(1/6)t}\left(\frac{1}{6}\,dt\right) \\ &= 60e^{(1/6)t} + c \\ v &= 60 + ce^{-(1/6)t} \end{aligned}$$

Initial condition: $v = 0$ when $t = 0$.

$0 = 60 + c$ or $c = -60$

Solution: $v = 60 - 60e^{-(1/6)t} = 60\left(1 - e^{-(1/6)t}\right)$.

21. Differentiating implicitly, we get $2x - 4y\dfrac{dy}{dx} = 0$ or $\dfrac{dy}{dx} = \dfrac{x}{2y}$.

Condition for orthogonal trajectories:

$$\begin{aligned} \frac{dy}{dx} &= -\frac{2y}{x} \\ \frac{dy}{y} &= -\frac{2\,dx}{x} \\ \ln y &= -2\ln x + \ln k \\ \ln y + 2\ln x &= \ln k \\ \ln y + \ln x^2 &= \ln k, \text{ whence } x^2y = k. \end{aligned}$$

Chapter 12

Higher-Order Linear Differential Equations

12.1 Higher-Order Homogeneous Differential Equations

1.
$$\begin{aligned} m^2 - 13m + 42 &= 0 \\ (m-6)(m-7) &= 0 \\ m &= 6,\ 7 \end{aligned}$$
$y = c_1 e^{6x} + c_2 e^{7x}$

3.
$$\begin{aligned} 6m^2 - m - 2 &= 0 \\ (3m-2)(2m+1) &= 0 \\ m &= \frac{2}{3},\ -\frac{1}{2} \end{aligned}$$
$y = c_1 e^{2x/3} + c_2 e^{-x/2}$

5.
$$\begin{aligned} 4m^2 + 7m - 2 &= 0 \\ (4m-1)(m+2) &= 0 \\ m &= \frac{1}{4},\ -2 \end{aligned}$$
$y = c_1 e^{x/4} + c_2 e^{-2x}$

7. $m^2 - m - 1 = 0$; by the quadratic formula,

$$\begin{aligned} m &= \frac{-(-1) \pm \sqrt{(-1)^2 - 4(1)(-1)}}{2 \cdot 1} = \frac{1 \pm \sqrt{5}}{2} \\ &= \frac{1}{2} \pm \frac{\sqrt{5}}{2}. \end{aligned}$$

$$\begin{aligned} y &= c_1 e^{\left(1/2+\sqrt{5}/2\right)x} + c_2 e^{\left(1/2-\sqrt{5}/2\right)x} \\ &= c_1 e^{(1/2)x} e^{\left(\sqrt{5}/2\right)x} + c_2 e^{(1/2)x} e^{-\left(\sqrt{5}/2\right)x} \\ &= e^{x/2}\left(c_1 e^{\sqrt{5}x/2} + c_2 e^{-\sqrt{5}x/2}\right) \end{aligned}$$

9.
$$\begin{aligned} 2m^2 - 3m + 1 &= 0 \\ (2m-1)(m-1) &= 0 \\ m &= 1,\ \frac{1}{2} \end{aligned}$$
$y = c_1 e^x + c_2 e^{x/2}$

11. $m^2 + m - 1 = 0$

$$m = \frac{-1 \pm \sqrt{1^2 - 4(1)(-1)}}{2} = \frac{-1 \pm \sqrt{5}}{2} = -\frac{1}{2} \pm \frac{\sqrt{5}}{2}$$

$$\begin{aligned} y &= c_1 e^{\left(-1/2+\sqrt{5}/2\right)x} + c_2 e^{\left(-1/2-\sqrt{5}/2\right)x} \\ &= c_1 e^{(-1/2)x} e^{\left(\sqrt{5}/2\right)x} + c_2 e^{(-1/2)x} e^{-\left(\sqrt{5}/2\right)x} \\ &= e^{-x/2}\left(c_1 e^{\sqrt{5}\,x/2} + c_2 e^{-\sqrt{5}\,x/2}\right) \end{aligned}$$

13. $m^2 - 2m - 2 = 0$

$$\begin{aligned} m &= \frac{-(-2) \pm \sqrt{(-2)^2 - 4(1)(-2)}}{2 \cdot 1} \\ &= \frac{2 \pm \sqrt{4+8}}{2} = \frac{2 \pm 2\sqrt{3}}{2} = \frac{2(1 \pm \sqrt{3})}{2} = 1 \pm \sqrt{3} \end{aligned}$$

$$\begin{aligned} y &= c_1 e^{\left(1+\sqrt{3}\right)x} + c_2 e^{\left(1-\sqrt{3}\right)x} \\ y &= c_1 e^{x+\sqrt{3}\,x} + c_2 e^{x-\sqrt{3}\,x} \\ y &= c_1 e^x e^{\sqrt{3}\,x} + c_2 e^x e^{-\sqrt{3}\,x} \\ y &= e^x\left(c_1 e^{\sqrt{3}\,x} + c_2 e^{-\sqrt{3}\,x}\right) \end{aligned}$$

15. $m^2 - 4m - 2 = 0$

$$\begin{aligned} m &= \frac{-(-4) \pm \sqrt{(-4)^2 - 4(1)(-2)}}{2} = \frac{4 \pm \sqrt{24}}{2} = \frac{4 \pm \sqrt{4 \cdot 6}}{2} \\ &= \frac{4 \pm 2\sqrt{6}}{2} = \frac{2\left(2 \pm \sqrt{6}\right)}{2} = 2 \pm \sqrt{6} \end{aligned}$$

$$\begin{aligned} y &= c_1 e^{\left(2+\sqrt{6}\right)x} + c_2 e^{\left(2-\sqrt{6}\right)x} \\ &= c_1 e^{2x} e^{\sqrt{6}\,x} + c_2 e^{2x} e^{-\sqrt{6}\,x} \\ &= e^{2x}\left(c_1 e^{\sqrt{6}\,x} + c_2 e^{-\sqrt{6}\,x}\right) \end{aligned}$$

17. $m^2 + 6m - 6 = 0$

$$\begin{aligned} m &= \frac{-6 \pm \sqrt{6^2 - 4(1)(-6)}}{2 \cdot 1} \\ &= \frac{-6 \pm \sqrt{36+24}}{2} = \frac{-6 \pm \sqrt{4 \cdot 15}}{2} = \frac{-6 \pm 2\sqrt{15}}{2} = \frac{2(-3 \pm \sqrt{15})}{2} \\ &= -3 \pm \sqrt{15} \end{aligned}$$

$$\begin{aligned} y &= c_1 e^{\left(-3+\sqrt{15}\right)x} + c_2 e^{\left(-3-\sqrt{15}\right)x} \\ &= c_1 e^{-3x} e^{\sqrt{15}\,x} + c_2 e^{-3x} e^{-\sqrt{15}\,x} \\ &= e^{-3x}\left(c_1 e^{\sqrt{15}\,x} + c_2 e^{-\sqrt{15}\,x}\right) \end{aligned}$$

19. $2m^2 + 4m + 1 = 0$

$$\begin{aligned} m &= \frac{-4 \pm \sqrt{4^2 - 4(2)(1)}}{2 \cdot 2} = \frac{-4 \pm \sqrt{8}}{4} = \frac{-4 \pm \sqrt{4 \cdot 2}}{4} \\ &= \frac{-4 \pm 2\sqrt{2}}{4} = \frac{2\left(-2 \pm \sqrt{2}\right)}{4} = \frac{-2 \pm \sqrt{2}}{2} = -1 \pm \frac{\sqrt{2}}{2} \end{aligned}$$

$$\begin{aligned} y &= c_1 e^{\left(-1+\sqrt{2}/2\right)x} + c_2 e^{\left(-1-\sqrt{2}/2\right)x} \\ &= c_1 e^{-x} e^{\left(\sqrt{2}/2\right)x} + c_2 e^{-x} e^{-\left(\sqrt{2}/2\right)x} \\ &= e^{-x}\left(c_1 e^{\left(\sqrt{2}/2\right)x} + c_2 e^{-\left(\sqrt{2}/2\right)x}\right) \end{aligned}$$

21. $m^3 - 7m + 6 = 0$. Possible rational roots: ± 1, ± 2, ± 3, ± 6.

$1 + 0 - 7 + 6)\underline{1}$

$\underline{\quad +1 + 1 - 6}$

$1 + 1 - 6 + 0 \qquad x = 1$ is a root.

$$\begin{aligned} m^2 + m - 6 &= 0 \\ (m-2)(m+3) &= 0 \\ m &= 2,\ -3 \qquad \text{(remaining roots)} \end{aligned}$$

$y = c_1 e^{x} + c_2 e^{2x} + c_3 e^{-3x}$

23. $m^3 - m^2 - 4m - 2 = 0$. Possible rational roots: ± 1, ± 2.

$1 - 1 - 4 - 2)\underline{-1}$

$\underline{\quad -1 + 2 + 2}$

$1 - 2 - 2 + 0 \qquad x = -1$ is a root.

$m^2 - 2m - 2 = 0$

$$\begin{aligned} m &= \frac{-(-2) \pm \sqrt{(-2)^2 - 4(1)(-2)}}{2 \cdot 1} \\ &= \frac{2 \pm \sqrt{4+8}}{2} = \frac{2 \pm 2\sqrt{3}}{2} = 1 \pm \sqrt{3} \end{aligned}$$

So the roots are: -1, $1 \pm \sqrt{3}$.

$y = c_1 e^{-x} + c_2 e^{(1+\sqrt{3})x} + c_2 e^{(1-\sqrt{3})x}$

25. $m^2 - 9 = 0$; $m^2 = 9$; $m = \pm 3$

$$\begin{aligned} y &= c_1 e^{3x} + c_2 e^{-3x} \\ Dy = y' &= 3c_1 e^{3x} - 3c_2 e^{-3x} \end{aligned}$$

Now substitute the initial conditions:

$0 = c_1 + c_2$	$y = 0$ when $x = 0$
$6 = 3c_1 - 3c_2$	$Dy = 6$ when $x = 0$
$0 = c_1 + c_2$	first equation
$2 = c_1 - c_2$	second equation
$2 = 2c_1$	adding

$c_1 = 1$ and $c_2 = -1$.

Substituting in the general solution: $y = e^{3x} - e^{-3x}$.

27.
$$\begin{aligned} m^2 - m - 2 &= 0 \\ (m+1)(m-2) &= 0 \\ m &= -1, 2 \\ y &= c_1 e^{-x} + c_2 e^{2x} \\ Dy = y' &= -c_1 e^{-x} + 2c_2 e^{2x} \\ 2 &= c_1 + c_2 && y = 2 \text{ when } x = 0 \\ 3 &= -c_1 + 2c_2 && Dy = 3 \text{ when } x = 0 \\ 5 &= 3c_2 && \text{adding} \end{aligned}$$

$c_2 = \frac{5}{3}$ and $c_1 = 2 - c_2 = 2 - \frac{5}{3} = \frac{1}{3}$

Thus $y = \frac{1}{3}e^{-x} + \frac{5}{3}e^{2x}$.

29.
$$\begin{aligned} 3m^2 - m - 2 &= 0 \\ (3m+2)(m-1) &= 0 \\ m &= 1, -\frac{2}{3} \\ y &= c_1 e^{x} + c_2 e^{-(2/3)x} \\ Dy = y' &= c_1 e^{x} - \frac{2}{3} c_2 e^{-(2/3)x} \\ -2 &= c_1 + c_2 && y = -2 \text{ when } x = 0 \\ 3 &= c_1 - \frac{2}{3}c_2 && Dy = 3 \text{ when } x = 0 \\ -5 &= \frac{5}{3}c_2 && \text{subtracting} \\ c_2 &= -3 \text{ and } c_1 = -2 + 3 = 1 \end{aligned}$$

Subsituting in the general solution, $y = e^{x} - 3e^{-(2/3)x}$.

12.2 Auxiliary Equations with Repeating or Complex Roots

1.
$$\begin{aligned} m^2 + 6m + 9 &= 0 \\ (m+3)^2 &= 0 \\ m &= -3, -3 && \text{(repeating root)} \end{aligned}$$
$y = c_1 e^{-3x} + c_2 x e^{-3x}$

3. $4m^2 - 4m + 1 = 0$
$$\begin{aligned} (2m-1)^2 &= 0 \\ m &= \frac{1}{2}, \frac{1}{2} && \text{(repeating root)} \end{aligned}$$
$y = c_1 e^{(1/2)x} + c_2 x e^{(1/2)x}$

5.
$$\begin{aligned} 9m^2 + 12m + 4 &= 0 \\ (3m+2)^2 &= 0 \\ m &= -\frac{2}{3}, -\frac{2}{3} && \text{(repeating root)} \end{aligned}$$
$y = c_1 e^{-(2/3)x} + c_2 x e^{-(2/3)x}$

7.
$$\begin{aligned} 4m^2 - 20m + 25 &= 0 \\ (2m-5)^2 &= 0 \\ m &= \frac{5}{2}, \frac{5}{2} \qquad \text{(repeating root)} \end{aligned}$$
$y = c_1 e^{(5/2)x} + c_2 x e^{(5/2)x}$

9. $m^2 - 4m + 5 = 0$

$$\begin{aligned} m &= \frac{-(-4) \pm \sqrt{(-4)^2 - 4(1)(5)}}{2 \cdot 1} = \frac{4 \pm \sqrt{16-20}}{2} \\ &= \frac{4 \pm \sqrt{-4}}{2} = \frac{4 \pm 2\mathrm{j}}{4} = \frac{2(2 \pm \mathrm{j})}{2} = 2 \pm \mathrm{j} \end{aligned}$$

$y = e^{2x}(c_1 \cos x + c_2 \sin x)$

11. $m^2 + 4m + 8 = 0$

$$m = \frac{-4 \pm \sqrt{4^2 - 4(1)(8)}}{2} = \frac{-4 \pm \sqrt{-16}}{2} = \frac{-4 \pm 4\mathrm{j}}{2} = -2 \pm 2\mathrm{j}$$

$y = e^{-2x}(c_1 \cos 2x + c_2 \sin 2x)$

13. $2m^2 - 2m + 1 = 0$

$$m = \frac{2 \pm \sqrt{4-8}}{4} = \frac{2 \pm \sqrt{-4}}{4} = \frac{2 \pm 2\mathrm{j}}{4} = \frac{1}{2} \pm \frac{1}{2}\mathrm{j}$$

$$y = e^{(1/2)x}\left(c_1 \cos \frac{1}{2}x + c_2 \sin \frac{1}{2}x\right)$$

15. $m^2 - 3m + 5 = 0$

$$m = \frac{-(-3) \pm \sqrt{(-3)^2 - 4(1)(5)}}{2} = \frac{3 \pm \sqrt{-11}}{2} = \frac{3 \pm \sqrt{11}\,\mathrm{j}}{2} = \frac{3}{2} \pm \frac{\sqrt{11}}{2}\mathrm{j}$$

$$y = e^{3x/2}\left(c_1 \cos \frac{\sqrt{11}}{2}x + c_2 \sin \frac{\sqrt{11}}{2}x\right)$$

17. $2m^2 + 4m - 1 = 0$

$$\begin{aligned} m &= \frac{-4 \pm \sqrt{4^2 - 4(2)(-1)}}{2 \cdot 2} = \frac{-4 \pm \sqrt{24}}{4} = \frac{-4 \pm \sqrt{4 \cdot 6}}{4} \\ &= \frac{-4 \pm 2\sqrt{6}}{4} = \frac{2\left(-2 \pm \sqrt{6}\right)}{4} = \frac{-2 \pm \sqrt{6}}{2} = -1 \pm \frac{\sqrt{6}}{2} \end{aligned}$$

$$\begin{aligned} y &= c_1 e^{\left(-1+\sqrt{6}/2\right)x} + c_2 e^{\left(-1-\sqrt{6}/2\right)x} \\ &= c_1 e^{-x} e^{\left(\sqrt{6}/2\right)x} + c_2 e^{-x} e^{-\left(\sqrt{6}/2\right)x} \\ &= e^{-x}\left(c_1 e^{\left(\sqrt{6}/2\right)x} + c_2 e^{-\left(\sqrt{6}/2\right)x}\right) \end{aligned}$$

19. $m^2 + 25 = 0,\ m = \pm 5j$
$y = c_1 \cos 5x + c_2 \sin 5x$

21. $m^2 - 25 = 0,\ m = \pm 5$
$y = c_1 e^{5x} + c_2 e^{-5x}$

23.
$$\begin{aligned} m^2 - 6m + 9 &= 0 \\ (m-3)^2 &= 0 \\ m &= 3,\ 3 \end{aligned}$$
$y = c_1 e^{3x} + c_2 x e^{3x}$

25. $2m^2 - 4m + 5 = 0$
$$m = \frac{4 \pm \sqrt{16 - 40}}{4} = \frac{4 \pm \sqrt{-24}}{4} = \frac{4 \pm \sqrt{4 \cdot (-6)}}{4} = \frac{4 \pm 2\sqrt{6}\,\mathrm{j}}{4} = 1 \pm \frac{1}{2}\sqrt{6}\,\mathrm{j}$$
$$y = \mathrm{e}^{x}\left(c_1 \cos \frac{1}{2}\sqrt{6}\,x + c_2 \sin \frac{1}{2}\sqrt{6}\,x\right)$$

27. $m^2 - 2m - 4 = 0$
$$\begin{aligned} m &= \frac{2 \pm \sqrt{4 + 16}}{2} = \frac{2 \pm \sqrt{20}}{2} = \frac{2 \pm 2\sqrt{5}}{2} \\ &= \frac{2\left(1 \pm \sqrt{5}\right)}{2} = 1 \pm \sqrt{5} \end{aligned}$$
$$\begin{aligned} y &= c_1 \mathrm{e}^{(1+\sqrt{5})x} + c_2 \mathrm{e}^{(1-\sqrt{5})x} \\ y &= \mathrm{e}^{x}\left(c_1 \mathrm{e}^{\sqrt{5}x} + c_2 \mathrm{e}^{-\sqrt{5}x}\right) \end{aligned}$$

29.
$$\begin{aligned} 3m^3 - 2m^2 + m &= 0 \\ m(3m^2 - 2m + 1) &= 0 \\ m &= 0, \\ m &= \frac{2 \pm \sqrt{4 - 12}}{6} = \frac{2 \pm \sqrt{2(4)(-1)}}{6} = \frac{2 \pm 2\sqrt{2}\,\mathrm{j}}{6} \\ &= \frac{1}{3} \pm \frac{1}{3}\sqrt{2}\,\mathrm{j} \\ y &= c_1 \mathrm{e}^{0x} + \mathrm{e}^{(1/3)x}\left(c_2 \cos \frac{\sqrt{2}}{3}x + c_3 \sin \frac{\sqrt{2}}{3}x\right) \\ &= c_1 + \mathrm{e}^{x/3}\left(c_2 \cos \frac{\sqrt{2}}{3}\,x + c_3 \sin \frac{\sqrt{2}}{3}x\right) \end{aligned}$$

31.
$$\begin{aligned} m^3 - 4m^2 + 4m &= 0 \\ m(m^2 - 4m + 4) &= 0 \\ m(m-2)^2 &= 0 \\ m &= 0,\ 2,\ 2 \end{aligned}$$
$y = c_1 \mathrm{e}^{0x} + c_2 \mathrm{e}^{2x} + c_3 x \mathrm{e}^{2x} = c_1 + c_2 \mathrm{e}^{2x} + c_3 x \mathrm{e}^{2x}$

33.
$$\begin{aligned} m^4 + 2m^3 &= 0 \\ m^3(m+2) &= 0 \\ m &= 0,\ 0,\ 0,\ -2 \end{aligned}$$
Here we have a triple root and a single root:
$$\begin{aligned} y &= c_1 \mathrm{e}^{0x} + c_2 x \mathrm{e}^{0x} + c_3 x^2 \mathrm{e}^{0x} + c_4 \mathrm{e}^{-2x} \\ &= c_1 + c_2 x + c_3 x^2 + c_4 \mathrm{e}^{-2x} \end{aligned}$$

35.
$$\begin{aligned} m^3 - 9m = m(m^2 - 9) &= 0 \\ m &= 0,\ \pm 3 \end{aligned}$$
$$\begin{aligned} y &= c_1 \mathrm{e}^{0x} + c_2 \mathrm{e}^{-3x} + c_3 \mathrm{e}^{3x} \\ y &= c_1 + c_2 \mathrm{e}^{-3x} + c_3 \mathrm{e}^{3x} \end{aligned}$$

37. $$\begin{aligned}(m-2)^4 &= 0\\ m &= 2,\ 2,\ 2,\ 2\end{aligned}$$
$y = c_1 e^{2x} + c_2 x e^{2x} + c_3 x^2 e^{2x} + c_4 x^3 e^{2x}$

39. $$\begin{aligned}m^2+1 &= 0\\ m &= \pm j\end{aligned}$$
$y = c_1 \cos x + c_2 \sin x$
Substituting $(0,0)$ and $\left(\frac{\pi}{2},1\right)$, respectively, we get
(1) $0 = c_1 \cdot 1 + c_2 \cdot 0$ or $c_1 = 0$; $\quad y=0,\ x=0$
(2) $1 = c_1 \cdot 0 + c_2 \cdot 1$ or $c_2 = 1$. $\quad y=1,\ x=\frac{\pi}{2}$
Thus $y = \sin x$ is the solution.

41. $m^2 - 2m + 2 = 0$

$$m = \frac{-(-2) \pm \sqrt{(-2)^2 - 4(1)(2)}}{2} = \frac{2 \pm \sqrt{-4}}{2} = \frac{2 \pm 2j}{2} = 1 \pm j$$

$y = e^x(c_1 \cos x + c_2 \sin x)$
By the product rule,
$Dy = y' = e^x(-c_1 \sin x + c_2 \cos x) + e^x(c_1 \cos x + c_2 \sin x)$
Now substitute the initial conditions:
$0 = 1(c_1+0) \qquad y=0$ when $x=0$
$-1 = 1(0+c_2) + 1(c_1+0) \qquad Dy=-1$ when $x=0$
From the first equation, $c_1 = 0$. Substituting in the second equation, we obtain
$-1 = 1(c_2) + 1(0)$ or $c_2 = -1$.
Finally, we substitute $c_1 = 0$ and $c_2 = -1$ in the general solution:
$y = e^x\left[0 + (-1)\sin x\right] = -e^x \sin x$.

43.
$$\begin{aligned}m^4 + 18m^2 + 81 &= 0\\ (m^2+9)^2 = (m^2+9)(m^2+9) &= 0\\ m &= 3j,\ 3j,\ -3j,\ -3j\end{aligned}$$
$$\begin{aligned}y &= c_1 e^{3jx} + c_2 x e^{3jx} + c_3 e^{-3jx} + c_4 x e^{-3jx}\\ &= c_1(\cos 3x + j\sin 3x) + c_2 x(\cos 3x + j \sin 3x)\\ &\quad + c_3(\cos 3x - j\sin 3x) + c_4 x(\cos 3x - j\sin 3x)\\ &= (c_1+c_3)\cos 3x + j(c_1-c_3)\sin 3x + (c_2+c_4)x\cos 3x + j(c_2-c_4)x\sin 3x\end{aligned}$$
So y has the following form:
$y = c_1 \cos 3x + c_2 \sin 3x + c_3 x \cos 3x + c_4 x \sin 3x$

45. $-1,\ \frac{1}{2} \pm \frac{\sqrt{2}}{2}j$

47. $6,\ -4,\ -4$

49. $-1,\ -1,\ 1 \pm \frac{\sqrt{5}}{2}j$

12.3 Nonhomogeneous Equations

1. $(D^2 - 6D + 9)y = e^x$

y_c:
$$\begin{aligned} m^2 - 6m + 9 &= 0 \\ (m-3)^2 &= 0 \\ m &= 3,\ 3 \end{aligned}$$

$y_c = c_1 e^{3x} + c_2 x e^{3x}$

y_p : $y_p = Ae^x$; $y_p' = Ae^x$; $y_p'' = Ae^x$

Substituting in $(D^2 - 6D + 9)y = e^x$:

$$\begin{aligned} y_p'' - 6y_p' + 9y_p &= e^x \\ Ae^x - 6Ae^x + 9Ae^x &= e^x \\ 4Ae^x &= e^x \\ A &= \frac{1}{4}. \end{aligned}$$

So $y_p = Ae^x = \frac{1}{4}e^x$ and $y = y_c + y_p = c_1 e^{3x} + c_2 x e^{3x} + \frac{1}{4}e^x$.

3. $(D^2 - 6D + 9)y = 9x$

y_c : $y_c = c_1 e^{3x} + c_2 x e^{3x}$ (same as in Exercise 1)

y_p : $y_p = Ax + B$; $y_p' = A$; $y_p'' = 0$

Substituting in $(D^2 - 6D + 9)y = 9x$:

$$\begin{aligned} y_p'' - 6y_p' + 9y_p &= 9x \\ 0 - 6A + 9(Ax + B) &= 9x \\ 9Ax - 6A + 9B &= 9x + 0. \end{aligned}$$

Now compare coefficients:

$$\begin{aligned} 9A &= 9 && x\text{-coefficients} \\ -6A + 9B &= 0. && \text{constants} \end{aligned}$$

We obtain $A = 1$ and $B = \frac{6}{9}A = \frac{2}{3}$. So $y_p = x + \frac{2}{3}$ and $y = y_c + y_p = c_1 e^{3x} + c_2 x e^{3x} + x + \frac{2}{3}$.

5. $(D^2 - D - 2)y = 2x^2$

y_c :
$$\begin{aligned} m^2 - m - 2 = (m-2)(m+1) &= 0 \\ m &= 2,\ -1 \end{aligned}$$

$y_c = c_1 e^{2x} + c_2 e^{-x}$

y_p : $y_p = Ax^2 + Bx + C$, $y_p' = 2Ax + B$, $y_p'' = 2A$

Substituting in $(D^2 - D - 2)y = 2x^2$, we get

$$\begin{aligned} 2A - (2Ax + B) - 2(Ax^2 + Bx + C) &= 2x^2 \\ 2A - 2Ax - B - 2Ax^2 - 2Bx - 2C &= 2x^2 \\ -2Ax^2 + (-2A - 2B)x + (2A - B - 2C) &= 2x^2 + 0x + 0. \end{aligned}$$

$$\begin{aligned} -2A &= 2 && x^2\text{-coefficients} \\ -2A - 2B &= 0 && x\text{-coefficients} \\ 2A - B - 2C &= 0 && \text{constants} \end{aligned}$$

We obtain: $A = -1$, $B = 1$, and $C = -\frac{3}{2}$.

So $y_p = Ax^2 + Bx + C = -x^2 + x - \frac{3}{2}$ and $y = y_c + y_p = c_1 e^{2x} + c_2 e^{-x} - x^2 + x - \frac{3}{2}$.

7. $(2D^2 + D - 3)y = 2x - 3x^2$

$$y_c: \quad \begin{aligned} 2m^2 + m - 3 &= 0 \\ (2m+3)(m-1) &= 0 \\ m &= -\frac{3}{2}, 1 \end{aligned} \qquad y_c = c_1 e^{-(3/2)x} + c_2 e^x$$

y_p: $y_p = Ax^2 + Bx + C$; $y_p' = 2Ax + B$; $y_p'' = 2A$

Substituting in $(2D^2 + D - 3)y = 2x - 3x^2$ we get

$$\begin{aligned} 4A + (2Ax + B) - 3(Ax^2 + Bx + C) &= 2x - 3x^2 \\ -3Ax^2 + (2A - 3B)x + (4A + B - 3C) &= -3x^2 + 2x + 0. \end{aligned}$$

Comparing coefficients:

$$\begin{aligned} -3A &= -3 && x^2\text{-coefficients} \\ 2A - 3B &= 2 && x\text{-coefficients} \\ 4A + B - 3C &= 0. && \text{constants} \end{aligned}$$

We obtain: $A = 1$, $B = 0$, and $C = \frac{4}{3}$. So $y_p = x^2 + \frac{4}{3}$ and $y = c_1 e^{-(3/2)x} + c_2 e^x + x^2 + \dfrac{4}{3}$

9. $(D^2 - D + 2)y = 4e^{3x}$

$$y_c: \quad \begin{aligned} m^2 - m + 2 &= 0 \\ m &= \frac{-(-1) \pm \sqrt{(-1)^2 - 4(1)(2)}}{2 \cdot 1} \\ &= \frac{1 \pm \sqrt{-7}}{2} = \frac{1 \pm \sqrt{7}\,\mathrm{j}}{2} = \frac{1}{2} \pm \frac{\sqrt{7}}{2}\mathrm{j} \end{aligned}$$

$$y_c = e^{x/2}\left(c_1 \cos\frac{\sqrt{7}}{2}x + c_2 \sin\frac{\sqrt{7}}{2}x\right)$$

y_p : $y_p = Ae^{3x}$; $y_p' = 3Ae^{3x}$; $y_p'' = 9Ae^{3x}$

Substituting in $(D^2 - D + 2)y = 4e^{3x}$:

$$\begin{aligned} y_p'' - y_p' + 2y_p &= 4e^{3x} \\ 9Ae^{3x} - 3Ae^{3x} + 2Ae^{3x} &= 4e^{3x} \\ 8Ae^{3x} &= 4e^{3x} \\ A &= \frac{1}{2}. \end{aligned}$$

So $y_p = \dfrac{1}{2}e^{3x}$ and $y = e^{x/2}\left(c_1 \cos\dfrac{\sqrt{7}}{2}x + c_2 \sin\dfrac{\sqrt{7}}{2}x\right) + \dfrac{1}{2}e^{3x}$.

11. $\left(D^2 + 3D - 10\right)y = 15e^{5x}$

$$y_c: \quad \begin{aligned} m^2 + 3m - 10 &= 0 \\ (m+5)(m-2) &= 0 \\ m &= -5, 2 \end{aligned}$$

$y_c = c_1 e^{-5x} + c_2 e^{2x}$

y_p : $y_p = Ae^{5x}$, $y_p' = 5Ae^{5x}$, $y_p'' = 25e^{5x}$

Substituting in $\left(D^2 + 3D - 10\right)y = 15e^{5x}$, we get

$$\begin{aligned} 25Ae^{5x} + 3\left(5Ae^{5x}\right) - 10Ae^{5x} &= 15e^{5x} \\ 30A = 15 \text{ and } A &= \frac{1}{2} \end{aligned}$$

So $y_p = \dfrac{1}{2}e^{5x}$ and $y = c_1 e^{-5x} + c_2 e^{2x} + \dfrac{1}{2}e^{5x}$.

13. $\left(D^2 - D + 2\right) y = 42e^{4x}$

$$\begin{aligned} y_c:\ m^2 - m + 2 &= 0 \\ m &= \frac{1 \pm \sqrt{1-8}}{2} = \frac{1}{2} \pm \frac{1}{2}\sqrt{7}j \\ y_c &= e^{(1/2)x}\left(c_1 \cos\frac{1}{2}\sqrt{7}x + c_2 \sin\frac{1}{2}\sqrt{7}x\right) \end{aligned}$$

$y_p:\ y_p = Ae^{4x},\ y_p' = 4Ae^{4x},\ y_p'' = 16Ae^{4x}$

Substituting in $\left(D^2 - D + 2\right) y = 42e^{4x}$, we get

$$\begin{aligned} 16Ae^{4x} - 4Ae^{4x} + 2Ae^{4x} &= 42e^{4x} \\ 14A = 42 \text{ and } A &= 3 \end{aligned}$$

So $y_p = 3e^{4x}$ and $y = e^{(1/2)x}\left(c_1 \cos\frac{1}{2}\sqrt{7}x + c_2 \sin\frac{1}{2}\sqrt{7}x\right) + 3e^{4x}$.

15. $(D^2 - 6D + 9)y = 9\cos 3x$

$y_c : m^2 - 6m + 9 = 0;\ (m-3)^2 = 0;\ m = 3,\ 3$

$y_c = c_1e^{3x} + c_2xe^{3x}$

$y_p : y_p = A\cos 3x + B\sin 3x;\ y_p' = -3A\sin 3x + 3B\cos 3x;\ y_p'' = -9A\cos 3x - 9B\sin 3x$

Substituting in $(D^2 - 6D + 9)y = 9\cos 3x$,

$$\begin{aligned} y_p'' - 6y_p' + 9y_p &= 9\cos 3x \\ -9A\cos 3x - 9B\sin 3x - 6(-3A\sin 3x + 3B\cos 3x) & \\ +9(A\cos 3x + B\sin 3x) &= 9\cos 3x \\ -18B\cos 3x + 18A\sin 3x &= 9\cos 3x + 0\sin 3x \end{aligned}$$

So $A = 0$ and $B = -\frac{1}{2}$, and $y_p = -\frac{1}{2}\sin 3x$.

$y = c_1e^{3x} + c_2xe^{3x} - \frac{1}{2}\sin 3x$

17. $\left(D^2 - 2D + 2\right) y = 5\sin 2x$

$y_p:\ y_p = A\cos 2x + B\sin 2x,\ y_p' = -2A\sin 2x + 2B\cos 2x,\ y_p'' = -4A\cos 2x - 4B\sin 2x$

Substituting,

$$-4A\cos 2x - 4B\sin 2x - 2(-2A\sin 2x + 2B\cos 2x) + 2(A\cos 2x + B\sin 2x) = 5\sin 2x$$

$$\begin{aligned} -2A - 4B &= 0 \\ 4A - 2B &= 5 \\ \hline A + 2B &= 0 && \text{dividing by } -2 \\ 4A - 2B &= 5 \\ \hline 5A &= 5 && \text{adding} \\ A &= 1 \text{ and } B = -\frac{1}{2} \end{aligned}$$

So $y_p = \cos 2x - \frac{1}{2}\sin 2x$ and $y = y_c + y_p$.

19. $\left(D^2 - 2D - 4\right) y = 20\cos 2x$

$$\begin{aligned} y_c:\ m^2 - 2m - 4 &= 0 \\ m &= \frac{2 \pm \sqrt{4+16}}{2} = \frac{2 \pm 2\sqrt{5}}{2} = 1 \pm \sqrt{5} \end{aligned}$$

$y_c = e^x\left(c_1e^{\sqrt{5}x} + c_2e^{-\sqrt{5}x}\right)$

$y_p:\ y_p = A\cos 2x + B\sin 2x,\ y_p' = -2A\sin 2x + 2B\cos 2x,\ y_p'' = -4A\cos 2x - 4B\sin 2x$

Subsituting in $\left(D^2 - 2D - 4\right) y = 20\cos 2x$, we get

$-4A\cos 2x - 4B\sin 2x - 2\left(-2A\sin 2x + 2B\cos 2x\right) - 4A\cos 2x - 4B\sin 2x = 20\cos 2x$

$(-8A - 4B)\cos 2x + (4A - 8B)\sin 2x = 20\cos 2x + 0\sin 2x$

Comparing coefficients,

$$\begin{array}{rcrcl} -8A & - & 4B & = & 20 \\ 4A & - & 8B & = & 0 \\ \hline -8A & - & 4B & = & 20 \\ 8A & - & 16B & = & 0 \quad \text{multiplying by 2} \\ \hline & & -20B & = & 20 \\ & & B & = & -1; \quad \text{also } A = -2 \end{array}$$

So $y_p = -2\cos 2x - \sin 2x$ and $y = \mathrm{e}^x\left(c_1\mathrm{e}^{\sqrt{5}x} + c_2\mathrm{e}^{-\sqrt{5}x}\right) - 2\cos 2x - \sin 2x$.

21. y_c :

$$\begin{aligned} m^2 + 5m + 6 &= 0 \\ (m+2)(m+3) &= 0 \\ m &= -2,\ -3 \end{aligned}$$

$y_c = c_1\mathrm{e}^{-2x} + c_2\mathrm{e}^{-3x}$

$y_p:\ y_p = A\cos x + B\sin x;\ y_p' = -A\sin x + B\cos x;\ y_p'' = -A\cos x - B\sin x$

Substituting in the equation $(D^2 + 5D + 6)y = 4\cos x + 6\sin x$, we get

$$\begin{aligned} (-A\cos x - B\sin x) + 5(-A\sin x + B\cos x) + 6(A\cos x + B\sin x) &= 4\cos x + 6\sin x \\ (5A + 5B)\cos x + (-5A + 5B)\sin x &= 4\cos x + 6\sin x. \end{aligned}$$

Comparing coefficients:

$$\begin{array}{rcl} 5A + 5B & = & 4 \\ -5A + 5B & = & 6 \\ \hline 10B & = & 10, \quad B = 1 \end{array}$$

$5A + 5 = 4,\ A = -\dfrac{1}{5}$ first equation

So $y_p = -\dfrac{1}{5}\cos x + \sin x$ and $y = c_1\mathrm{e}^{-2x} + c_2\mathrm{e}^{-3x} - \dfrac{1}{5}\cos x + \sin x$.

23. $\left(2D^2 + 3D - 2\right)y = 2x^2 + 1$

$$\begin{aligned} y_c:\ 2m^2 + 3m - 2 &= 0 \\ (2m-1)(m+2) &= 0 \\ m &= \frac{1}{2}, -2 \end{aligned}$$

$y_c = c_1\mathrm{e}^{(1/2)x} + c_2\mathrm{e}^{-2x}$

$y_p:\ y_p = Ax^2 + Bx + C;\ y_p' = 2Ax + B;\ y_p'' = 2A$

Substituting in $\left(2D^2 + 3D - 2\right)y = 2x^2 + 1$, we get

$4A + 6Ax + 3B - 2Ax^2 - 2Bx - 2C = 2x^2 + 1$

$-2Ax^2 + (6A - 2B)x + (4A + 3B - 2C) = 2x^2 + 0x + 1$

$$\begin{aligned} -2A &= 2 && x^2\text{-coefficients} \\ 6A - 2B &= 0 && x\text{-coefficients} \\ 4A + 3B - 2C &= 1 && \text{constants} \end{aligned}$$

We obtain: $A = -1,\ B = -3,\ C = -7$.

So $y_p = -x^2 - 3x - 7$ and $y_c = c_1\mathrm{e}^{(1/2)x} + c_2\mathrm{e}^{-2x} - x^2 - 3x - 7$.

25. $(D^2 - 2D + 1)y = 3\mathrm{e}^{2x}$

$$\begin{aligned} y_c:\ m^2 - 2m + 1 &= 0 \\ (m-1)^2 &= 0 \\ m &= 1,\ 1 \end{aligned}$$

$y_c = c_1\mathrm{e}^x + c_2x\mathrm{e}^x$

$y_p : y_p = A\mathrm{e}^{2x};\ y_p' = 2A\mathrm{e}^{2x};\ y_p'' = 4A\mathrm{e}^{2x}$

Substituting in $(D^2 - 2D + 1)y = 3e^{2x}$, we get

$$\begin{aligned} 4Ae^{2x} - 2(2Ae^{2x}) + Ae^{2x} &= 3e^{2x} \\ Ae^{2x} &= 3e^{2x} \text{ and } A = 3. \end{aligned}$$

So $y_p = 3e^{2x}$ and $y = c_1e^x + c_2xe^x + 3e^{2x}$.

27. $y_c:\ m^3 - 2m^2 - m + 2 = 0$

Possible rational roots: $\pm1,\ \pm2$; one root is $m = 1$.

$$\begin{array}{l} 1 - 2 - 1 + 2)\underline{1} \\ \quad 1 - 1 - 2 \\ \hline 1 - 1 - 2 + 0 \end{array}$$

Since the resulting factor is $m^2 - m - 2 = (m-2)(m+1)$, the other roots are 2 and -1.

$y_c = c_1e^x + c_2e^{-x} + c_3e^{2x}$

$y_p : y_p = Ae^{3x};\ y_p' = 3Ae^{3x};\ y_p'' = 9Ae^{3x};\ y_p''' = 27Ae^{3x}$

Substituting in the equation: $27Ae^{3x} - 2\left(9Ae^{3x}\right) - 3Ae^{3x} + 2Ae^{3x} = 8e^{3x}$.

$8A = 8,\ A = 1$ and $y_p = e^{3x}$.

$y = c_1e^x + c_2e^{-x} + c_3e^{2x} + e^{3x}$

29. $(D^2 - 2D + 5)y = 4xe^x$

$y_c : m^2 - 2m + 5 = 0$

$$m = \frac{-(-2) \pm \sqrt{(-2)^2 - 4(1)(5)}}{2} = \frac{2 \pm \sqrt{-16}}{2} = 1 \pm 2j$$

$y_c = e^x(c_1 \cos 2x + c_2 \sin 2x)$

$y_p : y_p = Axe^x + Be^x;$

$y_p' = Axe^x + Ae^x + Be^x$ product rule

$y_p'' = Axe^x + 2Ae^x + Be^x$

Substituting in $(D^2 - 2D + 5)y = 4xe^x$,

$(Axe^x + 2Ae^x + Be^x) - 2(Axe^x + Ae^x + Be^x) + 5(Axe^x + Be^x) = 4xe^x$.

Collecting terms, $4Axe^x + 4Be^x = 4xe^x$.

We obtain $A = 1$ and $B = 0$, so that $y_p = xe^x$ and $y = e^x(c_1 \cos 2x + c_2 \sin 2x) + xe^x$.

31. $y_c : m^2 + 9 = 0$

$m = \pm 3j$

$y_c = c_1 \cos 3x + c_2 \sin 3x$

$y_p:\ y_p = Ae^{3x};\ y_p' = 3Ae^{3x};\ y_p'' = 9Ae^{3x}$

Substituting in the equation:

$9Ae^{3x} + 9Ae^{3x} = 9e^{3x}$

$$18A = 9,\ A = \frac{1}{2} \text{ and } y_p = \frac{1}{2}e^{3x}.$$

$y = c_1 \cos 3x + c_2 \sin 3x + \frac{1}{2}e^{3x}$

$Dy = -3c_1 \sin 3x + 3c_2 \cos 3x + \frac{3}{2}e^{3x}$

Substituting in the given conditions, we have

(1) $1 = c_1 \cdot 1 + c_2 \cdot 0 + \frac{1}{2}$ $\quad y = 1,\ x = 0$

(2) $\frac{3}{2} = -3c_1 \cdot 0 + 3c_2 \cdot 1 + \frac{3}{2}$ $\quad Dy = \frac{3}{2},\ x = 0$

$c_1 = \frac{1}{2},\ c_2 = 0$

The solution is $y = \frac{1}{2}\cos 3x + \frac{1}{2}e^{3x} = \frac{1}{2}\left(\cos 3x + e^{3x}\right)$.

33. $(D^2+1)y = 6\cos 2x$

$y_c:\ m^2+1=0;\ m=\pm \mathrm{j}$

$y_c = c_1\cos x + c_2 \sin x$

$y_p:\ y_p = A\cos 2x + B\sin 2x;\ y_p' = -2A\sin 2x + 2B\cos 2x;\ y_p'' = -4A\cos 2x - 4B\sin 2x$

Substituting,

$$(-4A\cos 2x - 4B\sin 2x) + (A\cos 2x + B\sin 2x) = 6\cos 2x$$
$$-3A\cos 2x - 3B\sin 2x = 6\cos 2x.$$

We obtain $A=-2$ and $B=0$, so that $y_p = -2\cos 2x$ and

$$y = c_1\cos x + c_2\sin x - 2\cos 2x$$

$Dy = y' = -c_1\sin x + c_2\cos x + 4\sin 2x.$

Substituting the initial conditions,

$3 = c_1 + 0 - 2;\qquad y=3$ when $x=0$

$1 = 0 + c_2 + 0.\qquad Dy = 1$ when $x=0$

So $c_1=5$ and $c_2=1$; the solution is $y = 5\cos x + \sin x - 2\cos 2x$.

35. $(D^2+2D+5)y = 10\cos x$

$y_c: m^2+2m+5=0$

$$m = \frac{-2\pm\sqrt{4-20}}{2} = \frac{-2\pm 4\mathrm{j}}{2} = -1\pm 2\mathrm{j}$$

$y_c = \mathrm{e}^{-x}(c_1\cos 2x + c_2\sin 2x)$

$y_p: y_p = A\cos x + B\sin x;\ y_p' = -A\sin x + B\cos x;\ y_p'' = -A\cos x - B\sin x$

Substituting in the given equation:

$$(-A\cos x - B\sin x) + 2(-A\sin x + B\cos x) + 5(A\cos x + B\sin x) = 10\cos x$$
$$(4A+2B)\cos x + (-2A+4B)\sin x = 10\cos x$$
$$(4A+2B)\cos x + (-2A+4B)\sin x = 10\cos x + 0\sin x.$$

$$\begin{array}{r} 4A+2B=10 \\ -2A+4B=0 \\ \hline 4A+2B=10 \\ -4A+8B=0 \\ \hline 10B=10, B=1 \end{array}$$

$-2A+4=0,\ A=2\qquad y_p = 2\cos x + \sin x$

General solution:

$$y = \mathrm{e}^{-x}(c_1\cos 2x + c_2\sin 2x) + 2\cos x + \sin x$$

$Dy = \mathrm{e}^{-x}(-2c_1\sin 2x + 2c_2\cos 2x) - \mathrm{e}^{-x}(c_1\cos 2x + c_2\sin 2x) - 2\sin x + \cos x.$

Now substitute the given conditions:

$5 = c_1 + 2\qquad x=0,\ y=5$

$6 = 2c_2 - c_1 + 1\qquad x=0,\ Dy=6$

Thus $c_1=3$ and $c_2=4$, so that the solution is

$$y = \mathrm{e}^{-x}(3\cos 2x + 4\sin 2x) + 2\cos x + \sin x.$$

37. $(D^2+2)y = 2\mathrm{e}^{2x}+2$

$y_p = A\mathrm{e}^{2x} + B;\ y_p' = 2A\mathrm{e}^{2x};\ y_p'' = 4A\mathrm{e}^{2x}$

Substituting:

$$4A\mathrm{e}^{2x} + 2(A\mathrm{e}^{2x}+B) = 2\mathrm{e}^{2x}+2$$
$$6A\mathrm{e}^{2x} + 2B = 2\mathrm{e}^{2x}+2$$
$$A = \frac{1}{3},\ B=1.$$

$$y = y_c + \frac{1}{3}\mathrm{e}^{2x} + 1$$

39. $(D^2 - D - 4)y = x + 2e^{3x}$
$y_p = Ax + B + Ce^{3x};\ y_p' = A + 3Ce^{3x};\ y_p'' = 9Ce^{3x}$
Substituting:
$$9Ce^{3x} - (A + 3Ce^{3x}) - 4(Ax + B + Ce^{3x}) = x + 2e^{3x}$$
$$-4Ax + (-A - 4B) + 2Ce^{3x} = x + 0 + 2e^{3x}.$$
$-4A = 1$ x-coefficients
$-A - 4B = 0$ constants
$2C = 2$ coefficients of e^{3x}
We obtain: $A = -\frac{1}{4},\ B = \frac{1}{16}$, and $C = 1$.
$y = y_c - \frac{1}{4}x + \frac{1}{16} + e^{3x}$.

41. $(D^2 + 3)y = 2x + \cos x$
$y_p = Ax + B + C\cos x + D\sin x;\ y_p' = A - C\sin x + D\cos x;\ y_p'' = -C\cos x - D\sin x$
Substituting:
$$(-C\cos x - D\sin x) + 3(Ax + B + C\cos x + D\sin x) = 2x + \cos x$$
$$3Ax + 3B + 2C\cos x + 2D\sin x = 2x + \cos x$$
$$= 2x + 0 + \cos x + 0\sin x.$$
$A = \frac{2}{3},\ B = 0,\ C = \frac{1}{2},\ D = 0$
$y = y_c + \frac{2}{3}x + \frac{1}{2}\cos x$

43. $(D^2 - D - 3)y = 6xe^x$
$\frac{d}{dx}(6xe^x) = 6xe^x + 6e^x$
Since the right side is a linear combination of y_p and its derivatives, the form of y_p is the following:
$y_p = Axe^x + Be^x;\ y_p' = Axe^x + Ae^x + Be^x;\ y_p'' = Axe^x + 2Ae^x + Be^x$.
Substituting:
$$Axe^x + 2Ae^x + Be^x - (Axe^x + Ae^x + Be^x) - 3(Axe^x + Be^x) = 6xe^x$$
$$-3Axe^x + (A - 3B)e^x = 6xe^x + 0e^x.$$
We obtain: $A = -2,\ A - 3B = 0$, so $B = -\frac{2}{3}$.
$y = y_c - 2xe^x - \frac{2}{3}e^x$

45. $(D^2 - 4)y = 8e^{2x}$
$y_c:\ m^2 - 4 = 0,\ m = \pm 2$
$y_c = c_1e^{-2x} + c_2e^{2x}$
Since the right side of the given equation, $8e^{2x}$, is a term in y_c, we need the annihilator. (In other words, y_p is not of the form Ae^{2x}.)
Right side $= 8e^{2x}$
$m' = 2$
$m' - 2 = 0$
Annihilator: $D - 2$.

Applying the annihilator to the equation, we get

$(D-2)(D^2-4)y=(D-2)(8e^{2x})=0$

$(m-2)(m^2-4)=0$

$(m-2)(m-2)(m+2)=0$

$m=2,\ 2,\ -2.$

So y has the form: $y=c_1e^{-2x}+c_2e^{2x}+c_3xe^{2x}$.

Since $y_c=c_1e^{-2x}+c_2e^{2x}$, we conclude that

$y_p=Axe^{2x}$;

$y_p'=2Axe^{2x}+Ae^{2x}$;

$y_p''=4Axe^{2x}+2Ae^{2x}+2Ae^{2x}=4Axe^{2x}+4Ae^{2x}$.

Substituting in the given equation, we get

$(4Axe^{2x}+4Ae^{2x})-4Axe^{2x}=8e^{2x}$

$4Ae^{2x}=8e^{2x}$

$A=2$ and $y_p=2xe^{2x}$.

$y=y_c+y_p=c_1e^{-2x}+c_2e^{2x}+2xe^{2x}$

47. $(D^2-D-6)y=10e^{-2x}$

$y_c:\quad m^2-m-6=0$

$(m-3)(m+2)=0$

$m=3,\ -2$

$y_c=c_1e^{3x}+c_2e^{-2x}$

Since the right side of the given equation, $10e^{-2x}$, is of the form c_2e^{-2x}, y_p cannot be Ae^{-2x}.

So we need the annihilator:

Right side$=10e^{-2x}$

$m'=-2$

$m'+2=0$

Annihilator: $D+2$.

Applying the annihilator to the given equation, we get

$(D+2)(D^2-D-6)y=(D+2)(10e^{-2x})=0$

$(m+2)(m+2)(m-3)=0$

$m=-2,\ -2,\ 3.$

So y has the form: $y=c_1e^{3x}+c_2e^{-2x}+c_3xe^{-2x}$.

Since $y_c=c_1e^{3x}+c_2e^{-2x}$, we conclude that

$y_p=Axe^{-2x}$; $y_p'=-2Axe^{-2x}+Ae^{-2x}$; $y_p''=4Axe^{-2x}-2Ae^{-2x}-2Ae^{-2x}=4Axe^{-2x}-4Ae^{-2x}$.

Substituting in $(D^2-D-6)y=10e^{-2x}$, we get

$(4Axe^{-2x}-4Ae^{-2x})-(-2Axe^{-2x}+Ae^{-2x})-6Axe^{-2x}=10e^{-2x}$

$-5Ae^{-2x}=10e^{-2x}$ and $A=-2$.

So $y_p=-2xe^{-2x}$ and $y=c_1e^{3x}+c_2e^{-2x}-2xe^{-2x}$.

49. $(D^2+3D-28)y=11e^{4x}$

$y_c: m^2+3m-28=(m+7)(m-4)=0$

$m=4,\ -7$

$y_c=c_1e^{-7x}+c_2e^{4x}$

(Note that the right side, $11e^{4x}$, is one of the terms in y_c.)

$y_p:$ Right side$=11e^{4x}$

$m'=4$

$m'-4=0$

Annihilator: $D-4$.

Applying the annihilator to the equation, we get

$(D-4)(D^2+3D-28)y=(D-4)(11e^{4x})=0$

$(m-4)(m-4)(m+7)=0$

$m=4,\ 4,\ -7.$

So y has the following form: $y=c_1e^{-7x}+c_2e^{4x}+c_3xe^{4x}$.

Since $y_c=c_1e^{-7x}+c_2e^{4x}$, we conclude that

$y_p=Axe^{4x};\ y_p'=4Axe^{4x}+Ae^{4x};$

$y_p''=16Axe^{4x}+4Ae^{4x}+4Ae^{4x}=16Axe^{4x}+8Ae^{4x}$.

Substituting in the given equation,

$(16Axe^{4x}+8Ae^{4x})+3(4Axe^{4x}+Ae^{4x})-28Axe^{4x}=11e^{4x}$

$16Axe^{4x}+8Ae^{4x}+12Axe^{4x}+3Ae^{4x}-28Axe^{4x}=11e^{4x}$

$11Ae^{4x}=11e^{4x}$

$A=1.$

So $y_p=xe^{4x}$ and $y=y_c+y_p=c_1e^{-7x}+c_2e^{4x}+xe^{4x}$.

51. $(D^2-1)y=2e^x$

$y_c:\ m^2-1=0;\ m=\pm1$

$y_c=c_1e^{-x}+c_2e^x$

(Observe that the right side is one of the terms in y_c.)

$y_p:$ Right side$=2e^x$

$m'=1$

$m'-1=0$

Annihilator: $D-1$.

Applying the annihilator to the equation, we get

$(D-1)(D^2-1)y=(D-1)(2e^x)=0$

$(m-1)(m-1)(m+1)=0$

$m=1,\ 1,\ -1.$

So y has the form: $y=c_1e^{-x}+c_2e^x+c_3xe^x$.

Since $y_c=c_1e^{-x}+c_2e^x$, we conclude that

$y_p=Axe^x;\ y_p'=Axe^x+Ae^x;\ y_p''=Axe^x+2Ae^x$.

Substituting in the given equation, we get

$Axe^x+2Ae^x-Axe^x=2e^x$

$A=1.$

So $y_p=xe^x$ and $y=c_1e^{-x}+c_2e^x+xe^x=c_1e^{-x}+(c_2+x)e^x$.

53. $(D^2-4)y=4+e^{2x}$

$y_c:\ m^2-4=0,\ m=\pm2$

$y_c=c_1e^{-2x}+c_2e^{2x}$

$y_p:$ Right side$=4+e^{2x}$

$m'=0,\ 2$

$m'(m'-2)=0$

Annihilator: $D(D-2)$.

Applying the annihilator to the equation:

$$D(D-2)(D^2-4)y = D(D-2)(4+e^{2x}) = 0$$

$$m(m-2)(m^2-4) = 0$$

$$m = 0,\ 2,\ 2,\ -2.$$

So y has the form: $y = c_1 e^{-2x} + c_2 e^{2x} + c_3 x e^{2x} + c_4 e^{0x}$.

Since $y_c = c_1 e^{-2x} + c_2 e^{2x}$, we conclude that

$y_p = Axe^{2x} + B$; $y_p' = 2Axe^{2x} + Ae^{2x}$; $y_p'' = 4Axe^{2x} + 2Ae^{2x} + 2Ae^{2x} = 4Axe^{2x} + 4Ae^{2x}$.

Substituting in the given equation,

$$(4Axe^{2x} + 4Ae^{2x}) - 4(Axe^{2x} + B) = 4 + e^{2x}$$

$$4Axe^{2x} + 4Ae^{2x} - 4Axe^{2x} - 4B = 4 + e^{2x}$$

$$-4B + 4Ae^{2x} = 4 + e^{2x}$$

$$-4B = 4$$

$$4A = 1.$$

So $A = \frac{1}{4}$ and $B = -1$ and $y_p = \frac{1}{4} xe^{2x} - 1$.

$$y = y_c + y_p = c_1 e^{-2x} + c_2 e^{2x} + \frac{1}{4} xe^{2x} - 1$$

55. $(D^2+1)y = 2\sin x$

$y_c:\ m^2 + 1 = 0;\ m = \pm \mathrm{j}$

$y_c = c_1 \cos x + c_2 \sin x$

Since the right side is one of the terms in y_c, we need the annihilator.

$y_p:$ Right side $= 2\sin x$

$$m' = \pm \mathrm{j}$$

$$(m' - \mathrm{j})(m' + \mathrm{j}) = 0$$

$$(m')^2 + 1 = 0$$

Annihilator: $D^2 + 1$.

Applying the annihilator to the given equation, we get

$$(D^2+1)(D^2+1)y = (D^2+1)(2\sin x) = 0$$

$$(m^2+1)(m^2+1) = 0$$

$$m = \pm \mathrm{j},\ \pm \mathrm{j}.$$

By Exercise 43, Section 12.2, $y = c_1 \cos x + c_2 \sin x + c_3 x \cos x + c_4 x \sin x$.

It follows that

$$y_p = Ax\cos x + Bx\sin x;$$

$$y_p' = -Ax\sin x + A\cos x + Bx\cos x + B\sin x;$$

$$y_p'' = -Ax\cos x - A\sin x - A\sin x - Bx\sin x + B\cos x + B\cos x$$

$$= -Ax\cos x - 2A\sin x - Bx\sin x + 2B\cos x.$$

Substituting in the given equation:

$$-Ax\cos x - 2A\sin x - Bx\sin x + 2B\cos x + Ax\cos x + Bx\sin x = 2\sin x$$

$$-2A\sin x + 2B\cos x = 2\sin x.$$

We obtain $A = -1$ and $B = 0$, so that $y_p = -x\cos x$ and $y = c_1\cos x + c_2 \sin x - x\cos x$.

12.4 Applications of Second-Order Equations

1. By Hooke's law

$$\begin{aligned} F &= kx \\ 4 &= k\cdot\frac{1}{2} \text{ or } k=8 && \text{spring constant} \\ \text{mass} &= \frac{4}{32}=\frac{1}{8}\text{ slug} && \text{mass} \\ m\frac{d^2x}{dt^2}+kx &= 0 && b=0 \\ \frac{1}{8}\frac{d^2x}{dt^2}+8x &= 0 \\ \frac{d^2x}{dt^2}+64x &= 0 && \text{equation} \end{aligned}$$

Conditions: (1) If $t=0,\ x=\frac{1}{4}$ ft initial position

(2) If $t=0,\ \frac{dx}{dt}=0$ intital velocity

Auxiliary equation: $m^2+64=0,\ m=\pm 8\text{j}$.

Thus
$$\begin{aligned} x(t) &= c_1\cos 8t+c_2\sin 8t; \\ x'(t) &= -8c_1\sin 8t+8c_2\cos 8t. \end{aligned}$$

From the initial conditions:
$$\begin{aligned} \frac{1}{4} &= c_1+c_2(0) \quad \text{or} \quad c_1 = \frac{1}{4}; \\ 0 &= 0+8c_2 \quad \text{or} \quad c_2 = 0. \end{aligned}$$

Substituting c_1 and c_2, we get $x(t)=\frac{1}{4}\cos 8t$.

3. By Hooke's law,

$$\begin{aligned} F &= kx \\ 12 &= k\cdot 2 \text{ or } k=6 && \text{spring constant} \\ \text{mass: } \frac{12}{32} &= \frac{3}{8}\text{ slug} && \text{mass} \\ \frac{3}{8}\frac{d^2x}{dt^2}+6x &= 0 && b=0 \\ \frac{d^2x}{dt^2}+16x &= 0 && \text{equation} \end{aligned}$$

Initial conditions: (1) If $t=0,\ x=\frac{8}{12}=\frac{2}{3}$ ft initial position

(2) If $t=0,\ \frac{dx}{dt}=3$ ft/s initial velocity

Auxiliary equation: $m^2+16=0$ or $m=\pm 4\text{j}$.

Thus
$$\begin{aligned} x(t) &= c_1\cos 4t+c_2\sin 4t; \\ x'(t) &= -4c_1\sin 4t+4c_2\cos 4t. \end{aligned}$$

From initial conditions:
$$\begin{aligned} \frac{2}{3} &= c_1+c_2(0) \quad \text{or} \quad c_1 = \frac{2}{3}; \\ 3 &= 0+4c_2 \quad \text{or} \quad c_2 = \frac{3}{4}. \end{aligned}$$

Substituting c_1 and c_2: $x(t)=\frac{2}{3}\cos 4t+\frac{3}{4}\sin 4t$.

5. From Exercise 3, $\frac{d^2x}{dt^2}+16x=0$.

Initial conditions: (1) If $t=0,\ x=\frac{2}{3}$ ft initial position

(2) If $t=0,\ \frac{dx}{dt}=-4$ ft/s initial velocity

From $m^2 + 16 = 0$, we get $m = \pm 4j$ and

$$\begin{aligned} x(t) &= c_1 \cos 4t + c_2 \sin 4t; \\ x'(t) &= -4c_1 \sin 4t + 4c_2 \cos 4t. \\ \frac{2}{3} &= c_1 + 0 \qquad x = \frac{2}{3} \text{ when } t = 0 \\ -4 &= 0 + 4c_2 \qquad \frac{dx}{dt} = -4 \text{ when } t = 0 \end{aligned}$$

So $c_1 = \frac{2}{3}$ and $c_2 = -1$ and $x(t) = \frac{2}{3}\cos 4t - \sin 4t$.

7. From Hooke's law:

$$\begin{aligned} F &= kx \\ 4 &= k \cdot \frac{1}{2} \text{ or } k = 8 \qquad \text{spring constant} \\ m = \frac{4}{32} &= \frac{1}{8} \text{ slug} \qquad \text{mass} \\ m\frac{d^2x}{dt^2} + kx &= f(t) \qquad b = 0 \\ \frac{1}{8}\frac{d^2x}{dt^2} + 8x &= \frac{1}{4}\cos 6t \\ \frac{d^2x}{dt^2} + 64x &= 2\cos 6t \qquad \text{multiplying by 8} \end{aligned}$$

Conditions: (1) If $t = 0$, $x = \frac{1}{4}$ ft — intitial position

(2) If $t = 0$, $\frac{dx}{dt} = 0$ — initial velocity

Auxiliary equation:

$$\begin{aligned} x_c: \ m^2 + 64 &= 0, \ m = \pm 8j \\ x_c &= c_1 \cos 8t + c_2 \sin 8t \end{aligned}$$

x_p : $x_p = A\cos 6t + B\sin 6t$; $x_p' = -6A\sin 6t + 6B\cos 6t$; $x_p'' = -36A\cos 6t - 36B\sin 6t$

Substituting, $(-36A\cos 6t - 36B\sin 6t) + 64\,(A\cos 6t + B\sin 6t) = 2\cos 6t$

$28A\cos 6t + 28B\sin 6t = 2\cos 6t + 0\sin 6t$

We obtain: $A = \frac{2}{28} = \frac{1}{14}$ and $B = 0$. So $x_p = \frac{1}{14}\cos 6t$ and

$$\begin{aligned} x(t) &= c_1 \cos 8t + c_2 \sin 8t + \frac{1}{14}\cos 6t \\ x'(t) &= -8c_1 \sin 8t + 8c_2 \cos 8t - \frac{6}{14}\sin 6t. \end{aligned}$$

From the initial conditions,

$$\begin{aligned} \frac{1}{4} &= c_1 + 0 + \frac{1}{14}; \qquad x = \frac{1}{4} \text{ when } t = 0 \\ 0 &= 0 + 8c_2 - 0: \qquad \frac{dx}{dt} = 0 \text{ when } t = 0 \end{aligned}$$

So $c_2 = 0$ and $c_1 = \frac{1}{4} - \frac{1}{14} = \frac{5}{28}$. The solution is $x(t) = \frac{5}{28}\cos 8t + \frac{1}{14}\cos 6t$.

9. From Hooke's law:

$$\begin{aligned} F &= kx \\ 4 &= k \cdot 2 \text{ or } k = 2 \qquad \text{spring constant} \\ \text{mass: } \frac{4}{32} &= \frac{1}{8} \text{ slug} \qquad \text{mass} \\ \frac{1}{8}\frac{d^2x}{dt^2} + \frac{1}{8}\frac{dx}{dt} + 2x &= 0 \qquad b = \frac{1}{8} \\ \frac{d^2x}{dt^2} + \frac{dx}{dt} + 16x &= 0 \qquad \text{equation} \end{aligned}$$

Initial conditions: (1) If $t = 0$, $x = \frac{1}{2}$ ft   initial position

(2) If $t = 0$, $\frac{dx}{dt} = 0$ initial velocity

Auxiliary equation:

$$\begin{aligned} m^2 + m + 16 &= 0 \\ m &= \frac{-1 \pm \sqrt{1-64}}{2} = \frac{-1 \pm \sqrt{63}\,\mathrm{j}}{2} \\ &= \frac{-1 \pm 3\sqrt{7}\,\mathrm{j}}{2} = -\frac{1}{2} \pm \frac{3}{2}\sqrt{7}\,\mathrm{j}. \end{aligned}$$

Thus $x(t) = \mathrm{e}^{-t/2}\left(c_1 \cos\frac{3}{2}\sqrt{7}\,t + c_2 \sin\frac{3}{2}\sqrt{7}\,t\right)$.

By the product rule,

$$x'(t) = \mathrm{e}^{-t/2}\left(-\frac{3}{2}\sqrt{7}\,c_1 \sin\frac{3}{2}\sqrt{7}\,t + \frac{3}{2}\sqrt{7}\,c_2 \cos\frac{3}{2}\sqrt{7}\,t\right) - \frac{1}{2}\mathrm{e}^{-t/2}\left(c_1 \cos\frac{3}{2}\sqrt{7}\,t + c_2 \sin\frac{3}{2}\sqrt{7}\,t\right).$$

In the first equation, using $t = 0$ and $x = \frac{1}{2}$:

$\frac{1}{2} = 1(c_1 + c_2 \cdot 0)$, so that $c_1 = \frac{1}{2}$.

In the second equation, let $t = 0$ and $x'(t) = 0$:

$0 = 1\left(0 + \frac{3}{2}\sqrt{7}c_2\right) - \frac{1}{2}(c_1 + 0)$, so that $0 = \frac{3}{2}\sqrt{7}\,c_2 - \frac{1}{2}c_1$. Since $c_1 = \frac{1}{2}$,

$\frac{3}{2}\sqrt{7}\,c_2 - \frac{1}{4} = 0$ and $c_2 = \frac{1}{6\sqrt{7}}$.

It follows that $x(t) = \mathrm{e}^{-t/2}\left(\frac{1}{2}\cos\frac{3}{2}\sqrt{7}\,t + \frac{1}{6\sqrt{7}}\sin\frac{3}{2}\sqrt{7}\,t\right)$.

11. From Hooke's Law, $F = kx$ $m = \frac{5}{32}$ slug $b = 0$

$$\begin{aligned} 5 &= k \cdot \frac{1}{2} \\ k &= 10 \end{aligned}$$

Equation: $\frac{5}{32}\frac{d^2x}{dt^2} + 10x = \frac{5}{8}\sin 3t$

Multiplying by $\frac{32}{5}$: $\frac{d^2x}{dt^2} + 64x = 4\sin 3t$

$$\begin{aligned} x_c:\ m^2 + 64 &= 0,\ m = \pm 8j \\ x_c &= c_1 \cos 8t + c_2 \sin 8t \end{aligned}$$

$x_p: x_p = A\cos 3t + B\sin 3t$; $x_p' = -3A\sin 3t + 3B\cos 3t$; $x_p'' = -9A\cos 3t - 9B\sin 3t$

$-9A\cos 3t - 9B\sin 3t + 64A\cos 3t + 64B\sin 3t = 4\sin 3t$

$55A\cos 3t + 55B\sin 3t = 0\cos 3t + 4\sin 3t$

We obtain: $A = 0$ and $B = \frac{4}{55}$.

Since $x_p = \frac{4}{55}\sin 3t$, we get $x(t) = c_1\cos 8t + c_2\sin 8t + \frac{4}{55}\sin 3t$.

(1) When $t = 0$, $x = -\frac{1}{3}$ ft   initial position

(2) When $t = 0$, $\frac{dx}{dt} = 0$ initial velocity

$x'(t) = -8c_1\sin 8t + 8c_2\cos 8t + \frac{12}{55}\cos 3t$

From the initial conditions,

$$\begin{aligned} x(0) &= -\frac{1}{3} = c_1 + 0 + 0,\ \text{or } c_1 = -\frac{1}{3} \\ x'(0) &= 0 = 0 + 8c_2 + \frac{12}{55} \text{ and } c_2 = -\frac{3}{110} \end{aligned}$$

The solution is $x(t) = -\frac{1}{3}\cos 8t - \frac{3}{110}\sin 8t + \frac{4}{55}\sin 3t$.

13. Given: $k = 3$, $m = \frac{8}{32} = \frac{1}{4}$ slug, and $b = \frac{1}{4}$.

$$\begin{aligned} m\frac{d^2x}{dt^2} + b\frac{dx}{dt} + kx &= 0 \\ \frac{1}{4}\frac{d^2x}{dt^2} + \frac{1}{4}\frac{dx}{dt} + 3x &= 0 \\ \frac{d^2x}{dt^2} + \frac{dx}{dt} + 12x &= 0 \end{aligned}$$

(1) If $t = 0$, $x = -\frac{3}{12} = -\frac{1}{4}$ ft   initital position

(2) If $t = 0$, $\frac{dx}{dt} = 0$ initial velocity

Auxiliary equation:

$$\begin{aligned} m^2 + m + 12 &= 0 \\ m &= \frac{-1 \pm \sqrt{1^2 - 4(1)(12)}}{2} = \frac{-1 \pm \sqrt{-47}}{2} = -\frac{1}{2} \pm \frac{\sqrt{47}}{2}\mathrm{j}. \end{aligned}$$

Thus $x(t) = \mathrm{e}^{-t/2}\left(c_1 \cos\frac{\sqrt{47}}{2}t + c_2 \sin\frac{\sqrt{47}}{2}t\right)$.

By the product rule,

$$x'(t) = \mathrm{e}^{-t/2}\left(-\frac{\sqrt{47}}{2}c_1 \sin\frac{\sqrt{47}}{2}t + \frac{\sqrt{47}}{2}c_2 \cos\frac{\sqrt{47}}{2}t\right) - \frac{1}{2}\mathrm{e}^{-t/2}\left(c_1 \cos\frac{\sqrt{47}}{2}t + c_2 \sin\frac{\sqrt{47}}{2}t\right).$$

In the first equation, let $t = 0$ and $x = -\frac{1}{4}$:

$-\frac{1}{4} = 1(c_1 + 0)$ or $c_1 = -\frac{1}{4}$.

In the second equation, let $t = 0$ and $x'(t) = 0$:

$$\begin{aligned} 0 &= 1\left(0 + \frac{\sqrt{47}}{2}c_2\right) - \frac{1}{2}(c_1 + 0) \\ 0 &= \frac{\sqrt{47}}{2}c_2 - \frac{1}{2}c_1 \\ 0 &= \frac{\sqrt{47}}{2}c_2 - \frac{1}{2}\left(-\frac{1}{4}\right) \text{ and } c_2 = -\frac{1}{4\sqrt{47}}. \end{aligned}$$

The solution is $x(t) = \mathrm{e}^{-t/2}\left(-\frac{1}{4}\cos\frac{\sqrt{47}}{2}t - \frac{1}{4\sqrt{47}}\sin\frac{\sqrt{47}}{2}t\right)$.

15. We are given that $k = 10$, $b = \frac{5}{8}$, and $m = \frac{10}{32} = \frac{5}{16}$.

$$\begin{aligned} \frac{5}{16}\frac{d^2x}{dt^2} + \frac{5}{8}\frac{dx}{dt} + 10x &= 0 \\ \frac{d^2x}{dt^2} + 2\frac{dx}{dt} + 32x &= 0 \quad \text{equation} \end{aligned}$$

(1) If $t = 0$, $x = \frac{1}{2}$ ft   initial position

(2) If $t = 0$, $\frac{dx}{dt} = -3$ ft/s   initial velocity

Auxiliary equation:

$$\begin{aligned} m^2 + 2m + 32 &= 0 \\ m &= \frac{-2 \pm \sqrt{4 - 4(32)}}{2} = \frac{-2 \pm \sqrt{-124}}{2} = \frac{-2 \pm 2\sqrt{31}\,\mathrm{j}}{2} = -1 \pm \sqrt{31}\,\mathrm{j}. \end{aligned}$$

Thus $x(t) = \mathrm{e}^{-t}\left(c_1 \cos\sqrt{31}\,t + c_2 \sin\sqrt{31}\,t\right)$.

By the product rule,

$$x'(t) = \mathrm{e}^{-t}\left(-\sqrt{31}\,c_1 \sin\sqrt{31}\,t + \sqrt{31}\,c_2 \cos\sqrt{31}\,t\right) - \mathrm{e}^{-t}\left(c_1 \cos\sqrt{31}\,t + c_2 \sin\sqrt{31}\,t\right).$$

In the first equation, let $t = 0$ and $x = \frac{1}{2}$:
$\frac{1}{2} = 1(c_1 + 0)$ or $c_1 = \frac{1}{2}$.
In the second equation, let $t = 0$ and $x'(t) = -3$:
$-3 = 1\left(0 + \sqrt{31}\, c_2\right) - 1(c_1 + 0)$.
Since $c_1 = \frac{1}{2}$, we get $-3 = \sqrt{31}\, c_2 - \frac{1}{2}$, or $c_2 = -\frac{5}{2\sqrt{31}}$.
Finally, $x(t) = \mathrm{e}^{-t}\left(\frac{1}{2}\cos\sqrt{31}\, t - \frac{5}{2\sqrt{31}}\sin\sqrt{31}\, t\right)$.

17. By Hooke's law,

$$\begin{aligned} F &= kx \\ 2 &= k \cdot \frac{1}{2} \text{ or } k = 4 && \text{spring constant} \\ m &= \frac{2}{32} = \frac{1}{16}\text{ slug} && \text{mass} \end{aligned}$$

damping force: $1\frac{dx}{dx}$ so that $b = 1$.

$$\begin{aligned} \frac{1}{16}\frac{d^2x}{dt^2} + 1\frac{dx}{dt} + 4x &= 2\sin 8t \\ \frac{d^2x}{dt^2} + 16\frac{dx}{dt} + 64x &= 32\sin 8t \end{aligned}$$

Initial conditions: (1) If $t = 0$, $x = \frac{3}{12} = \frac{1}{4}$ ft   initial position
(2) If $t = 0$, $\frac{dx}{dt} = 0$ initial velocity

$$\begin{aligned} x_c:\ m^2 + 16m + 64 &= 0 \\ (m+8)^2 &= 0 \\ m &= -8,\ -8 \end{aligned}$$

$x_c = c_1\mathrm{e}^{-8t} + c_2 t\mathrm{e}^{-8t}$
$x_p : x_p = A\cos 8t + B\sin 8t$; $x_p' = -8A\sin 8t + 8B\cos 8t$; $x_p'' = -64A\cos 8t - 64B\sin 8t$
Substituting,

$$(-64A\cos 8t - 64B\sin 8t) + 16(-8A\sin 8t + 8B\cos 8t) + 64(A\cos 8t + B\sin 8t) = 32\sin 8t.$$

Collecting terms, $-128A\sin 8t + 128B\cos 8t = 32\sin 8t + 0\cos 8t$.
We obtain: $A = -\frac{32}{128} = -\frac{1}{4}$ and $B = 0$. So $x_p = -\frac{1}{4}\cos 8t$ and

$$\begin{aligned} x(t) &= c_1\mathrm{e}^{-8t} + c_2 t\mathrm{e}^{-8t} - \frac{1}{4}\cos 8t; \\ x'(t) &= -8c_1\mathrm{e}^{-8t} - 8c_2 t\mathrm{e}^{-8t} + c_2\mathrm{e}^{-8t} + 2\sin 8t. \end{aligned}$$

Substituting the initial conditions,

$$\begin{aligned} \frac{1}{4} &= c_1 + 0 - \frac{1}{4} && x = \frac{1}{4} \text{ when } t = 0 \\ 0 &= -8c_1 + 0 + c_2 + 0. && \frac{dx}{dt} = 0 \text{ when } t = 0 \end{aligned}$$

We obtain: $c_1 = \frac{1}{2}$ and $0 = -8\left(\frac{1}{2}\right) + c_2$ or $c_2 = 4$; thus
$x(t) = \frac{1}{2}\mathrm{e}^{-8t} + 4t\mathrm{e}^{-8t} - \frac{1}{4}\cos 8t$.

19. $k = 2$ (given), $m = \frac{8}{32} = \frac{1}{4}$ slug, $b = \frac{1}{2}$

$$\begin{aligned}\frac{1}{4}\frac{d^2x}{dt^2} + \frac{1}{2}\frac{dx}{dt} + 2x &= \cos 4t \\ \frac{d^2x}{dt^2} + 2\frac{dx}{dt} + 8x &= 4\cos 4t\end{aligned}$$

Initial position: $x(0) = \frac{1}{2}$
Initial velocity: $x'(0) = 0$

$$\begin{aligned}x_c:\ m^2 + 2m + 8 &= 0 \\ m &= \frac{-2 \pm \sqrt{4-32}}{2} = \frac{-2 \pm 2\sqrt{7}j}{2} = -1 \pm \sqrt{7}j \\ x_c &= e^{-t}\left(c_1 \cos\sqrt{7}t + c_2 \sin\sqrt{7}t\right)\end{aligned}$$

$x_p:\ x_p = A\cos 4t + B\sin 4t;\ x_p' = -4A\sin 4t + 4B\cos 4t;\ x_p'' = -16A\cos 4t - 16B\sin 4t$
Substituting, we get
$(-16A\cos 4t - 16B\sin 4t) + 2(-4A\sin 4t + 4B\cos 4t) + 8A\cos 4t + 8B\sin 4t = 4\cos 4t$
$(-8A + 8B)\cos 4t + (-8A - 8B)\sin 4t = 4\cos 4t + 0\sin 4t$
Comparing coefficients:

$$\begin{aligned}-8A + 8B &= 4 \\ -8A - 8B &= 0 \\ \hline -16A &= 4,\quad A = -\frac{1}{4} \text{ and } B = \frac{1}{4}\end{aligned}$$

Next, we write $x(t)$ and $x'(t)$:

$$\begin{aligned}x(t) &= e^{-t}\left(c_1\cos\sqrt{7}t + c_2\sin\sqrt{7}t\right) - \frac{1}{4}\cos 4t + \frac{1}{4}\sin 4t \\ x'(t) &= e^{-t}\left(-c_1\sqrt{7}\sin\sqrt{7}t + c_2\sqrt{7}\cos\sqrt{7}t\right) - e^{-t}\left(c_1\cos\sqrt{7}t + c_2\sin\sqrt{7}t\right) + \sin 4t + \cos 4t\end{aligned}$$

From the initial conditions,
$x(0) = \frac{1}{2} = c_1 - \frac{1}{4}$; hence $c_1 = \frac{3}{4}$
$x'(0) = 0 = c_2\sqrt{7} - c_1 + 1 = c_2\sqrt{7} - \frac{3}{4} + 1$; hence $c_2 = -\frac{1}{4\sqrt{7}}$.
The solution is $x(t) = e^{-t}\left(\frac{3}{4}\cos\sqrt{7}t - \frac{1}{4\sqrt{7}}\sin\sqrt{7}t\right) - \frac{1}{4}\cos 4t + \frac{1}{4}\sin 4t$.

21. From $F = kx$, $k = 200$. We are also given that $m = 2.0\,\text{kg}$, $b = 4$, and $f(t) = 20\sin 5t$.

$$\begin{aligned}m\frac{d^2x}{dt^2} + b\frac{dx}{dt} + kx &= f(t) \\ 2.0\frac{d^2x}{dt^2} + 4\frac{dx}{dt} + 200x &= 20\sin 5t \\ &\text{and} \\ \frac{d^2x}{dt^2} + 2\frac{dx}{dt} + 100x &= 10\sin 5t \qquad (*)\end{aligned}$$

$$\begin{aligned}x_c:\ m^2 + 2m + 100 &= 0 \\ m &= \frac{-2 \pm \sqrt{4-400}}{2} = \frac{-2 \pm \sqrt{-396}}{2} \\ &= -1 \pm \frac{1}{2}\sqrt{396}\,\mathrm{j} = -1 \pm \sqrt{99}\,\mathrm{j}\end{aligned}$$

$x_c = e^{-t}\left(c_1\cos\sqrt{99}\,t + c_2\sin\sqrt{99}\,t\right)$

x_p : $x_p = A\cos 5t + B\sin 5t$; $x_p' = -5A\sin 5t + 5B\cos 5t$; $x_p'' = -25A\cos 5t - 25B\sin 5t$

Substituting in equation (*):

$$\begin{aligned}(-25A\cos 5t - 25B\sin 5t) + 2(-5A\sin 5t + 5B\cos 5t)&\\ +100(A\cos 5t + B\sin 5t) &= 10\sin 5t\\ (75A + 10B)\cos 5t + (-10A + 75B)\sin 5t &= 0\cos 5t + 10\sin 5t\\ 75A + 10B &= 0\\ -10A + 75B &= 10\end{aligned}$$

$$A = \frac{\begin{vmatrix} 0 & 10 \\ 10 & 75 \end{vmatrix}}{\begin{vmatrix} 75 & 10 \\ -10 & 75 \end{vmatrix}} = \frac{0-100}{75^2+10^2} = \frac{-100}{5725} = -0.01747$$

$$B = \frac{\begin{vmatrix} 75 & 0 \\ -10 & 10 \end{vmatrix}}{\begin{vmatrix} 75 & 10 \\ -10 & 75 \end{vmatrix}} = \frac{750}{5725} = 0.131$$

General solution:

$$\begin{aligned}x(t) &= \mathrm{e}^{-t}\left(c_1\cos\sqrt{99}\,t + c_2\sin\sqrt{99}\,t\right) - 0.01747\cos 5t + 0.131\sin 5t;\\ x'(t) &= \mathrm{e}^{-t}\left(-\sqrt{99}\,c_1\sin\sqrt{99}\,t + \sqrt{99}\,c_2\cos\sqrt{99}\,t\right) - \mathrm{e}^{-t}\left(c_1\cos\sqrt{99}\,t + c_2\sin\sqrt{99}\,t\right)\\ &\quad +0.0874\sin 5t + 0.655\cos 5t.\end{aligned}$$

Initial conditions: if $t = 0$, $x = 0.25$ and $\dfrac{dx}{dt} = 0$.

$$\begin{aligned}(1)\quad 0.25 &= c_1 - 0.01747, \text{ or } c_1 = 0.267\\ (2)\quad 0 &= \sqrt{99}\,c_2 - c_1 + 0.655\\ 0 &= \sqrt{99}\,c_2 - 0.267 + 0.655, \text{ or } c_2 = -0.0390\end{aligned}$$

$$\begin{aligned}\text{Thus }\ x_c &= \mathrm{e}^{-t}\left(0.267\cos\sqrt{99}\,t - 0.0390\sin\sqrt{99}\,t\right);\\ x_p &= -0.01747\cos 5t + 0.131\sin 5t.\end{aligned}$$

Using two significant digits,

$$\begin{aligned}x_c &= \mathrm{e}^{-1.0t}\left(0.27\cos 9.9t - 0.039\sin 9.9t\right);\\ x_p &= -0.017\cos 5t + 0.13\sin 5t.\end{aligned}$$

23.
$$\begin{aligned}L\frac{d^2q}{dt^2} + \frac{1}{C}q &= 0\\ \frac{d^2q}{dt^2} + \frac{1}{1.0\times 10^{-4}}q &= 0\\ \frac{d^2q}{dt^2} + 10^4 q &= 0\end{aligned}$$

Initial conditions: (1) If $t = 0$, $q = 0$.

(2) If $t = 0$, $i = \dfrac{dq}{dt} = 10$.

Auxiliary equation: $m^2 + 10^4 = 0$; $m = \pm 10^2\mathrm{j} = \pm 100\mathrm{j}$.

$$\begin{aligned}q(t) &= c_1\cos 100t + c_2\sin 100t\\ q'(t) &= -100c_1\sin 100t + 100c_2\cos 100t\\ 0 &= c_1 + 0 \text{ or } c_1 = 0\\ 10 &= 0 + 100c_2 \text{ or } c_2 = \frac{1}{10}\\ q(t) &= \frac{1}{10}\sin 100t\\ i &= \frac{dq}{dt} = 10\cos 100t\end{aligned}$$

25.
$$\begin{aligned} L\frac{d^2q}{dt^2} + \frac{1}{C}q &= \mathrm{e}(t) \\ 0.5\frac{d^2q}{dt^2} + \frac{1}{8\times 10^{-4}}q &= 50\sin 100t \\ \frac{d^2q}{dt^2} + 2500q &= 100\sin 100t \qquad (*) \end{aligned}$$

$$\begin{aligned} q_c: \quad q_c &= c_1\cos 50t + c_2\sin 50t \\ q_p: \quad q_p &= A\cos 100t + B\sin 100t \\ q_p' &= -100A\sin 100t + 100B\cos 100t \\ q_p'' &= -10000A\cos 100t - 10000B\sin 100t \end{aligned}$$

Substituting in equation (*):

$$\begin{aligned} (-10000A\cos 100t - 10000B\sin 100t) + 2500(A\cos 100t + B\sin 100t) &= 100\sin 100t \\ (-10000+2500)B\sin 100t + (-10000+2500)A\cos 100t &= 100\sin 100t. \end{aligned}$$

$$\begin{aligned} -7500B &= 100,\ B = -\frac{1}{75} \\ A &= 0 \end{aligned}$$

General solution:
$$\begin{aligned} q &= c_1\cos 50t + c_2\sin 50t - \left(\frac{1}{75}\right)\sin 100t; \\ \frac{dq}{dt} &= -50c_1\sin 50t + 50c_2\cos 50t - \left(\frac{4}{3}\right)\cos 100t. \end{aligned}$$

Initial conditions: if $t = 0$, then $q = 0$ and $\dfrac{dq}{dt} = 0$.

$$\begin{aligned} (1)\quad 0 &= c_1 \\ (2)\quad 0 &= 50c_2 - \frac{4}{3},\ \text{or } c_2 = \frac{2}{75} \end{aligned}$$

Thus $q(t) = \dfrac{2}{75}\sin 50t - \dfrac{1}{75}\sin 100t$.

27. $\dfrac{d^2q}{dt} + 10\dfrac{dq}{dt} + 100q = 50\cos 10t$, $q(0) = q'(0) = 0$

$$\begin{aligned} q_c: \ m^2 + 10m + 100 &= 0 \\ m &= \frac{-10\pm\sqrt{100-400}}{2} = \frac{-10\pm 10\sqrt{3}j}{2} = -5\pm 5\sqrt{3}j \\ q_c &= \mathrm{e}^{-5t}\left(c_1\cos 5\sqrt{3}t + c_2\sin 5\sqrt{3}t\right) \end{aligned}$$

$q_p: \ q_p = A\cos 10t + B\sin 10t;\ q_p' = -10A\sin 10t + 10B\cos t;\ q_p'' = -100A\cos 10t - 100B\sin 10t$

Substituting,

$-100A\cos 10t - 100B\sin 10t - 100A\sin 10t + 100B\cos 10t + 100A\cos 10t + 100B\sin 10t = 50\cos 10t$

$-100A\sin 10t + 100B\cos 10t = 50\cos 10t$

We obtain: $A = 0$ and $B = \dfrac{1}{2}$. Next we write $q(t)$ and $q'(t)$:

$$\begin{aligned} q(t) &= \mathrm{e}^{-5t}\left(c_1\cos 5\sqrt{3}t + c_2\sin 5\sqrt{3}t\right) + \frac{1}{2}\sin 10t \\ q'(t) &= \mathrm{e}^{-5t}\left[c_1\left(-5\sqrt{3}\right)\sin 5\sqrt{3}t + c_2\left(5\sqrt{3}\right)\cos 5\sqrt{3}t\right] - 5\mathrm{e}^{-5t}\left(c_1\cos 5\sqrt{3}t + c_2\sin 5\sqrt{3}t\right) \\ &\quad +5\cos 10t \end{aligned}$$

From the initial conditions, $q(0) = 0 = c_1$ or $c_1 = 0$

$q'(0) = 0 = 1\left[0 + c_2\left(5\sqrt{3}\right)\right] - 5\left(c_1 + 0\right) + 5$. Since $c_1 = 0$, we get $c_2 = -\dfrac{1}{\sqrt{3}}$.

Solution: $q(t) = -\dfrac{1}{\sqrt{3}}\mathrm{e}^{-5t}\sin 5\sqrt{3}t + \dfrac{1}{2}\sin 10t$.

Chapter 12 Review

1. $D^4y = 0,\ m^4 = 0$, so that $m = 0,\ 0,\ 0,\ 0.$
   $y = c_1 + c_2x + c_3x^2 + c_4x^3$

3. $(D^2 + 4)y = 0$
   $m^2 + 4 = 0;\ m^2 = -4;\ m = \pm 2\mathrm{j}$
   $y = c_1 \cos 2x + c_2 \sin 2x$

5. $(D^2 - 2D - 2)y = 0;$

$$\begin{aligned} m^2 - 2m - 2 &= 0 \\ m &= \frac{2 \pm \sqrt{4+8}}{2} = \frac{2 \pm 2\sqrt{3}}{2} = 1 \pm \sqrt{3} \end{aligned}$$

$$y = c_1 e^{(1+\sqrt{3})x} + c_2 e^{(1-\sqrt{3})x} = e^x \left(c_1 e^{\sqrt{3}x} + c_2 e^{-\sqrt{3}x} \right)$$

7. $(3D^2 - D + 1)y = 0$

$$\begin{aligned} 3m^2 - m + 1 &= 0 \\ m &= \frac{-(-1) \pm \sqrt{(-1)^2 - 4(3)(1)}}{2 \cdot 3} = \frac{1 \pm \sqrt{-11}}{6} = \frac{1}{6} \pm \frac{\sqrt{11}}{6}\mathrm{j} \end{aligned}$$

$$y = e^{(1/6)x} \left(c_1 \cos \frac{\sqrt{11}}{6}x + c_2 \sin \frac{\sqrt{11}}{6}x \right)$$

9.
$$\begin{aligned} (m-2)^2(m^2+1) &= 0 \\ m &= 2,\ 2,\ \pm\mathrm{j} \end{aligned}$$
$$y = c_1 e^{2x} + c_2 x e^{2x} + c_3 \cos x + c_4 \sin x$$

11. $2m^2 - m + 1 = 0$

$$m = \frac{-(-1) \pm \sqrt{(-1)^2 - 4(2)(1)}}{2 \cdot 2} = \frac{1 \pm \sqrt{-7}}{4} = \frac{1}{4} \pm \frac{\sqrt{7}}{4}\mathrm{j}$$

$$y = e^{(1/4)x} \left(c_1 \cos \frac{\sqrt{7}}{4}x + c_2 \sin \frac{\sqrt{7}}{4}x \right)$$

13. $(3D^2 - 2D - 2)y = 0$

$$\begin{aligned} 3m^2 - 2m - 2 &= 0 \\ m &= \frac{2 \pm \sqrt{4+24}}{6} = \frac{2 \pm 2\sqrt{7}}{6} = \frac{1}{3} \pm \frac{1}{3}\sqrt{7} \end{aligned}$$

$$y = c_1 e^{(1/3+\sqrt{7}/3)x} + c_2 e^{(1/3-\sqrt{7}/3)x} = e^{(1/3)x} \left(c_1 e^{(\sqrt{7}/3)x} + c_2 e^{-(\sqrt{7}/3)x} \right)$$

15. $(D^2 - 3D - 4)y = 6e^x$

y_c:
$$\begin{aligned} m^2 - 3m - 4 &= 0 \\ (m-4)(m+1) &= 0 \\ m &= 4, -1 \end{aligned}$$

$y_c = c_1e^{4x} + c_2e^{-x}$

y_p: $y_p = Ae^x$; $y_p' = Ae^x$; $y_p'' = Ae^x$

Substituting in $(D^2 - 3D - 4)y = 6e^x$, we get
$$\begin{aligned} Ae^x - 3Ae^x - 4Ae^x &= 6e^x \\ -6Ae^x &= 6e^x \\ A &= -1. \end{aligned}$$

So $y_p = -e^x$ and $y = y_c - e^x$.

17. $(D^2 - 3D - 4)y = 2\sin x$

y_c:
$$\begin{aligned} m^2 - 3m - 4 &= 0 \\ (m-4)(m+1) &= 0 \\ m &= 4, -1 \end{aligned}$$

$y_c = c_1e^{4x} + c_2e^{-x}$

y_p: $y_p = A\cos x + B\sin x$; $y_p' = -A\sin x + B\cos x$; $y_p'' = -A\cos x - B\sin x$

Substituting in the equation:
$$\begin{aligned} (-A\cos x - B\sin x) - 3(-A\sin x + B\cos x) - 4(A\cos x + B\sin x) &= 2\sin x \\ (-5A - 3B)\cos x + (3A - 5B)\sin x &= 0\cos x + 2\sin x \\ -5A - 3B &= 0 \\ 3A - 5B &= 2. \end{aligned}$$

Solution: $B = -\dfrac{5}{17}$, $A = \dfrac{3}{17}$.

Thus $y = c_1e^{4x} + c_2e^{-x} + \dfrac{3}{17}\cos x - \dfrac{5}{17}\sin x$.

19. $(D^2 - 3D - 4)y = 10xe^{-x}$

y_c:
$$\begin{aligned} m^2 - 3m - 4 &= 0 \\ (m-4)(m+1) &= 0 \\ m &= 4, -1 \end{aligned}$$

$y_c = c_1e^{4x} + c_2e^{-x}$

(Observe that the root $m = -1$ of the auxiliary equation coincides with a root associated with the annihilator.)

y_p:
$$\begin{aligned} \text{Right side} &= 10xe^{-x} \\ m' &= -1, -1 \qquad \text{repeating root} \\ (m'+1)^2 &= 0 \end{aligned}$$

Annihilator: $(D+1)^2$.

Now apply the annihilator to the given equation:
$$\begin{aligned} (D+1)^2(D^2 - 3D - 4)y &= (D+1)^2(10xe^{-x}) = 0 \\ (m+1)^2(m+1)(m-4) &= 0 \\ m &= 4, -1, -1, -1. \end{aligned}$$

It follows that y has the form: $y = c_1e^{4x} + c_2e^{-x} + c_3xe^{-x} + c_4x^2e^{-x}$.

Since $y_c = c_1 e^{4x} + c_2 e^{-x}$, y_p must have the following form:

$$\begin{aligned} y_p &= Axe^{-x} + Bx^2e^{-x}; \\ y_p' &= -Axe^{-x} + Ae^{-x} - Bx^2e^{-x} + 2Bxe^{-x}; \\ y_p'' &= Axe^{-x} - Ae^{-x} - Ae^{-x} + Bx^2e^{-x} - 2Bxe^{-x} - 2Bxe^{-x} + 2Be^{-x} \\ &= Axe^{-x} - 2Ae^{-x} + Bx^2e^{-x} - 4Bxe^{-x} + 2Be^{-x}. \end{aligned}$$

Substituting in $(D^2 - 3D - 4)y = 10xe^{-x}$, we get

$(Axe^{-x} - 2Ae^{-x} + Bx^2e^{-x} - 4Bxe^{-x} + 2Be^{-x})$

$-3(-Axe^{-x} + Ae^{-x} - Bx^2e^{-x} + 2Bxe^{-x}) - 4(Axe^{-x} + Bx^2e^{-x}) = 10xe^{-x}$.

After collecting terms, we get $(-5A + 2B)e^{-x} - 10Bxe^{-x} = 10xe^{-x}$.

We obtain:
$$\begin{aligned} -10B &= 10 \\ -5A + 2B &= 0. \end{aligned}$$

Thus $B = -1$ and $A = -\frac{2}{5}$; so that $y_p = -\frac{2}{5}xe^{-x} - x^2e^{-x}$.

21. $(D^2 - 2D - 3)y = 0$

$$\begin{aligned} m^2 - 2m - 3 &= 0 \\ (m-3)(m+1) &= 0 \\ m &= 3,\ -1 \end{aligned}$$

$$\begin{aligned} y &= c_1e^{3x} + c_2e^{-x} \\ Dy &= 3c_1e^{3x} - c_2e^{-x} \end{aligned}$$

If $x = 0$, then $y = 0$ and $Dy = -4$:

$$\begin{aligned} (1)\quad 0 &= c_1 + c_2 \\ (2)\quad -4 &= 3c_1 - c_2 \\ \hline -4 &= 4c_1 \end{aligned}$$

or $c_1 = -1$; $c_2 = 1$.

Thus $y = e^{-x} - e^{3x}$.

23.
$$\begin{aligned} 0.100\frac{d^2q}{dt^2} + 40.0\frac{dq}{dt} + \frac{1}{2.00 \times 10^{-3}}q &= 100.0\cos 20.0t \\ \frac{d^2q}{dt^2} + 400\frac{dq}{dt} + 5000q &= 1000\cos 20t \end{aligned}$$

$$\begin{aligned} q_p &= A\cos 20t + B\sin 20t; \\ q_p' &= -20A\sin 20t + 20B\cos 20t; \\ q_p'' &= -400A\cos 20t - 400B\sin 20t \end{aligned}$$

Substituting,

$$\begin{aligned} (-400A\cos 20t - 400B\sin 20t) + 400(-20A\sin 20t + 20B\cos 20t) & \\ +5000(A\cos 20t + B\sin 20t) &= 1000\cos 20t \\ (4600A + 8000B)\cos 20t + (-8000A + 4600B)\sin 20t &= 1000\cos 20t \end{aligned}$$

$$\begin{aligned} 4600A + 8000B &= 1000 \\ -8000A + 4600B &= 0 \\ \hline 46A + 80B &= 10 \\ -80A + 46B &= 0 \end{aligned}$$

$$A = \frac{\begin{vmatrix} 10 & 80 \\ 0 & 46 \end{vmatrix}}{\begin{vmatrix} 46 & 80 \\ -80 & 46 \end{vmatrix}} = \frac{10(46)}{46^2 + 80^2} = \frac{460}{8516} = 0.054$$

$$B = \frac{\begin{vmatrix} 46 & 10 \\ -80 & 0 \end{vmatrix}}{\begin{vmatrix} 46 & 80 \\ -80 & 46 \end{vmatrix}} = \frac{0 + 10(80)}{8516} = 0.0939$$

$q(t) = 0.054\cos 20.0t + 0.0939\sin 20.0t$

$i(t) = q'(t) = -1.08\sin 20.0t + 1.88\cos 20.0t$

25. By Hooke's law,

$$\begin{aligned} F &= kx \\ 5 &= k \cdot \frac{1}{2} \text{ or } k = 10 && \text{spring constant} \\ \text{mass} &= \frac{5}{32} \text{ slug} && \text{mass} \\ m\frac{d^2x}{dt^2} + kx &= f(t) && b = 0 \\ \frac{5}{32}\frac{d^2x}{dt^2} + 10x &= \frac{1}{8}\cos 4t && f(t) = \frac{1}{8}\cos 4t \\ (*) \quad \frac{d^2x}{dt^2} + 64x &= \frac{4}{5}\cos 4t && \text{multiplying by } \frac{32}{5} \end{aligned}$$

$x_c:\ m^2 + 64 = 0,\ m = \pm 8\text{j}$

$x_c = c_1\cos 8t + c_2\sin 8t$

$x_p:\ x_p = A\cos 4t + B\sin 4t;\ x_p' = -4A\sin 4t + 4B\cos 4t;\ x_p'' = -16A\cos 4t - 16B\sin 4t$

Substituting in (*):

$$(-16A\cos 4t - 16B\sin 4t) + 64(A\cos 4t + B\sin 4t) = \frac{4}{5}\cos 4t$$

or

$$48A\cos 4t + 48B\sin 4t = \frac{4}{5}\cos 4t.$$

It follows that $B = 0$ and $A = \frac{4}{5} \cdot \frac{1}{48} = \frac{1}{60}$.

General solution:

$$\begin{aligned} x(t) &= x_c + x_p = c_1\cos 8t + c_2\sin 8t + \frac{1}{60}\cos 4t; \\ x'(t) &= -8c_1\sin 8t + 8c_2\cos 8t - \frac{1}{15}\sin 4t. \end{aligned}$$

Initial conditions: (1) If $t = 0$, $x = \frac{1}{4}$ ft — initial position

(2) If $t = 0$, $\frac{dx}{dt} = -5$ ft/s — initial velocity

Substituting:

$\frac{1}{4} = c_1 + 0 + \frac{1}{60}$, or $c_1 = \frac{1}{4} - \frac{1}{60} = \frac{15}{60} - \frac{1}{60} = \frac{14}{60} = \frac{7}{30}$;

$-5 = 0 + 8c_2 + 0$, or $c_2 = -\frac{5}{8}$.

Substituting the values of c_1 and c_2, we get

$x(t) = \frac{7}{30}\cos 8t - \frac{5}{8}\sin 8t + \frac{1}{60}\cos 4t.$

Chapter 13

The Laplace Transform

Sections 13.1-13.3

1. Transform 5:
$\mathcal{L}\{\sin at\} = \int_0^\infty e^{-st} \sin at\, dt$
Integration by parts: $u = e^{-st}$, $dv = \sin at\, dt$; $du = -se^{-st}$, $v = -\frac{1}{a}\cos at$

$$\lim_{b\to\infty}\left(-\frac{1}{a}e^{-st}\cos at\Big|_0^b\right) - \frac{s}{a}\int_0^\infty e^{-st}\cos at\, dt$$

$$= \lim_{b\to\infty}\left(-\frac{1}{a}e^{-sb}\cos ab + \frac{1}{a}\right) - \frac{s}{a}\int_0^\infty e^{-st}\cos at\, dt = 0 + \frac{1}{a} - \frac{s}{a}\int_0^\infty e^{-st}\cos at\, dt$$

Integrating by parts again: $u = e^{-st}$, $dv = \cos at\, dt$; $du = -se^{-st}$, $v = \frac{1}{a}\sin at$

$$\begin{aligned}\int_0^\infty e^{-st}\sin at\, dt &= \frac{1}{a} - \frac{s}{a}\left[\frac{1}{a}e^{-st}\sin at\Big|_0^\infty + \frac{s}{a}\int_0^\infty e^{-st}\sin at\, dt\right]\\ &= \frac{1}{a} - \frac{s}{a}(0) - \frac{s^2}{a^2}\int_0^\infty e^{-st}\sin at\, dt\end{aligned}$$

Solving for the integral,

$$\begin{aligned}\left(1+\frac{s^2}{a^2}\right)\int_0^\infty e^{-st}\sin at\, dt &= \frac{1}{a}\\ \int_0^\infty e^{-st}\sin at\, dt &= \frac{1}{a}\cdot\frac{a^2}{s^2+a^2} = \frac{a}{s^2+a^2}.\end{aligned}$$

3. $f(t) = 2 + 3e^{-t} = 2\cdot 1 + 3e^{-t}$
$F(s) = 2\mathcal{L}\{1\} + 3\mathcal{L}\{e^{-t}\} = 2\cdot\frac{1}{s} + 3\frac{1}{s+1}$ (by transforms 1 and 4, respectively) or
$F(s) = \frac{2}{s}\frac{s+1}{s+1} + \frac{3}{s+1}\frac{s}{s} = \frac{2s+2+3s}{s(s+1)} = \frac{5s+2}{s(s+1)}$.

5. $f(t) = t + \cos 2t$
$F(s) = \frac{1}{s^2} + \frac{s}{s^2+4}$ by transforms 2 and 6, respectively.
Thus, $F(s) = \frac{s^2+4+s^3}{s^2(s^2+4)} = \frac{s^3+s^2+4}{s^2(s^2+4)}$.

7. $f(t) = e^{2t}\sin 5t$. By transform 7 with $a = 2$ and $b = 5$:
$F(s) = \frac{5}{(s-2)^2+5^2} = \frac{5}{(s-2)^2+25}$.

9. $f(t) = t^3 e^{-4t}$. By transform 9 with $n = 3$ and $a = -4$:
$F(s) = \dfrac{3!}{(s+4)^4} = \dfrac{6}{(s+4)^4}.$

11. $f(t) = 2t^4 e^{-t}$. By transform 9 with $n = 4$ and $a = -1$:
$F(s) = 2 \cdot \dfrac{4!}{(s+1)^{4+1}} = \dfrac{48}{(s+1)^5}.$

13. $f(t) = 4 - 5\sin 2t$. By transforms 1 and 5, respectively:
$F(s) = \dfrac{4}{s} - 5\dfrac{2}{s^2+4} = \dfrac{4}{s} - \dfrac{10}{s^2+4}.$

15. $F(s) = \dfrac{10}{s^2+4} = 5\dfrac{2}{s^2+2^2}$
$f(t) = 5\sin 2t$ by transform 5.

17. $F(s) = \dfrac{1}{(s+2)^5}$. Transform 9 with $n = 4$ and $a = -2$. The form needs to be adjusted: insert 4! and place $\dfrac{1}{4!}$ in front.
$F(s) = \dfrac{1}{4!}\dfrac{4!}{(s+2)^5} = \dfrac{1}{24}\dfrac{4!}{(s+2)^5}.$
It follows that $f(t) = \dfrac{1}{24} t^4 e^{-2t}$.

19. $F(s) = \dfrac{4}{(s^2+4)^2}$. This form can be made to fit transform 12 with $a = 2$: the numerator must be $2a^3 = 2 \cdot 2^3 = 16$. One way to make this adjustment is to insert the factor 4 and place $\dfrac{1}{4}$ in front. So

$$\begin{aligned} F(s) &= \frac{1}{4}\frac{4 \cdot 4}{(s+4)^2} = \frac{1}{4}\frac{2 \cdot 2^3}{(s+2^2)^2}; \\ f(t) &= \frac{1}{4}(\sin 2t - 2t\cos 2t). \end{aligned}$$

21. $F(s) = \dfrac{2s}{(s^2+9)^2}$. This form can be made to fit transform 13 with $a = 3$: insert the required 3 and place $\dfrac{1}{3}$ in front. $F(s) = \dfrac{1}{3}\dfrac{2 \cdot 3s}{(s^2+3^2)^2}$ so $f(t) = \dfrac{1}{3} t \sin 3t$.

23. $F(s) = \dfrac{3s}{s^2+6} = 3\dfrac{s}{s^2+6}.$
$f(t) = 3\cos\sqrt{6}\,t$ by transform 6 with $a = \sqrt{6}$.

25. $F(s) = \dfrac{3}{s^2+6} = \dfrac{3}{\sqrt{6}}\dfrac{\sqrt{6}}{s^2+6}.$
$f(t) = \dfrac{3}{\sqrt{6}}\sin\sqrt{6}t$ by transform 5 with $a = \sqrt{6}$.

27. $F(s) = \dfrac{2s+4}{(s+2)^2+4} = 2\dfrac{s+2}{(s+2)^2+2^2}$
$f(t) = 2e^{-2t}\cos 2t$ by transform 8 with $a = -2$ and $b = 2$.

29. $F(s) = \dfrac{1}{(s+3)^2+5}$. Transform 7 with $a = -3$ and $b = \sqrt{5}$. To fit this form, we need to insert $\sqrt{5}$ and place $\dfrac{1}{\sqrt{5}}$ in front:

$$F(s) = \frac{1}{\sqrt{5}} \frac{\sqrt{5}}{(s+3)^2+5} \text{ and } f(t) = \frac{1}{\sqrt{5}} e^{-3t} \sin \sqrt{5}t.$$

31. Here we need to complete the square in the denominator and use transform 7:

$$\begin{aligned} F(s) &= \frac{\sqrt{10}}{s^2-2s+11} = \frac{\sqrt{10}}{(s-1)^2+10}; \\ f(t) &= e^t \sin \sqrt{10}\,t. \qquad (a = 1,\ b = \sqrt{10}) \end{aligned}$$

33. Here we complete the square in the denominator and then use transforms 8 and 7 (after splitting the fraction).

$$\begin{aligned} \mathcal{L}^{-1}\left\{\frac{s}{s^2-6s+10}\right\} &= \mathcal{L}^{-1}\left\{\frac{s}{(s-3)^2+1}\right\} = \mathcal{L}^{-1}\left\{\frac{(s-\underline{3}+\underline{3})}{(s-3)^2+1}\right\} \\ &= \mathcal{L}^{-1}\left\{\frac{s-3}{(s-3)^2+1}\right\} + \mathcal{L}^{-1}\left\{\frac{3}{(s-3)^2+1}\right\} \\ &= \mathcal{L}^{-1}\left\{\frac{s-3}{(s-3)^2+1}\right\} + 3\mathcal{L}^{-1}\left\{\frac{1}{(s-3)^2+1}\right\} \\ &= e^{3t}\cos t + 3e^{3t}\sin t = e^{3t}(\cos t + 3\sin t) \end{aligned}$$

35. $$\begin{aligned} F(s) &= \frac{s}{s^2-2s+6} = \frac{s}{(s-1)^2+5} = \frac{(s-1)+1}{(s-1)^2+5} \\ &= \frac{s-1}{(s-1)^2+5} + \frac{1}{(s-1)^2+5} = \frac{s-1}{(s-1)^2+5} + \frac{1}{\sqrt{5}}\frac{\sqrt{5}}{(s-1)^2+5} \end{aligned}$$

$f(t) = e^t \cos \sqrt{5}\,t + \dfrac{1}{\sqrt{5}} e^t \sin \sqrt{5}\,t$ by transforms 8 and 7, respectively. Factoring e^t, we also have $f(t) = e^t \left(\cos \sqrt{5}\,t + \dfrac{1}{\sqrt{5}} \sin \sqrt{5}\,t\right)$.

37. $$\begin{aligned} \mathcal{L}^{-1}\left\{\frac{\frac{1}{3}s-1}{s^2-4s+9}\right\} &= \frac{1}{3}\mathcal{L}^{-1}\left\{\frac{s-3}{(s-2)^2+5}\right\} \\ &= \frac{1}{3}\mathcal{L}^{-1}\left\{\frac{(s-2)-1}{(s-2)^2+5}\right\} \\ &= \frac{1}{3}\mathcal{L}^{-1}\left\{\frac{s-2}{(s-2)^2+5} - \frac{1}{\sqrt{5}}\frac{\sqrt{5}}{(s-2)^2+5}\right\} \\ &= \frac{1}{3}\left(e^{2t}\cos\sqrt{5}t - \frac{1}{\sqrt{5}}e^{2t}\sin\sqrt{5}t\right) \\ &= \frac{1}{3}e^{2t}\cos\sqrt{5}t - \frac{1}{3\sqrt{5}}e^{2t}\sin\sqrt{5}t \end{aligned}$$

by transforms 8 and 7, respectively.

39. $\dfrac{1}{s(s+1)}$. Rule I, distinct linear factors:

$$\begin{aligned} \frac{A}{s} + \frac{B}{s+1} &= \frac{A(s+1)+Bs}{s(s+1)} \\ A(s+1)+Bs &= 1. \\ \text{Let } s = 0: \quad A(1)+0 &= 1 \\ A &= 1. \\ \text{Let } s = -1: \quad 0 + B(-1) &= 1 \\ B &= -1. \end{aligned}$$

$\mathcal{L}^{-1}\left\{\dfrac{1}{s(s+1)}\right\} = \mathcal{L}^{-1}\left\{\dfrac{1}{s} - \dfrac{1}{s+1}\right\} = 1 - e^{-t}$ by transforms 1 and 4, respectively.

41. $\dfrac{2s+1}{(s-2)(s+3)}$. Distinct linear factors:

$$\begin{aligned}\frac{2s+1}{(s-2)(s+3)} &= \frac{A}{s-2}+\frac{B}{s+3}=\frac{A(s+3)+B(s-2)}{(s-2)(s+3)}\\ A(s+3)+B(s-2) &= 2s+1.\end{aligned}$$

Let $s=-3:\ 0+B(-5) = -5.$ Let $s=2:\ A(5)+0 = 5$

$B = 1. \qquad A = 1.$

$$\mathcal{L}^{-1}\left\{\frac{2s+1}{(s-2)(s+3)}\right\}=\mathcal{L}^{-1}\left\{\frac{1}{s-2}+\frac{1}{s+3}\right\}=\mathrm{e}^{2t}+\mathrm{e}^{-3t}$$ by transform 4.

43. $\dfrac{s^2}{(s-2)(s+2)(s-4)}$. Rule I, distinct linear factors:

$$\frac{A}{s-2}+\frac{B}{s+2}+\frac{C}{s-4}=\frac{A(s+2)(s-4)+B(s-2)(s-4)+C(s-2)(s+2)}{(s-2)(s+2)(s-4)}$$

$A(s+2)(s-4)+B(s-2)(s-4)+C(s-2)(s+2)=s^2.$

$$\begin{aligned}\text{Let } s=-2:\ 0+B(-4)(-6)+0 &= (-2)^2\\ 24B &= 4\\ B &= \frac{1}{6}.\\ \text{Let } s=4:\ 0+0+C(2)(6) &= 4^2\\ 12C &= 16\\ C &= \frac{4}{3}.\\ \text{Let } s=2:\ A(4)(-2)+0+0 &= 2^2\\ -8A &= 4\\ A &= -\frac{1}{2}.\end{aligned}$$

$$\mathcal{L}^{-1}\left\{-\frac{1}{2}\frac{1}{s-2}+\frac{1}{6}\frac{1}{s+2}+\frac{4}{3}\frac{1}{s-4}\right\}=-\frac{1}{2}\mathrm{e}^{2t}+\frac{1}{6}\mathrm{e}^{-2t}+\frac{4}{3}\mathrm{e}^{4t}$$ by transform 4.

45. $\dfrac{3s^2}{(s+2)^2(s-1)}$. Rule I, linear factors, one repeating and one distinct:

$$\begin{aligned}\frac{3s^2}{(s+2)^2(s-1)} &= \frac{A}{s+2}+\frac{B}{(s+2)^2}+\frac{C}{s-1}\\ &= \frac{A}{s+2}\frac{(s+2)(s-1)}{(s+2)(s-1)}+\frac{B}{(s+2)^2}\frac{s-1}{s-1}+\frac{C}{s-1}\frac{(s+2)^2}{(s+2)^2}\\ &= \frac{A(s+2)(s-1)+B(s-1)+C(s+2)^2}{(s+2)^2(s-1)}\end{aligned}$$

$A(s+2)(s-1)+B(s-1)+C(s+2)^2=3s^2.$

Let $s=-2:\ 0+B(-3)+0=12$ or $B=-4$. Let $s=1:\ 0+0+C(3)^2=3$ or $C=\dfrac{1}{3}$.

Since there are only two distinct factors, we seem to have run out of values to substitute. So we use the values already obtained for B and C and then let $s=$ any value:

$A(s+2)(s-1)-4(s-1)+\dfrac{1}{3}(s+2)^2=3s^2.$

Let $s=$ any value (such as $s=0$):

$$\begin{aligned}A(2)(-1)-4(-1)+\frac{1}{3}(2)^2 &= 0\\ -2A+4+\frac{4}{3} &= 0\\ A &= \frac{8}{3}.\end{aligned}$$

$F(s)=\dfrac{8}{3}\dfrac{1}{s+2}+\dfrac{-4}{(s+2)^2}+\dfrac{1}{3}\dfrac{1}{s-1}$;

$f(t)=\dfrac{8}{3}\mathrm{e}^{-2t}-4t\mathrm{e}^{-2t}+\dfrac{1}{3}\mathrm{e}^{t}$ by transforms 4 and 9, respectively.

47. $\dfrac{9s}{(s+2)^2(s-1)}$. Rule I, repeating linear factors:

$$\begin{aligned}\frac{9s}{(s+2)^2(s-2)} &= \frac{A}{s+2}+\frac{B}{(s+2)^2}+\frac{C}{s-1}\\ &= \frac{A}{s+2}\cdot\frac{(s+2)(s-1)}{(s+2)(s-1)}+\frac{B}{(s+2)^2}\cdot\frac{s-1}{s-1}+\frac{C}{s-1}\frac{(s+2)^2}{(s+2)^2}.\end{aligned}$$

Equating numerators: $A(s+2)(s-1)+B(s-1)+C(s+2)^2=9s$.

We start by substituting convenient values.

Let $s=-2:\ B(-3)=-18 \qquad B=6$.

Let $s=1:\ C(9)=9 \qquad C=1$.

Because of the repeating factor, we have run out of values. So let's use the values already found,

$A(s+2)(s-1)+6(s-1)+1(s+2)^2=9s$

and let $s=$ any value (such as $s=0$) to find A:

$$\begin{aligned}A(2)(-1)+6(-1)+2^2 &= 0\\ A &= -1.\end{aligned}$$

We now have

$$\begin{aligned}F(s) &= \frac{-1}{s+2}+\frac{6}{(s+2)^2}+\frac{1}{s-1}\\ f(t) &= -\mathrm{e}^{-2t}+6t\mathrm{e}^{-2t}+\mathrm{e}^{t}\end{aligned}$$

by transforms 4 and 9, respectively.

49. $\dfrac{2}{(s^2+1)(s-2)}$. (Distinct factors, one linear, one quadratic) By Rules I and II,

$$\begin{aligned}\frac{2}{(s^2+1)(s-1)}=\frac{As+B}{s^2+1}+\frac{C}{s-1} &= \frac{(As+B)(s-1)+C\left(s^2+1\right)}{(s^2+1)(s-1)}\\ (As+B)(s-1)+C\left(s^2+1\right) &= 2\end{aligned}$$

Let $s=1$: $C(2)=2$ and $C=1$. We have no more real values to substitute, since s^2+1 is positive for all real s. So we use the value for C already found, i.e.,

$$(As+B)(s-1)+1\left(s^2+1\right)=2,$$

and let $s=0$ to find B:

$$B(-1)+1=2 \text{ and } B=-1.$$

We now have $(As-1)(s-1)+1(s^2+1)=2$. To find A, we let $s=$ any value (such as $s=2$).

$$\begin{aligned}s=2:\ (2A-1)(1)+5 &= 2 \qquad\qquad C=1;\ B=-1\\ 2A-1 &= 2-5=-3\\ 2A &= -2\\ A &= -1\end{aligned}$$

$F(s)=\dfrac{-s-1}{s^2+1}+\dfrac{1}{s-1}=\dfrac{1}{s-1}-\dfrac{s}{s^2+1}-\dfrac{1}{s^2+1}$

$f(t)=\mathrm{e}^{t}-\cos t-\sin t$

51. $\dfrac{1}{(s+1)(s^2+1)}$. (Distinct factors, one linear, one quadratic.) By Rules I and II:

$$\begin{aligned}
\frac{A}{s+1}+\frac{Bs+C}{s^2+1} &= \frac{A(s^2+1)+(Bs+C)(s+1)}{(s+1)(s^2+1)}\\
A(s^2+1)+(Bs+C)(s+1) &= 1.\\
\text{Let } s=-1:\ A(2)+0 &= 1\\
A &= \frac{1}{2}.
\end{aligned}$$

We have no more real values to substitute, since s^2+1 is positive for all real s. So we use the value for A already found

$\frac{1}{2}(s^2+1)+(Bs+C)(s+1)=1$

and let $s=0$, yielding C:

$\frac{1}{2}(1)+C(1)=1$ and $C=\frac{1}{2}$.

We now have $\frac{1}{2}(s^2+1)+\left(Bs+\frac{1}{2}\right)(s+1)=1$. To find B, let $s=$ any value (such as $s=1$).

$$\begin{aligned}
\text{Let } s=1:\ \frac{1}{2}(2)+\left(B+\frac{1}{2}\right)(2) &= 1 \qquad A=\frac{1}{2},\ C=\frac{1}{2}\\
1+\left(B+\frac{1}{2}\right)(2) &= 1\\
\left(B+\frac{1}{2}\right)(2) &= 0\\
B &= -\frac{1}{2}.
\end{aligned}$$

$$\mathcal{L}^{-1}\left\{\frac{1}{2}\frac{1}{s+1}-\frac{1}{2}\frac{s}{s^2+1}+\frac{1}{2}\frac{1}{s^2+1}\right\}=\frac{1}{2}e^{-t}-\frac{1}{2}\cos t+\frac{1}{2}\sin t$$

53. $\dfrac{s}{(s+1)(s^2+1)}$. Distinct factors, one linear and one quadratic:

$$\begin{aligned}
\frac{s}{(s+1)(s^2+1)} &= \frac{A}{s+1}+\frac{Bs+C}{s^2+1}=\frac{A(s^2+1)+(Bs+C)(s+1)}{(s+1)(s^2+1)}\\
A(s^2+1)+(Bs+C)(s+1) &= s.
\end{aligned}$$

Let $s=-1:\ A(2)+0=-1$ or $A=-\frac{1}{2}$.

We have no more real values to substitute since s^2+1 is positive for all real s. So we use the value for A already found, that is,

$-\frac{1}{2}(s^2+1)+(Bs+C)(s+1)=s$

and let $s=0$ to find C:

$-\frac{1}{2}(1)+C(1)=0$ or $C=\frac{1}{2}$.

We now have $-\frac{1}{2}(s^2+1)+\left(Bs+\frac{1}{2}\right)(s+1)=s$.

To find B, let $s=$ any value (such as $s=1$):

$$\begin{aligned}
-\frac{1}{2}(2)+\left(B+\frac{1}{2}\right)(2) &= 1\\
-1+2B+1 &= 1\\
B &= \frac{1}{2}.
\end{aligned}$$

$$\begin{aligned}
F(s) &= -\frac{1}{2}\frac{1}{s+1}+\frac{\frac{1}{2}s+\frac{1}{2}}{s^2+1}=-\frac{1}{2}\frac{1}{s+1}+\frac{1}{2}\frac{s}{s^2+1}+\frac{1}{2}\frac{1}{s^2+1};\\
f(t) &= -\frac{1}{2}e^{-t}+\frac{1}{2}\cos t+\frac{1}{2}\sin t.
\end{aligned}$$

55. $\dfrac{20s}{(s-1)(s+2)(s^2+4)}$. (Two linear factors, one quadratic.)

$$\frac{A}{s-1}+\frac{B}{s+2}+\frac{Cs+D}{s^2+4}=\frac{A(s+2)(s^2+4)+B(s-1)(s^2+4)+(Cs+D)(s-1)(s-2)}{(s-1)(s+2)(s^2+4)}$$

$$A(s+2)(s^2+4)+B(s-1)(s^2+4)+(Cs+D)(s-1)(s+2)=20s$$

$\underline{s=-2}: B(-3)(8)=-40$, and $B=\dfrac{5}{3}$

$\underline{s=1}: A(3)(5)=20$, and $A=\dfrac{4}{3}$

Substituting the values of A and B, we get

$$\frac{4}{3}(s+2)(s^2+4)+\frac{5}{3}(s-1)(s^2+4)+(Cs+D)(s-1)(s+2)=20s$$

$\underline{s=0}: \dfrac{4}{3}(8)+\dfrac{5}{3}(-4)+D(-2)=0$, and $D=2$

Now let $D=2$ and let $s=$ any value (such as $s=-1$):

$$\frac{4}{3}(1)(5)+\frac{5}{3}(-2)(5)+(-C+2)(-2)(1)=-20$$

Solving for C, we find that $C=-3$.

$F(s)=\dfrac{4}{3}\dfrac{1}{s-1}+\dfrac{5}{3}\dfrac{1}{s+2}+\dfrac{-3s}{s^2+4}+\dfrac{2}{s^2+4}$

$f(t)=\dfrac{4}{3}e^{t}+\dfrac{5}{3}e^{-2t}-3\cos 2t+\sin 2t$

57. $\dfrac{s}{(s+1)^2}$. Repeating linear factor:

$$\frac{s}{(s+1)^2}=\frac{A}{s+1}+\frac{B}{(s+1)^2}=\frac{A(s+1)+B}{(s+1)^2}$$

$$A(s+1)+B = s.$$

Let $s=-1:\ 0+B=-1$ or $B=-1$.

Using this value, we get $A(s+1)-1=s$. Now let $s=$ any value (such as $s=0$)

$s=0:\ A(1)-1=0$ or $A=1$.

$$F(s) = \frac{1}{s+1}+\frac{-1}{(s+1)^2};$$

$$f(t) = e^{-t}-te^{-t}.$$

A simple alternative is to write

$\dfrac{s}{(s+1)^2}=\dfrac{s+1-1}{(s+1)^2}=\dfrac{s+1}{(s+1)^2}-\dfrac{1}{(s+1)^2}=\dfrac{1}{s+1}-\dfrac{1}{(s+1)^2}$.

59.
$$\frac{2s^2+2s+1}{(s^2+2s+2)(s-1)} = \frac{As+B}{s^2+2s+2}+\frac{C}{s-1} = \frac{(As+B)(s-1)+C(s^2+2s+2)}{(s^2+2s+2)(s-1)}$$

$$(As+B)(s-1)+C(s^2+2s+2) = 2s^2+2s+1.$$

Let $s=1:\ 0+C(5) = 5$

$C = 1.$

Because of the quadratic factor, let us use the value of C already found,

$(As+B)(s-1)+1(s^2+2s+2)=2s^2+2s+1$

and let $s=0$ to find B.

Let $s=0:\ B(-1)+1(2) = 1$ (since $C=1$)

$B = 1.$

$$\begin{aligned}\text{Let } s=2:\ (2A+1)(1)+1(10) &= 13 \qquad B=1,\ C=1\\ 2A+1 &= 3\\ A &= 1.\end{aligned}$$

$$\mathcal{L}^{-1}\left\{\frac{s+1}{(s+1)^2+1}+\frac{1}{s-1}\right\}=\mathrm{e}^{-t}\cos t+\mathrm{e}^{t}$$

61.
$$\begin{aligned}\frac{1}{(s^2+4s+7)(s+4)} &= \frac{As+B}{s^2+4s+7}+\frac{C}{s+4}\\ &= \frac{(As+B)(s+4)+C(s^2+4s+7)}{(s^2+4s+7)(s+4)}\\ (As+B)(s+4)+C(s^2+4s+7) &= 1.\end{aligned}$$

Let $s=-4:\ 0+C(7)=1$ or $C=\dfrac{1}{7}$.

Because of the quadratic factor, let us use the value of C already found and then let $s=0$ to determine B:

$(As+B)(s+4)+\dfrac{1}{7}(s^2+4s+7)=1.$

$s=0:\ B(4)+\dfrac{1}{7}(7)=1$ or $B=0$.

We now have $(As+0)(s+4)+\dfrac{1}{7}(s^2+4s+7)=1.$

To find A, let $s=$ any value (such as $s=1$):

$A(5)+\dfrac{1}{7}(12)=1$ or $A=-\dfrac{1}{7}$.

$$\begin{aligned}F(s) &= -\frac{1}{7}\frac{s}{s^2+4s+7}+\frac{1}{7}\frac{1}{s+4}\\ &= -\frac{1}{7}\frac{s}{(s+2)^2+3}+\frac{1}{7}\frac{1}{s+4}\\ &= -\frac{1}{7}\frac{s+2-2}{(s+2)^2+3}+\frac{1}{7}\frac{1}{s+4}\\ &= -\frac{1}{7}\left(\frac{s+2}{(s+2)^2+3}-2\frac{1}{(s+2)^2+3}\right)+\frac{1}{7}\frac{1}{s+4}\\ &= -\frac{1}{7}\left(\frac{s+2}{(s+2)^2+3}-\frac{2}{\sqrt{3}}\frac{\sqrt{3}}{(s+2)^2+3}\right)+\frac{1}{7}\frac{1}{s+4}\\ &= -\frac{1}{7}\frac{s+2}{(s+2)^2+3}+\frac{2}{7\sqrt{3}}\frac{\sqrt{3}}{(s+2)^2+3}+\frac{1}{7}\frac{1}{s+4}\\ &= \frac{1}{7}\frac{1}{s+4}-\frac{1}{7}\frac{s+2}{(s+2)^2+3}+\frac{2\sqrt{3}}{21}\frac{\sqrt{3}}{(s+2)^2+3};\\ f(t) &= \frac{1}{7}\mathrm{e}^{-4t}-\frac{1}{7}\mathrm{e}^{-2t}\cos\sqrt{3}\,t+\frac{2\sqrt{3}}{21}\mathrm{e}^{-2t}\sin\sqrt{3}\,t.\end{aligned}$$

13.4 Solution of Linear Equations by Laplace Transforms

1. $y'-2y=0,\ y(0)=1$

Step 1. $sY(s)-y(0)-2Y(s)=0$

Step 2. $sY(s)-1-2Y(s)=0$ since $y(0)=1$

Step 3.
$$\begin{aligned}sY(s)-2Y(s) &= 1\\ (s-2)Y(s) &= 1 && \text{factoring } Y(s)\\ Y(s) &= \frac{1}{s-2} && \text{dividing by } s-2\end{aligned}$$

Step 4. $y=\mathrm{e}^{2t}$ by transform 4

3. $2y' + y = 0,\ y(0) = -2$

Step 1. $2\,[sY(s) - y(0)] + Y(s) = 0$

Step 2. $2\,[sY(s) + 2] + Y(s) = 0$ $\qquad y(0) = -2$

Step 3.
$$\begin{aligned} 2sY(s) + 4 + Y(s) &= 0 \\ 2sY(s) + Y(s) &= -4 \\ (2s+1)Y(s) &= -4 \qquad \text{factoring } Y(s) \\ Y(s) &= \frac{-4}{2s+1} = -\frac{2}{s+\frac{1}{2}} \end{aligned}$$

Step 4. $y = -2e^{-(1/2)t}$ $\qquad$ by transform 4

5. $y' - 2y = 4,\ y(0) = 0$

Step 1. $sY(s) - y(0) - 2Y(s) = \dfrac{4}{s}$ $\qquad$ by (13.11)

Step 2. $sY(s) - 0 - 2Y(s) = \dfrac{4}{s}$ $\qquad y(0) = 0$

Step 3.
$$\begin{aligned} Y(s)(s-2) &= \frac{4}{s} \qquad \text{factoring } Y(s) \\ Y(s) &= \frac{4}{s(s-2)} \qquad \text{dividing by } s-2 \end{aligned}$$

Step 4. $Y(s) = \dfrac{A}{s} + \dfrac{B}{s-2} = \dfrac{A(s-2) + Bs}{s(s-2)}$

$A(s-2) + Bs = 4$

$s = 2:\ 0 + 2B = 4$ or $B = 2$.

$s = 0:\ A(-2) + 0 = 4$ or $A = -2$.

$$\begin{aligned} Y(s) &= \frac{-2}{s} + \frac{2}{s-1} \\ y &= 2e^{t} - 2 \end{aligned}$$

7. $y' - 2y = e^{2t},\ y(0) = 0$

Step 1. $sY(s) - y(0) - 2Y(s) = \dfrac{1}{s-2}$

Step 2. $sY(s) - 2Y(s) = \dfrac{1}{s-2}$ $\qquad y(0) = 0$

Step 3.
$$\begin{aligned} (s-2)Y(s) &= \frac{1}{s-2} \qquad \text{factoring } Y(s) \\ Y(s) &= \frac{1}{(s-2)^2} \qquad \text{dividing by } s-2 \end{aligned}$$

Step 4. $y = te^{2t}$ $\qquad$ by transform 9

9. $y' + 4y = te^{-4t},\ y(0) = 3$

Step 1. $sY(s) - y(0) + 4Y(s) = \dfrac{1}{(s+4)^2}$ $\qquad$ transform 9

Step 2. $sY(s) - 3 + 4Y(s) = \dfrac{1}{(s+4)^2}$ $\qquad y(0) = 3$

Step 3.
$$\begin{aligned} sY(s) + 4Y(s) &= \frac{1}{(s+4)^2} + 3 \qquad \text{solving for } Y(s) \\ Y(s) &= \frac{1}{(s+4)^3} + \frac{3}{s+4} \end{aligned}$$

Step 4. $Y(s) = \dfrac{1}{2!}\dfrac{2!}{(s+4)^3} + \dfrac{3}{s+4}$

$y = \dfrac{1}{2}t^2e^{-4t} + 3e^{-4t}$ (by transforms 9 and 4, respectively).

11. $y'' + 9y = 0,\ y(0) = 2,\ y'(0) = 0$

Step 1. $s^2Y(s) - sy(0) - y'(0) + 9Y(s) = 0$ by (13.12)

Step 2. $s^2Y(s) - 2s - 0 + 9Y(s) = 0$ $y(0) = 2,\ y'(0) = 0$

Step 3.
$$\begin{aligned} s^2Y(s) + 9Y(s) &= 2s \\ (s^2+9)Y(s) &= 2s \\ Y(s) &= \frac{2s}{s^2+9} \end{aligned}$$

Step 4. $y = 2\cos 3t$

13. $y'' + 9y = 0,\ y(0) = 1,\ y'(0) = -2$

Step 1. $s^2Y(s) - sy(0) - y'(0) + 9Y(s) = 0$ by (13.12)

Step 2. $s^2Y(s) - s + 2 + 9Y(s) = 0$ $y(0) = 1,\ y'(0) = -2$

Step 3.
$$\begin{aligned} s^2Y(s) + 9Y(s) &= s - 2 \\ (s^2+9)Y(s) &= s - 2 \\ Y(s) &= \frac{s-2}{s^2+9} \end{aligned}$$

Step 4. Find the inverse transform:

$$\begin{aligned} Y(s) &= \frac{s-2}{s^2+9} = \frac{s}{s^2+9} - \frac{2}{s^2+9} \\ &= \frac{s}{s^2+9} - 2\frac{1}{s^2+9} \\ &= \frac{s}{s^2+9} - \frac{2}{3}\frac{3}{s^2+9} \\ y &= \cos 3t - \frac{2}{3}\sin 3t \end{aligned}$$

15. $y'' + y = 2\sin t,\ y(0) = 0,\ y'(0) = 0$

Step 1. $s^2Y(s) - sy(0) - y'(0) + Y(s) = \dfrac{2}{s^2+1}$ by (13.12)

Step 2. $s^2Y(s) + Y(s) = \dfrac{2}{s^2+1}$

Step 3.
$$\begin{aligned} Y(s)(s^2+1) &= \frac{2}{s^2+1} \\ Y(s) &= \frac{2}{(s^2+1)^2} \end{aligned}$$

Step 4. By transform 12 with $a = 1$, $y = \sin t - t\cos t$.

17. $y'' + 4y = 2\cos 2t,\ y(0) = -2,\ y'(0) = 0$

Step 1. $s^2Y(s) - sy(0) - y'(0) + 4Y(s) = \dfrac{2s}{s^2+4}$

Step 2. $s^2Y(s) + 2s - 0 + 4Y(s) = \dfrac{2s}{s^2+4}$

Step 3.
$$\begin{aligned} s^2Y(s) + 4Y(s) &= -2s + \frac{2s}{s^2+4} \\ Y(s)(s^2+4) &= -2s + \frac{2s}{s^2+4} \\ Y(s) &= -\frac{2s}{s^2+4} + \frac{2s}{(s^2+4)^2} \end{aligned}$$

Step 4. (Transforms 6 and 13 with $a = 2$.) For the second term on the right, we need to insert a 2 and place $\frac{1}{2}$ in front:

$$\begin{aligned} Y(s) &= -\frac{2s}{s^2+4} + \frac{1}{2}\frac{2\cdot 2s}{(s^2+4)^2} \\ y &= -2\cos 2t + \frac{1}{2}t\sin 2t. \end{aligned}$$

19. $y'' - 4y' + 10y = 0,\ y(0) = -3,\ y'(0) = 0$

Step 1. $s^2Y(s) - sy(0) - y'(0) - 4[sY(s) - y(0)] + 10Y(s) = 0$

Step 2. $s^2Y(s) + 3s - 4sY(s) - 12 + 10Y(s) = 0$

Step 3.
$$\begin{aligned} s^2Y(s) - 4sY(s) + 10Y(s) &= -3s + 12 \\ Y(s)(s^2 - 4s + 10) &= -3s + 12 \\ Y(s) &= \frac{-3s+12}{s^2 - 4s + 10} \end{aligned}$$

Step 4. To find the inverse transform, we need to complete the square in the denominator and use transforms 7 and 8:

$$\begin{aligned} Y(s) &= -3\frac{s-4}{(s-2)^2+6} \\ &= -3\frac{s-2-2}{(s-2)^2+6} \\ &= -3\left[\frac{s-2}{(s-2)^2+6} - 2\frac{1}{(s-2)^2+6}\right] \\ &= -3\left[\frac{s-2}{(s-2)^2+6} - \frac{2}{\sqrt{6}}\frac{\sqrt{6}}{(s-2)^2+6}\right] \\ &= -3\frac{s-2}{(s-2)^2+6} + \frac{6}{\sqrt{6}}\frac{\sqrt{6}}{(s-2)^2+6}; \\ y &= -3e^{2t}\cos\sqrt{6}\,t + \frac{6}{\sqrt{6}}e^{2t}\sin\sqrt{6}\,t \\ &= -3e^{2t}\cos\sqrt{6}\,t + \sqrt{6}e^{2t}\sin\sqrt{6}\,t \\ &= e^{2t}\left(-3\cos\sqrt{6}\,t + \sqrt{6}\sin\sqrt{6}\,t\right). \end{aligned}$$

21. $y'' + 4y' + 9y = 0,\ y(0) = 2,\ y'(0) = 0$

Step 1. $s^2Y(s) - sy(0) - y'(0) + 4[sY(s) - y(0)] + 9Y(s) = 0$

Step 2. $s^2Y(s) - 2s - 0 + 4sY(s) - 8 + 9Y(s) = 0$

Step 3.
$$\begin{aligned} Y(s) &= \frac{2s+8}{s^2+4s+9} = 2\frac{s+4}{(s+2)^2+5} \\ &= 2\frac{(s+2)+2}{(s+2)^2+5} = 2\frac{s+2}{(s+2)^2+5} + 4\frac{1}{(s+2)^2+5} \\ &= 2\frac{s+2}{(s+2)^2+5} + \frac{4}{\sqrt{5}}\frac{\sqrt{5}}{(s+2)^2+5} \end{aligned}$$

Step 4. $y = 2e^{-2t}\cos\sqrt{5}t + \frac{4}{\sqrt{5}}e^{-2t}\sin\sqrt{5}t$ transforms 8 and 7

23. $y'' - 4y' + 4y = e^{3t},\ y(0) = 0,\ y'(0) = -2$

Step 1. $s^2Y(s) - sy(0) - y'(0) - 4[sY(s) - y(0)] + 4Y(s) = \frac{1}{s-3}$

Step 2. $s^2Y(s) + 2 - 4sY(s) + 4Y(s) = \frac{1}{s-3}$

Step 3.
$$\begin{aligned} s^2Y(s) - 4sY(s) + 4Y(s) &= -2 + \frac{1}{s-3} \\ Y(s)(s^2 - 4s + 4) &= -2 + \frac{1}{s-3} \\ Y(s)(s-2)^2 &= -2 + \frac{1}{s-3} \\ Y(s) &= -\frac{2}{(s-2)^2} + \frac{1}{(s-3)(s-2)^2} \end{aligned}$$

Step 4. Find the inverse transform. The second fraction on the right needs to be expanded by Rule I (repeating linear factors):

$$\begin{aligned}\frac{1}{(s-3)(s-2)^2} &= \frac{A}{s-3}+\frac{B}{s-2}+\frac{C}{(s-2)^2}\\ &= \frac{A}{s-3}\cdot\frac{(s-2)^2}{(s-2)^2}+\frac{B}{s-2}\frac{(s-2)(s-3)}{(s-2)(s-3)}+\frac{C}{(s-2)^2}\frac{s-3}{s-3}\end{aligned}$$

Equating numerators: $A(s-2)^2+B(s-2)(s-3)+C(s-3)=1$.

Let $s=2:\ -C=1$ or $C=-1$.

Let $s=3:\ A(1)=1$ or $A=1$.

Because of the repeating factor, we have run out of values to substitute. So let's use the values of A and C already found,

$1(s-2)^2+B(s-2)(s-3)+(-1)(s-3)=1$

and let $s=$ any value (such as $s=0$) to find B.

Let $s=0:\ (-2)^2+B(-2)(-3)+(-1)(-3) = 1$

$$B = -1.$$

We now have

$$\begin{aligned}Y(s) &= -\frac{2}{(s-2)^2}+\frac{1}{s-3}+\frac{-1}{s-2}+\frac{-1}{(s-2)^2}\\ &= -\frac{3}{(s-2)^2}-\frac{1}{s-2}+\frac{1}{s-3}.\end{aligned}$$

Step 4. $y=-3te^{2t}-e^{2t}+e^{3t}$ by transforms 9 and 4, respectively.

25. $y''-6y'+9y=12t^2e^{3t},\ y(0)=y'(0)=0$

Step 1. $s^2Y(s)-sy(0)-y'(0)-6[sY(s)-y(0)]+9Y(s)=12\frac{2!}{(s-3)^3}$ (by transform 9)

Step 2. $s^2Y(s)-6sY(s)+9Y(s)=\frac{24}{(s-3)^3}$

Step 3.
$$\begin{aligned}Y(s)(s^2-6s+9) &= \frac{24}{(s-3)^3}\\ Y(s)(s-3)^2 &= \frac{24}{(s-3)^3}\\ Y(s) &= \frac{24}{(s-3)^5}\end{aligned}$$

Step 4. By transform 9 with $n=4$,

$Y(s)=\frac{4!}{(s-3)^5}$ and $y=t^4e^{3t}$.

27. $y'-y=\cos 2t,\ y(0)=0$

Step 1. $sY(s)-sy(0)-Y(s)=\frac{s}{s^2+4}$

Step 2. $sY(s)-Y(s)=\frac{s}{s^2+4}$ $\quad y(0)=0$

Step 3.
$$\begin{aligned}Y(s)(s-1) &= \frac{s}{s^2+4}\\ Y(s) &= \frac{s}{(s-1)(s^2+4)}\end{aligned}$$

Step 4.
$$\begin{aligned}Y(s) &= \frac{s}{(s-1)(s^2+4)}=\frac{A}{s-1}+\frac{Bs+C}{s^2+4}\\ &= \frac{A(s^2+4)+(Bs+C)(s-1)}{(s-1)(s^2+4)}\end{aligned}$$

$$A(s^2+4)+(Bs+C)(s-1) = s$$

$s=1:\ A(5)+0=1$ or $A=\frac{1}{5}$.

$\frac{1}{5}(s^2+4)+(Bs+C)(s-1)=s \qquad A=\frac{1}{5}$

Now let $s = 0$ to find C:
$\frac{1}{5}(4) + C(-1) = 0$ or $C = \frac{4}{5}$.
Using this value, we get
$\frac{1}{5}(s^2+4) + \left(Bs + \frac{4}{5}\right)(s-1) = s$.
To find B, let $s =$ any value (such as $s = -1$):

$$\begin{aligned} \frac{1}{5}(5) + \left(-B + \frac{4}{5}\right)(-2) &= -1 \\ 1 + 2B - \frac{8}{5} &= -1 \\ B &= -\frac{1}{5} \end{aligned}$$

$$\begin{aligned} Y(s) &= \frac{1}{5}\frac{1}{s-1} - \frac{1}{5}\frac{s}{s^2+4} + \frac{4}{5}\frac{1}{s^2+4} = \frac{1}{5}\frac{1}{s-1} - \frac{1}{5}\frac{s}{s^2+4} + \frac{2}{5}\frac{2}{s^2+4}. \\ y &= \frac{1}{5}e^t - \frac{1}{5}\cos 2t + \frac{2}{5}\sin 2t. \end{aligned}$$

29. $y'' + y = 4e^t,\ y(0) = y'(0) = 0$

Step 1. $s^2Y(s) - sy(0) - y'(0) + Y(s) = \dfrac{4}{s-1}$

Step 2. $s^2Y(s) + Y(s) = \dfrac{4}{s-1}$

Step 3. $Y(s) = \dfrac{4}{(s-1)(s^2+1)}$

Step 4. $Y(s) = \dfrac{A}{s-1} + \dfrac{Bs+C}{s^2+1} = \dfrac{A(s^2+1) + (Bs+C)(s-1)}{(s-1)(s^2+1)}$

$A(s^2+1) + (Bs+C)(s-1) = 4$

$s = 1:\ A(2) + 0 = 4$ or $A = 2$.

$2(s^2+1) + (Bs+C)(s-1) = 4$ (since $A = 2$).

Now let $s = 0$ to find C: $2(1) + C(-1) = 4$ or $C = -2$.

Using this value, we now have: $2(s^2+1) + (Bs-2)(s-1) = 4$.

Finally, to find B, we let $s =$ any value (such as $s = -1$): $2(2) + (-B-2)(-2) = 4$ or $B = -2$.

$$\begin{aligned} \text{So } Y(s) &= \frac{2}{s-1} + \frac{-2s-2}{s^2+1} = 2\left(\frac{1}{s-1} - \frac{s}{s^2+1} - \frac{1}{s^2+1}\right); \\ y &= 2\left(e^t - \cos t - \sin t\right). \end{aligned}$$

31. $y'' + 4y = 10e^{-t},\ y(0) = 0,\ y'(0) = 1$

Step 1. $s^2Y(s) - sy(0) - y'(0) + 4Y(s) = \dfrac{10}{s+1}$

Step 2. $s^2Y(s) - 1 + 4Y(s) = \dfrac{10}{s+1}$

Step 3.
$$\begin{aligned} s^2Y(s) + 4Y(s) &= 1 + \frac{10}{s+1} \\ \left(s^2+4\right)Y(s) &= 1 + \frac{10}{s+1} \\ Y(s) &= \frac{1}{s^2+4} + \frac{10}{(s+1)\left(s^2+4\right)} \end{aligned}$$

Step 4. For the second fraction:

$$\frac{A}{s+1} + \frac{Bs+C}{s^2+4} = \frac{A\left(s^2+4\right) + (Bs+C)(s+1)}{(s+1)\left(s^2+4\right)}$$

$$A\left(s^2+4\right) + (Bs+C)(s+1) = 10$$

$$\underline{s=-1}: \quad A(5) = 10, \text{ and } A = 2$$

$$2\left(s^2+4\right)+(Bs+C)(s+1) = 10 \qquad A=2$$

$$\underline{s=0}: \quad 2(4)+C(1) = 10, \text{ and } C=2$$

$$2\left(s^2+4\right)+(Bs+2)(s+1) = 10 \qquad C=2,\ A=2$$

$\underline{s = \text{any}}$: (such as $s=1$)

$$2(5)+(B+2)(2) = 10, \text{ and } B=-2$$

$$\frac{10}{(s+1)\left(s^2+4\right)} = \frac{2}{s+1}+\frac{-2s+2}{s^2+4}$$

As a result,

$$Y(s) = \frac{1}{s^2+4}+\frac{2}{s+1}+\frac{-2s+2}{s^2+4} = \frac{2}{s+1}-\frac{2s}{s^2+4}+\frac{3}{s^2+4}$$

$$y = 2e^{-t}-2\cos 2t+\frac{3}{2}\sin 2t$$

33. $y''-4y=3\cos t,\ y(0)=y'(0)=0$

$$\begin{aligned}
s^2Y(s)-sy(0)-y'(0)-4Y(s) &= \frac{3s}{s^2+1}\\
(s^2-4)Y(s) &= \frac{3s}{s^2+1}\\
Y(s) &= \frac{3s}{(s^2-4)(s^2+1)}\\
Y(s) &= \frac{3s}{(s-2)(s+2)(s^2+1)}
\end{aligned}$$

$$\begin{aligned}
\frac{3s}{(s-2)(s+2)(s^2+1)} &= \frac{A}{s-2}+\frac{B}{s+2}+\frac{Cs+D}{s^2+1}\\
&= \frac{A(s+2)(s^2+1)+B(s-2)(s^2+1)+(Cs+D)(s-2)(s+2)}{(s-2)(s+2)(s^2+1)}
\end{aligned}$$

$$A(s+2)(s^2+1)+B(s-2)(s^2+1)+(Cs+D)(s-2)(s+2)=3s$$

Let $s=2$: $\quad A(4)(5)+0+0 = 6$

$$A = \frac{3}{10}.$$

Let $s=-2$: $\quad 0+B(-4)(5)+0 = -6$

$$B = \frac{3}{10}.$$

Now use the values of A and B already found. Then let $s=0$ to find D.

Let $s=0$: $\quad \frac{3}{10}(2)(1)+\frac{3}{10}(-2)(1)+D(-2)(2) = 0 \qquad A=B=\frac{3}{10}$

$$D = 0.$$

Let $s=1$: $\quad \frac{3}{10}(3)(2)+\frac{3}{10}(-1)(2)+C(-1)(3) = 3$

$$\begin{aligned}
\frac{9}{5}-\frac{3}{5}-3C &= 3\\
-3C &= \frac{15}{5}-\frac{9}{5}+\frac{3}{5}=\frac{9}{5}\\
C &= -\frac{3}{5}.
\end{aligned}$$

$$\begin{aligned}
Y(s) &= \frac{3}{10}\frac{1}{s-2}+\frac{3}{10}\frac{1}{s+2}-\frac{3}{5}\frac{s}{s^2+1}\\
y &= \frac{3}{10}e^{2t}+\frac{3}{10}e^{-2t}-\frac{3}{5}\cos t.
\end{aligned}$$

35. $y'' + 2y' + 5y = 8e^t,\ y(0) = 0,\ y'(0) = 0$

Step 1. $s^2Y(s) - sy(0) - y'(0) + 2(sY(s) - y(0)) + 5Y(s) = \dfrac{8}{s-1}$

Step 2. $s^2Y(s) + 2sY(s) + 5Y(s) = \dfrac{8}{s-1}$

Step 3.
$$\begin{aligned} Y(s)(s^2+2s+5) &= \frac{8}{s-1} \\ Y(s) &= \frac{8}{(s-1)(s^2+2s+5)} \end{aligned}$$

Step 4. $\dfrac{A}{s-1} + \dfrac{Bs+C}{s^2+2s+5}$

$A(s^2+2s+5) + (Bs+C)(s-1) = 8$

$s = 1:\ A(8) = 8$ or $A = 1$.

$1(s^2+2s+5) + (Bs+C)(s-1) = 8$ since $A = 1$

Now let $s = 0$ to find C:

$1(5) + C(-1) = 8$ or $C = -3$.

Using this value, we now have: $1(s^2+2s+5) + (Bs-3)(s-1) = 8$.

Finally, to find B, we let $s =$ any value (such as $s = -1$):

$4 + (-B-3)(-2) = 8$ or $B = -1$.

$$\begin{aligned} \text{So } Y(s) &= \frac{1}{s-1} + \frac{-s-3}{s^2+2s+5} \\ &= \frac{1}{s-1} - \frac{s+3}{(s+1)^2+4} = \frac{1}{s-1} - \frac{s+1+2}{(s+1)^2+4} \\ &= \frac{1}{s-1} - \frac{s+1}{(s+1)^2+4} - \frac{2}{(s+1)^2+4}; \\ y &= e^t - e^{-t}\cos 2t - e^{-t}\sin 2t. \end{aligned}$$

37.
$$\begin{aligned} L\frac{d^2q}{dt^2} + R\frac{dq}{dt} + \frac{1}{C}q &= e(t) \\ 0.1\frac{d^2q}{dt^2} + 6.0\frac{dq}{dt} + \frac{1}{0.02}q &= 6.0 \end{aligned}$$

Omitting final zeros, we get

$$\begin{aligned} 0.1\frac{d^2q}{dt^2} + 6\frac{dq}{dt} + 50q &= 6 \\ \frac{d^2q}{dt^2} + 60\frac{dq}{dt} + 500q &= 60. \end{aligned}$$

$$\begin{aligned} s^2Q(s) - sq(0) - q'(0) + 60(sQ(s) - q(0)) + 500Q(s) &= \frac{60}{s} \\ s^2Q(s) + 60sQ(s) + 500Q(s) &= \frac{60}{s} \end{aligned}$$

$$\begin{aligned} Q(s) &= \frac{60}{s(s^2+60s+500)} = \frac{60}{s(s+50)(s+10)} \\ &= \frac{A}{s} + \frac{B}{s+50} + \frac{C}{s+10} = \frac{3}{25}\frac{1}{s} + \frac{3}{100}\frac{1}{s+50} - \frac{3}{20}\frac{1}{s+10} \\ q(t) &= \frac{3}{25} + \frac{3}{100}e^{-50t} - \frac{3}{20}e^{-10t} \\ &= 0.12 + 0.03e^{-50t} - 0.15e^{-10t}. \end{aligned}$$

39.
$$
\begin{aligned}
0.2\frac{d^2q}{dt^2}+\frac{1}{0.05}q &= 20.8\mathrm{e}^{-2t}\\
\frac{d^2q}{dt^2}+100q &= 104\mathrm{e}^{-2t}\\
s^2Q(s)+100Q(s) &= \frac{104}{s+2}\\
Q(s) &= \frac{104}{(s+2)(s^2+100)}=\frac{1}{s+2}-\frac{s}{s^2+100}+\frac{2}{s^2+100}\\
q(t) &= \mathrm{e}^{-2t}-\cos 10t+\frac{1}{5}\sin 10t
\end{aligned}
$$

41. (Exercise 1)

$F=kx$

$4=k\cdot\frac{1}{2}$ or $k=8$; mass$=\frac{4}{32}=\frac{1}{8}$ slug.

Equation: $\frac{1}{8}\frac{d^2x}{dt^2}+8x=0$.

$\frac{d^2x}{dt^2}+64x=0,\ x(0)=\frac{1}{4},\ x'(0)=0.$

$$
\begin{aligned}
s^2X(s)-sx(0)-x'(0)+64X(s) &= 0\\
X(s)(s^2+64) &= \frac{1}{4}s\\
X(s) &= \frac{1}{4}\frac{s}{s^2+64}\\
x(t) &= \frac{1}{4}\cos 8t.
\end{aligned}
$$

(Exercise 9)

$F=kx$

$4=k\cdot 2$ or $k=2$; mass$=\frac{4}{32}=\frac{1}{8}$ slug; $b=\frac{1}{8}$.

Equation: $\frac{1}{8}\frac{d^2x}{dt^2}+\frac{1}{8}\frac{dx}{dt}+2x=0$.

$\frac{d^2x}{dt^2}+\frac{dx}{dt}+16x=0,\ x(0)=\frac{1}{2},\ x'(0)=0.$

$$
\begin{aligned}
s^2X(s)-sx(0)-x'(0)+sX(s)-x(0)+16X(s) &= 0\\
s^2X(s)-\frac{1}{2}s-0+sX(s)-\frac{1}{2}+16X(s) &= 0\\
X(s)(s^2+s+16) &= \frac{1}{2}s+\frac{1}{2}
\end{aligned}
$$

$$
\begin{aligned}
X(s) &= \frac{1}{2}\frac{s+1}{s^2+s+16}=\frac{1}{2}\frac{s+1}{\left(s+\frac{1}{2}\right)^2+\frac{63}{4}}\\
&= \frac{1}{2}\frac{s+\frac{1}{2}+\frac{1}{2}}{\left(s+\frac{1}{2}\right)^2+\frac{63}{4}}\\
&= \frac{1}{2}\frac{s+\frac{1}{2}}{\left(s+\frac{1}{2}\right)^2+\frac{63}{4}}+\frac{1}{4}\frac{1}{\left(s+\frac{1}{2}\right)^2+\frac{63}{4}}\\
&= \frac{1}{2}\frac{s+\frac{1}{2}}{\left(s+\frac{1}{2}\right)^2+\frac{63}{4}}+\frac{1}{4}\frac{2}{\sqrt{63}}\frac{\sqrt{63}/2}{\left(s+\frac{1}{2}\right)^2+\frac{63}{4}}\\
x(t) &= \frac{1}{2}\mathrm{e}^{-t/2}\cos\frac{\sqrt{63}}{2}t+\frac{1}{2\sqrt{63}}\mathrm{e}^{-t/2}\sin\frac{\sqrt{63}}{2}t
\end{aligned}
$$

or

$$
x(t) = \mathrm{e}^{-t/2}\left(\frac{1}{2}\cos\frac{3}{2}\sqrt{7}\,t+\frac{1}{6\sqrt{7}}\sin\frac{3}{2}\sqrt{7}\,t\right).
$$

43. $y'' + 4y = 5\cos 3t,\ y(0) = y'(0) = 0$

$$\begin{aligned}
s^2Y(s) - sy(0) - y'(0) + 4Y(s) &= \frac{5s}{s^2+9} \\
Y(s)(s^2+4) &= \frac{5s}{s^2+9} \\
Y(s) &= \frac{5s}{(s^2+9)(s^2+4)} = \frac{s}{s^2+4} - \frac{s}{s^2+9} \\
y &= \cos 2t - \cos 3t.
\end{aligned}$$

45. $y'' + 9y = 2\sin 2t,\ y(0) = 0,\ y'(0) = 2$

$$\begin{aligned}
s^2Y(s) - sy(0) - y'(0) + 9Y(s) &= \frac{4}{s^2+4} \\
s^2Y(s) - 2 + 9Y(s) &= \frac{4}{s^2+4}
\end{aligned}$$

$$\begin{aligned}
Y(s) &= \frac{2}{s^2+9} + \frac{4}{(s^2+4)(s^2+9)} = \frac{2}{s^2+9} + \frac{4}{5}\frac{1}{s^2+4} - \frac{4}{5}\frac{1}{s^2+9} \\
&= \frac{6}{5}\frac{1}{s^2+9} + \frac{4}{5}\frac{1}{s^2+4} = \frac{2}{5}\frac{3}{s^2+9} + \frac{2}{5}\frac{2}{s^2+4}. \\
y &= \frac{2}{5}\sin 3t + \frac{2}{5}\sin 2t.
\end{aligned}$$

47. $y'' + 4y' + 6y = 3\cos\sqrt{6}\,t,\ y(0) = -2,\ y'(0) = 0$

$$\begin{aligned}
s^2Y(s) - sy(0) - y'(0) + 4[sY(s) - y(0)] + 6Y(s) &= \frac{3s}{s^2+6} \\
s^2Y(s) + 2s + 4sY(s) + 8 + 6Y(s) &= \frac{3s}{s^2+6} \\
(s^2+4s+6)Y(s) &= -2s - 8 + \frac{3s}{s^2+6}
\end{aligned}$$

$$\begin{aligned}
Y(s) &= \frac{-2s-8}{s^2+4s+6} + \frac{3s}{(s^2+6)(s^2+4s+6)} \\
&= \frac{-2s-8}{s^2+4s+6} + \frac{\frac{3}{4}}{s^2+6} - \frac{\frac{3}{4}}{s^2+4s+6} \\
&= \frac{-2s - \frac{32}{4} - \frac{3}{4}}{s^2+4s+6} + \frac{\frac{3}{4}}{s^2+6} \\
&= -2\cdot\frac{s+\frac{35}{8}}{(s+2)^2+2} + \frac{\frac{3}{4}}{s^2+6} \\
&= -2\left[\frac{s+2-2+\frac{35}{8}}{(s+2)^2+2}\right] + \frac{\frac{3}{4}}{s^2+6} \\
&= -2\left[\frac{s+2}{(s+2)^2+2} + \frac{\frac{19}{8}}{(s+2)^2+2}\right] + \frac{\frac{3}{4}}{s^2+6} \\
&= -2\cdot\frac{s+2}{(s+2)^2+2} - \frac{19}{4}\frac{1}{\sqrt{2}}\frac{\sqrt{2}}{(s+2)^2+2} + \frac{\frac{3}{4}}{s^2+6}. \\
y &= -2e^{-2t}\cos\sqrt{2}\,t - \frac{19}{4\sqrt{2}}e^{-2t}\sin\sqrt{2}\,t + \frac{3}{4\sqrt{6}}\sin\sqrt{6}\,t.
\end{aligned}$$

49. $F = kx$

$12 = k\cdot 2$ or $k = 6$; mass$= \frac{12}{32} = \frac{3}{8}$ slug; $b = 0$.

$$\frac{3}{8}\frac{d^2x}{dt^2} + 6x = 12\sin t$$

$$\frac{d^2x}{dt^2} + 16x = 32\sin t,\ x(0) = 1,\ x'(0) = 0$$

$$\begin{aligned} s^2X(s) - sx(0) - x'(0) + 16X(s) &= \frac{32}{s^2+1} \\ s^2X(s) - s + 16X(s) &= \frac{32}{s^2+1} \end{aligned}$$

$$\begin{aligned} X(s) &= \frac{s}{s^2+16} + \frac{32}{(s^2+16)(s^2+1)} = \frac{s}{s^2+16} + \frac{32}{15}\frac{1}{s^2+1} - \frac{32}{15}\frac{1}{s^2+16} \\ &= \frac{s}{s^2+16} + \frac{32}{15}\frac{1}{s^2+1} - \frac{8}{15}\frac{4}{s^2+16}. \\ x(t) &= \cos 4t + \frac{32}{15}\sin t - \frac{8}{15}\sin 4t. \end{aligned}$$

51. $0.100\dfrac{d^2q}{dt^2} + 10.0\dfrac{dq}{dt} + \dfrac{1}{1.00\times 10^{-3}}q = 10.0\sin 20.0t$
Multiplying by 10 and leaving out final zeros:
$\dfrac{d^2q}{dt^2} + 100\dfrac{dq}{dt} + 10000q = 100\sin 20t.$

$$\begin{aligned} Q(s) &= \frac{2000}{(s^2+400)(s^2+100s+10000)} \\ &= \frac{\frac{5}{2404}s}{s^2+100s+10000} + \frac{\frac{5}{601}}{s^2+100s+10000} - \frac{\frac{5}{2404}s}{s^2+400} + \frac{\frac{120}{601}}{s^2+400} \end{aligned}$$

The last two terms lead to: $\dfrac{-5}{2404}\cos 20.0t + \dfrac{6}{601}\sin 20.0t.$
The first two terms need to be rewritten as follows:

$$\begin{aligned} \frac{5}{2404}\frac{s+\frac{2404}{5}\cdot\frac{5}{601}}{(s+50)^2+7500} &= \frac{5}{2404}\frac{s+4}{(s+50)^2+7500} = \frac{5}{2404}\frac{s+50-50+4}{(s+50)^2+7500} \\ &= \frac{5}{2404}\frac{s+50}{(s+50)^2+7500} + \frac{5}{2404}(-46)\frac{1}{\sqrt{7500}}\frac{\sqrt{7500}}{(s+50)^2+7500}. \end{aligned}$$

Inverse transform: $0.00208\mathrm{e}^{-50t}\cos 50\sqrt{3}\,t - 0.00110\mathrm{e}^{-50t}\sin 50\sqrt{3}\,t.$

Chapter 13 Review

1.
$$\begin{aligned} f(t) &= 2\mathrm{e}^{-3t} \\ F(s) &= 2\frac{1}{s+3} = \frac{2}{s+3} \qquad \text{by transform 4} \end{aligned}$$

3. $f(t) = 2t^3 + \sin 3t$
$F(s) = 2\mathcal{L}\{t^3\} + \mathcal{L}\{\sin 3t\} = 2\cdot\dfrac{3!}{s^{3+1}} + \dfrac{3}{s^2+3^2}$ by transforms 3 and 5, respectively, or
$F(s) = \dfrac{12}{s^4} + \dfrac{3}{s^2+9}.$

5. $f(t) = 2t - \sin 2t$
$F(s) = \dfrac{2}{s^2} - \dfrac{2}{s^2+4}$ by transforms 2 and 5, respectively.
Thus $F(s) = \dfrac{2(s^2+4)-2s^2}{s^2(s^2+4)} = \dfrac{8}{s^2(s^2+4)}.$

7. $F(s) = \dfrac{s-2}{(s-2)^2+5}$. By transform 8 with $a=2$ and $b=\sqrt{5}$,
$f(t) = \mathrm{e}^{2t}\cos\sqrt{5}\,t.$

9.
$$\begin{aligned}F(s) &= \frac{s}{s^2-2s+5} = \frac{s}{(s-1)^2+4} = \frac{(s-1)+1}{(s-1)^2+4} \\ &= \frac{s-1}{(s-1)^2+4} + \frac{1}{2}\frac{2}{(s-1)^2+4} \\ f(t) &= e^t\cos 2t + \frac{1}{2}e^t \sin 2t \qquad \text{transforms 8 and 7, respectively} \\ &= e^t\left(\cos 2t + \frac{1}{2}\sin 2t\right).\end{aligned}$$

11. $F(s) = \dfrac{s}{(s+1)(s-2)} = \dfrac{A}{s+1} + \dfrac{B}{s-2} = \dfrac{A(s-2)+B(s+1)}{(s+1)(s-2)}$

$A(s-2)+B(s+1) = s$

Let $s=2:\ 0+B(3)=2$ or $B=\dfrac{2}{3}$. Let $s=-1:\ A(-3)+0=-1$ or $A=\dfrac{1}{3}$.

$F(s) = \dfrac{1}{3}\dfrac{1}{s+1} + \dfrac{2}{3}\dfrac{1}{s-2}$.

$f(t) = \dfrac{1}{3}e^{-t} + \dfrac{2}{3}e^{2t}$.

13.
$$\begin{aligned}F(s) &= \frac{1}{(s+2)(s-3)(s-4)} = \frac{A}{s+2} + \frac{B}{s-3} + \frac{C}{s-4} \\ &= \frac{A(s-3)(s-4)+B(s+2)(s-4)+C(s+2)(s-3)}{(s+2)(s-3)(s-4)}\end{aligned}$$

$A(s-3)(s-4)+B(s+2)(s-4)+C(s+2)(s-3)=1$

Let $s=3:\ B(5)(-1) = 1.\quad B = -\dfrac{1}{5}.$ Let $s=4:\ C(6)(1) = 1.\quad C = \dfrac{1}{6}.$

Let $s=-2:\ A(-5)(-6) = 1.\quad A = \dfrac{1}{30}.$

$F(s) = \dfrac{1}{30}\dfrac{1}{s+2} - \dfrac{1}{5}\dfrac{1}{s-3} + \dfrac{1}{6}\dfrac{1}{s-4}$.

$f(t) = \dfrac{1}{30}e^{-2t} - \dfrac{1}{5}e^{3t} + \dfrac{1}{6}e^{4t}$.

15. $y' + 2y = 0,\ y(0) = 1$

Step 1. $sY(s) - y(0) + 2Y(s) = 0$

Step 2. $sY(s) - 1 + 2Y(s) = 0 \qquad y(0)=1$

Step 3. $Y(s)(s+2) = 1$

$Y(s) = \dfrac{1}{s+2}$

Step 4. $y = e^{-2t}$

17. $y' + 2y = te^{-2t},\ y(0) = -1$

$$\begin{aligned}
sY(s) - y(0) + 2Y(s) &= \frac{1}{(s+2)^2} && \text{by transform 9} \\
sY(s) + 1 + 2Y(s) &= \frac{1}{(s+2)^2} && y(0) = -1 \\
(s+2)Y(s) &= \frac{1}{(s+2)^2} - 1 \\
Y(s) &= \frac{1}{(s+2)^3} - \frac{1}{s+2} \\
&= \frac{1}{2}\frac{2!}{(s+2)^3} - \frac{1}{s+2}. \\
y &= \frac{1}{2}t^2 e^{-2t} - e^{-2t} \\
&= e^{-2t}\left(\frac{1}{2}t^2 - 1\right) && \text{by transforms 9 and 4, respectively}
\end{aligned}$$

19. $y'' - 2y' - 3y = 0,\ y(0) = 0,\ y'(0) = -4$

Step 1. $s^2Y(s) - sy(0) - y'(0) - 2[sY(s) - y(0)] - 3Y(s) = 0$

Step 2. $s^2Y(s) - 0 - (-4) - 2sY(s) + 0 - 3Y(s) = 0$

Step 3. $s^2Y(s) - 2sY(s) - 3Y(s) = -4$

$$Y(s) = \frac{-4}{s^2 - 2s - 3} = \frac{-4}{(s+1)(s-3)}$$

Step 4. $Y(s) = \dfrac{A}{s+1} + \dfrac{B}{s-3} = \dfrac{A(s-3) + B(s+1)}{(s+1)(s-3)}$

$A(s-3) + B(s+1) = -4$

Let $s = 3:\ 0 + B(4) = -4$ or $B = -1$.

Let $s = -1:\ A(-4) + 0 = -4$ or $A = 1$.

$Y(s) = \dfrac{1}{s+1} - \dfrac{1}{s-3}$

$y = e^{-t} - e^{3t}$

21. $y'' + 2y' + 5y = 0,\ y(0) = 1,\ y'(0) = 0$

$$\begin{aligned}
s^2Y(s) - sy(0) - y'(0) + 2[sY(s) - y(0)] + 5Y(s) &= 0 \\
s^2Y(s) - s + 2[sY(s) - 1] + 5Y(s) &= 0 \\
s^2Y(s) - s + 2sY(s) - 2 + 5Y(s) &= 0 \\
(s^2 + 2s + 5)Y(s) &= s + 2 \\
Y(s) &= \frac{s+2}{s^2 + 2s + 5}
\end{aligned}$$

$$\begin{aligned}
Y(s) &= \frac{s+2}{(s+1)^2 + 4} = \frac{(s+1) + 1}{(s+1)^2 + 4} \\
&= \frac{s+1}{(s+1)^2 + 4} + \frac{1}{2}\frac{2}{(s+1)^2 + 4} \\
y &= e^{-t}\cos 2t + \frac{1}{2}e^{-t}\sin 2t = e^{-t}\left(\cos 2t + \frac{1}{2}\sin 2t\right)
\end{aligned}$$

23. $y'' + 2y' + 5y = 3e^{-2t},\ y(0) = 1,\ y'(0) = 1$

Step 1. $s^2Y(s) - sy(0) - y'(0) + 2[sY(s) - y(0)] + 5Y(s) = \dfrac{3}{s+2}$

Step 2. $s^2Y(s) - s - 1 + 2sY(s) - 2 + 5Y(s) = \dfrac{3}{s+2}$

$s^2Y(s) + 2sY(s) + 5Y(s) = s + 3 + \dfrac{3}{s+2}$

$Y(s) = \dfrac{s+3}{s^2 + 2s + 5} + \dfrac{3}{(s+2)(s^2 + 2s + 5)}$

We decompose the second fraction:

$$\frac{3}{(s+2)(s^2+2s+5)} = \frac{A}{s+2} + \frac{Bs+C}{s^2+2s+5}$$

$A(s^2+2s+5) + (Bs+C)(s+2) = 3$

Let $s = -2$: $A(5) = 3$ or $A = \frac{3}{5}$.

So $\frac{3}{5}(s^2+2s+5) + (Bs+C)(s+2) = 3$.

Now let $s = 0$ to find C:

$\frac{3}{5}(5) + C(2) = 3$ or $C = 0$.

Using this value, we get: $\frac{3}{5}(s^2+2s+5) + (Bs+0)(s+2) = 3$.

To find B, let $s =$ any value (such as $s = 1$):

$\frac{3}{5}(8) + B(3) = 3$ or $B = -\frac{3}{5}$.

We now have

$$\begin{aligned}
Y(s) &= \frac{s+3}{s^2+2s+5} + \frac{3}{5}\frac{1}{s+2} - \frac{3}{5}\frac{s}{s^2+2s+5} \\
&= \frac{3}{5}\frac{1}{s+2} + \frac{(2/5)s+3}{s^2+2s+5} \\
&= \frac{3}{5}\frac{1}{s+2} + \frac{2}{5}\frac{s+15/2}{(s+1)^2+4} \\
&= \frac{3}{5}\frac{1}{s+2} + \frac{2}{5}\frac{s+1-1+15/2}{(s+1)^2+4} \\
&= \frac{3}{5}\frac{1}{s+2} + \frac{2}{5}\frac{s+1}{(s+1)^2+4} + \frac{13}{5}\frac{1}{(s+1)^2+4} \\
&= \frac{3}{5}\frac{1}{s+2} + \frac{2}{5}\frac{s+1}{(s+1)^2+4} + \frac{13}{10}\frac{2}{(s+1)^2+4}; \\
y &= \frac{3}{5}e^{-2t} + \frac{2}{5}e^{-t}\cos 2t + \frac{13}{10}e^{-t}\sin 2t
\end{aligned}$$

25. By Hooke's law,

$$\begin{aligned}
F &= kx \\
4 &= k\cdot\frac{1}{2} \text{ or } k = 8 \\
\text{mass: } \frac{4}{32} &= \frac{1}{8} \text{ slug.}
\end{aligned}$$

Equation: $\frac{1}{8}\frac{d^2x}{dt^2} + 2\frac{dx}{dt} + 8x = 0$ since $b = 2$

and

$\frac{d^2x}{dt^2} + 16\frac{dx}{dt} + 64x = 0.$

Initial conditions: (1) If $t = 0$, $x = 0$ initial position

(2) If $t = 0$, $\frac{dx}{dt} = -4$ initial velocity

$$\begin{aligned}
s^2X(s) - sx(0) - x'(0) + 16[sX(s) - x(0)] + 64X(s) &= 0 \\
s^2X(s) - (-4) + 16sX(s) + 64X(s) &= 0 \\
X(s) = \frac{-4}{s^2+16s+64} &= -\frac{4}{(s+8)^2}
\end{aligned}$$

$x(t) = -4te^{-8t}$ by transform 9.

In the decades since its beginnings as a single acr Berry Farms has flourished under the care and pa of the Krause family. With their unwavering commitment to ethical and sustainable agriculture, the Krauses have grown their Fraser Valley estate into 200 acres devoted to mixed produce, four kitchens, a market, a bakery, a waffle bar, an ice-cream shop, and an award-winning winery. Year after year, close to half a million visitors make their way to Krause Berry Farms to revel in a day of family fun and feast on tasty home cooking, leaving with baskets of berries and lifelong memories.

Now, some of their most beloved recipes are available for all in *The Krause Berry Farms Cookbook.* Inside this book you'll find recipes for every meal, snack, and celebration that a cook or baker of any skill level could tackle, including:

- **Breakfast and Brunch**: Wake up with Krause Berry Farms' Sensational Cinnamon Buns, Auntie Dona's Overnight Brunch Eggs, or their famous Waffles.
- **Soups:** Enjoy a hearty Apple, Cheese, and Cider Soup, Roasted Corn Chowder, or French Onion Soup.
- **Main Meals:** Feast on wonderful spreads with loved ones such as Corn on the Cob, Six Ways, Cheesy Chicken Enchiladas, Farmers Eat Quiche, or have a pizza party with their popular Roasted Corn Pizza.
- **Breads and Muffins:** Munch on Hearty Multigrain Bread or Blueberry Muffins.
- **Cookies and Bars:** Satisfy your sweet tooth with Berry Farm Cookies, Gingerbread People, Gluten-Friendly Almond Crescents, or Colourful Nanaimo Bars.
- **Cakes and Desserts:** Celebrate a special occasion with Berry Merry Christmas Fruitcake or Peach Cobbler.
- **Pies and Tarts:** Bring a piece of the farm to your table with Krause Berry Farms' famous Strawberry Custard Pie, Apple Pie, and Pecan Butter Tarts.
- **Drinks and Snacks:** Serve up an Old-Fashioned Berry Milkshake, Sangria, or Hot Berry Mulled Wine and snack on Fresh-Cut Strawberry Salsa or Roasted Casa Corn Dip.

With something for everyone (even your four-legged friends), this book celebrates fresh-off-the-farm goodness and will inspire you to create with your own local, seasonal produce. Paired with gorgeous photography and charming farm and family tales, *The Krause Berry Farms Cookbook* transports the warm and fun-loving spirit of Krause Berry Farms right to your kitchen.

Apron Strings are Powerful Things

. . . Tie One On!

MARKET

The Krause Berry Farms Cookbook

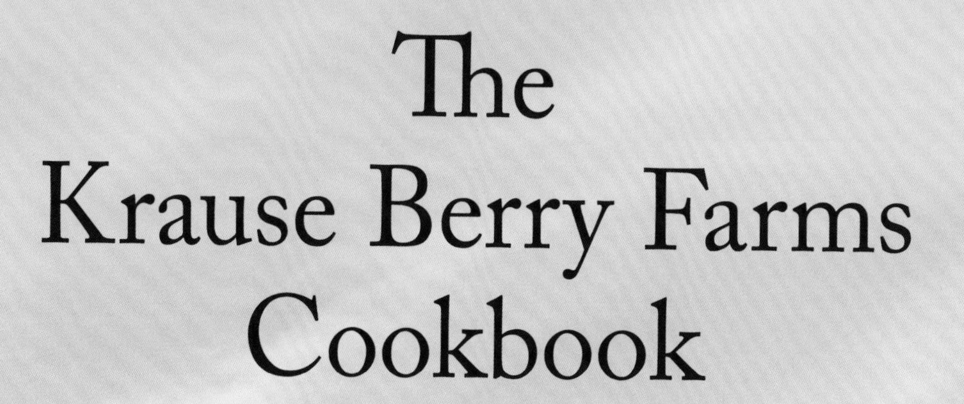

Sweet and Savoury Recipes from the Fraser Valley's Famous Farm and Bakery

Sandee Krause

Photography by Heather Cameron

appetite
by RANDOM HOUSE

Library and Archives Canada Cataloguing in Publication
is available upon request.
ISBN: 978-0-525-61190-5
eBook ISBN: 978-0-525-61191-2

Photography by Heather Cameron
Cover and book design by Matthew Flute

Printed in China

Published in Canada by Appetite by Random House®,
a division of Penguin Random House Canada Limited

www.penguinrandomhouse.ca

10 9 8 7 6 5 4 3 2 1

In Honour and Memory of My Son
Tanner Joseph Mikesh
(September 22, 1985–March 4, 2018)

Thank you for 32 years of being a dedicated and adventurous taste-tester. For writing the funniest birthday menus to include lobster and beer when you were only 15. For requesting out-of-the-box cakes that stretched my abilities but helped me gain new skills and confidence. For all the times we made serious omelets together like the pros before your baseball games—at the age of 12, you were just as serious about your food as you were about achieving your dream of playing in the Little League World Series representing Canada. For all the cookies, cakes, and gingerbread houses we made and decorated together. For valuing our family dinners and for all the times I heard you say, "May I please be excused, that was a very nice dinner."

Breaking bread with family and friends was a big part of your life. Watching you grow into a foodie, trying new places, and then bringing me and our family to the ones that passed your taste-test added such sweet spice to some of our best memories of you. Thank you for caring about your health and making healthy food choices. But most of all, thank you for the way you raised your eyebrows, flashing that mischievous glint of your eye, letting us all know when you were ready to indulge. We continue to *MARCH FORTH* on this food journey together, just as you would have liked.

. . .

This book is also dedicated to my husband, Alf; my other children, Sarah, son-in-law Ryan, grandchildren Kale, Emme and Brynn, Grayden, daughter-in-law Jayme, grand children River, and Casey, Jared and partner Caleigh, Holly, Tracey, son-in-law Chirag and my strong support system and dear friends Heather, Ann, Pauline, Tammy, and their families.

Contents

Cakes and Desserts

Pies and Tarts

Drinks and Snacks

Staples

KRAUSE
BERRY FARMS
&
Since 1974
ESTATE WINERY
Sandee

Welcome

Welcome to my world. Krause Berry Farms and Estate Winery—located in the southwestern corner of British Columbia within the Fraser Valley—was born in 1974, when a young man named Alf Krause planted 1 acre of strawberries. If someone had told him back then that one day he and his future wife would farm over 200 acres and be known throughout British Columbia and beyond not only for their top-quality berries and vegetables but also for their unwavering commitment to ethical and sustainable farming practices, he would have grinned. And if someone had told him that one day, he and his wife would host then prime minister Stephen Harper and his wife, Laureen Harper, on their farm, I'm sure he would have politely but firmly assured them that that was unlikely, to say the least. As for me, I love to bake, cook, serve, share, and see smiles. I'm blessed to be able to experience those loves every day on our farm and to add some sunshine to other people's lives at the same time.

The Fraser Valley hasn't always been my home. I was born in San Diego, California, to parents of Irish, English, and German descent. We moved as a family every few years, with stops including Oregon, Colorado, and Vancouver, BC, where I decided to stay while my parents and siblings moved on to Quebec, Florida, Texas, and then full circle back to California.

I was very fortunate to be born into a family with diverse backgrounds and interests. My father was in the computer industry, and my parents owned an antique store and two yarn and needlework stores. They often talked business, and I absorbed their dinner-table conversations like a sponge. I was fascinated by their conversations and decided to study business after I graduated from high school. My mother's mother, Veronica, had been a military wife and had regaled me with stories of her travels and how much she had enjoyed them. She encouraged me to travel, telling me that once I tasted the travel bug, I would be hooked. When I was still studying business, she suggested I apply to an airline to be a flight attendant—or air hostess, as they were called in those days. And that's how I found myself working for Air Canada in 1976.

Initially I moved to Montreal, which was hosting the Summer Olympics that year, but I later relocated to Winnipeg, then Calgary, and finally back to Vancouver. During those years, I travelled to many cities in North America and Europe. Since then, I have visited every state in the United States and all the Canadian provinces and territories—and I'm still travelling, whether it's by plane, boat, train, bus, bicycle, or on foot. I recently enjoyed walking over 600 miles through Spain along the beautiful Camino Frances to Santiago. I have also completed half of the South Coast Path in England and 100 miles of The Way of Saint James in France.

Some of the things I appreciate most about travelling are discovering new foods, getting to know the hands that grow and prepare the local food, and developing relationships around food. As soon as I've checked into where I'm staying, I head out to find the local bakery. I believe that bakeries are a window to the soul of any location. They are how I fondly remember the destinations. Restaurants are almost like classrooms, as they help me learn about cultures. I recall the restaurants and cities where I've tasted my favourite and most memorable eggplant parmesan, chicken Kiev, hot pot, paella, falafels, German pancakes, crepes, bagels, trifle—the list goes on and on. In fact, let me be honest here: I have planned more trips than I can remember based solely on visiting a certain bakery or restaurant. Luckily, Alf, my husband, enjoys food as much as I do, and he is always happy to travel to those special places with me.

My farmer husband is one of the few local boys born and raised within 5 miles of his family's homestead. He jokes that his boots got stuck in the mud—and are we ever glad! His parents were German and immigrated to Canada from modern-day Ukraine in 1948. They had an agrarian background and dreamed of farming in their new homeland. Eventually they were able to purchase a small plot of land with the intent of growing berries. Shortly after they realized that the soil was not suitable for growing, they changed directions and moved into raising poultry. Alf did farm chores and learned farm management skills and a good work ethic.

Although Alf decided poultry farming was not for him, he was still drawn to farming. During his college years, he often recalled watching his parents and extended family members picking berries on his parents' farm when he was young. The memories prompted him to take a chance: he leased 1 acre of land and began growing strawberries. His harvest was a success by any standard, and it confirmed that growing berries was what he wanted to do. He gradually expanded his acreage over the years, along with the number of crops he grew, and he adopted the practice of crop rotation to reintroduce important nutrients into the soil. Meanwhile, he continued to study science and business at BCIT. After he received his diploma in agriculture business, he became a full-time farmer, growing crops as well as working with all of the provincial berry associations, and attending and speaking at national conventions. Quickly establishing a solid reputation among his peers, he became president of the BC Raspberry

Growers Association (now called The Raspberry Industry Development Council) and the Lower Mainland Horticultural Improvement Association (LMHIA) for a number of years. He is currently the vice president of the BC Strawberry Growers Association and a director of LMHIA. He has always supported and worked closely with university research and breeding programs, providing acreage to grow experimental berry varieties. Alf remains a firm believer in the value of crop rotation and actively supports crop-protection initiatives with Agriculture and Agri-Food Canada. What keeps Alf going? He loves the cycle of farm life. The sprouting of seeds in spring, that new growth developing into the abundance and busyness and exhaustion of harvest, followed by the calm of the onset of winter. Winter re-energizes him, giving him time to research different varieties of berries, bugs, and developments in farming in general, along with sparking new ideas for projects and planning for the new year. He's constantly on the go.

Alf and I are a partnership in every sense of the word. Much of our story seems unlikely—that is a tale for another book—but I like to think of us as a match made in heaven. Somehow, we managed, as so many married couples do, to strengthen our marriage among the love and chaos! Our entrepreneurial spirits and gifts, combined with our work ethic, have guided us in the development of our business. Family is the most important thing to us. We were able to keep our family involved in our farming business, offering our children a way to earn their pocket money while building life skills, while we worked our responsibilities around their activities: horses, baseball, basketball, football, volleyball, soccer, dance, and music.

As the farm continued to grow, our family of six children continued to be involved in the many different areas of the farm and to look after the guests who return year after year, along with the tourists who flock to the farm by the busload. We regularly welcome over 350,000 guests—some familiar faces, some new—from all over the world every year. And yes, we welcome the occasional celebrity, which always causes a flutter of excitement! The boys, Grayden, Tanner, and Jared, gravitated to the work of growing and the building aspects of the farm, while the girls, Sarah, Holly, and Tracey, worked in the farm market, making and selling the products on the main farm and serving at events. Our son-in-law Ryan helped in the fields and for farm events over the years as well. When we gathered for dinner, we shared stories from our workday. One evening when our youngest, Tracey, was 13, she told all of us that she felt left out because she was too young to work alongside her siblings. The next year, we gave her the job of cleaning the washrooms. She was thrilled to feel part of our dinnertime sharing. She did such a great job that she graduated to the bakery counter the following year—where she created 50 flavours of fudge—and then moved into working in the winery along with our son-in-law Chirag. All the kids have benefited from being involved in one way or another, building life skills along the way. We couldn't feel prouder of them and their contributions to the farm.

We have hosted many events on the farm throughout our 50 years, but the visit by Stephen and Laureen Harper was an exceptional one. We never realized the amount of security that goes into keeping leaders safe! Within all that security, our family was offered the privilege of a private meeting with the prime minister. He addressed each one of our children individually, inquiring about their personal interests. This experience left an indelible impression on them. He then asked Alf and me about our farm, the food we grow, and the guests we serve. It was a memorable 25 minutes, to say the least.

The Farm

Alf's single acre of strawberries is now 200 acres of mixed produce, a market, a bakery, four kitchens, and a winery. We started off as a U-pick farm. People would come to buy berries to make their own jam. Times change, though, and we started to see our U-pick trade slow down as people turned to store-bought jams and jellies. We knew we had much more to give than being a U-pick location, so we decided that we'd offer our guests more of an experience when they visited the farm. I started to make jam, jellies, berry shortcakes and pies in the farm kitchen to sell fresh on the farm. The pies rapidly became popular throughout the region, and from there we gradually expanded the products we sold and introduced sit-down options.

These days our primary crops are strawberries, blueberries, raspberries, blackberries, tayberries, and sweet corn. We also grow goji berries, broccoli, cauliflower, Brussels sprouts, garlic, asparagus, artichokes, dill, squash, pumpkins, gourds, and a variety of flowers on a smaller scale. Alf enjoys getting out into the fields on his tractor (as pictured here) to see these crops up close and monitor changes over the seasons. It also allows him to reflect and think of new management and crop care ideas.

We sell our prepicked fresh produce in the market. And when I say "fresh," I mean it: during harvest season, the produce comes straight from the field to the berry counter throughout the day, every day. If you want to get down and dirty and really experience where your food comes from, we also offer U-pick options. And because we know nothing tastes better in winter than a hint of summer, we individually quick freeze (IQF) our berries so people can enjoy them all year round.

We also have created something that many guests haven't encountered anywhere else: a U-pick flower garden. I had the idea for this garden when I was travelling in the eastern United States. I saw the most amazing garden—a true blaze of colour and shapes—and instantly knew that I wanted something similar to share with our guests at the farm. Thanks to the magic of my good friend Kelly, who is a landscape designer, guests are able to pick their own flowers, vase included, or simply stroll through it or sit in its beauty. It radiates calm, making it a beautiful spot to catch your breath and just take a few moments away from the world outside. In August each year we host a butterfly release as a fundraiser for our local Langley hospice. Up to 400 butterflies are released at one time. It is breathtaking to see this number of butterflies released to commemorate the lives of loved ones.

Next to the flower garden is a labour of love that we call the Potager Garden, which was also designed by Kelly. It has symmetrical and geometrical beds planted in patterns, and groups of flowers, fruits, vegetables, and herbs. You will find figs, jalapeños, green, yellow, and red bell peppers, zucchini, tomatoes, tomatillos, eggplant, peas, beans, basil, cilantro, parsley, sage, lemon balm,

Alf and I opened our farm gates to share with our community the connection of where their food comes from so they can experience the local growing cycle and enjoy fresh foods that are grown in their area.

Food is always better when shared!

calendula . . . the list goes on. These are harvested for food boards and boxes, which we make in the winery's kitchen (page 259) and are easy to pick from any angle without having to do much more than reach in and snip.

The Potager Garden is a good example of the ancient art of companion planting, which my grandmother Sarah's sister Thelma introduced me to. Companion planting is a widely practised and extremely helpful tool for growing healthy crops and minimizing chemical intervention. It's essentially the practice of "buddying-up" flowers and vegetables by planting them alongside each other to help keep them healthy and free of disease and insect infestation. The Indigenous peoples who lived in North America were practising companion planting long before the Europeans arrived. Probably the most famous example of traditional companion planting in North America is the Three Sisters: beans, corn, and squash. The beans use the corn to support them as they climb skyward and in return add some nutritious nitrogen to the soil. When the leaves of the squash plants fall to the ground, they turn into mulch, which helps to retain the water in the soil. Ultimately, all three crops help each other thrive. At the farm, we have planted alfalfa strips down the aisles of our strawberries to attract bugs to the alfalfa instead of to the strawberries.

You may wonder why we opted to have a barley maze rather than the more traditional corn maze. It's a good question, and the answer lies in Alf's commitment to responsible farming. Barley is a good cover crop. It enriches the soil and prepares it for next year's planting, usually strawberries in our case. The barley maze lets our guests enjoy some fun while the soil enjoys some welcome nourishment.

In our early farming days, we would close the farm gate after Labour Day. Over the years, our guests convinced us to open for Christmas. That turned out to be a great decision. We aren't as busy on the farm then as we are at harvest times, so we had time to really get to know our guests and to share stories with them. This was a particularly lovely time for Alf. Let me just say that Alf does not fit the stereotype of the quiet farmer. He is not a man of few words. He loves to talk, whether it's sharing stories or passing on his farming knowledge, so he really enjoys the time we share with our guests. Following the success of our Christmas opening, we branched out into having a pumpkin patch and barley maze. One of the best things about extending our season was that we were able to offer year-round work to staff who previously worked for us only for part of the year.

In 2012, we decided to build a winery. This was never part of our plan, but after numerous conversations with various winery owners—and many guests who wanted us to host their weddings and other special events—we realized that we had both the berries and the people we would need to make a success of our own winery.

Our farm is a warm, welcoming farm environment that offers our guests a fun and memorable experience, and I wanted to carry that feeling over into the winery, but I struggled with how to achieve that. I reached out to various marketing

PEPPERMINT
SWISS CHARD

businesses, all of whom offered to rebrand us, but I didn't want to be rebranded. I wanted us to be who we already were. We liked who we were! In a moment of frustration, I blurted out to Alf, "Who are we, anyway?" He looked at me and calmly said, "I'm the farmer and you're the cowgirl." "That's it!" I shouted. "We are Farmers to Boot." I immediately knew I had found the theme for our winery.

One of the first things on my to-do list was to find a boot maker who could hand-make a pair of boots for each of us. After about three months of searching, I finally found Nevena Christi from Rocketbuster Boots. I sketched out a design on a napkin for the toe of my boots to look like a slice of pie and the sides of the boots to be decorated to tell the story of our farm, and sent it to her. The strawberry spurs added the finishing touch. Alf's boots tell the story of our brother-and-sister horse team, Mac and Bess, tilling the soil in previous years, along with the tool we use today, Alf's favourite tractor. The toe on his boot shows his favourite berries, while the back spine is a corn cob, illustrating his love of corn. Once completed, they were photographed and became the artwork for our wine labels. When we're not wearing them, they sit on a shelf behind the tasting bar for guests to enjoy. (You can see a photograph of them on page 258.)

While the boots were in progress, I gathered up saddles to use for seating at the tasting bar. The final touch was a chandelier. Unable to find just the right one, I decided to make my own. I took myself off to the feed store, where I purchased a large stainless-steel water trough. Alf and I placed it in the middle of the berry field, filled it with sand for safety, and shot the heck out of it with a 12-gauge shotgun. We hung it upside down and inserted a row of chandelier lights that shine through the bullet holes. Our guests love gathering around the harvest table enjoying farm food and wine, surrounded by these western-themed details.

Since opening the winery, we have received many awards. Our most prestigious was the All-Canadian Best Fruit Wine of the Year in 2016. We are known for our Sparkling Raspberry and Sparkling Blueberry wines, our Blackberry Portoe (a port-style wine), and our Cassis Dessert and Strawberry Dessert wines, all of which have received double golds, meaning all the judges voted them gold in the competition. I have also been a judge myself. I was honoured to be asked to be one of 24 judges at the NorthWest Wine Summit in Portland, Oregon, in 2016. The judges are a mix of winemakers and people who are involved in the manufacture and sale of wine, all of whom are passionate about wine. The tests are blind, with the wines arriving at the judges' tables without any distinguishing information to identify where they are made. The wines are rated on colour, clarity, aroma, bouquet, taste, and the judge's enjoyment of the wine. In this competition, wines were submitted from Oregon, Washington, Montana, British Columbia, and Alberta. On the second day of the competition, the wines that scored the highest on the first day were presented to us again for a final scoring. Hmmm . . . I thought I tasted our Sparkling Blueberry Wine, and indeed I had. It won the Crystal Rose award, which is a particularly prestigious honour. It's the highest award of all double-gold wines in its category (non-grape wine). You can imagine how thrilled Alf and I were to win it.

Even if you've never visited our farm, you might have still seen it. It's been a location for over 35 movies and a few TV series. We keep a list of movie and TV titles on our website. Be sure to check it out! The first movie that was filmed on the farm was Hallmark's *Pumpkin Pie Wars*, in 2017. I didn't know much about the film industry at that time, but being an animal lover, I was very excited when the contact person told me the circus would arrive at 5:00 on Monday morning. Imagine my disappointment when I arose early only to learn that "circus" is movie lingo for all the trucks and other equipment that come on site on filming day. There wasn't a tiger or zebra to be seen!

It's fun to receive phone calls from family, friends, and guests inquiring if a show they just watched was filmed at the farm. Alf and I have been caught off guard a few times ourselves. On more than one occasion, we've been watching TV on a winter's evening when something very familiar catches our eyes. We'll do a rewind, only to see our winery performing as a restaurant, for example. Our Farmhouse Kitchen has even posed as a location in Italy. Our farm manager, Ann, told us once that she saw someone walking from the Harvest Kitchen up the ramp into The Porch, only to emerge in Fort Langley, which is approximately 12 miles away. Many of the movies filmed in summer are Christmas movies, and it's always fun for our farm guests to see "snow" on the ground in the middle of August. Usually, life goes on as normal for us during filming, but occasionally we do need to scale back our daily activity to accommodate a film crew.

With all the filming, we have met many wonderful producers, directors, and actors. While she was at the farm enjoying our berry waffles recently, country music legend and actor Reba McEntire spotted our winery chandelier and fell in love with it! We have had many performers, both local and famous, on our centre stage, which we call the KB Corral. Located between the winery and the market, it's a lively place where guests can listen to great music, snap their fingers, and tap their

The Kitchens

I have eight kitchens!

My number-one kitchen, the one that's closest to my heart, is my Farmhouse Kitchen (page 189), where I cook and bake with my family and friends.

In 2000, in a little closet at the small berry barn, I began baking pies and shortcakes to sell. I outgrew the space in one season, so we built a new little building we called The Porch (page 248), where I served milkshakes and smoothies to the guests who came to pick berries.

After 2 years, I needed much more space, so we built a large industrial kitchen that I named the Harvest Kitchen (page 239). This is where all our volume production takes place. I then needed a separate area where we could create our specialty baking products, so the Specialty Foods Kitchen (page 198) was wedged between The Porch and Harvest Kitchen.

Because we live on the farm, we really needed a break by that stage from all the swirling activity around us, somewhere to escape with our family. I soon found I had yet another kitchen to cook and bake in for family and friends. This one I named my Relax Kitchen (page 50). Two years later, Alf and I took a leap and made the decision to build a winery, and another kitchen was born: the Winery Kitchen (page 259).

Escaping became more important now than ever, but it wasn't easy, so we transformed a small barn in a hidden wooded area of our farm into a cottage. I now have the Cottage Kitchen (page 163), only a 5-minute walk away from the farm and not open to guests. I can make us a quick lunch together, and I can have tea parties with my kids, grandkids, and friends.

Just when I thought I had enough kitchens, my son Tanner presented me with the gift of a 1970 vintage trailer, which I immediately named Sweetie Pie (page 65). So now I also have a kitchen on wheels.

toes while they wait to enjoy their berry waffles. On busy days in the summer season, we open a second stage out in the drive-in waffle field to entertain our guests while they tailgate and enjoy their berry waffles and the fresh farm air.

Of course, our journey hasn't been free of bumps in the road. What journey ever is? Like all farmers, we've experienced setbacks over the years: the recession in the early 1980s, when interest rates in excess of 20% limited our growth; the berry-processing pricing staying stagnant while farming expenses rose each year; and, of course, COVID, which affected us just as it affected everyone else. We struggled to find harvesting crew, which meant many crops were left on the vine, and we had to deal with supply-chain shortages and soaring prices. The heat wave of 2021 singed the tops of our raspberries and blueberries, melting them on the vines and affecting the yields for 2022. All this to say, farming is a gamble, and each year we remain hopeful and enthusiastic for every farmer's dream: a bumper crop!

The Produce

The farm, the winery, my home, and my kitchen on wheels are my world. They enable me to combine my business sense, creativity, and passion for food and fun. That being said, none of it would be possible if I didn't have the quality fruit and vegetables that my husband, Alf, grows.

Alf's decision to farm responsibly and follow crop-rotation practices is one of the best he's ever made. It means we have super-healthy soil, which results in healthy, nutritious produce.

WHAT WE PLANT AND WHY WE PLANT IT

Here are some of the things that Alf considers when deciding what berry varieties to plant:

- Will it thrive in our growing zone, zone 8, which is a cool and wet climate?
- What is its taste profile? (We don't want berries that aren't going to taste good!)
- What size is it? (Larger is better because of visual appeal and picking efficiency.)
- When is it harvested and will that clash with our other harvest schedules?
- What are the labour requirements to grow and harvest it?
- How susceptible is it to disease and pests?

He considers the same factors when he's deciding what vegetables to grow, but vegetables are also a fashion item. So, in addition to asking himself the six basic questions above, Alf checks on ever-changing food trends with chefs who work with local food so that we can deliver what they need. Fortunately, we have a great friend, Master Chef Wolfgang Schmelcher, as a fabulous resource.

We grow mostly berries, not only because they were Alf's chosen crop when he started farming, but also because berries generally have a higher market value. Land prices are high in the Fraser Valley, so we needed to grow a higher-valued crop to cover the costs associated with the land.

To ensure the quality and higher production of our berries, we rotate them with vegetable crops to keep our soils healthy by breaking disease, insect, and weed cycles and return some nutrients back to the soil. The longevity of the berries we grow varies. Strawberries are on a 2- to 3-year rotation, while raspberries and blackberries are 5–15 years and blueberries can be up to 30 years. As part of the rotation process, a vegetable crop, such as corn, pumpkins, or cucumbers, is planted along with a cover crop, such as barley, fall rye, mustard, or forage turnips, for 2–4 years. This allows the soil to recover and enrich its nutrients before we rotate a berry crop back into it.

GETTING THE MOST FROM BERRIES

Berries benefit from a little extra care once you get them home. Here are a few tips to help you maintain the freshness of our berries and get the most out of them.

PICKING

All berries, no matter what type, should be left on the vine until they are ripe and ready to pick. How can you tell when they're ready? They will have a rich uniform colour and shape. The ones that are half coloured and half green should be left on the vine to ripen. Berries do not ripen after they are picked. Gently move the leaves around to find hidden ripe berries, working back and forth from different angles to find berries you didn't see the first pass through. You'll be surprised!

Strawberries are ready to pick when they're a bright, shiny red and they feel firm to the touch. To pick them, twist the berry at its stem, leaving the green leaves (calyx) in place at the top of strawberries if you plan on eating them fresh or decorating with them. If you plan on making jam or freezing them, use both hands to twist the entire strawberry loose from the calyx, leaving it on the vine. That will make your home prep easier and faster. If I plan on dipping strawberries in chocolate, I take scissors into the row and cut the stem 2–3 inches above the strawberry.

Raspberries are ready to pick when their colour deepens to a rich red. They will release from their stem when gently pulled, leaving the core on the vine. If a raspberry doesn't release when you do this, move on to the next one so the first berry can continue to ripen on the vine.

Blueberries are ready to pick when the entire berry is a deep blue. To pick, gently roll the blueberry between your thumb and finger and let it fall into your palm.

Blackberries are ready to pick when their colour turns a rich, deep, shiny black and their drupelets feel plump and full. With a little twist, the berry should release from the stem; if it doesn't, leave it to ripen.

CLEANING

Always sort through your berries before eating or storing them. Discard any that are brown or have begun to mould. You might find some stray bugs or worms from the soil when you do this.

Moisture will cause berries to brown and decay, which reduces their shelf life. Therefore, do not wash them until right before you're going to eat them or work with them. A rinse and a very thorough drying with paper towel are strongly recommended. Strawberries and blueberries can be rolled in paper towel to remove excess moisture, while raspberries and blackberries should be gently patted to remove excess moisture.

STORING

Freshly picked berries from the farm need to be put in the refrigerator as soon as you arrive home. Leaving them in their original container is the best option because the less you handle them before you're ready to use them, the better. There is one exception to this rule: upon inspection, if you encounter a decaying berry, remove it immediately so the decay doesn't spread to the other berries. Don't clean berries until you're ready to eat them (this point bears repeating). When you're ready to eat them or use them, see the cleaning section for tips on how to clean them. If the berries have their stem on, leave it on until you're ready to clean them. This will help to extend their shelf life. Most berries will keep in the refrigerator for 3–5 days. The longevity of any berry depends on various factors, including the weather it was picked in and its ripeness. Blueberries generally have the longest shelf life.

You can also freeze berries to enjoy them in the dark days of winter. We freeze our berries using a method called IQF, which stands for individually quick frozen. This is a simple process that you can use at home:

1. Sort, clean, and dry your berries. Ice crystals will form on berries that have any moisture on them, which adds more water to them when thawed, so pay close attention to drying them thoroughly.
2. Spread out the dry berries in a single layer on a clean, dry baking sheet that will fit into your freezer. Place the baking sheet into your freezer for at least 2 hours and up to 24 hours.
3. Remove the baking sheet and tap it on a counter to loosen the berries. Place the frozen berries in a freezer bag, removing as much air as possible, or in a freezer container.

Using this method, the berries will be frozen as individual berries instead of in undesirable clumps, making them easier to measure out and use. You can keep frozen berries in the freezer for up to 6 months. Thaw frozen berries in the refrigerator overnight or run cold water over them in a bowl, removing them as soon as they are thawed. If you're using them for baking, lay the thawed berries on a paper towel–lined baking sheet and pat them dry.

The People

The farm would not be the success it is without our amazing staff, some of whom have been with us for literally decades (pages 280–83). These staff are the backbone of our farm. We've even had several generations of various families work with us across the years.

We have a mix of full-time staff, part-time year-round, and seasonal full- and part-time of all ages who come from all parts of North America to help with harvest. In the early years, we hired a crew of six summer students who used their wages to help pay for their university education. Every year since, the number of students we hire has increased. We now hire just over 200 staff for our harvest season. One of my favourite parts of my job is reading the cover letters and résumés that flood in from students in early March. I will never tire of reading opening lines like "I've been coming to your farm since I was little and have been counting the years until I am finally old enough to work there" or "I have grown up on your milkshakes."

Alf and I feel great pride that we have been able to provide jobs in our community, and we chuckle when we think of how many kids, aside from our own, we have helped put through university. I still blush when these grown kids visit the farm to tell me it was the best summer job they could have had! We've also contributed in a small way to romantic encounters. I feel honoured when a former member of our staff drops by to introduce their spouse, whom they met while they were working with us—and, of course, I love, love, love meeting the children of former employees.

Our staff is very closely involved in almost every aspect of the farm and how it runs. One of the more fun aspects of this involvement is taste-testing. When we're developing a new product, I leave out samples for staff to try. Alongside the product, I leave feedback forms where they can write what they thought about the product's flavour, texture, appearance, etc. Staff also have the chance to develop new recipes and products in the off-season, when we're not rushed off our feet.

Like people in many industries, farmers and people in the hospitality industry are always working at least one season ahead. At Krause Berry Farms and Estate Winery, for example, we start to think about Christmas in January. Although I source local products throughout the year for the farm's market, I set aside January to plan for our farm's market—that basically means I look carefully at what we bought and sold the previous year, and when we bought and sold it. I work closely with our farm manager, Ann, our market manager, Leanne, and our numbers wizard, Tammy, to gather all the information I need before Ann and I head off to the Las Vegas Winter Market to hunt for products we think our guests will like. We spend four days scouring the market before coming back to the farm to begin ordering everything for the year. From there we move forward to planning for staffing and scheduling before the summer season begins. I then enjoy a little time to play in one of the eight kitchens, experimenting with different recipe ideas for the food areas around the farm before life becomes utterly and wonderfully chaotic.

The Animals

Alf never did change his mind about raising chickens, but animals still play a large part in our lives.

When my dad was growing up, he rode in rodeos in the Oregon outback. In my youth, he took my siblings and me to little rodeos as well as the legendary Pendleton Round-Up. He instilled a real love of horses in me, and he gifted me with his Stetson 3X Beaver cowboy hat the year we opened the winery (seen on page 128). My first horse was a little black and white pinto Welsh pony named Prince. We rode miles and miles together in all kinds of weather. If a blustery snowstorm hit when my parents were out for the evening, I would sneak him inside to keep him warm.

Over the years, Alf and I had quarter horses and three teams of draft horses on the farm. The drafts had many different jobs, but the one they loved the most was pulling the wagon around the farm fields to show our guests what we grow and how we grow it. They have since retired, but they live on through the wine labels we had made to commemorate their service to our farm. These days, we enjoy saddling up our Tennessee Walkers, Tucker and Mosaic, to ride around the farm checking the fields. After a ride, we reward them with Horse Muffin Treats (page 129).

Dogs were also a big part of my childhood. I have fond memories of Jojo, a cocker spaniel, and Lady, a Weimaraner. As an adult I had the pleasure of owning Puka Shell, a Samoyed; Tex, a husky; Chad, a boxer; RB (for Raspberry), a golden retriever; Rowan, an Akbash Maremma; and Biscuit, our very long-lived and -loved Jack Russell. We now live the love of dogs through our kids' dogs: Oliver, a Chesapeake Bay retriever, and Poppy and Willow, Moyen poodles, and Derek, a miniature Dachshund. Maybe you've even seen our daughter Holly's Oliver in the Winery Dogs of BC calendar, jumping up for one of my delicious dog treats (page 146)!

As much as Alf and I love animals, we don't allow animals to come onto the farm for hygienic reasons, except, of course, service dogs.

JAKE AND JERRY
Raspberry Apple
KRAUSE BERRY FARMS
& ESTATE WINERY
Pear
LUCY AND LILY
Tayberry
KRAUSE BERRY FARMS
& ESTATE WINERY

MENU

About This Book

Truth be told, it didn't occur to me to create this cookbook until a few years back, when Robert McCullough, the publisher of Appetite by Random House, approached me during a visit to the farm. He pointed out that not only was the story of the farm and everyone who works on it a piece of living history, but also guests had been asking me for years for the recipes for their favourite dishes and treats. His comments made me realize that I could offer guests a piece of the farm to take home—a true Fraser Valley farm-to-fork cookbook containing a collection of my most-loved recipes bound together with a glimpse or two into my life on a family-operated, working destination farm. The idea sat on a back burner behind the corn chowder and strawberry jam for many months. I had it on a medium-to-low simmer, giving it the occasional stir and adding the occasional pinch of seasoning.

And then I met Heather Cameron. By this time I had the idea for the basic concept of a book, a wealth of recipes and stories, and a host of people I knew I could call upon to be recipe testers. But one vital ingredient was missing: the visuals. When I met Heather, a magazine stylist, story producer, author, and photographer, in 2015, I knew I had found my missing ingredient. Since Heather joined us, she has followed me through eight consecutive seasons, capturing in real time the reality of life on the farm, the food, and the kitchens where the magic happens. She has also taken on all the farm's social media and advertising and developed the website and online store. She works with natural light and angles, and her photos are an authentic reflection of who we are and what we do—as I'm sure you'll realize as you page through this beautiful book.

With my missing ingredient safely in place, I decided to take Robert's advice and start writing. I was driven by my love for my family, the farm, our staff, the guests who return each year, the guests we have yet to meet, and the food we grow and produce. I set to work with gusto. Then, halfway through the writing of this cookbook, my world fell apart with the sudden death of my son Tanner. Life is never the same after your child dies. It changes you forever. I didn't feel

able to continue working on the book, but in quiet moments I began to hear Tanner encouraging me to keep going, reminding me that I wasn't a quitter. His love realigned me with my goals and gave me new inspiration and strength to continue writing. He, along with all the wonderful people who had been working with me on the book, continued to support me along the way.

Cooking is about so much more than just following a recipe (although that helps, of course). Please don't treat this cookbook as a textbook that you have to memorize or follow to the letter. Use what you have on hand, be brave with your ingredients and techniques, and honour your own tastes and style. And have fun! If you have your grandma's old aprons, wear them and celebrate all the meals those aprons have seen in their lifetimes. Wear a souvenir apron from your favourite place and recapture the joy of your travels. Turn the music up a notch or ten. Bake and cook with your kids and your grandkids. Love is the most important ingredient in any recipe. There is not a holiday dinner that goes by without one of my children reminding us of the year my daughter Holly, who was 11 at the time, was helping me prepare the dinner and noticed me taking the turkey's temperature. She asked me if it was sick. They all release the same chorus of laughter year after year. On my weekly babysitting days when three of my grandchildren at the time were young, I would bake with them, teaching them how to measure, crack eggs, mould, roll, and bake. My granddaughter Emme and I would don our strawberry aprons and she would really get into it, head and all! After she licked the spoon and the beaters, she went for the bowl! She learned how to wash the dishes with more bubbles than Mommy would have allowed, and after the cookies came out of the oven, we had not only clean dishes, but also a sparkling floor to boot. The smile on her face as her older brother, Kale, and her parents walked through the doors ready to devour her cookies is etched in my memory forever.

My purpose with this cookbook—and in life!—is to inspire people to be their creative selves. I hope you make my recipes your own. Write notes on the pages about substitutions or serving suggestions. Scribble down dates when you made a recipe for a special occasion, even if that occasion was one of mourning rather than celebrating. Put sticky notes on the pages to remind you about who expressed a particular liking for a dish or who suggested a possible substitution for an ingredient. Eventually you'll have your whole family history on the pages of your book. Try baking an 8-inch square cake in mini loaf pans or as muffins, or try baking regular muffins as mini muffins, adjusting all timings accordingly. Make mini cookies to use as checkers, like my granddaughters and I do. It makes for a very fun game and sweet time together!

About the Recipes

Just as Alf began with one strawberry, I began with one family recipe for pie passed down by my grandmother Sarah Greene. It was clearly a good recipe, as today we annually produce well over 100,000 pies based on it.

Choosing which recipes to include in this book was an exercise in itself. I was limited by space to include about 100 recipes, so I chose an assortment of farm and family favourites that I thought most of you would enjoy making. Some are very simple, perfect for the novice baker and cook, while others are a little more challenging and time-consuming, ideal for those with more experience—but all the finished products are worth the time and energy you'll invest in making them! I've even included a couple of recipes for our four-legged friends and a family-favourite recipe for baby teething cookies.

And remember, experienced bakers and cooks all started as novices, so even if you don't have much experience in the kitchen, don't be afraid to try all the recipes in this book!

Before you start to cook or bake, read the recipe all the way through to check that you have everything you need. Next, do all the measuring, peeling, chopping, dicing, and slicing so that everything is ready when you need to add it.

A Wink While You Cook, A Smile While You Eat

Throughout the recipes in this book, you'll find a **Wink** (or note) to help you make the recipe, take it up a notch, or make a variation. I've even included fun tidbits about how we make and serve these recipes at home and at the farm. I hope these winks *nudge* you to try them all and inspire you to create something new and add your own personal touch!

Equipment

I am a mother to six and a grandmother to five (and growing). I love my family and I love experiences. I strive to make each meal fun, unique, and memorable. I believe a food experience goes beyond the ingredients. It incorporates the apron (page 91), the bowls, the utensils, the cookware, the serving pieces, the environment, the aesthetics, and, most importantly, the people who work together in creating it.

If, like me, you are lucky enough to have a collection of seasoned pots, pans, and older dishes that belonged to your parents or grandparents, pull them out and use them to connect with the loved ones who baked and cooked before you and for you, and to allow you to continue the tradition of giving to others. If you don't have an inherited collection of equipment, treat your own equipment as a collection of blank pages waiting for you to write your story on them.

The Value of a Good Knife

In 1997, at my son Tanner's baseball game, his teammate's older brother came through the stands selling Cutco knives with a forever guarantee. It was his first job and I wanted to support his great sales spiel, so I agreed to buy one knife, a white-handled 7-inch trimmer with a serrated blade. When he presented me with the bill, I almost fell out of the stands. It cost $100! Well, that was the best $100 I could have spent on a knife. It was a really big deal in my house, and I laid down the law that no one but me was to use it, as it was so sharp. Within the first week, my oldest daughter, Sarah, decided to test its sharpness. She bears the little scar to this day.

I used that 7-inch knife for everything, and 10 years later I splurged and added a 4-inch paring knife, a 9¾-inch slicer, a 7-inch Santoku, a butcher knife, and a cheese knife to my collection. The last three knives are just plain luxury, but I consider the first three as essentials. And you know what? I have only professionally sharpened that first knife once!

I have many conveniences in my kitchens today that my grandmothers and mother did not, yet they were able to produce delicious, nutritious meals day after day. Don't feel you need to use the information below as a shopping list for stocking your kitchen (it's not a comprehensive list). Some of the items are very much according to taste rather than need. I'm sure you'll figure out what will be of most use to you in your kitchen.

POTS AND PANS

Let me say upfront that I have been very fortunate in terms of building my stock of kitchen equipment. My grandmother Sarah Greene gave me a set of Revere Ware copper-bottom pots, a cast-iron frying pan, and two Pyrex mixing bowls as a wedding gift in 1981. I still have all of them today and I use them daily, polishing the copper bottoms occasionally. I was also given my mother's and her mother's (my grandma Buehler's) Guardian Ware to add to my own collection (page 216). I am very particular about my pots and pans and the way I look after them. I hand-wash them all, along with my good knives, never putting them in the dishwasher.

For the recipes in this book, you'll need a 10-inch frying pan (preferably oven-proof and ideally cast iron), one small and medium pot, and one large heavy-bottomed pot. For the baking recipes, you'll need a selection of standard-sized baking pans and ideally at least two baking sheets (if you don't have two, you can bake in batches).

BOWLS

I have a collection of both glass and stainless-steel mixing bowls in all different sizes. Glass bowls can go into the microwave, provide great visuals, and look nice as serving bowls. Stainless-steel bowls are unbreakable, nest well, and come in very small to very large sizes. Both can go into the dishwasher. If you have one large, one medium, and one small bowl that are all heatproof, you'll be prepared for almost anything.

MIXERS

In my Harvest Kitchen, I use large mixers because I work with large volumes of ingredients, but in all my other kitchens, I use an electric hand mixer and a stand mixer. In my younger days, before I had children, I had only a hand mixer, so that's what I used. For mixing cookie dough and quick bread dough or batter, I found it was faster and easier to use a little muscle, a large spoon and a spatula rather than going through the extra step of setting up the hand mixer. It was all about efficiency when the kids were small. For cakes, though, I used the hand mixer, and there were always happy kids around to lick the beaters, bowls, and spatulas. For the recipes in this book, you can use either a hand mixer or a stand mixer when you see the words "electric mixer." If a stand mixer makes a difference, I've specified it.

KNIVES

Some say German, some say Japanese. I say it is truly a matter of preference, but be aware that when it comes to knives, you get what you pay for (see opposite page). If you're planning to splurge on kitchen equipment, splurge on your knives. You'll be pleased you did. Some chefs swear by their favourite knives,

To find a cast-iron pan, search second-hand and antique stores. You may even find one that is already seasoned. How can you tell? The inside will feel smooth as silk! A well-cared-for cast-iron piece is a treasure as well as a workhorse for cooking and baking. If you buy new, you must follow the seasoning instructions by washing the pan with soap and hot water, immediately followed by drying it in the oven or on the stove burner. When there is not a trace of water left, remove from heat and let it cool so it doesn't smoke when you drop a little oil into the pan. With a paper towel, rub the oil onto the bottom and sides of your piece. After you cook or bake in cast-iron pieces, keep them away from water until you have eaten and have the time to wash them in hot, soapy water. Rinse them and repeat the simple steps above after each use. If you take care of your cast iron in this way, it will return the "flavour" and become one of your most beloved pans.

carrying them from job to job. For the recipes in this book, a 3-inch paring knife, an 8-inch chef's knife, and a long serrated bread knife should see you through.

OVENS: CONVENTIONAL AND CONVECTION

While you probably aren't planning to change your oven any time soon, it's helpful to know a bit about the differences between conventional and convection ovens, not to mention that each oven is different.

Conventional and convection ovens both have heating elements on the top and bottom of the oven. Convection ovens also have a fan that helps to circulate the hot air throughout the oven, resulting in evenly distributed heat. Both can be gas or electric and both can have hot spots in different areas.

I work with both in my kitchens. I have found that conventional ovens work best for baking recipes that use yeast and where you want the final product to retain some moisture, such as breads, cakes, and muffins, without browning and crisping the tops. For me, convection ovens are the best for baking, roasting, and crisping, such as roasting corn, cooking enchiladas, and baking cookies.

If your convection oven doesn't reduce the baking temperature automatically, my general rule is to reduce the temperature by 25°F (for example, instead of baking something at 350°F, I'll bake it at 325°F). Just be sure to watch carefully with the oven light on to make sure the item doesn't overbake and check the item's doneness with a toothpick or knife—if it comes out clean, it's done. I don't have strict guidelines about shortening the cooking time, but, for example, if a cake normally bakes for 30 minutes and I'm baking it in a convection oven, I'll start checking at the 20-minute point.

If you're cooking something in a convection oven that needs to be covered, make sure your covering is well tucked under your dish so that it doesn't come loose and fly around in your oven. Also, never overcrowd a convection oven—the air won't be able to circulate, which will create hot spots.

Hot spots are exactly what they sound like: areas of the oven, usually the top, bottom, and sides, that are hotter than other parts, which means that your food cooks unevenly. They aren't unique to convection ovens. You can get them in conventional ovens as well. To find your oven's hot spots, heat the oven to 350°F and place one oven rack in the highest position and one in the lowest. Line them both with bread slices and note which slices turn brown and which slices stay pale. You'll then know how you can work around your oven's hot spots.

Get to know your oven. Once you do, you'll be able to work with it and manage any of its quirks. We have an oven in our Harvest Kitchen (page 239) that we call the Workhorse. We absolutely love her, but she has hot spots. However, because we are aware of them, we work with her and help her along by rotating and removing dishes and letting some cook a little longer if need be. That way, we're all happy with the finished product.

All ovens and stovetops have their quirks. The timings and temperatures that you see in the book are the ones we used when testing the recipes. Treat them as guidelines and adjust as you see fit.

OTHER USEFUL EQUIPMENT
Here are a few other pieces of equipment you may find useful.

Candy thermometer: This is a fairly inexpensive but very useful tool. It's indispensable for recipes like my Custard Sauce (page 276) to prevent overcooking and for any recipe that calls for scalded milk.

Cookie scoops: These will help you scoop out consistent amounts of cookie dough to keep your cookies a similar size. I recommend one small, #60 (2 tsp/¾ oz), scoop for a 2-inch cookie; one medium, #40 (1½ Tbsp/1½ oz), for a 3-inch cookie; and one large, #20 (3 Tbsp/approximately 3 oz) for a 4-inch cookie.

Flat skimmer: This is a round sieved tool for lifting poached eggs out of simmering water. I use it when I'm making Eggs Benedict (page 34).

French onion soup bowls: These are the best option for serving French Onion Soup (page 62). In a pinch, you could use any small ovenproof bowls.

Immersion blender: This tool makes it quick and easy to purée soups such as Cauliflower, Potato, and Leek Soup (page 58), and sauces such as Tomato Basil Sauce (page 275), and to smooth out any lumps in Custard Sauce (page 276).

Kitchen blowtorch: This is useful for browning meringue (see Baked Alaska, page 202) and putting a nice, toasted finish on Coquilles St. Jacques (page 92).

Loaf pans: Loaf pans come in different sizes, but I use a 5 × 9-inch one for Scrapple (page 38), Auntie Dona's Zucchini Bread (page 121), Grandma Greene's Blueberry Banana Bread (page 193), and many of the breads in the Bread and Muffins chapter (pages 110–29).

Perforated baking sheet: You'll need one of these for best results in my Roasted Corn recipe (page 73), but you'll find many uses for it, from baking bread to making pizza (page 77). The air can circulate for even cooking and gives a brown delicious crispy finish.

Piping bags and tips: Piping bags are triangular in shape and made of plastic or cloth. They tend to come in one size, but the tips come in different shapes and sizes, for various decorative techniques. A #2 tip is the most used type in our kitchens, so if you're only going to buy one tip, that's the one I would recommend. The Krause Berry Farms kitchen staff use that tip to decorate Strawberry Cookies (page 132), Gingerbread People (page 141), the Berry Merry Christmas Fruitcake (page 175), and Chocolate-Covered Strawberries (page 208).

Popsicle mould: These are useful for setting popsicles, like my Wine Popsicles (page 256). They come in fun shapes and sizes, which is perfect if you're throwing a theme party. Silicone moulds make it easy to remove your treats and they're easy to clean.

Probe thermometer: Use a digital instant-read probe thermometer to check internal temperatures when you are making the bread recipes in this book. And of course, this is the perfect tool for checking the temperature of meat and poultry.

Pudding mould: A pudding mould will distribute the heat evenly and keep the pudding moist. It's a must for my Steamed Berry Pudding (page 223).

Rolling pins: This is essential for rolling out the many doughs in this book. I have tried all types of rolling pins—marble, wood, nylon, and stainless steel—and have even used a wine bottle and a tall S'well thermos in a pinch. My favourite type of rolling pin is made of unfinished wood and has handles. I find it easier on my hands. I wash mine after each use, dry them thoroughly, and occasionally season with a little oil. I then dust with flour right before I use again. I know a lot of people who like to use a wooden dowel or French tapered pin, so go with what works best for you. (I have collected rolling pins for years and have over 80 on display at the bakery in the farm market as inspiration for our guests.)

Want the perfect pie slice? We love to use the "first out" pie spatula, which makes it easy to cut and lift out a slice without making a mess (except for custard pie, of course!).

Ingredients

In writing this cookbook, I took recipes used in our commercial kitchens that yield 100 servings—where I like to use Krause Berry Farms and Estate Winery produce and products as much as possible, of course—and adapted them for your home kitchen. This means that when you see one of our products listed in a recipe's ingredients list, it was tested with that particular product. Our guests visit from all over the world, so I understand that you might not have our products on hand when you get the urge to make a particular recipe. If that's the case, substitute with as close an equivalent as you can get. For example, if a recipe calls for our Gluten-Friendly Flour Blend (page 279), substitute your favourite gluten-free flour mix instead. Just keep in mind that if you use a substitute ingredient, you may not get the same result.

In general, use the freshest, best quality, most local ingredients you can lay your hands on. Sometimes, though, we just don't have the time to enjoy cooking or baking from scratch! I have certainly been in that position many times. I'll look at the recipe to see if I could substitute a jarred, canned, or frozen product to shorten the preparation time, such as tomatoes for tomato sauce or beans for chili. Usually I can, but I still try to ensure that the product is locally grown. When shopping, I read the labels on bagged, boxed, jarred, and canned items to find out where the ingredients were grown. If the ingredients were imported and only packaged locally, I try to avoid those products, unless I specifically want to use them and can't source

them any other way, such as spices like smoked Spanish paprika, Saigon cinnamon, nutmeg, or poppy seeds; grains like rice; and out-of-season items.

Using locally grown food supports not only your local farmers and your local economy but your food future as well. One of the lessons that the COVID-19 pandemic taught us was the value of supporting local food producers. They were there for us when other supply chains faltered. If you're able to grow some of your own produce, why not give it a try? Start small with a pot of herbs on your window ledge (or a small strawberry patch!).

BERRIES: FRESH VERSUS FROZEN

Whether fresh or frozen, berries retain their nutritional value. But you can't always use them interchangeably. Most berries are approximately 80% water, so when they're frozen, the water expands and breaks down the berries' cell structure. When thawed, they release more juice and become softer in texture, although the flavour remains the same. If a recipe calls for fresh berries, it should be made with fresh berries. Definitely, choose fresh berries over frozen for eating! And fresh is the best choice for using in decorative desserts, unbaked desserts, trifles, salads, fruit platters, charcuterie boards, and cocktails.

Pancakes and waffles are best made with fresh berries, but you can use frozen if that's all you have. A general rule of thumb is that if the recipe you're making has a short cooking time—for example, Blueberry Pancakes (page 45)—you will want to thaw the berries first. The best way to thaw berries whole is to put them in a sieve and run cold water over them until they separate. To help prevent steaks of colour in your batter, rinse them until any colour runs off, then pat them dry with a paper towel to remove the moisture so the batter cooks evenly and fold them in at the last minute.

If you want the juice from fresh berries, add just enough water to cover the bottom of the pot, bring to a boil on medium heat, then turn down to a simmer, stirring occasionally to help break them apart and not scorch. When they are soft enough to mash, line a fine-mesh sieve with cheesecloth, place over a large bowl, add the berries, and press against the sides and bottom to collect the juice. To cook frozen berries for juice, place them in a medium pot on medium heat without water and follow the method above.

Unthawed frozen berries are great in smoothies and milkshakes. A word of caution if you're thinking about using fresh blueberries in a smoothie or milkshake, blueberries are high in pectin and can curdle milk. I learned this lesson years ago when I ran out of frozen blueberries. I substituted fresh blueberries and my milkshake became gel-like and a greyish green!

Frozen berries are heavier than fresh, so if you're planning to use them in baked goods, toss them in a bowl with a little flour before adding them to the other ingredients. This will help distribute them evenly in the batter and prevent them from sinking to the bottom of your baking pan.

Sensational Cinnamon Buns (page 32)

Breakfast and Brunch

Sensational Cinnamon Buns

MAKES 15 BUNS

Dough
- 3½ cups all-purpose flour, divided
- ⅓ cup granulated sugar
- 1 Tbsp instant dry yeast
- 1 tsp table salt
- 1 cup 2% milk
- ⅓ cup salted butter, softened
- 1 large egg, at room temperature

Topping
- 1½ cups packed brown sugar
- ½ cup salted butter
- ⅓ cup light corn syrup
- ½–1 cup seasonal additions (optional, see Wink)

Filling
- ½ cup packed brown sugar
- 1½ tsp ground Saigon cinnamon
- 2 Tbsp salted butter
- ½–1 cup seasonal additions (optional, see Wink)

What do cinnamon buns and artichokes have in common? Their centres are the best part and worth eating to the middle to get to! I remember wondering as a child how my mother managed to put the cinnamon sugar so perfectly inside of her cinnamon buns. It wasn't until I was older and helped her make them that I understood. It's all about the roll! Even now that I'm older, I still like to eat these with my fingers. I even unroll them to eat the centre first. We like to match our cinnamon buns to the seasons, so in the summer they have fresh berries (of course!), in the fall we include apples and peaches, and in the winter we add dried berries. As we move into spring, we add yellow and black raisins in anticipation of the fresh fruit cycle beginning again. Frozen berries can be substituted for fresh. No matter the season, these are one of my favourite treats. See Wink for seasonal variations. (Note: You can make this recipe without a mixer using a bowl and spoon and a lot of elbow grease (just mix with a spoon until you cannot anymore) and knead for 8 to 10 minutes until smooth before continuing onto step 4.)

MAKE THE DOUGH

1. In a stand mixer fitted with the dough hook, combine 1¾ cups flour with the sugar, yeast, and salt on low speed for approximately 30 seconds.

2. Warm the milk in the microwave until lukewarm, approximately 30 seconds.

3. Add the milk, butter, and egg to the flour mixture. Beat for 1 minute on medium speed.

4. Add the remaining 1¾ cups flour and mix on low speed to make a soft dough. Turn the speed up to medium and knead the dough in the mixer for 2½ minutes. Shape into a ball. Transfer the dough to a large, lightly oiled bowl and cover with a clean tea towel or plastic wrap.

5. Let rise in a warm, draft-free area until doubled in size, approximately 1½ hours.

MAKE THE TOPPING

1. In a small microwave-safe mixing bowl, microwave the sugar and butter until melted, 40 seconds. Stir in the corn syrup.

2. Pour into a 9 × 13-inch baking pan with 2-inch-deep sides, and spread until evenly covering the baking pan. If you're using seasonal additions, scatter them evenly over the topping mixture. Set aside.

MAKE THE FILLING

1. In a small mixing bowl, whisk the sugar and cinnamon together until combined. Set aside.

2. Melt the butter separately in the microwave, approximately 20 seconds. Set aside.

ASSEMBLE THE BUNS FOR BAKING

1. Punch down the dough. Transfer the dough onto a lightly floured countertop. With a rolling pin lightly coated with flour, roll out into an 11 × 21-inch rectangle.

2. Brush the melted butter over the dough, then sprinkle with the cinnamon-sugar filling mixture. Sprinkle with any additional seasonal additions you want to use for the filling.

3. With the long side facing you, roll up the dough, rolling it away from you. When you're done rolling, ensure the long seam is on the bottom.

4. Using an electric knife or serrated knife, cut the dough parallel with the open ends of the roll into 15 evenly sized (approximately 1- to 1½-inch) pieces. Trim off the two ends for a clean edge. Place the rolls cut side down on either side in the prepared baking sheet on top of the topping, approximately 1 inch apart.

5. Cover in plastic wrap and let them rise in a draft-free area for 1½ hours until the buns are almost touching each other.

6. Preheat your oven to 350°F.

7. Bake the rolls, uncovered, for 10 minutes. Rotate the pan 180 degrees and bake until golden brown, fully cooked on the inside, and touching one another, for another 10–15 minutes.

8. Remove from the oven and let stand for 2 minutes in the pan, then invert onto a tray or plate.

9. For extra indulgence, let these cool completely and then top with cream cheese frosting (page 186).

10. These are best enjoyed the day you make them, but you can store them in an airtight container at room temperature for up to 2 days.

RECIPE PICTURED ON PAGE 30

WINK: *In spring, add ½–1 cup yellow or black raisins to the topping and another ½–1 cup to the filling. In summer, add 1 cup of your favourite fresh berries (sliced, if you're using strawberries) to the topping and another 1 cup to the filling. In fall, add 1 cup of peeled and diced apples or peaches to the topping and another 1 cup to the filling. And in winter, add 1 cup of your favourite dried berries for the filling.*

Eggs Benedict

To celebrate a 40th birthday one year, I felt inspired to host brunch for 50 friends. The first recipe that came to mind was this recipe, as it is so easy to streamline for everyone to enjoy their eggs Benny. You can make it classic like in this recipe or substitute the bacon with shrimp and crab for a West Coast Benny (like in this photo), or add an avocado for a California Benny. If you're brave, you can even turn up the heat by adding a few drops of hot sauce to the hollandaise when you add the butter! Speaking of butter, Hollandaise sauce requires a good-quality 100% butter with minimal water content; I like to use Kerrygold. This is one of my favourite breakfast dishes to make in my Relax Kitchen (page 50).

SERVES 2

Eggs Benedict

4 slices Canadian back bacon (see Wink)

2 English muffins (store-bought or homemade, page 126), split

Salted butter, for the muffins

1 Tbsp white vinegar

4 large eggs, at room temperature

Chopped chives (optional)

Hollandaise sauce

½ cup salted butter (see Wink)

3 large egg yolks

1 Tbsp fresh lemon juice

½ tsp kosher salt

MAKE THE EGGS BENEDICT

1. Preheat your oven to 200°F–250°F.

2. Warm a large frying pan on medium-high heat. Fry the bacon on each side until lightly browned. Drain on a paper towel. Keep warm in a baking dish in the oven.

3. Toast and butter the English muffins and keep them warm in the oven.

4. Add the vinegar to a large pot of water and bring to a boil on medium-high heat. Once boiling, carefully crack the eggs into a small bowl and slide each one into the boiling water (do not crowd them). When the eggs rise to the top, 2–3 minutes, use a flat skimmer to remove them carefully. Allow any water to drain back into the pot.

MAKE THE HOLLANDAISE

1. In a small pot, melt the butter on low heat.

2. In a blender, mix the egg yolks, lemon juice, and salt on low speed until fully mixed. With the blender running, slowly pour the melted butter through the hole in the top of the lid (see Wink). Mix until the butter is blended throughout the mixture.

ASSEMBLE AND SERVE

1. Arrange the English muffin halves on plates and top with the bacon and then the poached eggs.

2. Drizzle each egg with warm hollandaise sauce, chopped chives (if using), and serve.

WINK: *If you're serving a crowd, keep the toasted English muffins and cooked bacon warm in the oven, as described above. Make the hollandaise sauce and keep it warm on the stovetop. When you're ready to start serving, place warm bacon on a warmed muffin, slip an egg onto a muffin, and top with warm hollandaise sauce. Keep the plates coming! Leftover hollandaise sauce can be warmed gently in a saucepan on low or in the microwave for 50 seconds before serving.*

Auntie Dona's Overnight Brunch Eggs

Late every Christmas Eve, while sugar plums dance in the kids' heads, something stirs in the Krause kitchen, and it isn't a mouse. It's me, preparing for Christmas morning magic—or more specifically, ensuring this dish is made and refrigerated for Christmas brunch.

Auntie Dona, my dad's older sister, gave me this family recipe in 1982. Christmas morning can be magical for young kids but chaotic for parents. This dish made Christmas morning easy, as all the work was done the night before and it only needed to be popped into the oven, allowing the morning to be magical for parents as well. Hot, cheesy, and loaded with sausage and onions, it has everything you could want to get a good start on your Christmas morning. (Note: You need to start this recipe the night before you plan to eat it.)

SERVES 8

6 (½-inch-thick) slices sourdough bread
¾–1 lb (340–450 g) breakfast sausage (see Wink)
12 large eggs, at room temperature
1 cup 2% milk
1 medium yellow onion, diced
1 Tbsp Worcestershire sauce
½ tsp freshly ground black pepper
¼ tsp kosher salt
½ lb (225 g) shredded Monterey Jack cheese
½ lb (225 g) shredded Swiss cheese

1. Grease the bottom and sides of a 9 × 13-inch baking dish. Place the bread in the dish without overlapping (you may need to trim the bread slices to fit).

2. In a large frying pan on medium heat, fry the sausage, breaking it up and cooking it until browned. Drain off the excess oil and set aside.

3. In a large bowl, beat the eggs thoroughly with a fork.

4. Using a large spoon, mix in the milk, onions, Worcestershire sauce, pepper, and salt until everything is evenly distributed.

5. Pour the egg and onion mixture evenly over the bread slices.

6. In the mixing bowl you used for the eggs, mix the cheeses and cooked sausage together and layer over the egg and onion mixture.

7. Cover the dish with foil and refrigerate overnight.

8. Preheat your oven to 350°F. Let the dish sit for 15 minutes at room temperature before baking.

9. Bake the casserole, covered with the foil, on the middle rack in the oven for 40 minutes. Remove the foil and bake until the eggs are fully cooked, the cheese is browned, and a toothpick inserted into the centre comes out clean, approximately 10 minutes.

10. Store leftovers covered in foil in the refrigerator for up to 2 days.

WINK: *I love to use Jimmy Dean sausage. If you can't find it, you can use breakfast sausage or your preferred sausage. Just use the same weight of breakfast links and remove the meat from the sausage casings.*

Scrapple

Scrapple is a northern German dish. It's a mixture of cornmeal porridge and meat that's formed into a loaf, then sliced and fried. It's popular in the eastern United States because of the population's strong German and Dutch heritage. Longing for scrapple one day, I asked my mother to send me her recipe. Along with the recipe, she included a note with this story: She was making scrapple early one morning when the milkman tapped on her door. He was from Pennsylvania and had smelled the scrapple from down the road. He traced the smell back to Mom's house and told her the smell made him nostalgic. His West Coast wife had never heard of scrapple. My mother sent him on his way with one of the fresh loaves. The next morning, she found a baked cake and a thank-you note with her loaf pan in her milk box. My mother concluded her note to me by saying, "Enjoy the scrapple, enjoy the memories that come with it." I certainly do each time I make it. (Note: You need to start this recipe the day before you plan to enjoy it.)

MAKES TWO (5 × 9-INCH) LOAVES, EACH SERVES APPROXIMATELY 4

¾ lb (340 g) lean ground beef

½ lb (225 g) ground pork

3 cups water

1 large yellow onion, finely diced

1 cup + 2 Tbsp yellow cornmeal

1 cup cold water

1 Tbsp poultry seasoning

Salted butter, for frying

1. Crumble the ground beef and ground pork as small as possible and put them in a large pot with 3 cups water and the onions. Bring to a boil on medium-high heat, stirring and separating the meat constantly so there are no chunks, until there is no pink left in the meat, 5–10 minutes. (Chunks make cutting the scrapple difficult.) Turn the heat down to a low simmer.

2. Place the cornmeal and 1 cup cold water in a medium bowl. Stir until completely combined. Slowly add this to the simmering meat mixture, stirring constantly.

3. Add the poultry seasoning and turn the heat to medium-high to bring it back to a boil. Once the mixture is boiling again, keep it on medium-high heat and stir until it is thick and the boiling bubbles have turned into large popping eyes, 5–10 minutes. (Grandma Buehler used the term "large popping eyes" to describe the large bubbles that rise from the bottom of the pot and pop dramatically!)

4. Rinse two 5 × 9-inch loaf pans in cold water to aid the congealing process and pour the scrapple directly into the pans. Let cool at room temperature, then cover the pans with plastic wrap and refrigerate overnight.

5. Slice the chilled scrapple as thin as possible (¼–½ inch). Heat a large frying pan on medium-high heat and add 1 tsp butter. When the butter is melted, place 3–5 slices of scrapple in the pan and fry until they're dark golden brown. If the pan looks dry, add another 1 tsp butter. Carefully flip to the other side and fry until golden brown and crispy. Alternatively, slices of scrapple can be placed on a greased baking sheet and broiled on one side until the top is crispy, approximately 5 minutes, and then broiled on the other side until crispy, 3–5 minutes. Watch carefully so they do not burn.

6. Store wrapped in plastic wrap in the refrigerator for up to 4 days. Slice while cold and fry as directed above.

WINK: *I serve scrapple with ketchup and toast, the way my mother did, but my friend Brian, who lives in Virginia, serves it with a fried egg and a drizzle of maple syrup! Enjoy it for breakfast, lunch, or dinner.*

Healthy, Hearty Berry Granola

There are a few ways I like to start my day during harvest season. One is with a bowl of this granola, which we sell in the market, another is with a cup of yogurt topped with this granola. Harvest season can be very intense, and this granola is power-packed with the energy I need to keep going—it can take me from early rise to mid-afternoon if need be. It's not only a great breakfast but always a good snack with or without milk or yogurt!

MAKES APPROXIMATELY 17 CUPS

6 Tbsp salted butter
3 cups quick-cooking oats
1½ cups sweetened shredded coconut
¾ cup flaxseed
¾ cup wheat germ
¾ cup whole wheat flour
¾ cup dry-roasted pumpkin seeds
¾ cup dry-roasted sunflower seeds
¾ cup dry-roasted sesame seeds
¾ cup raw cashews
¾ cup unsalted almonds
¾ cup blanched almonds
6 Tbsp poppy seeds
6 Tbsp wheat bran
1 cup Krause Berry Farms syrup (any flavour)
1½ cups mixed dried berries (such as dried cranberries, blueberries, and cherries or raisins)

1. Preheat your oven to 300°F. Lightly grease two large baking sheets or line them with parchment paper.

2. In a small microwave-safe bowl, melt the butter in the microwave.

3. In a large bowl, mix together the oats, coconut, flaxseed, wheat germ, flour, pumpkin seeds, sunflower seeds, sesame seeds, cashews, both types of almonds, poppy seeds, wheat bran, and syrup. Mix in the melted butter.

4. Spread the mixture onto the prepared baking sheets. Place one sheet on the top rack of the oven and the other on the bottom.

5. Bake for 20 minutes, then take out of the oven and use a metal scraper or metal flipper to turn the granola over on each sheet. Return to the oven, placing the tray that was on the top rack on the bottom rack and the one that was on the bottom rack on the top rack. Bake for 20 more minutes. The granola will have dried out and darkened a bit and will no longer be sticky.

6. With a metal scraper or metal flipper, break up the granola into small pieces. Sprinkle with the dried berries and let cool on the baking sheets.

7. Once it's cool, toss it to break up any large chunks while transferring it into airtight containers. Store at room temperature for up to 4 months.

WINK: *This is great for kids' school snacks and packaged prettily for gifting.*

Berry German Pancake

When I lived in Portland, Oregon, our family would go downtown to a restaurant that was famous for the biggest and best German pancakes in the entire state. The 20-minute wait after ordering was always worth it. The presentation of this pancake is really "wow." The sides are crispy and come up over the pan, the pancakes are light and fluffy, and the look of the butter melting over the top with loads of fresh fruit is sure to make your mouth water. I bake mine for breakfasts and top them with fresh berries in a special German pancake pan and my cast-iron frying pan so I can bake two at a time. Either option will give you the characteristic puffed, raised sides you're aiming for.

SERVES 2

3 large eggs room, at temperature
3 Tbsp salted butter, divided, + more for serving
½ cup 2% milk
1 Tbsp granulated sugar
¼ tsp table salt
½ cup all-purpose flour
Fresh fruit (strawberries, raspberries, blueberries, or blackberries), for topping (optional)
Icing sugar, syrup, or whipped cream, for topping (optional)

WINK: *I use both a 10-inch cast-iron pan and a 10-inch German pancake pan designed specifically for this type of pancake. Both are ovenproof.*

1. Preheat your oven to 450°F.
2. In a medium bowl, lightly beat the eggs.
3. In a small microwave-safe bowl, melt 2 Tbsp butter in the microwave, approximately 45 seconds.
4. Stir the melted butter, milk, sugar, and salt into the bowl with beaten eggs. Add the flour one big spoonful (you don't need to be too precise) at a time, whisking after each addition until the flour is incorporated. Do not overmix.
5. Place the remaining 1 Tbsp butter in a 10-inch cast-iron frying pan (or another ovenproof frying pan—I like the 10-inch German pancake pan designed specifically for this type of pancake) and place in the oven to melt the butter. This will take only a few minutes, so watch carefully.
6. When the butter is melted, pour the batter into the hot frying pan.
7. Bake until the sides of the pancake have puffed up about 2 inches all the way around the pan and the pancake is golden brown, approximately 12–18 minutes. Switch on your oven light to help you see if this has happened. Once it's puffed up, carefully open the oven door, so as not to deflate the pancake, and insert a toothpick in the centre. If it comes out clean, the pancake is ready. If some batter is sticking to it, let it bake for another 2 minutes.
8. Cut the pancake in half and slide each half onto a serving plate. To serve, top with butter and fresh fruit (if using), and sprinkle with icing sugar or drizzle with syrup or whipped cream (if using). Serve immediately. The pancake will deflate soon after it comes out of the oven, so have everyone ready and waiting at the table to enjoy the visual beauty.

WINK: *If you are really hungry, double the recipe and use two pans, side by side in the oven and serve whole.*

Blueberry Pancakes

Pancakes are always a fun, lighthearted experience in the kitchen, but the addition of blueberries really elevates both the experience and the final dish, whether they're in the batter or on the plate. When we make blueberry pancakes, Alf plays Dean Brody's "The Kitchen Song," filling our kitchen with fun. It makes me laugh and we'll do a twirl or two, but has he convinced me yet?

MAKES 7 (4½-INCH) PANCAKES

¾ cup all-purpose flour
1 tsp baking powder
½ tsp baking soda
½ tsp table salt
1¼ cups buttermilk
1 large egg, at room temperature
3 Tbsp melted salted butter
1 cup fresh blueberries
1 Tbsp neutral-flavoured oil, for frying
Fresh blueberries, blackberries, and/or sliced strawberries, for serving
Krause Berry Farms and Estate Winery blueberry syrup, for serving

1. In a bowl, combine the flour, baking powder, baking soda, and salt and gently stir. Add the buttermilk and egg and whisk until smooth. Add the melted butter and continue whisking until well combined.

2. Cover and let rest for 30 minutes (or in the refrigerator overnight, but let it come to room temperature first), then add the blueberries and stir to distribute them evenly.

3. Heat a hot griddle or frying pan (I love using my cast-iron pan on medium-high heat). Add 1 tsp oil to the pan and then pour in about three 4-inch circles of batter. Cook until bubbles form. Gently lift and flip and continue cooking the other side for another few minutes for a perfect finish, lifting the edge to check if the bottom is golden brown. Continue with the remaining batter, adding more oil to the pan between batches.

4. Drizzle with syrup and top with berries to serve. These are best enjoyed the day you make them.

When Life Throws You a Curveball . . .

Hit a home run . . . with waffles

Several years ago, we were on the countdown to strawberry harvest, the farm gates were about to be opened, and we still hadn't received the final approvals for our new winery from the municipality. We were set to have a flood of people pouring through the gates, yet we were tied up in increasingly frustrating red tape. Our beautiful new winery was stocked full to the brim with amazing wines, but we couldn't open it, which meant we could not sell any of them or use the space we had set aside for the winery experience. Even worse, we couldn't offer this new experience to our guests. If we missed the harvest season, we would have to wait a full year to begin to recover from our losses and finally launch the experience we had worked so hard to prepare for our guests. We felt downhearted to say the least, but my feeling of helplessness in the hands of bureaucracy forced me to step back and look at what else I could offer to our guests in the meantime.

Thoughts of how I could use the space intended for guests to enjoy our wines rolled continuously around in my mind until one of my favourite breakfasts popped into my head—waffles! I grew up on waffles; my dad made them almost every weekend, and I love them. I immediately ordered four waffle makers and set up a few tables to serve waffles hot off the irons, drizzled with our berry syrup, heaped with fresh quality berries from our fields, and topped with whipped cream. I opened from 9 a.m. until noon each day, and our guests loved it! I managed for all of about 1 week, and then to keep up with demand, I ordered four more waffle makers—then four more. I couldn't manage 12 waffle makers alone, so I enlisted my three daughters to help me as the lineups got longer. Knowing efficiency is one of my four core values, my business-minded daughter Tracey pointed out ways to improve my process—by cutting back on wasted motion, for instance—and before long we had a well-oiled operation. Today we have 16 waffle makers and 12 staff serving approximately 1,000 guests hot berry waffles from 9 a.m. to 4 p.m., 7 days a week during our busy harvest season. The red tape was removed 3 months after I served the first waffle, and ever since then, wine and waffles have sat side by side in perfect harmony for our guests to enjoy.

My biggest lesson: When it seems like something isn't going the way I think it should, step back and wonder if maybe it's for a reason—a better one! Thank you, Dad and waffles. I'm grateful!

Waffles

Guests regularly tell us that they happily drive 1 hour and then wait another hour to enjoy one of our waffles. I thought I'd share some of our guests' comments about why they go to such lengths to buy our waffles: "The waffles are a perfect balance of sweet." "They are light and fluffy, with a hint of vanilla filling the air and making them smell so good." "They are homemade and fresh, everything about them is quality, and you get to look out into the fields where the berries were just picked." "They are big, have a mountain of fresh berries on top with their own syrup and swirled with whipped cream; they are a taste of the country." Two of my all-time favourite comments came from a young girl who said, "I would bring my boyfriend, if I had one, for a berry waffle to impress him," and a grown, burly man who told me with tear-filled eyes, "They are nostalgia on a plate. As soon as I have mine in front of me, I think of my grandma's kitchen and, with one melt-in-my-mouth bite, I am transported back in time."

What can I say except "The customer is always right!"

MAKES 4 WAFFLES (DEPENDING ON SIZE OF WAFFLE MAKER)

Waffles

1½ cups all-purpose flour
1 Tbsp granulated sugar
1 Tbsp baking powder
2 Tbsp malt powder (see Wink)
½ tsp table salt
1½ cups 2% milk
½ cup melted salted butter
2 large eggs, at room temperature
1 tsp pure vanilla extract

Toppings

Syrup (I recommend using one that matches your berry flavour)
1½ cups fresh berries, washed (strawberries should be sliced as well)
Whipped cream (homemade or from a can)

1. In a medium bowl, gently whisk together the flour, sugar, baking powder, malt powder (if using), and salt.

2. In a large bowl, whisk together the milk, melted butter, eggs, and vanilla. Using your whisk, slowly add the dry ingredients to the wet ingredients and mix until incorporated. Let rest for 30 minutes.

3. Preheat your waffle maker as per the manufacturer's instructions.

4. When your waffle maker is ready, lightly spray the upper and lower plates with cooking spray. Stir the batter and pour about ⅔ cup of batter into each prepared waffle mould. Cook your waffles according to the waffle maker's operating instructions, until golden brown, 3–4 minutes.

5. Once the waffles are cooked, place each one onto a plate and drizzle with syrup in a spiral motion from the centre to the outside. Then mound the berries on top of the centre of the waffle and finish with whipped cream in the same spiral motion.

6. Leftover waffles that have not been topped can be stored in freezer bags in the freezer for up to 3 months. Reheat them at 350°F for about 10 minutes in a toaster oven to enjoy them as if they were freshly made.

WINK: *Malt powder gives a toasty, sweet caramel flavour and a crunchy outside texture to waffles. If you can't find it in the baking aisle of your local supermarket, it's easy to find online.*

Vegan Gluten-Friendly Waffles

When we began making and serving our gluten-friendly waffles (see page 46), our vegan guests felt left out, so back to the kitchen I went to create a version for both the vegans and the gluten-free guests. This recipe is the one we make and serve at our farm waffle bar so that everyone, vegan or gluten-free, is happy!

MAKES 8 WAFFLES (DEPENDING ON SIZE OF WAFFLE MAKER)

Waffles
- 2 cups Gluten-Friendly Flour Blend (page 279)
- 2 Tbsp granulated sugar
- 2 Tbsp baking powder
- 1½ tsp table salt
- ¾ tsp xanthan gum
- 1½ cups coconut milk
- ¾ cup applesauce
- ½ cup salted vegan butter, melted
- ½ tsp pure vanilla extract

Toppings
- Syrup (I recommend using one that matches your berry flavour)
- 1½ cups fresh berries, washed (strawberries should be sliced as well)
- Coconut whipped topping

1. In a medium bowl, gently whisk to combine the flour blend, sugar, baking powder, salt, and xanthan gum.

2. In a large bowl, whisk the coconut milk, applesauce, melted butter, and vanilla. Using your whisk, gradually add the dry ingredients to the wet ingredients and mix until blended. Let rest for 30 minutes.

3. Preheat your waffle maker as per the manufacturer's instructions.

4. When your waffle maker has preheated, lightly spray the upper and lower plates with cooking spray. Pour about ⅔ cup of batter into each prepared waffle mould. Cook your waffles according to the waffle maker's operating instructions, until golden brown.

5. Once the waffles are cooked, place each one onto a plate and drizzle with syrup in a spiral motion from the centre to the outside. Mound the berries on top of the centre of the waffle and finish with whipped cream in the same spiral motion.

6. Leftover waffles that have not been topped can be stored in freezer bags in the freezer for up to 3 months. Reheat them at 350°F for 10 minutes in a toaster oven to enjoy them as if they were freshly made.

WINK: *If I feel like making dessert waffles, or if I have leftover waffle batter, I'll finish cooking the batter and, once cooled, dip the quarters in chocolate, sprinkles, coconut, or wherever my imagination takes me using whatever I have in my pantry. It's also fun to set up a dessert bar, where friends and families can dip and decorate to their own delight.*

Relax Kitchen

Yes, you read that correctly. I do take some time now and again to relax. This kitchen isn't on the farm. It's in our desert house in southern California, my home away from home, where I gather with family and friends or escape with Alf.

My son-in-law Ryan made me beautiful wooden letters as a gift for the desert house (and as a hint). They read, "RELAX." I placed them high on a prominent window ledge I see right when I walk through the front door as a reminder to do just that. That simple word has served us well over the years and continues to do so.

Things are laid back here. It's about family, good friends, good wine, and really good food—so, similar to life on the farm but without the work. Some of my favourite dishes to cook and enjoy here are Eggs Benedict on homemade English muffins (page 126) with freshly made Hollandaise sauce (page 34) for breakfast, Berry Lemonade (page 251) and Roasted Casa Corn Dip (page 263) with fresh tortilla chips to snack on by the pool, and a bottle of wine and some warm Berry Crisp (page 220) to watch the sun go down. Nothing is fussy here—it's all about comfort food that makes us smile.

Cream of Asparagus Soup (page 54)

Soups

Cream of Asparagus Soup

Asparagus is our first crop in the spring. The staff in the Harvest Kitchen (page 239) look forward to this crop so they can begin the first major fresh production of the year—cream of asparagus soup and pickled asparagus. If you want to double down on the asparagus, serve this alongside my Asparagus Cheese Bread (page 110).

SERVES 6–8

½ cup salted butter
5 cups water
2 lb (900 g) chopped, blanched asparagus (see Wink, page 111)
8 cups half-and-half (10%) cream, divided
¾ cup all-purpose flour
1 Tbsp kosher salt

1. In a large soup pot on medium heat, melt the butter.
2. Add the water, increase the heat to high, and bring to a boil.
3. Add the asparagus and return to a boil, still on high heat.
4. Pour 4 cups of the cream into a separate pot on your work surface (not over heat).
5. Sift the flour and salt together. Using a whisk, gradually stir the flour mixture into the pot with the cream.
6. Whisking constantly, pour the cream and flour mixture through a sieve (to remove any lumps) into the pot with the asparagus.
7. Once the cream and flour mixture is blended into the butter and water, whisk in the remaining 4 cups cream until combined.
8. Place the pot on medium-high heat and bring the soup to a boil. Turn the heat down to low and let simmer for 15 minutes, stirring occasionally.
9. Store in an airtight container in the refrigerator for up to 2 days or in the freezer for up to 6 months.

RECIPE PICTURED ON PAGE 52

Cream of Wild Rice Soup

Years ago, I discovered a wild rice soup in a little café when I was travelling from Vancouver, BC, to Salem, Oregon, where my grandmother Sarah lived. Stopping off in a little family restaurant in Chehalis for a big hot, hearty bowl along the way (both directions) was the highlight of the drive! I loved how the creamy texture of the soup was paired with the chewy texture of the wild rice along with the smoky taste of the ham.

SERVES 4–6

1 cup uncooked wild rice
3 cups water
1 tsp kosher salt, divided
½ cup salted butter
1 cup diced yellow onions
2 cups sliced button mushrooms
1 medium carrot, thinly sliced
½ cup thinly sliced celery
¾ cup all-purpose flour
6½ cups vegetable or chicken broth
1 cup cubed smoked ham
½ tsp dry mustard
½ tsp Spanish smoked paprika
2 cups half-and-half (10%) cream
⅔ cup Krause Berry Farms and Estate Winery Blackberry Portoe

1. Rinse the rice in a fine-mesh sieve under cold water and drain. Place the rice in a pot and add the water and ½ tsp salt. Cook on medium-high heat until the water comes to a full boil, then turn the heat down to simmer and cover. Continue simmering until some of the rice kernels have opened, approximately 45 minutes. Drain and set aside.

2. In a large cast-iron pan, melt the butter on medium heat. Add the onions and sauté until golden, approximately 5 minutes. Add the mushrooms, carrots, and celery. Cook, stirring often, until softened, approximately 2 minutes.

3. Sprinkle in the flour, then, stirring constantly, gradually add the broth. Stir until it starts to thicken, 5–10 minutes. Stir in the cooked rice, ham, dry mustard, paprika, and remaining ½ tsp salt. Turn the heat down to low and stir in the cream and Portoe. Allow to simmer for 5–10 minutes, just to finish thickening, stirring occasionally to prevent separation.

4. Store in an airtight container in the refrigerator for up to 4 days or in the freezer for up to 4 months.

RECIPE PICTURED ON PAGE 64

Apple, Cheese, and Cider Soup

This fall soup is rich and creamy with a mild cheese flavour and a bit of tang from the apples. Try garnishing it with chopped chives, green onions, bacon bits, and a teaspoon of freshly grated cheese. I like to use Krause Berry Farms and Estate Winery Oaked Apple Wine in this recipe, but use *your* favourite cider or white wine. We source our apples from farms in the Fraser Valley and the Okanagan. Honeycrisp are delicious in this, but again, use your preferred varietal to make this soup yours. Pair this with the Cheese Bread in a Can (page 117) for a satisfying lunch.

SERVES 4–6

- 1 cup salted butter, divided
- ⅔ cup all-purpose flour
- 3 cups whole milk
- 1 cup half-and-half (10%) cream
- 1 tsp kosher salt
- 2 cups shredded Colby cheese (or your favourite mild cheese) + more for optional garnish
- 1¼ cups peeled and diced apples
- ¾ cup diced carrots
- ¾ cup diced celery
- ¾ cup diced yellow onions
- 1–2 cups vegetable broth
- 1 cup cider or white wine (optional)
- ¼ cup chopped fresh chives, for garnish (optional)
- ¼ cup chopped green onions, for garnish (optional)
- ¼ cup bacon bits, for garnish (optional)

1. In a large saucepan on low heat, melt ½ cup of the butter. With a whisk, stir in the flour to incorporate. Don't let it start to brown. Slowly whisk in the milk, cream, and salt. Continue to whisk for a couple of minutes to blend. Add the cheese and continue whisking until melted and the sauce has thickened and come to a low boil. Let sit on a low simmer while you prepare the other ingredients, whisking occasionally.

2. In a deep, large frying pan on medium-high heat, melt the remaining ½ cup butter. Add the apples, carrots, celery, and onions and sauté until lightly browned and just fork-tender, 8–10 minutes.

3. If you'd like to keep the soup alcohol-free, add 2 cups vegetable broth to the vegetables and let cook for 5 minutes, using a wooden spoon to dislodge any pieces of vegetables from the bottom of the pan. Turn the heat off. If you're using the cider or wine, add 1 cup broth and the cider or wine to the vegetables and cook as above.

4. Using a ladle, slowly add the vegetables and broth mixture to the simmering cheese sauce, whisking constantly as you ladle them in to fully incorporate.

5. Let simmer for 5–10 minutes, stirring occasionally, to allow the flavours to meld before serving. If you like, garnish each portion with a quarter of the shredded cheese, chives, green onions, and bacon bits.

6. Store in an airtight container in the refrigerator for up to 2 days or in the freezer for up to 6 months.

WINK: *Over the holiday season, I like to use red and green bell peppers instead of apples. They keep the hint of sweetness while giving the soup a festive look. When warming up leftovers, stir in ½ cup vegetable broth, cider, or wine to thin it back to its regular consistency.*

Cauliflower, Potato, and Leek Soup

Your stash of frozen Roasted Garlic (page 270) will come in handy for this recipe (see Wink)! I once had a call from the mom of one of Jared's high school friends asking for this recipe. She said her son, Devon, raved about it being creamy, hearty, and chunky, and he found it rich and comforting after playing an intense basketball game, and he wanted her to make it. Always happy to share!

SERVES 4–6

3 cups cubed cauliflower (1–2 heads)

3 cups trimmed and sliced leeks (3–4 large leeks, white and light green parts only)

2 large russet potatoes, peeled and chopped into 1-inch pieces

1 medium yellow onion, thinly sliced

6 cloves garlic

2–4 Tbsp extra virgin olive oil

Kosher salt and freshly ground black pepper

8 cups vegetable or chicken broth

¼ cup chopped fresh flat-leaf parsley

1. Preheat your oven to 400°F. Line a baking sheet with parchment paper.
2. In a bowl, combine the cauliflower, leeks, potatoes, onions, garlic, and oil. Toss well with 2 Tbsp of olive oil to coat the vegetables. Add more oil as needed to evenly coat the vegetables (you don't want them to be soggy). Transfer the vegetables to the prepared baking sheet and roast until tender, 30–40 minutes. Transfer all the potatoes and ¼ cup of the cauliflower to a medium bowl, season with salt and pepper to taste, and set aside.
3. Place the remaining vegetables in a large soup pot and add the broth. On high heat, bring to a boil. Reduce the heat to low, cover, and simmer gently until the vegetables are tender, approximately 20–25 minutes. Allow to cool.
4. Blend with an immersion blender or transfer to a stand blender (in batches if necessary), and blend until smooth.
5. If blended in a stand blender, pour the soup back into the pot, then add the roasted potatoes and reserved roasted cauliflower pieces, and continue to cook for 15 minutes.
6. Ladle into serving bowls and garnish with the parsley and caramelized onions.
7. Store in an airtight container in the refrigerator for up to 2 days or in the freezer for up to 6 months.

WINK: *If you have Roasted Garlic (page 270) in the freezer, pop 6 cloves (no need to thaw first) into the pot in step 3. I also love to garnish with 5 Tbsp of Caramelized Onions (page 62)*

Roasted Corn Chowder

Each August as the corn is picked from our fields, the Harvest Kitchen (page 239) crew prepares large cooking kettles to make this hearty corn chowder. This soup is creamy, crunchy, and sweet from the roasted corn, and the green onions offer a little zing of both colour and taste. This chowder pairs beautifully with our Asparagus Cheese Bread (page 110) for a complete meal.

SERVES 6–8

½ cup salted butter
⅔ cup chopped green onions (approximately 2 green onions)
5 cups water
1 Tbsp kosher salt
5 cups Roasted Corn (page 73)
⅔ cup all-purpose flour
8 cups half-and-half (10%) cream, divided

1. In a large soup pot on medium heat, melt the butter. Add the green onions, reserving a few for garnish, and simmer until tender, approximately 3 minutes.
2. Add the water and salt. Bring to a boil, then add the roasted corn.
3. In a medium bowl, whisk together the flour and 1½ cups cream.
4. Add the cream and flour mixture to the corn soup. Whisk together for 1 minute, until blended. Return the soup to a boil. Stir in the remaining 6½ cups cream. Bring to a boil again, then lower heat to a simmer. The soup will look liquidy but not runny. Let cook, stirring occasionally, for 15–30 minutes, just to let the flavours meld. Serve immediately, garnished with the reserved chopped green onions.
5. Store in an airtight container in the refrigerator for up to 2 days or in the freezer for up to 3 months.

French Onion Soup

From the time they were little, my children have always written their own birthday menus. They were encouraged to have an appetizer, a main dish, a dessert, and a special drink. When my youngest daughter, Tracey, was turning 10, this French onion soup appeared on her menu. It is a cozy, comforting soup that is hearty, sweet, and salty and covered in hot, bubbly, crispy cheese. It uses croutons instead of the traditional slice of bread, as they perfectly come apart with the cheese with each bite, don't sink so fast or get soggy. At the time, her oldest sister, Sarah, was dating a young man named Ryan (now Sarah's husband) who didn't care for onions. When Sarah saw our family's favourite French onion soup on Tracey's menu, she tried everything to get Tracey to take it off, knowing Ryan would be attending the birthday celebration. Do you think she was successful? (Note: For an authentic look, use French onion soup bowls for this recipe.)

SERVES 6–8

- ⅓ cup salted butter
- 1½ lb (680 g) thinly sliced yellow onions (approximately 2 large yellow onions)
- ¾ tsp kosher salt
- ½ tsp granulated sugar
- 8 cups vegetable, beef or chicken broth
- ⅔ cup Krause Berry Farms and Estate Winery Oaked Apple Wine
- 2 cups store-bought croutons
- ½ cup shredded Swiss or Gruyère cheese
- 3–6 tsp shredded parmesan
- Chopped green onions, for garnish

1. In a large (3-quart/approximately 11-inch-diameter) frying pan, a 3-quart Dutch oven, or a large cast-iron pot, melt the butter on low heat.

2. Turn the heat to medium-low. Add the onions and salt, stirring until they begin to look translucent, approximately 10 minutes.

3. Cover and continue cooking until the onions are limp and fully translucent, stirring occasionally, approximately 10 minutes.

4. Uncover and sprinkle with sugar. Continue cooking on medium-low, still uncovered, stirring occasionally, until the onions are tender and caramelized, 10–12 minutes (see Wink).

5. Pour the broth and wine over the onions, cover again, and simmer, stirring occasionally, approximately 10 minutes.

6. Preheat your oven to broil.

7. Divide the soup between your French onion soup bowls, top with the croutons, and sprinkle each with roughly 4 tsp Swiss cheese and ½–1 tsp parmesan. Broil until the cheese is lightly browned. Sprinkle with green onions, if desired.

WINK: *If you want your onions to be a little more caramelized for toppings, turn the heat up to medium-high at the end of step 4 for 2–3 minutes. I like to add a few of these to the tops of each individual serving. They're also great for topping baked potatoes, steaks, hummus, savoury Pierogies (page 101), Pesto, German and Hawaiian pizzas (pages 78–79), and German, Hawaiian, and Italian (Pesto) corn cobs (pages 69–72).*

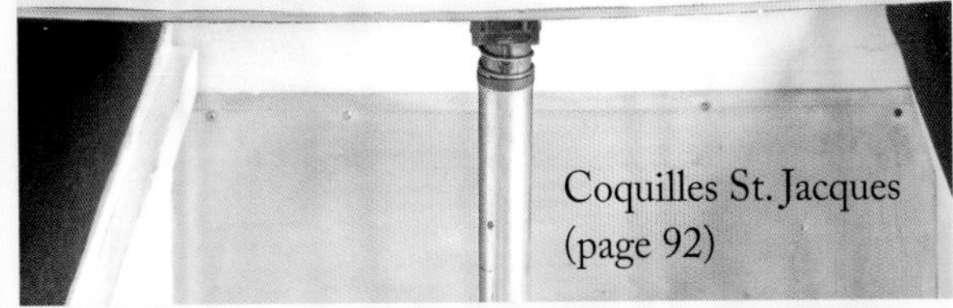

Coquilles St. Jacques (page 92)

Cream of Wild Rice Soup (page 55)

Garden Chili (page 88)

Sweetie Pie Kitchen

I enjoy cooking and baking anywhere, even on the road in my 1970 vintage Travelaire trailer, named Sweetie Pie. It was a Mother's Day gift from my son Tanner and is parked in a secluded area on our home farm, known as the Ponderosa (page 189). Alf and I camp right there by the pond when we are unable to get away but want to enjoy the feeling of camping. When we are lucky enough to get a couple of days off in a row, we'll pull her not far from home to a campsite to enjoy the crackling fire, some comfort food, and each other before heading back to the farm. Some dishes I make at home and package up for us to enjoy on the road: Garden Chili (page 88), Cream of Wild Rice Soup (page 55), and Coquilles St. Jacques (page 92) are all big favourites. I warm them up on the campfire or in Sweetie Pie's oven. Other favourites, such as Blueberry Pancakes (page 45) and Zucchini Pancakes (page 87), are easy to prepare right inside the trailer and enjoyed around the campfire.

Hawaiian Pizza (page 79)

Main Meals

Corn on the Cob, Six Ways

MAKES 4 COBS

Over the years, Alf has grown many varieties of corn, starting with standard (su) varieties such as Jubilee. He has changed to growing a combination of sugar-enhanced (se) and supersweet (sh2) all of which are hybrid varieties. They all have their own names, and I never tire of hearing them: Vision (an all yellow corn), Awesome, and Kandy Korn are the varieties we grow now.

They can be yellow, bi-coloured, and white in colour. My favourite is the yellow, and Alf's is the bi-colour, commonly known as peaches and cream. He brings some home almost every day during the first 2 weeks of harvest. We love it! Luckily, there are many different ways to cook and dress corn on the cob. We rotate through a variety of preparations and toppings, so we never get bored of eating corn.

How much time I have usually determines how I cook the cobs. Here are my main methods:

Microwaved: Some days you have to make hay while the sun shines. If you're in a hurry, you can peel off a few of the outer layers of the cob until it's free from any field dirt and then pop it in the microwave for 3 minutes on high. Serve with butter, salt, and pepper.

Boiled: If you have the time, fill a pot with water to boil while you strip all the leaves off each cob. Remove any of the silk threads that are still attached. Carefully drop the corn cobs into the boiling water and cook until tender, approximately 5 minutes, then drain. Serve with butter, salt, and pepper.

Barbecued (1): Fill your sink with water and place the corn cobs in their husks to soak for at least 1 hour, but no longer than 3 hours. When you're ready to cook them, drain them, pat them dry, and place them right onto the barbecue on high heat for approximately 25 minutes, rotating them every 3 minutes. They will grill and steam at the same time, which will give a lovely smoky flavour and tenderness.

Barbecued (2): Strip the corn cobs, discarding all the leaves and silk as you would if you were boiling them. Fill your sink with water and place the corn cobs in the water to soak for at least 1 hour, but no longer than 3 hours. After draining, pat them dry and place them right on the barbecue on high heat for approximately 25 minutes. The corn will char in places, so constantly rotate the cobs around the grill so that they all have a bit of time on the lower-heat areas on the barbecue and don't burn.

Roasting: I love to roast corn and freeze it to use in all sorts of recipes. See page 73 for my Roasted Corn recipe.

Once the corn is cooked, it's time to top it. I've listed a few of our go-to toppings (see right). My favourite is Tex-Mex, while Alf prefers Hawaiian.

Each of these topping recipes will make enough to top about 4 cobs. Add the toppings to hot, freshly cooked cobs for maximum deliciousness. The toppings can be prepped and stored in an airtight container in the refrigerator for up to 2 days.

GREEK

½ cup tzatziki (page 77)
¼ cup chopped red onions
½ cup diced tomatoes
1 tsp dried oregano or 1 Tbsp chopped fresh oregano
¾ cup crumbled or grated feta
Sprinkle of sliced pitted Greek olives

1. Place hot cooked corn cobs on individual plates and brush each one with ⅛ cup tzatziki. Sprinkle with onions, tomatoes, oregano, feta, and olives.

GERMAN

4 large, thin slices corned beef, each cut into 4 pieces, heated
¼ cup hot mustard
½ cup sauerkraut, heated

1. Place hot cooked corn cobs on individual plates. Wrap each cob in a corned beef slice and drizzle each with hot mustard. Pat heated sauerkraut on top.

ITALIAN (PESTO SAUCE)

½ cup Pesto (page 275)
½ cup shredded parmesan
½ cup chopped fresh basil leaves

1. Place hot cooked corn cobs on individual plates, brush with pesto, and sprinkle with parmesan and basil leaves.

WINK: *I like to add a drizzle of chipotle aioli or a few caramelized onions to Italian (Pesto Sauce), German, and Hawaiian corn cobs. See Wink in French Onion Soup (page 62).*

ITALIAN (TOMATO BASIL SAUCE)

½ cup Tomato Basil Sauce (page 275)
24 slices pepperoni, fried
½ cup shredded parmesan
½ cup sliced pitted green olives

1. Place hot cooked corn cobs on individual plates. Brush each one with tomato basil sauce. Top with 6 fried pepperoni slices, and sprinkle green olives and cheese over top.

RECIPES PICTURED ON PAGE 70–71

Left to right, Greek (page 69), German (page 69), Italian (Pesto Sauce) (page 69), Italian (Tomato Basil Sauce) (page 69), Hawaiian (page 72), Tex-Mex (page 72)

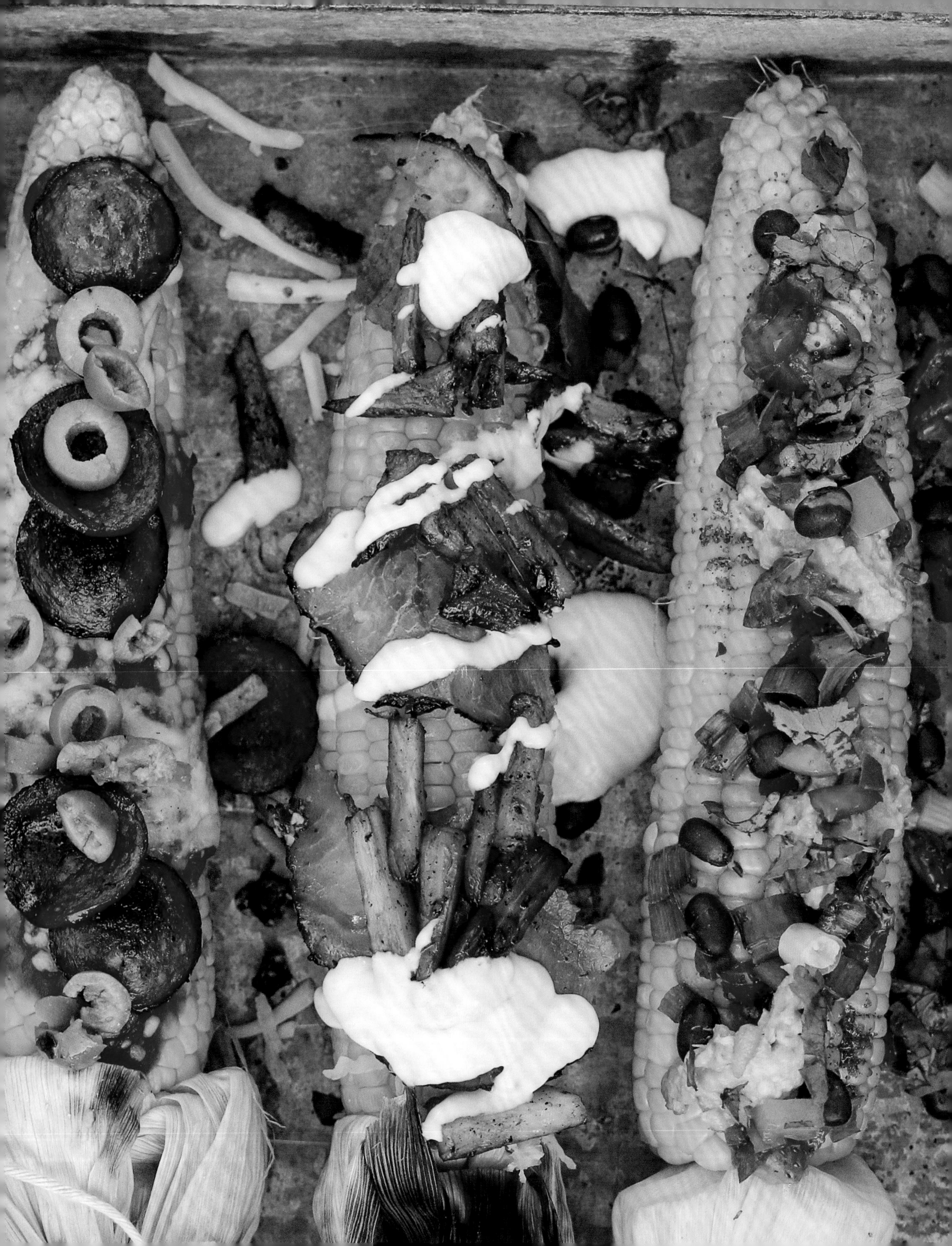

HAWAIIAN

Extra virgin olive oil
4 pineapple rings
4 large slices Black Forest ham, each cut into 3 lengths, heated

1. Preheat your barbecue to 400°F, or leave it on after you've finished barbecuing your corn cobs.

2. Brush a little oil on the pineapple rings. Place on the grill and roast until caramelized, 3 minutes. Remove from the grill. Cut the pineapple rings into small chunks.

3. Place hot cooked corn cobs on individual plates and wrap each in 3 lengths of heated ham. Top with roasted pineapple chunks.

RECIPES PICTURED ON PAGE 70–71

TEX-MEX

¼ cup jarred Estate Winery Roasted Corn Salsa
½ cup canned black beans, drained, rinsed and heated
½ cup canned navy beans, drained, rinsed and heated
½ cup chopped tomatoes
2 green onions, chopped
2 Hass avocados, mashed
4 Tbsp finely chopped fresh cilantro
1 tsp chipotle chili powder, to garnish

1. Place hot cooked corn cobs on individual plates and brush with corn salsa. Add heated beans, tomatoes, cheese, and green onions. Top with mashed avocados and finish with cilantro and chili powder.

RECIPES PICTURED ON PAGE 70–71

WINK: *If you prefer to batch-cook the beans, see Garden Chili (page 88) for instructions. You can freeze cooked beans by letting them cool and then placing them in portions in freezer bags for up to 5 months. Be sure to lie them flat in the freezer so they freeze evenly.*

Roasted Corn

Roasting is one of my favourite ways of preparing corn. It is hands-on but so worth it, as the caramelized flavour enhances the natural sweetness of the corn. It is easily frozen and thawed, making it convenient to use in many recipes, like my KBF Roasted Corn Pizza (page 74), Roasted Corn Chowder (page 61), and Roasted Casa Corn Dip (page 263). (Note: You need a perforated baking sheet for this.)

MAKES APPROXIMATELY 10 CUPS

15–20 cobs of corn
¼ cup extra virgin olive oil

1. Preheat your oven to 400°F.

2. Bring a large pot of water to a boil on high heat. You'll need enough to cover the cobs completely. Strip the leaves and silk from the cobs, then carefully drop the cobs into the boiling water to cook. Take them out when the water hits the boiling point again. (You can do this in batches if you don't have a pot big enough to hold all the cobs.)

3. Let the corn cool enough to touch. Using an electric or serrated knife, remove the kernels from the cobs. To do this, hold a cob vertically over a baking sheet and carefully cut the sides from top to bottom as close as you can to the centre of the cob, cutting slow and steady.

4. Spread the kernels out on a baking sheet and drizzle with just enough olive oil to coat. You don't want to drench the corn. Toss to evenly coat the kernels. Roast the corn for 20 minutes, tossing at the 10-minute mark. The kernels will be an array of golden-brown jewels.

5. Place a perforated baking sheet over a rimmed baking sheet and place the roasted corn on top. This will let any liquids easily drain off the corn.

6. Spread the roasted corn in an even layer on the sheet and then freeze, uncovered, for a couple of hours. Transfer to an airtight container or freezer bag and store in the freezer for up to 6 months. If you plan on using the corn soon, you can store it in an airtight container in the refrigerator for up to 3 days.

KBF Roasted Corn Pizza

Both my ears were ringing. My husband was asking me to create recipes for our corn harvest and our guests were asking for savoury choices. The cob literally hit me over the head, and our famous Roasted Corn Pizza was created. In our Harvest Kitchen (page 239) we process 10,000 lb of corn annually, which translates to approximately 50,000 cobs. Guests loved our new pizza and constantly encouraged me to continue producing volumes for the freezer for them to take home and enjoy. Now it's a matter of keeping up with the demand.

MAKES ONE (10-INCH) PIZZA

- 1 par-baked 10-inch Pizza Shell (page 274)
- 1 cup shredded cheddar
- 1 cup shredded mozzarella
- ¼ cup Tomato Basil Sauce (page 275)
- 2 cups Roasted Corn (page 73)
- 1 Tbsp diced red bell peppers
- 1 Tbsp diced orange bell peppers
- 1 Tbsp diced green onions
- 4 freshly cooked, frozen, or canned artichoke hearts, quartered

1. Lightly grease a 10-inch pizza pan.

2. If you're baking the pizza now, preheat your oven to 350°F and place the pizza shell on the prepared pan. If you're prepping this ahead of time to freeze until needed, see Wink.

3. In a medium bowl, toss both cheeses together until evenly mixed.

4. Spread the sauce evenly across the pizza shell, leaving ½ inch bare around the edge. Then sprinkle with 2 Tbsp cheese followed by, in this order, the corn, bell peppers, green onions, artichoke hearts, and remaining shredded cheeses.

5. Bake until the bottom and side crusts are golden brown and the cheese is melted and bubbly, approximately 10 minutes. Remove the pizza from the oven and slice it in quarters (or however you prefer) and continue baking for 10 minutes of until the centre is cooked all the way through. The reason for this is the generous amount of corn used.

WINK: *If you want to freeze the pizza, place the pizza shell on a 10-inch round piece of stiff cardboard. Build the pizza, following the steps above, and then carefully slide it into a large freezer bag and freeze for up to 2 months. To cook your frozen pizza, thaw as directed on page 274 and bake as directed in step 5.*

Pizza Party

There are so many possibilities when it comes to pizza. I've given you a variety of delicious pizza options to get you started on your own pizza flavour journey. You can use my KBF Roasted Corn Pizza recipe (page 74) as your template for putting them all together.

EACH RECIPE MAKES ONE (10-INCH) PIZZA

BENNY

Base
1 par-baked 10-inch Pizza Shell (page 274)
¼ cup Hollandaise Sauce (page 34)

Toppings
7 slices round Canadian back bacon
½ cup shredded mozzarella
7 poached large eggs (see Eggs Benedict recipe, page 34, for method)
¼ cup warm Hollandaise Sauce (page 34)
7 fresh spinach leaves

1. Follow steps 1–2 on page 74.

2. Cover the pizza shell with the sauce, leaving ½ inch bare around the edge. Top with the bacon and cheese.

3. Bake the pizza according to the instructions on page 74.

4. Place a poached egg on each bacon round, drizzle with the warm hollandaise sauce, and garnish with fresh spinach.

GREEK

Base
1 par-baked 10-inch Pizza Shell (page 274)

Tzatziki sauce (makes ¾ cup)
½ cup plain Greek yogurt
½ cup peeled and grated cucumber
1 tsp dried or 1 Tbsp chopped fresh oregano
1 tsp fresh lemon juice
Kosher salt and freshly ground black pepper, to taste

Toppings
½ cup quartered tomatoes
¼ cup chopped white onions
¼ cup whole pitted black olives
8 red onion rings
¾ cup crumbled feta, divided
1 tsp dried oregano
½ cup peeled and diced cucumber

1. Follow steps 1–2 on page 74.

2. In a small bowl, combine the yogurt, cucumber, oregano, lemon juice, salt, and pepper. Mix until blended, approximately 1 minute.

3. Cover the pizza shell with all but 2 Tbsp tzatziki sauce, leaving ½ inch bare around the edge. Top with the tomatoes, white onions, olives, red onion rings, and ½ cup feta, then sprinkle with the dried oregano.

4. Bake the pizza according to the instructions on page 74.

5. Garnish with the cucumbers and remaining feta.

ITALIAN PEPPERONI

Base
1 par-baked 10-inch Pizza Shell (page 274)
¼ cup Tomato Basil Sauce (page 275)

Toppings
1 cup shredded mozzarella, divided
24 slices (2-inch-diameter) pepperoni
¼ cup black or green pitted olives (or both)
¼ cup shredded parmesan
12 arugula leaves, for garnish

1. Follow steps 1–2 on page 74.
2. Cover the pizza shell with the sauce, leaving ½ inch bare around the edge. Top with ½ cup mozzarella, followed by the pepperoni slices and olives. Add the remaining ½ cup mozzarella and the parmesan.
3. Bake the pizza according to the instructions on page 74.
4. Garnish with the arugula.

ITALIAN PESTO

Base
1 par-baked 10-inch Pizza Shell (page 274)
¼ cup Pesto (page 275)

Toppings
1 cup shredded mozzarella, divided
6 cloves Roasted Garlic (page 270)
⅛ cup toasted pine nuts
5 fresh basil leaves, for garnish

1. Follow steps 1–2 on page 74.
2. Cover the pizza shell with the sauce, leaving ½ inch bare around the edge. Top with ½ cup mozzarella, followed by the garlic and pine nuts. Top with remaining ½ cup cheese.
3. Bake the pizza according to the instructions on page 74.
4. Garnish with the basil.

GERMAN

Base
1 par-baked 10-inch Pizza Shell (page 274)
¼ cup Tomato Basil Sauce (page 275)

Toppings
5 large slices corned beef
½ cup sauerkraut
3 heaping Tbsp diced white onions
1½ cups mixed shredded cheddar and mozzarella
2 white onion rings
2 slices red cabbage
Hot mustard, for drizzling

1. Follow steps 1–2 on page 74.
2. Cover the pizza shell with the sauce, leaving ½ inch bare around the edge. Top with corned beef, sauerkraut, onions, and cheese.
3. Bake the pizza according to the instructions on page 74.
4. Place 2 onion rings for eyes, a slice of cabbage to form a smiley face, and another slice of cabbage to form a nose. Drizzle with hot mustard.

HAWAIIAN

Base
1 par-baked 10-inch Pizza Shell (page 274)
¼ cup Tomato Basil Sauce (page 275)

Toppings
1¼ cups shredded mozzarella, divided
5 slices Black Forest ham
½ cup pineapple chunks
6 slices lime
1 tsp extra virgin olive oil
2 stems fresh chives, thinly sliced, for garnish

WINK: *I like to add a few caramelized onions to Pesto, German, and Hawaiian pizzas; see Wink in French Onion Soup (page 62).*

1. Follow steps 1–2 on page 74.

2. Cover the pizza shell with the sauce, leaving ½ inch bare around the edge. Sprinkle with ½ cup cheese, then top with the ham, pineapple, and remaining ¾ cup mozzarella.

3. Bake the pizza according to the instructions on page 74.

4. Turn your oven to 450°F and line a baking sheet with foil. Brush the lime slices with the olive oil and place on the baking sheet. Bake for 3–5 minutes, until the undersides of the fruit are a caramel colour.

5. Place the roasted lime slices on the pizza and sprinkle with the chives.

RECIPE PICTURED ON PAGE 66

SEAFOOD

Base
1 par-baked 10-inch Pizza Shell (page 274)
¼ cup cocktail or tartar sauce

Toppings
½ cup sliced button mushrooms
¼ cup fresh shelled shrimp
¼ cup fresh shelled crab
½ cup shredded parmesan
5 fresh basil leaves

1. Follow steps 1–2 on page 74.

2. Cover the pizza shell with the sauce, leaving ½ inch bare around the edge. Top with the mushrooms, shrimp, crab, and parmesan.

3. Bake the pizza according to the instructions on page 74.

4. Garnish with the basil leaves.

MEXICAN

Base

1 par-baked 10-inch Pizza Shell (page 274)

¼ cup Tomato Basil Sauce (page 275) or Krause Berry Farms and Estate Winery Roasted Corn Salsa

Toppings

¼ cup canned black beans, drained and rinsed

¼ cup canned white beans, drained and rinsed

½ cup Roasted Corn (page 73)

3 Tbsp diced red onions

¼ cup sliced pitted black olives

1 cup mixed shredded cheddar and Monterey Jack cheese

1 tsp chipotle chili powder, plus more for garnish

1 Tbsp finely sliced green onion, for garnish

¼ cup diced tomatoes, for garnish

¼ cup Guacamole (page 264), for serving

1. Follow steps 1–2 on page 74.

2. Cover the pizza shell with the sauce, leaving ½ inch bare around the edge. Top with the black beans, white beans, corn, onions, and olives. Sprinkle with the cheese and chipotle.

3. Bake the pizza according to the instructions on page 74.

4. Garnish with the green onions and tomatoes and serve with a spoonful of guacamole in the centre of the pizza and sprinkle with chipotle chili powder.

WINK: *If you prefer to batch-cook the beans, see Garden Chili (on page 88) for instructions. You can freeze cooked beans by letting them cool and then placing them in portions in freezer bags for up to 5 months. Be sure to lay them flat in the freezer so they freeze evenly.*

Stuffed Pattypan Squash

I love making risotto year-round, but I especially love making it when the pattypans are harvested in the late summer. This squash is moist and light, and the risotto filling adds a touch of heartiness and comfort.

SERVES 6
(4 PER PERSON)

- 2½ cups vegetable broth
- 2½ cups water
- ½ tsp kosher salt
- 2 Tbsp salted butter (see Wink)
- 1 cup sliced button mushrooms (see Wink)
- ¾ cup finely chopped white or yellow onions
- 1½ cups arborio rice
- 3 Tbsp Krause Berry Farms and Estate Winery Oaked Apple Wine
- ¾ cup shredded parmesan, divided (optional, see Wink)
- Kosher salt and freshly ground black pepper
- 24 pattypan squash
- Extra virgin olive oil, for oiling the squash
- 1 green onion, finely diced up to the whites, for garnish

WINK: *Instead of mushrooms, you can stir 1 cup chopped, steamed asparagus (1-inch pieces) or 1 cup steamed fresh or frozen peas into your risotto. This recipe can easily be made vegan by using salted vegan butter and vegan (or no) parmesan.*

1. In a large pot on medium heat, bring the broth, water, and salt to a boil, then turn the heat down to low and keep at a low simmer.

2. In a large skillet, melt the butter on medium heat. Add the mushrooms and onions and sauté until translucent.

3. Add the rice to the pan and stir for 1 minute to coat the rice in the butter. Add the wine, stirring until all the liquid has evaporated, approximately 2 minutes.

4. Transfer the broth mixture to a large liquid measuring cup so you can track how much you're adding. Add ½ cup of the warm broth to the rice, stirring until it is absorbed. The rice should bubble gently in this process. Continue adding broth ½ cup at a time, letting each addition be absorbed, until you have just under 1 cup left.

5. Add ¾ cup of the remaining warm broth to the rice, stirring as before. Remove from heat and stir in ½ cup parmesan and the salt and pepper to taste.

6. Preheat your oven to 375°F. Lightly spray a rimmed baking sheet with cooking spray.

7. Level the bottom of each squash. Carefully cut a circle around the stem, enough to create a slight hollow for your filling while leaving a ½-inch rim all around. Gently lift the squash lid off by its handle and set aside. Scoop out the seeds and the pulp from the pattypans. Lightly oil the inside of each squash.

8. Turn each squash upside down on the prepared baking sheet and bake for 15 minutes. The flesh will be half baked. Oil the exposed flesh of the squash tops. Gently lift them by their handles onto the baking sheet, placing them alongside the whole squash. Fill each squash with mushroom risotto and continue baking, uncovered, until the flesh of the squash is fork-tender, approximately 20 minutes.

9. Remove the squash from the oven and sprinkle green onions over the risotto as garnish. Before serving, place the tops back on top of each squash.

10. If making the pattypans as a side dish, cut the recipe in half and reduce the number of pattypans to 12.

11. Store in an airtight container in the refrigerator for up to 2 days.

Roasted Succotash Salad

This hearty salad is a colourful blend of our summer vegetable crops, like corn and bell peppers, and power-packed edamame for protein. I love to serve it with Asparagus Cheese Bread (page 110).

SERVES 4

1½ cups fresh or frozen shelled edamame beans

½ tsp kosher salt

1 Tbsp canola oil

½ cup chopped red bell peppers

¼ cup chopped red onions

2 cloves garlic, minced

2 cups Roasted Corn (page 73)

3 Tbsp water

2 Tbsp rice vinegar

2 Tbsp chopped fresh flat-leaf parsley

2 Tbsp chopped fresh basil or 1 Tbsp dried basil, plus more for garnish

Kosher salt and freshly ground black pepper

1. Place the edamame beans in a large saucepan with enough water to cover them. Add the salt and give it a stir to mix the salt into the water. Cook the edamame on medium heat until tender, approximately 4 minutes. Drain and rinse them well.

2. Heat the oil in a large non-stick skillet on medium heat. Add the bell peppers, onions, and garlic. Cook, stirring frequently, until the vegetables start to soften, approximately 2 minutes. Stir in the drained edamame beans, corn, and water and continue to cook, stirring frequently, until the water has evaporated, approximately 4 minutes.

3. Remove from heat. Stir in the vinegar, parsley, and basil, and season with salt and pepper.

4. Garnish with basil.

5. This is best served immediately, but you can store leftovers in an airtight container in the refrigerator for up to 2 days.

Zucchini Pancakes

We grow zucchini in our Potager Garden (page 4) just for this recipe and my Auntie Dona's Zucchini Bread (page 121). Harvest begins at the end of June, and it always puts me in the mood for these savoury, crispy pancakes. They are a great vegetarian (but not vegan) meal on their own or make a delicious appetizer.

SERVES 2 AS AN ENTRÉE OR 4 AS AN APPY

1½ cups grated zucchini
¼ cup Roasted Corn (page 73) (see Wink)
2 Tbsp sliced green onions
2 Tbsp chopped fresh mint leaves or ½ tsp dried mint
2½ Tbsp all-purpose flour or Gluten-Friendly Flour Blend (page 279)
½ tsp kosher salt
½ tsp freshly ground black pepper
2 large eggs, room temperature, beaten
2 Tbsp salted butter, divided
½ cup sour cream or plain Greek yogurt, for serving
2 green onions, finely chopped, for garnish

1. Put the grated zucchini in a colander and gently squeeze it to remove any excess water.

2. In a large bowl, mix together the zucchini, corn, sliced green onions, mint, flour, salt, and pepper. Add the eggs and mix again to blend well. Let rest for 20 minutes.

3. Heat a frying pan on medium heat and melt 1 Tbsp butter. Drop ¼ cup of the pancake batter into the hot pan. Spread the batter evenly to form a circle. Fry until browned on both sides, 2–3 minutes per side. Repeat with the remaining batter, adding the remaining 1 Tbsp butter if needed.

4. Serve with sour cream or yogurt and garnish with green onions.

5. Store leftovers in an airtight container in the refrigerator for up to 2 days. Warm gently in a toaster oven for about 10 minutes to reheat.

WINK: *Let the frozen corn sit for 15 minutes to thaw. Pat dry before adding.*

Garden Chili

I put this chili on the menu at the Winery and The Porch in the fall. I also make it for my family as the temperature drops. Thicker than a soup and hearty like a stew, it gives us all a nice feeling of warmth and comfort, especially for lunch when we're up on the ski slopes. This is a great batch-cooking recipe to add to your repertoire. It's perfect for larger gatherings or packing up and storing in your freezer to pull out for a nourishing dinner on a cold winter's eve. (Note: You need to start this recipe the night before you plan to enjoy it.)

SERVES 24

- 1 cup dried chickpeas (see Wink)
- 1 cup dried black beans (see Wink)
- 1 cup dried red kidney beans (see Wink)
- ½ cup canola oil
- 4 cups diced white onions
- 2 Tbsp minced garlic
- 1 cup pineapple juice
- 1 Tbsp fresh lemon juice
- 2 tsp red wine vinegar
- 2 cups Tomato Basil Sauce (page 275) (see Wink)
- 3 (each 28 oz/796 ml) cans crushed tomatoes
- 1 (28 oz/796 ml) can diced tomatoes
- ¼ cup mild chili powder
- ½ tsp chipotle chili powder
- 2 tsp kosher salt
- 2 heaping tsp granulated sugar
- ¾ tsp dried basil
- ¾ tsp granulated garlic
- ¾ tsp smoked sweet Spanish paprika
- ¾ tsp ground coriander
- ¾ tsp freshly ground black pepper
- ¾ tsp ground cumin
- 2 Tbsp smoked salt
- 3 bay leaves
- 3 cups frozen Roasted Corn (page 73) or frozen corn kernels
- 1 cup fresh or frozen shelled edamame beans
- 1 cup shredded cheddar cheese, for garnish (optional)

1. Fill three bowls each with 4 cups cold water. Add one type of dried beans to each bowl and leave to soak, uncovered, in the refrigerator overnight.
2. The next day, preheat your oven to 350°F.
3. Drain the beans and place them in a 9 × 13-inch ovenproof glass pan that is 2 inches deep. Pour 6 cups boiling water over the beans.
4. Cover the pan with foil and cook the beans for 2½–3 hours until tender, stirring at the 1½-hour mark.
5. Once the beans have finished cooking, in a large pot (at least 4 gallons), heat the oil on medium-high heat. Add the onions and garlic and sauté until the onions are tender.
6. Add the pineapple juice, lemon juice, and red wine vinegar. Stir to combine. Add the tomato sauce and the crushed and diced tomatoes and stir to combine.
7. Add the mild and chipotle chili powders, salt, sugar, basil, garlic, paprika, coriander, pepper, cumin, smoked salt, and bay leaves and stir until well mixed. Add the cooked beans, corn, and edamame and mix well.
8. While stirring, bring to a rolling boil on medium-high heat. Turn the heat down to low and simmer, uncovered, for 1 hour, stirring occasionally. The chili will have reduced slightly. Serve with shredded cheddar cheese, if using.
9. When cooled, store in an airtight container in the refrigerator for up to 3 days or in the freezer for up to 4 months.

WINK: *If you're pressed for time, you can use 3 cups canned beans for each type of bean (9 cups beans in total) instead of cooking the beans from scratch. Add them to the chili as directed for the cooked beans. Always drain and rinse canned beans thoroughly before using. You can also substitute 2 cups canned tomato sauce in place of the Tomato Basil Sauce.*

EAT.

Apron Strings Are Powerful Things

I have a vast collection of aprons. I choose my apron according to what I am planning on making. If I am set to make any of Grandma Greene's recipes, I like to wear one of the many aprons she gave me. She made them on her old Singer treadle sewing machine and patched her favourites by hand, many times over. I feel they all have her personality stitched right into their seams. If I'm making Coquilles St. Jacques (page 92), I tie on the apron I brought back from Paris and enjoy my many memories of the city and the cooking classes I've taken there. If I am missing my children, I don the canvas and leather apron they gave me one year for Christmas.

I know I'm not alone in my love of aprons. I like to keep the farm market well stocked with a wide variety of them. They sell out almost as fast as I bring them in. Many of our regular guests have confided in me that they count on those aprons for their special gift giving. And if you're ever passing by in early summer, check to see if we're hosting one of our occasional apron fashion events, when the staff and their kids (like Lily (pictured left) and Esther (pictured right)) walk across the stage in the KB Corral wearing the latest apron designs.

Coquilles St. Jacques

The first Coquilles St. Jacques I enjoyed wasn't in France, as you might expect, but in a little French restaurant in Vancouver's Gastown district in the '80s. I developed my own version, substituting the scallops in the original recipe with halibut, baby shrimp, and crab claws—three family favourites. I served it in seashells for a bit of fun for the kids, but eventually they became chipped and then they were nowhere to be found. One day my daughter Holly sent me a photo from Craigslist of a set just like the ones she remembered when she was growing up. She bought them and brought them to the Farmhouse Kitchen, and I taught her how to make this creamy, rich, delicious dish. I buy the seafood from our local seafood store for the freshest and best results. If you don't live near the ocean, frozen, jarred, or tinned will work. Always check the labels to ensure you are purchasing from as close to your home as you can.

SERVES 6–8

- ¾ cup salted butter, divided
- 1½ cups diced yellow onions
- 4 cloves Roasted Garlic (page 270), flesh squeezed out and mashed
- 2 cups sliced button mushrooms
- 1½ lb (680 g) halibut, cut into cubes
- 1¼ cups fresh baby shrimp
- Meat from 6 whole crab claws
- ½ cup all-purpose flour
- 1 cup 2% milk
- 1 cup Krause Berry Farms and Estate Winery Apple Wine
- ½ cup sour cream
- 1 Tbsp finely grated lemon zest
- ½ tsp kosher salt
- ½ tsp freshly ground black pepper
- ½ cup shredded parmesan

1. In a large frying pan, melt 2 Tbsp butter on medium heat. Sauté the onions until translucent, stirring occasionally, then mix in the garlic. Set aside in a small dish.

2. Add another 2 Tbsp butter to the frying pan, still on medium heat, and sauté the mushrooms, stirring occasionally, until lightly browned. Set aside in a separate dish.

3. Add 1 Tbsp butter to the frying pan, still on medium heat. Cook the halibut until the pink almost disappears, flipping to cook all sides, approximately 2 minutes in total. Add the shrimp and cook, stirring, until the shrimp are pink and have curled, approximately 1 minute. Add the crabmeat and stir for 1 minute more to separate some of its chunks. Set the halibut, shrimp, and crabmeat aside in a separate dish.

4. Add the remaining 7 Tbsp butter to the frying pan and melt on medium heat.

5. Add the flour, stirring constantly to make a thickened smooth paste (also called a roux). Continue stirring constantly, slowly add half the milk, half the wine, and half the sour cream. Repeat with the remaining milk, wine, and sour cream. Turn the heat down to low. Add the lemon zest, salt, and pepper and simmer for 5 minutes.

6. Return the onions, mushrooms, and seafood to the pan and stir well to combine. Transfer to ovenproof scallop seashell serving dishes or individual casserole dishes.

7. Set your oven to broil. Sprinkle the seafood mixture with the parmesan and broil until golden brown, 2–4 minutes. Watch carefully to avoid overbrowning or burning.

8. Store in an airtight container in the refrigerator up to overnight.

WINK: *This recipe can be made the day before you plan to eat it. Place the mixture in a bowl or individual seashell dishes and cover with plastic wrap. When you are ready to bake, bring the mixture to room temperature, fill your serving dishes (if you stored the mixture in a bowl), and reheat for 15 minutes at 350°F. Use your kitchen blowtorch or broil for the last few minutes to brown the cheese.*

Seafood-Stuffed Artichokes

Bless my husband's heart—before he met his California wife, he had never tasted an artichoke in any form or fashion. Love changed that, and he is now passionate not only about the artichokes he grows, but also about what he dips his artichoke leaves in. This dish is elegant and simple. The artichoke is stuffed with a rich, creamy, and flavourful filling that complements the earthy tones of the vegetable. In our family we have even been known to auction off our artichoke hearts to the highest bidder at the table! Who do you think usually ends up the highest bidder?

SERVES 3–4

3 large artichokes
1 cup mayonnaise
½ tsp kosher salt
½ tsp freshly ground black pepper
1½ cups shredded mozzarella, divided
1 cup fresh baby shrimp, divided
Chopped chives, for garnish

1. Wash the artichokes and trim off the stems and bottom leaves. Cut off any spikes on the tips of the leaves with a pair of scissors.

2. Place the artichokes in a pot of cold water and boil on high heat until the base of each artichoke is fork-tender, 15–20 minutes.

3. Once they're ready, use a pair of tongs to remove them from the water. (Pick them up so they are upside down. This allows the water to drain out the top of the artichokes instead of running down the tongs and potentially scalding you.) Lay the artichokes upside down as best as possible on a clean tea towel or paper towel to allow more water to drip out.

4. Allow the artichokes to cool enough to handle before cutting in half. The easiest approach is to cut them upside down from bottom to top.

5. Use a small spoon to remove the stringy edge at the inside bottom of the artichoke heart.

6. Preheat your oven to 350°F. Line a baking sheet with foil.

7. In a medium bowl, mix together the mayonnaise, salt, pepper, 1 cup cheese, and ½ cup shrimp.

8. Stuff this mixture into each artichoke half and place the artichokes, cut side up, on the prepared baking sheet.

9. Divide the remaining ½ cup shrimp evenly between the artichokes. Then top the shrimp with the remaining ½ cup cheese.

10. Bake until the cheese in the centre of the artichokes melts, 15–20 minutes, then broil for 3–5 minutes, until the cheese on top is completely melted, slightly browned, and bubbling.

11. Enjoy these straight from the oven, sprinkled with the chives. They don't store well.

WINK: *Artichokes are delicious cooked as above until step 3, then leave whole, drain, and serve with drawn butter or mayonnaise for dipping the leaves.*

Cheesy Chicken Enchiladas

My husband was one of the first farmers in our area to grow goji berries—which have been used in traditional medicine for thousands of years, and are packed with antioxidants and are a great source of vitamins and minerals, such as fibre, iron, and vitamins A and C. I experimented with them fresh, right off the bush, of course. They are fun to sprinkle on sauces, salads, pasta, and pizza. In this recipe, their bright red colour is a fantastic contrast with the light sauce, and their sweet/sour flavour is absorbed by both the sauce and the cheese. These enchiladas are a batch recipe, which means it's a definite crowd-pleaser (you can easily halve the recipe as well)! Serve with taco chips and either Krause Berry Farms and Estate Winery Roasted Corn Salsa, or the Fresh-Cut Strawberry Salsa (page 260).

SERVES 12–14

- 6 boneless, skinless chicken breasts
- 12 tomatillos, quartered
- 4 jalapeños, diced
- 1½ cups diced yellow onions
- 3 Tbsp Roasted Garlic (page 270) or minced fresh garlic
- 4 ripe Hass avocados, diced
- ½ cup water
- 8 cups sour cream
- 10 cups mixed shredded cheddar and Monterey Jack cheese, divided
- 24 Homemade Tortillas (page 271) or store-bought
- ¼ cup fresh goji berries when in season, or dried when not in season (see Wink)
- Guacamole (page 264), for serving

1. Place the chicken into a large pot with 12–14 cups water and bring to a boil on high heat. Turn the heat down to medium and simmer until the chicken is fully cooked and there is no pink remaining when cut through, approximately 25 minutes.

2. Remove the chicken from its cooking liquid and allow to cool in a medium bowl. Reserve ⅔ cup of the cooking liquid. When almost cooled, shred the chicken using a couple of forks. Set aside.

3. To make the sauce, in two batches, using a blender, combine the tomatillos, jalapeños, onions, garlic, and avocados, and ½ cup water and reserved cooking liquid. Blend until smooth, 2 minutes per batch. Pour the puréed mixture into a very large pot. Repeat with the second batch. Set the pot on medium heat. Add the sour cream. Stir until well blended. Remove from heat and set aside.

4. Preheat your oven to 350°F. Grease two 10 × 15-inch baking dishes.

5. Warm a frying pan on medium heat and add 1 Tbsp oil. Lightly fry the tortillas to soften them and make them pliable, otherwise they may crack when rolled. Remove excess oil from the tortillas by patting with paper towel. Continue adding oil a little at a time, as the tortillas will absorb it. It's good to wait until the oil is hot again before adding the next tortilla.

6. Spread 1 tsp sauce on each tortilla. Fill each tortilla with approximately ⅓ cup each of the shredded chicken and cheese, placing the fillings in a line down one side of the tortilla.

7. Roll up the tortillas by tucking one side over the chicken and cheese and leaving the ends open. Line them in three rows, seam sides down on the long side of the prepared baking dish until the pan is full. Pour the sauce over the top, making sure to cover all the enchiladas.

8. Sprinkle the remaining 2 cups cheese and the goji berries evenly over the top to finish. The goji berries add a bright pop of colour.

9. Bake for 35–45 minutes or until the cheese is melted, bubbly, and crispy on the edges. Serve warm.

10. These are best enjoyed the day or the day after you make them.

WINK: *If you don't have fresh goji berries, you can use the same volume of dried. Just soak them in warm water for 20 minutes, drain, and add in step 6. Other great add-ins are sliced black olives and pico de gallo in step 6.*

Tamale Pie

SERVES 6–8

- 1 Tbsp extra virgin olive oil
- 1 lb (450 g) lean ground beef
- 1 cup chopped yellow onions
- 1 large Anaheim pepper, seeded, crushed, and finely chopped
- 2 cups Roasted Corn (page 73)
- 1 (6 oz/170 g) can whole pitted black olives, drained
- 1 (750 ml) jar Krause Berry Farms and Estate Winery Roasted Corn Salsa
- 1 Tbsp chili powder
- 4 cups cold water
- 2 tsp kosher salt
- 1 cup + 3 Tbsp cornmeal
- 1½ cups shredded cheddar

This dish originated in the southwestern United States and is a classic American comfort food. Many families have their own version, and this is ours. This recipe has been handed down through the generations, and I like to serve it the same way my mother and her mother did—in their vintage 9-inch round Guardian Ware pot (page 216) with its glass lid, as shown.

Flavourful with a spicy ground-beef base and a crispy cornbread topping, this recipe also contains one of my favourite ingredients: black olives. When I was young, a bowl of black olives was always set on the table at extended family gatherings. When all was clear, my siblings and I would put them on our fingers, wiggle our hands around, and then quickly eat one right after the other before we were caught. This tradition carries on with my grandchildren.

1. Preheat your oven to 350°F.
2. In a large frying pan on medium-high heat, warm the oil. Add the ground beef and break it up with a spoon. Cook for 2 minutes, then add the onions and peppers and continue sautéing until the beef has browned.
3. Add the corn, olives, corn salsa, and chili powder and stir to distribute the seasoning evenly.
4. Turn the heat down to low and simmer for 5–10 minutes while you make the cornmeal.
5. In a medium pot on medium-high heat, combine the cold water and salt. Stir to dissolve. While stirring, sprinkle the cornmeal into the water 1 Tbsp at a time to prevent lumps. Stir until thickened, 5–10 minutes. The mixture will be thick but pourable.
6. Pour two-thirds of the cornmeal mixture into a greased 9-inch round casserole dish, spreading it around the bottom and up all sides of the dish.
7. Pour the meat mixture into the cornmeal-lined casserole dish and then cover with the remaining cornmeal, ensuring the cornmeal goes right to the edge of the dish to create a seal with the edge of the cornmeal.
8. Bake until the cornmeal is golden brown and crisp, 20–30 minutes.
9. Sprinkle the cheese evenly over the top and return to the oven until the cheese is melted and browned, approximately 5–7 minutes.
10. Store leftovers in an airtight container in the refrigerator for up to 3 days.

Pierogies

Before Alf and I met, he had never had a savoury pierogi and I had never had a sweet one. One time when we were dating, he was going to a family gathering and was asked to bring a dish, but he had gotten behind in his day. I saved him by making him my savoury pierogies. As the host removed the foil after baking and set them on the table, he along with others noticed something was very, very different about the pierogies! With a smile, Alf confided that, yes indeed, there was someone special in his life. We still laugh about the pierogie dish that gave us away! After Alf and I were married, I enjoyed many of my mother-in-law Katie's, tasty sweet berry pierogies. I asked her if she would teach me how to make them so we could share her delicious pierogies with our guests. She loved the idea and came into the Harvest Kitchen (page 239) to teach us all how to make them. We continue to make and sell her pierogies to this day.

MAKES 20–25 PIEROGIES

Dough
- 1½ cups cold water
- ¾ cup salted butter, cubed, at room temperature
- 2 large eggs, at room temperature
- 1½ tsp table salt
- 5½ cups all-purpose flour, divided, + more as needed

Filling for sweet pierogies
- 1 cup berries of your choice (blueberries, raspberries, and blackberries can all be added whole; strawberries should be chopped into manageable pieces) (see Wink)

Filling for savoury pierogies
- 1 cup cottage cheese or mashed potatoes
- ½ cup cooked diced bacon, for sprinkling
- ½ cup shredded cheese, for sprinkling
- ½ cup caramelized onions (see Wink, page 62), for sprinkling
- ¼ cup diced green onions, for sprinkling
- ¼ cup sour cream, for serving

WINK: *When I'm frying sweet pierogies, I like to break open a few to release the juice, which then caramelizes to make a sauce.*

MAKE THE DOUGH

1. In the bowl of a stand mixer fitted with the dough hook or whisk attachment, mix the water, butter, eggs, and salt until the eggs are fully incorporated. You might see pieces of butter at this point.

2. Add 3 cups flour and mix on low speed. The butter will mix in when the flour is added.

3. Add the remaining 2½ cups flour and mix on low speed until thoroughly combined. The dough should be soft. If it's too sticky, add 1 Tbsp more flour at a time until you have a soft dough.

4. On a clean surface dusted with flour, roll the dough to an ⅛-inch thickness.

5. Using a 3-inch round cutter, cut out circles. With floured fingers, stretch out each piece of cut dough to make it a 4-inch round.

FILL THE PIEROGIES

1. For sweet pierogies: Place berries of your choosing (3 for whole berries, 5–6 for blueberries, and/or 1½ tablesoons chopped strawberries) in the middle of each circle of dough.

2. For savoury pierogies: Place approximately 2½ tsp of your filling of choice in the middle of each circle of dough.

3. Use a finger to spread a bit of water along the edge of half the circle before folding and pinching the edges together to make sure they are well sealed. (The water acts like glue.)

4. If you wish to cook the fresh pierogies now, follow the instructions below. If you want to freeze them, place them on a baking sheet lined with parchment paper and transfer to the freezer. When they're completely frozen, package them in freezer bags and return to the freezer for up to 6 months. Follow step 3 below to cook them.

Recipe continues

COOK THE PIEROGIES

1. For fresh sweet pierogies: Bring a large pot of water to boil. Drop the pierogies in the boiling water and cook until they float to the surface plus 1 minute more. Drain on a paper towel. Now pan-fry them in butter or light-flavoured oil (see Wink) and sprinkle with sugar while they're still hot. Serve with whipped cream, ice cream, Custard Sauce (page 276), your favourite syrup, or berries.

2. For fresh savoury pierogies: Bring a large pot of water to boil. Drop the pierogies in the boiling water and cook until they float to the surface plus 1 minute more. Drain on a paper towel. (If you like pierogies a little crisper, you can also pan-fry these in butter or oil after boiling.) Sprinkle the toppings over top to garnish. Serve with sour cream on the side with a sprinkling of green onions.

3. For frozen pierogies: To cook from frozen, cook as above, but add an additional minute for the savoury or sweet pierogies to float to the top of the boiling water.

Asparagus Grilled Cheese

This grilled cheese is another family favourite. Pick asparagus from the garden, steam, and presto, you have a crunchy, cheesy, tart-sweet, melt-in-your-mouth sandwich. I love to use the loaf version of the Cheesy Potato Bread (page 118) for this. My Multi-Purpose Gluten-Friendly Bread (page 114) is also a delicious choice.

MAKES 1 SANDWICH

1 Tbsp salted butter
2 slices bread of your choice
2 thick slices cheddar
4 asparagus spears, fried or steamed until tender

WINK: *In the fall, I replace the asparagus with ½ small apple, sliced, and ¼ cup of Caramelized Onion (see Wink on page 62).*

1. Butter one side of each slice of bread.
2. Place a slice of cheese on the unbuttered side of one of the slices of bread. Lay the asparagus spears on top of the cheese. Place the second slice of cheese on top of the asparagus spears and top with the second slice of bread, buttered side up.
3. Heat a cast-iron frying pan on medium heat. Place the sandwich in the pan and cook until the bread is toasted and golden and the cheese is melted, approximately 3 minutes per side. If you'd prefer to use a panini press rather than a frying pan, cut a piece of parchment paper big enough to wrap around the sandwich, wrap the sandwich, and grill for 3 minutes.
4. This is best enjoyed immediately.

Farmers Eat Quiche

When I was in my mid-20s, my friend Bisenia served this delicious quiche at a flight attendant luncheon. I had never tasted quiche before, and I immediately loved it. It also became one of my family's favourites for special-occasion brunches. One day in the Harvest Kitchen (page 239), I spotted bins and bins of freshly harvested asparagus on one side of the room and mounds of pie dough on the other. That was when Farmers Eat Quiche was born. This dish has since become a mainstay of our winery restaurant and in our market store. My family does not have to wait now for a special occasion to enjoy it!

MAKES TWO (10-INCH) QUICHES

- 1 batch Whole Wheat Pie Dough (page 273)
- 1 cup chopped, blanched asparagus (see Wink, page 111)
- 4 Tbsp chopped green onions
- 1½ cups shredded cheddar
- 1½ cups shredded mozzarella
- 3 Tbsp whole wheat flour
- ¼ tsp table salt
- ⅛ tsp freshly ground black pepper
- 6 large eggs, at room temperature
- ½ cup half-and-half (10%) cream

1. Follow steps 1–3 on page 273 to prepare the dough. Divide the dough into two portions.
2. Follow step 4 on page 273 to roll out each portion of dough into a circle a bit bigger than a 10-inch diameter. Carefully drape each portion of dough into a 10-inch pie plate and gently press down on it evenly, leaving the sides to hang over the pie plate slightly. Use a knife to trim any excess dough off the rim of the pie plate and press with a fork to stamp all around the pie.
3. Preheat your oven to 350°F.
4. Divide the asparagus and green onions between the shells.
5. In a small mixing bowl, mix together the cheeses, flour, salt, and pepper.
6. Divide the cheese mixture between the shells.
7. In the mixing bowl you used for the cheese mixture, whisk the eggs and cream until combined, then divide between the unbaked shells.
8. Bake until a knife inserted in the centre comes out clean, approximately 40–45 minutes. If the knife does not come out clean, bake for another 5 minutes, then test again. Repeat until the knife comes out clean.
9. The quiches are best served warm but can be enjoyed cold or rewarmed. Wrap in plastic wrap or foil and store in the refrigerator for up to 2 days, or double-wrap in plastic wrap and store in the freezer for up to 3 months. Thaw in the refrigerator overnight, and then heat in a 350°F oven for 20 minutes.

Asparagus Cheese Bread (page 110)

Breads and Muffins

Asparagus Cheese Bread

At our farm, asparagus is in season from April through May. When it's in season, we are all-hands-on-deck in the Harvest Kitchen (page 239) as we try to make the most of this classic taste of spring. We pickle some of the spears, blanch for Cream of Asparagus Soup (page 54) sell others fresh, and the luckiest ones end up in this delicious bread. One of the highlights of spring for me is eating a slice of this fresh bread warm with a schmear of butter or using it to make a gooey grilled cheese sandwich (page 105). Bliss.

MAKES ONE (10-INCH) LOAF

Bread

1 cup lukewarm water
1 Tbsp instant yeast
1 Tbsp granulated sugar
½ Tbsp table salt
1 Tbsp extra virgin olive oil
3 cups all-purpose flour + more as needed
¼ cup finely grated parmesan
⅓ cup chopped, blanched asparagus (approximately 2 medium spears) (see Wink)

Herbed olive oil

1 Tbsp extra virgin olive oil
½ tsp dried rosemary
½ tsp dried basil
½ tsp dried oregano

Toppings

¼ cup shredded cheddar
3–4 (each 2–3 inches long) asparagus tips

MAKE THE BREAD

1. In a stand mixer fitted with the dough hook, mix the lukewarm water, yeast, sugar, salt, and olive oil on low speed until combined.

2. Add the flour and parmesan. Knead on low speed until the dough feels elastic and smooth, approximately 3 minutes. If mixing by hand, knead the dough very thoroughly to make sure all flour is incorporated into the dough. Although mixing by hand will take a little longer, the same results are achievable.

3. After the dough has been thoroughly kneaded, add the asparagus. You may need to add a little more flour, about 1 Tbsp at a time, if the dough is too sticky after adding the asparagus. (The dough should peel away from the side of the bowl and should not come off in bits on your hands.)

4. Cover the bowl with plastic wrap or a clean light tea towel and set aside in a draft-free spot until doubled in size, approximately 30–45 minutes.

5. Line a baking sheet with parchment paper. Once the dough has doubled in size, form it into a 10-inch-long loaf on the baking sheet. Cover the dough with the tea towel and let rise until the dough has doubled in size again, approximately 1 hour.

MAKE THE HERBED OLIVE OIL

1. While the dough is rising, in a small mixing bowl, whisk together the oil, rosemary, basil, and oregano.

BAKE THE BREAD

1. Preheat your oven to 350°F.

2. Once the loaf of dough has doubled in size, lightly brush all the herbed olive oil over the top; avoid pressing on the dough to prevent it from collapsing.

3. Bake the bread until golden or the middle of the bread registers 200°F on a thermometer, approximately 25 minutes.

4. Top the bread with the cheddar and asparagus tips. Return to the centre rack of the oven and bake until the cheddar is bubbly and melted, approximately 7 minutes.

5. Remove from the baking sheet and transfer the baked bread to a wire rack to cool.

6. Serve warm or at room temperature, or cool completely (approximately 1½ hours) and store in an airtight container in the refrigerator for up to 2 days or wrapped in plastic wrap in the freezer for up to 1 month.

RECIPE PICTURED ON PAGE 108

WINK: *To blanch the asparagus, trim the woody ends and chop the remaining asparagus into 1-inch pieces. Fill a large bowl with cold water and ice. Set aside. Fill a medium saucepan with water and bring to a boil on high heat. Add the asparagus, return the water to a boil, and maintain a boil on high heat to cook for 3 minutes. Using a slotted spoon, remove the asparagus from the boiling water and quickly plunge into the ice water bath. Let cool in the ice water for 5 minutes. Transfer the asparagus to a baking sheet and let air-dry, or blot dry with paper towel if you're in a rush. You can freeze the blanched asparagus to use all year. Place the dried asparagus pieces on a parchment-lined baking sheet and individually quick freeze until solid. Transfer to a freezer bag or freezer-safe container and freeze for up to 6 months.*

GINGE

Hearty Multigrain Bread

One day while rolling out pie shells in the Harvest Kitchen (page 239), Pauline, one of our earliest staff members and now wonderful friend, surprised me when she told me that she makes all of her own bread. She said she tried to introduce store-bought bread to her family after her twins were born, but they wouldn't have any part of it. Once I tasted hers, I knew why. I loved the idea of adding her soft, hardy, and flavourful bread to our menu. It quickly became a farm favourite and we now offer it year-round. Some of our guests live quite a distance from the farm, so they order loaves ahead. We also keep extra in the freezer, as we sell out daily. This bread tastes delicious with any of our soups (pages 53–63) or our Asparagus Grilled Cheese (page 105). This recipe makes three loaves, so you can enjoy one fresh and keep the others in the freezer or share one with a friend.

MAKES THREE (5 × 9-INCH) LOAVES

- ⅔ cup Potato Slurry (page 279)
- ⅓ cup granulated sugar
- 1 large egg, at room temperature
- 2 tsp table salt
- 2¼ cups warm (110°F) water
- 1½ Tbsp melted salted butter
- 1 Tbsp instant dry yeast
- 3 cups whole wheat flour
- ½ cup cracked rye
- ½ cup flaxseed
- ½ cup cracked wheat
- 4 cups all-purpose flour

1. Preheat your oven to 350°F. Lightly spray three 5 × 9- inch loaf pans with cooking spray.
2. In a stand mixer fitted with the dough hook, mix the potato slurry, sugar, egg, salt, warm water, and butter at low speed for 1 minute to blend. Sprinkle the yeast over top. Mix to combine.
3. In a large bowl, whisk together the whole wheat flour, rye, flaxseed, and cracked wheat. Add this multigrain mixture to the wet ingredients and mix on low speed to combine.
4. With the mixer still running on low speed, gradually add the all-purpose flour. Keep mixing until the dough peels away from the sides of the bowl.
5. Cut the dough into three equal portions. Shape each piece into a loaf and place in the prepared pans.
6. Cover each loaf with plastic wrap or a clean tea towel, then place them in a warm, draft-free spot to rise for 30 minutes. The loaves should double in size. If they haven't, let rest for up to 30 minutes longer.
7. Bake until the tops of the loaves are golden brown, approximately 35 minutes. Use a probe thermometer to take the internal temperature; it should be 210°F–220°F. Let cool in the pan for 5 minutes before turning out onto the rack.
8. Wrap the loaves individually in plastic wrap and store in the refrigerator for up to 3 days or in the freezer for up to 4 months.

WINK: *You can use this recipe to make hamburger buns instead. Just grease a 9 × 13-inch cake pan instead of a loaf pan. Cut the dough into 12 evenly sized, bun-shaped pieces, place them in the greased cake pan, loosely cover with a clean tea towel or plastic wrap, and let rise until doubled in size, 20–30 minutes. Bake as directed above.*

Multi-Purpose Gluten-Friendly Bread

Those of us who eat gluten-free products, whether it's because of an allergy or simply a preference, now have many more options than we did even a few years ago. This is my recipe for gluten-friendly bread. It's simple to make and tastes delicious. It can be used in many recipes in this book, such as Auntie Dona's Overnight Brunch Eggs (page 37), Asparagus Grilled Cheese (page 105), French Onion Soup (page 62), or in place of the English muffins in Eggs Benedict (page 34). It's also perfect toasted to accompany any of the soups (pages 53–63) or the Garden Chili (page 88). See right for more variations (I don't call it multi-purpose for nothing!).

MAKES ONE (5 × 9-INCH) LOAF

- 1½ cups 2% milk
- 2 Tbsp granulated sugar
- 1½ tsp instant yeast
- 3 cups Gluten-Friendly Flour Blend (page 279) + more for dusting
- 4 tsp baking powder
- 1½ tsp xanthan gum
- 1 tsp table salt
- ¼ cup extra virgin olive oil
- 2 tsp fresh lemon juice
- 2 large eggs, at room temperature

1. In a heatproof measuring cup, warm the milk for 30-second increments in the microwave to 100°F. It should be warm to the touch but not burning hot as the milk needs to be moderately warm to fully activate the yeast.

2. Stir the sugar into the milk, followed by the yeast. Set the yeast mixture aside and let it activate for approximately 10 minutes.

3. In a medium mixing bowl, whisk together the flour blend, baking powder, xanthan gum, and salt.

4. In a stand mixer fitted with the paddle attachment, mix the oil, lemon juice, and eggs on low speed for 10 seconds to combine.

5. Add the proofed yeast mixture to the lemon juice mixture and continue to mix on low speed for just 5 seconds to combine.

6. With the mixer running on low speed, slowly add the dry mix to the yeast mixture. Continue to mix until fully incorporated, approximately 5 minutes, then beat the mixture on medium-high until it is thick and sticky, but still wet, approximately 3 minutes.

7. Grease a 5 × 9-inch loaf pan with non-stick cooking spray or olive oil and dust with gluten-friendly flour blend. Using a rubber spatula, scrape the dough mixture into the prepared loaf pan and smooth the top of the loaf with the spatula or wet fingers. Loosely cover with a clean tea towel or plastic wrap and set in a warm place until doubled in size, approximately 20–40 minutes depending on your room temperature. It is important not to overproof the bread. Don't let it rise above your loaf pan, just let it rise until it crests the top. Gluten-free breads do not maintain their structure and will flow over the pan or collapse if left to over-rise.

8. Preheat your oven to 375°F.

9. Bake the bread until golden, approximately 35 minutes. If the crust is darkening too quickly, you can tent it with foil and continue baking until done. The bread is finished baking when it has an internal temperature of 210°F–220°F. Use a probe thermometer to check the temperature.

10. Remove the loaf pan from the oven and let cool completely. To remove the loaf from the pan, loosen its edges with a knife, carefully turn the loaf pan on its side to release the loaf, and transfer the loaf to a cooling rack.

11. Store any leftovers in an airtight container at room temperature for up to 2 days or in the refrigerator for up to 1 week, but be aware that gluten-free bread tends to dry out quickly. It's best to use fresh or to freeze for up to 3 months in the freezer for lasting freshness. If it's stale try the breadcrumbs variation below.

This Recipe Isn't Just for Bread!

- **Breadcrumbs:** If leftover bread has staled, make breadcrumbs by toasting in the oven at 350°F for 10–20 minutes, then grinding to the desired coarseness in a blender or food processor.
- **Pizza Dough:** This dough also makes an excellent pizza base. Place the dough in a greased 9 × 13-inch cake pan or two 10-inch round cake pans instead of a loaf pan. Let rise until doubled in size, 20–30 minutes, and then bake as directed above. Let cool in the pan for 3–5 minutes and then transfer to a wire rack to cool completely. If you're using a rectangular pan, cut the dough into two 6½ × 9-inch rectangles. Add your favourite pizza toppings (see pages 77–80 for some ideas) to the crusts and bake at 350°F for 10–15 minutes.

Cheese Bread in a Can

This bread pairs very well with the Apple, Cheese, and Cider Soup (page 57)—or, in fact, any of the soups in this book. As you can see, it can be made in any can, from a simple coffee can to a beautiful brass wine chiller! My grandmother Sarah used Folgers coffee cans for many things in and around her farm, including for baking bread. She was raised during the Great Depression and I'm sure she found the can to be a cost-saving vessel to bake her bread in.

MAKES ONE (5 × 9-INCH) LOAF OR 2 COFFEE-CAN LOAVES

¼ cup warm (approximately 110°F) water
1 package active dry yeast (2¼ tsp)
2 Tbsp granulated sugar
2⅓ cups all-purpose flour
½ tsp freshly ground black pepper
1 tsp table salt
¼ tsp baking soda
1 large egg, at room temperature
1 cup sour cream, at room temperature
1 cup shredded cheddar

1. Grease two 6-inch-high × 5-inch-diameter metal coffee cans or copper cylinders, or a 5 × 9-inch loaf pan.

2. Warm the bowl of a stand mixer with hot water. Dry with a tea towel. Pour in the warmed water, yeast, and 1 tsp of the sugar. Let sit until it bubbles, approximately 10 minutes.

3. Meanwhile, in a medium bowl, stir together ⅓ cup of the flour, salt and pepper, baking soda, and the remaining sugar.

4. In a small bowl, add the egg and sour cream.

5. Affix the mixer with the dough hook. Add the flour mixture and egg mixture to the bowl of the stand mixer. Mix for 30 seconds on low, then beat on high until incorporated, approximately 1 minute, scraping down the sides of the bowl as required. Stir in the remaining flour and add the cheese, then mix thoroughly on medium speed for 1 minute.

6. Using floured hands, transfer the dough to the prepared coffee cans or loaf pan. Cover with a clean tea towel and let rise in a warm, draft-free place for 1 hour. (The dough will rise slightly but it will not double.)

7. Preheat your oven to 350°F.

8. Bake the bread until it is golden brown and a knife inserted in the middle comes out clean, approximately 40–45 minutes.

9. Let cool in the pan for 2 minutes. Remove to a wire rack to cool completely. This bread tastes great the next day, as the cheese keeps it moist. It's also delicious warmed in the oven or toasted.

10. Store wrapped in plastic wrap in the refrigerator for up to 3 days or in the freezer for up to 3 months.

WINK: *If you want to change it up a bit, you can add ¼ cup diced green onions and ¼ cup more cheese when you're adding the cheese in step 4. Or try adding ¼ cup of either Roasted Garlic (page 270) or diced apples.*

Cheesy Potato Bread

Knowing how I like to make individualized foods with a personal touch, Pauline came up with the idea to serve this light and airy bread, with delicious pockets of cheese, in these cute and small jars. Everyone loves them fresh and warm right out of the oven. Make sure you have butter ready! My Herb Butter (page 268) and Garlic Parmesan Butter (page 268) go exceptionally well and will have your mouth watering.

MAKES 12 (1-CUP WIDE-MOUTH) JARS

¼ cup + 2 tsp granulated sugar, divided

1½ cup warm (approximately 110°F) water, divided

1 Tbsp active dry yeast

1½ Tbsp melted salted butter

1 large egg, at room temperature

½ cup Potato Slurry (page 279), at room temperature

5½ cups all-purpose flour

1½ tsp table salt

2 cups shredded cheddar, divided

Flaked sea salt, for garnish (optional)

Rosemary, for garnish (optional)

1. Stir 2 tsp of the sugar into ½ cup of the warm water, followed by the yeast. Set the yeast mixture aside and let it proof for approximately 10 minutes. The yeast will bubble and smell yeasty.

2. In a stand mixer fitted with the dough hook, mix the remaining 1 cup of warm water, butter, and egg for 1 minute on low speed until just combined. Add the remaining ¼ cup sugar, potato slurry, and salt, then the proofed yeast mixture and mix for another couple of seconds on low speed until just incorporated.

3. With the mixer still running on low speed, slowly add the flour and salt to the yeast mixture until fully incorporated, approximately 5 minutes. Turn the mixer speed to medium-high and beat the mixture for approximately 3 minutes. Your dough should be thick and elastic. With the mixer still running on medium speed, add 1 cup cheese a little at a time; as the dough gets drier, it will form into a ball.

4. Loosely cover the dough with a clean tea towel or plastic wrap and place in a warm, draft-free place to proof 45 minutes–1 hour. The dough should double in size before you put it into jars.

5. Grease or spray twelve 1-cup wide-mouth heatproof glass jars with non-stick cooking spray or olive oil and dust with flour.

6. Using your hands, divide the dough into equal portions and place into the prepared jars. Wet your fingers and use them to smooth the top of the loaves. Loosely cover the jars with the tea towel or plastic wrap and place in a warm, draft-free place to proof. It is important not to overproof the bread. Just let it rise until the mixture is approximately double in size (20–30 minutes, depending on your room temperature). Watch it carefully and don't let it rise above your jar. It should just crest the top.

7. Preheat your oven to 350°F.

8. Bake until golden, 20–25 minutes. If the crust is browning too quickly, you can tent it with foil and continue baking until done. The bread is finished baking when it's golden and has an internal temperature of 210°F–220°F. Use a probe thermometer to check the temperature.

9. Remove the bread from the oven and sprinkle the remaining 1 cup cheese evenly over the loaves. Return to the oven until the cheese is melted, approximately 5 minutes.

10. Remove the jars from the oven and let cool for 3–5 minutes before removing and transferring the bread to a cooling rack.

11. Wrap the loaves individually in plastic wrap stored in the refrigerator for up to 3 days or in the freezer for up to 4 months.

WINK: *You can omit the cheese for plain potato bread, if you prefer. This recipe will also make two 5 × 9-inch loaves or 12 buns. Follow the steps in the Wink on page 113.*

AUSE
FARMS
WINER

Auntie Dona's Zucchini Bread

My dad's older sister, Auntie Dona, grew a beautiful garden filled with many vegetables, including zucchini. This is her recipe for zucchini bread. She believed that freezing the loaves and then thawing them to warm or toast them gives them a fuller flavour than they have the day they're baked. I agree—and I'm sure you will too. My children grew up with this bread, never knowing they were eating vegetables along with the plump blueberries or chocolate chips tucked in!

MAKES TWO (5 × 9-INCH) LOAVES

3 cups all-purpose flour
3 tsp ground Saigon cinnamon
1 tsp table salt
1 tsp baking soda
¼ tsp baking powder
3 large eggs, at room temperature
2 cups granulated sugar
1 cup vegetable oil
2 cups grated zucchini
3 tsp pure vanilla extract
1 cup dark chocolate chips or blueberries (optional)

1. Preheat your oven to 350°F. Grease and flour two 5 × 9-inch loaf pans.

2. In a medium bowl, sift together the flour, cinnamon, salt, baking soda, and baking powder. Set aside.

3. In a stand mixer fitted with the paddle attachment, beat the eggs, sugar, and oil together until well mixed. With a spoon, stir in the zucchini and vanilla to combine. Add the dry ingredients to the wet ingredients and mix by hand until the flour is fully incorporated. Stir in the chocolate chips or blueberries (if using) to evenly distribute.

4. Pour the batter into the prepared loaf pans. Bake until golden brown on top and a toothpick inserted in the centre of each loaf comes out clean, approximately 50–60 minutes.

5. Cool for 15 minutes before releasing the loaves from the pans. Run a knife along the inside edges before inverting onto a rack to cool.

6. Store in plastic wrap at room temperature or in the refrigerator for up to 3 days, or double-wrapped in plastic wrap and foil in the freezer for up to 3 months. Let thaw at room temperature for approximately 1 hour.

WINK: *I love toasting a slice of this to enjoy with a bit of butter.* *(See page 268 for some delicious butter variations.)*

Blueberry Muffins

When my husband finally convinced me to hire someone to help me in the kitchen, my first choice was Pauline Schroeder, whom he had known for 10 years. The farm, especially the Harvest Kitchen (page 239), would not be what it is today without Pauline and her many talents. Managing a kitchen this size, where all the different crops are brought in from the fields, is not for the faint of heart. She managed it seamlessly for 13 years until her retirement in 2018. Pauline still pops in now and then when we're super busy to help and we are so grateful!

Pauline's blueberry muffins graced the table at many Friday-morning managers' meetings and special events, until we decided to sell them in the bakery for our guests to enjoy. Now we can enjoy them every day!

MAKES 12 MUFFINS

- ¾ cup granulated sugar
- ¾ cup 2% milk
- ½ cup vegetable oil
- 1 large egg, at room temperature
- 1 tsp pure vanilla extract
- 2 cups all-purpose flour + extra for coating (coating optional)
- 2½ tsp + pinch of baking powder
- 1½ cups fresh or frozen blueberries
- ¼ cup brown sugar, for sprinkling (optional)
- 1 tsp ground Saigon cinnamon, for sprinkling (optional)

1. Preheat your oven to 350°F. Line a standard 12-cup muffin pan with paper liners or lightly spray with cooking spray.
2. In a stand mixer, combine the sugar, milk, oil, egg, and vanilla and mix on low speed, or place the ingredients in a large bowl and mix with a spatula.
3. In a medium bowl, combine the flour and baking powder, using a dry whisk to gently stir in the baking powder. This distributes the baking powder evenly.
4. Add the dry ingredients to the wet ingredients and mix until just combined. The flour should be fully incorporated at this point. Do not overmix.
5. If using frozen berries, coat them lightly with a bit of flour, so they aren't too juicy, and then add them to the batter. If using fresh berries, add them to the batter without any extra flour. Fold in the berries until just combined.
6. Fill each muffin cup about two-thirds full of batter. In a small bowl, mix together the brown sugar and cinnamon, if using, and sprinkle on top of each muffin.
7. Bake until golden brown and a toothpick comes out clean when inserted into the centre, approximately 20 minutes.
8. Wrap the muffins individually in plastic wrap and store in the refrigerator for up to 3 days or in the freezer for up to 4 months.

WINK: *For an extra treat, top these with cream cheese frosting (page 186).*

Cheesy Corn Muffins

Nothing goes better with Garden Chili (page 88) than moist cheesy corn muffins. This classic combination provides a hearty comfort meal. During corn season, I make sure to freeze a good amount of roasted corn so that I have plenty on hand to make these in the winter.

MAKES 12 MUFFINS

- 1 cup cornmeal
- 1 cup all-purpose flour or Gluten-Friendly Flour Blend (page 279)
- 2 Tbsp granulated sugar
- 1 Tbsp baking powder
- ½ tsp table salt
- 1 cup sour cream
- ½ cup 2% milk
- 2 large eggs room temperature
- ⅓ cup extra virgin olive oil
- 2 cups fresh or frozen, thawed and drained Roasted Corn (page 73)
- ¾ cup chopped green onions
- ¾ cup shredded cheddar

1. Preheat your oven to 350°F. Lightly spray a 12-cup muffin pan with cooking spray.

2. In a large bowl, gently whisk together the cornmeal, flour, sugar, baking powder, and salt.

3. In a medium bowl, use a fork to stir together the sour cream, milk, eggs, and oil.

4. Add the wet ingredients to the dry ingredients and stir with the fork until the flour is no longer visible. Fold in the corn, green onions, and cheese.

5. Fill the muffin cups two-thirds full. Bake until lightly golden and a toothpick inserted in the centre of a muffin comes out clean, 15–18 minutes. Let them cool in the pan for 3–5 minutes before removing them.

6. These are best eaten freshly baked, but can be wrapped individually in plastic wrap and stored in the refrigerator for up to 3 days or in an airtight container in the freezer for up to 2 months. To reheat from frozen, place them on a parchment-lined baking sheet and warm in a 400°F oven for 10 minutes (they will taste freshly baked with a crispy finish!).

English Muffins

Why make English muffins yourself? Because, like other yeast breads, when fresh they offer a taste you just can't buy. Their heady aroma fills your home with a nostalgic warmth.

MAKES 14–18 MUFFINS

1 cup 2% milk
¼ cup room temperature water
2 Tbsp granulated sugar
1 Tbsp instant dry yeast
½ cup warm (110°F) water
½ cup melted salted butter, lukewarm
4 cups all-purpose flour, divided, + more as needed
1 tsp table salt
Cornmeal, for sprinkling
Olive oil, for cooking

1. In a small saucepan, warm the milk and room temperature water on medium heat. Bring to a simmer, then remove from heat. Mix in the sugar, stirring until dissolved. Let cool until lukewarm.

2. In a small bowl, dissolve the yeast in the warm water and let stand for 10 minutes.

3. In the bowl of a stand mixer fitted with the dough hook, place the following in this order: milk mixture, yeast mixture, butter, and 3 cups flour. Mix on low speed until smooth.

4. Gradually add the salt and remaining 1 cup flour. You may need to add more flour, ¼ cup at a time, to reach the required soft dough. This will take approximately 1½ minutes of mixing.

5. Remove the dough from the mixer and knead for a couple of minutes on a lightly floured surface. Form the dough into a ball. Place the dough in a greased bowl, cover with plastic wrap or a clean tea towel, and let rise in a warm, draft-free spot until doubled in size, approximately 1 hour.

6. Lay out an approximately 15 × 22-inch piece of parchment paper and place on a baking sheet. Sprinkle the parchment with cornmeal and set aside.

7. Punch down the dough in the bowl. On a clean surface dusted with flour, roll out the dough to about ½ inch thick. Cut out rounds with a 3½-inch cookie cutter for smaller English muffins or 4½-inch cookie for medium-size English muffins. Set the rounds on the prepared baking sheet. Gather the remaining dough, roll it out again, and continue cutting out more rounds until the dough is all cut into rounds. Dust the tops of the muffins with some cornmeal. Cover with the plastic wrap or tea towel and let rise until doubled in size, approximately 45 minutes.

8. Grease a griddle or a cast-iron frying pan with olive oil. Place on medium-low heat and cook the muffins on each side until the bottoms are crisp and browned, approximately 10 minutes on each side.

9. If you plan on serving these not long after cooking them, you can keep the muffins in the oven at 200°F until needed. Otherwise, you can toast them in the oven before serving.

10. Store in an airtight container in the refrigerator for up to 3 days or in the freezer for up to 3 months.

WINK: *Spread these with any of the butters on page 268 to switch up the taste for a sweet or savoury snack.*

Horse Muffin Treats

Horses have been such a large part of my life, I couldn't write a cookbook without including a special treat for them. My daughters Sarah and Holly share the same love of horses as I do. I was introduced to horse muffins at a horse show we attended together many years ago. When we saw how much the horses loved muffins, we created our own recipe to make at home. Tucker (pictured left) and Mosaic (pictured right) love them too!

MAKES 36 MINI MUFFINS

1 cup quick-cooking rolled oats
½ cup oat milk
1 large egg, at room temperature
½ cup packed brown sugar
¼ cup shredded carrots
¼ cup diced apples, skin on
¼ cup flaxseed
1 cup whole wheat flour
2 tsp baking powder
1 tsp baking soda
1 tsp fenugreek powder

1. In a small bowl, soak the oats in the oat milk for 30 minutes.

2. Preheat your oven to 250°F. Grease the cups of a 36-cup mini-muffin pan.

3. In a large bowl, beat together the egg and brown sugar. Add the carrots, apples, flaxseed, and oat mixture. Stir until well combined.

4. In a small bowl, mix together the flour, baking powder, baking soda, and fenugreek powder.

5. Add the oat milk mixture to the dry ingredients and stir until just combined.

6. Fill the muffin cups two-thirds full and bake for 30 minutes, then lower the temperature to 225°F for 1 hour. There's no need to test these muffins with a toothpick the way you'd check muffins for humans. They'll be nice and crunchy and ready for your horse to enjoy when cool.

7. Store in a tin with a tight-fitting lid in the barn for up to 1 week or in a freezer bag in the freezer for up to 4 months. Thaw completely before feeding your horse the treat.

Strawberry Cookies (page 132)

Cookies and Bars

Strawberry Cookies

MAKES 24 COOKIES AND 24 MINI COOKIES

Since 2005, my friend Lee, from Vista D'oro Farms & Winery, has organized a cookie-tin sale to raise money for BC Children's Hospital in Vancouver. Her friends with bakeries make a special cookie for her tins. Here at Krause Berry Farms and Estate Winery we love contributing these sweet, delicate, and beautiful little cookies bursting with the flavour of strawberries. We chose strawberries to represent the history of our farm. We are all so happy to donate our time to something so meaningful. (Note: You'll need strawberry-shaped cookie cutters for this recipe. If you don't have any, these will taste equally delicious in other shapes!)

Cookies

1 batch Sugar Cookies dough (page 167)

1 cup Krause Berry Farms and Estate Winery Strawberry Jam

1 cup icing sugar

Green royal icing

1 cup Meringue Powder Royal Icing (page 278)

Leaf-green food colouring

MAKE THE COOKIES

1. Preheat your oven to 350°F. Spray three baking sheets with cooking spray or line them with parchment.

2. Dust a clean work surface with flour and roll out half of the dough to ¼ inch thick for the tops of the cookies. Using a 3-inch strawberry cookie cutter, cut out 24 tops.

3. Place the cookies on a prepared baking sheet ½ inch apart, then cut out the inside of each cookie with a 2-inch strawberry cookie cutter. (They hold their shape better when cut on the baking sheet.) Place the small strawberry shapes on a prepared baking sheet and set aside.

4. Roll out the other half of the dough to ¼ inch thick for the bottoms of the cookies. Use the 3-inch strawberry cookie cutter to cut out 24 cookies. Place them on the third prepared baking sheet ½ inch apart.

5. Bake both trays of 3-inch cookies until the edges are beginning to brown, 4–5 minutes. Transfer to a cooling rack to cool completely.

6. Meanwhile, bake the 2-inch cookies until beginning to brown, approximately 4–5 minutes.

7. You can assemble and decorate the 3-inch cookies now, using the instructions below, or freeze them in an airtight container for up to 3 months to assemble and decorate later. Thaw frozen cookies for 1 hour before assembling.

8. Spread the jam on the top of the full cookies, which will now be the bottoms, leaving a ¼-inch margin around the outside of each one. You'll need approximately 2 tsp jam for each cookie.

9. Lay the top cookies, the ones with the centre cut out of them, in a row and cover the leaf part with a strip of parchment paper.

10. Place the icing sugar in a fine-mesh sieve and sprinkle the sugar on these cut-out cookies.

11. Lay the top cookies over the bottom cookies.

12. If you like, you can sandwich together the 2-inch cookies and sprinkle them with icing sugar as well.

MAKE THE GREEN ROYAL ICING

1. Place the icing in a mixing bowl and add green food colouring, a little at a time, until a nice leafy colour appears. Using an electric mixer (a stand mixer or hand-held mixer would both work for this) blend well on medium speed for 2 minutes.

2. Use an angled spatula to place the icing into a piping bag fitted with a #2 round tip, standing the piping bag in a tall cup and folding the top of the bag over the rim of the cup for easy filling. Twist the top of the piping bag closed and secure with a clip or elastic. If not using a #2 round tip, snip a small tip off the end of the piping bag and squeeze to push the icing down.

3. Carefully pipe the green icing over the leaves of each cookie. Allow the icing to set completely before indulging.

4. Store in an airtight container at room temperature or in the refrigerator for up to 1 week.

RECIPE PICTURED ON PAGE 130

Berry Farm Cookies

What does every kid want waiting on the kitchen counter when they arrive home from school? A big plate of warm chocolate chip cookies and a glass of milk. And if you're a berry farmer's kid, like ours were, you would find dried berries mixed in with the chocolate chips! By request, we now stock the dough for these cookies in the freezer section of our market for our guests to take home and bake their own fresh cookies. I use a blend of dried berries mixed in equal parts. I typically combine dried blueberries, dried cranberries, and dried cherries. You can make this recipe entirely by hand or in a stand mixer fitted with the paddle attachment.

MAKES 30 COOKIES

1 cup salted butter, softened
½ cup packed brown sugar
1 cup granulated sugar
2 large eggs, at room temperature
1 tsp pure vanilla extract
2 cups + 2 Tbsp all-purpose flour
1 tsp baking soda
½ tsp table salt
¾ cup mixed dried berries, chopped (larger berries coarsely chopped) (see Wink)
½ cup white chocolate chips or chunks
½ cup dark chocolate chips or chunks

1. Preheat your oven to 375°F. Line two or three baking sheets (enough to fit 30 cookies 1 inch apart) with parchment paper.

2. Using a stand mixer fitted with the paddle attachment, or in a bowl using a spatula, beat the butter until completely smooth.

3. Add both sugars and cream for 1–2 minutes. The mixture should lighten in colour.

4. Add both eggs and the vanilla and mix until incorporated.

5. Add the flour, baking soda, and salt. Mix until the flour is incorporated and a thick, sticky dough forms, 3–5 minutes.

6. Mix in the berry mix and chocolate chips until evenly distributed.

7. Scoop 2 Tbsp of dough per cookie onto the prepared baking sheets, 1 inch apart. You should be able to fit 12 cookies on each sheet.

8. You can bake these one or two sheets at a time, depending on your oven's hot spots (see page 24). To bake one sheet, place it on the middle rack and bake the cookies until they are chewy on the inside and crispy on the outside, 8–11 minutes. Do not overbake. To bake two sheets, place one sheet on the top rack and one on the bottom rack of your oven. Bake as directed but switch the top and bottom sheets halfway through the baking time.

9. Immediately transfer the cookies to a wire rack or on a piece of parchment paper.

10. Store in an airtight container at room temperature for up to 1 week or in the freezer for up to 6 months.

RECIPE ALSO PICTURED ON PAGE 153

WINK: *For an easy dessert that all ages can make and enjoy, try portioning out the desired amount of dough in step 7, and lightly pat down the dough until it is ¾-inch thick and evenly fits into a seasoned cast-iron skillet (preferably 3½-inch if you have it). Bake in a 375°F oven until golden brown, the edges are slightly crispy, and the middle is still soft and chewy, approximately 15–20 minutes, depending on the size of the skillet. Serve warm in the skillet with a scoop of vanilla ice cream to add a nice finish. If you don't want to use berries, you can substitute them with ¾ cup pecan pieces or add an extra ¾ cup chocolate chips.*

Bird's Nest Cookies

To me, this classic Christmas cookie tastes just like being home for the holidays. These were made only at Christmas in my house when I was growing up. They remain one of my very favourites to this day and can be found in the farm's bakery showcase only at Christmastime. My grandmother Sarah Greene would make these by the dozens, but the recipe halves easily, so you don't need to make a full batch.

MAKES 40–45 COOKIES

1 cup salted butter, at room temperature
1 cup vegetable shortening, at room temperature
1 cup packed brown sugar
4 large eggs, at room temperature, separated
1 tsp pure vanilla extract
4 cups all-purpose flour
1 tsp table salt
2 cups quick-cooking rolled oats
1 (250 ml) jar of Krause Berry Farms and Estate Winery Blueberry or Raspberry Jam

1. Preheat your oven to 325°F. Line two baking sheets with parchment paper.

2. In a stand mixer fitted with the paddle attachment, cream the butter, shortening, and sugar on medium-high speed until light and fluffy. Add the egg yolks one at a time, mixing well between additions, and then add the vanilla. Beat until well combined. Turn the speed down to low and gradually add the flour, about ½ cup at a time, and the salt. Mix until well combined. Chill for 45 minutes.

3. In a large bowl, lightly beat the egg whites using a handheld whisk. You want to just loosen and start to mix them together. Set aside.

4. Place the oats in another large bowl.

5. Using a #40 cookie scoop (1½ Tbsp/1½ oz), scoop out a ball of dough. Place the dough ball into the beaten egg whites, letting any excess drip back into the bowl, and then into the oats. Put the coated ball on a prepared baking sheet. Repeat with the remaining dough, spacing the dipped balls 2 inches apart on the baking sheets.

6. Using the handle of a wooden spoon, gently press down into the centre of each cookie, making a small indent.

7. Bake for 7 minutes, with one pan on the top rack and the other on the bottom rack. Remove the trays from the oven and use the handle of your wooden spoon to re-establish the indents. Return the pans to the oven, switching their positions on the racks, until golden brown and the cookie edges feel firm but soft to the touch, approximately 7 minutes. Remove from the oven and fill each indent with about 1 tsp of jam while still hot.

8. Store in an airtight container at room temperature for up to 5 days or in the freezer for up to 2 months.

Blueberry White Chocolate Biscotti

These biscotti are a popular item in our gift baskets and boxes because of their long shelf life. So many guests have told us how special they felt when they received a couple of cellophaned and bowed biscotti as a gift. These crunchy treats can be flavoured in many ways (see Wink for some ideas). Enjoy with a cup of coffee or tea, with a few dunks between sips, or with a glass of sweet dessert wine to wind down after a busy day.

MAKES 45 BISCOTTI

1 cup dried blueberries
1 cup salted butter, softened
2 cups granulated sugar
6 large eggs, at room temperature
2 Tbsp amaretto extract
6½ cups all-purpose flour
2 Tbsp baking powder
1 cup white chocolate chunks
1 cup white chocolate melts

1. Immerse the blueberries in hot (just boiled) water to soak for 30 minutes. Drain well.
2. Preheat your oven to 300°F. Line a baking sheet with parchment paper.
3. In a stand mixer fitted with the paddle attachment, cream the butter and sugar on medium speed until light and fluffy, 3–5 minutes.
4. Add all the eggs and amaretto extract and continue to mix on medium speed until well combined, approximately 3 minutes.
5. In a medium mixing bowl, sift the flour and baking powder. With the mixer running on low speed, gradually add the flour mixture to the butter mixture until fully combined, approximately 5 minutes.
6. Using a spatula, fold in the white chocolate chunks until evenly distributed.
7. Stir in blueberries with the same spatula; this will keep them from turning the batter too blue.
8. On the prepared baking sheet, divide the dough into three even pieces and form each piece into a ½ × 6-inch log.
9. Bake the biscotti until golden brown, approximately 30 minutes.
10. Remove from the oven and increase the oven temperature to 325°F.
11. Cool the biscotti until just able to handle, then cut crosswise into ½-inch slices. Place the slices back on the baking sheet.
12. Bake for 15 minutes, flip the biscotti, and continue to bake until lightly toasted, 5–15 minutes. Remove from the oven and let cool.
13. When the biscotti are completely cooled, place the white chocolate melts in a microwave-safe bowl. Microwave on low in 30-second intervals, stirring between each interval, until smooth and fully melted, then drizzle over one of the cut sides of the biscotti. Or, for extra indulgence, dunk these on one side in the melted chocolate.
14. Let the biscotti sit to allow the chocolate to harden, 20–30 minutes.
15. Store in an airtight container at room temperature for up to 2 weeks or in the freezer for up to 3 months.

WINK: *Biscotti can be flavoured with 2 Tbsp vanilla, almond, lemon, or anise extract in place of the listed amaretto. In step 7, you can add 1–2 tsp seeds of your choice, ½–1 cup nuts, other soaked, finely chopped dried fruits instead of blueberries, or 1 Tbsp lemon or orange zest.*

For Santa
Reindeer too,!!

Gingerbread People

I moved to Vancouver as a 17-year-old and lived in Kitsilano near the beach. I often walked along Yew Street, past a little bakery called Elsie's Bakery. As Christmas season approached that first year, I began to see little decorated gingerbread people and houses in the window. I had become a regular and asked Elsie if I could buy a gingerbread slab to cut out and decorate my own cookies and house. She couldn't resist my cost-saving approach and said yes. For the next 16 years, I ordered gingerbread slabs from Elsie, increasing the number of slabs as my family grew and our gingerbread creations went from gingerbread people and small houses to replicas of our home, school, and church. Elsie retired and sold her bakery to a young family, with a request that they make gingerbread for me each year. I moved to Langley and continued to order from the new owner until I decided I needed to (finally) develop my own recipe, which our guests love. It's fun to watch people decorate—their imaginations go wild! And it's equally delightful that Elsie and her family would visit the farm over the years!

MAKES APPROXIMATELY 35 GINGERBREAD PEOPLE

- 1½ cups salted butter
- 1½ cups packed brown sugar
- 1½ cups fancy molasses
- 3 large egg yolks, at room temperature
- 6 cups all-purpose flour
- 1½ tsp baking powder
- 1½ tsp baking soda
- 1 Tbsp ground ginger
- 2 tsp ground cloves
- 1½ tsp ground Saigon cinnamon
- 1½ tsp ground nutmeg
- 1 tsp table salt
- 1 batch Meringue Powder Royal Icing (page 278)
- 1 (7 oz /200 g) bag plain chocolate M&Ms

1. Preheat your oven to 325°F. Line two baking sheets with parchment paper.
2. In a stand mixer fitted with the paddle attachment, cream the butter and sugar until smooth. Add the molasses and egg yolks, mixing until well combined.
3. In a large bowl, gently whisk together the flour, baking powder, baking soda, ginger, cloves, cinnamon, nutmeg, and salt until well combined.
4. Gradually add the dry ingredients to the wet ingredients, about ¼ cup at a time, mixing between additions until just combined. Do not overmix. Divide the dough into six balls, flatten, and wrap in plastic wrap. Refrigerate for 1 hour.
5. On a clean surface dusted with flour, roll the dough to a ¼-inch thickness and use cookie cutters to cut the people shapes. Place the cut-out cookies on the prepared baking sheets, 2 inches apart.
6. Bake in batches until slightly soft and light brown, 6–8 minutes. Do not overbake. Allow the cookies to cool completely before decorating.
7. To decorate, use an angled spatula to place ½ cup royal icing into a piping bag fitted with a #2 round tip. Stand the piping bag in a tall cup and fold the top of the bag over the rim of the cup for easy filling. Twist the top of the piping bag closed and secure with a clip or elastic. If not using a #2 round tip, snip a small tip off the end of the piping bag and squeeze to push the icing down. Carefully pipe the icing to outline each cookie. Pipe two dots for the eyes and two or three dots for buttons, then cover each dot with an M&M. Refill piping bag as necessary. Any leftover icing can be left in the piping bag with a damp cloth over the tip and sealed in a baggie for a few days.
8. Store in an airtight container at room temperature or in the refrigerator for up to 1 week, or in the freezer for up to 3 months.

Chocolate "Gingerbread" People

This recipe is for those who don't like the spice of gingerbread but would still like to make "gingerbread" people—such as my kids and grandchildren, who prefer the taste of chocolate. We are a family-friendly farm and we want to offer gingerbread people for all ages. We also sell them in kits to take home for a decorating activity.

MAKES APPROXIMATELY 18 "GINGERBREAD" PEOPLE

1 cup salted butter
1½ cups granulated sugar
2 large eggs, at room temperature
1 tsp pure vanilla extract
2½ cups all-purpose flour
½ cup black cocoa powder (see page 201)
1 tsp baking powder
½ tsp table salt
1 (10 oz/283 g) bag plain chocolate M&Ms
½ batch Meringue Powder Royal Icing (page 278)

1. Preheat your oven to 350°F. Line two baking sheets with parchment paper.

2. In a large bowl with an electric mixer, cream the butter and sugar together.

3. Crack the eggs into a small bowl, add the vanilla, and whisk together. Add this egg mixture to the butter mixture and beat until light and fluffy.

4. In a medium bowl, whisk together the flour, cocoa powder, baking powder, and salt.

5. With the mixer running on low speed, gradually add the dry ingredients to the wet ingredients until the flour is fully incorporated.

6. On a clean surface dusted with flour, roll the dough out to a ½-inch thickness. Using a gingerbread-person cookie cutter, cut out your people. Place the cookies on the prepared baking sheets ½ inch apart, or they may end up holding hands! Reroll the dough up to two more times to get as many cookies as possible.

7. Bake until a toothpick inserted in the centre of a cookie comes out clean and the edges have darkened slightly, approximately 5 minutes. Once done, allow to cool before enjoying or decorating.

8. To decorate, use an angled spatula to place ½ cup royal icing into a piping bag fitted with a #2 round tip, standing the piping bag in a tall cup and folding the top of the bag over the rim of the cup for easy filling. Twist the top of the piping bag closed and secure with a clip or elastic. If not using a #2 round tip, snip a small tip off the end of the piping bag and squeeze to push the icing down. Carefully pipe the icing to outline each cookie. Pipe two dots for the eyes and two or three dots for buttons, then cover each dot with an M&M. Refill your piping bag as needed. Any leftover icing can be left in the piping bag with a damp cloth over the tip and sealed in a baggie for a few days.

9. Store decorated cookies in an airtight container for up to 1 week or in the freezer for up to 3 months.

WINK: *There are great instructions on how to fill and use piping bags online for beginners.*

Peter was not very well during the evening.
His mother put him to bed,
and made some camomile tea;
'One table-spoonful to be taken at bed-time.'

Cookies for Babies

My dad gave me his mother's recipe for baby teething cookies. The date on the recipe was 1956. It makes me smile to think of all the babies who have enjoyed these cookies through many generations! Don't worry about the sweeteners in these cookies. There's only a tiny amount in each one—and store-bought cookies will have much more sugar in them. I make them with a teddy bear cutter and serve them with a teddy bear I made for my youngest daughter, Tracey.

MAKES 48 (2-INCH) COOKIES OR 24 (4-INCH) COOKIES

¼ cup vegetable shortening, at room temperature
¼ cup lightly packed light or dark brown sugar
2 Tbsp medium-dark molasses
2 Tbsp Krause Berry Farms and Estate Winery Blueberry Syrup
2 large egg yolks room temperature, lightly beaten
2 tsp pure vanilla extract
1¾ cups all-purpose flour
3 tsp baking powder
½ tsp table salt

1. Preheat your oven to 350°F. Line a baking sheet with parchment paper or grease it with cooking spray.

2. In a large bowl with an electric mixer or by hand, cream the shortening and sugar together.

3. Add the molasses, blueberry syrup, egg yolks, and vanilla. Stir to combine.

4. Sift the flour, baking powder, and salt into a medium bowl. Mix the flour mixture ⅓ cup at a time into the sugar mixture, incorporating each addition before adding the next.

5. With clean hands, gather the dough together and work it into a ball. Divide the ball in half. Working on a lightly floured surface, roll each half out to a ¼-inch thickness.

6. Cut into your preferred shapes and place 1 inch apart on the prepared baking sheet.

7. Bake until golden brown, 10–12 minutes.

8. Store in an airtight container at room temperature for up to 5 days or in the freezer for up to 1 month. These can be given cold, not frozen, to an 8- to 12-month-old baby under supervision.

Give a Dog a Bone

Dogs have always been a large part of my life (see page 16). I couldn't possibly write a cookbook without including a special treat for our furry friends. We're very careful about what we feed our dogs, but we think they deserve a treat now and again, just like we do. Kale (pictured here), my daughter Sarah's son, with his family's puppy, Poppy. Kale has enjoyed baking with me since he was old enough to hold a spoon, and now Poppy, and his dog friends, get to enjoy his efforts in the kitchen!

MAKES 30 (3-INCH) DOG-BONE-SHAPED BISCUITS

- 1 tsp instant beef or chicken bouillon
- ½ cup just-boiled water
- 1 cup whole wheat flour
- 1 cup all-purpose flour
- ½ cup non-fat milk powder
- 1 Tbsp packed brown sugar
- ⅓ cup extra virgin olive oil
- 1 large egg room temperature
- ¼ cup fried and crumbled bacon pieces (4 strips)
- ¼ cup grated cheddar

1. Preheat your oven to 325°F. Grease a baking sheet or line it with parchment paper.
2. In a small bowl, mix the bouillon into the water until dissolved; set aside.
3. In a large bowl, stir the flours, milk powder, and sugar until well mixed. Add the oil, egg and dissolved bouillon, and mix to combine. Now add the bacon and cheese, mixing until well combined and a dough has formed.
4. Flour your work surface. Using your hands, form the dough into a ball and pat it down on the work surface.
5. Using a lightly floured rolling pin, roll out the dough to ⅛ inch thick. It's helpful to pat the edges as they separate, for a uniform circle.
6. Use any cookie cutter to cut out your cookies (mine are dog-bone-shaped). Place the cut-outs approximately ½ inch apart on the prepared baking sheet.
7. Bake until the biscuits are firm, 30–40 minutes.
8. Let cool completely on a wire rack and then give a dog a bone!
9. Store in an airtight container in the refrigerator for up to 4 days or in the freezer for up to 4 months. Thaw completely before serving.

WINK: *These dog bones are a great gift to welcome a new dog or to treat dogs that come to visit. I carry them in my pocket on walks to give to furry friends along the trail, providing their owner approves.*

Dog Bones
for
Oliver
From: Kale

BERRY FARMS & ESTATE

Gluten-Friendly Almond Crescents

Many years back, we had a wonderful staff member, Rita, who worked with us in the Harvest Kitchen (page 239). She made these little crescents for a friend who followed a gluten-free diet, and brought some in for us to taste. They became our first gluten-friendly option for guests. They are a dainty, buttery, two-bite cookie with an almond flavour, dusted in a delicate layer of icing sugar.

MAKES 32 CRESCENTS

- 2 cups almond meal
- 1 cup brown rice flour
- ½ cup sorghum flour
- ¼ cup cornstarch
- ¼ cup potato starch
- ¼ cup tapioca starch
- ½ tsp xanthan gum
- 1 cup salted butter
- ½ cup icing sugar + ½ cup for rolling (optional)
- 2½ tsp almond extract
- 1 cup dark chocolate chips, melted, for drizzling (optional)

1. Preheat your oven to 325°F. Grease a baking sheet or line it with parchment paper.
2. In a large bowl, combine the almond meal, rice flour, sorghum flour, cornstarch, potato starch, tapioca starch, and xanthan gum, and gently whisk to combine.
3. In a medium bowl with an electric mixer, cream the butter and ½ cup icing sugar until fluffy. Add the almond extract and mix well.
4. Add about one-quarter of the flour mixture to the creamed butter mixture and mix to combine. Continue adding the flour in that increment, mixing after each addition, until the dough is firm.
5. Divide the dough into 32 evenly sized pieces, rolling and shaping the portions into crescent shapes with your hands. Place the crescents on the prepared baking sheet. You can place them close together, as they don't spread.
6. Bake until golden, approximately 15 minutes.
7. If you wish, while the crescents are still warm but easy to handle, roll them in approximately 1 tsp icing sugar. Or let them cool completely and then drizzle with melted dark chocolate.
8. Store in an airtight container for up to 1 week or in the freezer for up to 3 months.

COOKIES

COOKIES

Cookie Jars on Parade

When you're standing in our farm market, be sure to look up. My collection of cookie jars is on display on the ceiling rail. My cookie-jar obsession began when I was tall enough to see over the counter in my grandmother Sarah Greene's kitchen. That was when I first met Smiley, who now sits on the third shelf, third jar to the right in the photo—the keeper of Grandma Greene's yummy cookies.

Smiley is a collector's piece from the pottery company Shawnee and was made in the 1940s in Zanesville, Ohio. Grandma Greene's favourite café in town was called The Chalet. The owner displayed her collection of cookie jars on a ceiling rail in the café. I knew that one day I would display cookie jars on a shelf in my kitchen too. On one occasion, I asked the owner if I could purchase a little pink kitten, an antique McCoy, that sat on her rail, but she explained that it was very special to her. Twenty-five years later, I found one just like it and gifted it to my cat-loving daughter Tracey.

The first cookie jar I bought was Cinderella, for my daughter Sarah. She was purchased in Martha's Vineyard outside Boston, Massachusetts. She sits on the second shelf, third to the right (pictured here in the top-right photo). Since then, I have collected cookie jars for all my children, my grandchildren, and the farm! Along with my family collection, I have 30 and growing strawberry-themed jars and dozens of jars that feature other berries, vegetables, ice cream cones, and other images that relate to the farm. My collection became too large to display in my Farmhouse Kitchen, so I now share it on the ceiling rail of the farm market for all to enjoy.

Hearty Farm Cookies

I stash these cookies in a drawer when I need to keep them from being eaten by the hollow-legged humans who roam my kitchen. They are packed with fibre-rich ingredients, making for a satisfying, chewy, and hearty snack. When my friends stop by for tea, I pull them from the drawer for us to enjoy. Of course, I'll have to think of a new hiding place now . . .

MAKES 24 COOKIES

1 cup salted butter, softened
2 cups packed brown sugar
2 cups granulated sugar
4 large eggs, at room temperature
1½ tsp pure vanilla extract
2½ cups all-purpose flour
1 tsp baking soda
½ tsp table salt
3 cups quick-cooking rolled oats
1 cup sweetened shredded coconut
½ cup dried blueberries
½ cup dried cranberries
½ cup wheat germ
¼ cup flaxseed
¼ cup wheat bran

1. Preheat your oven to 350°F. Line two baking sheets with parchment paper.
2. In a large bowl with an electric mixer, beat the butter until light and fluffy. Add both sugars and beat until fluffy once again. Add the eggs and vanilla, then cream until fluffy again.
3. In a separate large bowl, stir together the flour, baking soda, and salt.
4. Add the flour mixture to the creamed butter mixture in three increments, combining well with the electric mixer on medium speed between additions.
5. In a medium bowl, mix the oats, coconut, dried blueberries, dried cranberries, wheat germ, flaxseed, and wheat bran. Use a spatula to fold this mixture into the cookie dough.
6. Use a #40 cookie scoop (1½ Tbsp/1½ oz) to scoop the dough onto the baking sheets, leaving 1 inch between cookies.
7. Bake the cookies, one sheet on the top rack and one on the bottom, switching positions halfway through baking, until golden brown and a toothpick inserted comes out clean, approximately 8–11 minutes. Do not overbake. They should be chewy on the inside and crispy on the outside.
8. Store in an airtight container at room temperature for up to 1 week, or in the freezer for up to 3 months.

WINK: *You can also freeze these cookies unbaked. Place the balls of dough on a parchment-lined baking sheet and put in the freezer until solid, approximately 2 hours. Transfer to a freezer bag or container and return to the freezer for up to 3 months. Thaw at room temperature for 1 hour before baking as directed above. You'll have a stash on hand to bake up whenever the notion strikes!*

L–R: Hearty Farm Cookies (opposite), Berry Farm Cookies (page 134), and even more Hearty Farm Cookies!

Linzer Cookies

One year when planning for our Christmas season, I fondly remembered my trips to Austria, where I would peer into the bakery windows and see a beautiful display of Linzer cookies. They are an elegant buttery cookie, filled with sweet farm jelly and finished with icing sugar, making them a perfect cookie to enjoy with tea or to package in Christmas tins for gift giving.

MAKES 15 COOKIES AND 15 MINI COOKIES

1 batch Sugar Cookies dough (page 167)

1 (9 oz/260 ml) jar Krause Berry Farms raspberry jelly

1 cup icing sugar, for dusting

1. Preheat your oven to 350°F. Line two baking sheets with parchment paper.

2. Dust a clean work surface with flour and roll out the dough to ⅛ inch thick. It's important to make sure the thickness is consistent.

3. Cut out rounds with a 3-inch round cookie cutter, rerolling the dough as needed to make 30 dough rounds.

4. Place half of the circles onto one of the prepared baking sheets about 1 inch apart. Bake until the edges are beginning to brown, 6–7 minutes.

5. Place the other half of the circles onto a second baking sheet. Press out the centre with the cookie-cutter shape of your choice—it just needs to fit inside the circle and leave the edges intact. You'll see from the photo that we use a 2-inch six-pointed star or heart for this. Place the cut-out pieces on the baking sheet to make mini cookies.

6. Bake the second half of the cookies until the edges are beginning to brown, 4–5 minutes. The mini cookies will be ready just before the regular ones. Baked, unassembled cookies can be frozen for up to 3 months and can be taken out to assemble and enjoy at any time. Allow them to thaw for 1 hour before assembling.

7. When the freshly baked cookies are cool or frozen ones are thawed, spread the jelly on the bottom side of the full-circle cookies, leaving a ¼-inch margin on the outer rim of the cookie. You'll need approximately 1 Tbsp jelly for each cookie.

8. Place the icing sugar in a fine-mesh sieve and sprinkle the sugar on the cut-out cookies. Gently tap these on the counter before placing the cut-out cookies on top of the bottoms with the jelly.

9. Store assembled cookies in an airtight container at room temperature or in the refrigerator for up to 1 week.

WINK: *When making jelly, you must first cook down the berries until soft (see page 228), then strain through a sieve with cheesecloth to remove the pulp and seeds, leaving the juice. When making jams, you use whole berries, mashed. When you leave larger berry pieces, you have what's called preserves. We use jelly for these cookies to create the look of a stained-glass window.*

Macaroons

MAKES 24 MACAROONS

1¾ cups unsweetened shredded coconut

½ cup + 2 Tbsp sweetened condensed milk

2 tsp pure vanilla extract

1 large egg white, at room temperature

¼ tsp table salt

¼ cup blue, pink, purple, or dark chocolate candy melts

I have a love thing for coconut. Maybe it's because I grew up watching *Gilligan's Island*, but it's more likely due to the winter of 1967. My mother (pictured on page 195) was crowned Mrs. Oregon that year and participated in the Mrs. America Pageant. It was one of the happiest years of my childhood, and for a very good reason. One category of the pageant was homemaking, which included baking. Among the many items my mother perfected for the contest was a fluffy white bunny cake with a coconut frosting. She made many practice cakes, improving them each time, and I was the beneficiary of each one. The judges must have loved that coconut-frosted cake as much as I did because she came in third out of 50 women.

I wanted so badly to make the bunny cake at the farm, but the market and bakery aren't open at Easter and the chances of a bunny cake selling in the summer are nil, so I settled on coconut macaroons with berry-flavoured icing drizzled on top. These bite-sized cookies are soft and chewy on the inside, crispy on the outside, and bursting with the flavour of sweet coconut and chocolate.

1. Line a baking sheet with parchment paper or foil and spray lightly with a non-stick cooking spray.

2. In a large bowl, place the coconut, condensed milk, vanilla, egg white, and salt. Using a spatula, mix to combine, 2–3 minutes. The mixture will be scoopable (not too hard, not too soft).

3. Using a large spoon, scoop and form a heaping tablespoon of the mixture into pyramid-shaped mounds and place on the prepared baking sheets about 1 inch apart. The unbaked coconut mounds can be frozen at this point. Freeze them on the baking sheets for at least 1 hour, then transfer to an airtight container and return to the freezer for up to 3 months.

4. Preheat your oven to 300°F, if baking freshly made macaroons, or to 325°F, if baking from frozen.

5. Bake the macaroons until golden brown on the top and the bottom and chewy in the middle, approximately 15–20 minutes. Watch them carefully, though, as they can burn quickly. Let cool completely on the baking sheet.

6. Place the candy melts in a large microwave-safe bowl and melt in the microwave, heating in 30-second intervals and stirring for at least 30 seconds between each interval to ensure the candy melts do not burn.

7. Using a spoon, drizzle the melted candy melts over the macaroons. Let sit until the candy melts have set.

8. Store in an airtight container at room temperature or in the refrigerator for up to 1 week.

WINK: *It is easy to flavour these cookies like an Almond Joy candy bar by replacing the vanilla with almond flavouring. Make an indent in the top centre before putting in the oven. Then, when cooled, drizzle with dark chocolate and place an almond in the indent. For Easter, drizzle with pastel-coloured chocolate as shown in the photo and place a malted chocolate egg in the indent.*

Reindeer Cookies

My kids and grandkids love this version of a peanut butter cookie, and so does Santa! The pretzel antlers make a nice salty contrast with the sweet candies. Definitely a recipe to make with the kids in your life!

MAKES 32 COOKIES

1 cup salted butter, at room temperature
1 cup smooth peanut butter
1 cup granulated sugar
1 cup packed brown sugar
2 large eggs, at room temperature
1 tsp pure vanilla extract
2¾ cups all-purpose flour
1 tsp baking powder
1 tsp baking soda
64 brown Smarties
32 red Smarties
½ cup milk or dark chocolate chips
32 hard knot-shaped pretzels

1. Preheat your oven to 350°F. Line two baking sheets with parchment paper.

2. In a large bowl and using a large spoon or an electric mixer, cream the butter, peanut butter, and both sugars until light and fluffy.

3. Add both eggs and the vanilla. Mix until fully blended.

4. In a medium bowl, gently whisk together the flour, baking powder, and baking soda. Gradually add the dry ingredients to the butter mixture, approximately ¼ cup at a time, and mix to combine between additions.

5. Once smooth and well mixed, portion the dough out into ¼-cup balls and place them approximately 3 inches apart on the prepared baking sheets.

6. Flatten the balls by pressing down on them with another baking sheet until the dough is ¼ inch thick.

7. Bake until lightly browned, approximately 6 minutes.

8. While the cookies are still warm and fresh out of the oven, press 2 brown Smarties and a red Smartie into each one for the eyes and nose. The cookies will firm up when cooled.

9. Melt the chocolate in the microwave at 30-second intervals until smooth and silky, stirring between each interval. Break each pretzel in half to form two antlers. Dip the bottom of the pretzel half in the melted chocolate to glue the pretzel to the top of the cookie above the eyes, as seen in the photo.

10. Store in an airtight container at room temperature for up to 5 days or in the freezer for up to 2 months.

Shortbread Rounds

Many of my recipes in this book, and many of the recipes we use at the farm, have been passed down through the generations in my family—from my grandmothers, mother and father, mother-in-law, and aunts. Others have been passed on to me through friends and acquaintances. This shortbread recipe was given to me as a Christmas gift from my friend Donna, and her gift continues to give. Even after 30 years, this recipe remains one of my favourites. This rich, buttery shortbread is crunchy on the outside and melt-in-your-mouth perfect on the inside. We bake it into large rounds, wrapping some with cellophane and a bow and cutting others into slices for those who want to indulge right on the spot.

MAKES THREE (8-INCH) ROUNDS OR 24 WEDGES

4 cups salted butter, at room temperature
4 cups granulated sugar
10½ cups all-purpose flour
Approximately 6 Tbsp granulated sugar, for topping (optional)
1 cup melted dark chocolate, for topping (optional)

1. Preheat your oven to 300°F. Spray or grease three 8-inch round cake pans.

2. In a large bowl with an electric mixer, cream the butter and sugar until light and fluffy.

3. Gradually add the flour, about ½ cup at a time, to the creamed butter mixture and mix on low speed until all the flour is blended in.

4. Divide the dough into three. Spread each dough portion evenly into the pans, pressing down firmly, ensuring it goes right to the edges of the pan.

5. Score each round to make eight wedges by inserting a paring knife ¼ inch into the top of each round. This will make cutting easier once cooled.

6. Bake until lightly browned around the edges and a toothpick inserted into the centre comes out clean, approximately 40 minutes.

7. The shortbread rounds are delicious as is, but you can decorate them with either sugar or chocolate if you wish.

8. If you're using sugar, sprinkle the shortbread round with just enough, approximately 2 Tbsp, to cover the top right after you take the shortbread out of the oven. Cool completely in the pan (approximately 1 hour), then carefully flip the rounds onto a baking sheet. At this point, any excess sugar will fall off. You can cut the rounds into wedges, following the scored lines, or leave them as large rounds.

9. If you're using chocolate, let the shortbread rounds cool completely in the pan (approximately 1 hour), then carefully flip them onto a baking sheet. Drizzle the melted chocolate over top and let the shortbread sit to allow the chocolate to set. You can cut the rounds into wedges, following the scored lines, or leave them as large rounds.

10. No matter what shape and decorating option you choose for your rounds, they will freeze beautifully for up to 3 months if wrapped in plastic wrap once cooled and placed in an airtight container. Thaw for 1 hour before eating or before wrapping as a gift.

WINK: *These make wonderful holiday gifts. At the farm, we sell them wrapped in cellophane as sugared rounds, chocolate-drizzled rounds, and sugared or drizzled wedges.*

Snickerdoodles (page 164)

Strawberry Rhubarb Loaves (page 197)

Carrot Cake (page 186)

Carrot Cake (page 186)

Cottage Kitchen

In the season of magnificent chaos—otherwise known as summer—Alf and I escape the farm and head north across the berry fields into the woods. There awaits a tiny white cottage with a tiny white kitchen, a big-shaded porch, and a swing made for two. Everyone needs a little time alone.

The Cottage Kitchen is where I cook and bake for tea parties and outside harvest dinners for friends and family. As a mother to six and grandmother to five, it's also where I enjoy sharing my cooking tips and recipes with my young granddaughters and grandsons and revisit traditions with my daughters. We make our favourites, like Grandma Greene's Blueberry Banana Bread (page 193) and Grandma Buehler's Poppy Seed Cake (page 190).

One of my favourite things to do for tea parties is to make mini versions of baked goods with my loved ones. For example, as pictured here, I enjoyed making Snickerdoodles (page 164) with my grandaughters Brynn (left) and Emme (right), the Strawberry Rhubarb Loaves (page 197) with my daughter Sarah (their mother), the mini cupcake version of my Carrot Cake (page 186) with my daughter Holly, and the Poppyseed Cake (page 190) and the large version of the Carrot Cake (page 186) with Tracey. Then my grandchildren often run off to play checkers with the mini snickerdoodles we made together after filling up on tarts, sandwiches, and cakes. The sound of their laughter is music to my ears.

Snickerdoodles

I am grateful I was able to babysit my three grandchildren a couple of days a week before they entered elementary school while my daughter Sarah and her husband, Ryan, were at work as teachers. My grandchildren are the icing on the cake of my life. We donned our aprons, sifted, cracked, stirred, tasted, rolled, sprinkled, baked . . . and ate. We enjoyed licking spoons, spatulas, and bowls for hours as we prepared treats to taste and share with Mommy and Daddy when they arrived home. I love sharing the baking (and eating!) of so many of my childhood-favourite recipes with them, including these snickerdoodles. I also love being creative with add-ins, such as pink and purple sprinkles and sparkles, to delight my grandchildren.

MAKES 12 LARGE OR 20 MEDIUM COOKIES

- ¼ cup salted butter, at room temperature
- ¼ cup vegetable shortening, at room temperature
- ¾ cup + 1 Tbsp granulated sugar, divided
- 1 large egg, at room temperature
- ¼ tsp pure vanilla extract
- 1⅓ cups all-purpose flour
- 1 tsp cream of tartar
- ½ tsp baking soda
- ¼ tsp table salt
- 1 tsp ground Saigon cinnamon

1. Line two baking sheets with parchment paper.
2. In a large bowl with an electric mixer, beat the butter and shortening until well combined. With the mixer running on low speed, slowly beat in ¾ cup sugar until combined and light yellow.
3. Beat in the egg, then add the vanilla.
4. In a separate bowl, combine the flour, cream of tartar, baking soda, and salt. Add to the wet ingredients in three increments, mixing well to incorporate the flour between additions. Chill the dough for 15 minutes in the refrigerator.
5. Meanwhile, place the remaining 1 Tbsp sugar in a small bowl, add the cinnamon, and mix.
6. Preheat your oven to 400°F.
7. Scoop out the dough with a #60 (2 tsp/¾ oz) cookie scoop (see Wink). Using two teaspoons, toss each ball in the cinnamon-sugar mixture to coat well, then place them 2 inches apart on the prepared baking sheets. Lightly press each cookie with your fingers; this will prevent the centres from puffing while baking.
8. Bake until the cookies form light cracks on their surfaces, 7–9 minutes. Do not overbake.
9. Let cool for a minute or two and then remove to racks to cool completely.
10. Store in an airtight container at room temperature for up to 5 days or in the freezer for up to 2 months

WINK: *Make one dough ball the right size and place it where you can compare it to all the others so that the balls are a consistent size. This gives young bakers a visual guide for forming their cookie balls.*

SWEETIE
PIE
I'M
YOURS
XOXO
BE
MINE
KISSES

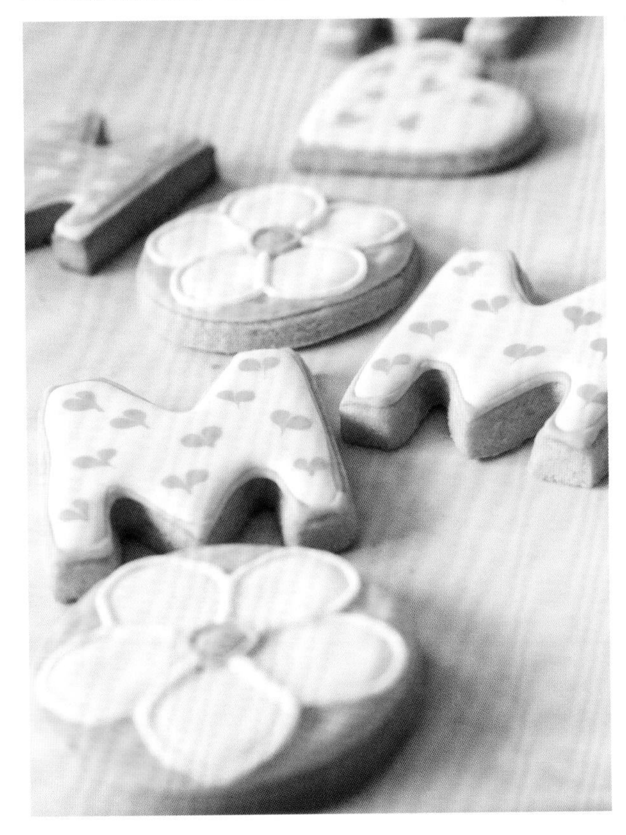

Sugar Cookies

On a visit to California in the early 1980s, my mother took me to a farm called the Nut Tree (long since closed, unfortunately). She told me the drive would be well worth it, as she knew how much I loved cookies. When we walked in the front door, I stood frozen looking up at a display wall of what seemed to be hundreds of cookies decorated in a vast assortment of designs and colours. How would I possibly choose just one cookie? On trips home to California, I worked in a drive to the farm when it fit the agenda, to stare in awe at the cookie wall, eventually choosing one. The Nut Tree's cookies were the inspiration for the decorated ones we make at our farm today. They take patience and a steady hand, but the results are masterpieces!

MAKES 18–24 COOKIES, DEPENDING ON THE SIZE OF YOUR CUTTER

1 cup salted butter, at room temperature

1½ cups granulated sugar

2 large eggs, at room temperature

1 tsp pure vanilla extract

3 cups all-purpose flour

1 tsp baking powder

½ tsp table salt

1. Preheat your oven to 350°F. Line two baking sheets with parchment paper or grease with cooking spray, shortening, or butter.

2. In a large bowl with an electric mixer, cream the butter and sugar together until light and fluffy. Add the eggs one at a time, beating well after each addition. Add the vanilla and mix until fully incorporated.

3. In a medium bowl, mix the flour, baking powder, and salt.

4. Add the dry ingredients to the butter mixture. With the electric mixer on low speed or by hand with a spatula, mix until fully incorporated.

5. Dust a clean work surface with a little flour and roll out the dough to ¼–⅜ inch thick. These puff up when baked, so don't be tempted to make them any thicker.

6. Use your preferred cookie cutters to cut out your cookies. Dip the cookie cutter into flour to prevent the dough from sticking to it. Excess dough can be rerolled up to three times before it gets too worked, which will cause shrinkage when baking.

7. Lay the cookies out on the prepared baking sheets. Bake with one sheet on the top rack and one on the bottom, swapping positions halfway through baking, until the edges are golden brown, approximately 10 minutes.

8. Let the cookies cool before you decorate them according to your preference. You can also freeze the cookies before decorating them in an airtight container for up to 3 months. Just let them come to room temperature a bit first.

9. Store in an airtight container at room temperature for up to 1 week.

WINK: *You'll see from the photo that Krause Berry Farms cookies are decorated in a variety of ways. Don't feel you need to copy what we've done. Use our Meringue Powder Royal Icing (page 278) and let your imagination run wild. I could write an entire book about decorating cookies, but for now I suggest you explore some of the amazing tutorials available online.*

Colourful Nanaimo Bars

Nanaimo bars originated on Vancouver Island in the city of Nanaimo, BC. I imagine many lunch boxes contained these hearty and delicious treats as the fishermen set off from the harbour for their daily catch. This recipe was introduced to our Specialty Foods Kitchen (page 198) by Leslie, who came to us from Vancouver Island. We customized her recipe by using the crops we grow to add beautiful colour and serious new flavours! (Note: You'll need a candy thermometer for this recipe. See page 25.)

MAKES 16 BARS

Bottom layer

1 cup salted butter

⅓ cup granulated sugar

2 large eggs, at room temperature, lightly whisked

1¾ cups graham crumbs

⅓ cup cocoa powder

2 cups sliced almonds

1¼ cups sweetened flaked coconut

Middle layer

1¼ cups salted butter

½ cup Krause Berry Farms and Estate Winery Jam (any flavour)

2¼ cups icing sugar

Top layer

2 cups dark chocolate chips

1 Tbsp + 1 tsp salted butter

MAKE THE BOTTOM LAYER

1. Line an 8-inch square baking pan with parchment paper leaving an overhang for lifting.

2. In a double boiler on medium heat, melt the butter and sugar.

3. Once melted, add the whisked eggs and continue whisking on medium heat until the temperature of the mixture reaches 70°F. Remove from heat.

4. Add the graham crumbs, cocoa, almonds, and coconut and mix until incorporated.

5. Spread the mixture into the prepared pan and chill, uncovered, for 30 minutes.

MAKE THE MIDDLE LAYER

1. Meanwhile, in a medium bowl with an electric mixer, beat the butter and jam together on high speed.

2. Beat in the icing sugar until emulsified.

3. Once the bottom layer has finished chilling, spread the middle layer smoothly over the bottom. Return to the refrigerator to chill, uncovered, for another 30 minutes.

MAKE THE TOP LAYER

1. In a double boiler on low to medium heat, melt the chocolate and butter together, stirring until well combined.

2. Once the second layer has finished chilling, pour the chocolate over the second layer and spread evenly. Chill, uncovered, overnight.

3. Using the two end of hanging parchment paper, carefully lift the bars out of the baking pan. Score the chocolate layer using a hot knife, cleaning the knife between scores. Now cut through all three layers into 15 bars (or your desired size).

4. Store in an airtight container in the refrigerator for up to 3 days or in the freezer for up to 3 months.

Berry Cream Cake (page 172)

Cakes and Desserts

Berry Cream Cake

Although we are known for our pies, we wanted to have a signature berry cake so we could offer both to our guests! As luck would have it, I met fourth-generation Master Pastry Chef Marco Röpke at a Christmas party many years ago. As I described my perfect cake to complement our berries, his eyes lit up and he said he would love to share his passion and knowledge by coming to the Harvest Kitchen (page 239) and showing us how to make this perfect cake. Marco owns and teaches at the Pastry Training Centre of Vancouver, which offers serious and intense hands-on courses for both professionals and home chefs. What a privilege it was to learn from him.

We bake this cake on two large baking sheets for specialty cakes such as wedding, anniversary, and birthday cakes. I've amended the recipe to make a smaller cake, but if you'd like to make it the way we do, just double the recipe and use two baking sheets instead of one. You can also cut the cake to make two smaller two-layer cakes, if you like. The photo shows this cake filled with fresh berries and whipped cream, glazed with berry syrup, and topped with seasonal berries, but we decorate it to match the occasion that it's celebrating.

MAKES ONE (8 × 12-INCH) TWO-LAYER CAKE, TRIMMED

Cake

- 6 large eggs, at room temperature, separated
- 1 cup granulated sugar, divided
- ⅓ tsp finely grated lemon zest
- 2 cups cake or pastry flour, sifted
- 4 cups fresh berries (if using strawberries, use 2 cups whole berries + 2 cups sliced), divided

Berry soaking syrup

- 2 cups fresh or frozen berries (strawberries, blueberries, raspberries, or blackberries)
- ⅛ cup granulated sugar
- ⅛ cup boiling water
- ¼ cup cold water
- 2 drops fresh lemon juice

Whipped cream

- 2 cups whipping (35%) cream
- ¼ cup icing sugar

MAKE THE CAKE

1. Preheat your oven to 350°F. Line a 13 × 18-inch baking sheet with parchment paper.
2. In a stand mixer fitted with the whisk attachment, whip the egg whites with ¼ cup of the sugar on high speed. With the mixer running, gradually add the remaining sugar. Once all of the sugar is incorporated, whisk for another 2 minutes or until medium peaks form.
3. In a large mixing bowl, use a spatula to combine the egg yolks and lemon zest.
4. Using a clean spatula, gently fold the whipped egg whites into the egg yolk mixture. Gradually fold in the flour until completely mixed.
5. Spread the batter evenly in the prepared baking sheet. Bake until golden and a toothpick inserted in the centre comes out clean, approximately 13 minutes.
6. Let the cake cool in the pan on a wire rack for 20 minutes, then turn it out onto the rack and flip it over so it's parchment side down. Let cool completely, approximately 15 minutes. Cut the cake in half horizontally for 1 cake or cut each half lengthwise, making 4 pieces.

MAKE THE SOAKING SYRUP

1. Place the fresh or frozen berries in a blender and blend until smooth.
2. Place the sugar in a small bowl and add the boiling water. Stir until the sugar is dissolved. Stir in ¾ cup berry purée, cold water, and lemon juice until combined.

MAKE THE WHIPPED CREAM

1. Place a large mixing bowl in the freezer for 15 minutes.

2. Remove the bowl from the freezer and pour in the whipping cream. Using an electric mixer, whip the cream until it reaches soft peaks. Add the icing sugar and continue to whip until stiff peaks form.

ASSEMBLE THE CAKE

1. Place one cake half, parchment side up, on a serving tray. Peel off the parchment paper.

2. Brush the cake with half of the soaking syrup.

3. Spread one-third of the whipped cream over the surface of the cake and sprinkle 2 cups of fresh berries over top. Spread half of the remaining whipped cream over the berries. Make sure to spread the berries and cream right to the edge of the cake so it is level and even when cut.

4. Place the second cake on top of the first and peel off the parchment paper.

5. Brush the top of the cake with the remaining syrup. Spread the remaining whipped cream on top of the cake and garnish with the remaining berries in your desired design.

6. Using a serrated knife, carefully trim ½ inch off all sides of the cake for a uniform presentation.

7. This cake is best served fresh, but it will keep in the refrigerator for up to 3 days. Place toothpicks around the edges of the cake and drape plastic wrap over the toothpicks, ensuring you don't disturb the berries and whipped cream.

WINK: *As the baker, you have the perk of nibbling on the cake trimmings once you've assembled the main cake!*

WINERY
KRAUSE
FARMS & ESTATE

Berry Merry Christmas Fruitcake

This recipe is the ultimate in fruitcakes and comes with a 100% satisfaction guarantee! If you haven't cared for fruitcake in the past, please take the time to make this one—you won't be disappointed. Thanks go to my friend Alfred Voss, a retired pastry chef, for this recipe. He popped into the Harvest Kitchen (page 239) one day and out came this beautiful fruitcake! It uses different ingredients than you might find on traditional store bought fruitcakes, because we wanted to bust the myths people may have about fruitcakes. It also uses quality ingredients, and I like to present them wrapped and bowed for gift giving for the holidays. I shipped it to my mother in California for years. She kept it in her refrigerator and sliced the tiniest bit off throughout the year, savouring its richness while awaiting the next annual shipment. Don't forget to keep one hidden somewhere safe for yourself to indulge in throughout the holidays!

MAKES TWO (4½ × 5-INCH) CAKES OR ONE (5 × 9-INCH) CAKE

Cake

- 1 cup raisins
- 1 cup chopped mixed dried orange and lemon peels
- ⅔ cup dried blueberries
- ⅔ cup dried cranberries
- ½ cup chopped dried cherries (not maraschino cherries)
- 1 cup chopped walnuts
- 1 cup chopped pecans
- 1 Tbsp + 1 cup all-purpose flour, divided
- 1 cup packed brown sugar
- 2 large eggs, at room temperature
- ¼ cup salted butter, softened
- ½ tsp pure vanilla extract
- ⅓ cup Simple Syrup (page 177)
- ¼ cup brandy or rum (optional)
- ⅛ cup blackstrap molasses
- ½ tsp baking powder
- ⅛ tsp baking soda
- ⅛ tsp table salt
- ¼ tsp ground Saigon cinnamon
- ⅛ tsp ground allspice
- ⅛ tsp ground mace
- 1/16 tsp ground cloves

For decorating

- 3 Tbsp water + more for the double boiler
- 1 Tbsp envelope gelatin
- ½ Tbsp vegetable shortening
- ½ Tbsp clear vanilla extract

MAKE THE CAKE

1. In a large bowl, combine the raisins, orange and lemon peels, dried blueberries, dried cranberries, dried cherries, walnuts, and pecans. Sprinkle with 1 Tbsp flour, toss to coat the ingredients, and set aside.

2. Preheat your oven to 250°F and grease a 5 × 9-inch loaf pan.

3. In a medium bowl, mix together the sugar, eggs, butter, and vanilla until creamy. Add the simple syrup, brandy (if using), and molasses. Combine well.

4. In a small bowl, mix together the remaining 1 cup flour and the baking powder, baking soda, salt, cinnamon, allspice, mace, and cloves. Add these dry ingredients to the wet ingredients gradually until well combined. The batter will be quite thick and clumpy, almost like a dough. Lastly, add the dried fruits and nuts mixture to the batter.

5. Pour the batter into the prepared loaf pan. With a spatula, press the batter into the pan, making sure to push it into the corners. Make a small indent in the top centre of the dough. This will help it bake flat instead of puffing up in the centre.

6. Bake until you see the cake is slightly darker than golden (but not dark brown), has slightly pulled away from the pan, and a toothpick inserted in the centre of the cake comes out clean, approximately 3 hours.

7. I cut the loaf in half and decorate each half like a little present, as shown in the photo, but you can leave the cake uncovered if you prefer.

DECORATE THE CAKE

1. To decorate as in the photo, first make the fondant. Fill the bottom of a double boiler with approximately 2 inches of water and bring to a boil. Place the top of the double boiler over the bottom part. Add 3 Tbsp water and then the gelatin. Turn the heat to medium-high. Stir with a spatula until the mixture is smooth and clear, approximately 2 minutes.

Recipe continues

¼ cup white corn syrup
5 cups icing sugar + ¼ cup for hands, work surface, and rolling pin
2 (6-inch square) cake boards
1 cup Meringue Powder Royal Icing (page 278)
74 inches ½-inch-wide ribbon
36 silver dragées

2. Add the shortening and continue to stir with a spatula for 1 more minute, just to let it blend a little. Add the vanilla and corn syrup. Continue stirring for 3 minutes, until the mixture is clear. You will see little oil bubbles on the surface. Remove from heat.

3. Sift in the icing sugar in four or five additions by pressing the sugar against the sides of the sieve into the gelatin mixture, stirring in the sugar with the spatula after each addition.

4. Rub some icing sugar into your hands, gather the mixture into a ball, and knead until it is smooth and pliable, approximately 5 minutes. Sprinkle your work surface and rolling pin with icing sugar. Cut the ball in half and gently roll out each half to ¼ inch thick and approximately 9 inches square.

5. Centre each cake on a 6-inch square cake board. Place a piece of fondant on each cake to completely cover the top and sides. Cut away excess with clean scissors where the cake meets the board.

6. Use an angled spatula to place the royal icing into a piping bag fitted with a #2 round tip, standing the piping bag in a tall cup and folding the top of the bag over the rim of the cup for easy filling. Twist the top of the piping bag closed and secure with a clip or elastic. If not using a #2 round tip, snip a small tip off the end of the piping bag and squeeze to push the royal icing down.

7. Cut the ribbon into four 9-inch lengths and two 19-inch lengths. Place one 9-inch ribbon across the centre of the cake, letting it drape down to touch the base. Secure it to the cake with a dot of royal icing. Place another 9-inch length across the cake perpendicular to the first ribbon, making a cross. Secure it to the cake with a dot of royal icing. Repeat with the other cake.

8. Pipe dots of royal icing to decorate around the base of each cake and three dots on all four sides of each. Then pipe decorative curls on the top of each cake as shown. Place silver dragées in the centre of the decorative curls as shown. Place more silver dragées in the centre of the piped dots on the four sides of each cake.

9. Tie two bows with the remaining 19-inch ribbon lengths. Dab royal icing on the back of the centre of the bows and place one bow on the centre of each cake.

10. Because these cakes are sealed in fondant, they will keep in an airtight container on either the counter, in the refrigerator, or in the freezer, for up to 1 year. If you don't cover them in fondant and royal icing, they will keep in an airtight container in the refrigerator for up to 6 months or in the freezer for up to 1 year. Wrap any leftover fondant pieces in plastic and place in a sealed bag. Store at room temperature in a dry location for up to 2 months.

WINK: *Using white shortening, clear vanilla, and white corn syrup will keep your fondant looking white.*

Christmas Stollen

My friend and retired chef Alfred Voss created this farm-favourite recipe. I wanted a stollen that was rich and authentic. Alfred is not only German but also an incredibly skilled baker and pastry chef. My kitchen became a classroom the day he came in to perfect the berry stollen with his authentic techniques, our farm-grown berries, and both of our German heritages. This stollen has a special place in my heart. I make it every Christmas and slice it on Grandma Greene's beautiful antique blue, wooden, round breadboard with its matching knife. It speaks to her memory and our German heritage. (Note: You need to start this recipe the day before you plan to bake it.)

MAKES 4 LOAVES

Fruit and nut filling
1½ cups dried blueberries
1¾ cups dried cranberries
¾ cup chopped dried cherries
1 cup slivered almonds

Simple syrup
⅓ cup boiling water
⅓ cup granulated sugar

Marzipan
1½ cups almond meal
⅔ cup superfine sugar
1 large egg white, at room temperature

Dough
4 cups all-purpose flour, divided, + more as needed
1 tsp instant yeast
½ cup 2% milk
1 cup salted butter, at room temperature
½ cup granulated sugar
½ tsp table salt
½ tsp ground Saigon cinnamon
½ tsp ground cardamom
½ tsp ground coriander
⅛ tsp ground fennel
⅛ tsp ground anise
⅛ tsp ground nutmeg
⅛ tsp baking powder
1½ tsp pure vanilla extract
1 tsp lemon extract

Topping
½ cup salted butter, melted
½ cup granulated sugar
¼ cup icing sugar

MAKE THE FRUIT AND NUT FILLING

1. In a large bowl, combine the dried blueberries, dried cranberries, dried cherries, and almonds. Pour the simple syrup over top, cover the bowl, and soak the fruit overnight at room temperature.

MAKE THE SIMPLE SYRUP

1. In a glass measuring cup or small bowl, whisk the water and sugar until the sugar has dissolved. Set aside and let cool.

MAKE THE MARZIPAN

1. In a medium bowl, mix the almond meal, sugar, and egg white until they bind together and the mixture is smooth. Separate it into four 5-inch lengths, each about ½ inch in diameter.

2. Cover with plastic wrap and set aside at room temperature until ready to use. If you make the marzipan ahead of time, refrigerate it and then bring it to room temperature before using.

MAKE THE DOUGH

1. Place 1½ cups flour in the bowl of a stand mixer fitted with the dough hook. Add the yeast and milk and mix on low speed until well combined. The batter will be stiff. Cover the bowl with parchment paper or a clean tea towel and allow to rest for 30 minutes.

2. After resting, add the remaining 2½ cups flour and the butter, sugar, salt, cinnamon, cardamom, coriander, fennel, anise, nutmeg, baking powder, vanilla, and lemon extract. With the dough hook, mix for 4–5 minutes on low speed or until the flour is fully incorporated. Let rest for another 20 minutes, uncovered this time.

3. Gradually add the soaked fruit and nut mixture into the dough until incorporated. Add 1 Tbsp flour at a time if the dough is too wet. It should be soft but not sticky.

Recipe continues

4. Divide the dough into four evenly sized balls. Knead each gently on a clean, floured surface to form 6-inch rounds. Transfer the rounds to a greased baking sheet and let rest for 30 minutes, covered with parchment or a tea towel.

5. Once rested, shape the dough rounds into ovals and place back onto the greased baking sheet. Make a depression with a small rolling pin just off centre along the length of one side of the oval. Place 1 length of marzipan into the depression, fold the dough over, and seal it around the marzipan. Repeat with the remaining ovals and marzipan. Let rest, uncovered, for 30 minutes.

6. Preheat your oven to 325°F.

7. Bake until browned on the bottom, approximately 50 minutes.

8. Remove the stollen from the oven to cool. You'll see bits and pieces of over-cooked fruits. Just pick them off and discard them. While the stollen are cooling on the baking sheet but still warm, brush the top of each cake with 1 Tbsp melted butter. Cool completely, approximately 1 hour.

9. Reheat the remaining melted butter. Brush it along the tops of the stollen, then sprinkle with granulated sugar and generously dust with icing sugar.

10. Wrap tightly in plastic wrap and store at room temperature for up to 2 weeks. You can also wrap it tightly in plastic, place it in a freezer bag, and freeze for up to 2 months.

WINK: *Wrapped in cellophane with bows, these make wonderful gifts!*

Bread

Berry Shortcakes

MAKES SIX (4-INCH) SHORTCAKES

Shortcakes
2 large eggs, at room temperature
⅔ cup buttermilk
½ cup vegetable shortening
¼ cup granulated sugar
3 cups all-purpose flour
2 tsp baking powder
1 tsp table salt
3½ cups fresh berries, divided (strawberries need to be quartered before measuring)
2 tsp ground Saigon cinnamon

Icing
1 cup icing sugar
2 Tbsp water

We describe these as sweet biscuits bursting with the summer flavour of berries. When guests step up to the bakery counter in our market and start gesturing with their hands to describe our shortcakes, we know a friend must have sent them and they have forgotten what they are called. When we first started making these, Pauline and I baked and sold them from The Porch (page 248). We got to know what our regular guests liked to order, and as we watched them making their way to The Porch we would prepare their order, so it was ready by the time they got to the serving window. On one of those earlier days 20-odd years ago, there was a couple who came each Friday before lunch. One day the man told us his wife had lost 50 lb while eating our shortcakes. Trying to remain calm as we each collected our wild thoughts, we asked how this could be possible. He glanced at her and proudly said she had committed to losing weight and had joined Weight Watchers. He went on to explain how she would weigh in each Friday morning, and if she had met her weekly weight-loss goal, she would be able to enjoy a treat of her choice. Each week during her 50-lb weight loss, she had chosen to come to our farm and enjoy a berry shortcake. We have never heard a story quite like that one again, and we felt very honoured to be the reward for her hard-earned achievement!

MAKE THE SHORTCAKES

1. Preheat your oven to 350°F. Spray a baking sheet with cooking spray or line it with parchment paper.

2. In a stand mixer fitted with the paddle attachment, mix the eggs, buttermilk, shortening, and sugar on low speed until smooth and fully incorporated, approximately 2 minutes.

3. Add the flour, baking powder, and salt and mix until well combined and smooth.

4. Weigh out six 1.7-oz balls for the bottom and six 2.2-oz balls for the top. If you don't have a food scale: the smaller balls are approximately 3½ Tbsp and the larger are approximately 4½ Tbsp.

5. Using your fingers, spread the six smaller balls into 4-inch circles (use a 4-inch cookie cutter for guidance) and place them 1 inch apart on the prepared baking sheet.

6. Pile ½ cup berries on top of each round of dough, leaving a ½-inch rim of dough bare around the edge.

7. Press each of the remaining six dough balls into a 5-inch round with your fingers or a rolling pin. Place them on top of the berries.

8. Fold the edges of the top rounds underneath the bottom rounds and press tightly to seal and prevent leaking.

9. Sprinkle cinnamon on top of each shortcake.

10. Bake until golden brown, approximately 25 minutes. Allow to cool for 10 minutes.

MAKE THE ICING

1. In a small bowl, whisk together the icing sugar and water until combined. Using a fork, drizzle the icing over the top of the shortcakes. They are best served when they are still a little warm.

2. Store these in an airtight container in the refrigerator for up to 2 days or in the freezer for up to 1 month.

Berry Upside-Down Cake

Upon graduating high school at 17, I received as a gift a large orange suitcase and left home to find my way in the world. Knowing how much I loved this cake, my mom gave me her recipe for it before I ventured off into the unknown. It was the first cake I made when I was on my own and I continue to make it today. It is like a best friend to me. One of the things I love most about it is that I can make it with berries and then move on to plums and peaches as they come into season. The fruit caramelizes on the bottom or top (depending on which way you want to look at it), making for a flavourful, soft, moist cake. It never fails to be enjoyed at family gatherings—including by Oliver, our grand-dog. This doubles as a coffee cake for breakfast if there's any left!

MAKES ONE (9-INCH) CAKE

½ + ⅓ cup salted butter, at room temperature
¾ cup packed brown sugar
1½ cups fresh berries or peeled and sliced soft fruit
2 cups cake flour
1 Tbsp baking powder
¼ tsp table salt
1 cup granulated sugar
3 medium eggs, at room temperature, well beaten
2 tsp pure vanilla extract
1 cup 2% milk

1. Preheat your oven to 350°F.

2. In a 9-inch round pan, melt ⅓ cup of the butter in the oven, approximately 2–3 minutes.

3. Sprinkle the brown sugar evenly over the melted butter. Arrange the berries or fruit over the sugar mixture.

4. In a medium bowl, mix together the flour, baking powder, and salt. Set aside.

5. With an electric mixer, cream the sugar and the rest of the butter together. Add the eggs and vanilla, mixing until creamed.

6. Add about half the flour mixture and mix to incorporate. Add about half the milk and mix to incorporate. Repeat with the remaining flour mixture and milk.

7. Pour the batter over the berries or fruit.

8. Bake until the cake is golden brown and a knife inserted into the centre comes out clean, approximately 65 minutes. Immediately invert onto a serving plate by putting a serving plate upside down on top of the cake pan, carefully flipping both the pan and plate over, and then gently removing the cake pan.

9. This is best served immediately, but you can store it wrapped in plastic wrap at room temperature or in the refrigerator for up to 3 days, or in the freezer for up to 3 months.

WINK: *This is delicious served warm with whipped cream or ice cream!*

Black Forest Cake

Half of the recipe for this cake was already being made in our Harvest Kitchen (page 239) for our delicious Cherry Pie (page 235), so it made perfect sense to make the other half to create not only a tasty cake but a beautiful one as well. Okay, I'll confess we'd been making our Cherry Pie for quite some time before I realized that I could use some of it to make this delicious cake. The rich chocolate cake and sour cherry filling combine to create a sweet-tart flavour that is balanced by whipped cream.

MAKES ONE (8-INCH) THREE-LAYER CAKE

- 2¼ cups all-purpose flour
- 1½ cups granulated sugar
- ⅔ cup black cocoa powder (see page 201)
- 1⅛ tsp baking soda
- 1 tsp table salt
- ¼ tsp baking powder
- 1¼ cups water
- ¾ cup vegetable shortening, at room temperature
- 2 large eggs, at room temperature
- 2 large egg yolks, at room temperature
- 5 cups whipping (35%) cream
- 3 cups Cherry Pie filling (page 235)
- 1½ cups shaved dark chocolate
- Sprig fresh mint (optional)

1. Preheat your oven to 350°F. Spray three 8-inch cake pans with cooking spray. Place a large bowl in the freezer for whipping the cream.
2. In a separate large bowl, combine the flour, sugar, cocoa, baking soda, salt, baking powder, water, shortening, eggs, and egg yolks. With an electric mixer, beat on low speed for 1 minute and then on high speed for 2 minutes, scraping down the sides of the bowl as required. The batter should be smooth and dark.
3. Pour the batter into the prepared pans. Bake until a toothpick inserted in the centre comes out clean, approximately 35 minutes. The cakes will look quite dark from the black cocoa. Allow to cool completely in the pans.
4. While the cakes cool, place the cream in the bowl you placed in the freezer and whip with an electric mixer until stiff peaks form. Place in the refrigerator until ready to use. You may need to whip it up again when you're ready to use just to stiffen it back up.
5. Trim away the rounded tops of the cakes to level each cake. I usually freeze the cakes first, as it makes it easier to level them.
6. Place the first cake layer on a serving plate. Evenly spread 1¼ cups cherry filling on the cake layer and then 1 cup whipped cream. Place the second cake on top and repeat.
7. Place the third layer on top and place a 3- to 4-inch round ring (a cookie cutter is ideal), light teacup, or glass in the centre of the top layer (this will allow a place for the cherry filling that tops the cake).
8. Starting at the bottom and working in an upward motion with a flat spatula, cover all the sides with the remaining whipped cream. Also cover the top of the cake around the ring.
9. Sprinkle the shaved chocolate on the sides and top of the cake by letting it fall down the side of your palm (see Wink). You may need to press the shavings into the sides if they aren't sticking well.
10. Remove the ring, cup, or glass from the top of the cake. Place the remaining ½ cup cherry filling inside the saved space. If using mint, pluck the leaves from the sprig and place them on top.
11. Place toothpicks around the perimeter of the cake and drape plastic wrap over the toothpicks. Store in the refrigerator until you're ready to serve, up to 2 days.

WINK: *To make cleanup easier when you're adding the shaved chocolate, place the entire plated cake on top of waxed or parchment paper before you sprinkle the chocolate on.*

Carrot Cake

This recipe was given to me over 40 years ago by a firefighter who brought it to a summer pool potluck dinner. The recipe I remember, his name I do not. (But if you're reading this, please do get in touch so I can thank you.) It is one of my all-time favourites, and I feel pretty confident that it will be the best carrot cake you will ever taste!

SERVES 12

Carrot cake

2 cups all-purpose flour
2 tsp baking powder
2 tsp baking soda
2 tsp ground Saigon cinnamon
2 tsp ground nutmeg
1 tsp table salt
4 large eggs, at room temperature
2 cups granulated sugar
1 cup vegetable oil
2 tsp pure vanilla extract
4 cups grated carrots (approximately 12 carrots) + more for garnish
1 cup diced fresh pineapple or canned, drained pineapple tidbits
¾ cup dark raisins

Cream cheese frosting

1 (8 oz/226 g) package firm cream cheese, cut in quarters, at room temperature
3½–4 cups icing sugar
2 tsp salted butter, at room temperature
2 tsp pure vanilla extract

WINK: *The cream cheese frosting is also delicious on Grandma Buehler's Poppy Seed Cake (page 190) and Sensational Cinnamon Buns (page 32).*

MAKE THE CAKE

1. Preheat your oven to 350°F. Grease a Bundt pan or line the cups of a 12-cup muffin pan with cake liners.

2. In a medium bowl, sift together the flour, baking powder, baking soda, cinnamon, nutmeg, and salt. Set aside.

3. In a large bowl with an electric mixer, beat the eggs, then add the sugar, oil, and vanilla on high speed. Turn the speed to low. With the mixer still running, add the flour mixture in three increments, scraping the sides of the bowl with each addition.

4. With a large spoon, fold in the carrots, pineapple, and raisins.

5. Spoon the batter into the prepared Bundt pan or muffin pan.

6. If you're using a Bundt pan, bake until lightly browned and a toothpick inserted into the cake comes out clean, 50–55 minutes. Allow the cake to cool completely in the pan before transferring to a serving plate.

7. If you're making muffins, bake until lightly browned and a toothpick inserted into a muffin comes out clean, 18–20 minutes. For mini-muffins, bake for 10–12 minutes. For larger muffins, bake for 25–30 minutes. Allow the muffins to cool in the pan before removing them.

MAKE THE CREAM CHEESE FROSTING

1. In a large bowl, use an electric mixer on medium speed to beat the cream cheese with 3½ cups of the icing sugar and the butter until it holds its shape, approximately 3 minutes. Add more icing sugar, about ¼ cup at a time, if needed. Add the vanilla and beat until smooth.

2. Spread or pipe the frosting on the cooled cake or muffins. Garnish with some grated carrot.

3. Store in an airtight container in the refrigerator for up to 5 days. This also keeps in the freezer for up to 2 months, just pop the whole cake into the freezer, uncovered, for a minimum of 2 hours. Remove from the freezer, wrap in aluminum foil, then place in a large freezer bag. When you take it out to thaw, tent the foil so it doesn't stick to the frosting.

FARMGIRL

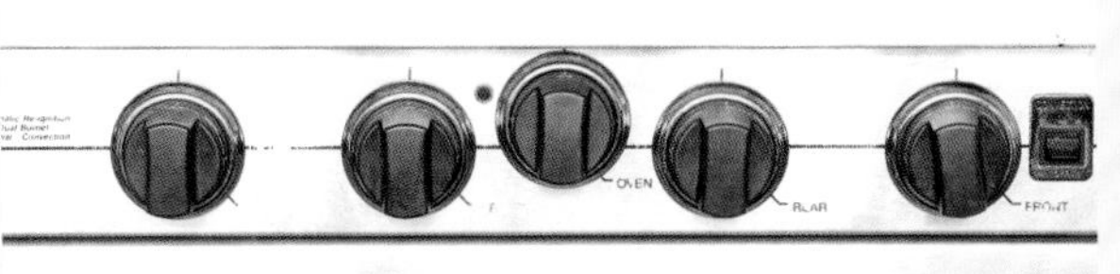

Farmhouse Kitchen

This kitchen is in my farmhouse on the 10 acres where I live with my husband, Alf (pictured here), which we call the Ponderosa. It was named the Ponderosa for two reasons: it has a large pond, and we both grew up watching and loving the western TV series *The Ponderosa.*

Our home is not connected to the farm but it's where we grow 4 acres of corn to supplement the 20 acres we grow at the farm. It's where we start seedlings, protected in a glass house known as The Orangery. The kitchen remains the hub where I serve my family and friends, wear my sentimental or fun aprons (page 91), pull out my mother's and grandmother's pots, pans, and bowls, listen to my favourite playlist, and get down to doing what I love, surrounded by those I love. I simply adore having a cooking day with one of the kids, like making pasta with Tracey (as shown here).

Many of the ingredients I cook with in the Farmhouse Kitchen are grown in the farm's fields or in the large French-inspired Potager Garden (page 4) and brought fresh to the Ponderosa.

Grandma Buehler's Poppy Seed Cake

My grandma Veronica Buehler would make this cake when my family visited her in San Diego. It has become my family's all-time favourite. It's been shipped far and wide, sending love and comfort to my children who have travelled, worked, and lived around the world. I usually make three at a time: one for today, one for tomorrow, and one to either give away or freeze. Each time I bake this I am reminded of my grandmother and mother, and it always brings a smile and an occasional tear. This cake is very moist and buttery and has a little crunch from the poppy seeds. It's also very flexible when it comes to serving, as it can be finished in many ways: plain, drizzled with butter, iced, or made into mini or regular muffins.

MAKES 1 BUNDT CAKE

Cake

4 large eggs, at room temperature, separated
1 cup salted butter
1½ cups granulated sugar
1 tsp pure vanilla extract
2 cups all-purpose flour
1 tsp baking soda
½ tsp table salt
1 cup sour cream
¾ cup poppy seeds

Glaze

1 cup icing sugar
2–3 Tbsp fresh or bottled lemon juice (see Wink)

MAKE THE CAKE

1. Preheat your oven to 350°F. Grease and flour a Bundt pan.

2. In a medium bowl with an electric mixer, beat the egg whites until soft peaks form, 1–2 minutes. Set aside.

3. In a stand mixer fitted with the paddle attachment, cream the butter and sugar on medium-high speed until light and fluffy, approximately 1 minute, With the mixer still running on medium-high, add the egg yolks one at a time, making sure each one is incorporated before adding the next. Add the vanilla and mix until fully blended.

4. In a large bowl, sift together the flour, baking soda, and salt.

5. Add approximately one-third of the flour mixture to the butter mixture and mix to incorporate. Mix in half of the sour cream. Repeat this with half of the remaining flour, the rest of the sour cream, and the final batch of flour until they are all fully incorporated.

6. With a spoon, fold the egg white mixture into the batter to combine, followed by the poppy seeds.

7. Pour the batter into the prepared Bundt pan. Bake until golden brown and a toothpick inserted into the centre comes out clean, approximately 1 hour.

8. Let cool in the pan on a cooling rack. If you want to eat it while it's still warm (as we love to do!), turn it out of the pan 20–30 minutes after it's finished baking. If you want to ice it or eat it with ice cream, turn it out of the pan after 1 hour and let cool completely on the rack before proceeding.

MAKE THE GLAZE

1. In a medium bowl, whisk together the icing sugar and lemon juice. Start with 2 Tbsp lemon juice and gradually add up to 1 Tbsp more to reach your preferred consistency.

2. Drizzle the glaze over the cooled cake.

3. Store in an airtight container at room temperature or in the refrigerator for up to 4 days, or in the freezer for up to 2 months.

WINK: *To make this cake vegan and gluten-friendly, substitute 1 cup applesauce for the 4 eggs, 1 cup vegan butter for the butter, 2 cups Gluten-Friendly Flour Blend (page 279) (or your favourite gluten-free flour) for the all-purpose flour, and 1 cup dairy-free sour cream for the sour cream. Follow the method above, but skip step 2 and add the applesauce in step 3, where it says to add the egg yolks. Add the applesauce about 1 Tbsp at a time until fully incorporated. Skip adding the egg whites in step 6. The cake will be a little denser and darker, but it will be equally delicious and moist.*

Grandma Greene's Blueberry Banana Bread

This is one of Grandma Greene's recipes and one of my all-time favourites. I used to hide bananas from my kids so they would ripen for this recipe. I love the taste combination of banana and blueberries and how they complement this bread's soft centre and crunchy top. Slicing, toasting, and buttering is my routine until the last slice is gone!

MAKES ONE (5 × 9-INCH) LOAF

- 1½ tsp baking soda
- ½ cup 2% milk
- 2 cups granulated sugar
- 1½ cups butter, at room temperature
- 2 large eggs
- 3 cups all-purpose flour
- ⅛ tsp table salt
- 3 large ripe bananas
- 1 tsp fresh lemon juice
- 1 cup fresh or frozen blueberries

1. Preheat your oven to 350°F. Grease well a 5 × 9-inch loaf pan. If your pan has a large lip, cut one 4 × 15-inch and one 8 × 15-inch piece of parchment and grease. Line the pan with the prepared parchment, letting the parchment overhang along the sides. This will make it easier to unmold the bread.
2. In a small bowl, dissolve the baking soda in the milk. Let sit for 5 minutes.
3. In a large bowl and with an electric mixer, cream the sugar, butter, and eggs on low speed until smooth. In a small bowl, mix the flour and the salt. With the mixer running on low speed, gradually add one-third of the flour mixture and one-third of the milk mixture, alternating until all is incorporated.
4. Place the bananas on a large plate and use a fork to mash them. Drizzle the lemon juice over the bananas and mix them together with the fork. Add the bananas to the batter with a spoon and mix until combined.
5. Using a large spoon, gently fold the blueberries into the batter.
6. Spoon the batter into the prepared loaf pan and bake until golden brown and a knife inserted in the centre of the loaf comes out clean, approximately 1½ hours. If the edges are starting to look crisp in the final 15–30 minutes of baking, then cover with a tented foil and continue to keep an eye on the edges.
7. Enjoy warm or let cool. Store in plastic wrap on the counter or in the refrigerator for up to 3 days, or wrapped in plastic wrap and then in a freezer bag in the freezer for up to 3 months.

WINK: *If you prefer smaller loaves or muffins, this recipe will make 2 small loaves, 12 regular muffins, or 36 mini muffins. Grease small loaf or muffin pans or line them with liners, and bake at 350°F until golden brown and a toothpick comes out clean, 20–25 minutes for small loaf pans, 15–20 for regular muffins, and 10–12 minutes for mini muffins. They will have the same shelf life and storage as the large loaf. This is also great with frozen strawberries in place of the blueberries. Just microwave 1 cup of frozen strawberries for 3–4 minutes until they start to break down. Strain the juice (cool and add to lemonade!) and bake as directed. This will result in a beautiful light pink loaf.*

Mrs. Oregon's Strawberry Cake

My mother, Veronica (known at the farm as Baby Grandma), was third runner-up in the Mrs. America pageant in 1967, representing Oregon. She had a lot of fun presenting her homemaking skills and we benefited from all the baking and cooking recipes she perfected before the final competition. She made this cake to represent her strawberry-growing state, and the judges loved it. Light and bursting with flavour, it makes a very refreshing summer dessert.

MAKES ONE (9-INCH) TWO-LAYER CAKE

Cake

1 cup salted butter, softened
2 cups granulated sugar
4 large eggs, at room temperature, separated
1½ tsp pure vanilla extract
2⅔ cups sifted cake flour
1½ tsp baking powder
¾ tsp table salt, divided
1¼–1½ cups mashed frozen strawberries, semi-thawed with their juice

Filling and topping

2½ cups whipping (35%) cream
1 tsp pure vanilla extract
½ cup icing sugar
1 cup sliced fresh strawberries
10 whole strawberries, stems removed, for garnish

MAKE THE CAKE

1. Preheat your oven to 350°F. Grease two 9-inch round cake pans and line them with parchment paper. Set one oven rack in the highest position and one in the lowest.

2. In a large bowl, use an electric mixer to beat the butter with the sugar until fluffy. Add the egg yolks one at a time, beating well between additions. Mix in the vanilla.

3. In a medium bowl, combine the flour, baking powder, and ½ tsp salt.

4. Place the mashed strawberries in a large bowl. With a spatula, gradually mix the flour mixture into the mashed berries.

5. Add the flour and berry mixture to the butter and sugar mixture in three additions, beating well between additions until both mixtures are blended together.

6. In a medium bowl, use an electric mixer and clean beaters to beat the egg whites with the remaining ¼ tsp salt until stiff peaks form.

7. With a spatula, fold the egg whites into the batter and divide evenly between the prepared baking pans.

8. Place one pan on the top rack and one on the bottom. Switch their positions carefully halfway through the baking time to ensure an even bake and to prevent the cake from falling in the middle. Bake until the top and edges have turned a golden brown and a toothpick inserted into the centre comes out clean, 30–35 minutes. The top of the cake will be pale pink and will spring back when lightly pressed.

9. Let cool in the pans for 10 minutes, then use the flat edge of a knife to carefully release the cakes from the side of the pans and invert onto a cooling rack to cool completely.

MAKE THE FILLING AND TOPPING

1. Place a medium bowl, beaters, and spatula in the refrigerator for 15 minutes.

2. Fit your mixer with the chilled beaters. Place the cream, vanilla, and icing sugar in the chilled bowl and whip until stiff peaks form, 5–8 minutes.

3. Level the top of both cakes with a serrated knife.

4. Place 1 cake on a cake stand or serving plate. Top with a layer of approximately 1 cup whipped cream, then a layer of fresh sliced strawberries, and then one more layer of 1 cup whipped cream.

5. Place the second cake on top of the fillings and use your chilled spatula to cover the entire cake with the remaining whipped cream, creating waves around the sides and the top. Garnish with the whole strawberries.

6. Store in an airtight container in the refrigerator for up to 2 days.

Strawberry Rhubarb Loaves

When we begin to see the rhubarb brought off the fields and into the kitchen, it signals us to get ready. Strawberries will follow closely behind and then the countdown to our busy harvest season on the farm begins. I love to make this recipe as these two crops overlap. It has a lovely sweet and tart taste that makes for a perfect kickoff to the summer! The lemon glaze adds a bit of colour and texture to the loaf.

MAKES TWO (5 × 9-INCH) LOAVES

2½ cups all-purpose flour
1 tsp baking powder
1 tsp baking soda
½ tsp table salt
1 cup 2% milk (see Wink)
2 tsp white vinegar
1¼ cups packed brown sugar
½ cup vegetable oil
1 large egg, at room temperature
2 tsp pure vanilla extract
1 cup sliced strawberries
1 cup chopped rhubarb
½ cup icing sugar
1 Tbsp fresh lemon juice + more as needed

1. Preheat your oven to 350°F. Spray two 5 × 9-inch loaf pans with cooking spray.

2. In a small bowl, gently whisk together the flour, baking powder, baking soda, and salt. Set aside.

3. In a small bowl, mix together the milk and vinegar and let sit for 5 minutes to make sour milk.

4. In a large bowl, combine the sugar, oil, egg, vanilla, and sour milk. Stir well by hand or with an electric mixer at low speed. Stir in the strawberries and rhubarb. Add the dry ingredients to the wet ingredients and mix just until incorporated. Do not overmix.

5. Spoon the batter into the prepared loaf pans. Bake until golden brown and a toothpick inserted in the centre comes out clean, 30–35 minutes.

6. Let cool slightly in the pans, then remove from the pans to cool completely on wire racks.

7. While the loaves are cooling, make the glaze. In a small bowl, whisk the icing sugar with the lemon juice until smooth. Add another 1 tsp lemon juice at a time if the icing sugar has not fully dissolved. Once the loaves are cool, apply the glaze to the tops with a pastry brush.

8. Wrap in plastic wrap and store at room temperature for up to 3 days or in the refrigerator for up to 4 days, or double-wrap in plastic wrap, place in an airtight container, and store in the freezer for up to 2 months.

WINK: *If you have buttermilk, you can use it in place of the milk and omit the vinegar. Instead of topping with the glaze after you bake the loaves, you can combine ¼ cup brown sugar and 1 tsp cinnamon in a small bowl, then sprinkle the cinnamon sugar on top of the loaves* ***before*** *you bake them.*

Specialty Foods Kitchen

I don't like to see anything go to waste, so when I saw a little spare space between The Porch and the Harvest Kitchen, I quickly claimed it as a new work-space and named it the Specialty Foods Kitchen. It has a viewing window just like the Harvest Kitchen (page 239), inviting our guests to be part of what our bakers make in this kitchen daily.

We use the Specialty Foods Kitchen to produce all the specialty, small-batch bakery products that guests find in our bakery showcase and in The Porch (page 248)—for example, our specialty cakes, decorated cookies, chocolate barks, creamy fudge, dipped and drizzled strawberries, and waffle cones—all of which are easier to decorate when the tables don't jiggle and aren't moved daily in the hustle and bustle of the busy Harvest Kitchen! These products are time-intensive to make, and the techniques require a great deal of skill. Our guests literally buy these tasty treats faster than we can make them.

It's
THE MOST
WONDER
FUL
time of the
YEAR

Blackout Ice Cream Sandwiches

We truly have so much fun with our food, and these ice cream sandwiches are an excellent example. The cookies use black cocoa, which tastes much like dark chocolate and gives chocolate products a darker, almost black finish. If you can't find it, you can substitute the darkest cocoa on the grocery shelf, but keep in mind that the cookies may not have the same deep black colour or rich chocolate flavour. We fill them with our fresh farm berry ice cream—raspberry, strawberry, blueberry, or blackberry. We then sandwich the ice cream between these tasty, rich, and crisp black cookies. They hold the ice cream together firmly and make a nice contrast in colour and texture with the ice cream as it melts. The fun continues by dipping them in chocolate or rolling them in pistachios, sprinkles, or coloured confetti. For different flavours, try substituting other cookies for the cookies in this recipe. Hearty Farm Cookies (page 152) and Berry Farm Cookies (page 134) are both delicious options.

MAKES APPROXIMATELY 12 (3-INCH) SANDWICHES

1 cup salted butter
1½ cups granulated sugar
¼ cup black cocoa powder or dark cocoa powder
2 large eggs, at room temperature
2 tsp pure vanilla extract
3½ cups all-purpose flour
1 tsp baking powder
3 cups ice cream of your choice
½ cup sprinkles (optional)

1. Preheat your oven to 350°F. Line two baking sheets with parchment paper. Place one oven rack on the top position and one on the bottom position.

2. In a stand mixer fitted with the paddle attachment, cream the butter and sugar. Add the cocoa powder, eggs, and vanilla. Mix to combine well.

3. In a medium bowl, mix together the flour and baking powder.

4. Gradually add the flour mixture, about ¼ cup at a time, to the butter mixture, and mix until blended and the dough holds the shape of a smooth ball. Refrigerate for 30 minutes.

5. On a clean surface, roll out the dough into a circle by moving the roller away from the centre evenly in all directions, until you reach an ⅛-inch thickness. Using a 3-inch round cookie cutter, cut out cookies and place them 2 inches apart on the prepared baking sheets. Reroll the dough up to two times to cut as many cookies as you can. You should get about 24.

6. Bake until a toothpick inserted in a cookie comes out clean, approximately 8 minutes. Switch the position of the baking racks at the 4-minute point.

7. Let the cookies cool completely on the sheets.

8. Slightly soften the ice cream by letting it warm at room temperature for approximately 30 minutes. Scoop ¼ cup of ice cream on the underside of one cookie and top with another cookie. Press lightly to join them. Run a spatula or knife around the cookie to level the ice cream. Roll each ice cream sandwich in ½ tsp sprinkles, if desired.

9. Place the finished sandwiches on a plate or baking sheet, cover with plastic wrap, and place in the freezer for at least 2 hours. Before serving, let sit for 5–10 minutes. If you intend to keep them long-term in the freezer, wrap them in plastic wrap and place them in an airtight container. They'll keep for up to 3 months.

WINK: *In 1898, a bakery called Ebinger's in Brooklyn, New York, created a black cookie. During the Second World War, it was coined the blackout cookie in recognition of the mandatory blackouts to protect the Brooklyn Navy Yard.*

Baked Alaska

This is a WOW dessert! With a base of rich dark chocolate cake, it's packed with strawberry ice cream, coated with soft meringue, and finished by torching the meringue to create golden-brown peaks. Even better, it's not just very impressive but also very simple—and you can make it ahead of time and freeze it. Right before you're ready to serve, all that's left to do is whip your meringue, cover, and torch. Be creative by substituting different flavours of ice cream and have lots of fun with your presentation by swirling your meringue into different shapes and patterns before torching.

SERVES 4–6

Cake

4 cups ice cream of your choice (I like berry ice cream for the colour contrast)
2¼ cups cake flour
1¾ cups granulated sugar
⅔ cup black cocoa powder + more for sprinkling (see Wink)
1 tsp table salt
1¼ tsp baking soda
¼ tsp baking powder
1¼ cups water
¾ cup salted butter, at room temperature
2 large eggs, at room temperature
2 large egg yolks, at room temperature
1 tsp pure vanilla extract

Meringue

6 large egg whites, at room temperature
½ tsp cream of tartar
1 cup granulated sugar

MAKE THE CAKE

1. Line a 2½-quart, 7- to 7½-inch-diameter bowl with plastic wrap. Bring your ice cream to room temperature for approximately 30 minutes, just so it's soft enough to spoon into the bowl. Fill the bowl just over half full with the ice cream. Cover with aluminum foil and freeze for at least 3 hours.

2. Preheat your oven to 350°F. Spray the bottom and sides of two 8-inch round baking pans with cooking spray and line the bottoms with parchment. Sprinkle the sides and bottom of the pan with a little cocoa powder (a great way to keep the cake looking dark).

3. In the bowl of a stand mixer fitted with the paddle attachment, place the flour, sugar, cocoa powder, salt, baking soda, baking powder, water, butter, eggs, egg yolks, and vanilla. Mix on low speed for 1 minute and then on high speed until well blended, scraping down the sides of the bowl often with a spatula.

4. Pour the batter into the prepared pans and bake until a toothpick inserted into the centre comes out clean, 25–30 minutes.

5. Let the cakes cool in the pans for 10 minutes, then turn out onto a rack to cool completely.

6. Line a flat ovenproof plate or baking sheet with aluminum foil, making sure you have enough of an overhang to come up and over the cake once the ice cream has been placed on top. Place the cake onto the foil. Invert the bowl of ice cream onto the cake. Trim the cake so it's even with the circumference of the ice cream. Cover the ice cream and cake with plastic wrap. Now bring up the foil and secure it. Transfer to the freezer for at least 3 hours or up to overnight.

MAKE THE MERINGUE

1. Make the meringue right before you're ready to serve it so it holds the peaks beautifully. If you make it the morning of or the night before, it becomes watery and will not hold the peaks.

2. Preheat your oven to 500°F.

Recipe continues

3. Place the egg whites and cream of tartar into a medium bowl. Beat with an electric mixer on high speed as you gradually add the sugar. Continue to beat until stiff peaks form, 5–7 minutes. Be careful not to overbeat.

4. Carefully remove the foil and plastic wrap from the frozen cake.

5. Quickly cover the ice cream dome with the meringue, swirling with a spatula and ensuring it creates a seal between the foil-lined ovenproof plate and the cake.

6. Place on the centre rack of the oven for 2 minutes. Watch carefully. You want the meringue to brown and not burn! You can also use a kitchen blowtorch to brown the meringue if you prefer.

7. Trim the foil with a sharp knife to the edge of the meringue and serve.

8. This is best enjoyed the day you make it.

WINK: *Black cocoa powder is a cocoa powder that has been treated with an alkalizing agent to reduce the natural acidity of cocoa, giving it a less bitter taste and a darker, almost black colour. In this cake, it adds a rich flavour and creates a striking contrast to the pink ice cream and white meringue. Black cocoa is readily available in most specialty food stores.*

Berry S'mores

There are many different ways to build a s'more. I first learned to make a basic s'more when I was a 10-year-old Camp Fire Girl. Boy, have my s'mores come a long way since then. These days, I lay out graham crackers and bowls of add-ons for everyone to build their own. The berry marshmallows in this recipe give the s'mores a sweet berry flavour while providing a soft pink colour to the filling as it oozes out from between the crispy golden grahams.

MAKES 24 S'MORES

- 1 batch Honey Graham Crackers (recipe below)
- 24 squares milk or dark chocolate from chocolate bars
- 1 batch Berry Marshmallows (recipe below)

1. Campfire method: Lay out a 10-inch square piece of foil. Place a graham cracker down first, then a square of chocolate and a marshmallow, and top with a second graham cracker. Bring the foil up from the sides, leaving space between the foil and the s'more, and twist the top to seal. Repeat with remaining grahams, chocolate, and marshmallows to make 24 s'mores. Place on the side of a campfire to allow the chocolate and marshmallow to melt. Carefully open the foil to check on it after a few minutes. Reseal and return it to the fire if it's not quite melted.

2. Oven method: Preheat your oven to 300°F and line two baking sheets with aluminum foil. Place half of the graham crackers on the prepared baking sheets. Build the s'mores as described above. Cover loosely with foil and bake until the marshmallow is oozing from the sides of the crackers, 5–10 minutes.

WINK: *Use the extra marshmallows to top mugs of hot chocolate. You can also try adding Nutella, caramel spread, peanut butter, or your favourite chocolate bar, broken into pieces, to personalize your s'mores.*

HONEY GRAHAM CRACKERS

These crispy, sweet, caramel-flavoured crackers are the muscle men holding the warm, gooey, and chocolatey s'mores together! They are also perfect for a late-night dunk in milk and an unwind talk before bed. I don't want to give away any secrets, but I know someone, long from being a kid, who comes over to dunk and talk. It's not easy to break this comfort routine with these snappy little cracker cookies, and am I ever glad! (Note: You need to start this recipe a few hours before, or the night before, you plan to eat these.)

MAKES 24 CRACKERS

- 2 cups all-purpose flour
- ¾ cup whole wheat pastry flour
- ¾ cup packed brown sugar
- 1 tsp baking soda
- 1 tsp ground Saigon cinnamon
- ½ tsp table salt
- ¼ cup wheat germ
- ½ cup cold unsalted butter, cut into 10–12 chunks
- ⅓ cup clover honey
- 3 Tbsp 2% milk
- 1 Tbsp pure vanilla extract
- ½ tsp caramel flavouring or extract

1. Place both flours, wheat germ, and the sugar, baking soda, cinnamon, and salt in the bowl of a stand mixer fitted with the paddle attachment. Mix on low to combine the ingredients well. Add the butter and mix on low speed until the mixture resembles coarse crumbs.

2. In a small bowl, whisk together the honey, milk, vanilla, and caramel flavouring. Add this to the flour mixture and pulse several times or mix on low until the dough comes together. Scrape the sides and bottom of the bowl with a spatula a few times to ensure it is evenly mixed. The dough resembles a cookie dough.

3. Turn the dough out onto a lightly floured work surface. Pat the dough into a 4 × 8-inch rectangle, approximately 1¼ inches thick.

Recipe continues

4. Cut this rectangle into four equal sections, each one 1 inch wide and 2 inches long. Wrap each section in plastic wrap and refrigerate for 2 hours or overnight.

5. Line two baking sheets with parchment paper.

6. Stand each section of dough on its long side and slice through each piece to get ¼-inch slices. Place the slices on the prepared baking sheets flat side down and, using a knife, score a line horizontally across each cracker to form a break line, being careful not to cut through. Use a fork to prick the dough with three evenly spaced dotted rows down each cracker. Repeat with the remaining pieces of dough. Refrigerate the dough again for 30 minutes before baking, this will help them keep their shape.

7. Preheat your oven to 300°F.

8. Bake the crackers in the centre of the oven for 10 minutes. (You can bake the trays side by side.) Turn the baking sheet 180 degrees and bake for another 10–12 minutes or until golden brown with crisp edges but soft to the touch. Allow to cool completely until firm.

9. Store in an airtight container at room temperature for up to 3 weeks.

BERRY MARSHMALLOWS

Pillowy soft and tender, and bursting with the flavour of fresh berries, when these are added to graham crackers and chocolate for s'mores they turn into a rich, gooey, melt-in-your-mouth treat. They're also super delicious on their own or as an indulgent topping for hot chocolate. Make ahead the day before.

MAKES 24 (1½-INCH) CUBES

- ⅓–⅔ cup cornstarch, for dusting
- ¼ cup Krause Berry Farms berry juice or 1 cup fresh or frozen berries
- ½ cup + 6 tsp water, divided
- 2 envelopes gelatin (2 Tbsp)
- 1½ cups granulated sugar
- ½ cup + 2 Tbsp light corn syrup

1. Line an 8-inch square baking pan with parchment paper, making sure you have at least 1 inch of overhang on all sides, and dust liberally with cornstarch.

2. If using fresh or frozen berries, in a saucepan over medium-low heat, cook the berries for approximately 10 minutes, stirring occasionally. Blend the softened berries in a blender or food processer. Otherwise, move to the next step.

3. In the bowl of a stand mixer fitted with the whisk attachment, combine ¼ cup + 3 tsp of the water and ½ cup berry juice. Sprinkle the gelatin on top and stir briefly to distribute. Let the gelatin sit and dissolve, a maximum of 4 minutes.

4. In a medium saucepan, combine the remaining ¼ cup + 3 tsp water, sugar, and corn syrup. Heat over medium heat, stirring constantly, until the sugar is dissolved. Turn the heat to high. When the mixture starts to boil, cover it for approximately 3 minutes to allow any crystals that have formed to be washed down from the sides of the pan. Remove the lid, insert a candy thermometer, and continue to cook over high heat, without stirring, until the mixture reaches 240°F–244°F, then remove from heat immediately. Do not overcook or the marshmallows will be tough.

5. Turn the stand mixer on low speed and slowly pour the hot syrup into the mixer bowl. Be careful, as the syrup is extremely hot. Gradually increase the speed of the mixer until it is running on high. With the whisk attachment, whip the marshmallow mixture for 10–15 minutes, until it is stiff and shiny. You can tell it's done when you stop the mixer and lift the beater; the marshmallow will slowly drip back down into the bowl in a thick, shiny stream.

6. Pour the marshmallow mixture into the prepared pan and smooth the top with a spatula. Allow it to sit and firm up, uncovered, at room temperature for at least 10 hours before cutting.

7. Lift the marshmallows out of the pan. Dust the top of the marshmallows with the cornstarch, flip them over onto a cutting board, and carefully remove the parchment paper. Dust a sharp knife or scissors with cornstarch and then cut the marshmallows into squares, dusting all sides with a light coating of cornstarch.

8. Store in an airtight container at room temperature for up to 3 months.

WINK: *I like to score before slicing through the rectangles of dough to ensure I have even slices. If you prefer a soft graham, bake as above. If you prefer a snappy graham cracker, add a few more minutes to the baking time.*

Chocolate-Covered Strawberries

Twice a year, our field manager, Nanaki, calls to let us know when he is able to pick our strawberries with the stem left on for our guests. These are very special calls. He is a busy guy, and it's a time-consuming task to cut the strawberries with about a 2-inch stem so we have something to hold onto when dipping them. We eagerly await the calls and almost get silly when we receive them, thinking up all the different ways we can dip and sprinkle these long-stemmed strawberries to make our guests smile.

MAKES 18–24 STRAWBERRIES

1 cup milk chocolate chips
3–4 tsp coconut oil
18–24 strawberries, stems on
3 Tbsp white chocolate chips, for drizzling (optional)
3 Tbsp rainbow sprinkles

1. Line a large baking sheet with parchment or waxed paper.

2. Place the milk chocolate chips and 3 tsp coconut oil in a large microwave-safe bowl. Place in the microwave and heat in 30-second intervals until melted, stirring for at least 30 seconds between each interval to ensure the chocolate does not burn.

3. Holding the stem of a strawberry securely, dip the berry into the melted chocolate to just below the leaves, keeping it there for approximately 10 seconds before pulling it up. Let the chocolate drip off, then place it on the lined baking sheet.

4. If using the white chocolate chips, melt them with 1 tsp coconut oil in the same way you melted the milk chocolate chips in step 2. Transfer the melted chocolate to a piping bag fitted with a #2 round tip, standing the piping bag in a tall cup with the top folded down over the rim of the cup for easy filling. Use an angled spatula to scrape all of the chocolate into the piping bag. When ready to decorate, twist the top of the piping bag closed and secure it with a clip or elastic. If you're not using a #2 round tip, snip a small tip off the end of the piping bag and squeeze to push the chocolate out. Drizzle or pipe the chocolate in a spiral starting from the centre and moving outward.

5. Pinch sprinkles with your fingertips and then sprinkle over the piped icing. Repeat for all strawberries.

6. Let the strawberries sit at room temperature for 2 hours to let the chocolate set.

7. Chocolate-dipped strawberries are best served the day they are made, but you can store them in an airtight container in the refrigerator for up to 2 days.

WINK: *If you prefer, you can use dark, semisweet, or even white chocolate chips for dipping the strawberries. If you use white for dipping, use milk, dark, or semisweet for your drizzle (if you're using one), for contrast. Mix and match sprinkles with flowers in pinks and purples. If adding flowers, stars, or any other shapes, add them right before the dipped berries are almost set by placing them with your finger. If placed too soon, they may not hold securely.*

Wine Sundae

This wine sundae was born one extra-hot summer's eve when Alf and I were sipping our glasses of wine. We decided to turn our wine into dessert. A trip back inside to the freezer did the trick. Instant cool-down! We felt so much better, although I'm not sure if it was from the wine or the cold ice cream. Whatever the case, we unanimously agreed they tasted great together! This adult version of an ice cream sundae combines berries and ice cream, making it a refreshing summer dessert.

MAKES 1 SUNDAE

1 large scoop vanilla ice cream
½ cup fresh or frozen mixed berries (such as strawberries, raspberries, blueberries, and blackberries) + 1 for decorating
¼ cup Krause Berry Farms and Estate Winery dessert wine (any flavour)
A couple spoonfuls of real whipped cream

1. Place the ice cream in a margarita glass. Allow it to be higher in the centre than on the edges of the glass, leaving room for the wine to sit around the edge of the inside of the glass. Circle the berries around the inside of the glass on top of the ice cream. Pour the wine over the ice cream and berries.

2. Garnish with the whipped cream and a single berry on top. Serve with a spoon immediately.

WINK: *These are fun to serve in martini glasses or mason jars.*

Raspberry Mousse

Smooth, creamy, and light, this mousse imparts a strong flavour of raspberries, signalling that summer is finally here. When I look out my upstairs office window and see the welcoming reds of the raspberries on the vines, my mouth begins to water and I think of this mousse. As you can see from the photo, part of the fun of this recipe is choosing which serving vessels to use! There are so many different choices, from clear glasses of many shapes and sizes to assorted delicate teacups.

SERVES 4–6

2¼ cups fresh raspberries
2 Tbsp cornstarch
¾ cup fresh orange juice (approximately 3 oranges)
3 large eggs, at room temperature, separated
1 cup castor sugar (see Wink)
3 cups whipping (35%) cream, divided
20 fresh raspberries, for garnish
Fresh mint leaves, for garnish

1. In a blender, purée the raspberries. Put them in a sieve placed over a large bowl and press on them with the back of a spoon to release the purée juice into the bowl, leaving the pulp and seeds in the sieve to discard.

2. In a small bowl, stir the cornstarch into the orange juice until the cornstarch is dissolved. Add this mixture to the raspberry juice, stir to combine, and set aside.

3. Place the egg yolks in a large mixing bowl and the whites in a medium bowl.

4. Add the sugar to the egg yolks and beat with an electric mixer until they are pale yellow and have a glossy shine, approximately 2 minutes. Fold in the raspberry purée mixture with a spatula and transfer the mixture to a large saucepan. Heat the mixture over the medium-low heat, stirring constantly, until it is thick enough to coat the back of a spoon, and the white and pink colours have blended into a pale pink, approximately 6–9 minutes. Remove from heat and let it cool completely, stirring occasionally (so it doesn't form a skin), at room temperature for approximately 1 hour.

5. Chill a clean medium bowl and beaters for 30 minutes. In the cold bowl with the cold beaters, whip 1 cup whipping cream on medium-high speed with an electric mixer until you begin to see ridges form, approximately 1½ minutes. Turn the speed down to low and continue whipping until soft peaks form, approximately 4 minutes.

6. With the electric mixer and clean beaters, beat the egg whites on low until soft peaks form, approximately 3½ minutes, then turn the speed up to high and beat until peaks stand up straight, approximately 2 minutes.

7. Gently fold the whipped cream into the raspberry mixture until there are no streaks. Now gently fold in the egg whites until the colour is evenly pink. The mousse will be thickened but still pourable at this point. It will firm up in the refrigerator.

8. Divide the mousse evenly between individual 6- to 8-oz cups. Cover each cup with plastic wrap. Chill in the refrigerator for a minimum of 4 hours and up to 6 hours.

9. Remove the mousse cups from the refrigerator 15 minutes before you're ready to serve. Meanwhile, whip the remaining 2 cups whipping cream as above. Serve the mousse with a generous dollop of whipped cream and garnish with raspberries and a few mint leaves.

10. Store in the refrigerator for up to 2 days.

WINK: *If you can't find castor sugar in the store, you can make your own very easily. Put 1 cup granulated sugar in a blender or food processor fitted with the steel blade and process until finely ground. The texture should be midway between that of granulated sugar and icing sugar.*

Mixed Berry Cheesecake

When I was in my mid-20s, I was gifted a cookbook entitled *Mama Never Cooked Like This*, written by the owner of the Lazy Gourmet, and now culinary legend, chef Susan Mendelson. I had never made a cheesecake before but had begun to see them at parties and on menus in restaurants. I made Susan Mendelson's cheesecake that year, and I'm so glad I did! I have tried other cheesecake recipes over the years, but I always go back to hers. It is by far the best! Thank you, Susan, for letting me use your recipe in this book. I fill it, top it, and decorate it in many, many different ways, but the base is always the same because you can't make the best any better!

MAKES ONE (9-INCH) CHEESECAKE

Crust

1⅓ cups graham crumbs or chocolate cookie crumbs

¼ cup brown sugar

⅓ cup salted butter, melted

Cheesecake filling

1 lb (500g) package cream cheese, cubed, at room temperature

⅔ cup granulated sugar

½ cup sour cream

3 medium eggs, at room temperature

2 Tbsp fresh lemon juice

Topping

1 cup sour cream

4 Tbsp granulated sugar

2 Tbsp fresh lemon juice

For serving

1 cup fresh seasonal berries

Grated lemon peel

Mint leaves

MAKE THE CRUST

1. Preheat your oven to 350°F.

2. In a medium bowl, mix the graham crumbs and sugar. Mix in the melted butter and then press into the bottom of a 9-inch springform pan.

3. Bake until set, approximately 5 minutes. Set aside. Leave the oven on.

MAKE THE FILLING

1. In a large mixing bowl, combine the cream cheese, sugar, sour cream, eggs, and lemon juice. Using an electric mixer, mix on medium speed until the filling is smooth, approximately 4–5 minutes.

2. Pour over the par-baked crust and bake until the centre is almost set and jiggles only slightly, 30–35 minutes.

MAKE THE TOPPING

1. While the cheesecake is baking, in a small bowl, mix together the sour cream, sugar, and lemon juice. You can keep this in the refrigerator until you're ready to use it.

2. When the cheesecake is finished baking, spread the topping over the cheesecake and return to the oven until the topping has set, approximately 5 minutes. Remove from the oven and allow to cool at room temperature before placing, uncovered, in the refrigerator.

3. Remove from the refrigerator 1 hour before you plan to serve. Run a flat knife around the inside of the springform pan before you remove the outside ring. Decorate with the berries, mint leaves, and grated lemon peel and place back into the refrigerator for up to 1 hour before serving.

4. Wrap in plastic and store in the refrigerator for up to 4 days or in the freezer for up to 1 month.

WINK: *You can use this recipe to make a cherry almond cheesecake. When the cheesecake has cooled, spread 1 cup Cherry Pie filling (page 235) on top. If you're using cherry filling that was previously frozen, warm it on low heat for approximately 30 minutes to remove the excess moisture from the thawing process. Let cool before spreading on top of the cheesecake.*

Guardian Ware

I had the good fortune of being given my grandmother's and then my mother's Guardian Ware cookware, which added onto my own growing collection. Guardian Ware was made by the Century Metalcraft Corporation of Chicago. In 1938 the company moved to Los Angeles, where it launched a line of strong aluminum cookware called Guardian Service, which they proclaimed was "the world's finest combination cooking and tableware service equipment." Guardian Ware is one of three cookware products from that line.

The cookware was designed to be sold by independent salesmen at house parties, an early version of Tupperware and Pampered Chef parties—they would even prepare a meal to demonstrate the power of the cookware. The salesmen tended to be handsome, strapping men sent in to "rescue" housewives with the modern, versatile kitchen equipment designed to produce healthy, cost-efficient meals.

The key to Guardian Ware is the construction of its high-domed lid. The lids were designed so that the food's moisture would condense while cooking and then drip back down on the food, minimizing the need for added fat. The flat bases of the cookware ensured even heating, and there were even some triangular pieces that sat on top of each other. These pieces were known as the "economy trio" because they allowed you to cook various parts of a meal on a single burner. Each piece bears the Guardian Service Knight, the defender of nutrition, health, and vitality.

Production paused during the Second World War to accommodate production in the aerospace industry. After the war, Century resumed production of Guardian Ware. A devastating fire in 1956 brought an end to the company. Today, Guardian Ware pieces are highly sought-after collectibles and loved by their users.

Peach Cobbler

My grandma Buehler was from a large Irish family, and many of her recipes stretched the dollar while providing warmth. She created this recipe after she immigrated to the eastern United States, where peaches grow in abundance. It is an example of a dish that is inexpensive to make yet is tasty and heart-warming. The biscuits are crunchy on the top and moist in the middle. Because they float on top of the sweet peaches, the bottoms have a soft and caramelized texture and taste. We buy our peaches from a farmer in the Okanagan who comes to our farm for berries to take back home with him. It's a sweet deal for both of us!

SERVES 6–8

½ cup melted salted butter
4–5 cups sliced fresh peaches or canned sliced peaches, well drained
1¼ cups all-purpose flour
1 cup granulated sugar
2 tsp baking powder
¾ cup 2% milk

1. Preheat your oven to 325°F.
2. Pour the melted butter into a 9-inch round baking dish, swirling to coat the bottom and sides.
3. Gently arrange the peaches on top of the butter. It's fine if they overlap.
4. In a medium bowl, gently whisk together the flour, sugar, and baking powder. Stir in the milk with a spoon until everything is fully incorporated and you have a soft dough.
5. Using floured hands, pinch pieces from the dough and place on top of the peaches. Leave a little space for them to rise and gently touch while still allowing the juices from the peaches to bubble around them. (You'll see from the photo that the dough spreads and the pieces socialize with each other.)
6. Bake until the dough pieces have spread and are golden brown, and the peaches are caramelized underneath and around the biscuits.
7. To serve, scoop up a spoonful of peaches topped with the biscuits.
8. Cover with a lid or plastic wrap and store in the refrigerator for up to 2 days.

WINK: *The peaches can be exchanged for the same volume of any single berry or a mixture of berries for a lovely berry cobbler instead.*

Berry Crisp

This easy dessert of sweet berries sandwiched between a thick oat base and a crispy crumble topping tastes delicious warm or cold. It's delightful all year round and pairs wonderfully with a scoop of creamy vanilla ice cream. I make these in mason jars for my children so they each have their own individual dessert (see Wink). I continued the tradition my mother created of personalizing whatever she could to let us know we were special. We like to pack these up in Sweetie Pie (page 65), my vintage camper trailer, when we go camping!

MAKES ONE (9 × 13-INCH) DESSERT

Crisp base and topping
2 cups all-purpose flour
1 cup old-fashioned oats
½ cup graham crumbs
2 cups packed brown sugar
2 tsp ground Saigon cinnamon
1 cup salted butter
2 tsp pure vanilla extract

Berry filling
10 cups fresh or frozen berries of your choice
4 Tbsp cornstarch
2 cups water
1 cup granulated sugar
2 tsp pure vanilla extract

MAKE THE CRISP BASE AND TOPPING

1. Lightly grease the bottom and sides of a 9 × 13-inch baking dish.

2. In a large bowl, use a spoon to combine the flour, oats, graham crumbs, sugar, and cinnamon.

3. In a microwave-safe bowl, melt the butter in the microwave, approximately 40 seconds. Mix in the vanilla.

4. Add the butter mixture to the dry ingredients and mix until combined.

5. Preheat your oven to 350°F.

6. Spread 2¼ cups of the mixture into the prepared baking dish, pressing it into an even base.

MAKE THE FILLING

1. Distribute the mixed berries evenly over the base of the baking dish.

2. In a large saucepan, combine the cornstarch, water, sugar, and vanilla. Heat, stirring constantly, on medium heat until the fruit mixture thickens, is smooth, and is no longer cloudy, approximately 3 minutes.

3. Let the filling cool for 5 minutes, then layer it over the berries in your baking dish. Sprinkle the remaining crisp base and topping over the filling and press down gently.

4. Bake until the filling is bubbly and peeking through and the topping is golden brown, 35–45 minutes for a baking dish. Let sit for a few minutes before eating warm.

5. Store leftovers in the refrigerator, covered in plastic wrap, for up to 3–4 days.

WINK: *This recipe halves and even quarters well, so you can adapt it easily for smaller groups. For individual servings, I make this in 24 (1-cup) ovenproof mason jars. Divide 2¼ cups of the crumble mixture between the jars (no need to grease them), then fill with the berry or apple pie filling and top with more oat mixture and some caramel sauce (your favourite homemade recipe or store-bought will work). Bake for 15 minutes, then let sit for a few minutes before serving. This recipe also makes a delicious pie. Simply divide the filling between two 10-inch pie shells (use the pastry recipe on page 273), add the topping, and bake as directed for 30 minutes. Why not give you one more option?! Substitute rhubarb for half the berries and increase the sugar to 2 cups. Yum!*

Steamed Berry Pudding

What happens when you grow up with parents who had an antique store? You get hooked on antiquing, of course. On an antiquing outing one day, I found a pudding mould that was crying out from the shelf to be filled. Luckily, I had the perfect recipe, thanks to Grandma Greene's treasured recipe collection. This pudding comes out very moist and is an elegant dessert to celebrate the holidays with. There's really no substitute for a pudding mould for this recipe. You can find them online, in stores, and at antique outlets. I promise, it will be worth the investment.

SERVES 8

- 2 cups fresh cranberries
- 1½ cups sifted all-purpose flour, divided
- 2 tsp baking powder
- ¼ tsp table salt
- ¼ cup salted butter
- ½ cup granulated sugar
- 2 large eggs, at room temperature
- ½ cup 2% milk
- Custard Sauce (page 276) or Lemon Sauce (page 278), for serving

1. Preheat your oven to 350°F. Grease a 5-inch-high × 7-inch-diameter pudding mould with butter.

2. In a small bowl, toss the cranberries in 2 Tbsp flour.

3. In a large bowl, combine the remaining sifted flour, the baking powder, and the salt. Set aside.

4. In another large bowl with an electric mixer, beat the butter and then add the sugar. Beat the eggs in a separate bowl and add to mixture and cream until light and fluffy, 1–2 minutes.

5. Still using the electric mixer, mix in the milk in three additions, alternating with the flour in three additions, until all milk and flour are mixed into the batter. Fold in the cranberries with a spatula.

6. Pour the batter into the prepared mould. Cover tightly with the mould lid and place the mould into a pan large and deep enough to fit the mould. Fill the pan with water to halfway up the pudding mould. Place in the oven to steam for 2 hours. Add more boiling water while steaming if needed, making sure the water doesn't completely evaporate during the steaming process.

7. After 2 hours, check if the pudding is done by carefully removing the lid and poking a skewer through the centre. If the pudding is ready, the skewer will come out clean. If it does not come out clean, place the lid back on the mould, return it to the oven, and steam for 10–15 minutes longer.

8. Once the pudding is cooked, remove it from the oven and the pan of water and set the mould on a cooling rack for 15 minutes before taking off the lid and gently flipping the mould onto a plate. Carefully remove the mould to reveal the pudding. Top with custard sauce or lemon sauce and slice to serve.

9. Store leftovers in an airtight container in the refrigerator for up to 2 days.

WINK: *As an alternative, brush ¼ cup Krause Berry Farms and Estate Winery berry syrup or your favourite berry syrup on top of the cooked pudding to create a shiny finish.*

Grandma Greene's Strawberry Dessert Pie

This is the pie to make when you want a light summer dessert, as it's made without a crust. When our strawberry crop is ready to pick, my mind goes straight to celebrating the berries' arrival with this pie. It's a real throwback to the days when cream desserts were found on kitchen tables at family gatherings.

MAKES ONE (10-INCH) PIE

¼ cup 2% milk
1 tsp pure vanilla extract
¼ tsp table salt
35 regular marshmallows
2 cups whipping (35%) cream
2 cups sliced strawberries
30–35 whole, stemmed strawberries, for decorating

1. Put a mixing bowl and beaters in the refrigerator to chill.

2. In a double boiler on medium heat, heat the milk until warm to the touch. Add the vanilla and salt, mixing well. Add the marshmallows and cook, stirring occasionally, until almost melted, approximately 9 minutes. Do not overcook. The marshmallows should be smooth and thick. Let cool at room temperature, for approximately 3 minutes. Do not chill.

3. Take the bowl and beaters out of the refrigerator and beat the cream on medium speed until soft peaks form. Turn the speed up to high and beat until stiff peaks form.

4. With a spatula, fold half of the whipped cream into the melted and cooled marshmallows. Fold in the remaining whipped cream.

5. Pour a thin layer (about one-quarter) of the marshmallow mixture into a 10-inch glass pie plate. Use half of the sliced strawberries to line the bottom and sides of the pie dish, over top of the marshmallow mixture. Pour another thin layer of the marshmallow mixture on top of the strawberries. Layer the remaining sliced strawberries over top, and top with a final thin layer of the marshmallow mixture.

6. Arrange the whole strawberries on top of the pie.

7. Use an angled spatula to place the remaining marshmallow mixture into a piping bag fitted with a #1A (½-inch round piping tip), standing the piping bag in a tall cup and folding the top of the bag over the rim of the cup for easy filling. Twist the top of the piping bag closed and secure with a clip or elastic. Pipe "puffs" around the outside rim of the pie plate and in between the strawberries.

8. Refrigerate, uncovered, for at least 8 hours. This is best eaten on the day it's made.

WINK: *If you would like to serve this pie in a crust, use the pastry recipe on page 272. Simply bake it and allow it to cool completely before adding the filling. Or you could use your favourite baked graham crust recipe. Remove the pie from the refrigerator approximately 20 minutes before serving to remove the chill from the pastry.*

Strawberry Custard Pie (page 228)

Pies and Tarts

Strawberry Custard Pie

This is our most famous pie. It's so famous that we needed to create a "pie jail" to lock them all up. They sit in a cooler with a huge lock and chain on the door, waiting for the people who have ordered them online to pick them up. Our Harvest Kitchen staff works around the clock in harvest season to make over 400 pies on weekend days, ensuring that everyone who comes to Krause Berry Farms and Estate Winery looking to buy a custard pie will be able to leave with one—along with a smile on their face—even if they didn't order online in advance.

Nothing beats sweet local strawberries (or any berries from the fields!) paired with thick creamy custard and topped with strawberry glaze. It makes for a tasty, refreshing, and impressive dessert. Be warned, though: this pie can be messy to cut and plate, although no one really seems to mind. See Wink for how to make individual servings.

MAKES ONE (10-INCH) SINGLE-CRUST PIE

Base

¼ batch Grandma Greene's Pie Dough (page 272), chilled

Custard

3 Tbsp cornstarch

¼ tsp table salt

⅓ cup granulated sugar

5 large egg yolks, at room temperature

1 cup 2% milk

1 cup half-and-half (10%) cream

1 tsp pure vanilla extract

6 cups whole fresh strawberries, hulled (see Wink)

Strawberry glaze

5 cups frozen whole strawberries (see Wink)

3 tsp cornstarch

¼ cup granulated sugar

2 Tbsp fresh lemon juice

3 Tbsp water

MAKE THE BASE

1. Preheat your oven to 400°F.

2. Follow steps 1–3 on page 272 to prepare the dough.

3. Follow step 4 on page 272 to roll out one portion of dough into a circle a bit bigger than a 10-inch diameter. Carefully drape the dough into a 10-inch pie plate and gently press down on it evenly, leaving the sides to hang over the pie plate slightly.

4. Using a knife, trim any excess dough off the rim, and prick all around the pie shell with a fork.

5. Bake until golden, approximately 10 minutes. Remove from the oven and set aside.

MAKE THE CUSTARD

1. In a small bowl, whisk together the cornstarch, salt, and sugar.

2. In a separate large bowl, gently beat the egg yolks until smooth. Set aside.

3. In a double boiler, combine the milk and cream and set on medium heat. Scald to 180°F. Clip a candy thermometer to the side of the pot to monitor the temperature. Pour the scalded milk into a glass measuring cup for easy pouring. Now pour the milk mixture into the eggs in a very slow, steady stream, whisking constantly to avoid cooking the eggs. Continue to whisk until the two mixtures are combined, approximately 1 minute. Return to the double boiler on medium-low heat and whisk in the dry ingredients. Clip the thermometer to your pot again to ensure the temperature stays at 170°F–173°F. Whisk constantly until smooth and creamy, approximately 3 minutes.

4. Remove the custard from heat, let cool for 2 minutes, then mix in the vanilla. Cover with plastic wrap, making sure it touches the surface to prevent a skin from forming.

5. The custard can be refrigerated for up to 3 days. Do not freeze.

WINK: *This pie can be made into lovely individual desserts and take-home gifts. Break up the baked pie shell and layer on the bottom of jars (I recommend using 2-cup wide-mouthed jars). Distribute evenly the cooled custard and berries of your choice and finish with cooled glaze. Close the jars with their lids and pop in the refrigerator. They're best served the day they're made but can be kept in the refrigerator for 1 day. Depending on the size of jar you choose, you will get 2 to 6 desserts. Using two-cup wide-mouthed jars will yield 6 servings. Then use leftover berry juice and pour it into lemonade for a refreshing drink!*

MAKE THE GLAZE

1. While the custard cools, start on the glaze. In a medium pot, cook the frozen strawberries on medium-high heat, stirring occasionally, until soft, approximately 10 minutes.

2. Line a fine-mesh sieve with cheesecloth and place over a medium bowl. Push the cooked strawberries through to collect their juice (the amount you collect will depend on the size and ripeness of the berries). Save 1 cup juice for the glaze. (Place the rest in freezer bags and freeze for another use.)

3. In a large saucepan on medium heat, stir the strawberry juice with the cornstarch until the cornstarch is dissolved. The mixture will look cloudy. Add the sugar and continue stirring constantly until thickened, dark, and clear, approximately 5 minutes. Remove from heat and mix in the lemon juice. If you need to thin the glaze, add 1 Tbsp of water at a time until you reach the desired thickness between pourable and spreadable. The glaze should be thick enough to coat the back of a spoon. It will thicken further as it cools.

ASSEMBLE THE PIE

1. When the custard has cooled, spoon it into the cooled baked pie shell, levelling the top. Add the hulled fresh strawberries on top of the custard in a circular pattern, starting at the outer edges of the pie shell and continuing inward, ensuring that the custard does not come above the pie shell. Continue building up strawberry over strawberry in a circular motion.

2. Fill a large spoon with glaze. Beginning at the middle of the pie, let the glaze fall off the spoon and trickle a little down the sides of the mound of berries, then work around the entire pie from the bottom of the berries in an upward motion to prevent the glaze and berries from falling.

3. Place the pie in the refrigerator, uncovered, for at least 30 minutes to set the glaze.

4. This pie is best served on the same day it's made, but it can be refrigerated for 2 days. It can't be frozen.

RECIPE PICTURED ON PAGE 226

Apple Pie

One day, a guest told me our pies had saved his marriage. "How so?" I asked. His family claimed that his mother's pies were the best. His new wife made a darn good pie. So good, in fact, that her family claimed hers was the best. He felt loyalty to his mother *and* his new wife. A friend told him about our pies, and he brought them to the next family dinner. Everyone agreed that they were the best and a truce was had . . . and they all lived happily ever after!

MAKES ONE (10-INCH) DOUBLE-CRUST PIE

½ batch Grandma Greene's Pie Dough (page 272)
1½ lb (340 g) peeled and sliced apples (approximately 6 medium apples, such as Golden Delicious)
¾ cup granulated sugar
1½ tsp ground Saigon cinnamon
2 Tbsp cornstarch
2 Tbsp water
1 egg (optional)

1. Follow steps 1–3 on page 272 to prepare the dough. Divide the dough into two portions.

2. Follow step 4 on page 272 to roll out one portion of dough into a circle a bit bigger than a 10-inch diameter. Carefully drape the dough into a 10-inch pie plate and gently press down on it evenly, leaving the sides to hang over the pie plate slightly.

3. In a large saucepan, combine the apples, sugar, and cinnamon. Cook on medium heat until the apples become fork-tender and hold their sliced form, stirring occasionally. The time will vary depending on the variety of apples used but allow at least 5–10 minutes.

4. In a medium bowl, whisk the cornstarch and water together thoroughly until smooth. Slowly add the cornstarch slurry to the cooked apples. It will look cloudy at first, but keep stirring until the liquid has thickened and looks glossy, then remove from heat. Allow to cool slightly for 15–20 minutes.

5. Pour the cooled apple mixture into the prepared pie shell. Don't let it spill over the edge of the pastry. Spread the apples evenly across the pie shell.

6. Follow step 4 on page 272 to roll out the second portion of pie dough to an 11-inch-diameter circle. Place it on top of the apples, and pinch the edges of the bottom and top crusts together. Pinch the outer edge of dough by rolling up the sides of the dough all the way around the pie to seal, then pinch the dough between your index finger and thumb all around the pie. Poke the top crust with a fork (to allow the steam to release when baking) in the design of your choice. If desired, whisk the egg to make an egg wash for the crust. Brush with egg wash.

7. For best results, loosely cover the pie with plastic wrap or foil and freeze the pie for a couple of hours. This will help keep the decorative effect of the pinches intact. (You can also freeze the pie to bake at a later date, just wrap it in plastic wrap and store it in the freezer for up to 3 months.)

8. If baking the pie from frozen, preheat your oven to 425°F and bake for 15 minutes. Turn down the oven to 375°F and bake until the crust is crisp at the edges and golden brown on top, 50–60 minutes. If baking it from fresh, preheat your oven to 350°F and bake for 35–45 minutes (the filling may bubble through the holes in the top).

9. Let cool for at least 2 hours. This makes it easier to cut and hold a wedge shape. The pie can then be covered in foil or plastic wrap and refrigerated for up to 3 days or wrapped in plastic wrap and frozen for up to 3 months.

WINK: *Instead of using a top crust, top the pie with 2 cups berry crisp topping (page 220). Spread the topping evenly over the cooked apples, pressing it down around the edges of the pastry shell. Bake the pie at 375°F for 40 minutes. The topping should be golden brown. You can also make individual servings by baking the bottom pie crust, breaking it into pieces, and dividing it between six 2-cup mason or other decorative jars. Layer the filling on top of the pastry and top with baked crisp topping (page 220), whipped cream, ice cream, or your favourite caramel topping.*

Bumbleberry Pie

My Grandma Greene and I shared a special relationship, and we had lots of fun over the years tasting pie when we ate at different restaurants during our travels. We kept searching for a better pie than hers, but we both agreed we couldn't find it. We spent many occasions baking and chasing one another around our kitchens with long-handled tongs getting in a loving pinch now and then. Krause pies are filled with all the love my grandma and I had for one another, and I'm very sure that's why they taste so wonderful! Use your favourite berries and play around with proportions. I like to use 1 cup raspberries, 1 cup strawberries, 1 cup blackberries, and 2 cups blueberries.

MAKES ONE (10-INCH) DOUBLE-CRUST PIE

½ batch Grandma Greene's Pie Dough (page 272), chilled

5 cups mixed berries, fresh or frozen and thawed

½ cup granulated sugar (see Wink)

⅓ cup cornstarch

⅓ cup water

1 egg (optional)

WINK: *As a long-time borderline diabetic, I felt disappointed at missing out on pie at the holiday dinners, so I created a sugar-free variation of this recipe. It tastes just the same as our regular bumbleberry pie, with a nice and balanced fruit flavour. Just substitute ⅓ cup of your preferred powder sweetener for the sugar in the filling. Any of the pies can be made in this way, although the flavour may change slightly.*

1. Follow steps 1–3 on page 272 to prepare the dough. Divide the dough into two portions.

2. Follow step 4 on page 272 to roll out one portion of dough into a circle a bit bigger than a 10-inch diameter. Carefully drape the dough into a 10-inch pie plate and gently press down on it evenly, leaving the sides to hang over the pie plate slightly.

3. Put the berries in a large saucepan on medium heat and bring to a boil. Add the sugar and stir well until dissolved, approximately 1 minute.

4. In a medium bowl, whisk together the cornstarch and water thoroughly until smooth.

5. Once the berries and sugar have returned to a boil, add the cornstarch slurry. The mixture will look cloudy. Stir constantly until the liquid has thickened and becomes dark and clear, 1–2 minutes.

6. Allow the berry mixture to cool for approximately 45 minutes before pouring into the prepared pie shell, spreading it so it sits even with the top of the pie shell and leaving a little room around the edges.

7. Follow step 4 on page 272 to roll out the second portion of pie dough to an 11-inch-diameter circle. Place it on top of the berries and pinch the edges of the bottom and top crusts together. Pinch the outer edge of dough by rolling up the sides of the dough all the way around the pie to seal, then pinch the dough between the sides of your index finger and thumb all around the pie. Poke the top crust with a fork (to allow the steam to release when baking) in the design of your choice. If desired, whisk the egg to make an egg wash for the crust. Brush with egg wash.

8. For best results, loosely cover the pie with plastic wrap or aluminum foil and pop the pie in the freezer for a couple of hours before baking. This will help to keep the decorative effect of the pinches intact. (You can also freeze the pie to bake at a later date, just wrap it in plastic wrap and store it in the freezer for up to 3 months.)

9. If you're baking the pie from frozen, preheat your oven to 425°F and bake for 15 minutes. Turn down the oven to 350°F and bake until the crust is crisp

around the edges and golden brown on top, 50–60 minutes. If you're baking it from fresh, preheat your oven to 375°F and bake for 35–45 minutes (the filling may bubble through the holes in the top).

10. Let cool for at least 2 hours. This makes it easier to cut and hold a wedge shape. When cooled, the pie can be covered in foil or plastic wrap and refrigerated for up to 3 days or wrapped in plastic wrap and frozen for up to 3 months.

Cherry Pie

Anytime anyone asks me what my favourite pie is, I lower my voice, look around me to make sure my husband isn't near, and say . . . "Cherry, because of the sweet and tart marriage of flavours!" Followed by, "Please don't mention this to my strawberry-growing husband!" But I agreed to tell all to Robert at Appetite, so I guess my secret is out now. I'm certain Alf and I will have a conversation about it, but I'm even more certain he will still love me!

MAKES ONE (10-INCH) DOUBLE-CRUST PIE

- ½ batch Grandma Greene's Pie Dough (page 272), chilled
- 1¾ lb (790 g) fresh or frozen and thawed pitted dark sour cherries
- ½ cup granulated sugar
- ¼ cup cornstarch
- ¼ cup water
- 1 egg (optional)

1. Follow steps 1–3 on page 272 to prepare the dough. Divide the dough into two portions.

2. Follow step 4 on page 272 to roll out one portion of dough into a circle a bit bigger than a 10-inch diameter. Carefully drape the dough into a 10-inch pie plate and gently press down on it evenly, leaving the sides to hang over the pie plate slightly.

3. Put the cherries in a large saucepan on medium heat and bring to a boil. Add the sugar and stir well until dissolved, approximately 1 minute. The sugar will temporarily reduce the boil to a simmer.

4. In a medium bowl, whisk together the cornstarch and water until smooth.

5. Bring the cherries and sugar back to a good boil, then add the cornstarch slurry. The mixture will look cloudy. Stir constantly until the liquid has thickened and becomes dark and clear. Allow to cool for 2 hours.

6. Pour the cooled filling into the prepared pie shell, spreading it so it sits evenly with the top of the pie shell, leaving a little room around the edge.

7. Follow step 4 on page 272 to roll out the second portion of pie dough to an 11-inch-diameter circle. Place it on top of the cherries and pinch the edges of the bottom and top crusts together. Pinch the outer edge of dough by rolling up the sides of the dough all the way around the pie to seal, then pinch the dough between your index finger and thumb all around the pie. Poke the top crust with a fork (to allow the steam to release when baking) in the design of your choice. If desired, whisk the egg to make an egg wash for the crust. Brush with egg wash.

8. For best results, loosely cover the pie with plastic wrap or aluminum foil and pop the pie in the freezer for a couple of hours before baking. This will help to keep the decorative effect of the pinches intact. (You can also freeze the pie to bake at a later date. See Wink.)

9. If baking the pie from frozen, preheat your oven to 425°F and bake for 15 minutes. Turn down the oven to 375°F and bake until the crust is crisp at the edges and golden brown on top, 50–60 minutes. If baking it from fresh, preheat your oven to 350°F and bake for 35–45 minutes (the filling may bubble through the holes in the top).

10. Let cool for at least 2 hours; this makes it easier to cut and hold a wedge shape. When cooled, the pie can be covered in aluminum foil or plastic wrap and refrigerated for up to 3 days or wrapped in plastic wrap and frozen for up to 3 months.

WINK: *To freeze the pie to bake later, wrap it in plastic wrap and store it in the freezer for up to 3 months. Bake as directed.*

Pumpkin Pie

Our family celebrates Easter, Thanksgiving, and Christmas, and pumpkin pie always makes an appearance. Pumpkin signifies the end of harvest for the year, and is a welcome marker for a farm family. It serves as a reminder for all we are thankful for, including Grandma Greene's delicious pumpkin pie, and it evokes memories of our holidays spent together.

MAKES TWO (10-INCH) SINGLE-CRUST PIES

½ batch Grandma Greene's Pie Dough (page 272), chilled

4 large eggs, at room temperature

1 (28 oz/796 ml) can pure pumpkin purée (not canned pumpkin pie filling)

2 cups granulated sugar

¾ tsp ground Saigon cinnamon

½ tsp ground nutmeg

½ tsp ground ginger

½ tsp table salt

1 (12 oz/354 ml) can evaporated milk

1. Follow steps 1–3 on page 272 to prepare the dough. Divide the dough into two portions.

2. Follow step 4 on page 272 to roll out each portion of dough into a circle a bit bigger than a 10-inch diameter. Carefully drape each piece of dough into a 10-inch pie plate and gently press down on it evenly, leaving the sides to hang over the pie plate slightly. Using a knife, trim any excess dough off the rim, and prick all around the pie shell with a fork. You can use the trimmings to make decorations (see Wink).

3. Preheat your oven to 350°F.

4. In a stand mixer fitted with the whisk attachment, whisk together the eggs and pumpkin for 1 minute.

5. In a separate medium bowl, mix together the sugar, cinnamon, nutmeg, ginger, and salt.

6. Add the sugar and spice mixture to the pumpkin mixture and continue whisking for another 30 seconds, using a spatula to scrape the sides of the bowl. To prevent dark spots on the pie, mix another 15 seconds to ensure the spices are mixed well.

7. Add the evaporated milk to the mixture. Beat for another 30 seconds until just mixed.

8. Pour the filling into the two prepared unbaked pie shells.

9. Bake until golden and the filling is not jiggly in the centre, approximately 45 minutes–1 hour. Insert a toothpick; if it comes out clean, the pie is done. Overcooking can cause cracking.

10. Turn the oven off and leave the pies in for 5–10 more minutes. This bakes the pies thoroughly without burning them. Watch carefully to make sure the pies don't burn.

11. Let cool for at least 2 hours. This makes it easier to cut and hold a wedge shape. When cooled, the pie can be covered in foil or plastic wrap and refrigerated for up to 3 days or wrapped in plastic wrap and frozen for up to 4 months. To serve from frozen, thaw overnight in the refrigerator.

WINK: *If you want to make just 1 pie, you can pop half the filling in a freezer bag or container and freeze it, or freeze the entire baked pie, for up to 3 months. Don't freeze this pie before you bake it. If you want to make decorations, roll out the trimmings and use cookie cutters to cut out desired shapes. You can also shape them by hand. Bake for a few minutes at 350°F until brown. Decorate the pie with your new creations once it has cooled a bit after coming out of the oven, otherwise they may sink.*

Harvest Kitchen

The Harvest Kitchen is the farm's main industrial kitchen, so named because it's where all the freshly harvested produce comes to straight from the field. It's located right in the centre of the farm. Freshly picked produce is brought from the fields directly to this kitchen to be cleaned before being turned into baked goods or frozen or processed into jarred and bottled products. This is where our very first product, strawberry jam, was born. Since then, of course, we've expanded significantly. On a Saturday in harvest season, our fresh Strawberry Custard Pies (page 228) initially sold out at 12 pies, and now they sell out at 400+.

Krause Berry Farms and Estate Winery may have started from a single acre of strawberries, but these days, this kitchen produces over 100,000 pies, 14,000 bottles of syrup, and 39,000 jars of jam and jelly. To produce these and other farm-made products, we need 12,000 lb of lard and 4,200 lb of butter, 20,000 lb of sugar, 15,000 lb of cornstarch, 3,300 lb of glucose, 34,000 lb of flour, over 200,000 lb of berries and fruits, 56,000 cobs of corn, and 40,000 lb of apples. It takes around 400 apple trees to produce this volume of apples! The Fraser Valley, where we farm, has a very limited number of apple trees, so we're thankful to have connected with farms a few hours away in the Okanagan Valley so we can purchase apples as locally and directly as we can. Fourteen full-time and thirteen part-time kitchen staff dedicate days to peeling enough apples to make the volume we require for our apple pies and crumbles. Needless to say, this kitchen literally never closes.

This kitchen has a large viewing window, so our guests are able to watch our kitchen staff work their magic in real time. Guests can even press a buzzer to speak to staff if they have questions.

Strawberry Rhubarb Pie

When I was a child, I would watch my aunts and uncles eat strawberry rhubarb pie, which, of course, my grandmother Sarah made for Easter, Thanksgiving, and Christmas holiday dinners. I was very confused when I listened to them rave about how delicious it was and how much they loved it. From the kids' table, I wondered how they could possibly love it. Really? Rhubarb? I thought I had even overheard them say in the garden that rhubarb was poisonous, so I never took a chance. Well, they lived, and I have since learned that they must have been talking about the leaves of the rhubarb. Like them, I now enjoy how the tartness of the rhubarb and the sweetness of the strawberry combine to make a delicious pie.

MAKES ONE (10-INCH) DOUBLE-CRUST PIE

½ batch Grandma Greene's Pie Dough (page 272), chilled
2½ cups chopped rhubarb
2½ cups whole strawberries, hulled
¾ cup granulated sugar
3 Tbsp cornstarch
3 Tbsp water
1 egg (optional)

1. Follow steps 1–3 on page 272 to prepare the dough. Divide the dough into two portions.

2. Follow step 4 on page 272 to roll out one portion of dough into a circle a bit bigger than a 10-inch diameter. Carefully drape the dough into a 10-inch pie plate and gently press down on it evenly, leaving the sides to hang over the pie plate slightly.

3. Put the rhubarb in a large saucepan on medium heat and bring to a boil, stirring occasionally, for 10 minutes. Once it's boiling, add the strawberries. Return to a boil and add the sugar, stirring well.

4. In a medium bowl, whisk together the cornstarch and water thoroughly until smooth.

5. Once the berries and sugar have returned to a good boil, add the cornstarch slurry. The mixture will look cloudy. Stir constantly until the mixture has thickened and becomes dark.

6. Allow the filling to cool for 45 minutes before pouring into the prepared pie shell, leaving room around the edge. Spread it evenly around the crust.

7. Follow step 4 on page 272 to roll out the second portion of pie dough to an 11-inch-diameter circle. Place it on top of the strawberry rhubarb mixture and pinch the edges of the bottom and top crusts together. Pinch the outer edge of dough by rolling up the sides of the dough all the way around the pie to seal, then pinch the dough between your index finger and thumb all around the pie. Poke the top crust with a fork (to allow the steam to release when baking) in the design of your choice. If desired, whisk the egg to make an egg wash for the crust. Brush with egg wash.

8. For best results, loosely cover the pie with plastic wrap or aluminum foil and pop the pie in the freezer for a couple of hours before baking. This will help to keep the decorative effect of the pinches intact. (You can also freeze the pie to bake at a later date. See Wink.)

WINK: *To freeze the pie to bake later, wrap it in plastic wrap and store it in the freezer for up to 3 months. Bake as directed above.*

9. If you're baking the pie from frozen, preheat your oven to 425°F and bake for 15 minutes. Turn down the oven to 375°F and bake until the crust is crisp around the edges and golden brown on top, 50–60 minutes. If you're baking it from fresh, preheat your oven to 350°F and bake for 35–45 minutes. The filling will look bubbly and may poke through some of the holes in the top of the pie.

10. Let cool for at least 2 hours; this makes it easier to cut and hold a wedge shape. When cooled, the pie can be covered in aluminum foil or plastic wrap and refrigerated for up to 3 days or wrapped in plastic wrap and frozen for up to 3 months.

Pecan Butter Tarts

Creamy and caramelly with a pecan crunch, these little tarts make a perfect and delightful bite. They are very rich, so don't be tempted to eat too many in one sitting. We reserve these treats for the holidays.

MAKES 24 TARTS

2 cups granulated sugar
1½ cups light corn syrup
1 cup salted butter, cubed
½ tsp table salt
4 large eggs
1 tsp pure vanilla extract
1⅔ cups chopped pecan pieces (see Wink)
24 store-bought frozen 3-inch unsweetened tart shells

1. Preheat your oven to 325°F.

2. In a medium pot on medium heat, mix together the sugar, syrup, butter, and salt. Stir frequently until the mixture reaches a boil. Remove from heat.

3. In a large bowl, beat the eggs and vanilla.

4. Pour the hot mixture very slowly into the beaten eggs while whisking constantly. It's crucial to do this extremely slowly, as you do not want to cook the eggs.

5. Distribute the pecans evenly among the tart shells.

6. Pour the filling into the shells. Carefully transfer the shells to a baking sheet. Bake until the tops of the tarts are golden brown, approximately 30 minutes.

7. Cool completely before serving.

8. Store in an airtight container at room temperature for up to 4 days or in the refrigerator for up to 1 week. The tarts can also be frozen in freezer bags for up to 3 months. Let them thaw in the refrigerator before eating.

WINK: *You can replace the chopped pecan pieces with dried blueberries for an extra sweet flavour and a chewy finish.*

Pie in the Sky

Our pies have travelled many a distance, the farthest being all the way to the Netherlands. One of our staff, Grace, who worked in the Harvest Kitchen (page 239), had relatives visiting from the Netherlands. During their visit she baked many of the pies from the farm for dessert. They continued to tell her how much they loved them. A few years later as she prepared to visit them in the Netherlands, she decided she would bring them a taste of the pies they had so enjoyed. She took a frozen pie in its box, wrapped it in plastic wrap, and tucked it into her suitcase, hoping it would stay cold in the belly of the aircraft. Her plane took off without any delays and arrived on time 9½ hours later; she then had another hour's drive to her destination. She excitedly unpacked her suitcase to find the blueberry pie . . . still frozen. Her family was suitably impressed—they baked it and enjoyed it for dessert that night.

Old-Fashioned Berry Milkshake (page 253)

Drinks and Snacks

The Porch

This kitchen was my second farm kitchen—and what a difference it made in my life! My first kitchen was a converted closet with a sink, refrigerator, and small tabletop oven where I made a dozen berry shortcakes and a dozen pies each day. The Porch is an enclosed space with French windows on two sides that open up toward the ceiling to create a serving hatch. It's a versatile space that's used for cooking (it's where we make our milkshakes, smoothies, doughnuts, and fresh-cut farm fries, for example), prepping, and serving as required. It is connected to the Specialty Foods and Harvest Kitchens along its back wall, so we can bring products, such as berry shortcakes, cookies, pies, and corn pizza, from those kitchens into The Porch with ease.

I, along with our guests, was happy to move over to the newly revamped Porch, where I was able to produce more shortcakes and pies and see more happy faces. Since then, it has continued to provide our guests with the many products whose ingredients come fresh from our kitchens and fields. This is an open-air space where our guests love to look out into our fields, perched up on a tractor stool, while enjoying the many delicious farm-grown foods we serve.

Berry Lemonade

I enjoy making big batches of this lemonade when I'm in my Relax Kitchen (page 50) in California, where we have large lemon trees and can pick the lemons right off the tree. It's very refreshing to sip poolside, and I love having it on hand to serve to guests when they stop in. It takes a little time, but it is well worth it. The berries add flavour and beautiful colours.

MAKES 18 CUPS

- 1½–3 cups granulated sugar (depending on personal taste), divided
- 16 cups water, divided
- 2 cups fresh lemon juice (approximately 12 large lemons)
- 1 cup fresh mint leaves
- 1 cup freshly picked berries (mixed or just 1 variety)

1. Place 1½ cups sugar in a large pot and add 8 cups water. Stir on medium heat until the sugar has dissolved, approximately 5 minutes. Remove from heat and let cool.
2. Add the remaining 8 cups water and the lemon juice, stirring until well combined.
3. Pour into a large pitcher and add the mint leaves and berries.
4. Adjust the sweetness to your taste by adding more sugar, ¼ cup at a time.
5. Pour over ice to serve.
6. Enjoy immediately.

WINK: *Try adding blueberry or raspberry juice or wine to create your own spin on this lemonade. For an extra-cold drink, add a few frozen berries along with the fresh ones.*

Hot Berry Mulled Wine

Coming home after a long day out in the cold, crisp air, you step inside and catch the scent of sweet spice coming from your hot wine mulling in the slow cooker. The spiced aroma is just a hint of what's to come when you take the first sip and taste this wintery wine. This is great for a cozy night in by the fire with a book or a fun night playing cards with friends. We also sell the ingredients for berry mulled wine boxed up as kits, ready for our guests to take home to enjoy.

MAKES APPROXIMATELY 6 CUPS

- 1 (750 ml) bottle Krause Berry Farms and Estate Winery Blueberry or Blackberry wine
- 1 (147 ml) bottle Krause Berry Farms and Estate Winery berry syrup (any flavour)
- 4 cups apple cider
- 1 cup fresh or frozen berries Krause Berry Farms and Estate Winery (strawberries, blueberries, raspberries, or blackberries)
- 1 stick cinnamon
- 1–2 whole cloves
- 1 star anise

1. In a large pot, combine the wine, syrup, apple cider, berries, cinnamon stick, cloves, and star anise. Warm on medium heat until close to boiling, then reduce the heat to low. Cover and let simmer for 15 minutes.
2. Ladle the mulled wine into cups and enjoy.
3. Leftovers can be strained, kept in the refrigerator for 2 days, and reheated on the stove over low heat.

Old-Fashioned Berry Milkshake and Smoothie

These were our first offerings up at The Porch in the early years. I always think of them as a couple, which is why I've presented them together on this page. They are very refreshing and rewarding after a warm summer day of berry picking! For best results, use an ice cream or syrup flavour that matches the flavour of the berries you're using.

MAKES 1 MILKSHAKE

MILKSHAKE

- 1½ cups or 3 large scoops Krause Berry Farms and Estate Winery ice cream
- 1 cup fresh or frozen and partially thawed Krause Berry Farms and Estate Winery berries, plus extra to garnish (garnish optional) (see Wink)
- ½ cup 2% milk
- 1 Tbsp Krause Berry Farms and Estate Winery berry syrup
- Mint leaves, for garnish (optional)

1. Place all the ingredients in a blender and blend until smooth.
2. Pour the mixture in a glass. Top with berries and a few mint leaves, if using.
3. Enjoy immediately.

WINK: *If you're using blueberries in your milkshake, I suggest using only frozen. The natural pectin in fresh blueberries makes them gel with the milk, which will change the texture and turn your shake grey. Any other berry can be used fresh or frozen.*

SMOOTHIE

MAKES 1 SMOOTHIE

- 1 cup fresh or frozen and partially thawed Krause Berry Farms and Estate Winery berries, plus extra for garnish (garnish optional)
- ¼ cup cold water
- 1 cup crushed ice
- 2 Tbsp Krause Berry Farms and Estate Winery berry syrup (syrup flavour to match berries)
- Mint leaves, for garnish (optional)

1. Place all the ingredients in a blender and blend until smooth.
2. Pour the mixture in a glass. Top with berries and a few mint leaves, if using.
3. Enjoy immediately.

Margaritas

When I'm visiting my second home in California, I look forward to sitting down in one of my favourite Mexican cafés and ordering guacamole, salsa, and a margarita to start. My family and I all love Mexican cuisine—we were raised on it and probably ate it a couple of times a week. San Diego's proximity to the Mexican border and the abundance of ingredients nearby made it possible to incorporate this delicious cuisine into the dishes my grandmother taught my mom, and which she then taught me. Nothing quite hits the spot and quenches the thirst on a hot summer's day like this drink!

MAKES 4 MARGARITAS

1 (10 oz/295 ml) carton frozen pink lemonade concentrate
2 cups frozen strawberries
½ cup water
4 oz tequila

Optional Garnishes
Salt, for rim
4 lime wedges
4 fresh strawberries
4 hibiscus flowers

1. In a blender, combine the concentrate, frozen strawberries, water, and tequila and blend until smooth.

2. Rim the glass with salt, if using. Pour the margarita into glasses. Decorate the rim of each glass with a lime wedge and a partially sliced fresh strawberry or a hibiscus flower.

3. Enjoy immediately.

WINK: *Try this with frozen limeade concentrate for a vibrant green margarita.*

Sangria

Sangria is a refreshing summertime drink. Originating in Spain, it is made with wine and fruits and served in large pitchers for sharing. Sangria is typically served with a long-handled wooden spoon, which is also used for stirring in the fruits before pouring the next glass. This recipe, which we serve in our Winery Kitchen (page 259), is made with Krause Berry Farms and Estate Winery fresh berries—unlike traditional sangria, which is made with oranges, lemons, and limes. Ours is sweet, fruity, and delicious! (Note: You will need to start this recipe the night before you plan to drink it.)

SERVES 4

24 fresh or frozen blueberries
12 fresh or frozen raspberries
12 fresh or frozen blackberries
3 fresh or frozen strawberries, sliced
1 bottle (750 ml) Krause Berry Farms and Estate Winery berry wine

1. Place the berries in a small bowl and lightly toss.

2. Place ½ cup of the berry mixture in a large pitcher. Add the wine and stir well with a long-handled spoon.

3. Cover both the pitcher and the bowl of remaining berries tightly with plastic wrap and place them in the refrigerator overnight.

4. Strain the sangria through a sieve into a fresh pitcher.

5. Pour the sangria into four wine glasses and add 2 Tbsp of the reserved berry mix to each glass.

Wine Popsicles

Thanks to Heather (also our incredible photographer of this very cookbook!) for teaching me this fun and very easy way to make wine popsicles for adults in the summer. Our fruit wines make excellent wine popsicles and can be mixed or used individually. You can use any combination of berry wines, syrups, and berries! Have fun with mixing and matching.

MAKES 16–20 POPSICLES

2 cups Krause Berry Farms and Estate Winery berry wine (flavour of your choice)

⅓–½ cup Krause Berry Farms and Estate Winery berry syrup (flavour of your choice)

Fresh or frozen berries (if using strawberries, slice them)

1. Pour the wine into a medium bowl and then add the berry syrup to sweeten to your liking. Start with ⅓ cup and taste. If you wish, add more berry syrup, 1 Tbsp at a time, until it tastes just how you like it. Stir well.
2. Fill your popsicle moulds about halfway with this wine mixture.
3. Drop in some berries, pour in more wine mixture, then add more berries to top it off. Freeze according to the popsicle mould instructions. Once the popsicles are frozen, I suggest wrapping each popsicle in foil and transferring them to a labelled freezer bag.
4. Store in a freezer bag in the freezer for up to 3 months.

WINK: *Be careful. The littles will want these too! Substitute juice for them and use a different mould. Make sure to label them so as not to confuse the two, especially if you store them side by side!*

KRAUSE

Estate Winery and Kitchen

We grow all the berries we use for our berry wines, many of which are award-winning (page 8). We make table wines from all four of our berries, plus apples and pears, ranging from sweet to dry; five sparkling wines with our berries; six dessert wines; and two beautiful port-style wines from our blueberries and blackberries. Look for our boots (page 258) on some of the labels.

Keep an eye out for our winery manager, Ted (pictured here). He looks just like a barkeep from the Old West, with his magnificent moustache, satin vest, and bowler hat. He has an encyclopedic knowledge of wines and pairings and he's always happy to share that knowledge with guests.

The Estate Winery is a gathering place that will make you feel like you're in a cowboy movie. We use it to host parties and celebrations, but it's also somewhere you can come alone and still feel comfortable. When we are hosting a special event at the farm or in the fields, we invite a guest chef for the evening. In addition, our Winery Kitchen serves guest favourites such as corn pizza, nachos, krauseadillas, garden paninis, garden boxes, winearitas, and sangrias (page 255).

If you pop into the winery, don't forget to say hello to Slim. He's a wooden cowboy with a special question written on the mirror located at the end of the hallway for all our guests. (You'll have to come by to find out what that question is!)

Fresh-Cut Strawberry Salsa

When I want to make anyone in my family—including myself—smile, this is my go-to. Fresh-cut salsa paired with a freshly made guacamole (page 264) and chips brings a smile to any face! This salsa is refreshing in the summer heat and tastes all the sweeter because it's only available for a very short time. We sell this in the refrigerated section of our market and it is a favourite of our guests. (Note: This recipe can easily be halved.)

MAKES APPROXIMATELY 9½ CUPS

Salsa

4 Roma tomatoes, diced

3 cups chopped fresh strawberries

1½ cups finely chopped green, yellow, and red bell peppers

½ cup minced green onions

2 jalapeños, seeded and finely diced (see Wink)

2 Tbsp minced fresh cilantro

4 cloves garlic, minced (approximately 1⅓ Tbsp)

Dressing

1 Tbsp extra virgin olive oil

1 Tbsp fresh lime or lemon juice

½ tsp kosher salt

½ tsp freshly ground black pepper

1 bag of tortilla chips (or homemade, page 271), for serving

MAKE THE SALSA

1. In a large bowl, mix together the tomatoes, strawberries, peppers, green onions, jalapeños, cilantro, and garlic.

MAKE THE DRESSING

1. In a small bowl, whisk together the oil, juice, salt, and pepper. Pour this dressing over the salsa and mix until the salsa is evenly coated.

SERVE!

1. Serve the salsa with tortilla chips.

2. Strawberries will get mushy within 2 days of being cut. Keep the salsa and dressing separate, and combine them only when you're ready to eat the salsa. If you do this, they'll each keep separately in an airtight container in the refrigerator for up to 2 days.

WINK: *Wear rubber gloves when handling jalapeños to avoid a burning sensation on your hands!*

Roasted Casa Corn Dip

This recipe was born out of love for my husband, our mutual love of the corn he grows, and my love for Mexican cuisine. Alf's face lights up when he sees I've made something with his corn, or in fact any of the produce he grows. I know no better way of showing him how much his efforts are appreciated.

When the mood to make this recipe strikes, I just go to my freezer and pull one of the large bags of roasted corn that I've made during harvest season, making this a quick dip to assemble. Some of the corn kernels pop with the freshness of summer, while others impart a sweet, smoky, caramelized taste. The lime brings out the best of all these flavours, and the soft cheese adds a beautiful contrast to the flavours and textures of the other ingredients.

MAKES 3½ CUPS

1 cup mayonnaise
¼ cup vegetable broth
¼ cup crumbled cotija or feta
2 Tbsp fresh lime juice
1 Tbsp medium hot sauce
2 Tbsp chopped fresh cilantro
1 tsp kosher salt
1 tsp freshly ground black pepper
Pinch of chili powder or tagine, for finishing
3 cups Roasted Corn (page 73)
Corn tortilla chips (store-bought or homemade, page 271), for serving

1. In a large serving bowl, combine the mayonnaise, broth, cheese, lime juice, hot sauce, cilantro, salt, pepper, and chili powder.
2. In a medium pot on medium-low heat, warm the roasted corn. Add the warm corn to the mayonnaise mixture and stir together.
3. Place in a bowl and finish with a pinch of chili powder.
4. Serve the dip warm with tortilla chips.
5. Store in an airtight container in the refrigerator for up to 2 days. Reheat before serving.

WINK: *This corn dip is best made with roasted corn. It just will not taste the same if you make it from canned or frozen corn. Serving it in a special bowl, such as the one you see in the photograph, which my dad made from wood in 1936, adds another layer of warmth.*

Guacamole, Two Ways

URUAPAN MICHOACÁN GUACAMOLE

MAKES APPROXIMATELY 3 CUPS

While travelling in Mexico several years ago, I met a woman from Uruapan Michoacán, west of Mexico City—also known as the Avocado Capital of the World. We had quite the conversation about guacamole, which resulted in her sharing her special recipe with me. I explained to her that when I go to a Mexican restaurant, I first order guacamole and salsa. If I love them both, I'll continue ordering a main course. If I don't, I'll search for another Mexican restaurant, because the quality of those two dishes will let me know if I want to stay. She understood because she was passionate about guacamole, even more than I am. This Uruapan Michoacán guacamole is bright green, thick, creamy, and crunchy, with a little kick from the jalapeños.

3 ripe Hass avocados, pitted, peeled, and chopped
Juice of ½ medium lemon
¼ cup finely diced red onions
½ jalapeño, stemmed, deseeded, and finely diced (see Wink on page 260)
1 Tbsp finely diced green bell peppers
½ tsp kosher salt
½ tsp freshly ground black pepper
¼ cup diced Roma tomatoes
1 Tbsp chopped fresh cilantro, for garnish (optional)

1. Place the avocados on a large plate and squeeze the lemon juice over them.

2. With a fork, mash until fairly smooth, but leave some small chunks.

3. Add the red onions, jalapeños, green peppers, salt, and pepper. Mix with the fork to blend. Garnish with the tomatoes and cilantro.

4. This is best made right before serving.

APIZACO TLAXCALA GUACAMOLE

MAKES APPROXIMATELY 3 CUPS

This second recipe is completely different from the first one, not least because it contains tomatillos. It originates from Apizaco, Tlaxcala, east of Mexico City, and was shared with me by Gilberto, who has worked with us since 2010. This guacamole has a sauce-like texture and is tangy and lighter than the other guacamole. When I tasted it for the first time, I'm embarrassed to say that I almost ate it like a chilled soup, it was sooooo delicious!

1 red onion
6 tomatillos, stemmed and halved
1½ jalapeños, stemmed, deseeded, and chopped (see Wink on page 260)
½ clove garlic
Pinch of kosher salt
8 pitted, peeled, and chopped large ripe Hass avocados

1. Slice the red onion into rounds. Reserve a single slice from the centre for garnish, if desired, and dice the rest.

2. In a blender, combine the tomatillos, jalapeños, diced onions, garlic, and salt and blend until perfectly smooth. Pour the purée into a medium bowl and add the avocados, crushing them with a fork into the sauce until they are well incorporated, leaving some smaller chunks.

3. Place in a serving bowl and top with the optional red onion slice for a pop of colour.

4. Store in an airtight container in the refrigerator for up to 3 days.

WINK: *If I have some fresh cilantro on hand, I like to add a couple of sprigs as a garnish.*

Roasted Garlic (page 270)

Staples

Butters with Dad

My dad, Dale (pictured here), made butter from his mother's cows' cream from the time he was little. He churned, moulded, and packed it not only for the family but also to sell. He rode his handcrafted delivery bicycle through his town of Paisley in the Oregon outback, dropping off butter and buttermilk to the neighbours and the town's only mercantile. This was an essential activity for their pioneer family. And while it wasn't essential for me, I felt it was still important to learn these skills. My grandmother gifted me her wooden moulds, press, and stamps, and my dad taught me how to use each of them. Try these butters, instead of regular butter, on any of the breads in this book.

EACH RECIPE MAKES APPROXIMATELY ¾ CUP (1 CUP OF BERRY BUTTER)

HERB BUTTER

¾ cup salted butter, at room temperature
3 tsp chopped fresh chives
2 tsp chopped fresh flat-leaf parsley
¾ tsp minced garlic
¼ tsp freshly ground black pepper

GARLIC PARMESAN BUTTER

¾ cup salted butter, at room temperature
3 Tbsp shredded parmesan
3 tsp garlic powder
¼ tsp freshly ground black pepper

JALAPEÑO BUTTER

¾ cup salted butter, at room temperature
3 tsp lemon zest
1¼ tsp minced jalapeño
1 tsp minced garlic

LAVENDER BUTTER

¾ cup salted butter, at room temperature
4 tsp lemon zest
2 tsp edible lavender

BERRY BUTTER

1 cup salted butter, at room temperature
¼ cup icing sugar
⅓ cup sliced fresh strawberries, raspberries, blackberries, or blueberries (combo or just one)

1. For each flavour of butter, place all the ingredients in a bowl or mould. With a hand mixer on medium speed, mix until well blended, 30 seconds to 1 minute.

2. Store the butter in an airtight container in the refrigerator for up to 5 days or in the freezer for up to 4 months.

Roasted Garlic

10 heads garlic
¼ cup extra virgin olive oil

There are so many ways to use roasted garlic! Each summer during harvest I spend an entire day in my Farmhouse Kitchen roasting garlic heads. I then freeze them to use in many, many ways throughout the year. Cheesy Chicken Enchiladas (page 97) and Italian Pesto Pizza (page 78) are two of my favourite recipes for using them. I love the way the garlic caramelizes when roasted and how it adds a mild sweet and smoky flavour to almost any dish. Another of my favourite ways to enjoy roasted garlic is beautifully simple: right after it comes out of the oven, I spread it on bread, fresh or toasted. Any of the breads in this book—such as Cheesy Potato Bread (page 118) or Asparagus Cheese Bread (page 110)—are perfect for this.

1. Preheat your oven to 400°F.

2. Rinse the whole garlic heads. Remove some of the papery skin. Slice off just enough of the top of each head of garlic to expose the cloves (approximately ¼ inch).

3. Oil your clean hands and rub each garlic head with some of the oil, just to give each one a light coating. Place the garlic heads, cut side up, in an oiled cast-iron frying pan, casserole dish, or baking sheet. Drizzle the tops lightly with the remaining oil.

4. Roast until the garlic heads have caramelized. The cloves will be fork-tender and some cloves may poke out of the head. This will take 35–40 minutes. Allow to cool slightly before lightly squeezing to remove the whole roasted cloves to serve.

5. If you're planning to use the garlic within the next few days, place the cloves in an airtight container and refrigerate for up to 4 days. You can also freeze the garlic by spreading the roasted garlic cloves out on a baking sheet lined with parchment paper and then flash-freezing them. When frozen, approximately 1 hour, transfer to a freezer bag or airtight container and return to the freezer for up to 1 year. (This will be a time saver for you all year when any recipe calls for roasted garlic.) Frozen cloves thaw within minutes, and for some recipes you can add them still frozen.

RECIPE PICTURED ON PAGE 266

WINK: *During harvest season, I roast approximately 40 garlic heads on a baking sheet in a single session. If I run out of my annual supply, I will roast 9 at a time, for a perfectly snug fit in my cast-iron pan, until we harvest garlic in season. My mother and grandmothers cooked with cast iron, and when I married, I inherited my grandmother Sarah's cast-iron pans, perfectly seasoned. I prefer them over any pan and use them for small garlic batches, as shown. When I'm cooking in them, my mind often wanders to memories of her doing the same, years ago.*

Homemade Tortillas

These round, flat, thin discs made from unleavened corn are the pride of Mexico. They are soft, chewy, and crispy all at the same time, and bursting with the flavour of corn. They use instant corn masa flour, which is available in most grocery stores.

MAKES APPROXIMATELY 16 TORTILLAS

2 cups instant yellow or white corn masa flour (such as MASECA)
1½ cup water
Vegetable oil, for frying
Shredded cheese (optional)

1. In a large bowl, stir the masa flour and water with a spoon until it begins to come together, then knead 5–7 times with your hands until the dough feels smooth.

2. Form into a round and cover with a tea towel or plastic wrap and let the dough rest at room temperature for 15 minutes. This dough doesn't rise while it rests.

3. Cut the dough round into quarters. Take each quarter and cut in half horizontally. Then cut again in half. You should have 16 wedges. Roll each wedge into a ball and keep covered with the damp tea towel.

4. Cut open the sides of a resealable bag and cut off the sealing top. Place 1 ball in the middle of the bag, cover with the other side of the bag and press down to create a flat 5-inch circle. Repeat with the other balls.

5. In a frying pan on medium heat, dry cook a tortilla for 30 seconds at a time, flipping the tortilla over 3 times. Fold them in half to keep them warm and hold their shape. If you are serving right away, you can sprinkle with cheese, if using, to melt in the tortilla fold. Alternatively, you can serve by frying the tortillas in 1 tsp vegetable oil on medium heat for 30 seconds per side, using a fork to flip them. Press each tortilla between paper towels to remove excess oil. Then sprinkle with cheese as directed.

6. Let cool and store wrapped in plastic wrap or in a resealable bag in the refrigerator for up to 5 days.

WINK: *Gabriel, a long-time staff member who is from Mexico, taught me to put a tortilla in the prepared pan, sprinkle shredded cheese on top as the underside cooks, and fold it in half. ¡Delicioso! If you have extra tortillas, you can turn them into chips. Just cut the tortilla into quarters and fry with oil on medium heat for 10–15 seconds on each side. Pat with paper towel to remove excess oil. Sprinkle with salt or seasoning salt (you can shake them in a bag to coat).*

Grandma Greene's Pie Dough

I have my wonderful grandmother Sarah Greene to thank for this pie dough recipe, which she lovingly made for every holiday meal. She was known in her family and in her town of Paisley, Oregon, as the best pie maker there was. She learned her pie-making skills from her mother, Ella May Vernon, on an old wood stove, with no access to electricity. Sarah was the oldest of nine children and began cooking and baking during the hay harvest in the Oregon outback at the age of 14. She married at 16, raised a family of four, and worked in the kitchen of the town's one and only hotel. Folks rode into town on horseback, in wagons, and occasionally in vehicles to have a piece of her pie!

MAKES ENOUGH DOUGH FOR FOUR (10-INCH) SINGLE-CRUST PIES OR TWO (10-INCH) DOUBLE-CRUST PIES

4 cups all-purpose flour
1½ tsp table salt
1 Tbsp granulated sugar
1 cup lard or vegetable shortening, cubed, at room temperature
1 Tbsp white vinegar
½ cup ice-cold water
1 large egg, at room temperature

1. Sift the flour and salt into a large bowl. Whisk in the sugar.

2. Using your hands, mix the shortening into the flour mixture until crumbly. Some of the lard will still be visible in pea-sized pieces.

3. Add the vinegar, water, and egg. Mix with a fork until combined, then use your hands to form the dough into a ball. Cut the ball into quarters, cover with plastic wrap, and chill for 15 minutes before rolling out the dough. Each quarter can be kept wrapped in plastic wrap and stored in the refrigerator for up to 2 days (let it sit at room temperature for about 15 minutes before rolling it) or in freezer bags in the freezer for up to 3 months. Just thaw frozen pie dough in the refrigerator overnight before baking, then let thaw at room temperature for 30 minutes. It should still be cool and pliable for rolling.

4. To use the dough, sprinkle a little flour on a clean work surface, your rolling pin, and your hands. Working with one portion of dough at a time, begin by using your hands to press the dough in the centre to flatten. Switch to the rolling pin and, working from the centre of the dough, press gently while rolling away from you, switching the direction of your pin at the halfway point. Still working from the centre, press gently while rolling the pin toward you now, switching the direction of your pin until a full circle forms. If using the dough for a bottom crust, roll the circle to a bit bigger than a 10-inch diameter. If using the dough for a top crust, roll it to an 11-inch diameter.

5. Carefully drape the dough in your pie plate and gently press down evenly, leaving the sides to hang over the pie plate slightly.

6. If the pie shells are for single-crust pies, with a baked filling, use a fork and poke all around the inside and bottom of the shell. If the filling is unbaked, you don't need to poke the dough. For a double-crust pie, you don't need to poke the dough either.

7. For best results, freeze pie shells or double-crust pies before baking. They will hold their pie pinch much better than if you bake them right after you make them.

8. Fill and bake the pastry shells according to your pie recipe instructions.

Whole Wheat Pie Dough

This recipe was created for a savoury pie, such as Farmers Eat Quiche (page 106). It is nicely flaky and tender, with a slight nutty taste and a light brown colour.

MAKES ENOUGH DOUGH FOR TWO (10-INCH) PIE SHELLS

1 cup pastry flour
1 cup whole wheat flour
1 Tbsp granulated sugar
1 tsp table salt
1 cup lard or vegetable shortening, cubed, at room temperature
½ cup ice water

1. In a stand mixer fitted with the paddle attachment, whisk the flours, sugar, and salt until combined.

2. Add the cubed lard and mix until the lard and dry ingredients start to come together and break into smaller pieces, then gradually add the water and mix until blended. No chunks of lard should be visible.

3. Divide the dough into two equal portions. Wrap in plastic wrap and chill for 30 minutes, and up to overnight, before using. Let sit at room temperature for 30–40 minutes before rolling.

4. To use the dough, sprinkle a little flour on a clean work surface, your rolling pin, and your hands. Working with one portion of dough at a time, begin by using your hands to press the dough in the centre to flatten. Switch to the rolling pin and, working from the centre of the dough, press gently while rolling away from you, switching the direction of your pin at the halfway point. Still working from the centre, press gently while rolling the pin toward you now, switching the direction of your pin until a full circle forms. If using the dough for a bottom crust, roll the circle to a bit bigger than a 10-inch diameter. If using the dough for a top crust, roll it to an 11-inch diameter.

5. Carefully drape the dough in your pie plate and gently press down evenly, leaving the sides to hang over the pie plate slightly.

6. If the pie shells are for single-crust pies, with a baked filling, use a fork and poke all around the inside and bottom of the shell. If the filling is unbaked, you don't need to poke the dough. For a double-crust pie, you don't need to poke the dough either.

7. For best results, freeze pie shells or double-crust pies before baking. They will hold their pie pinch much better than if you bake them right after you make them. Thaw frozen pie dough in the refrigerator overnight before baking, then let thaw at room temperature for 30 minutes. It should still be cool and pliable for rolling.

8. Fill and bake the pastry shells according to your pie recipe instructions.

WINK: *Make this gluten-free by replacing both flours with the Gluten-Friendly Flour Blend (page 279) and adding 3 tsp xanthan gum to the dry mixture.*

Pizza Shell

MAKES TWO (10-INCH) OR FOUR (6-INCH) ROUND PIZZA SHELLS

Just when I thought I had enough ovens, my husband came home with one more. This time he said it was for him, not me. Alf loves pizza and he loves to barbecue, so he put those two thoughts together and purchased a very fun outdoor barbecue oven. This inspired me to make a variety of tasty, fun, and whimsical pizzas that I knew my family would love. We're all about from scratch here at the farm, so of course we had to create a pizza dough recipe. Pauline and I had fun creating a pizza shell that would be able to hold the hefty amount of roasted corn we wanted on top. This is the recipe that passed the test! It's the perfect pizza shell to hold any of your favourite toppings—not too thick and not too thin, while crisping perfectly. I've shared my favourite pizza combinations on pages 74–80.

1 cup warm (approximately 110°F) water
1 Tbsp granulated sugar
1 Tbsp instant yeast
2 Tbsp extra virgin olive oil
¼ tsp table salt
2¼ cups all-purpose flour

1. Preheat your oven to 375°F. Spray two 10-inch perforated pizza pans with cooking spray.

2. In a large bowl, mix together the water, sugar, yeast, oil, and salt until well incorporated. Add the flour and knead by hand. Once the flour is incorporated, continue kneading until smooth, approximately 2 minutes. If using a stand mixer fitted with the dough hook, start on low speed, and once the flour is incorporated, turn the speed up to medium for 2 minutes or until the dough peels away from the hook and the sides of the bowl, and feels smooth.

3. Cover the dough with a clean tea towel or plastic wrap and let rise in a draft-free place until doubled in size, 20–50 minutes, depending on the temperature of your house.

4. Divide the dough in half and spread onto the prepared pizza pans. Poke with a fork to prevent bubbling.

5. Par-bake until set but not browned, approximately 10 minutes.

6. The pizza shells can be used in the recipes on pages 74–80 or can be frozen after par-baking, with or without toppings. Store in freezer bags in the freezer for up to 2 months. Thaw shells with no toppings on them for 15 minutes at room temperature before using them according to your recipe directions. Thaw shells with toppings on them overnight in the refrigerator or for a minimum of 2 hours at room temperature and bake at 350°F for 10–15 minutes. For the KBF Roasted Corn Pizza, continue as directed in step 5 on page 74.

WINK: *If you prefer, you can make four 6-inch pizza shells for a personalized experience. Follow the steps above but divide the dough into four, use 6-inch pizza pans, and reduce the par-baking time to 7 minutes.*

Pizza Sauces

Here are two tasty pizza sauces to experiment with at your next Pizza Party (page 77).

PESTO

MAKES APPROXIMATELY 3 CUPS

- 1⅔ cups extra virgin olive oil, divided, plus more to taste
- 4½ cups packed fresh basil leaves
- 1 cup shredded parmesan
- ¾ cup toasted pine nuts (see Wink)
- 5 cloves garlic, chopped (approximately 1⅔ Tbsp)
- ½ tsp kosher salt
- ¼ tsp freshly ground black pepper

I love this deep, rich, earthy flavour of the pesto. When we harvest garlic and basil in the summer, I make sure to double this recipe and freeze the extra to enjoy on pizzas and pastas throughout the year!

1. Place 1 cup olive oil in a food processor fitted with the steel blade. Add the basil leaves, parmesan, pine nuts, garlic, salt, and pepper.

2. Pulse the ingredients, stopping occasionally to push the ingredients down into the food processor, until the basil is broken down and the pine nuts are blended.

3. Add the remaining ⅔ cup olive oil in a slow, steady stream until reaching your desired thickness. You want a thicker, spreadable pesto for pizzas, but you can go a little thinner for a pasta sauce by adding a little more oil, 1 tsp at a time, until you have your desired consistency.

4. Store in an airtight container in the refrigerator for up to 1 week. To freeze, pour the pesto into ice cube trays until frozen, transfer the frozen pesto cubes to airtight containers or freezer bags, and return to the freezer for up to 6 months.

WINK: *To toast the pine nuts, place them in a dry cast-iron pan on medium heat, stirring carefully until golden brown, approximately 2½ minutes.*

TOMATO BASIL SAUCE

MAKES APPROXIMATELY 2 CUPS

- 2½ Tbsp extra virgin olive oil, divided
- 6 cloves garlic, smashed and then chopped (see Wink)
- 8 large garden tomatoes of your choice, diced
- ½ cup chopped fresh basil
- ¼ cup stemmed and chopped fresh oregano
- ¼ cup Bumbleberry Wine
- ¼ cup balsamic vinegar
- 1 tsp granulated sugar
- ⅛ tsp smoked Spanish paprika
- Kosher salt and freshly ground black pepper
- 1 (6 oz/170 g) can tomato paste

This is a great recipe for using in-season garden tomatoes. You can make a double batch to freeze, letting you bring back flavourful warm summer memories from your garden any time of year. Otherwise, you will run out—guaranteed!

1. In a medium or large pot on medium heat, heat ½ Tbsp oil. Add the garlic and sauté until slightly browned, approximately 1–2 minutes.

2. Add the remaining 2 Tbsp oil, the tomatoes, basil, oregano, wine, and balsamic vinegar, sugar, and smoked Spanish paprika. Stir well and bring to a boil for 5 minutes on medium-high heat, stirring constantly. Reduce the heat to low and allow the sauce to simmer, uncovered, stirring occasionally for 1 hour. If you prefer a thicker sauce, let it simmer, stirring occasionally, for another 30 minutes.

3. Let cool and blend with an immersion blender for 3 minutes. If you're concerned about seeds, pour into a sieve over a clean pot on medium heat. Gradually mix in the tomato paste. Season with salt and pepper to taste.

4. Store in an airtight container in the refrigerator for up to 4 days or in the freezer for up to 6 months.

WINK: *Mashing the garlic before you chop it releases more flavour.*

Custard Sauce

This custard sauce is creamy and velvety soft, with a hint of vanilla. It's a perfect finish to my traditional Steamed Berry Pudding (page 223). Try pouring it over a bowl of berries or Sensational Cinnamon Buns (page 32). Alf loves pouring it over sweet Pierogies (page 101), Waffles (page 48), and Blueberry Pancakes (page 45)—but not all at once!

MAKES 2 CUPS

- 1 cup whipping (35%) cream
- ⅔ cup whole milk
- ⅓ cup granulated sugar, divided
- 6 large egg yolks, at room temperature
- ¾ tsp pure vanilla extract
- Whipped cream, for garnish (optional)
- Lemon zest, for garnish (optional)

1. Fill the bottom part of a double boiler with water and bring it to a simmer on medium heat. In the top part of the double boiler, whisk the whipping cream and milk with 1 tsp of the sugar until it reaches 180°F. Clip a candy thermometer to the side of the pot to monitor the temperature. Pour the scalded milk into a measuring cup for easier slow pouring. For best results, warm your measuring cup before adding the milk.

2. In a medium bowl, whisk the egg yolks with the remaining sugar until combined, 1 minute. Whisking continuously, pour the scalded milk mixture very slowly into the egg mixture to avoid scrambling the eggs. Continue whisking until both mixtures are combined.

3. Pour the mixture into the top of the double boiler. On medium-low heat, stir continuously with a wooden spoon until the temperature reaches 170°F–173°F, approximately 15 minutes. Do not allow the temperature to go above 173°F. Adjust the heat level of your burner if necessary. Going even slightly over this temperature will cause the eggs to begin to scramble. The mixture should be thick enough to coat the back of a wooden spoon and will have a silky-smooth consistency. (If you run your finger across the wooden spoon, you'll create a path in the sauce.)

4. Remove from heat as soon as you have the required consistency, as overcooking can cause the custard sauce to curdle.

5. Place a fine-mesh sieve over a large bowl and strain the custard through it. Let cool completely, then mix in the vanilla. Cover the bowl with plastic wrap, making sure it touches the custard to prevent it from forming a skin, and place in the refrigerator for up to 2 days. I do not recommend freezing custard.

6. Warm the custard sauce slightly before serving, either in a small pot on medium-low heat or in the microwave at 30-second intervals, stirring between each interval.

7. Top with whipped cream and sprinkle with lemon zest, if desired.

WINK: *You can substitute the vanilla for ¾ tsp fresh orange or lemon juice for an alternative flavour.*

STRAWBERRY
U-PICK

Lemon Sauce

This sauce has a soft, smooth texture with a sweet pucker finish, making it the perfect way to add a touch of brightness to my Steamed Berry Pudding (page 223). Try pouring it over Grandma Buehler's Poppy Seed Cake instead of the glaze (page 190) for a smooth, tangy twist.

MAKES 2 CUPS

1½ Tbsp cornstarch
⅛ tsp table salt
¼ tsp cold water
1½ cups boiling water
1 Tbsp salted butter
¾ cup granulated sugar
Pinch of ground nutmeg
Zest and juice of ½ lemon

1. In a large bowl, combine the cornstarch and salt. Mix in the cold water until the cornstarch is smooth.

2. Pour the boiling water over the cornstarch mixture, stirring constantly for 3–5 minutes to prevent lumps. Once smooth, set aside.

3. In a medium bowl with an electric mixer, cream the butter and sugar together, then add to the cornstarch mixture. Combine well with a whisk.

4. Add the nutmeg, lemon zest, and lemon juice.

5. Transfer to a medium saucepan and cook on medium heat, stirring constantly, until smooth and thickened, 1–2 minutes.

6. Remove from heat and strain through a fine-mesh sieve to remove the lemon zest. Serve warm or cool.

7. Store in an airtight container in the refrigerator for up to 3 days. Warm slightly to serve, either in a small pot over medium-low or in the microwave in 30-second intervals, stirring between intervals.

WINK: *To make vanilla sauce, omit the nutmeg, lemon juice, and lemon zest, and use 1½ tsp pure vanilla extract.*

Meringue Powder Royal Icing

We use royal icing to decorate all our seasonal cookies, as shown in the photograph on page 166. When you're colouring the icing, add only a tiny drop at a time until you have the shade you want. It's a good idea to keep notes about how much you add so that you can recreate the colour. Space doesn't allow for a tutorial on cookie-decorating techniques, but there are many, many resources online.

MAKES 2 CUPS

3½ cups icing sugar
¼ cup meringue powder
Approximately 2 cups almost-boiling water
Food colouring (optional)

1. In the bowl of a stand mixer fitted with the whisk attachment, combine the icing sugar and meringue powder on low speed. With the mixer still running on low, add the hot water by the tablespoon until the sugar is moistened. You need just enough water to melt the icing sugar and meringue powder.

2. Mix on high speed for 10 minutes, until completely smooth.

3. Add food colouring, if desired. Add more water or icing sugar to get the consistency you want, depending on whether you'll be outlining or flooding your cookies. For outlining, it should be stiff but smooth. For flooding, you want it runnier so it will spread to the outline.

WINK: *This icing, once dry, has a matte finish to it.*

4. Store in an airtight bag or container at room temperature until you're ready to use it. Icing will last for up to 2 weeks. You might need to give it a bit of a mix before you use it. If it separates or dries out, discard it.

Gluten-Friendly Flour Blend

In 2012, I was advised to remove wheat from my diet to reduce the inflammation in my joints. My arthritis has since improved immensely. I have replaced regular flour with this blend in many recipes in this book, so I'm able to continue enjoying my favourite recipes! This recipe was created after many, many trials with Jody, a staff member who worked in the Specialty Foods Kitchen (page 198). We call it "gluten-friendly" rather than "gluten-free" because it isn't made in a certified gluten-free environment. Its texture is smooth but it's not as powdery as regular flour. To make it vegan, use coconut milk powder instead.

MAKES 3¾ CUPS

2½ cups white rice flour
½ cup brown rice flour
½ cup potato starch
¼ cup tapioca starch
2 Tbsp non-fat milk powder or 2 Tbsp coconut milk powder

1. In a large bowl, gently whisk together all ingredients until well combined.
2. Store in an airtight container in your pantry for up to 8 months.

WINK: *When you use this flour blend in my Whole Wheat Pie Dough (page 273), you'll need to add xanthan gum as noted in that recipe's Wink. It helps to bind the ingredients together in place of the missing gluten.*

Potato Slurry

Use this as a thickener, such as in the Cheesy Potato Bread (page 118) and Hearty Multigrain Bread (page 113).

MAKES ½–⅔ CUP

1 medium russet potato, peeled and cut into 8 pieces
¼–⅓ cup water

1. Place the potatoes in a small pot and cover with the water. Put the lid on and bring to a boil on high heat, then turn down the heat and keep at a low boil for 15 minutes.
2. Break up any pieces with a potato masher and cook on low heat for another 5 minutes. It should be a smooth consistency similar to applesauce. Potatoes vary in size, so if the slurry doesn't measure ½–⅔ cup, you can add a little water to make up the difference.
3. Let cool for 10 minutes before using.
4. Store in an airtight container in the refrigerator for up to 1 day. Stir well to restore its smooth consistency before using.

With Gratitude

THANK YOU, THANK YOU, THANK YOU.
I am sincerely **GRATEFUL** to many!

This book would not have been possible without the many wonderful and dedicated people I work with. I couldn't be prouder of my family and the farm family.

Thank you, Alf, for the way you care for our family and our community, for the meticulous way you take care of our soil, and for the way you always make the right choices for the soil and for the future of our children and grandchildren and the generations to come. Thank you for the wonder you still show after 50 years of seeing the first little green sprouts of new life emerging from the soil each spring. I have never met anyone who works as hard as you. Thank you for taking care of our family and our farm family with such dedication to a bigger picture. And thank you for encouraging me to be creative, never saying no to my next idea, and for trusting me.

Thank you to my children, Sarah, Grayden, Tanner, Jared, Holly, and Tracey, for filling out the endless testing sheets asking for your opinions on taste, texture, and presentation over all the years for every new product I've created. You are and always have been my best testers. Thank you for your detailed honesty, your politeness about every creation, and your willingness to work in the kitchen with me, have fun, make memories, and try new things. You are my pride, joy, and inspiration, and my most valuable teachers in life.

Thank you to my grandchildren, Kale, Emme, Brynn, River, and our most recent addition, Casey, for putting life into perspective for me.

Thank you, Heather, for your amazing photography talents and for sharing your incredible gifts with me and the farm. Your contributions to this book are captured in every beautiful image portrayed. What began as a come-and-take-photos-of-the-food-I-created arrangement for a year turned into an incredible journey going on 8 years. While snapping me and everything I create in all eight kitchens and in every area throughout the farm, you have become part of our farm family and captured not only the image but an authentic view of farming as it really happens, giving the reader a firsthand glimpse of the life of our farm, the crops, and our products. Thank you for all the encouragement you have given me, the knowledge you have shared with me throughout this journey, and your kind words of comfort during the times I could hardly breathe.

Thank you to Ann, our farm manager and my everything! The support you have given to me and to the farm for the past 18 years has been incredible. I could not do what I do without your knowledge, endurance, and professionalism. You have led with a passionate heart and commitment to excellent care to Alf and I, our staff, our guests, and our products. You dot my i's, cross my t's, and know exactly what I would say or do, giving me the opportunity to continue to create and grow our business. You are never afraid to help me make the difficult choices and are always there to help me reach my goals of self-improvement. Thank you for the stability you provide and for being the strength for the farm when I was weak. But most importantly, thank you for the gift of your loyal friendship of 30 years.

Thank you, Pauline, for your kitchen leadership over the past 20 years. The Harvest Kitchen wouldn't be what it is today without your dedication to the farm and your unfailing good humour and willingness to take whatever fresh crops Alf brought in from the fields and turn them into the many Krause products we all enjoy today. You have mentored many young people throughout the years, and there is no doubt that they owe many of their baking skills to you. Thank you for being one of the hardest-working women I know and for always thinking of the farm, staff, wholesalers, and guests. Thank you for having one of the kindest and most caring hearts I have ever had the good fortune to encounter, one that allows you to be thoughtful enough to provide a bucket of joy in the shape of a homemade cookie each week for a year for a grieving family. Thank you, too, for the many years of friendship we have shared, for allowing me to be creative and continue to create, and, most of all, for all the fun times we had over the years as we

worked side by side to reach our production goals, from early mornings to late, late nights.

Thank you to our field manager, Nanaki. I feel privileged to have watched you grow from the age of 15 and to have been a part of your life and your family's life for the past 27 years. I appreciate everything you have contributed to our farm, fields, and staff. Thank you for coordinating all the field workers each season and the dedicated way you assist Alf in taking such excellent care of all that goes into making operations run smoothly. What you manage each and every day is the starting point for what I am able to do on a daily basis in my kitchens. Your contributions are outstanding in every way!

Thank you to Paul, our crop care manager, for 27 years of helping Alf with field prep, planting, irrigating, tractor work, and maintaining our fields at the farm. Your dedication to your work is unwavering. I look forward to and enjoy seeing you on the tractor out in the fields, working so passionately and with such pride in the quality of our crops.

Thank you to our farm mechanic, Balwant, for 25 years of keeping all the tractors, trucks, buses, vans, farm equipment, and machinery operating safely, which enables the work to stay on schedule throughout the year. Thank you for your willingness to not only fix the farm equipment but also keep our personal vehicles operating smoothly as well.

Thank you, Gabriel, for supporting Ann and me and for keeping the farm up and running for our guests to enjoy. Your ability to fix anything, move anything, maintain the continuous rotation throughout the four seasons, and even flip waffles on busy days is remarkable. Your dedication to ensuring our vision remains a reality and your ability to see the larger picture are so valued.

Office staff

Thank you, Tammy, for looking after all the important details of accounting and payroll with such a big heart, and for remembering the name of every supplier I have purchased from for the past 10 years. I don't know how you do it with such precision, but I am extremely grateful for it. Your dedication to each and every detail of your work allows all of us to perform ours better.

Thank you, Elizabeth and Kimmy, for your assistance and support to Ann and all the duties of operational support in our office, from phones to filing, label making, training, staffing, scheduling, jumping on a cash register, and helping flip waffles on busy days. Your flexibility is so valued.

Jeff, thank you for your professionalism with all our farm events, from the smallest to the largest. We always know we can count on you to deliver the very best experience to our guests.

Thank you, Selin, for your dedication to our production schedule goals, our inventory and its ever-changing numbers, and our many suppliers.

Norene, thank you for your loyalty to this cookbook and my vision. Keeping track of all the details of the recipes, the testers, the stories, the Winks, and the ever-changing Excel spreadsheets was so valuable.

Thank you, Michaela, for allowing me to depend on you to give me the answer to any question I ask, almost instantly. For helping me to break down large projects into bite-sized pieces and then to build them back together again. Your knowledge and confidence, and ability to multitask and perform at a level far beyond your years, amazes me. For your willingness to contribute in any area where we may need help, I know I can always count on you.

Market staff

Thank you Leanne and Cheryl, for the dedication you give to our sparkling retail market, our orders, and your wonderful, ever-changing, amazing seasonal displays.

Thank you, Sherry, for your incredible attention to detail in our bakery, for all the endless packaging and labelling you do to make everything look beautiful, and for operating a smooth online store.

Kitchen staff

Thank you, Amanda, for your fearless desire over the years to become our kitchen manager, adoring and respecting Pauline's leadership, guidance, and experience, and valuing the entire team, I'm very proud of you. Thank you for thinking ahead and achieving the

production goals. You, rise to every challenge and operate an efficient and very fun kitchen.

Thank you, Salvador, Elias, and Joshua, for assisting Amanda and your incredible support. Salvador, thank you also for making many varieties of rich, creamy, and drool-worthy fudge. Thank you to Elias for your creative popcorn making. Thanks also to Carlos M for making the tastiest ice cream with our fresh berries. And, of course, thank you Carlos Z for the creation of all the flavourful beauty that comes out of our Specialty Foods Kitchen.

Harjinder, Liz, Daniela, Vanessa, Claudia, Jogie, and Miguel, thank you for your dough making, pie filling, jam, jelly, and syrup making, green bean and asparagus stuffing, and endurance on the endless corn-roasting days.

Winery staff

Thank you, Ted, for your role in managing the winery from its conception in 2012. Your knowledge about wines and your dedication to giving our guests excellent care has led to a very large and growing fan club of which you are the star, with your showmanship punctuated by your memorable moustache.

Sandra Lee, Toby, Thomas, and Christine, thank you for your contribution as winemakers and dedication to creating award-winning wines for us. Your knowledge and creativity continually contribute to our guests' excellent experience and enable us to provide outstanding products.

Other amazing staff

Thanks to Eniko, our professional and very talented gardener since 2016, for your passionate and excellent care in all the garden areas of the farm. Your knowledge combined with your hard work results in the beauty that surrounds us every day.

Thank you Cal and Sandy, for the joyful contributions you each make to our guests, welcoming them so warmly on the busy days and providing them with directional information with such fun and personal attention.

Paul, Lorene, and Nadine, thank you for 12 years of parking order and for preventing complete chaos during our harvest season, special events, and holidays.

Thank you, Audrey, our resident artist since 2001, for the many wonderful contributions you have made to the farm and for our friendship of 30 years. From signs to cookie and cake decorating, your artistic talent shines bright in all areas of the farm.

Virginia, thank you for your passionate heart in keeping our personal spaces and workspaces clean. You are invaluable to me with the amazing care you provide by creating a sense of peace and order in spaces for the farm and my family. Thank you for being my friend and providing this calming service for over 30 years.

Thanks to all the students who work with us each summer, enabling us to serve our community at every location throughout the farm.

Thank you, recipe testers: Pauline, Leanne, Norene, Selin, Kathy, Mary, Judy, and Michaela, for your willingness to take on shopping for ingredients, testing, filling out a very detailed result and comment form, and providing a snapshot and taste sample of each recipe.

It takes a dedicated team to manage our farming operation. I appreciate all the contributions each one of you makes, and I look forward to seeing you all continue to take such great pride in your responsibilities at our family farm.

And Other VIPs

Thank you to Robert McCullough from Appetite by Random House for asking me to write this book years ago and for your thoughtful patience and compassion through its completion. It was a great pleasure to work with you. Knowing I was in such capable and professional hands gave me peace of mind throughout the entire project. I am grateful for the opportunity you gave me to share our farm and these recipes with a larger community.

Just like running a farm, making a cookbook takes a professional and talented group of people to put it all together. Thank you to Rachel Brown, Lesley Cameron, Victoria Walsh, and Katherine Stopa, my

editors, for working closely with me to preserve the vision of family, food, farming, and a community that works with us to protect farming practices for local food consumption. Thank you to Lana Okerlund, Judy Phillips, and Lindsey King for copyediting and proofreading the book. Thank you to this book's designer, Matthew Flute, for creating a design that captures the authenticity of the farm and thoughtfully complements the photography. Thank you, Sean Tai for typesetting the book. Lastly, thank you to Michelle Arbus, Charlotte Nip, Aakanksha Malhotra, and Wayne Miller for promoting and selling the book. It's a privilege and an honour to have the entire Penguin Random House Canada team take seeds of the farm and our recipes and nurtured them as they grew into this beautiful cookbook for all to enjoy.

Finally, thank you to my grandmothers, my mother and father, my aunt, my mother-in-law and the many people I have had the pleasure of working with in their kitchens and mine, and the many chefs and bakers who have shared their knowledge with me over the years. Watching you in your element continues to be amazing and inspiring.

And to all the farmers around the world: you dedicate more time and energy to your work than many people realize, even as they enjoy the fruits of your labour. Thank you!

L–R: Brynn, Holly, Tracey, Tanner, me, Alf, Emme, Jared, Ryan, Grayden (behind), Kale (in front), and Sarah. Not pictured (but still part of the family!): Jayme, River, Casey, Caliegh, and Chirag.

Index

QUEEN
OF
THE KITCHEN
CLARK
C25
CLARK
MARKET
UPS